ENGINEERING MATERIALS SCIENCE:

Properties, Uses Degradation and Remediation

"Talking of education, people have now a-days" (he said) "got a strange opinion that every thing should be taught by lectures. Now, I cannot see that lectures can do so much good as reading the books from which the lectures are taken. I know nothing that can be best taught by lectures, except where experiments are to be shown. You may teach chymestry by lectures -- You might teach making of shoes by lectures!"

James Boswell: *Life of Dr Samuel Johnson, 1766*

Horwood Publishing
Chichester

ABOUT THE AUTHORS

Hugh McArthur BA(Cantab) MSc PhD

Dr Hugh McArthur, born in Africa, was educated at Cambridge University where he read Natural Science Tripos, specialising in metallurgy and corrosion. He spent 10 years in the corrosion laboratories in the Nuclear Power industry before joining Leicester Polytechnic (now De Montfort) in 1963. He gained an MSc from Leicester University and a PhD from Loughborough University in 1969 for his work on high strain fatigue. Dr McArthur is a corrosion consultant to the motor industry and has authored two books on motor vehicle corrosion. He was Principal Lecturer at De Montfort University and has spent 30 years specialising in electron microscopy and materials science teaching, research and consultancy to the construction industry. This is Dr McArthurs's second book on materials science.

Duncan Spalding BSc (Hons) PhD

Dr Duncan Spalding was educated at Leicester Polytechnic (now De Montfort University) where he read Building Surveying. He spent a number of years working in the building industry as a development surveyor, before moving to De Montfort University as a Research Fellow in 1993 where he gained a PhD in 1999 for his work on modelling sources of contaminant emission from building materials. He has undertaken externally funded research, and has published articles in this area: and is currently Senior Lecturer in Materials Science at Leicester School of Architecture in De Montfort University.

ENGINEERING MATERIALS SCIENCE

Properties, Uses, Degradation and Remediation

Hugh McArthur
Former Principal Lecturer
De Montfort University, Leicester *and*
Corrosion Consultant to Automotive and Steel
Manufacturing Industries, and British Steel

Duncan Spalding
Senior Lecturer in Materials Science
Leicester School of Architecture
De Montfort University, Leicester

Horwood Publishing
Chichester, U.K.

HORWOOD PUBLISHING LIMITED
International Publishers in Science and Technology
Coll House, Westergate, Chichester,
West Sussex PO20 3QL England

First published in 2004

British Library Cataloguing in Publication Data
A catalogue record of this book is available from the British Library

ISBN: 1-898563-11-X

Printed and bound by Antony Rowe Ltd, Eastbourne

TABLE OF CONTENTS

SOURCES OF INFORMATION

Specific references relating to the subject matter of each Chapter is given at the end of each Chapter. More general sources of information for construction and engineering materials are produced by the following sources: The Building Regulations 1991, 1995 and Approved Documents; Building Research Establishment (BRE) publications; Trade Association publications; Trade literature; Trade journals; British and European Standards available from the British Standards Institution (BSI); and British Board of Agrément certificates.

EUROPEAN STANDARDISATION

For most construction industry professionals, knowledge of the current work on European standardisation and its impact on British Standards (**BS**) and UK practice is scant. However, the impact of European standardisation has been profound, not least because British Standards for most construction products are being withdrawn and replaced by European Standards (**EN**). In many cases, the basic product has not changed, or changed significantly, although it now conforms to a new standard and may be given a different title.

European standardisation was implemented by the European Economic Community (**EEC**), now the European Union (**EU**). When the EEC agreed to create a single market within Europe for goods and services, they commissioned a review of barriers to trade. This review identified national standards as the most important technical barrier to trade. The EEC therefore set up the European Committee for Standardisation (**CEN**) to harmonise the national standards of member states within Europe.

Most national standards tend to have a combination of performance and prescriptive requirements. For example, British cement standards have performance requirements for strength and prescriptive requirements for constituents. Harmonisation of standards is extremely difficult when based on prescriptive requirements. Therefore the CEN requires harmonised European standards to be expressed, as far as possible, in terms of product performance. The preferred solution of the CEN for harmonisation was for the EU member states to adopt standards produced by the International Standards Organisation (**ISO**). The ISO produce standards intended to have worldwide validity and may be adopted by individual countries as national standards. For example, an ISO standard adopted by Britain would be prefixed BS ISO. However, it was apparent that ISO standards did not cover all the needs of the member states of the EU and, in some cases, the ISO standards were lower than the national standards of EU member states. Hence the European Committee for Standardisation (**CEN**) was established to produce standards for the EU where no suitable alternative standard existed. The CEN has international membership to reflect the views of EU member states and is split into various Technical Committees with expertise in the relevant areas.

In accordance with the fundamental requirement to limit barriers to trade, the CEN base the free market for products on six essential requirements: mechanical resistance and stability; safety in case of fire; hygiene, health and environment; safety in use; protection against noise; energy economy and heat retention. With respect to construction products, these requirements relate to structure. Thus, the main thrust of standardisation by the EU has been to harmonise these essential requirements. Such a standard is called a 'harmonised' European standard (**hEN**), which contains only clauses that relate to the essential requirements. Products satisfying just the six essential requirements have the freedom to be placed on the market anywhere within the EU and they can carry the **CE** mark if they have the designated level of attestation of conformity as decided by the EU (rather than the CEN). The levels of attestation range from full third party certification (for critical aspects of performance, e.g. safety) to declaration by the manufacturer (for non-critical aspects of performance, e.g. mechanical resistance). Thus a harmonised European standard, together with its designated level of attestation of conformity, provides the basis for awarding a CE mark. For some construction products (e.g, cement), third party attestation is essential because the final destination of the cement in unknown; it could be placed in a critical load-bearing situation, for example. The importance of CE marking can be judged from the UK Building Regulations, Regulation 7, which states

> '*An EC marked material can only be rejected by the Building Control Authority or Approved Inspector on the basis that its performance is not in accordance with its technical specification. The onus of proof in such cases is on the Building Control Authority or Approved Inspector, who must notify the Trading Standards Officer.*'

The CEN, rather than restricting itself to the production of 'harmonised' European Standards in which only the essential requirements are considered, has tended to write standards for users. Such standards are either published directly as a European Standards (**EN**) or as a Voluntary European Standard (**ENV**), which have similar status to a BSI 'Draft for Development' (**DD**), and exist alongside national standards. ENVs have a maximum life of three years before either being upgraded to a full European Standard or being withdrawn. In practice, it has often been possible to extend the life of an ENV for up to a further 2 years. Occasionally, preEuropean Standards (**prENs**) are published; these documents are drafts with no official status other than a committee paper. prEN status documents are usually used as consultative documents to gain the views and comments of interested parties prior to acceptance of full EN status.

When a European Standard (EN) is published, conflicting national standards, or conflicting parts of national standards, ordinarily have to be withdrawn within 6 months (but see below). Every CEN member will publish the European Standard with a national forward, but this may not change the content of the European Standard (even if that particular CEN member voted against it). This is to ensure that, for example, the British BS EN 196-1 is the same the same as the German DIN EN 196-1. European Standards are reviewed and, if necessary, amended, every 5 years. European Standards and Voluntary European Standards contain more than the six essential requirements and consequently conformity to these standards exceeds the legal requirements necessary for products to have free access to the EU market and carry the CE mark.

European Standards (ENs) and Voluntary European Standards (ENVs) are agreed by a complex majority voting system (no single CEN member has the power to stop a standard being adopted in the EU). CEN members (such as the British Standards Institution) can ask the CEN to produce a standard for a new product of one that is already standardised nationally. In the latter case, the member would propose that the European Standard is based on the national standard.

Increasingly, standardisation work is being instigated by the European Commission directly by the publication of Directives. When the commission publish a Directive, such as the Construction Products Directive (CPD) (in 1991), they then produce Interpretative Documents which identify the need for standardisation work required to make the Directive effective. The European Commission then asks the CEN to undertake this work, which will eventually result in a European Standard.

The CEN has the near impossible task of standardising thousands of products from conflicting national standards. Not all related standards can be produced at the same time and therefore the concept of a hierarchy of packages of standards was developed, as shown over for concrete and associated products.

At the lowest level, Level 1, a package is a product with its associated test methods. For example, EN 197-1 (common cements) and EN 196 (methods of testing cement) would form a Level 1 package. The next level, Level 2, would be concrete and the highest level, Level 3, would be the concrete design code (a Eurocode). In the package concept, the EU member states need not withdraw conflicting national standards until a package is complete (this overrides the 6-month withdrawal period outlined above).

The European Commission's objective is for the Eurocodes to establish as set of common technical rules for the design of buildings and civil engineering works which will ultimately replace differing rules in the various member states.

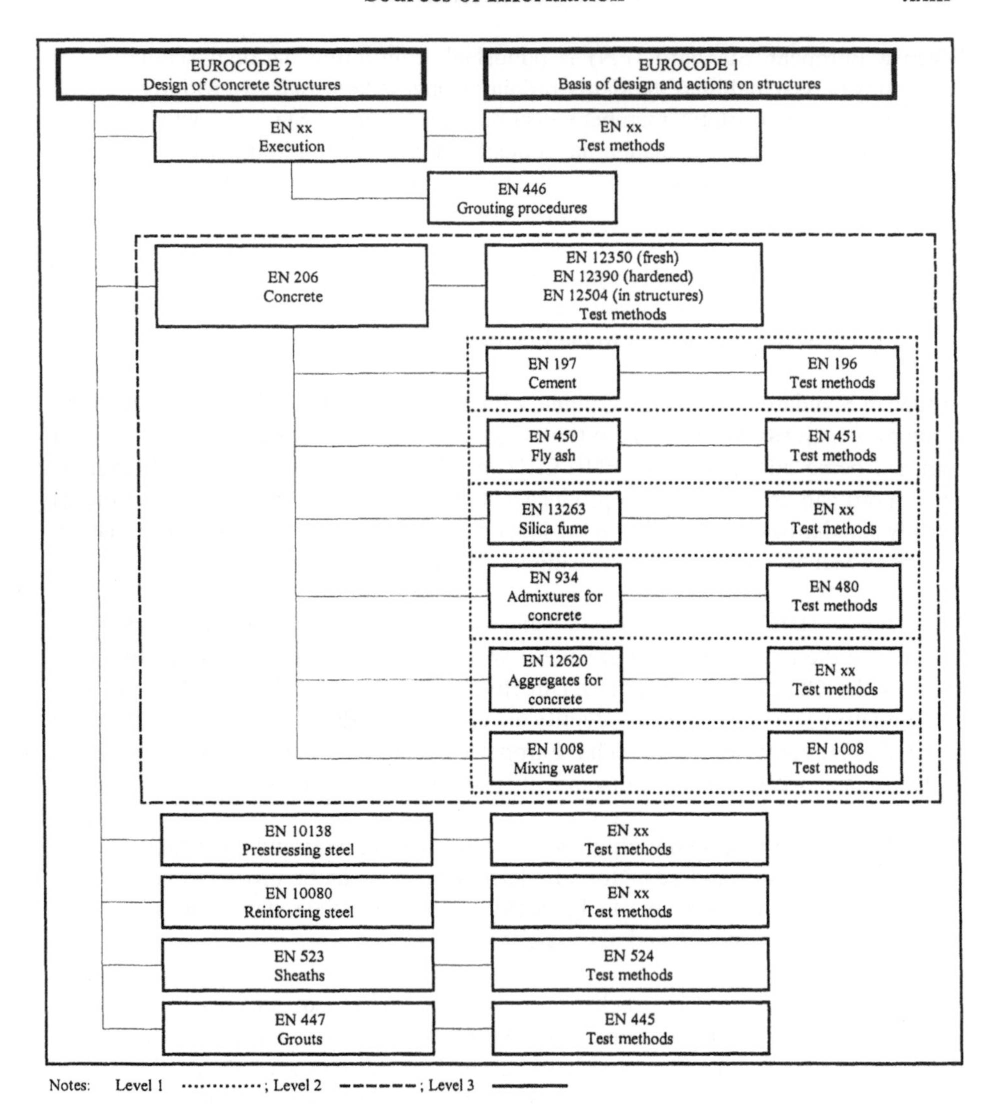

Hierarchy of European standards (for concrete and associated products)

Eurocodes are intended to serve as reference documents to be recognised by authorities of the member states for the following purposes

- as a means of compliance of building and civil engineering works with the essential requirements as set out in Council Directive 89/106/EEC (the Construction Products Directive), particularly essential requirement No. 1: Mechanical resistance and stability;
- as a basis for specifying contracts for the execution of construction works and related engineering services in the area of public works;
- as a framework for drawing up harmonised technical specifications for construction purposes.

In addition, the Eurocodes are foreseen to

- improve the functioning of the single European market for products' engineering services by removing barriers to trade;
- improve the competitiveness of the European construction industry and the professionals and industries connected to it, in countries outside the European Union.

The following structural Eurocodes, each generally consisting of a number of parts, will be released as ENs between 2000 and 2004; those not yet released exist as ENVs.

EN 1990 Eurocodes: Basis of Design (BS EN 1990: 2002)
EN 1991 Eurocode 1: Actions on structures (BS EN 1991-1.1: 2002. General actions. Densities, self-weight, imposed loads for buildings).
EN 1992 Eurocode 2: Design of concrete structures (BS EN 1992-1.1: 1992. General rules for buildings).
EN 1993 Eurocode 3: Design of steel structures (DD ENV 1993-1.1:1992. General rules and rules for buildings).
EN 1994 Eurocode 4: Design of composite steel and concrete structures (DD ENV 1994-1.1:1994. General rules and rules for buildings).
EN 1995 Eurocode 5: Design of timber structures (DD ENV 1995-1.1:1994. General rules and rules for buildings).
EN 1996 Eurocode 6: Design of masonry structures (DD ENV 1996-1.1:1996. General rules and rules for buildings - Rules for reinforced and unreinforced masonry).
EN 1997 Euronote 7: Geo technical design (DD ENV 1997-1: 1995. General rules).
EN 1998 Eurocode 8: Design of structures for earthquake resistance (DD ENV 1998-1.2: 1996. General rules - Rules for buildings).
EN 1999 Eurocode 9: Design of aluminium structures (DD ENV 1999-1.1:2000. General rules - General rules and rules for buildings).

AGRÉMENT CERTIFICATES

Agrément certificates are published in Britain by the British Board of Agrément (**BBA**) and in Europe by a European Union of Agrément (**UEA**). Agrément certificates generally relate to new products or processes not covered by a British or European Standard and are usually valid for a limited period (whilst appropriate British or other standards for the product or process are prepared). They are widely accepted by statutory bodies such as Local Authorities and therefore represent a means of obtaining relatively rapid approval for a product or system, hence encouraging innovation. The BBA focuses on the specification, manufacture and performance of a product and its suitability for use in buildings. Production is monitored by inspection, normally on a twice yearly basis throughout the period of validity of the certificate.

PREFACE

Engineering Materials: A guide to all branches of materials science is written primarily for undergraduate students of building construction disciplines, including civil and structural engineering, construction and architecture. The emphasis of the book is on materials defects and deterioration processes, and will therefore be of interest to those students and practitioners involved in identification and remediation of building material and component defects.

The process of scientific advancement requires a logical approach to problem-solving, termed **scientific modelling**. Science is built on models. In formulating a **model**, we construct a **hypothesis** that some event will occur as a consequence of certain actions. We then **test that hypothesis** (commonly by carrying out controlled experiments). If these tests show that the hypothesis is correct, it becomes a **theory**. A concise statement of the experimental results is known as a **law**. Over time, these 'laws' become the basis of our common sense, although not all theories survive the test of time and, indeed, not everyone wants these theories as the basis for their common sense. We hope the book will be of general interest to both the layman and craft worker by illustrating how this process of scientific modelling gives rise to traditional 'common sense' rules concerning the behaviour and uses of construction materials.

A main objective of the book is to develop an understanding of the behaviour of engineering materials from a knowledge of their structure, with particular emphasis on mechanisms of deterioration and degradation. The study of materials science has become an integral part of many diverse university courses, and encompasses the 'pure' disciplines of chemistry and physics. In our experience, many university students and practitioners in construction disciplines have little or no scientific knowledge. Science is therefore perceived to be a difficult subject for the student and practitioner alike, with a language and shorthand all of its own. To address these problems, the book is split into three parts. In the first part of the book, Chapters relate to fundamental chemistry, physics, and mechanical properties and testing. These Chapters provide the basis for understanding the behaviour of engineering materials developed in subsequent sections. In the second part of the book, we concentrate on the water, the primary agent of degradation and deterioration of building materials, components and the indoor environment. In the third section, we cover the manufacture, properties, uses and deterioration mechanisms of standard building materials, including masonry materials, cementitious materials, glass, metals, timber. Separate Chapters examine specific degradation problems relating to the corrosion of metals, and fire and fire resistance properties in detail. We believe that this approach allows the reader to appreciate the underlying scientific principles governing a degradation process, for example, and the consequent remedial action required to mitigate the problem. An extensive range of over 450 line illustrations and extensive cross-referencing both within and between Chapters has been included to aid this.

For the student and practitioner, we have included in the book relevant British and European Standards, all fully referenced to the text and current at the date of publication. However, as the process of European standardisation continues, additional standards will be published for new materials and existing British Standards will be replaced. The reader is encouraged to contact the British Standards Institution (BSI) for the most recent versions of standards.

It is not possible in a single book to comprehensively cover all construction and engineering materials currently available. We have therefore dealt primarily with the most widely used materials and mentioned, where relevant, notable recent advances in materials technology. We hope the book provides a sound foundation to the study of engineering materials and to the processes by which they deteriorate.

ACKNOWLEDGEMENTS

In writing a book of this nature, one accumulates indebtedness to a wide range of people, not least to authors of earlier books in the field. I am particularly indebted to RC Evans (Department of Mineralogy and Crystallography, Cambridge University) for the introduction given to me in my formative years, and to my wife for making available the luxury of time and an environment conducive to the long hours spent writing.

Hugh McArthur.

I wish to express my appreciation to colleagues in the Leicester School of Architecture for their encouragement throughout the development of this book, in particular to Professor Peter Swallow for his helpful advice and guidance, to Mr. Michael Ashley for lightening my University administrative responsibilities, and to Mrs Helen Monk for her continued laboratory support. I would also like to thank the various publishers who have given permission to reproduce figures and tables in the text. Finally, I wish to express my thanks to my parents, brother and sister for their unwavering support and to Caroline, who is always with me.

Duncan Spalding

Extracts from British Standards are reproduced with the permission of BSI under licence number 2003SK/0153 and are acknowledged with thanks. British Standards can be obtained from BSI Customer Services, 389 Chiswick High Road, London. W4 4AL. Tel. +44 (0)20 8996 9001. Figures 12.6, 12.10, 12.12a, 12.12b, and 14.10 are reproduced from Dinwoodie, J.M. (1994) Part 7. Timber. In: Illston, J.M. (ed.) *Construction Materials: Their Nature and Behaviour*, 2nd ed. London, E&FN Spon by permission of the BRE. Figure 12.13 and diagrams in Table 12.11 are reproduced from Bravery, A.F.; Berry, R.W.; Carey, J.K. and Cooper, D.E. (1992) *Recognising Wood Rot and Insect Damage in Buildings. BRE Report 232*. Garston, BRE by permission of the BRE. Tables 8.10, 8.11 and 8.12 are reproduced from BRE (2001) *Concrete in Aggressive Ground. Part 2: Specifying concrete and additional protective measures. BRE Special Digest 1*. Garston, BRE by permission of the BRE.

1

STRUCTURE OF MATERIALS

1.1 INTRODUCTION

When man first looked at a material, he decided that the fundamental building block of that material was the smallest unit to that it could be broken or **cleaved** (***Figure 1.1***). Hence, a crystal that grew from solution and had the external form of a cube (e.g, halide, or common salt) was thought to be composed of a fundamental unit that itself was cubic in shape. Likewise, all the geological crystals found in nature (e.g, calcite and mica) were considered to be composed of smaller crystals of the same shape all tightly packed together. Historically, before the advent of microscopes, scientists could grow very large crystals from solutions and their external faces always bore a constant angle with other crystals of the same material. For a time this seemed to be a very good basis on which to explain structure, as it was very easy to consider all materials as made up of smaller, miniature crystals with flat plane surfaces.

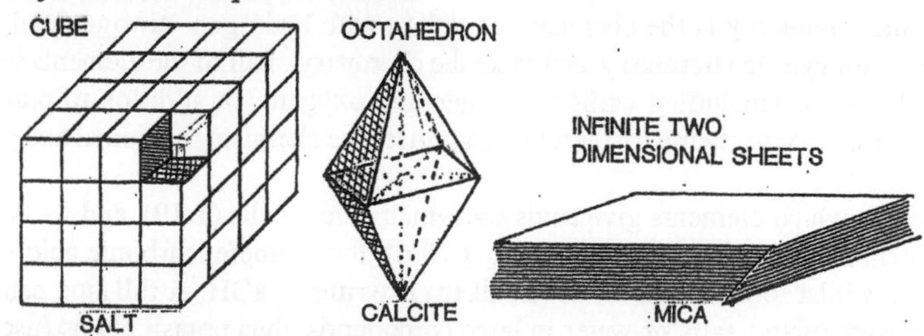

Figure 1.1 Examples of cleavage structures

It was soon realised, however, that this modelling approach had severe disadvantages. Some materials that crystallised as octahedrons (e.g, calcite) could not be stacked together to form a larger octahedron without leaving voids in the bigger crystal. This meant, according to the model, that the density of small crystals should be different to the density of larger crystals. Since both small and large crystals are composed of identical material, the model fails. Other materials are laminar in structure (e.g, mica) and hence comprise a fundamental unit that is an infinite two-dimensional sheet. This is clearly at variance with the model requirement to obtain the smallest unit to which the material can be cleaved.

sulfide (K_2S), calcium carbide (CaC_2). The ending **-ate** is used for compounds containing oxygen, e.g, potassium sulfate (K_2SO_4), calcium carbonate ($CaCO_3$) and **ions (1.11.5)** nitrate (NO_3^-), sulfate (SO_4^{2-}). The ending **-ite** is used for compounds containing less oxygen than those ending **-ate**, e.g, sodium nitrite ($NaNO_2$), potassium sulfite (K_2SO_3) and ions nitrite (NO_2^-), sulfite (SO_3^{2-}).

Note that there is nothing magical about these names; they merely enable us to better understand the composition of the compound. For example, the ending **-ade** is used in everyday language and is well known to refer to a fizzy drink (as in lemon**ade**).

1.2.1 Chemical Formulae

Using the symbol abbreviations, chemical elements and compounds can be written as **chemical formulae** that represent one molecule of the element or compound. For **elements that exist as molecules**, the chemical formula is the symbol of the element with the number of atoms in one molecule of the element written at the bottom right-hand corner of the symbol, e.g, H_2 means one molecule of hydrogen (containing two atoms of hydrogen), 2H means two separate atoms of hydrogen. For **compounds that exist as molecules**, the chemical formula consists of the symbols of the elements in the compound and shows the number of atoms of each element in one molecule of the compound, e.g, H_2O means two atoms of hydrogen and one atom of oxygen in each molecule of water; note that, by convention, the '1' (atom of oxygen) is never shown. Brackets are used in chemical formulae to denote multiples of groups of atoms, e.g, $Mg(OH)_2$ means one atom of magnesium (Mg), and two (OH) units, comprising a total of two atoms of oxygen (O) and two atoms of hydrogen (H).

1.2.2 Other Chemical Conventions

Organic chemistry is the chemistry of carbon with hydrogen, nitrogen and oxygen, whereas **inorganic chemistry** embraces the chemistry of all of the elements within the Periodic Table (including carbon, nitrogen and oxygen). In addition to plastics and adhesives, organic chemistry therefore comprises the chemistry of animal and plant life.

We know which elements give acids and which give alkalis **(1.19)**, and we write their chemical formulae to express this fact **(1.19.6)**; for example, carbonic acid is written H_2CO_3, whilst sodium hydroxide (an alkali) is written NaOH. A full stop can be used to separate mixed salts or water in large compounds, thus potash alums (used in fire protection, **14.7.3**) are represented $K_2SO_4.(NH_4)_2SO_4.24H_2O$ and not $K_2S_2N_2H_{56}O_{32}$, as the latter gives no indication of the chemical composition of the compound (there are exceptions to this general convention; for example, carbohydrates **(12.4.1)** are written as $C_6H_{12}O_6$ or $C_6(H_2O)_6$ but not as $C_6.6H_2O$. The solid sodium hydroxide is expressed by the chemical formula NaOH; however, when dissolved in a liquid, the compound is expressed in the **ionic form (1.12)** as Na^+ and OH^- **ions (1.4)**. As **covalently** bonded materials **(1.11)** cannot form ions, they are **insoluble**. In chemical equations, insoluble products precipitated out are represented with a downwards arrow, e.g, $CaCO_3\downarrow$, gases produced are represented with an upwards arrow, e.g, $CO_2\uparrow$, and $\rightleftharpoons$ means in equilibrium with.

1.3 ATOMIC THEORY

Chemists of the 18th century had to account for the relationship between the masses of reacting substances observed in nature, that always produced compounds containing elements in fixed proportions. For **example, calcium carbonate** (pure calcite or chalk) always consists of 40% calcium (Ca), 12% carbon (C) and 48% oxygen (O). The compounds formed by reactions were stable in the environment; they would not react further to produce new compounds unless exposed to other reactive compounds or elements. All the calculations on reacting masses and volumes of chemicals are based on the idea that each chemical element has a characteristic **atomic mass**. In 1808, the British chemist Dalton formulated his **atomic theory**. He postulated that all matter consists of **atoms** (from the Greek *atmos*=invisible), minute particles that cannot be created, or destroyed or split (this definition has since been modified following the nuclear fission (splitting of the atom) experiments of the early 20th century). Dalton theorised that all the atoms of an element are identical in every respect including, for example, their mass (but see **Isotopes, 1.3.3**). Hence the proportions of each element present in any compound can be calculated from the chemical formula of the compound using the atomic masses of each element present (recorded in the Periodic Table, **Table 1.2**). For example, calcium carbonate, $CaCO_3$ contains one calcium atom (atomic mass 40), one carbon atom (atomic mass 12) and 3 oxygen atoms (each having atomic mass 16). The overall atomic mass of calcium carbonate is therefore 100, and the percentage proportion by mass of calcium is 40%, carbon is 12% and oxygen is 48%.

Of course atomic theory must also be able to explain the crystal structures observed in nature. This is achieved by considering the atom as a sphere of constant radius, which has the advantage that there is a **constant ratio of free space** when equally sized spheres are packed as close as possible. The free space ratio is defined as

$$\frac{\text{Defined volume of space - Volume of atoms in a defined volume of space}}{\text{Defined volume of space}}$$

Unfortunately, however, this means that the modelling approach is more complex; for calcium carbonate, $CaCO_3$, we now have to fit one calcium (Ca) atom, one carbon (C) atom and three oxygen (O) atoms together and still form the crystal structure that results for this compound in nature. We consider how volumes of space are defined for solid structures in section **1.15.3** using **unit cells**. With atoms as spheres of constant radius packed as tightly as possible, the percentage free space can be calculated as 26% (**Table 1.12**). Chemical compounds comprise atoms and ions of different elements having different radii and therefore the ratio of free space will vary for different chemical compounds (**1.15.2**). Clearly spheres of different sizes can pack together to give a lower percentage of free space than spheres of the same size. In calcium carbonate, for example, the carbon (C) and oxygen (O) atoms combine to form **radicals** (CO_3^{2-}), which pack as closely as possible with the calcium ions (Ca^{2+}) to give stability.

Note that the volume of free space in a solid is **very** small; it cannot allow a water molecule (approximately 3 nm diameter) to enter, for example. However, in certain solid structures, chemically bound water, known as **water of crystallisation**, forms part of the solid structure (and therefore does **not** act like free water) (**1.18.5**). An example is gypsum, $CaSO_4.2H_2O$, which contains two molecules of water of crystallisation. Anhydrous gypsum, $CaSO_4$, is unstable in the environment and will react with water to become stable. This instability results from the large size difference between the small Ca^{2+} ion and large SO_4^{2-} ion; water of crystallisation effectively increases the size of the Ca^{2+} ion so that the compound is stable.

1.3.1 Structure of the Atom

In the late 19th century, scientists began to find evidence that atoms are composed of smaller particles. Crookes (1985) carried out experiments on the discharge of electricity of gases at low pressure and found that a beam of rays was given off by the cathode (the negative electrode). Crookes named these rays **cathode rays**. In 1898, Thompson studied the deflection of cathode rays in electric and magnetic fields, expressing his results as the ratio of charge/mass. Regardless of the gas and types of electrode used, the value of charge/mass was always the same, leading Thompson to deduce that these negatively charged particles were present in all matter. The particles were subsequently named **electrons** and were recognised as the particles of which electric current is composed and the means by which electric currents are conducted in metallic conductors, for example. Experiments by Millikan from 1909 to 1917 indicated that the mass of an electron is 1/1840 times that of a hydrogen atom (or a proton or neutron). The electron configuration of atoms is considered in section **1.3.2**. Electrons are the sub-atomic particles that govern the chemical reactivity of the element.

In 1911, Rutherford analysed the results of experiments on the scattering of charged helium nuclei (α-particles). The angular dependence of the scattering of these particles when passed through different materials could be explained only by postulating that the atom comprised a central positively charged **nucleus** surrounded by negative charges, with a diameter of the order 10^{-4} to 10^{-5} of that of an atom. In a chemical reaction, the atoms that make up the reactants enter into different combinations to form products, but the nuclei of the atoms remain unchanged; however, in certain reactions (nuclear reactions), rearrangement of sub-nuclear particles takes place, leading to the formation of new elements and the emission of radiation (**1.20.1**).

In 1913, Moseley suggested that the multiple charge on the nucleus arose from the presence of **protons**. The **number of protons** in the nucleus of an atom is called the **atomic number** (or **proton number**) of the element; this determines the **type of atom**, its **chemistry** and **nomenclature** and provides a means to classify chemical elements in to the Periodic Table (**Table 1.2**). Since atoms are electrically neutral, the number of electrons must be the same as the number of protons; hence the number of electrons increases with the atomic number (i.e., with the number of protons).

The atomic masses of elements are greater than the mass of the protons in the atom. To make up the extra mass, the existence of **neutrons** was postulated, particles with the same mass as the proton and zero charge. Chadwick finally established the presence of the neutron in 1934 by demonstrating that beryllium emitted uncharged particles when it was bombarded with helium nuclei (α-particles). The **number of protons plus**

neutrons is called the **atomic mass (number)** of the element. Neutrons account for the presence of **isotopes** of some elements (**1.3.3**).

Rutherford's model of the atom is rather similar to the modelling of the world. The central entity of the atom is the **nucleus** (equivalent to the Sun). The nucleus of the atom contains particles of mass number 1 that are positively charged. These are the **protons** (p). The atomic nucleus contains another particle of mass number 1 that has no charge. This is the **neutron** (n). The atom also contains particles of mass much smaller than the mass of the proton and neutron and of negative charge, the **electrons** (e). The electrons revolve around the nucleus in defined energy orbits (called **shells**) rather like the planets (Jupiter, Mars, etc.) revolve around the sun (***Figure 1.2***).

In the present day, scientists believe the structure of matter is a lot more complicated than merely being composed of protons, neutrons and electrons, and have proposed that atoms contain a range of sub-atomic particles (e.g, **quarks**, thought to be the 'building blocks' of protons and neutrons). ***Figure 1.2*** illustrates the relative size and properties of atomic and sub-atomic particles; the smallest living thing known to man is the virus (10^{-7} m in diameter) compared to the nucleus of an atom (10^{-14} m in diameter). Note the there is a large difference between the diameter of an atom (10^{-10} m) and that of the nucleus only (10^{-14} m), indicating that a large proportion of the space is occupied by the electrons. In this book we shall only be concerned with atoms, modelled as hard spheres approximately 10^{-10} m in diameter, composed of protons, neutrons and electrons.

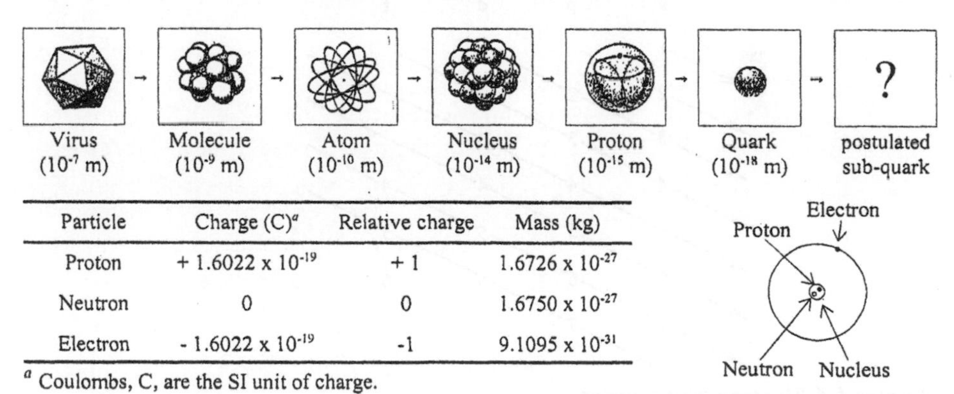

Particle	Charge (C)[a]	Relative charge	Mass (kg)
Proton	+ 1.6022×10^{-19}	+ 1	1.6726×10^{-27}
Neutron	0	0	1.6750×10^{-27}
Electron	- 1.6022×10^{-19}	-1	9.1095×10^{-31}

[a] Coulombs, C, are the SI unit of charge.

***Figure 1.2** Relative size, structure and properties of atomic and sub-atomic particles*

1.3.2 Electron Configuration of Atoms

The theory of the electron configuration of atoms is called **quantum theory**, and was developed in the early 1900s by Planck, Bohr, Sommerfield and others. A full description of quantum theory is outside the scope of this text and only a simplified explanation is included here. Electrons can be modelled as **particles** or **waves**. For our purposes, it is easier to visualise electron particles moving in an orbit around the nucleus of an atom. These electrons can be categorised according to the amount of energy they possess and the shape of their orbit. The term **shell** is used to classify the energy level.

An atom can contain up to 7 shells, numbered 1 to 7, where 1 represents the lowest energy level (nearest the nucleus). Each shell can be divided into **subshells**, which are assigned the letters s, p, d and f. The overall representation is shown in ***Figure 1.3***.

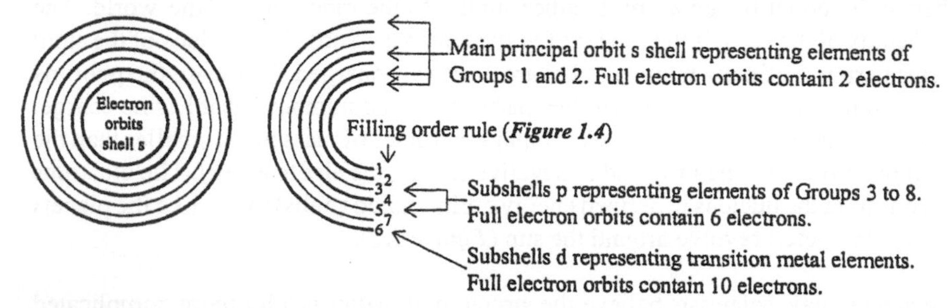

***Figure 1.3** Representation of electron orbits*

The maximum number of electrons accommodated within each subshell is two in the s subshell, six in the p subshell, ten in the d subshell and fourteen in the f subshell.

To calculate the arrangement of electrons in an atom, it must be appreciated that each added electron takes the position or level that **minimises the energy of the atom as a whole**. Thus shells and subshells are filled according to a specific order (***Figure 1.4***).

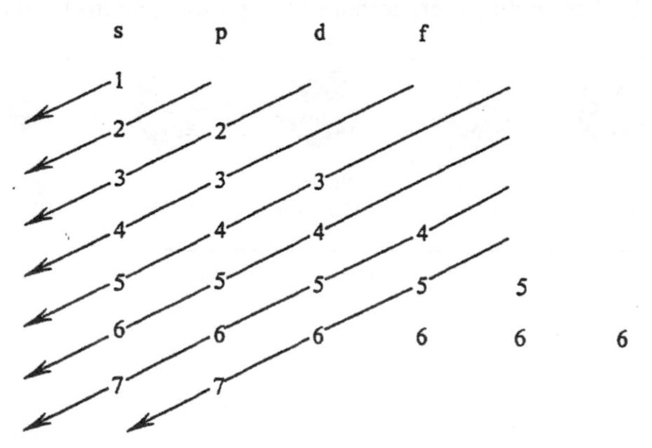

***Figure 1.4** Electron filling order*

In ***Figure 1.4***, 1 appears once, 2 appears twice, and so on. Thus shell 1 contains 1 subshell, shell 2 contains 2 subshells (s and p), and so on. The letters s, p, d and f read across and the numbers read down. To obtain the filling order, read diagonally downwards from right to left (following the arrows). Hence the filling order is 1s, 2s, 2p, 3s, 3p, 4s, 3d, 4p, 5s, 4d, 5p, 6s, 4f, 5d, 6p, 7s, etc. By convention, a number of electrons in each subshell ≥2 is denoted using a superscript. For example, the electron configuration of oxygen, which has 8 electrons (**Table 1.3**), would be $1s^2 2s^2 2p^4$.

Modelling electron orbits in shells and subshells must be able to explain the chemistry of materials observed in nature, including the reactivity of elements (**1.6**); differences between metals and non metals (**1.8.9**); the presence of the transition elements and the rare earth elements (**1.8.8**); the different alkalinity and acidity of elements, including elements that exhibit both acid and alkaline (**amphoteric**) properties (**1.5**).

1.3.3 Isotopes

Some elements have atomic masses that are not whole numbers; this is because they consist of a mixture of **isotopes.** Isotopes are variations of the same element, having the same number of protons but a different number of neutrons. Since chemical properties depend upon the number and structure of electrons, changes in atomic mass have little effect on chemical reactivity; therefore isotopes have the same chemical behaviour. Our modelling suggests that the lightest atom known, hydrogen (H), consists of one proton and one electron only (***Figure 1.5a***). To model deuterium (D) (***Figure 1.5b***), which is chemically the same as hydrogen, then we have one proton, **one neutron,** and one electron, hence deuterium is an isotope of hydrogen. Tritium (T) is another isotope of hydrogen (***Figure 1.5c***) and has the same chemical properties as hydrogen. Tritium is modelled as one proton, **two neutrons** and one electron (see also **1.20**).

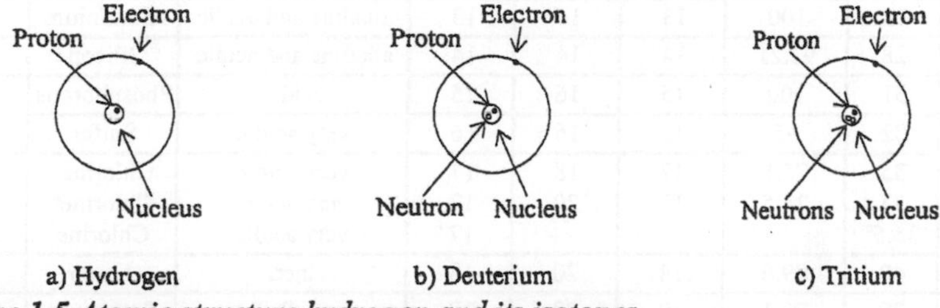

Figure 1.5 Atomic structure hydrogen and its isotopes

1.4 CHARACTERISTICS OF CHEMICAL ELEMENTS

Our modelling suggests that the lightest atom known, hydrogen (H), consists of one proton and one electron only. The next heaviest atom is helium (He), which has an atomic number of 2 and an atomic weight of 4. In our modelling this is accounted for by 2 protons and 2 neutrons (making up an atomic mass of 4) and 2 electrons to balance the charge of the 2 protons. This element, which is a gas, is chemically inert and so our model reflects this lack of chemical activity by stating that this is a **stable electron configuration** that cannot take part in any further chemical reaction. This "stability" is ascribed to the electron state ($1s^2$) (**1.3.2**).

The next heaviest atom is lithium (Li), which has an atomic number of 3 and an atomic mass of 6. In our modelling this is accounted for by 3 protons and 3 neutrons and 3 electrons to balance the 3 protons. Chemically this element, which is a solid and a metal, is very reactive and so our model reflects this by stating that this is an unstable electron configuration that takes a further part in some chemical reaction. This element has distinctive alkaline reactions (**1.8.1**). Subsequent elements can also be built up in a similar way (**Table 1.1**).

Table 1.1 List of chemical elements in order of increasing atomic mass

Atomic number	Atomic mass	Isotope abundance (%)	Number of			Chemical reactivity of oxide or element	Chemical name	Abbrev.
			Protons	Neutrons	Electrons			
1	1	99.98	1	0	1	Acid	Hydrogen	H
1	2	0.015	1	1	1	Acid	Deuterium[b]	D
1	3		1	2	1	Acid	Tritium[b]	T
2	4	100	2	2	2	Inert	Helium	He
3	6	92.58	3	3	3	Alkaline	Lithium	Li
4	7	100	4	3	4	less alkaline	Beryllium	Be
5	11	80.4	5	6	5	even less alkaline	Boron	B
6	12	98.89	6	6	6	slightly acidic	Carbon	C
7	14	99.63	7	7	7	more acidic	Nitrogen	N
8	16	99.76	8	8	8		Oxygen	O
9	19	100	9	10	9	very acidic	Fluorine	F
10	20	90.92	10	10	10	Inert	Neon	Ne
11	23	100	11	12	11	very alkaline	Sodium	Na[d]
12	24	78.7	12	12	12	less alkaline	Magnesium	Mg
13	27	100	13	14	13	alkaline and acidic	Aluminium	Al
14	28	92.21	14	14	14	alkaline and acidic	Silicon	Si
15	31	100	15	16	15	acidic	Phosphorous	P
16	32	95	16	16	16	very acidic	Sulfur	S
17	35	75.5	17	18	17	very acidic	Chlorine[c]	Cl
17	37	24.5	17	20	17	very acidic	Chlorine[c]	Cl
17	35.5[a]		[a]	[a]	17	very acidic	Chlorine	Cl
18	40	99.6	18	20	18	Inert	Argon	Ar
19	39	93.1	19	20	19	vary alkaline	Potassium	K[e]
20	40	96.97	20	20	20	less alkaline	Calcium	Ca

Notes: [a] Chlorine gas is composed of both isotopes in proportions 75% Cl^{35} and 25% Cl^{37} to give an average atomic mass of 35.5. Unlike hydrogen, isotopes of chlorine do not have separate names; [b] isotopes of hydrogen (they have the same chemistry as hydrogen, but are of different atomic mass); [c] isotopes of chlorine. [d] from the Latin *natrium*; [e] from the Latin *kalium*.

Table 1.1 lists the first 20 elements in order of the number of protons (atomic number). The list could be continued through the entire range of elèments up to an atomic number of 92 (for natural elements). Each element owes its individuality to the number of protons in the atom, whilst the chemical behaviour of the element is dependent upon its electron configuration. The important points to note in **Table 1.1** are

- Each atom differs from its neighbour by the **number of protons** it contains;
- The **atomic number** gives the **number of protons** within the atom. As the atom has no overall charge, the number of protons equals the number of electrons. The **atomic number** therefore also gives the **number of electrons** within the atom;
- A change in the number of protons gives a new **chemical element**. Note this is not so with a change in either the number of neutrons or electrons;
- The neutron is responsible for the number of **isotopes** and the **fractional atomic weights** (e.g, chlorine, **1.3.3**);
- A gradual change from alkalinity to acidity as the **number of protons increases**;
- An abrupt change from acidity to inertness to strong alkalinity at certain **atomic numbers** (for example, atomic numbers 2, 10 and 18). The inert gases separate the highly acidic elements from the highly alkaline elements (***Figure 1.6***).

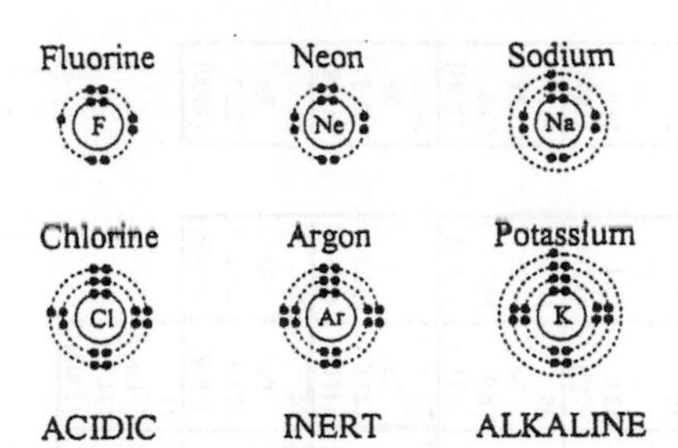

Figure 1.6 *Separation of highly acidic and alkaline elements by the inert gases*

In ***Figure 1.6***, the inert gases are made the centre of the figure and separate the highly acidic group 7 elements from the highly alkaline group 1 elements. As we will see (**1.8**), sodium (Na) and chlorine (Cl) are reactive elements; they react to produce a neutral salt, sodium chloride (NaCl). This is achieved through the formation of an **ionic bond** (**1.12**) in which the sodium atom donates its outer electron to the chlorine; sodium (Na^+) and chlorine (Cl^-) **ions** are formed (**1.18**). The sodium ion has the stable inert gas configuration of neon and is smaller than the sodium atom; the chlorine ion has the stable inert gas configuration of argon and is larger than the chlorine atom (**Table 1.5**).

Newlands (1864) is credited with first ordering chemical elements according to weight. With the knowledge of 63 elements known at that time, he noted that if these elements were tabulated in order of increasing atomic weight, the properties of the first seven elements reappeared in the second seven. He placed similar chemical elements beneath one another in vertical columns (**groups**) and named the classification the **law of octaves**. In 1869, the Russian scientist Mendeleev (1830-1907) developed a more elaborate and systematic representation of the elements, known as the **periodic law**, that included horizontal rows (**periods**). As some elements did not fit into this order, gaps were left for elements yet to be discovered. The fact that this ordering gave scientists new elements to look for was one of the most powerful applications of the periodic law. One new element was "eka silicon" (Greek *eka*=the same as), which we now call germanium (Ge), discovered in 1886. Others included gallium (Ga), scandium (Sc) and rhenium (Rh), which were not discovered until 20 years later. The period law eventually developed into the **Periodic Table** of elements as we know it today.

1.5 PERIODIC TABLE OF ELEMENTS

The Periodic Table of elements is shown in **Table 1.2** (over). The strongly alkaline elements (K, Ca) are to the left of the table (in groups 1 and 2) and the acidic elements (As, Se) are to the right (in groups 5, 6 and 7). The inert gases are all beneath one other in group 8 (He, Ne, etc.), whilst the metallic elements are grouped to the left and the non metals are grouped to the right. Between the metals and the non metals are a group of elements whose chemistry is both acidic and alkaline in nature (i.e., they are **amphoteric**) e.g, aluminium and silicon form both acidic oxides (alumin**ates** and silic**ates**) and alkali oxides (alumin**ium** and silic**on** compounds). Similar properties exist **diagonally** across the periodic table, e.g, Li, Mg and Ga have similar chemical properties, as do Be and Al, as do B, Si and As.

Table 1.2 Periodic Table of elements

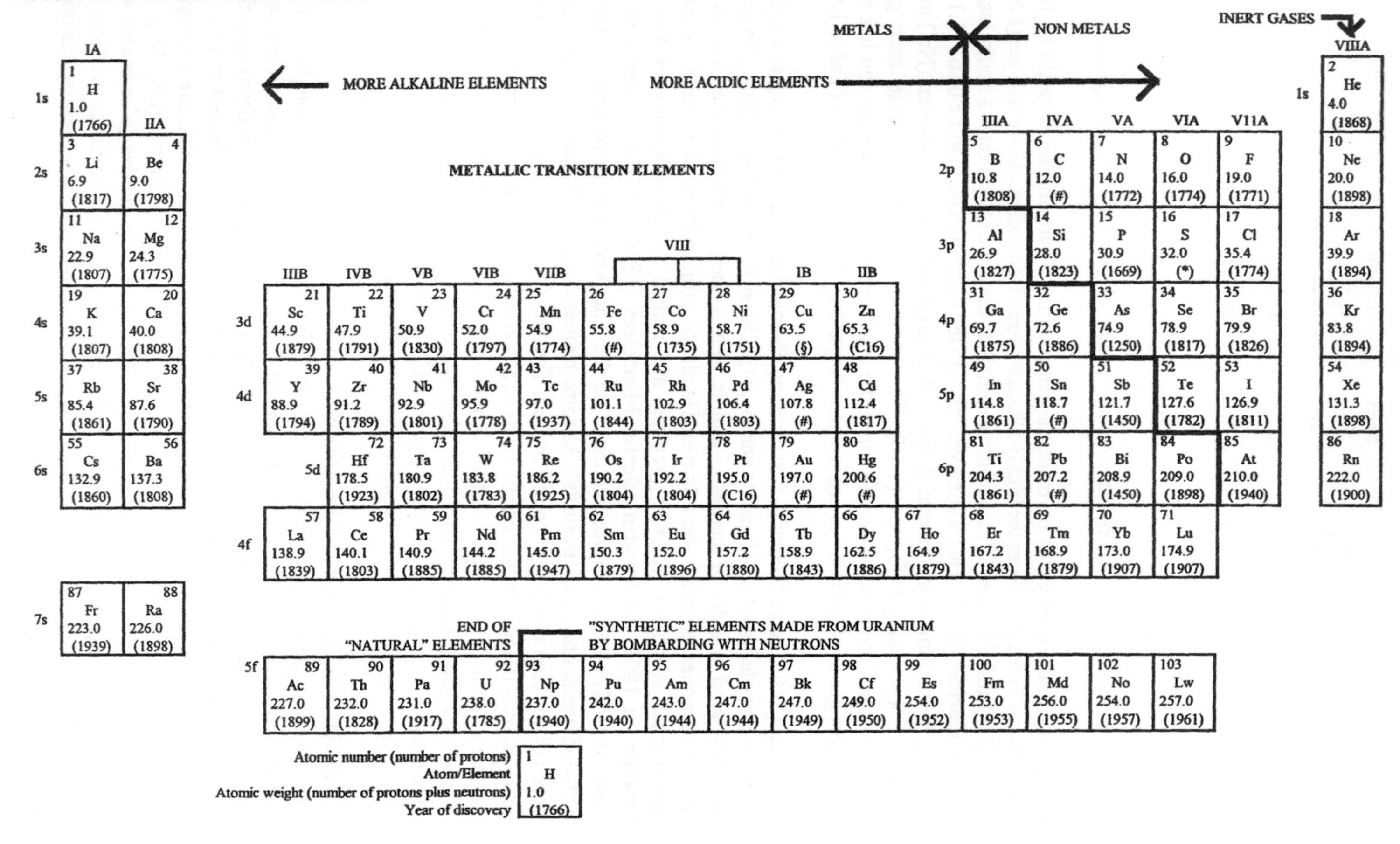

	IA	IIA		IIIB	IVB	VB	VIB	VIIB	VIII	VIII	VIII	IB	IIB		IIIA	IVA	VA	VIA	V11A		VIIIA
1s	1 H 1.0 (1766)																			1s	2 He 4.0 (1868)
2s	3 Li 6.9 (1817)	4 Be 9.0 (1798)												2p	5 B 10.8 (1808)	6 C 12.0 (#)	7 N 14.0 (1772)	8 O 16.0 (1774)	9 F 19.0 (1771)		10 Ne 20.0 (1898)
3s	11 Na 22.9 (1807)	12 Mg 24.3 (1775)												3p	13 Al 26.9 (1827)	14 Si 28.0 (1823)	15 P 30.9 (1669)	16 S 32.0 (*)	17 Cl 35.4 (1774)		18 Ar 39.9 (1894)
4s	19 K 39.1 (1807)	20 Ca 40.0 (1808)	3d	21 Sc 44.9 (1879)	22 Ti 47.9 (1791)	23 V 50.9 (1830)	24 Cr 52.0 (1797)	25 Mn 54.9 (1774)	26 Fe 55.8 (#)	27 Co 58.9 (1735)	28 Ni 58.7 (1751)	29 Cu 63.5 (§)	30 Zn 65.3 (C16)	4p	31 Ga 69.7 (1875)	32 Ge 72.6 (1886)	33 As 74.9 (1250)	34 Se 78.9 (1817)	35 Br 79.9 (1826)		36 Kr 83.8 (1894)
5s	37 Rb 85.4 (1861)	38 Sr 87.6 (1790)	4d	39 Y 88.9 (1794)	40 Zr 91.2 (1789)	41 Nb 92.9 (1801)	42 Mo 95.9 (1778)	43 Tc 97.0 (1937)	44 Ru 101.1 (1844)	45 Rh 102.9 (1803)	46 Pd 106.4 (1803)	47 Ag 107.8 (#)	48 Cd 112.4 (1817)	5p	49 In 114.8 (1861)	50 Sn 118.7 (#)	51 Sb 121.7 (1450)	52 Te 127.6 (1782)	53 I 126.9 (1811)		54 Xe 131.3 (1898)
6s	55 Cs 132.9 (1860)	56 Ba 137.3 (1808)	5d		72 Hf 178.5 (1923)	73 Ta 180.9 (1802)	74 W 183.8 (1783)	75 Re 186.2 (1925)	76 Os 190.2 (1804)	77 Ir 192.2 (1804)	78 Pt 195.0 (C16)	79 Au 197.0 (#)	80 Hg 200.6 (#)	6p	81 Ti 204.3 (1861)	82 Pb 207.2 (#)	83 Bi 208.9 (1450)	84 Po 209.0 (1898)	85 At 210.0 (1940)		86 Rn 222.0 (1900)
7s	87 Fr 223.0 (1939)	88 Ra 226.0 (1898)																			

4f	57 La 138.9 (1839)	58 Ce 140.1 (1803)	59 Pr 140.9 (1885)	60 Nd 144.2 (1885)	61 Pm 145.0 (1947)	62 Sm 150.3 (1879)	63 Eu 152.0 (1896)	64 Gd 157.2 (1880)	65 Tb 158.9 (1843)	66 Dy 162.5 (1886)	67 Ho 164.9 (1879)	68 Er 167.2 (1843)	69 Tm 168.9 (1879)	70 Yb 173.0 (1907)	71 Lu 174.9 (1907)

END OF "NATURAL" ELEMENTS — "SYNTHETIC" ELEMENTS MADE FROM URANIUM BY BOMBARDING WITH NEUTRONS

5f	89 Ac 227.0 (1899)	90 Th 232.0 (1828)	91 Pa 231.0 (1917)	92 U 238.0 (1785)	93 Np 237.0 (1940)	94 Pu 242.0 (1940)	95 Am 243.0 (1944)	96 Cm 247.0 (1944)	97 Bk 247.0 (1949)	98 Cf 249.0 (1950)	99 Es 254.0 (1952)	100 Fm 253.0 (1953)	101 Md 256.0 (1955)	102 No 254.0 (1957)	103 Lw 257.0 (1961)

Key	
Atomic number (number of protons)	1
Atom/Element	H
Atomic weight (number of protons plus neutrons)	1.0
Year of discovery	(1766)

Elements of different groups show different degrees of stability. Less stable elements will try to react with other elements to form a stable state. Since the only elements that are entirely stable in the environment are the inert gases of group 8, we model the instability of all other elements by comparing their electron configurations with those of the inert gases. In all cases, it is **only the outer electrons** that take part in chemical reactions and the inert gases have an stable electron configuration of 8 outer electrons.

The **valency** of an element is the number of atoms of hydrogen with which it will combine or which it will replace. In the case of elements that form **ionic bonds (1.12)**, the valency is an expression of the ease at which an element gains or loses an electron. In the case of elements that form **covalent bonds (1.11)**, the valency is the number of monovalent atoms (or half the number of divalent atoms, etc.) required as near neighbours to achieve a stable inert gas configuration. By definition, the valency of the inert gases is **zero**. The number of external electrons responsible for the valency is obtained from the group number

- for elements of groups 1 to 6 the valency is the group number;
- for elements of groups 5 to 8, the valency is 8 minus the group number (8-N).

Group 1 elements are monovalent (Li, etc.), as are Group 7 elements (8-7=1) (F, etc.). Group 2 elements are divalent (Mg, etc.), as are Group 6 elements (8-6=2) (O, etc.). Group 5 and 6 elements have two valency states. Group 5 elements have valencies of 5 and 3, e.g, nitrogen (N) can act with a valency of 5 to form nitric acid (HNO_3) or a valency of 3 (8-5=3) to form covalent bonds, as in ammonia (NH_3) and aluminium nitride (AlN). Group 6 elements valencies of 6 and 2. The characteristics of the elements in each group are considered in section **1.8.**

1.6 ELECTRON CONFIGURATION OF ELEMENTS

The electron configuration of the elements of the first three periods of the Periodic table is shown in **Table 1.3**. The electrons of hydrogen and helium (first period) fill the first shell (1), the electrons of the elements lithium to neon (second period) fill the second shell (2), the electrons of the elements sodium to argon (third period) contribute to, but do not fill, the third shell (3). The electrons of the first elements of the fourth period contribute to the fourth shell (4) rather than completing the third shell (in subshell 3d). Writing the electron configuration of argon (Ar) as $1s^22s^22p^63s^23p^6$ in accordance with the filling order (**1.3.2**), these elements are K (atomic number 19): (Ar)4s to Ca (atomic number 20): (Ar)$4s^2$. Once the 4s subshell is full, electrons of the next 10 elements enter the 3d subshell. These elements are Sc (atomic number 21): (Ar)$4s^23d$ to Zn (atomic number 30): (Ar)$4s^23d^{10}$. This means that the $4s^2$ electrons are acting as a 'cloak' whilst the inner 3d subshell is being filled. Since only outer electrons contribute to chemical reactivity, the additional electrons in the 3d subshell do not greatly affect the chemistry of the elements.

Thus the metals scandium (Sc) to zinc (Zn) are all similar, and form the first set of **transition metallic elements**. A second set of transition metals are included in the fifth period (yttrium, Y, to cadmium, Cd), where the 4d subshell fills after the 5s subshell. The sixth period includes the first set of **rare earth metals**, from lanthanum (La) to lutetium (Lu), which are even more similar to one another than the transition metals. The elements lutetium (Lu) to mercury (Hg) comprise a third transition series in which the 5d subshell fills after the 6s subshell. The seventh and final period starts at francium (Fr); the last naturally occurring element is uranium (U). Each subshell can be defined by the elements that start and complete it, i.e., $1s^{H\text{-}He}$; $2s^{Li\text{-}Be}$ $2p^{B\text{-}Ne}$; $3s^{Na\text{-}Mg}$ $3p^{Al\text{-}Ar}$ $3d^{Sc\text{-}Zn}$ $4s^{K\text{-}Ca}$ $4p^{Ga\text{-}Kr}$ $4d^{Y\text{-}Cd}$ $4f^{La\text{-}Yb}$; $5s^{Rb\text{-}Sr}$ $5p^{In\text{-}Xe}$ $5d^{Lu\text{-}Hg}$ $5f^{Th\text{-}U\text{-}Lr}$; $6s^{Cs\text{-}Ba}$ $6p^{Tl\text{-}Rn}$; $7s^{Fr\text{-}Ra}$.

Table 1.3 Electron configuration of elements in the Periodic table

Period	Group		Metallic transition elements																Electron configuration of inert gas (group 8)[a]							
	1	2											3	4	5	6	7	8	1s	2s	2p	3s	3p	3d	4s	4p
1	H																	He	2							
2	Li	B											B	C	N	O	F	Ne	2	2	6					
3	Na	Mg											Al	Si	P	S	Cl	Ar	2	2	6	2	6			
4	K	Ca	Sc	Ti	V	Cr	Mr	Fe	Co	Ni	Cu	Zn	Ga	Ce	As	Se	Br	Kr	2	2	6	2	6	10	2	6
Electron subshells filled in fourth period	4s		3d										4p													

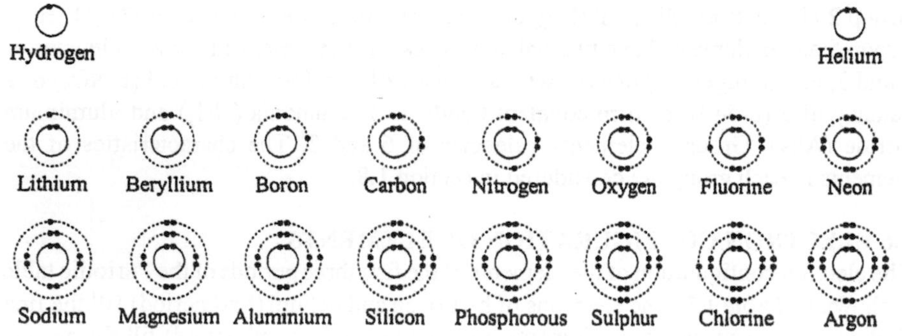

Hydrogen Helium

Lithium Beryllium Boron Carbon Nitrogen Oxygen Fluorine Neon

Sodium Magnesium Aluminium Silicon Phosphorous Sulphur Chlorine Argon

Notes: [a] subshell 4d and 4f not shown

1.7 TRENDS ACROSS A PERIOD IN THE PERIODIC TABLE

Bonding arrangements of elements, the nature (acid/alkali) of compounds formed and their solubility in relation to their position in the Periodic table is shown in **Table 1.4**.

The strength of the ionic bond decreases across the Periodic Table, as illustrated by the stability of compounds to heat. Taking carbonates as an example, **Table 1.4** illustrates that all Group 1 salts are stable and are not decomposed by heat, whilst Group 2 compounds are decomposed. Group 3 elements do not form salts with weak acids. There is a also general change from alkalinity to acidity from group 1 to group 8, i.e.,

Group 1	Group 2	Group 3	Group 4	...	Group 7	Group 8
very alkaline	alkaline	less alkaline	amphoteric	...	acidic	inert

Table 1.4 Characteristics of bonds and compounds formed in relation to position of elements in the Periodic Table

Period	Group 1	2	Metallic transition elements										3	4	5	6	7	8
1	H																	He
2	Li	B											B	C	N	O	F	Ne
3	Na	Mg											Al	Si	P	S	Cl	Ar
4	K	Ca	Sc	Ti	V	Cr	Mr	Fe	Co	Ni	Cu	Zn	Ga	Ce	As	Se	Br	Kr
	s block		d block										p block					

	s block	d block	p block
Summary of bonding in oxides, chlorides and hydrides	Ionic oxides, chlorides, hydrides	Oxides, chlorides ionic with large degree of covalent character, or covalent. Structures layered or macromolecular. Hydrides interstitial.	Covalent molecular chlorides, oxides, hydrides.
Summary of acidic and basic compounds	Oxides, hydrides give alkaline solutions. Chlorides give neutral solutions.	Oxides, hydroxides basic and insoluble. Some amphoteric. Chlorides hydrolysed to give acid solutions.	Oxides, chlorides and most hydrides are acidic (except NH_3, CH_4, etc.)
Summary of the solubility of salts and complex ion formation	Form cations. Group 1 salts all soluble. Group 2 salts vary. No complex salts.	Form variable valency cations. Oxides, hydroxides, many chlorides insoluble. Cations form complex ions.	Form anions. Anions act as electron donors in complex ion formation.
Summary of carbonates	Strong carbonates. Group 1[a] not heat decomposed; Group 2[b] decomposed	Weak, generally basic, carbonates formed.	No carbonates formed.[c]

Notes: [a] e.g, sodium carbonate, Na_2CO_3 (washing soda); [b] e.g, calcium carbonate, $CaCO_3$ (chalk) decomposed to lime (CaO) and carbon dioxide (CO_2) at 400°C; magnesium carbonate, $MgCO_3$ (dolomite) decomposed to magnesium oxide (MgO) and carbon dioxide (CO_2) at 400°C.; [c] the fact that aluminium carbonate does not exist is important where water proofing compounds (aluminium stearates) are decomposed by carbon dioxide to form an alumina gel and (calcium) stearate, the calcium being obtained from mortar or bricks (**5.8.1**)

Finally, covalent and ionic radii both decrease from left to right across any period, as shown in **Table 1.5** (over). In the second period (Li to F), the nuclear charge increases from 3 to 9. As the nuclear charge increases, it pulls the electrons of the outer shell 1 closer to the nucleus, and the radius of that shell decreases. The effect on the electrons of shell 2 is complicated by the fact that they are cloaked by shell 1 (**1.3.2**), so that the effective nuclear charge is less than the actual nuclear charge. For example, in lithium the outermost electron is attracted by the nucleus with a charge of +3 screened by two electrons. The net nuclear charge is closer to +1 than +3. In beryllium, the shell 2 electrons are attracted by a nucleus that has a charge of +4, and is screened by two electrons, which makes the effective charge closer to +2. Nevertheless, the effective nuclear charge increases from left to right across a period, causing a steady decrease in the atomic radius across the period. A comparison of ions with the same numerical charge (M^{2+}, etc.) shows that ionic radii follow the same pattern.

Table 1.5 Ionic and atomic radii (nm)

	1	2	3	4	5	6	7	8			0
1							H^{-} 0.154 H 0.046				He
2	Li 0.152 Li^{+} 0.078	Be 0.112 Be^{2+} 0.034	B 0.097	C 0.077 C^{4+} 0.02	N 0.071 N^{5+} .01-.02	O^{2-} 0.132 O 0.060	F^{-} 0.133				Ne 0.160
3	Na 0.186 Na^{+} 0.098	Mg 0.160 Mg^{2+} 0.078	Al 0.143 Al^{3+} 0.057	Si^{4-} 0.198 Si 0.117 Si^{4+} 0.039	P P^{6+} .03-.04	S^{2-} 0.174 S 0.104 S^{4+} 0.034	Cl^{-} 0.181 Cl 0.107				Ar 0.191
4	K 0.231 K^{+} 0.133 (NH_4^{+} 0.143)	Ca 0.196 Ca^{2+} 0.106	Sc 0.151 Sc^{3+} 0.069	Ti 0.146 Ti^{3+} 0.083 Ti^{4+} 0.064	V 0.130 V^{3+} 0.065 V^{4+} 0.061 V^{5+} ca.0.04	Cr 0.125 Cr^{3+} 0.064 Cr^{6+} .03-.04	Mn 0.118 Mn^{2+} 0.091 Mn^{3+} 0.070 Mn^{4+} 0.052	Fe 0.124 Fe^{2+} 0.083 Fe^{3+} 0.067	Co 0.125 Co^{2+} 0.082	Ni 0.124 Ni^{2+} 0.078	
	Cu 0.128 Cu^{+} 0.096	Zn 0.133 Zn^{2+} 0.083	Ga 0.122 Ga^{3+} 0.062	Ge 0.122 Ge^{4+} 0.044	As 0.125 As^{3+} 0.069 As^{5+} ca.0.04	Se^{2-} 0.191 Se 0.116 Se^{6+} .03-.04	Br^{-} 0.196 Br 0.119				Kr 0.201
5	Rb 0.243 Rb^{+} 0.149	Sr 0.215 Sr^{2+} 0.127	Y 0.181 Y^{3+} 0.106	Zr 0.156 Zr^{4+} 0.087	Nb 0.143 Nb^{4+} 0.069 Nb^{5+} 0.069	Mo 0.136 Mo^{4+} 0.068	Tc	Ru 0.133 Ru^{4+} 0.065	Rh 0.134 Rh^{4+} 0.065	Pd 0.137	
	Ag 0.144 Ag^{+} 0.133	Cd 0.149 Cd^{2+} 0.103	In 0.162 In^{3+} 0.092	Sn^{4-} 0.215 Sn 0.140 Sn^{4+} 0.074	Sb 0.145 Sb^{3+} 0.090	Te^{2-} 0.211 Te 0.143 Te^{4+} 0.089	I^{-} 0.220 I 0.136 I^{6+} 0.094				Xe 0.220
6	Cs 0.262 Cs^{+} 0.165	Ba 0.217 Ba^{2+} 0.143	La 0.186 La^{3+} 0.122	Ce 0.182 Ce^{3+} 0.118 Ce^{4+} 0.102	Pr 0.181 Pr^{3+} 0.116 Pr^{4+} 0.100	Nd 0.180 Nd^{3+} 0.115					
	Sm^{3+} 0.113	Eu^{3+} 0.113	Gd^{3+} 0.111	Tb^{3+} 0.109 Tb^{4+} 0.089	Dy^{3+} 0.107	Ho^{3+} 0.105	Er 0.186 Er^{3+} 0.104				
	Tm^{3+} 0.104	Yb^{3+} 0.100	Lu^{3+} 0.990	Hf 0.158 Hf^{4+} 0.084	Ta 0.143 Ta^{5+} 0.068	W 0.136 W^{4+} 0.068	Re	Os 0.135 Os^{4+} 0.067	Ir 0.135 Ir^{4+} 0.066	Pt 0.138	
	Au 0.144 Au^{+} 0.137	Hg 0.150 Ag^{3+} 0.112	Ti 0.170 Ti^{+} 0.149 Ti^{3+} 0.105	Pb^{4-} 0.215 Pb 0.175 Pb^{2+} 0.132 Pb^{4+} 0.084	Bi 0.155	Po	At				Rn
7	Fr	Ra	Ac	Th 0.180 Th^{4+} 0.110	Pa	U 0.138 U^{4+} 0.105					

Notes: The atomic radii are one half the distance of closest approach in the element. The ionic radii are those for six-fold coordination.

Trends in ionic radii and electronegativity in the Periodic Table are summarised in ***Figure 1.7***.

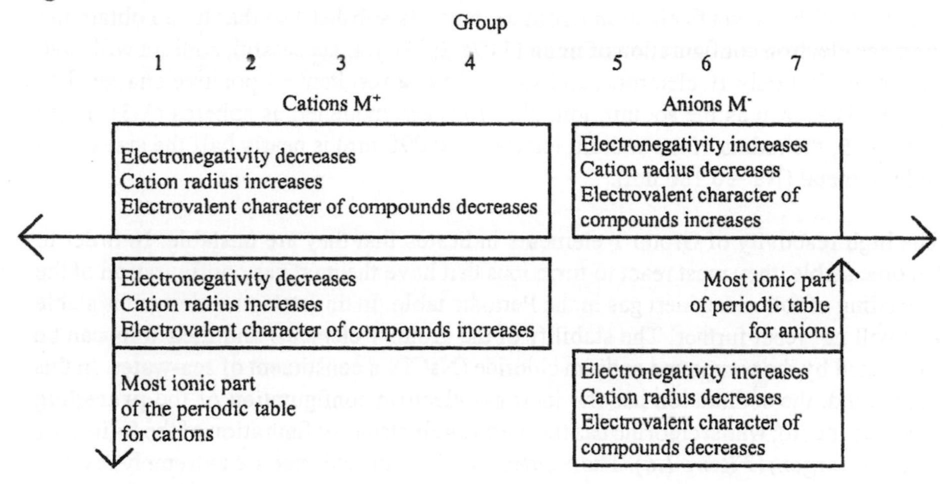

***Figure 1.7** Trends in ionic radii and electronegativity in the Periodic table*

Electropositive elements will show their most electrovalent behaviour at the top right corner of ***Figure 1.7***, where oxygen, fluorine and chlorine are located. The transition metals show a very small change in atomic radius across a series. In the first transition series, beginning with scandium, the size of the atoms is governed by the 3 shell, while the additional electrons are entering the two shell and are all of similar size. This similarity in size allows metals to form **substitutional solid solutions (1.16)**, important in producing **alloys (10.5.2)**.

1.8 TRENDS WITHIN A GROUP IN THE PERIODIC TABLE

In this section we consider the fundamental properties of elements within the Periodic table. To do this, we consider a number of elements as exemplars of each group. Those elements that have some application to the construction industry are denoted in bold typeface (i.e., **Na, K, Mg, Al, C, Si, Sn, Pd, N, O** and **S**).

1.8.1 Group 1 Elements (Li **Na K** Rb Cs)

Group 1 elements, sometimes called the **alkali metals**, are all monovalent (valency 1) (**1.5**). Denoting the element as M, they form M^+ ions. Group 1 elements are all metals. They form salts (carbonates, sulfates, nitrates, etc.) that are not decomposed by heat and only a few are insoluble. In general, group 1 elements become more metallic, more reactive and more alkaline as we go down the group. The increased reactivity of group 1 elements can be shown by comparing the heat emitted when sodium (Na) and potassium (K) react with water. Both reactions produce a lot of heat, but potassium produces sufficient heat to ignite the hydrogen liberated from the reaction, whereas sodium does not.

Sodium (Na) (electron configuration $1s^22s^22p^63s$)
The high reactivity of sodium is modelled by postulating that the sodium metal wants to get rid of the eleventh electron (from the outer 3s subshell) so that it can obtain the inert gas electron configuration of **neon** ($1s^22s^22p^63s^0$). If successful, sodium will have 11 protons but only 10 electrons, and so will have a resultant +1 positive charge. The product is known as the sodium ion, Na^+. Ions are modelled as spheres (**1.3**). From **Table 1.5**, the radius of the sodium ion ($Na^+ = 0.098$ nm) is nearly half the size of the sodium metal (Na = 0.186 nm).

The high reactivity of Group 1 elements indicates that they are unstable. In order to become stable, they must react to form ions that have the inert gas configuration of the preceding of following inert gas in the Periodic table. In this state, the element is stable and will not react further. The stability of the group 1 elements and their ions can be illustrated by the compound sodium chloride (NaCl), a constituent of sea-water. In this compound, the sodium ion has the inert gas electron configuration of the **preceding** inert gas (neon), whilst chlorine has the inert gas electron configuration of the **following** inert gas (argon) (***Figure 1.6***). The elements sodium and chlorine are extremely reactive (for example, sodium metal reacts vigorously with water and chlorine is highly toxic to humans), whereas when present as ions in the compound sodium chloride they are stable and form an essential component of the human diet. In general, elements present in the ionic form have completely different properties to those when present in the atomic form; the readiness of the atom to ionise is one of the parameters that gives rise to its reactivity. We can write the reaction for a metal element to produce its ions as

$$\underset{\text{(Metal)}}{M} \rightarrow \underset{\text{(ions)}}{M^{n+}} + \underset{\text{(electrons)}}{n\ \text{electrons (e)}}$$

where n is the valency (**1.5**) of the metal. Where the metal reacts with water, the metallic ions, M^{n+}, will be dissolved in the water. Hence for sodium reacting with water

$$\underset{\text{(sodium metal)}}{Na} \rightarrow \underset{\text{(sodium ions)}}{Na^{+}} + \underset{\text{(electrons)}}{1\ e}$$

We call the reaction of a metal with water to produce its ions an **anodic** reaction. In order for this reaction to proceed, the electrons produced by the anodic reaction must be consumed. This is achieved by a **cathodic** reaction, i.e.,

$$\underset{\text{(hydrogen ions)}}{2H^{+}} + \underset{\text{(2 electrons)}}{2\ e} \rightarrow \underset{\text{(hydrogen gas)}}{H_2}$$

Note that the hydrogen ions in the cathodic reaction are obtained from water (**1.19.1**)

$$\underset{\text{(water)}}{H_2O} \rightarrow \underset{\text{(hydrogen ions)}}{2H^{+}} + \underset{\text{(hydroxyl ions)}}{2OH^{-}}$$

Anodic and cathodic reactions are very important in the study of metallic corrosion, and we will return to these types of reactions in **Chapter 11**. We can write the overall reaction for sodium in water by combining the anodic and cathodic reactions as

Na	+	H_2O	→	Na^+	+	OH^-	+	H_2
(sodium metal)		(water)		(sodium ions)		(hydroxyl ions)		(hydrogen gas)

One important point about this representation of the reaction is that **matter cannot be created or destroyed**. This means that the number of elements on the right hand side of the chemical equation must equal the number of elements on the left hand side. We should always **balance** the chemical equation to ensure this. Thus

2Na	+	$2H_2O$	→	$2Na^+$	+	$2OH^-$	+	H_2
(sodium metal)		(water)		(sodium ions)		(hydroxyl ions)		(hydrogen gas)

In the above equation, the products of the reaction, $2Na^+ + 2OH^-$, are ionised. We can also write an equation for the products when not ionised, since these ions react to form a stable compound. This is the usual form in which chemical equations are expressed.

2Na	+	$2H_2O$	→	2NaOH	+	H_2
(sodium metal)		(water)		(sodium hydroxide)		(hydrogen gas)

Sodium hydroxide is soluble in water, and the aqueous solution is alkaline.

Potassium (K) (electron configuration $1s^2 2s^2 2p^6 3s^2 3p^6 4s^1$)
Potassium is a more highly reactive metal than sodium. To model this reactivity, we postulate that potassium metal wants to get rid of the outer nineteenth electron from the 4s subshell so it can obtain the inert gas configuration of **argon** ($1s^2 2s^2 2p^6 3s^2 3p^6 3d^0$). If successful, potassium will have 19 protons but only 18 electrons, and so will have a resultant +1 positive charge. The product is the potassium ion, K^+. Following the same procedure as for sodium (above), the overall balanced chemical equation for the reaction of potassium metal with water is

	2K	+	$2H_2O$	→	$2K^+$	+	$2OH^-$	+	H_2
or	2K	+	$2H_2O$	→		2KOH		+	H_2
	(potassium metal)		(water)			(potassium hydroxide)			(hydrogen gas)

Potassium hydroxide is soluble in water and the aqueous solution is **very** alkaline. By comparison with the reactions of sodium and potassium with water, elements become increasingly reactive and more alkaline as the group is descended. This is a characteristic of all groups to some extent.

1.8.2 Group 2 Elements (Be, **Mg**, **Ca**, Sr, **Ba**, Ra)

Group 2 elements, the **alkali earth metals**, are all divalent (valency 2), are all metals and form M^{2+} ions. Their salts (carbonates, sulfates, nitrates, etc.) are all decomposed by heat (indicating that they form weaker bonds than group 1 elements), and are not very soluble in water. In general, the group 2 elements become more reactive and more alkaline as we go down the group, but markedly less so than the group 1 elements. The group 2 elements all react with water, but not so readily as the group 1 elements.

Magnesium (Mg) (electron configuration $1s^2 2s^2 2p^6 3s^2$)
Magnesium is next to sodium in the first period, but is less reactive, evolving hydrogen

only slowly when placed in water. However, magnesium will react more quickly in acid solutions, when hydrogen is given off in abundance. The fact that magnesium is less reactive than sodium is modelled by postulating that the magnesium metal wants to get rid of the **two** outer electrons from the 3s subshell so that it can obtain the inert gas electron configuration of **neon** ($1s^22s^22p^63s^0$). More energy is required to remove two outer electrons (of magnesium) than is required to remove one outer electron (of sodium). As a result, magnesium is less reactive than sodium. If successful in removing the two outer electrons, magnesium will have 12 protons but only 10 electrons, and so will have a resultant +2 positive charge. The product is known as the magnesium ion, Mg^{2+}. The remaining 10 electrons are attracted more strongly to the 12 protons in the nucleus, and so the reduction in the radius (**Table 1.5**) of the magnesium atom (0.160 nm) to the magnesium ion (0.078 nm) is greater than that for sodium. As before, the overall balanced chemical equation for the reaction of magnesium metal with water is

$$Mg + 2H_2O \rightarrow Mg^{2+} + 2OH^- + H_2$$

or

$$\underset{\text{(magnesium metal)}}{Mg} + \underset{\text{(water)}}{2H_2O} \rightarrow \underset{\text{(magnesium hydroxide)}}{Mg(OH)_2} + \underset{\text{(hydrogen gas)}}{H_2}$$

The aqueous solution of magnesium hydroxide is only slightly alkaline. Note that the compound magnesium hydroxide, $Mg(OH)_2$, requires **two** hydroxyl (OH^-) ions to neutralise **one** magnesium (Mg^{2+}) ion.

1.8.3 Group 3 Elements (B, Al, Ga, In, Th)

With the exception of boron (B), group 3 elements are trivalent (valency 3), forming M^{3+} ions. With the exception of boron, they are all metals. Group 3 elements comprise the first of the acidic elements in the Periodic table. Oxides of boron are weakly acidic (some are used in eye washes), whilst the oxides of aluminium have both alkali and acidic properties (i.e., they are amphoteric).

Aluminium (Al) (electron configuration $1s^22s^22p^63s^23p$)

Aluminium is next to magnesium in the first period, but it is less reactive. This is modelled by postulating that the aluminium metal wants to get rid of the **three** outer electrons (from the 3s and 3p subshells) so that it can obtain the inert gas electron configuration of **neon** ($1s^22s^22p^6$). More energy is required to remove the three outer electrons of magnesium compared to that required to remove the outer electrons of sodium and magnesium. As a result, aluminium is less reactive than these elements. If successful in removing the three outer electrons, aluminium will have 13 protons but only 10 electrons, and so will have a resultant +3 positive charge. The product is known as the aluminium ion, Al^{3+}. The remaining 10 electrons are attracted more strongly to the 13 protons in the nucleus, and so the reduction in the radius (**Table 1.5**) of the aluminium atom (0.143 nm) to the aluminium ion (0.057 nm) is greater than that for both magnesium and sodium.

In general, the **smaller the radius of an ion, the more unstable the compound is forms.** This is particularly important for products based on the aluminium ion e.g, damp-proof compounds (**5.8.1**).

The true reactivity of aluminium is masked by the presence of a layer of aluminium oxide on its surface (termed a **passive oxide film**). When a fresh aluminium surface is exposed, it reacts immediately with water vapour in the air to form strands of aluminium hydroxide, $Al(OH)_3$. This is relatively unstable in air, and reacts further losing water to form aluminium oxide, Al_2O_3. Aluminium hydroxide is a gelatinous precipitate used in the dyeing industry, where it is termed a mordant (Latin: *mordere*=to bite) and helps the dye to 'bite' the cloth. Aluminium oxide and hydroxide are amphoteric. Aluminium oxide is insoluble in water (and hence protects aluminium from corrosion), but is soluble in alkaline environments. There are two important examples where this fact significant in the construction industry (**6.1**)

- cementitious products (**7.16**) hydrate to form an alkali (**1.19**), which will dissolve the passive aluminium oxide film, allowing the exposed aluminium metal to react with water and liberate hydrogen. Cementitious materials are mixed with powdered aluminium and water to create lightweight 'aerated' concrete blocks (e.g, Thermalite™). Note that the cementitious matrix is 'aerated' by hydrogen, not air;
- aluminium window frames are placed in wooden surrounds to protect the aluminium from the alkali mortar. Care should be taken to ensure that mortar splashes do not etch the aluminium window frames.

Aluminium, being a very weak alkali, does not form salts with weak acids (there is no aluminium carbonate). However, fairly concentrated solutions of hydrochloric and sulphuric acid will react with aluminium to form salts. Aluminium sulfate, $Al_2(SO_4)_3$, for example is used in fire extinguishers (**14.5.4**) and the sewage industry (**4.2.1**) to precipitate fines, as it very readily forms colloids (**2.9.2**). Aluminium can act as an acidic oxide (to form aluminates with other elements) and as an alkali oxide (producing, for example, aluminium silicates). The overall chemical equation for the reaction of aluminium metal with water is

$$\underset{\text{(aluminium metal)}}{2Al} + \underset{\text{(water)}}{6H_2O} \rightarrow 2Al^{3+} + 6OH^- + 3H_2$$

$$\text{or} \quad 2Al + 6H_2O \rightarrow \underset{\text{(aluminium hydroxide)}}{2Al(OH)_3} + \underset{\text{(hydrogen gas)}}{3H_2}$$

The reaction only takes place if the passive aluminium oxide film cannot be repaired. Note that the compound aluminium hydroxide, $Al(OH)_3$, requires **three** OH^- ions to neutralise **one** Al^{3+} ion. The compound $Al(OH)_3$ is more readily found as the oxide Al_2O_3. To illustrate this, we can rewrite the chemical formula for aluminium hydroxide as $Al_2O_3.3H_2O$, so that the balanced chemical equation becomes

$$\underset{\text{(aluminium metal)}}{2Al} + \underset{\text{(water)}}{6H_2O} \rightarrow \underset{\text{(aluminium hydroxide)}}{2Al_2O_3.3H_2O} + \underset{\text{(hydrogen gas)}}{3H_2}$$

Note also that $Al(OH)_3$ is the **alkaline** representation of the compound. If we want to express the acidic nature of the compound, we would represent the acid as H_3AlO_3 (**1.14.6**). If the acid were to lose **one** water molecule (H_2O), the chemical formula of the resulting acid would $HAlO_2$, as in the mineral **ghanite**, $Zn(AlO_2)_2$. If the acid were to lose **two** water molecules, the chemical formula of the resulting acid would be $H_2Al_2O_4$, as in the mineral **spinel**, $MgAl_2O_4$.

1.8.4 Group 4 Elements (C, Si, Ge, Sn, Pb)

The gradation in properties of the group 4 elements, from carbon (C) (a non-metallic element) to lead (Pb) (a metallic element) is much more marked than it is in other groups on the left or right of the Periodic table. The members of group 4 show valencies of 4 or 2. Denoting the electron configuration of all subshells except the outer s and p subshells as (core) (**Table 1.3**), all group 4 elements have the electron configuration (core)ns^2np^2. For the first elements of group 4 (C, Si), electrons are normally promoted from the s subshell to the p subshell so that these elements form covalent bonds. Elements lower down group 4 (Sn, Pb) sometimes show a valency of 2 less than the group valency through a failure of their s electrons to be promoted to the p subshell. Carbon and silicon form covalent, **macromolecular** structures (**1.17**).

Carbon (except for graphite) is a non-conductor, silicon and germanium as semi-conductors (used in the electronics industry), whilst tin and lead are metallic and will conduct electricity. **Table 1.6** summarises some of the physical properties of the group 4 elements; note that carbon has the highest melting and boiling point, which is related to the fact that carbon produces very strong **covalent** bonds (**1.11**). Stable **ionic** structures only form between a limited range of ionic radii (**1.18**). If carbon loses its electrons from the outer 2s and 2p subshells to obtain the stable electron configuration of **helium** ($1s^2$), the resulting carbon ion (C^{4+}) is too small (0.020 nm) to form a stable ionic structure. In a similar way, if carbon gains four electrons to fill the outer 2p subshell to obtain the inert gas electron configuration of **neon** ($1s^22s^22p^6$), the resulting ion would be too large to form a stable ionic structure. Carbon therefore **shares electrons** to form **covalent bonds** (**1.11**) and obtain a stable electron configuration.

Table 1.6 Some physical properties of group 4 elements

Element	Atomic number	Melting point (°C)	Atomic radii[a] (nm)	Electron configuration[b]															
				1s	2s	2p	3s	3p	3d	4s	4p	4d	4f	5s	5p	5d	5f	6s	6d
Carbon (C)	6	3750	C 0.077 C^{4+} 0.020	2	2	2													
Silicon (Si)	14	1421	Si^{4-} 0.198 Si 0.117 Si^{4+} 0.039	2	2	6	2	2											
Germanium (Ge)	32	937	Ge 0.122 Ge^{4+} 0.044	2	2	6	2	6	10	2	2								
Tin (Sn)	50	232	Sn^{4-} 0.215 Sn 0.140 Sn^{4+} 0.074	2	2	6	2	6	10	2	6	10		2	2				
Lead (Pb)	82	325	Pb^{4-} 0.215 Pb 0.175 Pb^{2+} 0.132 Pb^{4+} 0.084	2	2	6	2	6	10	2	6	10	14	2	6	10		2	2

Notes: 1 nm = 10^{-9} m; [a] covalent (atomic) and ionic radii from **Table 1.5**; [b] 6d and 6f subshells are not shown.

Passing down group 4, there is an increased tendency to use a valency of 2, i.e.,

- C, Si, Ge covalent compounds, almost exclusively 4 valent
- Sn covalent +4 and ionic +2 states are formed with almost equal ease; also ionic +4 state
- Pb mainly ionic +2 state; also covalent +4 state.

Carbon cannot form complex ions as the number of electrons in the outer (2) shell cannot exceed 8. Other elements in group 4 do form complex ions e.g, tin forms **stannates** and lead forms **plumbates**. Carbon forms gaseous oxides (CO and CO_2), in contrast to other elements in group 4. Group 4 elements do not react with water unless heated (for example, carbon reacts exothermically with oxygen and endothermically with steam at 450°C to produce hydrogen and carbon monoxide, as shown over).

The next group 4 elements down are germanium (Ge) (semi conductor industry), **tin (Sn) and lead (Pb)**, which are metals used extensively in the construction industry. Both tin and lead will readily lose four electrons and form a metallic ion e.g, Sn^{4+}, whilst at the same time follow the other group 4 elements (C, Si, Sn and Pb), forming compounds where they are acting in an acidic way (e.g, forming stannates and plumbates). Traditionally, lead has been used in plumbing, but for many purposes it has been replaced by copper and plastics. It form alloys with tin (solder). Lead compounds (plumbates) are used as pigments e.g, white basic lead carbonate, $Pb(OH)_2.PbCO_3$, and the orange pigment 'red lead', Pb_3O_4. In fact 'red lead' behaves as a mixed oxide $2PbO.PbO_2$, indicating its amphoteric character. Calcium plumbates are used in paints instead of poisonous red lead oxides. Red lead oxide is really lead plumbate $\mathbf{Pb_2}\boldsymbol{Pb}O_4$, and calcium plumbate is $Ca_2\boldsymbol{Pb}O_4$, where **Pb** is a small cation of radii 0.084 nm (**Table 1.5**) and the ***Pb*** atom is always found with the four covalently bonded oxygens as a PbO_4 radical (**1.19.2**).

Carbon (C) (electron configuration $1s^22s^22p^2$)
Carbon is present in all living beings; there are more compounds made from carbon (about four million) than for all other elements put together. When carbon forms four covalent bonds, it has a complete outer octet of electrons with the inert gas electron configuration of neon. It cannot expand this octet, as the empty 3rd electron shell is different in energy from the 2nd electron shell. This explains the lack of reactivity of many carbon compounds. With no lone pairs of electrons and a complete octet, it cannot give or accept electrons and therefore forms no complexes. Thus, unlike other elements in group 4, carbon does not form any compounds where the element is acting as an alkali i.e., the compound name starts with carbon, and is ionically bonded (**Table 1.4**).

Carbon has a unique way of linking up with other carbon atoms to form long chain and ring molecules and structures. For carbon to react with other carbon atoms rather than the atoms of other elements, it must form C-C bonds that are similar in strength to those of C to other elements, particularly C-O bonds. Since elements are exposed to the possibility of reaction with air and water in the environment, if C-O bonds are stronger than C-C bonds, then energy considerations will favour the formation of compounds containing C-O bonds rather than C-C-C chains. For silicon, for example, it is energetically more favourable to form a chain

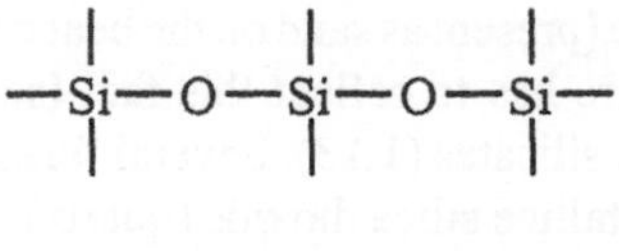

$$-\overset{|}{\underset{|}{Si}}-O-\overset{|}{\underset{|}{Si}}-O-\overset{|}{\underset{|}{Si}}-$$

while for carbon, energy considerations favour a chain

$$-\overset{|}{\underset{|}{C}}-\overset{|}{\underset{|}{C}}-\overset{|}{\underset{|}{C}}-\overset{|}{\underset{|}{C}}-\overset{|}{\underset{|}{C}}-$$

Carbon forms double and triple bonds between carbon atoms (e.g, in alkenes, etc.), with nitrogen atoms (in nitriles) and with oxygen atoms (e.g, in aldehydes, etc.) (**13.10**). Other elements in group 4 do not form corresponding compounds. In group 4, it is only for carbon that the C=O bond is more than twice as strong as the C-O bond. Carbon forms two double bonds in O=C=O, whereas for silicon the formation of four Si-O bonds is preferred on energy grounds.

Carbon combines with oxygen at about 450°C to produce heat and carbon monoxide

$$2C \quad + \quad O_2 \quad \rightarrow \quad 2CO \quad + \quad X \text{ joules}$$

Carbon combines with oxygen at about 500°C to produce heat and carbon dioxide

$$2C \quad + \quad O_2 \quad \rightarrow \quad CO_2 \quad + \quad X \text{ joules}$$

Carbon reacts with water (steam) to produce hydrogen and carbon monoxide

$$C \quad + \quad H_2O \quad \rightarrow \quad H_2 \quad + \quad CO \quad - \quad Y \text{ joules}$$

This reaction is produced in fires containing carbon fuels when water is used to extinguish them; the mixture of gases formed ($CO + H_2$) is combustible (**14.5.6**). Carbon dioxide is a stable, acidic gas and is present in the earth's atmosphere. The structure of the carbon dioxide has to reflect the fact that it is gas. For example, it has the same formula as silica, SiO_2, but this latter compound is a weather resistant solid (e.g, sand). The modelling of these structures is considered in section **1.17**. When dissolved in water, carbon dioxide produces a very weak acid, carbonic acid, H_2CO_3. Only about 0.4% of dissolved carbon dioxide is converted to carbonic acid, and when a solution is boiled nearly all the dissolved carbon dioxide is dispelled

$$H_2O \quad + \quad CO_2 \quad \rightleftharpoons \quad H_2CO_3$$

Carbonic acid reacts with calcium carbonate (limestones, chalk, etc) to produce calcium bicarbonate, $Ca(HCO_3)_2$, responsible for temporary water hardness (**4.3.1**).

Silicon (Si) (electron configuration $1s^22s^22p^63s^23p^2$)
Silicon is more metallic in character than carbon and, unlike carbon, does form silicon compounds (where the element acts as an alkali (e.g, silicon carbide). In contrast to carbon, silicon as the oxide is a crystalline solid structure (present as sand on the beach). It therefore does not dissolve in water, and its structure has to reflect this fact (see ***Figure 1.21***). Silicon also behaves as an acid and forms silicates (**1.13**). Several forms of silica dioxide, SiO_2, are known. The structure of crystalline silica dioxide (quartz) is

shown in ***Figure 1.21***. Silica dioxide melts at 1710°C to form a viscous liquid which, when cooled, forms silica glass used in the manufacture of laboratory glassware (**9.2**). The reaction of silica dioxide and molten **bases (1.19)** forms silicate ions. The silicate ion formed depends on the amounts of silica dioxide and base present e.g, the silicate ion may form a tetrahedral structure (SiO_4^{4-}), chains or sheets of extended $(SiO_3)_n^{2n-}$ and $(SiO_2)_n^{n-}$ ions, or be present in numerous other ionic states. This complexity results in a wide variety of silicates that exist as numerous 'polymeric' structures, considered further in the section on **mixed bonds (1.13)**

1.8.5 Group 5 Elements (N, P, As, Sb, Bi)

As with other groups, the character of the elements of group 5 change as we go down the group, in this case from non-metallic nitrogen (N), which exists as the diatomic gas N_2, to the metal bismuth (Bi). The oxides of nitrogen are acidic (except dinitrogen oxide, N_2O and nitrogen monoxide, NO), as are the oxides of phosphorous and arsenic. Antimony forms both amphoteric and acidic oxides, some of which are used as fire retardants (**14.7.3**). Bismuth oxide is alkaline. The elements of group 5 all accept 3 electrons to attain a full octet of electrons. All elements of group 5 except nitrogen show a covalency of 5 by promoting one of the s electrons to a d subshell. This cannot happen with nitrogen because it has no d subshells at energy levels comparable to the occupied shells, and this explains why nitrogen is a relatively unreactive element.

Nitrogen (N) (electron configuration $1s^2 2s^2 2p^3$)

Nitrogen is a relatively unreactive gas, constituting 78.06% by volume of dry air. Although unreactive in the free state, it is an essential element in all living things, being in present in proteins and nucleic acids. The balance of those reactions that remove nitrogen from the air and from soil and that put nitrogen into air and soil is termed the **nitrogen cycle**. Nitrogen is added to the soil by many means, for example direct absorption by soil bacteria, 'fixation' by bacteria present in the root structure of some plants, in rainwater (as nitric acid, HNO_3, in solution, formed by the action of lightning on atmospheric gases), by decay of organic matter by soil bacteria and by direct application of fertilisers (produced by the oxidation of ammonia, NH_3). Plants obtain nitrogen from the soil via their roots, which is subsequently incorporated into plant proteins, providing a food source for animals. The structure of nitric acid, HNO_3, and ammonia, NH_3, are shown in ***Figure 1.8***.

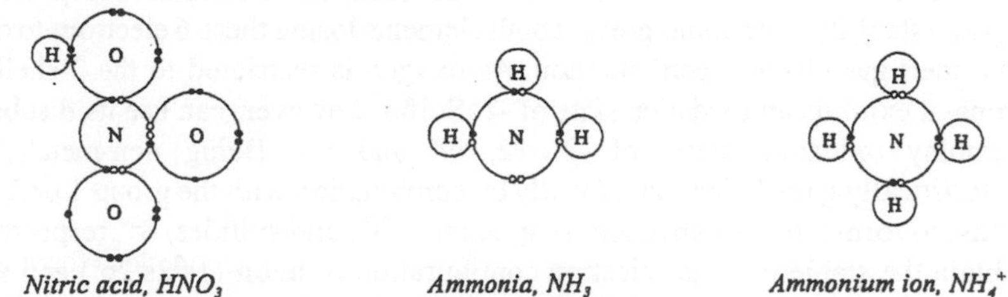

Figure 1.8 Structure of some nitrogen compounds

Nitric acid contains nitrogen and oxygen, which are bonded by the **covalent bond (1.11)** and hydrogen, which is bonded to the oxygen with an **ionic bond (1.12)**. The presence of covalent and ionic bonding in one compound is termed **mixed bonding** (**1.13**). Other compounds exhibit mixed bonding, for example sulphuric acid (H_2SO_4) (***Figure 1.27***) and phosphoric acid (H_3PO_4) (***Figure 1.31***). In each case, hydrogen is modelled as ionically bonded. This means that the hydrogen can be detected in aqueous solutions, whereas the sulfur, phosphorus and the nitrogen are never detected without the accompanying covalently bonded oxygen. Nitric acid forms salts called nitrates, NO_3^-, which are the covalently bonded units (PO_4^{2-}, SO_4^{2-}, in other acids) called **radicals** (**1.19.2**). ***Figure 1.8*** also illustrates another type of nitrogen radical, the ammonium ion (NH_4^+), derived from the gas ammonia NH_3, where the lone pair of electrons in the nitrogen atom enable it to accept a proton, for example when ammonia and hydrogen chloride react to form ammonium chloride (this reaction is used as a test for ammonia).

Phosphorous (P) (electron configuration $1s^22s^22p^63s^23p^3$)
In contrast to nitrogen, phosphorous is dangerously reactive and forms an acid when the oxide is mixed with water. Phosphorous is so reactive that it never occurs naturally as a free element. Phosphoric acid (H_3PO_4) (**1.19.6**), made by boiling phosphorous oxide (P_2O_5) in water (H_2O), is used to etch steel to key paint to the metal surface (**11.12.1**).

1.8.6 Group 6 Elements (O, S, Se, Te, Po)

Denoting all electron subshells excluding the outer s and p subshells as (core), all group 6 elements have the electron configuration (core) ns^2np^4. There is a transition in properties as we go down group 6, from non-metallic (oxygen, sulfur) to metallic (polonium). Oxygen (O) and sulfur (S) are unique in that they both can support life.

Oxygen (O) (electron configuration $1s^22s^22p^4$)
Sulfur (S) (electron configuration $1s^22s^22p^63s^23p^4$)
Oxygen constitutes 21% by volume of the air and 89% by mass of the earth's crust. It is the essential part of the air for animal respiration. Oxygen is sufficiently soluble in water to be able to support the respiration of marine animals. Oxygen is a colourless, diatomic gas (strictly speaking, we should refer to molecular oxygen, O_2, as dioxygen. Oxygen reacts readily directly with most other elements and forms compounds with all elements except the inert gases. Sulfur (S) is found as an element, as metal sulfide ores and as a number of sulfates e.g, calcium sulfate, $CaSO_4$, as in gypsum (**14.7.3**). Both oxygen and sulfur contain 6 electrons in their outer subshells (2nd and 3rd respectively). Energy and stability restrictions prevent both elements losing these 6 electrons to obtain a stable inert gas electron configuration. As oxygen is restricted to the 2 shell of 8 electrons, it exhibits an oxidation state of -2. Sulfur, however, can use its d subshells and employ oxidation states of -2, +2, +4 and +6. Being non-metals, they characteristically gain 2 electrons (usually by combination with the group 1 or 2 metal elements) to form large negative ions (e.g, oxides, O^{2-}, and sulfides, S^{2-}, respectively), and obtain the stable inert gas electron configuration of **neon** ($1s^22s^22p^6$) and **argon** ($1s^22s^22p^63s^23p^6$) respectively. The addition of two electrons nearly doubles the size of the elemental atoms (**Table 1.5**). These large negative ions determine the structure of the ionic compounds in which they appear (**1.18**) by controlling the size of the **unit cell** (**1.15.3**). These results are all characteristic of acidic elements.

Oxygen and sulfur will also form covalent bonds with hydrogen e.g, water (H_2O) in the liquid state and hydrogen disulphide (H_2S) in the gaseous state. Both will also form covalent bonds with carbon, producing compounds in the gaseous state e.g, carbon dioxide (CO_2) and carbon disulfide (CS_2). Sulfur shows an oxidation state of +4 in sulfur dioxide, SO_2

$$S + O_2 \rightarrow SO_2$$

and an oxidation state of +6 in sulfur trioxide, S_2O_6 (or SO_3)

$$2S + 3O_2 \rightarrow S_2O_6 \text{ (or } 2SO_3\text{)}$$

The hydroxide of sulfur can be written as $S(OH)_6$. In practice, however, it loses 2 molecules of water to become sulfuric acid, H_2SO_4 (***Figure 1.27***).

$$S(OH)_6 \underset{2H_2O \text{ removed as } 4OH^- \text{ ions}}{\rightarrow} H_2SO_4$$

Oxygen is responsible for the life support system of animals. In addition, sulfur supports micro-organisms in the complete absence of oxygen, such as in the deep active ocean trenches, at very high temperatures and pressures. In the spreading of the ocean plate, sulfurous gases are produced that sustain these organisms. They have a very large tolerance for extremes of alkalinity and acidity, from pH (**1.19.3**) 0 to 11, temperature (up to about 85°C) and pressure (up to about 1000 atmospheres). Similar micro-organisms have been found in the rubbish dumps where waste is deposited. Micro-organisms are classified according to whether they can exist in the absence of oxygen (**anaerobic**) or required the presence of oxygen (**aerobic**). The most important of these are **anaerobic bacteria** which reduce the **sulfate** radical (SO_4^{2-}) to the **sulfide** radical (S^{2-}), and are called sulfate-reducing bacteria.

$$SO_4^{2-} + 4H_2 \xrightarrow{\text{anaerobic bacteria}} S^{2-} + 4H_2O$$

(present in wet clays, boggy soils, and marshes)

The hydrogen required can be derived from cellulose, sugars or other organic products in the soil or compost. **Aerobic bacteria** oxidise sulfides to sulfates,

$$2S + 3O_2 + H_2O \xrightarrow{\text{aerobic bacteria}} H_2SO_4$$

(sulfur compounds)
(e.g, iron pyrites FeS)

which can cause damage to subsoil concrete (**8.19.1**). Overall, these bacteria can operate in a **cyclic** manner. When the ground is wet and boggy, the anaerobic bacteria flourish and when it dries, out the aerobic bacteria flourish. These bacteria can produce some awful smells. The sulfur equivalent of H_2O is H_2S, which is associated with putrefaction of rotten eggs; the odour is so powerful that it can be detected at concentrations of 10 ppm and causes death at concentrations of 1000 ppm. The sulfur equivalent of CO_2 is CS_2, which smells like addled cabbage. Thus complex reactions can take place in waste dumps and precautions may be required where structures are built on land fill sites; products containing the SO_4^{2-} radical, such as gypsum plaster, should not be used in hardcore. As reactions depend on the presence of water, the level of malodour often depends upon the water table level.

1.8.7 Group 7 Elements (F, Cl, Br, I)

Group 7 elements are referred to as **halogens**, a name derived from the Greek for salt-formers. They exist as diatomic molecules. The halogens form salts by accepting one electron to complete an octet of outer electrons, with the formation of a halide ion (e.g, F^-, Cl^-, Br^-, etc.). Halogens are never found in nature as free elements as they are far too reactive; they occur combined, principally with an alkali element as a salt. The high reactivity of fluorine gas can be seen by the fact that it reacts with gold, unlike all other elements. In addition, it will ignite explosively in contact with most metals, wood and rubber, whereas the other halogens will only combine on heating. Fluorine and chlorine oxidise water, forming hydrofluoric acid (HF) and hydrochloric acid (HCl) respectively. Hydrofluoric acid will attack glass, whereas hydrochloric acid can be stored in glass. Despite their reactivity, halogens form very stable molecular structures with non-metallic elements. Examples include organohalogens (or 'halons') used in fire protection (**14.5.5**), and poly(tetrafluoroethane) (PTFE), a plastic that resists attack from most chemicals, is an insulator (used as coating for electrical wiring) and has a low coefficient of friction (used as a coating for cooking pans and skis).

1.8.8 The Transition Metals

The transition metals are found mainly as sulfides and oxides. The least reactive (Cu, Au, Pt) are also found uncombined. The electron configuration of the transition metals is (argon)$4s^2 3d^n$. Once the 4s electrons have been removed, the 3d electrons may be removed. The difference in energy between the 3d and 4s electrons is much smaller than the difference the 3s and 3p electrons. A variety of oxidation states from +1 (Cu) to +7 (Mn) is possible. Some of these states are uncommon and unstable. In their lower oxidation states, elements form ionic compounds, but in their higher oxidation states they form covalent compounds. All transition metals have similar electrical and magnetic properties (dependent upon the number of electrons in the 3d subshell) and they have high melting and boiling points. Transition metals are commonly alloyed (**10.5.2**) to enhance performance (e.g, Cu and Ni in coinage metals, Cu and Zn in brass, Cu and Sn in bronze), are very dense (about 3000 to 8000 kg/m^3) and liberate hydrogen in contact with acids. All transition metals absorb hydrogen, which may cause embrittlement and early failure in some engineering alloys, notably steels (**10.14.1**).

Transition metals all form oxides, hydroxides or carbonates in air, some of which are stable and tenacious (e.g, zinc oxide, ZnO and aluminium oxide, Al_2O_3), whilst others flake away from the parent metal (e.g, iron oxide as rust, Fe_2O_3). The lower oxides (MO) are usually basic, amphoteric or both, whilst the higher oxides (M_2O_3) are acidic. Some transitional metal oxides are coloured. Colour results from the energy state of the electrons in the d subshell. In an isolated transition metal atom, the electrons in the five d subshells are all at the same energy level. In a complex transition metal ion, the electrons in the subshells differ slightly in energy, and electrons can jump from one d subshell to another if they are able to absorb energy. For most transition metal complexes, the frequency of light absorbed in these energy transitions is in the visible light spectrum, and therefore the ion appears coloured (the colour being complementary to the colours absorbed). For example, in Cu^{2+} and Zn^{2+} ions, with a d^{10} configuration, no electron jumps between d subshells can occur, and these ions are colourless. However, the complex ions produce blue and green chromium complexes, brown, pink and blue cobalt complexes, blood-red iron complexes (e.g, in haemoglobin, the oxygen-carrying compound in red blood cells) and blue and yellow copper complexes. Higher oxides are also coloured, for example $Fe_2O_3.nH_2O$ (rust) is brown and Fe_3O_4 is blue-black. Zinc products (e.g, galvanised steel) will react with weak acids to form salts of the hydrated zinc ion, $Zn(H_2O)_6^{2+}$, which reacts further to produce white zinc carbonate, $ZnCO_3.2Zn(OH)_2$ and zinc sulphide, ZnS salts (depending on the acid). These zinc salts are sometimes seen as run-off stains from galvanised steel materials that may stain adjoining materials. Copper is used as a roofing material because it weathers to acquire a coating (**patina**) of green basic copper carbonate, $CuCO_3.Cu(OH)_2.nH_2O$.

1.8.9 Metals and Non-metals

Table 1.7 summarises the chemical and physical properties of metals and non-metals; for the relation between these properties and electron configuration, see section **1.6**.

Table 1.7 Chemical and physical properties of metals and non-metals

Chemical Properties of Metals	Chemical Properties of Non Metals
Metallic oxides are bases and form alkalis if they are soluble in water (**Table 1.14**)	Non metal oxides are either neutral or form acidic oxides if they are soluble in water (**Table 1.14**)
Metals replace hydrogen in acids forming salts	Non metals do not form salts in this way
Metallic chlorides are ionic salts, dissolving in water to produce the Cl^- ion	Non metallic chlorides are covalent, do not usually produce electrolytes and are hydrolysed by water e.g, hydrochloric acid is a covalently bonded gas (HCl) which is hydrolysed by water $HCl - H^+ + Cl^-$
Metals do not usually form stable hydrides (H^-) (exceptions are K, Na and Ca)	Non metals form hydrides that are covalent and do not usually produce electrolytes in the absence of water (e.g, CH_4, NH_3). The 'hydrides' of Group 7 (e.g, HF) ionises to produce H^+ ions and not H^- ions
Physical Properties of Metals	**Physical Properties of Non Metals**
Good conductors of heat and electricity, capable of being shaped, ductile (may be drawn into wire) and have a lustre and may be polished.	Poor conductors of heat and electricity, incapable of being shaped, brittle and do not possess a lustre.
Some metals have high strengths and are very dense.	Non-metals have low strengths and are not dense.

1.9 CHEMICAL BONDS

As we have seen, the fundamental building block for chemical compounds is the atom, rather than the smallest unit to which the solid can be cleaved. The way atoms (or ions or molecules) are arranged in the solid state controls the properties of materials. When two or more atoms approach one another, they exert forces of attraction and repulsion. When these forces balance, a state of equilibrium is attained at which the atoms are at a certain definite distance apart. This distance is termed the **atomic radius**. When the attractive forces between the atoms is sufficiently strong so that a stable aggregation of atoms is formed, the atoms are said to be held together by **chemical bonds**. The arrangement of atoms in the solid state therefore depends on the relative radii of the atoms involved (**Table 1.5**) and also on the bonding between atoms. In a solid, each atom forms a bond with each of its neighbours so that each has a definite position in a three dimensional array of atoms, giving rise to various **crystalline** solid states.

Comparison of the atomic radii of atoms (**Table 1.5**) indicates that they do not alter very much with the number of protons and electrons in the nucleus, providing they are equal in number; for example, compare the atomic radii of uranium (U) atoms (atomic number 92), i.e., 0.138 nm, with that of lithium (Li) atoms (atomic number 3), i.e., 0.152 nm. The exception occurs where the number of electrons and protons are not equal, e.g, in the formation of ions; compare the radii of lithium (Li) atoms, i.e., 0.152 nm, with that of lithium ions (Li^+), i.e., 0.078 nm, half the radius of the neutral atom.

There are four classes of molecular bonds (although the boundaries between each bond type are not strictly defined)

- the **metallic bond**, formed only with the metallic elements from groups 1, 2, 3 and the transition elements in the pure (unalloyed) state (e.g, copper), or in the alloyed state (e.g, brass, an alloy formed between copper and zinc);
- the **covalent bond**, formed by the non-metallic elements usually those elements in groups 4, 5, 6, and 7;
- the **ionic bond**, formed when non-metallic elements (of groups 5, 6 and 7) combine with the metallic elements (of groups 1, 2 or 3). The elements in groups 1, 2 and 3 are **electropositive** elements that combine with **electronegative** elements in groups 5, 6 and 7 by the attraction of unlike charges;
- the **van der Waals force** (or **bond**), a residual bond that is therefore very weak and easily broken by mechanical or thermal means; it is present to some extent in all materials (from water to plastics).

Generally, we can state that

- two atoms of similar electronegativity form either a **metallic bond** or a **covalent bond**, according to whether they can release or accept electrons;
- when the electronegativities differ, the bond is partially **ionic**, the ionic character increasing with the difference in electronegativity;
- the **covalent bond** is **directional**, while the ionic bond is not. The degree of directionality changes with the bond character. Such influences have a large effect on the structure of a solid (**1.17**).

The **strength of the bond** is measured by the melting point of the bonded material giving, to a first approximation, the following order of bond strengths: covalent > ionic > metallic > van der Waals forces or bonds (the weakest bond of all).

1.10 METALLIC BONDS

The metallic elements lose some of their outer electrons (e.g sodium, **1.8.1**). These electrons are donated to form an electron cloud. The number of electrons lost (or donated) to this electron cloud depends upon the group that the element occupies. For example in brass, copper will donate one electron, whilst zinc will donate two electrons. This electron cloud is free to move through the three dimensional array of positively charged atoms under a small potential gradient; metals are therefore conductors of electricity. The structure of a metal is modelled in ***Figure 1.9*** as an array of positively charged centres representing the atoms that have given up their loosely bound electrons to an electron cloud. In ***Figure 1.9a***, the electron cloud is modelled as a sine wave travelling through the positive array of metallic atomic nuclei. The electron cloud is the bonding agent that cements all the positive charges into a three-dimensional array and gives a metal its characteristically good electrical properties. The positive charge on the array of positive centres is modelled as a first approximation as the **group in which the elements appear in the periodic table (Table 1.8).**

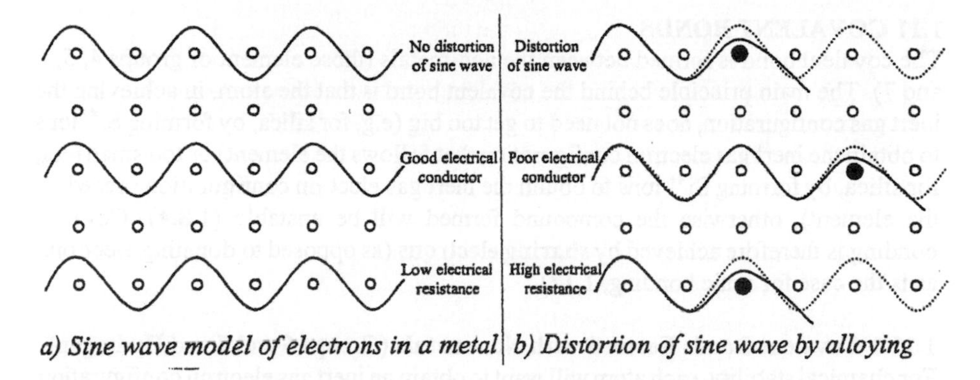

a) Sine wave model of electrons in a metal | *b) Distortion of sine wave by alloying*

***Figure 1.9** Structure of a metal showing sinusoidal wave model of electrons*

1.10.1 Electrical Resistance

Electrons can **only** travel easily in a metallic lattice where the metallic sites are all of the same. When a metal (A) from group 1 (e.g, copper, Cu) is alloyed with another metal (B) (which is often from a different group, e.g, zinc from group 2), the resulting brass alloy is said to be a **solid solution (1.16)**, where the metal (B) replaces the metal (A) in the lattice points of metal (A) structure. The number of electrons donated to this electron cloud (2 from metal (B) in this case) is different from the 'host' or base metal (A) (which donates 1 electron in this case). The electron cloud now suffers a resistance to its flow through the different positively charged atomic nuclei such that the cloud will stay longer around the +2 charged centres of metal (B). The electrical resistance arises because the alloy group 2 metal (B) (zinc) is in the higher group (group 2) and will therefore contribute two electrons to the electron cloud and so produce a +2 lattice site. If the flow of the electron cloud is modelled as a sinusoidal wave passing through the metal, then this sinusoidal wave is distorted as it passes the "foreign" higher charged centres of metal (B) under small potential differences, as shown diagrammatically in two dimensions in ***Figure 1.9b***; this is an example of **substitution solid solution (1.16)**.

As the number of electrons given to the electron cloud is dependent upon the group in which the element is located then, as a first approximation, those elements in group 1 (e.g, copper, Cu, silver, Ag and gold, Au) will not suffer in resistivity terms by alloying with each other, as they are all in group 1. It is only detrimental in terms of resistivity when alloying takes place across groups (e.g, group 1 elements with group 2 elements).

1.10.2 Summary: Properties of the Metallic Bond

- Non directional bond
- Metals are mechanically deformable and so atoms must "slide" over each other;
- Relatively weak bond and therefore metals have relatively low melting points;
- Metals are conductors of electricity (electrons must be free to move); alloys have a higher electrical resistance than the pure metals from which they are made;
- There is **no change in the radii** of the sphere (atom) upon forming the metallic bond (compare ionic case, **1.12**);
- The metallic bond is present in spot welding, brazing and soldering, etc.

1.11 COVALENT BONDS

The covalent bond is formed between the non-metals (those element of groups 4, 5, 6 and 7). The main principle behind the covalent bond is that the atom, in achieving the inert gas configuration, does not need to get too big (e.g, for silica, by forming Si^{4-} ions to obtain the inert gas electron configuration that follows the element) or too small (e.g, for silica, by forming Si^{4+} ions to obtain the inert gas electron configuration preceding the element), otherwise the compound formed will be unstable (**1.8.4**). Covalent bonding is therefore achieved by **sharing electrons** (as opposed to donating electrons, as is the case for ionic bonding, **1.12**).

1.11.1 Covalent Bonds Formed with Non-metals (Groups 6 and 7 and Hydrogen)

For chemical stability, each atom will want to obtain an inert gas electron configuration. Thus two atoms of the elements hydrogen, chlorine and fluorine combine to share their electrons and thus produce the diatomic (two atom) molecules shown in ***Figure 1.10***. Hydrogen cannot exist as atomic hydrogen H as it is far too reactive. Hydrogen gas exists as a diatomic molecule, written as H_2 (***Figure 1.10a***), where each hydrogen atom has obtained the inert gas electron configuration of the next inert gas (helium). Note that the hydrogen molecule is **not** charged, even though there are now two electrons in each shell. This is explained by the fact that the electrons are only half associated with each atom and so the diatomic molecule is overall electrically neutral. These electrons are usually further shortened as lines, one line denoting the sharing of **two** electrons (e.g, H-H). The hydrogen molecule separates the charge as (a positive nucleus):(electrons):(a positive nucleus). In some cases, for example where hydrogen is bonded to carbon, this creates a **dipole** that is responsible for the van der Waals bond or forces (***Figure 1.16***). Dipoles can arise from any covalent bond between unlike atoms (e.g, O-H, N-H, C-O, etc.), not only between carbon and hydrogen.

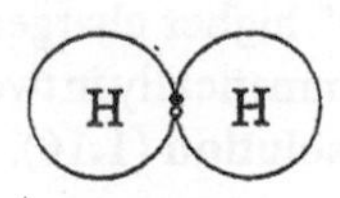

a) Hydrogen, H_2

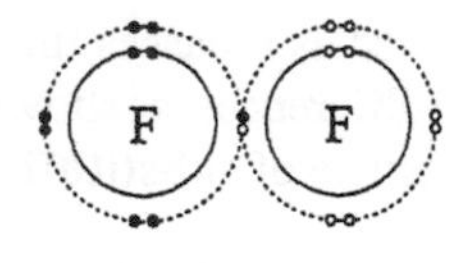

b) Fluorine, F_2

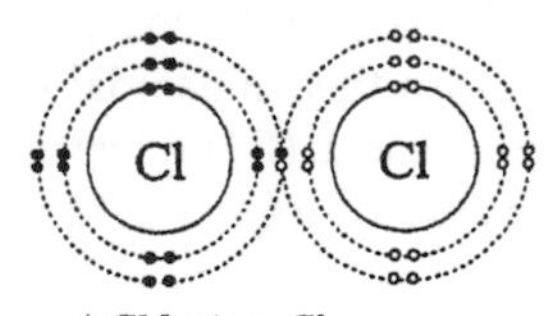

c) Chlorine, Cl_2

Figure 1.10 *Covalent bonding in diatomic molecules*

Figure 1.10b and 1.10c show the electron configurations of fluorine and chlorine. In fact these Figures show any group 7 element because, by convention, it is permissible to leave out the filled electron levels corresponding to the previous inert gas **as these inner electrons do not take part in any chemical reaction.** Two atoms of the element fluorine or chlorine (or any other element in group 7) therefore combine to share only one of their outer 7 electrons and thus produce a diatomic molecule. The sharing of the electrons produce a **molecule** that, in this case, comprises two atoms that each have two electrons in the outer electron shell. This therefore produces a **very directional bond** across the shared electrons, which are "fixed" at the point of sharing. This directionality is a characteristic of covalently bonded materials and is responsible for the arrangement of atoms in a covalently bonded solid (**1.17**). The sharing of **two** electrons is normally referred to as a **single bond** and is fairly difficult to break. It is usually represented by a single line. ***Figure 1.11*** shows three representations of the covalently bonded structure of a hydrocarbon, formed by the combination of hydrogen atoms with carbon atoms. ***Figure 1.11a*** shows the sharing of electrons between both the carbon and carbon, and between carbon and hydrogen making up the hydrocarbon. Each carbon electron is represented differently to facilitate identification. In ***Figure 1.11b***, the designated spatial orientation of the direction of these bonds is shown. Rotation around these directions is allowed (and is an important parameter in explaining the **glass transition temperatures** of plastics, **13.9.5**). In ***Figure 1.11c***, the electrons are usually further shortened as lines, one line denoting the sharing of two electrons (i.e., a **single bond**).

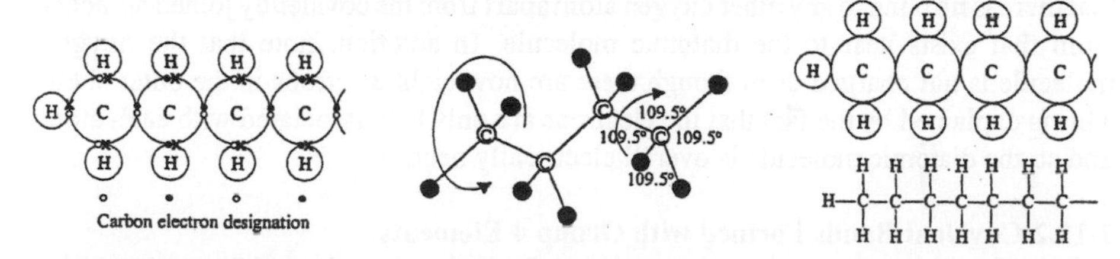

a) Electron designation *b) Spatial orientation* *c) General formula $C_nH_{(2n+2)}$*

***Figure 1.11** Representations of the structure of a covalently bonded hydrocarbon*

The sharing of **four** electrons is normally referred to as a **double bond** and is also not easily broken. It is usually represented by a double line. ***Figure 1.12*** shows the electron configuration of an atom of oxygen (O), and also the two molecules formed from oxygen, the oxygen (O_2) molecule and the ozone (O_3) molecule.

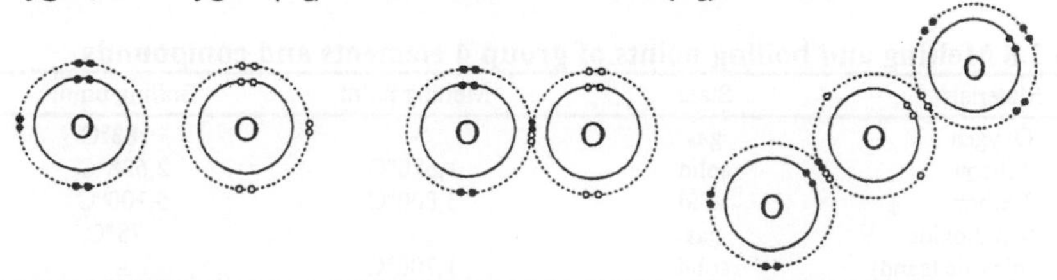

a) Atomic oxygen, O *b) Oxygen molecule, O_2* *c) Ozone, O_3*

***Figure 1.12** Structure of atomic and molecular (O_2 and O_3) oxygen*

By convention, it is permissible to leave out the filled electron levels (corresponding to the previous inert gas) as these inner electrons do not take part in any chemical reaction (this omission is shown in the ozone molecule, O_3, explained below).

As we have seen (**1.5**), the **number of nearest neighbours** within a structure is given by the number of electrons in the inert gas following the element **minus** the group in which the element occurs. Hence oxygen (O) (group 6) should have two nearest neighbours (8-6=2) on this model rule. This is obeyed in the combination of oxygen with other elements (e.g silica dioxide, ***Figure 1.13***) but not in combination with itself. Oxygen exists as two different molecular units (molecules) called oxygen (present in air) and ozone (present in the upper atmosphere). Ozone is the example that follows the above rule, but oxygen can also share more than two electrons, as in molecular oxygen (O_2). In molecular oxygen (O) (group 6) two electrons from each atom must be shared in order to form a stable inert gas configuration (of neon). This means that there are **four** shared electrons between each oxygen atom, i.e., a **double bond** (O=O), where the = sign represents a double bond. These diatomic (O_2) or triatomic (O_3) gases are only bonded in a directional bond across the shared electrons. They are **not** bonded in a three dimensional network because there are no spare electron spaces or electrons in the orbit, as all the atoms have the stable 8 electrons. These compounds are therefore gases. If the electrons are considered as discrete entities with their positions as shown in ***Figure 1.12***, then there is an electrical **dipole** (alternate positive and negative charges). The existence of this dipole is the 'bond model' required for the residual van der Waals forces (**1.14**) responsible for the weak intermolecular bonding of gases. Note that there is **no** bond to any other oxygen atom apart from the covalently joined adjacent atom that exists near to the diatomic molecule. In addition, note that the oxygen molecule is **not** charged even though there are now eight electrons in the outer shell. This is explained by the fact that the electrons are only half associated with each atom and so the diatomic molecule is overall electrically neutral.

1.11.2 Covalent Bonds Formed with Group 4 Elements

These elements can produce large three dimensional 'polymeric' structures by covalently bonding in a directional manner by having four nearest neighbours (in accordance with the (8-N) rule). Four electrons are shared from each of the four atoms that are spatially orientated around each other atom. Examples include carbon (diamond) and silica dioxide (sand) (considered in more detail in section **1.17**). In addition, carbon from group 4 forms covalent bonds with oxygen, as in carbon dioxide (CO_2). Some of the properties of group 4 elements are shown in **Table 1.8** below.

Table 1.8 Melting and boiling points of group 4 elements and compounds

Materials	State	Melting point	Boiling point
Oxygen	gas	-	- 183°C
Silicon	solid	1,680°C	2,628°C
Carbon	solid	3,800°C	5,100°C
Carbon dioxide	gas	-	- 78°C
Silicon dioxide (sand)	solid	1,700°C	-

Note that the group 4 elements carbon (C) and silicon (Si) both form the same oxides, XO_2, but their properties are quite different. From a knowledge of the position of these elements in the periodic table (**Table 1.2**) and their physical properties we would expect carbon dioxide (CO_2) to be a very high melting point solid, whereas it is in fact a gas; the modelling of the structure must reflect its properties (***Figure 1.13***). A covalently bonded material is a poor conductor of electricity because the electrons are "fixed" between each atom and cannot move away (under low potential differences). Covalently bonded materials are therefore insulators. Under high potentials, insulating properties can be broken down as electrons are stripped from their positions in the bond.

Figure 1.13a shows carbon dioxide, with carbon covalently bonded to two oxygen atoms. The carbon atom shares two of the outer 4 electrons with each oxygen, leaving the outer orbit completely full with 8 electrons. There are no available electrons to combine with another CO_2 unit. There is a very weak CO_2 - CO_2 attraction between the molecules; the distance between each CO_2 unit can be approximated as 12d, where d is the diameter of the carbon dioxide unit (**2.2.4** for derivation). ***Figure 1.13b*** shows the structure of silica dioxide. The SiO_2 is a large molecule and a solid. It does not ionise and thus is insoluble in water. It forms silicates (**1.17**), the most abundant group on earth. At the surface of the silicate unit there is an electron deficiency (arrowed), which is made up by an ionic bond formed by a donated electron; thus the surface of the silicate becomes negatively charged. In rivers, this negative charge prevents the silt from settling until the river reaches the sea, where the negative charge can be neutralised by the Na^+ ion in the sea water. The silt deposits to form the continental shelf. The construction industry uses this principle in the **silt or field settling test** (**8.6.2**), where the fine aggregate is shaken with salt water to make the silt deposit.

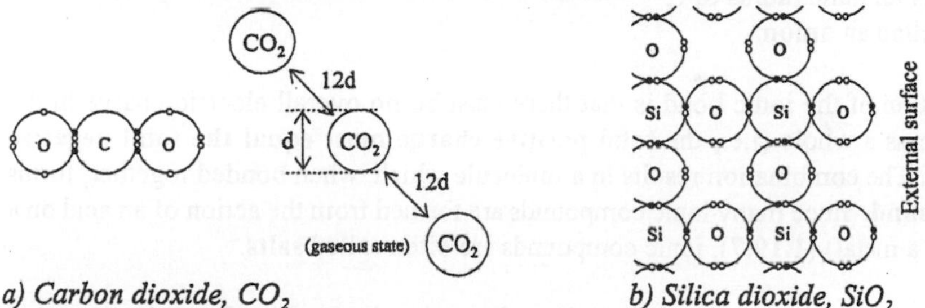

a) Carbon dioxide, CO_2 *b) Silica dioxide, SiO_2*

***Figure 1.13** Covalent bonding of some group 4 elements*

1.11.3 Summary: Properties of the Covalent Bond

- The covalent bond is a **very directional bond**;
- The covalent bond is a very strong bond (the strongest bond). **Wholly** covalently bonded materials are mechanically very strong and non-deformable, have very high melting points and are insoluble in water;
- Covalently bonded materials are poor conductors of electricity and are used as electrical insulators;
- The atomic radii of the elements involved in the covalent bond are not affected by the covalent bond;
- Ionic and covalent bonds are very similar and can exist in the same structural unit depending upon the properties being portrayed;
- Covalently bonded materials include diamond, timber, water, rubber and alcohol;
- The covalent bond is responsible for large macromolecules e.g, **cellulose** $(C_6H_{10}O_5)_n$, where n is greater than 2000; these large molecules give **anisotropic** properties to materials such as timber (**12.4.3**).

1.12 IONIC BONDS

The ionic bond is formed when a **non metal** (X) from either groups 5, 6 or 7 combines with a **metal** (A) from group 1, 2 or 3 to form charged ions. Ionic bonds give rise to compounds with the chemical formulae shown in **Table 1.9**.

Table 1.9 Chemical formulae of ionic compounds from different groups

Metal (A)	Non-metal (X)		
	Group 5 (anion X^{3-})	Group 6 (anion X^{2-})	Group 7 (anion X^{1-})
Group 1 (cation A^{1+})	A_3X	A_2X	**AX**
Group 2 (cation A^{2+})	A_3X_2	**AX**	AX_2
Group 3 (cation A^{3+})	**AX**	A_2X_3	AX_3

Metal (A) losses the outer electron(s) (the number lost depending upon the group) and
- becomes **positively** charged (the relative charge depending upon the group); and
- becomes **smaller in size** (nearly half the size of the neutral atom) (**Table 1.5**), having an ionic radius of $\mathbf{r_c}$

and is called a **cation. Non-metal** (X) gains electron(s) depending upon the (8-N) rule
- becoming **negatively** charged (the relative charge depending upon the group); and
- becoming **much larger** (sometimes twice the size of the neutral atom) (**Table 1.5**), having an ionic radius of $\mathbf{r_a}$

and is called an **anion.**

A condition of the ionic bond is that there must be **no** overall electric charge in the structure as a whole (i.e., the **total positive charge must equal the total negative charge**). The combination results in a **molecule** which, when bonded together, forms a **compound.** Since many ionic compounds are formed from the action of an acid on a base (or a metal) (**1.19.7**), ionic compounds are often called **salts.**

The ionic bond therefore produces a three-dimensional structure of **positive ions and negative ions.** It is the **difference in sizes of the ions** (or more accurately the **radius ratio** $\mathbf{r_c/r_a}$) and the **formula unit** that dictates the **structure** of the ionic salt (the structure of ionic compounds are considered in **1.18**). As an example, for a compound with the formula unit $\mathbf{A_2X_3}$, the structure must accommodate 2 atoms of element A and 3 atoms of element X, whilst still satisfying radius ratio conditions and packing efficiency requirements. As a further example, consider the $\mathbf{AX_2}$ formula typified by calcium fluoride (CaF_2) (**1.18.4**). If a group 2 metallic atom (e.g, Ca) combines with a group 7 atom (e.g, F), then the metallic atom loses 2 electrons and becomes the charged $\mathbf{Ca^{2+}}$ cation, whilst the non-metallic element gains only one electron (8-7=1) and becomes charged $\mathbf{F^-}$ anion. Therefore **two** of non-metallic anions are required to satisfy **one** divalent group 2 metallic cation. The resultant chemical formula is therefore $\mathbf{CaF_2}$ and **twice as many** $\mathbf{F^-}$ **cations** must be packed into the available space as $\mathbf{Ca^{2+}}$ **anions.**

1.12.1 Summary: The Properties of the Ionic Bond

- Non directional bonds. The structures formed are very dependent upon the number of near neighbours, which in turn is dictated by the radius ratio;
- Mechanically very weak and elastically non deformable. If ions are able to slide over

one another, two like charges may come together, resulting in mutual repulsion. Ionically bonded structures are therefore usually brittle and fracture without warning;

- Bond strength. Relatively medium to strong bond and therefore ionically bonded materials have average melting points;
- Ionically bonded materials are poor conductors of electricity in the solid state, but very good conductors in the liquid or molten state, where they are used in electroplating baths etc. In mass produced painting systems (e.g, electro-coating of motor vehicle bodies, windows, etc.) it is a requirement to make the covalently bonded paint film have an ionically bonded unit so that it ionises in water, can carry a charge and be plated out. The paint molecule is therefore be a big macromolecule so that good electrical efficiency can be achieved in that a large mass of paint can be deposited per unit charge (for solubilities of ionic compounds **Table 1.14**; for electrical resistivities **Table 1.16**; for the laws of electrolysis, see section **11.4**);
- The ions forming the **unit cell** (**1.15.3**) of the molecule are of different radii, but must be packed to give the smallest void space with no overall electrical charge. The ionic bond is not stable if the ions are either too large or too small (**1.18**);
- Examples of ionically bonded materials include natural stones, road de-icing salts, ceramic insulators on the spark plugs, etc. Processes involving charged ions include corrosion (**11.3**), which is enhanced by the presence of chloride ions (**8.21.5**).

1.13 MIXED BONDING

Nature does produce **mixed bonding** within the same molecular structure. Examples are **acids** (**1.19**); we can depict sulfuric acid (***Figure 1.28***) and phosphoric acid (***Figure 1.32***) as **covalently bonded**. However, when these acids dissolve in water, the hydrogen (H) becomes **ionically bonded** and forms **hydrogen (H^+) ions** (or, more correctly, the complex **oxonium ion H_3O^+**) (***Figure 1.30***). Generally, it is difficult for man to make use of materials with a mixed bond system. For example, **a covalently** bonded paint system adjacent to a **metallically** bonded metal presents a problem of adherence of the paint film to the metal surface (**11.12**). The model chosen to represent the compound under consideration will depend upon the properties being investigated. In complex structures e.g, the silicates (**1.17**), the ionic and covalent bonds can be considered as two extremes of the same phenomena of electron sharing and partition. Some simple structures containing OH^- groups that illustrate the concept of mixed bonding are shown later (**1.19.2**); here we consider a more complex mixed bond structure, aluminium nitride, formed from aluminium (group 3) and nitrogen (group 5) (***Figure 1.14***).

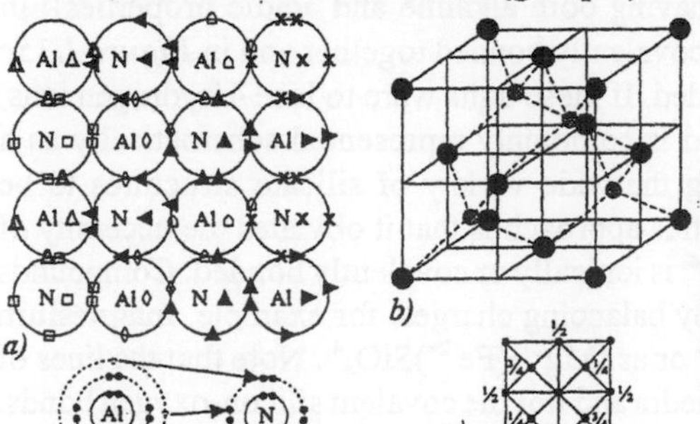

AlN is shown **covalently** bonded in ***Figure 1.14a***; each atom has a different electron representation to show where the electrons have originated. The three dimensional structure (***Figure 1.14b***) and plan structure (***Figure 1.14c***) are also shown, with the same lattice positions as ***Figure 1.14a***. In ***Figure 1.14d***. AlN is shown **ionically** bonded.

***Figure 1.14** Covalent and ionic bonding of aluminium nitride (AlN)*

However, another consideration is involved; this is the **packing efficiency (1.18)** of the positively and negatively charged units, which each have **different** radii. No two like charged ions can be adjacent to one another; this is important when the structural units are formed by **radicals (1.19.2)**. One example is compounds containing the sulfate ion (SO_4^{2-}) (considered further in section **1.18.5**), where the sulfur (S) to oxygen (O) bonding is **covalent**, whilst the whole SO_4^{2-} radical bonds with an **ionic** bond (**1.19.2**). Another example is the mixed bonding in the complex **silicates (1.17)**, which form a wide range of polymeric structures. ***Figure 1.15*** highlights the range of structural units formed by silicon and the chemical bonding.

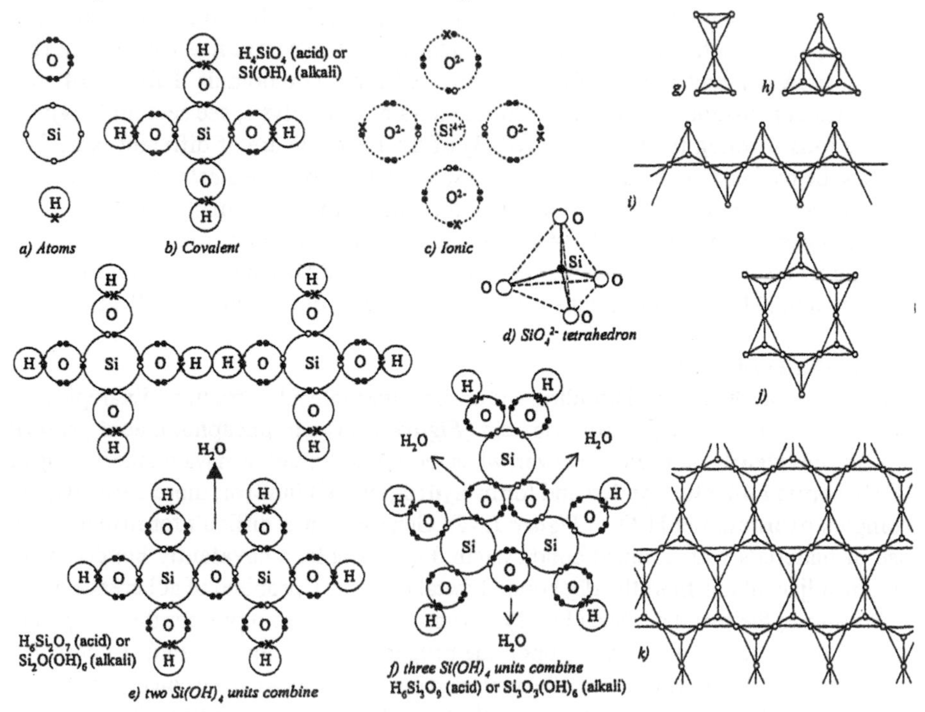

Figure 1.15 Structural units and chemical bonds formed by silicon

Figure 1.15a shows the electron representation of Si, O and H. Silicon produces hydroxides that are amphoteric (i.e., having both alkaline and acidic properties). In ***Figure 1.15b***, the atoms are all shown covalently bonded together and in ***Figure 1.15c*** the silicate unit is shown ionically bonded. If these units were to lose 4 hydrogen ions, the silicate (SiO_4^{4-}) unit is formed, and is commonly represented schematically as a **tetrahedron** (***Figure 1.15d***), enabling the wide variety of silicate structures to be simply represented. The advantage of this approach is that it obviates the necessity of discussing whether the structure $(SiO_4)^{4-}$ is ionically or covalently bonded. Compounds formed with this unit are determined by balancing charges; for example, **magnesium silicate** can be written as $(Mg^{2+})_2SiO_4^{4-}$ or as $(Mg^{2+})(Fe^{2+})SiO_4^{4-}$. Note that the lines of the tetrahedra are the edges of the tetrahedra and **not** the covalent silicon-oxygen bonds. In ***Figure 1.15e***, two silicic acid (H_4SiO_4) molecules are shown adjacent to one another.

These two molecules can be joined together to form one molecule ($H_6Si_2O_7$) through the loss of water. ***Figure 1.15g*** shows the tetrahedral representation of this unit ($Si_2O_7^{6-}$), which indicates that, for stability, a charge of six cations is required (e.g, $2Al^{3+}$ or $3Mg^{2+}$ ions) when this unit reacts. In ***Figure 1.15f***, three silicic acid (H_4SiO_4) molecules are shown combined together (on losing water, as before) to form the molecule ($H_6Si_3O_9$). The tetrahedral representation of this unit is also shown in ***Figure 1.15h***.

Silicic acid (H_4SiO_4) molecules can be joined up in this way to obtain

- **simple silicate structures** e.g, the linear silicates shown in ***Figure 1.15i***, which give rise to fibrous materials such as **asbestos**;
- **cyclic (ring) silicate structures** e.g, six silicic acid (H_4SiO_4) units can be joined to form a ring structure of the molecule $H_{12}Si_6O_{18}$ shown in ***Figure 1.15j***;
- **layered silicate structures**, e.g, **talc** and **mica** (***Figure 1.1*** illustrates the layered silicate structure of mica and shows how easily mica cleaves into thin sheets).

Each of these schematic silicate structures are held together by charged metallic ions. The layered silicate structures can absorb a large amount of water into the free space between the chains of Si_6O_{18}. Note that the more complex (cyclic) silicates can form large three dimensional polymeric structures (***Figure 1.15k***), which are essentially extensions of the structure of silica dioxide, ***Figure 1.13***). The OH^- compound of silicon is Si $(OH)_4$, which is not so stable and the two forms H_4SiO_4 and $Si(OH)_4$ coexist; thus silicon is said to be **amphoteric (1.5)**, having both acidic and alkaline properties. The formation of complex silicate structures can be summarised (***Figure 1.15k***) as

$2Si(OH)_4$	→	$Si_2O(OH)_6$	if an **alkali** (which it is not)
	water (H_2O) lost	$H_6Si_2O_7$	if an **acid** (which it is)
$3Si(OH)_4$	→	$Si_3O_3(OH)_6$	if an **alkali** (which it is not)
	water ($3H_2O$) lost	$H_6Si_3O_9$	if an **acid** (which it is)

$H_6Si_3O_9$ forms the mineral **Pectolite** ($Ca_2NaHSi_3O_9$) where Ca replaces 4 hydrogen ions, Na replaces one hydrogen ion leaving one hydrogen ion to be replaced.

$6Si(OH)_4$	→	$Si_6O_6(OH)_{12}$	if an **alkali**
	water ($6H_2O$) lost	$H_{12}Si_6O_{18}$	if an **acid**

The $H_{12}Si_6O_{18}$ unit gives rise to the mineral **Beryl** ($Be_3Al_2Si_6O_{18}$). Beryllium (Be) is in group 2 of the Periodic table and aluminium is in group 3; hence all 12 hydrogen ions in the acid $H_{12}Si_6O_{18}$ are replaced by Be (3 x 2+) and Al (2 x 3+) in the mineral Beryl. Note that, in the silicates, the silicon and oxygen atoms are bonded by the covalent bond allowing all the hydrogen to be manifested as H^+ ions.

1.14 VAN DER WAALS FORCES (OR 'BONDS')

The Van der Waals forces (or 'bonds') are residual bonds resulting from the separation of electrical charge. If the gases hydrogen (***Figure 1.10a***) or oxygen (***Figure 1.12***) are considered, it will be seen that the electrons in the covalent bond are now concentrated within the 'bond direction', so that a resultant (positive):(negative) charge separation occurs as a result of the nucleus being positively charged and the electron negatively charged. This charge arrangement is called an **electric dipole** and is present, for example, in the hydrocarbon chains that from **thermoplastics (13.6.1)** (***Figure 1.16***).

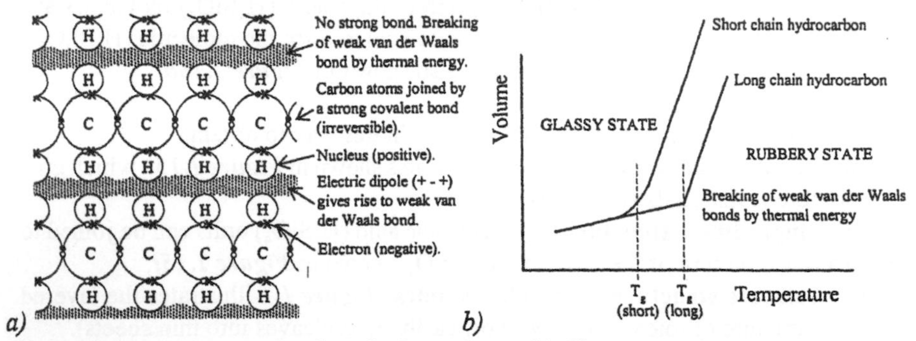

***Figure 1.16** a) Electric dipole in hydrocarbons; b) Glass transition temperature*

The bond **within** the hydrocarbon chain is the **covalent bond** and the 'bonds' **between** adjacent hydrocarbon chains are the **van der Waals** forces between the dipoles created by the hydrogen nucleus and the electrons forming the covalent bond **C-H**. The centre of each hydrogen atom is positively charged and the negative charge is fixed between the carbon and hydrogen atoms. It is breaking of the van der Waals 'bond' by thermal energy that causes the thermoplastic to melt; the temperature at which this readily happens is the **glass transition temperature (13.9.5)**, shown in ***Figure 1.16b*** (where the volume of the plastic is plotted against temperature). The low glass transition temperatures of thermoplastics reduce the energy required to form thermoplastic articles by moulding and cooling. However, due to the poor bond strength, some of the shorter chains are volatile and may be lost, making affected items brittle **(13.12.1)**. Plastics bonded by **covalent bonds** in three dimensions form a strong heat resistant; these plastics are known as **thermosetting plastics (13.6.2)**. Strong three dimensional bonding is required for high temperature applications, including repairs to motor vehicle tyres; adhesives **(13.14)** based on evaporation of solvents (producing van der Waals bonds) will fail at high temperature (and are illegal on motor vehicle tyres), so repairs are carried out by a process of **vulcanisation (13.10)**.

1.14.1 Summary: Properties of the van der Waals 'Bond'

- A directional force that can be described as a bond of varying strength;
- Mechanically very weak and elastically deformable; materials that contain the van der Waals bond (e.g, thermoplastics) have nonspecific melting points (although the size of the covalently bonded chain associated with the van der Waals bond usually overrides the mechanical properties; hence thermoplastics are almost universally used for low temperature electrical insulation, for example);
- Items manufactured from materials containing the van der Waals 'bond' are poor conductors of electricity in both the solid and liquid states;
- Transient van der Waals forces are often present between adjacent gas molecules;
- Certain adhesives **(13.15)** owe their properties to the van der Waals force or 'bond'.

1.15 STRUCTURE OF SOLIDS

The properties of a solid depend as much upon the arrangement of atoms as on the strength of the bonds between them. The packing arrangement of the atoms in a solid depends on the relative radii of the atoms involved and on the character of the bonds between them. Solids formed when the atoms are ordered in a regular three-dimensional array are termed **crystalline** solids. In other solids, the array of atoms is completely disordered, and these are termed **amorphous** solids. There are a range of solids between these two extremes in which the atoms show varying degrees of regularity; we refer to the **order** of the arrangement to distinguish the regularity of the atomic packing. For example, an atom in a diamond crystal (**1.17**) is always in the same position relative to neighbouring atoms, irrespective of where in the solid it lies. The diamond crystal has **long range order**. Some solids comprise different regions that exhibit long range order, called **crystallites** or **grains**, separated by distinct boundaries (**grain boundaries**). Although the orientation of the atoms differs from grain to grain, within each grain the atoms are arranged in a regular order; at each grain boundary there is a line of mismatch in the atomic arrangement. To distinguish solids where the regularity extends right through the material (e.g, the diamond crystal) from those that contain regularity in specific areas, we term the former **monocrystalline** and the latter **polycrystalline**. Many metals, some ceramics and many ionic and covalently bonded solids are polycrystalline. In some solids, regularity extends only over a few atoms, and we term this **short range order**. Many solids displaying short range order are **supercooled liquids**. Unlike other solids, these materials are not strictly in equilibrium, e.g, glass is a supercooled liquid which, over several hundred years or on exposure to shock, will crystallise from a structure showing short range order to one showing long range order.

1.15.1 Modelling Crystalline Solids

When chemical bonds are formed, the constituent atoms or ions form complete electron shells, which, as a first approximation, can be considered as rigid spheres (like billiard balls) (**1.3**). Adjacent atoms or ions in the structure are therefore modelled as a series of solid spheres stacked together in a regular three dimensional array. The form of this array in a given volume of space will depend on the relative size of the spheres and the way in which they are packed together. To get a feel for the modelling of the three dimensional array of atoms or ions in a solid, we first consider **close packed structures** of rigid spheres of identical radius. ***Figure 1.17a*** shows a number of spheres packed together **as close as possible** on a plane surface. Each sphere is surrounded by six nearest neighbours that just touch it. If we place two such planes in contact, each sphere in the second layer rests in the dimples formed between the spheres of the first layer. This is shown in ***Figure 1.17b***, where the centres of the spheres of the first layer are marked **A** and the centres of the spheres of the second layer are marked **B**. When we come to add a third layer, the centres of the spheres in it may **either** be placed in positions vertically above those in the first (lowest) layer, so that the sequence is ABABAB..., **or** they may be placed in the unoccupied dimples between the spheres of the second layer, denoted C, to give the sequence ABCABCABC.... In each case, the spheres of each layer are surrounded by 12 nearest neighbours (6 on the same plane, 3 in the plane below and 3 in the plane above). These two arrangements represent the **only** ways in which equally sized rigid spheres can be packed **as close as possible**, and hence produce the lowest amount of **free space** between the spheres. When modelling the three dimensional array of atoms in a solid, these two arrangements are known as **hexagonal close packed (hcp)** for the sequence ABABAB... and **face centred cubic (fcc)** for the sequence ABCABCABC....

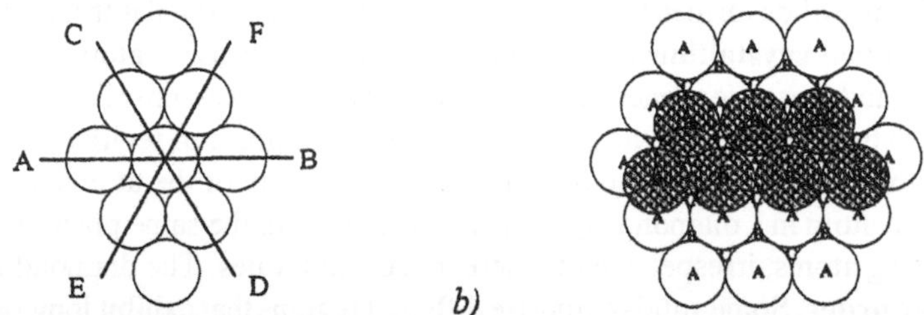

a) Close packed directions in a close packed plane of spheres; b) The B spheres (shaded) rest on the A spheres. A third close packed layer may then rest on the B spheres either directly above the A spheres or in C positions

***Figure 1.17** Close packing of identical spheres*

Although the fcc and hcp arrangements are the only ways in which rigid spheres of identical radius can be **closely packed** together, **looser packing** arrangements can appear in nature; these are termed **body centred cubic (bcc)** and **primitive**. The bcc arrangement results in each atom or ion having only 8 nearest neighbours whilst in the primitive arrangement each atom or ion has only 6 nearest neighbours (**1.15.3**).

1.15.2 Packing Arrangements of Solids

The previous sections identify four classes of packing arrangements that can exist for solids in nature with the atoms or ions considered as rigid spheres of identical radius

- the **hexagonal close packed (hcp)** (close packed; 12 nearest neighbours);
- the **face centred cubic (fcc)** (close packed; 12 nearest neighbours);
- the **body centred cubic (bcc)** (not close packed; 8 nearest neighbours);
- the **primitive** (not close packed; 6 nearest neighbours).

The close packed structures should be the preferred arrangement for solids with van der Waals bonds (**1.14**) or metallic bonds (**1.10**) because the atoms or ions have closed electron shells and are therefore approximated fairly well by the rigid sphere model. Many metals do crystallise as hcp (e.g, magnesium, zinc, cadmium, cobalt, titanium) or fcc (e.g, aluminium, copper, nickel, iron (above 910°C), lead, silver, gold) structures. Some metals exhibit the looser packing arrangements exemplified by the bcc (iron (below 910°C), chromium, molybdenum) and primitive arrangements, although the primitive arrangement is not of much importance since it is exhibited only by the metal polonium. Not all solids have atoms or ions that can be considered rigid spheres of identical radius. Other solids have bonding arrangements that dictate the arrangement of atoms in their structure. In ionic solids (**1.18**), the size of ions comprising the solid vary (**Table 1.5**) and they do not normally form close packed structures. Similarly, covalent solids (**1.17**) do not normally seek closest packing, as maintenance of the direction of their bonds overrides close packing considerations. Given the wide variety of possible atomic and ionic arrangements of solids, a model is required that accounts for all possible arrangements. This model is the **space lattice**.

1.15.3 Space Lattice and the Unit Cell

By 1848, the concept of a **space lattice** as the fundamental building block was well developed, superseding the cleaved structure model of solid materials (**1.1**). The space lattice considers solid structure as composed of atoms or ions of specific radii arranged in a regular three dimensional array (**1.15.1**). The space lattice is an infinite array of points in space, each representing an atom or ion in a solid, so arranged that each point divides the space into equal volumes with no space excluded. Every point, called a **lattice point**, has identical surroundings with every other point. The smallest volume that contains the full repetition of lattice points is called a **unit cell**. Identical unit cells must completely fill the space when they are packed face to face (like building blocks), thus forming the space lattice. Considering the three dimensional atomic arrangements geometrically, only a limited number of arrangements are possible whereby atoms and ions can fit together to give an extended space lattice. For example, a two dimensional space lattice using seven sided shapes (such as 50 pence pieces) or five sided shapes cannot be fitted together to form an extended two dimensional space lattice. As a result of similar considerations, Bravais found that there are only 14 different kinds of unit cell that can be fitted together to form an extended space lattice (***Figure 1.18***).

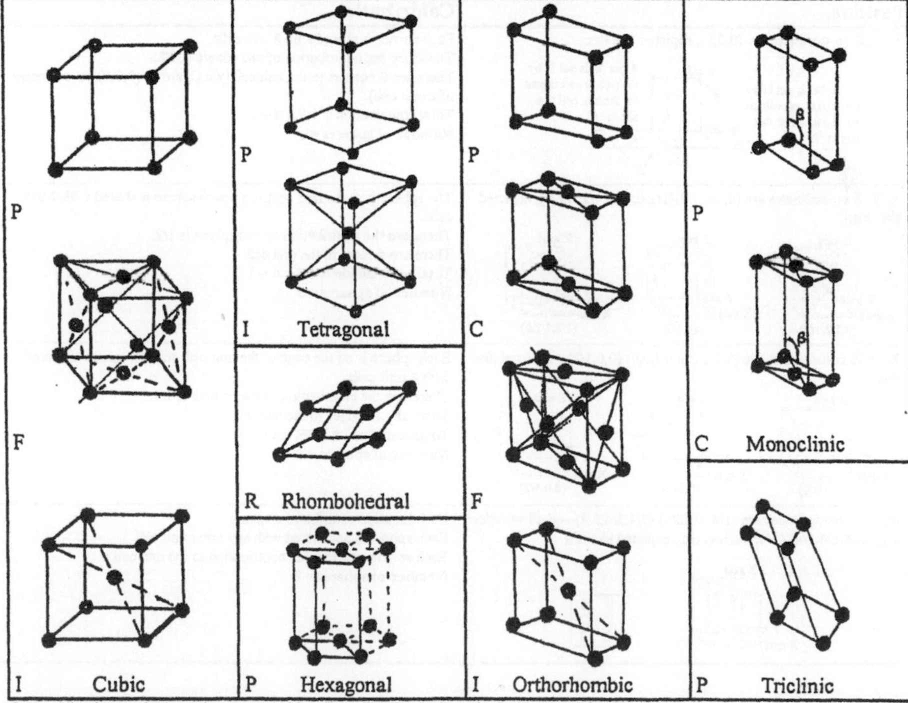

Notes: All points are lattice points. P = primitive. C = cell with a lattice point in the centre of two parallel faces. F = all face centred. I = cell with the lattice point in the centre of the interior. R = rhombohedral primitive cell.

Figure 1.18 The 14 Bravais lattices

Each lattice point (represented by a circle in ***Figure 1.18***) may accommodate more than a single atom so that more than 14 kinds of crystal structure can be formed. However, every crystal system is based on one of the 14 Bravais lattices. Accordingly, we can classify all crystalline solids into seven **crystal systems**, as shown in **Table 1.10** (which also identifies possible unit cells, i.e., hcp, fcc, bcc and primitive, available in each). Only the orthorhombic crystal system includes all unit cell types.

Table 1.10 Seven crystal systems for crystalline solids

Crystal System	Type of unit cell (***Figure 1.18***)	Axes	Axial Angles
Cubic	P F I	$a = b = c$	$\alpha = \beta = \gamma = 90°$
Tetragonal	P I	$a = b \neq c$	$\alpha = \beta = \gamma = 90°$
Orthorhombic	P F I C	$a \neq b \neq c$	$\alpha = \beta = \gamma = 90°$
Rhombohedral (trigonal)	R	$a = b = c$	$\alpha = \beta = \gamma \neq 90°$
Hexagonal	P	$a = b \neq c$	$\alpha = \beta = 90°, \gamma = 120°$
Monoclinic	P C	$a \neq b \neq c$	$\alpha = \gamma = 90° \neq \beta$
Triclinic	P	$a \neq b \neq c$	$\alpha \neq \beta \neq \gamma$

Lattice coordinates describe the position of a point on or within the unit cell. These coordinates are functions of the unit length of the axes X, Y and Z having their origin at the corner of the unit cell. The contribution of these points are shown in **Table 1.11**.

Table 1.11 Axis coordinates and number of atoms/ions contributing to the unit cell

Position	**Contribution**
X, Y, Z co-ordinates are (0,0,0), depicted like this: 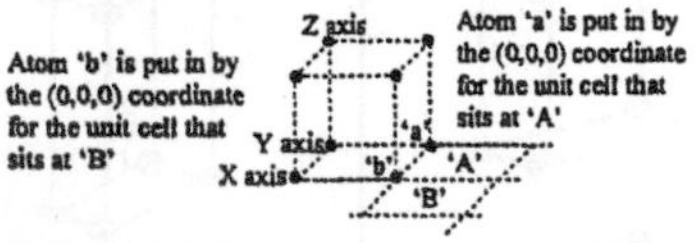 Atom 'b' is put in by the (0,0,0) coordinate for the unit cell that sits at 'B' Atom 'a' is put in by the (0,0,0) coordinate for the unit cell that sits at 'A'	Each corner is shared with 8 unit cells. Therefore the contribution of one sphere is 1/8. There are 8 corners to the unit cell (which are all (0,0,0) co-ordinates of a unit cell). Total contribution is 1/8 x 8 = 1. **Number of spheres =1**
X, Y, Z co-ordinates are (0,1/2,1/2)(1/2,0,1/2)(1/2,1/2,0), depicted like this: 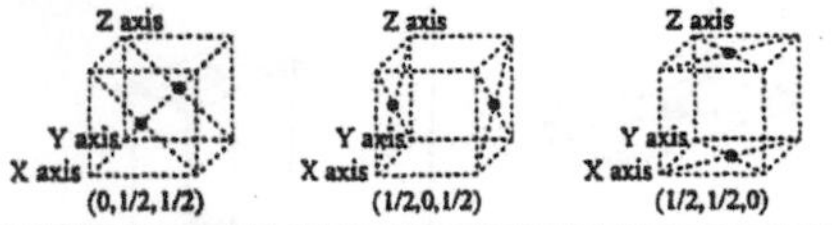 (0,1/2,1/2) (1/2,0,1/2) (1/2,1/2,0)	The sphere is at the face centre, so each sphere is shared with 2 unit cells. Therefore the contribution of one sphere is 1/2. There are 6 faces to the unit cell. Total contribution is 1/2 x 6 = 3. **Number of spheres = 3**
X, Y, Z co-ordinates are (0,1/2 ,0)(1/2,0,0) (0,0,1/2), depicted like this: 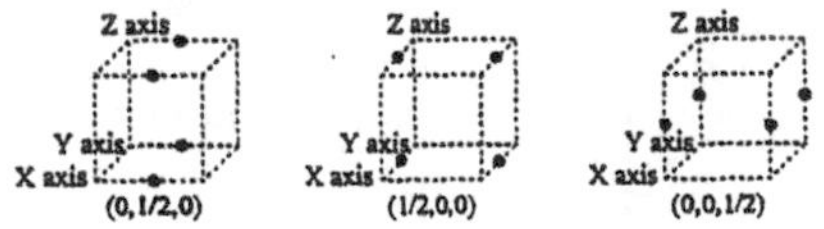 (0,1/2,0) (1/2,0,0) (0,0,1/2)	Each sphere is on the edge of the unit cell, so each sphere is shared with 4 unit cells. Therefore the contribution of one sphere is 1/4. There are 12 edges to the unit cell.. Total contribution is 1/4 x 12 = 3. **Number of spheres = 3**
X, Y, Z co-ordinates are (1/4,1/4,3/4) (3/4,3/4,3/4) and all variations e.g, (3/4,1/4,1/4) (1/4,3/4,1/4) etc., depicted like this 	Each location contributes a sphere. Each sphere is not shared with any other unit cell. Each sphere makes a full contribution to the unit cell. **Number of spheres = 8**

The packing efficiency (volume of space occupied) and the consequent free space of the atoms or ions in the unit cell can be calculated (**Table 1.12** for the cubic system only). As a first approximation, in order to form stable crystal structures, the packing efficiency must be high. It is for this reason that no metals (except polonium) have a primitive unit cell structure, as the free space is too high for stability.

Table 1.12 Percentage of free space in the cubic crystal system

Unit cell	Determination

Face centred cubic (fcc) structures (close packed structure)

a)

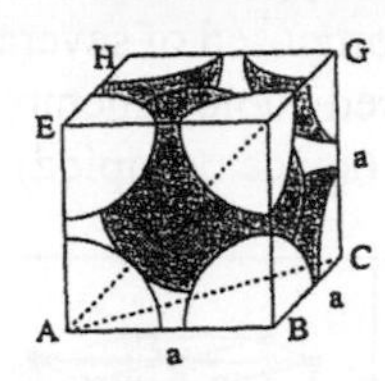

b) Face AEFB

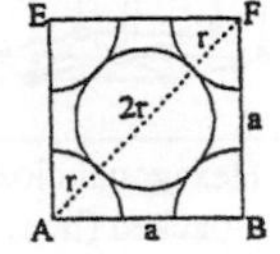

Number of atoms in the unit cell
From a), the 8 atoms at the corners of the cube are each shared by 8 unit cells, since the unit cells extend in three dimensions. However, each atom in the face is shared by only 2 adjacent unit cells. Thus the total number of atoms per unit cell is $[(8 \times 1/8) + (6 \times ½)] = 4$.
Volume of atoms per unit cell, V_a
$V_a = 4 \times (4/3\pi r^3) = 16/3\pi r^3$.
Volume of unit cell, V
From b), the constant, a, is related to the atomic radius, r, by
$AF^2 = a^2 + a^2$, so $(4r)^2 = 2a^2$, thus $a = 4r/\sqrt{2}$. Hence
$V = a^3 = \left[\frac{4r}{\sqrt{2}}\right]^3$
Packing efficiency
$V_a/V = (16/3\pi r^3)/(4r/\sqrt{2})^3 = (\pi\sqrt{2})/6 = 0.74$
(Void space $= 1 - 0.74 = 0.26$)

Body centred cubic (bcc) structures (not a close packed structure)

c)

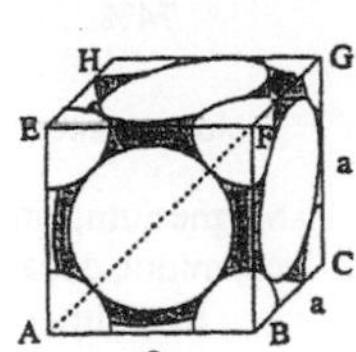

d) Face AEGC

Number of atoms in the unit cell
From c), the number of atoms per unit cell is 2, i.e., 8 atoms in the corners shared by 8 unit cells and 1 atom total within the unit cell. Hence $[(8 \times 1/8) + 1] = 2$ atoms per unit cell.
Volume of atoms per unit cell, V_a
$V_a = 2 \times (4/3\pi r^3) = 8/3\pi r^3$.
Volume of unit cell, V
From c) and d), the constant, a, is related to the atomic radius, r, by
$AC^2 = a^2 + a^2$, so $AG^2 = AC^2 + a^2 = a^2 + a^2 + a^2$. From d),
$AG^2 = (4r)^2$. Thus $(4r)^2 = 3a^2$, giving $r = (\sqrt{3}a)/4$. Hence

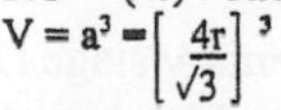

$V = a^3 = \left[\frac{4r}{\sqrt{3}}\right]^3$
Packing efficiency
$V_a/V = (8/3\pi r^3)/(4r/\sqrt{3})^3 = (\pi\sqrt{3})/8 = 0.68$
(Void space $= 1 - 0.68 = 0.32$)

Primitive lattice structures (not a close packed structure)

e)

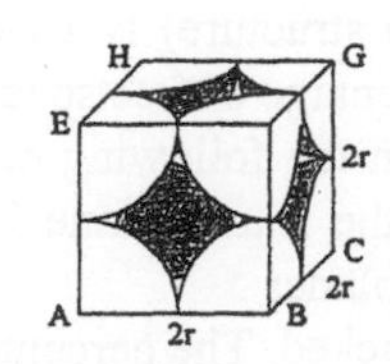

f) Face AEFB

Number of atoms in the unit cell
From e), the 8 atoms at the corners of the cube are each shared by 8 unit cells, since the unit cells extend in three dimensions. Thus the total number of atoms per unit cell is $[(8 \times 1/8)] = 1$.
Volume of atoms per unit cell, V_a
$V_a = 1 \times (4/3\pi r^3) = 4/3\pi r^3$.
Volume of unit cell, V
From f), the constant, a, is related to the atomic radius, r, by
$a = 2r$. Hence
$V = a^3 = (2r)^3$
Packing efficiency
$V_a/V = (4/3\pi r^3)/8r^3 = \pi/6 = 0.52$
(Void space $= 1 - 0.52 = 0.48$)

1.16 METALLIC STRUCTURES

The metal atoms are considered as hard **non deformable** spheres packed together so that each sphere is in contact with the adjacent spheres. The centres of the spheres are shown in ***Figure 1.19*** to represent the fundamental unit of the four types of lattices: primitive, bcc, fcc, and hcp lattices (**1.15.2**). Both fcc and hcp are composed of several close-packed planes stacked together. The primitive and body centred cubic structures are not close packed. The Figure also gives the packing efficiency (space occupied).

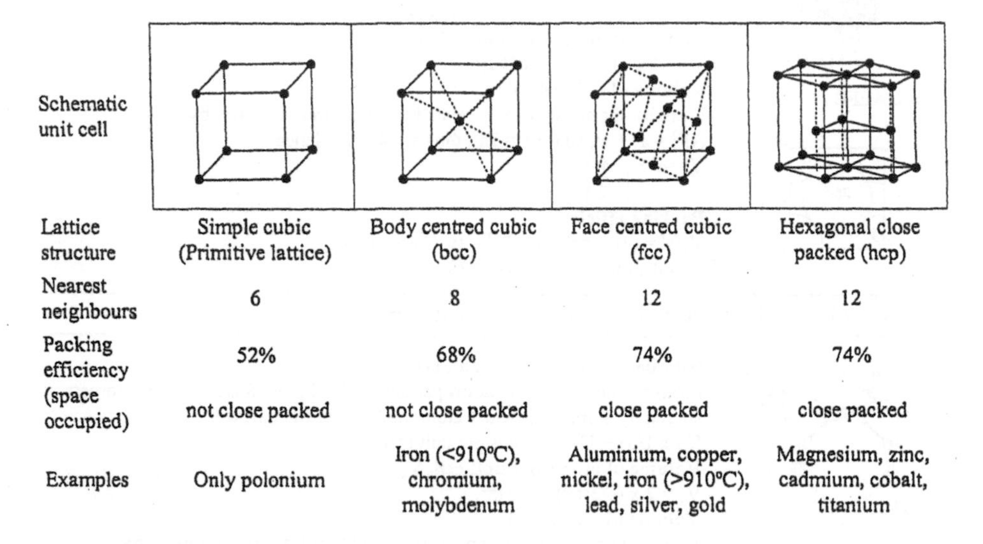

Schematic unit cell				
Lattice structure	Simple cubic (Primitive lattice)	Body centred cubic (bcc)	Face centred cubic (fcc)	Hexagonal close packed (hcp)
Nearest neighbours	6	8	12	12
Packing efficiency (space occupied)	52%	68%	74%	74%
	not close packed	not close packed	close packed	close packed
Examples	Only polonium	Iron (<910°C), chromium, molybdenum	Aluminium, copper, nickel, iron (>910°C), lead, silver, gold	Magnesium, zinc, cadmium, cobalt, titanium

***Figure 1.19** The structure (unit cells) for metals*

The three structures that concern us at this elementary stage (***Figure 1.19***) are

- the **hexagonal close packed structure (hcp)** which is a **close packed** structure of equally sized spheres or atoms. The percentage of free space is 26%. It is formed by closed packed planes being stacked in the following order **AB(C=A)ABAB...**, where (C=A) means that the third plane C is spatially over the first plane A and so is referred to as the A plane ;
- the **face centred cubic structure (fcc)** which (like the hcp structure) is a **close packed** structure of equally sized spheres or atoms. The percentage of free space is 26%. It is formed by closed packed planes being stacked in the following order **ABC(D=A)BCABC...** (*1.15.1*), where (D=A) means that the fourth plane D is spatially over the first plane A and so is referred to as the A plane;
- the **body centred cubic structure (bcc)** which is **not** close packed. The percentage of free space is 32%. This structural unit cannot be described as a series of close packed planes. The structures are shown above and because the **atoms are all metallic**, they are shown in the same way (as **filled in dots**).

As we have seen (**1.15.1**), a high packing efficiency (and low free space) are required for stable crystal structures. However, this simple theory does not explain why metals exhibit **different** crystal structures. The actual crystal structure formed is that which gives the crystal the least energy (this need not necessarily be a close packed structure). In addition, it should be noted that the energy differences between different structures is often very small. This means that a crystal structure that has the lowest energy at one temperature may not have the lowest energy at another, e.g, iron is bcc below 910°C and fcc above 910°C; this characteristic is termed **allotropy** (or **polymorphism**).

Very few metals are used in the pure state. They are nearly always alloyed with other elements to obtain better mechanical and/or enhanced durability. The alloying elements dissolve in the basis metal to form **solid solutions**, which are in many ways similar to liquid solutions (**2.9.1**). Some elements dissolve readily, some with more difficulty. For example, iron can only dissolve about 0.007% carbon (to make steels) at room temperature, but copper can dissolve about 30% zinc (to make brass) and in the copper-nickel system (Monel metal, cupronickels, etc.) there is complete solubility. There are two principle classes of solid solution. In one, small atoms (such as carbon and nitrogen) fit into the spaces between the larger atoms to form **interstitial solid solutions** (**10.5.2**). The solubility of these small atoms is generally limited, but the effects on the properties of the resultant alloy can be dramatic, In the other class, the dissolved atoms are similar in size to those of the basis metal and they simply replace some of the basis metal atoms to form **substitutional solid solutions** (**10.5.2**). The classic example of substitutional solid solutions is the copper-nickel system, where the sizes of the atoms of copper (Cu = 0.128 nm) and nickel (Ni = 0.124 nm) are similar (**Table 1.5**); they are also formed with copper and zinc (Zn = 0.133 nm) in the brass system.

The maximum ratio of atomic sizes of alloying elements that can be tolerated in substitutional solid solutions is about 15%. Some atoms are too big to fit comfortably into the lattice and some atoms are too small (though not small enough to achieve interstitial solubility. In either case, the energy of the lattice is increased and the surplus rejected, usually in the form of an **intermediate compound** (e.g, iron carbide, Fe_3C, in steel or copper aluminate, $CuAl_2$, in some aluminium alloys). Generally, such compounds are hard and brittle, and exert a reinforcing effect of the soft matrix. Some intermediate compounds are used in their own right, such as tungsten carbide for cutting tools and tantalum carbide in 'superalloys' that can operate at temperatures of 1200°C as, for example, in gas turbines. Intermediate compounds are examples of **phases**, simply a material or region of a material that has the same properties. Water and petrol are liquid phases, whereas ice is a solid phase, and a mixture of water and ice is a two-phase mixture. Phases in metals are examined in **10.6**.

1.17 COVALENT STRUCTURES

There are several simple structures formed by covalently bonded materials, as each atom has to be surrounded by (8-N) atoms in order to obtain the stable electron configuration. Therefore there are essentially different structures for each group that combine to form a three dimensional structure. A good example is carbon, a group 4 element that forms three completely different structures: diamond, graphite (***Figure 1.20***) and fullerenes. Graphite is formed quickly in the iron carbon phase diagram (**10.11**) and is soft and used as a lubricant. Conversely, diamond is the hardest known substance and is used in jewellery and diamond cutting equipment.

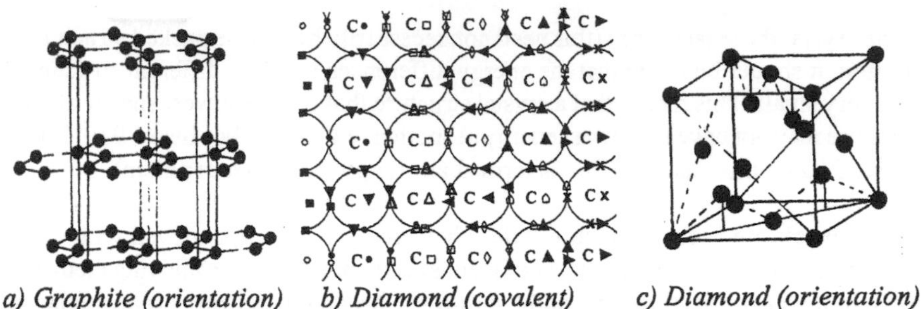

a) Graphite (orientation) *b) Diamond (covalent)* *c) Diamond (orientation)*
***Figure 1.20** Structure of covalently bonded carbon (graphite and diamond)*

The **graphite** structure is elongated in one direction (the C-C interatomic distance is 0.340 nm, compared to 0.142 nm within the plane and 0.154 nm in the diamond structure). The softness of graphite is ascribed to the atoms in one direction being further apart due to weak (van der Waals) forces or bonds (**1.14**), which means that the planes can accommodate occluded gases, water, oxygen etc. (These gases are removed in a vacuum, leading to rapid wear of the carbon brushes of high-flying airliners). Conversely, the **diamond** structure is modelled as a **three-dimensional** structure in which the carbon atoms are all equal distances apart (interatomic distance 0.154 nm), bonded by very strong **covalent bonds (1.11)**. This structure gives diamonds have very high melting (>3800°C) and boiling points (5100°C). Graphite may be changed into diamond at very high pressure (5×10^7 N/m^2) and temperatures (1500°C); this process is used industrially, but only a small amount of diamond is produced.

Covalent structures are also formed between silicon (Si) (group 4) and oxygen (O) (group 6) in the **silicates (1.13)**. Applying the (8-N) rule, each atom of silicon must have (8-4=) 4 nearest neighbours and each oxygen (8-6=) 2 nearest neighbours (***Figure 1.21***).

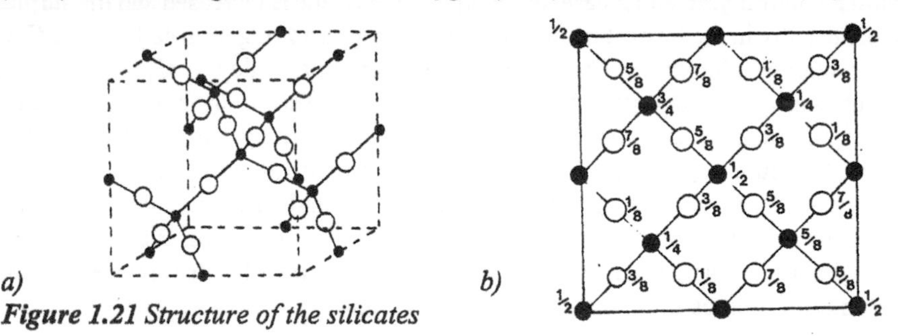

a) *b)*
***Figure 1.21** Structure of the silicates*

The scientist's model is shown as a three-dimensional repeating structure, whose unit cell is shown in ***Figure 1.21a*** (a plan view of the unit cell is shown in ***Figure 1.21b***; **Table 1.11** gives the ordinate system). The silicate tetrahedra, i.e., silica dioxide ('silica') (SiO_2), is covalently bond together in a close packed arrangement to give a large molecule (sand) with very little free space. The schematic three dimensional view of the structure of sand illustrates the bond direction of 109° within the tetrahedra. Given the ability of the silicates to join up into long chains and three dimensional structures, it is easy to see how a durable brick (**7.2**) is made from a very plastic clay, even though (in the unfired state) it will disperse in water. It is more difficult to see how glass (**9.1**) is modelled. In glass, the angles within each tetrahedra still have to be 109°, but they are not ordered and repeated regularly. The metallic cation additions to glass are dispersed within the disordered chains (***Figure 9.3***). Glass is an amorphous structure (**1.15**) and is formed if silica is cooled too quickly; this can occur in nature (amorphous silica from volcanoes is called **obsidian**) or during manufacturing processes (bricks become glassy if they are fired at too high a temperature).

1.18 IONIC STRUCTURES

A (pure) ionic structure is modelled as a three dimensional array of positively and negatively charged ions packed together so that no two similarly charged ions are adjacent. A unit cell (**1.15.3**) is arbitrarily taken from this array, where the corners of the cell are formed by like ions. Orthogonal axes are usually preferred by ionic structures, although in some of the more complex ionic structures, nonorthogonal axes are required. The ions of the different elements in the structure are of different chemistry (e.g, metals and non-metals) and are distinguished in the following Figures by open and closed circles to represent the lattice positions. Ionic structures are usually formed from the same number of cations (A^{n+}) and anions (X^{n-}), and are collectively known as **AX structures.**

The sizes of the ions are related through the ratio of the radius, r_c, of the cations over the radius, r_a, of the anions. This ratio is known as the **radius ratio**, and dictates the **packing efficiency** and **percentage free space** in the structure. Generally, stability is attained at a packing efficiency of about 0.6 to 0.74 (corresponding to a percentage free space of 40 to 26%), although the exact packing efficiency ranges for stability will depend on the type of unit cell (**1.15.3**). Note that the unit cell must accommodate the atoms in the formula unit of the compound, whilst still satisfying the packing density conditions. For example, in a compound with the formula unit of A_2X_3, the structure must accommodate 2 ions of element A and 3 ions of element X whilst still satisfying the radius ratio and packing density conditions for stability.

There are essentially three simple AX structures that concern us at this initial stage. These are normally designated by reference to **the number of nearest neighbours (or coordination number) in the structural unit for both ions A and X**, denoted in square brackets [ ]. Ionic structures possess anions and cations of different sizes such that **coordination polyhedrons** of anions are formed around every cation (and vice versa); the ionic structure will only be stable if the cation is in contact with each of its neighbours. If, for example, the cations are too small to touch the anions, the forces of attraction and repulsion between the ions do not balance and the structure is not stable. Thus the structural stability of ionic compounds is dictated by the difference in the sizes of the constituent anions and cations and the nature of the unit cell.

1.18.1 [8]:[8] Structures

[8]:[8] structures, exemplified by caesium chloride (CsCl) ($A = Cs^+, X = Cl^-$), are shown in ***Figure 1.22***. [8]:[8] structures are formed from the same number of anions and cations. Each ion in the structural unit is surrounded by 8 nearest neighbours.

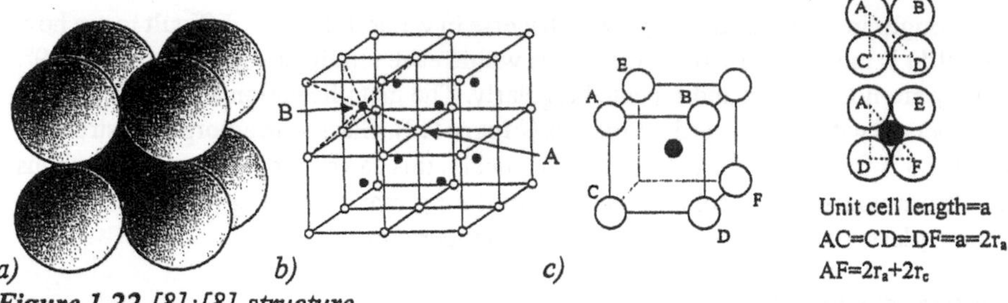

Figure 1.22 *[8]:[8] structure*

In ***Figure 1.22a***, the larger white spheres represent the anions (Cl^-) and the smaller black sphere in the centre of the cube represents the cations (Cs^+). Note we could have chosen the unit cell so that the corner positions were occupied by the smaller cations (Cs^+) with no effect on the modelling procedures. As ***Figure 1.22a*** is rather cumbersome, we reduce the size of the spheres to represent the [8]:[8] structure in three dimensions in ***Figure 1.22b*** as 8 units cells (with the anions at corner positions of each unit cell) and the single unit cell in ***Figure 1.22c***. In ***Figures 1.22b and c***, (○) represent the anions (Cl^-) and (●) represent the cations (Cs^+). ***Figure 1.22c*** can be used to derive the radius ratio for the [8]:[8] structure. From face ABCD, $AD^2 = AC^2 = CD^2 = (2r_a)^2 + (2r_a)^2$, hence $AD = 2\surd 2r_a$. From body AEDF, length $AF^2 = AD^2 + DF^2 = (2r_a)^2 + (2r_a)^2 + (2r_a)^2 = 3(2r_a)^2$. As $AF^2 = (2r_a + 2r_c)^2$, we can write $(2r_a + 2r_c)^2 = 3(2r_a)^2$, hence $2r_a + 2r_c = 2\surd 3r_a$, yielding $r_c = (\surd 3 - 1)r_a$. The theoretical radius ratio for an [8]:[8] structure, r_c/r_a is $(\surd 3 - 1) = 0.732$. [8]:[8] structures are only formed when the radius ratio is between 1.0 to 0.732.

Figure 1.22b illustrates that the anions (○) lie at the corners of **cubic** coordination polyhedron and a single cation (●) is at the body centre, with 8 anion nearest neighbours (as shown at A). Note that it is also apparent that each anion is surrounded by 8 cation nearest neighbours (B in ***Figure 1.22b***). From ***Figure 1.22b***, the 8 anions at the corners of the cube are each shared by 8 unit cells, since the unit cells extend in three dimensions. However, the cation in the centre of the unit cell is not shared by any other unit cell. Thus there are [1 x 1/8] = 1 anion per unit cell and 1 cation per unit cell. Hence the structure is referred to as an [8]:[8] structure (and since there are the same number of anions and cations, the chemical formula for caesium chloride is CsCl; there is one molecule of CsCl to the unit cell). The volume of unit cell, $V = a^3$. From face ABCD in ***Figure 1.22c***, $AC = a = 2r_a$, hence $r_a = a/2$. Since $AF = 2\surd 3r_a$, then $AF = a\surd 3$. From body AEDF, we know $AF = 2r_a + 2r_c$ and, since $r_a = a/2$, we can write $AF = a + 2r_c$. Hence $AF = a\surd 3 = a + 2r_c$, which yields $2r_c = a(\surd 3 - 1)$. As there is one anion and one cation per unit cell, the volume of space occupied by the ions, $V_a = (4/3)\pi r_a^3 + (4/3)\pi r_c^3$, so $2V_a = (\pi/3)(2r_a)^3 + (\pi/3)(2r_c)^3$. We can now substitute known terms into this relationship to give the volume of space occupied by the ions, $V_a = \{(\pi/3)a^3 + (\pi/3)(\surd 3 - 1)^3 a^3\}/2$. Hence the **packing efficiency** for the [8]:[8] structure,

$$\frac{V_a}{V} = \frac{(\pi/3)a^3 + (\pi/3)(\surd 3 - 1)^3 a^3}{2a^3} = 0.729$$

and the **free space** is 1 - 0.729 = 27%.

1.18.2 [6]:[6] Structures

[6]:[6] structures, exemplified by sodium chloride (NaCl) ($A = Na^+$, $X = Cl^-$), are shown in ***Figure 1.23***. [6]:[6] structures are formed from the same number of anions and cations. Each ion in their structural unit is surrounded by 6 nearest neighbours.

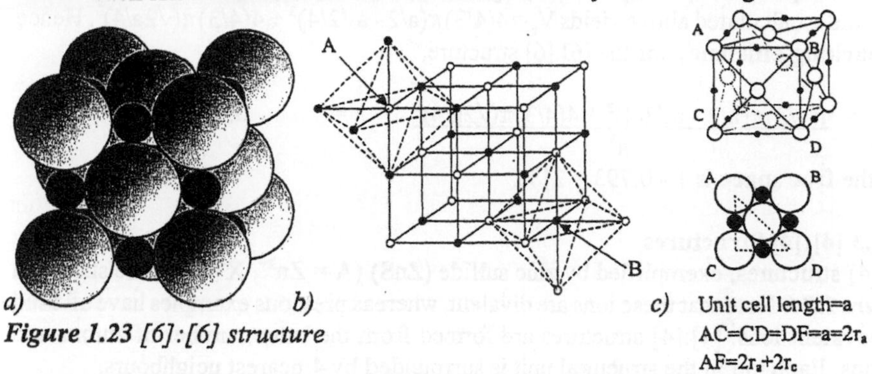

***Figure 1.23** [6]:[6] structure*

In ***Figure 1.23a***, the white spheres at the corners of the cube represent the anions (Cl^-) and the smaller dark spheres represent the cations (Na^+). Note that in this Figure, as before, we could have chosen the unit cell so that the corner positions were occupied by the smaller cations (Na^+) with no effect on the modelling procedures. In addition, note that, as in all ionic structures, the cations must be close enough to the anions to form bonds and so create the structure; the cations are in contact with the anions, just fitting the space between them. The three dimensional [6]:[6] structure is simplified in ***Figure 1.23b*** and the unit cell is shown in ***Figure 1.23c***. In ***Figures 1.23b and 1.23c***, (○) represent the anions (Cl^-) and (●) represent the cations (Na^+). From ***Figure 1.23*** it is apparent that the [6]:[6] structure is equivalent to two interpenetrating F Bravais lattices (**1.15.3**). ***Figure 1.23c*** can be used to derive the radius ratio for the [6]:[6] structure. From face ABCD, $AD^2 = AC^2 + CD^2$, thus $(4r_a)^2 = (2r_c + 2r_a)^2 + (2r_c + 2r_a)^2 = 2(2r_c + 2r_a)^2$. Hence $4r_a = 2\sqrt{2}(r_c + r_a)$, which simplifies to yield $r_c = (\sqrt{2} - 1)r_a$. The theoretical radius ratio for a [6]:[6] structure, r_c/r_a is $(\sqrt{2} - 1) = 0.414$. [6]:[6] structures are only formed when the radius ratio is between 0.732 to 0.414.

The anion (○) marked A in ***Figure 1.23b*** is surrounded by 6 cations (●) arranged towards the corners of a regular **octahedron** (by inspection, this is true for all anions). Likewise, the cation (●) marked B in ***Figure 1.23b*** is surrounded by 6 anions (○) arranged towards the corners of a regular **octahedron** (by inspection, this is true for all cations). In ***Figure 1.23b***, there are 8 anions at the corners of the unit cell, each shared by 8 unit cells. In addition, there are 6 anions in the centre of each face of the unit cell, each shared by 1 unit cell. Thus there are $[(8 \times 1/8) + (6 \times ½)] = 4$ anions per unit cell. The 12 cations at the cube edges are each shared by 4 unit cells. In addition, there is a cation at the centre of the unit cell that is not shared by any other unit cell. Thus there are $[(12 \times 1/4) + 1] = 4$ cations per unit cell. Since each ion is surrounded by 6 ions of the opposite type, this structure is referred to as a [6]:[6] structure (and since there are the same number of anions and cations, the chemical formula for sodium chloride is NaCl; there are four molecules of NaCl to the unit cell). The volume of unit cell, $V = a^3$. From face ABCD in ***Figure 1.23c*** and our previous calculations, we know that AD

$= 4r_a = \sqrt{2}(2r_a + 2r_c)$. Since $AC = CD = 2r_a + 2r_c = a$, we can state $4r_a = a\sqrt{2}$, so $r_a = a\sqrt{2}/4$. From our radius ratio calculations above we know that $r_c = r_a(\sqrt{2} - 1)$, thus $r_c = a\sqrt{2}(\sqrt{2} - 1)/4 = a/2 - a\sqrt{2}/4$. As there are four anions and four cations per unit cell, the volume of space occupied by the ions, $V_a = 4(4/3)\pi r_a^3 + 4(4/3)\pi r_c^3$. Substituting values for r_a and r_c calculated above yields $V_a = 4(4/3)\pi(a/2 - a\sqrt{2}/4)^3 + 4(4/3)\pi(\sqrt{2}a/4)^3$. Hence the **packing efficiency** for the [6]:[6] structure,

$$\frac{V_a}{V} = \frac{4(4/3)\pi(a/2 - a\sqrt{2}/4)^3 + 4(4/3)\pi(\sqrt{2}a/4)^3}{a^3} = 0.793$$

and the **free space** is 1 - 0.793 = 21%.

1.18.3 [4]:[4] Structures

[4]:[4] structures, exemplified by zinc sulfide (ZnS) ($A = Zn^{2+}$, $X = S^{2-}$), are shown in ***Figure 1.24***. Note that these ions are divalent, whereas previous examples have all used monovalent ions. [4]:[4] structures are formed from the same number of anions and cations. Each ion in the structural unit is surrounded by 4 nearest neighbours.

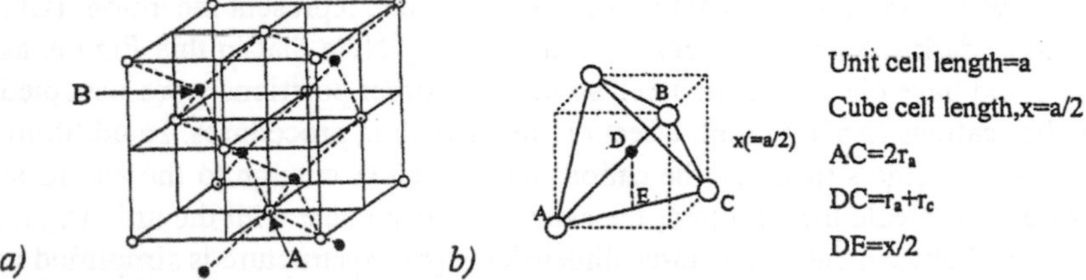

Figure 1.24 *[4]:[4] structure*

The [4]:[4] structure is represented in three dimensions in ***Figure 1.24***. In ***Figures 1.24a and 1.24b***, (○) represent the anions (S^{2-}) and (●) represent the cations (Zn^{2+}). Note that the anions (S^{2-}) are arranged precisely in the same way as the chlorine anions (Cl^-) in the [6]:[6] structure described above. In the [4]:[4] structure, however, it can be seen that the anion (○) marked A in ***Figure 1.24a*** is surrounded by 4 cations (●) arranged towards the corners of a regular **tetrahedron**. Likewise, by inspection, it can be seen that each cation (●) is surrounded by 4 anions (○) arranged towards the corners of a regular tetrahedron (B in ***Figure 1.24a***). Geometrically, it is most simple to consider the four coordinating anions occupying four of the eight corners of a regular cube (***Figure 1.24a***) whose sides are half the unit cell length, x = a/2; the resulting coordination polyhedron is then a regular tetrahedron (***Figure 1.24b***). Therefore $AC^2 = x^2 + x^2 = 2x^2$, thus $AC = x\sqrt{2}$. Since $AC = 2r_a$, we can write $x = 2r_a/\sqrt{2}$. Also, $DC^2 = (AC/2)^2 + DE^2 = (x\sqrt{2}/2)^2 + (x/2)^2 = 3x^2/4$, thus $DC = x\sqrt{3}/2$. We know that $DC = r_a + r_c$, so we can write $r_a + r_c = x\sqrt{3}/2 = (\sqrt{3}/2)(2r_a/\sqrt{2}) = r_a\sqrt{(3/2)}$. Hence $r_c = [\sqrt{(3/2)} - 1]r_a$ and the theoretical radius ratio for a [6]:[6] structure, r_c/r_a is $\sqrt{(3/2)} - 1 = 0.225$. [4]:[4] structures are only formed when the radius ratio is between 0.414 to 0.225.

In ***Figure 1.24a***, the 8 anions at the corners of the unit cell are each shared by 8 unit cells. In addition, there are 6 anions in the centre of each face of the unit cell, each shared by 1 unit cell. Thus there are [(8 x 1/8) + (6 x 1/2)] = 4 anions per unit cell. There are 4 cations within the unit cell that are not shared with any other unit cell. Thus there are 4 cations per unit cell. Since each ion is surrounded by 4 ions of the opposite type, this structure is referred to as a [4]:[4] structure (and since there are the same number of anions and cations, the chemical formula for zinc sulfide is ZnS). The volume of unit cell, $V = a^3$. From our previous calculations we know $x = 2r_a/\sqrt{2}$, so $r_a = x\sqrt{2}/2 = a\sqrt{2}/4$ (since $x = a/2$). In addition, we know $r_c = [\sqrt{(3/2)} - 1]r_a$, therefore $r_c = [\sqrt{(3/2)} - 1]a\sqrt{2}/4 = a(\sqrt{3} - \sqrt{2})/4$. As there are four anions and four cations per unit cell, the volume of space occupied by the ions, $V_a = 4(4/3)\pi r_a^3 + 4(4/3)\pi r_c^3$. Substituting values for r_a and r_c calculated above yields $V_a = 4(4/3)\pi\{(a\sqrt{2}/4)^3 + [a(\sqrt{3} - \sqrt{2})/4]^3\}$. Hence the **packing efficiency** for the [6]:[6] structure,

$$\frac{V_a}{V} = \frac{4(4/3)\pi\{(a\sqrt{2}/4)^3 + [a(\sqrt{3} - \sqrt{2})/4]^3\}}{a^3} = 0.749$$

and the **free space** is 1 - 0.749 = 25%.

A characteristic of ionic structures is that they have non directional bonds (**1.12**). The structure formed is very dependent upon the number of nearest neighbours, which in turn is dictated by the radius ratio. The radius ratios determined in this section are for idealised structures in which the anions and cations of the unit cell all just touch one another. In practice, ionic solids with intermediate structures can be formed (and therefore the radius ratios are quoted as ranges). ***Figure 1.25*** shows how the **packing efficiency** is altered for the different **radius ratios**.

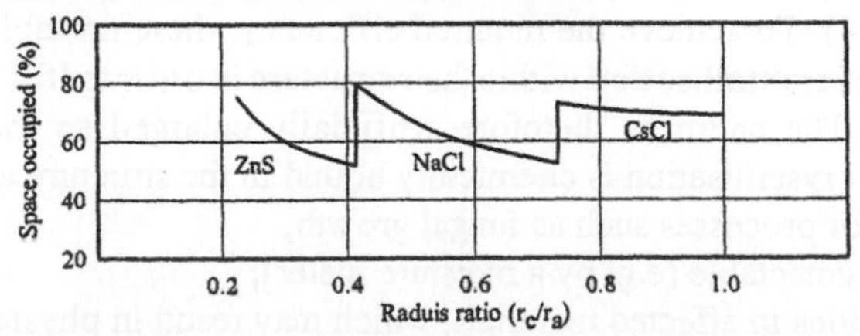

***Figure 1.25** The relationship of packing efficiency and radius ratio for ionic structures*

1.18.4 More Complex Ionic Structures

More complex ionic structures like A_2X, AX_2, AX_3 and ABX_3 are possible and each must have the correct number of ions/atoms in each unit cell. The atoms/ions have specific positions in the unit cell, as shown in ***Figure 1.26***.

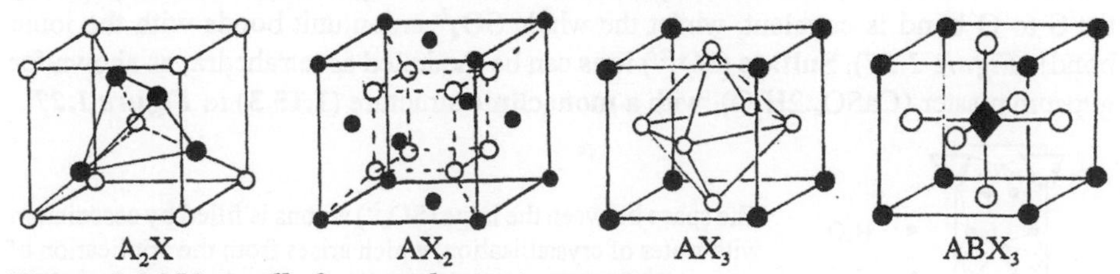

***Figure 1.26** Unit cells for complex ionic structures*

AX_2 structures, such as the [8]:[4] structure exemplified by the salt calcium fluoride (CaF_2) ($A = Ca^{2+}$, $X = F^-$), require twice as many anions (F^-) as cations (Ca^{2+}) packed in the three-dimensional array. In ***Figure 1.26***, (●) represent the anions (F^-) and

(○) represent the cations (Ca^{2+}). By inspection it can be seen that each anion (●) is surrounded by 4 cations (○) arranged **tetrahedrally**, whereas each cation (○) is surrounded by **8** anions (●) arranged towards the corners of a **cube**. Hence this structure is an [8]:[4] structure (and the chemical formula for calcium fluoride is CaF_2). In the ABX_3 structure, it would seem that there are more cations (○) (A) than anions (●) (X). In order to obtain the correct number of anions and cations for a single unit cell, we must consider the three-dimensional picture. Each corner cation (○) can be seen to be shared with 8 unit cells and therefore each contributes 1/8 of its volume to the unit cell. As there are 8 corners, the total contribution of the cations (A) is (8 x 1/8) 1 cation. Likewise, the anions (●) contribute only 1/2 to the unit cell, as each face is shared with 2 unit cells. As there are 6 faces, the total contribution is (1/2 x 6) 3 anions (X). The B ion (♦) is at the body centred position and is not shared with any other unit cell. Hence the formula unit is ABX_3. The remaining formula units can be obtained in a similar way.

If the radius ratio of an ionic structure alters (e.g, by **ionic substitution, 1.18.6**) then different structures are formed; the structural arrangement of the ions must reform if the cation is replaced.

1.18.5 Water of Crystallisation

Some anions (X^{n-}) are so large in comparison to the size of available cations (A^{n+}) that the packing efficiency of the compound will not come within the range required for stability (0.6 to 0.74). To achieve the required efficiency, these unstable compounds incorporate **water of crystallisation** within their structure in order to 'fill' the space left around the cation. The cation is therefore artificially enlarged so that stability is achieved. Water of crystallisation is **chemically** bound in the structure and therefore

- is not available for processes such as fungal growth;
- is not physically detectable (e.g, by a moisture meter);
- produces expansions in affected materials, which may result in physical damage;
- relatively high temperatures are required to remove chemically bound water of crystallisation from the structure (e.g, concrete 300°C; gypsum plaster 220°C).

There are various examples of construction materials containing water of crystallisation, some of which are highlighted below. In all cases, the ions are usually formed from a mixture of **covalently** and **ionically** bonded structures (e.g, in the sulfate SO_4^{2-} anion, the S to **O** bond is covalent, whilst the whole SO_4^{2-} anion unit bonds with the ionic bond) (***Figure 1.27***). **Sulfate (SO_4^{2-}) ions** can be modelled as tetrahedra, as shown for gypsum plaster ($CaSO_4.2H_2O$), with a **monoclinic** structure (**1.15.3**) in ***Figure 1.27***.

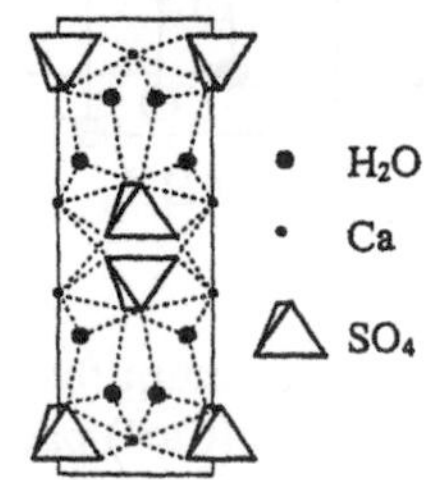

The space between the large (SO_4^{2-}) anions is filled by association with water of crystallisation, which arises from the application of mix water to gypsum plaster 'in the bag' (calcium sulfate hemihydrate, $CaSO_4.½H_2O$) to form $CaSO_4.2H_2O$ (**14.7.3**). The water of crystallisation is the cause of the expansion in setting plaster work, which hides shrinkage cracks in the substrate.

***Figure 1.27** The sulfate (SO_4^{2-}) ion in gypsum plaster ($CaSO_4.2H_2O$)*

Several other sulfates contain water of crystallisation, for example

- copper sulfate ($CuSO_4$.**5H$_2$O**), used in the anhydrous state for tests for free water;
- potash alums ($K_2SO_4.(NH_4)_2SO_4$.**12H$_2$O**, $K_2SO_4.Al_2(SO_4)_3$.**24H$_2$O**, etc.) (**14.7.3**);
- zeolites $(Na,Ca)_{4\text{ to }6}Al_6(Al,Si)_4Si_{26}O_{72}$.**24 H$_2$O** (**4.4.2**);
- a calcium sulfoaluminate known as **ettringite** ($3CaO.Al_2O_3.3CaSO_4$.**32H$_2$O**), formed when a constituent of cement (tri-calcium aluminate) reacts with water soluble sulfates. The reaction gives rise to **sulfate attack** within hardened cementitious materials that causes degradation of, for example, concrete and mortar beds (**8.19.2**) and, in some cases, converts the cementitious material back into its constituent parts (fine and coarse aggregate). Note that this reaction also occurs in the manufacture of cement when gypsum is added at the grinding stage (**7.15.1**). In all cases, it is the chemically combined water of crystallisation that produces the volume expansion and consequent degradation;
- in another form of the sulfate attack, the mineral **thaumasite** is formed in SRPC when sulfates and carbonates are present (**8.19.2**). In this case, the sulfates react with the calcium hydrate silicates (the bonding compounds in cement) in the presence of water to form the expansive sulfate compound $CaSiO_3.CaCO_3.CaSO_4$.**15H$_2$O**;
- sulfuric acid H_2SO_4 can be formed from the 6 valent sulfur (S^{6+}), combining with $6OH^-$ ions to form $S(OH)_6$ or H_6SO_6 (***Figure 1.28***). In this case, the anion formed (SO_6^{6-}) would be so large that few cations are big enough to fill the free space and so achieve stability. The required packing efficiency is therefore obtained by removal from the $S(OH)_6$ molecule of 2 water molecules (***Figure 1.28***) to leave sulfuric acid (H_2SO_4). The removal of water is equivalent to a **condensation reaction** that takes place when a thermosetting plastic is made (**13.5.2**). Sulfuric acid (H_2SO_4) is used as 12V battery acid; when it **ionises** (**1.12**), it loses the two H^+ ions so that the resultant unit (i.e., the SO_4^{2-} anion) becomes charged (2-).

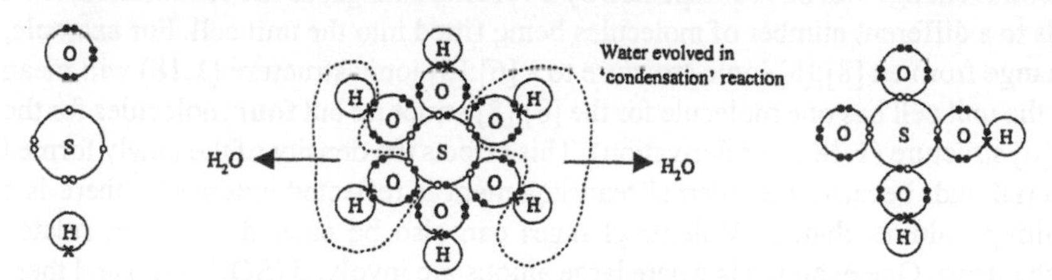

a) Atoms *b) Electron configuration of OH compound* *c) H_2SO_4*

Figure 1.28 *Formation of sulfuric acid (H_2SO_4) by a condensation type reaction*

In addition to sulfates, phosphates (PO_4^{3-}), aluminates ($Al_2O_4^{3-}$), borates ($B_2O_4^{3-}$ and $B_4O_7^{3-}$) and, to some extent, silicates (SiO_4^{4-}) are all associated with water of crystallisation. For example, Na_3PO_4.**12H$_2$O**, used in the polymer form for descaling boilers under the trade name **Calgon**^{TM} (an example of the phosphate radical joining up into larger units in the same way as the silicates is shown in ***Figure 1.15***). As with sulfates, these ions may cause degradation to cementitious products.

1.18.6 Ionic Substitution

As we have seen, models for the structure of an ionic material are dependent upon the packing efficiency of charged ions to form the unit cell, which can theoretically be repeated forever (**1.15.1**). **The dependence of the unit cell upon the radius ratio of the constituent ions has a profound effect upon durability, stability and melting points of the ionic compounds formed.** For example, the size of the cation determines

- the **packing efficiency** (**1.15.2**) of the ions and the stability of the ionic structure;
- the number of **nearest neighbours** (**coordination number**) in the structure (**1.18**);
- the **free (void) space** (**1.18**) in the structure.

Free (void) space is particularly important where the ions in the structure can be substituted, since the **valency** of the original ion may not be the same as that replacing it. For example, a (2+) cation replacing a (1+) cation in ionic AX structures must cause a vacant (1+) site for every **one** (2+) cation substituted if overall neutrality is to be maintained. This produces a **defect lattice** structure (***Figure 1.29***). In many instances, substitution may be yet more complex, with many ions of differing valency capable of substituting for the original ions in the structure; thus many new ionic structures can be formed. For example, assume that water containing both metallic (A) and non metallic (X) elements in the ionic form is running down the surface of an ionic material of composition BY. There will be random replacement of ions in the BY structure by a process termed **ion exchange**, a process is responsible for the degradation of many building materials. Ion exchange is progressive through the following stages

- the exchange of ions of the same valency, producing no structural changes (e.g, the substitution of Ca^{2+} ions with Mg^{2+} ions, as occurs in magnesium limestones);
- the exchange of ions of different valency causes a **defect lattice** to be formed (***Figure 1.29***) if there is insufficient ionic substitution to produce a **structural change** (a structural change only occurs when there is sufficient ionic substitution);
- further substitution may result in a **structural change** (**Table 1.13**);
- structural change can be accompanied by a **volume** change, as the substitution often leads to a different number of molecules being fitted into the unit cell. For example, a change from an [8]:[8] ionic structure to a [6]:[6] ionic structure (**1.18**) will mean that the unit cell has one molecule for the [8]:[8] structure but **four** molecules for the [6]:[6] structure (**1.18.2** for derivation). This affects the density of the newly formed material and, because the internal rearrangement is reflected externally, there is a resulting volume change. Volume changes can also be caused by other, related mechanisms. One example is where large anions are involved (SO_4^{2-}, etc.) and there no large cations to fill the free (void) space between the tetrahedra (***Figure 1.27***), so the cations become larger by associating with **water of crystallisation** (**1.18.5**).

As an example of these processes, consider a material whose structure is bonded by calcium carbonate. The possible mechanisms of erosion are replacement of the cations or anions, formation of a defect lattice that may remain stable (but if the substitution is high the lattice structure must change to accommodate the differently charged ions, which will have different radii because they have lost more electrons, **Table 1.5**) or a change from an insoluble compound into a soluble compound (**4.3.1**). The formation of a defect lattice is shown diagrammatically in ***Figure 1.29***. If mixed valency ions are considered as replacements for any one ion in the original ionic structure (shown here as a Ca^{2+} cation replacing a Na^{+} cation in the original sodium chloride ionic lattice), then for every one 2+ cation that substitutes for a 1+ cation, another 1+ cation must be lost from the original structure to keep overall electrical neutrality. This process produces a void space in the ionic structure (forming a **defect lattice** structure). Further substitution would eventually produce an [8]:[4] structure (***Figure 1.26***).

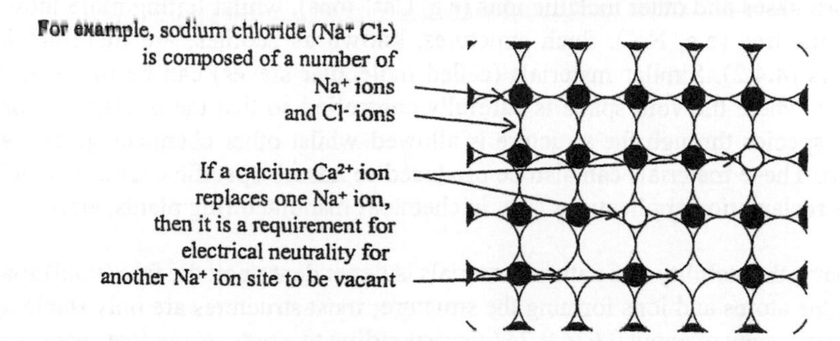

***Figure 1.29** Formation of a defect lattice in an ionic structure*

Formation of a defect lattice structure has far reaching implications for multivalent steel, since iron (Fe) has two valency states in the ionic form, one as Fe^{2+} (forming **ferrous** salts) and the other as Fe^{3+} (forming **ferric** salts). These salts produce a defect lattice structure that prevents the iron oxide film being protective to steel (see **11.15.2**). Ionic substitution resulting in structural change is summarised in **Table 1.13**.

Table 1.13 Structural changes due to ionic substitution

Original structure	Resulting structure	Conditions
BY [6] : [6]	**BY** [8]:[8]	if a **larger cation** is introduced to satisfy the radius ratio conditions
BY [6] : [6]	**BY** [4]:[4]	if a **smaller cation** is introduced to satisfy the radius ratio conditions
BY [6] : [6]	**A_2X** [8]:[4]	if the substituting ion is from a **different group** in the periodic table, then • the structure will change from [6]:[6] to [8]:[4] for an **anion** substitution
BY [6] : [6]	**AX_2** [4]:[8]	• the structure will change from [6]:[6] to [4]:[8] for a **cation** substitution In each case there is a chemical formula change (e.g, to **A_2X** or **AX_2**)

Note that, in nature, the structural units are a lot more complicated than those discussed here, but the same principles apply. Sometimes the defect lattice formed remains stable (as for example with the **zeolites** used in water softeners, **4.4.2**). However, if the level of substitution is high, the lattice structure must change to accommodate the differently charged ions. In addition, the substitutional ions may have different radii depending on the number of electrons lost in forming the ion (e.g, for multivalent ions Fe^{2+} and Fe^{3+}) (**Table 1.5** for comparison of ionic radii). The situation may be further complicated if sulfates diffuse into the structure, as most of the sulfate compounds are not stable unless associated with water of crystallisation (**1.18.5**). Some structures (e.g, the silicates, **1.17**) have both **ionic and covalent bonding** mechanisms in their structures (**1.13**). The covalent bond **does not** alter the size of radii of the atom, whereas the ionic bond **does** increase the size of the anion. Hence the largest atom(s) or the anion(s) govern the volumetric size of the fundamental building block of the structure, whilst the smaller atoms(s) or the cations must fill the remaining free (void) space. Ionic substitution can occur in silicates. The Si^{4+} ion can be replaced by an Al^{3+} ion in the structural unit (***Figure 1.21***) and still keep the tetrahedral shape, but the tetrahedron becomes negatively charged. In fact the silicate tetrahedra will form large open structures that will absorb gases and other metallic ions (e.g, Ca^{2+} ions), whilst letting more loosely bonded ions free (e.g, Na^{+}). Such structures, known as zeolites, are therefore ion exchangers (**4.4.2**). Similar materials (called molecular sieves) can be made in the laboratory, where the void space is carefully controlled so that the passage of some chemical species through the structure is allowed whilst other chemical species are prevented. These materials can also be produced to adsorb specific chemical species and have reclamation applications (e.g, in chemical manufacturing plants, etc.).

In summary, the stability of a natural materials is dependent upon the free (void) space between the atoms and ions forming the structure; most structures are only stable at a packing efficiency of about 0.6 to 0.74 (corresponding to a percentage free space of 40 to 26%). Outside this stability range, the atoms or ions must rearrange themselves to form a different structure and attain a stable free (void) space range. This rearrangement is usually initiated by the cation in the structure. A change in valency of either ion in an ionic structure will produce a defect lattice for small scale substitutions. From these simple considerations, a material is only stable if the ions comprising that structure have the correct radius ratio pertaining for that structure. In addition, note that on the atomic scale, a change in structure means a change in percentage free (void) space, which in turn means an expansion in the dimensions on the structural element. A change in the distance between the ions on the atomic scale will produce high lattice strains in a three-dimensional structure on the macro scale. As the ionic bonds become strained, there will be changes in the coefficient of expansion in the localities where ions have been substituted. Where some ions are large (e.g, the sulfate radical, SO_4^{2-}), there are no large cations to make the structure stable in terms of packing efficiency and so the structure is formed with water of crystallisation. Most sulfates have water of crystallisation and the crystallisation of these compounds produces large expansions (e.g when gypsum plaster in the bag, $CaSO_4.\frac{1}{2}H_2O$, transforms to gypsum plaster on the wall, $CaSO_4.2H_2O$). Finally, it is worth appreciating that the first formed compound is not always the most stable. Examples include the formation of unstable **graphite** in the manufacture of steels (**10.11**), which occurs more quickly than the formation of stable **cementite**; the initial formation of **ettringite** in the hydration of cement particles (**7.15.1**) and its subsequent transformation to stable **monosulfates**; and hydrocarbons, which produce soot and not diamonds when they burn!

1.19 ACIDS AND ALKALIS (BASES)

Alkalis or **bases** are terms that essentially mean the same thing. *A **base** is a compound that will react with an acid to give a **salt** and **water** only*. Bases may be oxides or hydroxides, and the majority are insoluble in water (**Table 1.14**). *An **alkali** is a base that is soluble in water*; examples include sodium hydroxide (NaOH), potassium hydroxide (KOH), ammonium hydroxide (NH_4OH), and the slightly soluble calcium hydroxide ($Ca(OH)_2$) and barium hydroxide ($Ba(OH)_2$). Alkali elements or ions are capable of taking the place of a hydrogen ion from an acid (HX) to form a compound (salt, AX) and produce water. When a base reacts with an acid, the reaction is called neutralisation. *An* ***acid*** *is a solution of a compound that contains hydrogen ions as its* ***only*** *positive ions* (hence $NaHSO_4$ is **not** an acid, since it contains both positive hydrogen (H^+) and sodium (Na^+) ions). The salts formed in these reactions can be soluble or insoluble (**Table 1.14**) (compare groups of elements in this table with their position in the Periodic table, **Table 1.2**).

Table 1.14 Solubility of salts of common acids

Salt	Exceptions	Slightly soluble	Comments
Soluble in cold water			
All nitrates All hydrogen carbonates All potassium, sodium and ammonium salts			Exist in the form of ions when in water.
All chlorides	Silver chloride, mercury chloride, lead chloride.		
All sulfates	Barium sulfate, lead sulfate	Calcium sulfate	Insoluble materials do not produce ions in water.
Insoluble in water			
All carbonates	Carbonates of Na, K and ammonium.		
All hydroxides	Hydroxides of Na, K and ammonium.	Barium and calcium	

If the salts formed are soluble, they dissolve in water to form **electrolytes**. Electrolytes are compounds that, when mixed with water, **dissociate (ionise)** to form ions (A^{n+} and X^{n-}). Acids are electrolytes, liberating hydrogen (H^+) ions in solution. The fraction of molecules that dissociate to form ions is called the **degree of dissociation (ionisation)**. As the concentration of a solution of a weak electrolyte decreases, the degree of dissociation increases. For example, for a weak acid AX, an equilibrium is set up between the undissociated molecules AX and the ions A^+ and X^-

$$\underset{\text{solid (not dissociated)}}{AX} + H_2O \rightleftharpoons \underset{\text{(in liquid)}}{H^+} + \underset{\text{(in liquid)}}{X^-} \quad \text{(dissociated/ionised)}$$

The equilibrium constant is called the (acid) **dissociation** (or **ionisation**) **constant**, k_a. Denoting concentration with square [ ] brackets,

$$k_a = \frac{[A^+] \times [X^-]}{[AX]}$$

and is temperature dependent. The value of the dissociation constant k_a tells us how strong an acid is and how vigorously it will take part in reactions that are typical of acids. The degree of dissociation for some common acids is shown **Table 1.15**. The product $k_a \times [AX]$ ($= [A^+] \times [X^-]$) is known as the **solubility product** (and is a constant). If a very soluble compound BX in solution is added to a slightly soluble compound AX in solution, then the compound AX can be precipitated out so that the solution is devoid of A^+ ions, i.e., $k_a \times [AX] = [A^+] \times ([X^-]_{from AX} + [X^-]_{from BX})$; the contribution of $[X^-]_{from BX}$ completely swamps the contribution from $[X^-]_{from AX}$.

Table 1.15 Degree of dissociation for some common acids

pH	Dissociation (ionisation)	Notes
1	Hydrochloric acid HCl (gas) $\rightarrow Cl^- + H^+$	Fully dissociated when dissolved in water (we cannot detect the molecules HCl and HNO in water)
1	Nitric acid $HNO_3 \rightarrow NO_3^- + H^+$	
1.5	Phosphoric acid $H_3PO_4 \rightleftharpoons (H_2PO_4)^- + H^+$; or $H_3PO_4 \rightleftharpoons (H_1PO_4)^{2-} + 2H^+$; or $H_3PO_4 \rightleftharpoons PO_3^{3-} + 3H^+$	Not fully dissociated. The pH is dependent upon the alkali group, sodium forming a more alkaline substance than calcium (**Table 1.19**). Fully dissociated
2.9	Acetic acid (vinegar) $CH_3COOH \rightleftharpoons \mathbf{CH_3}COO^- + H^+$ Stearates (soaps, waterproofing compounds) $C_{17}H_{35}COOH \rightleftharpoons C_{17}\mathbf{H_{35}}COO^- + H^+$	Partially dissociated (we can detect the molecules CH_3COOH and $C_{17}\mathbf{H_{35}}COOH$ in water). Note the hydrogen in bold does not contribute to the acidity.
5.2	Carbonic acid $H_2O + CO_2$ (gas) $\rightleftharpoons H_2CO_3$ $H_2CO_3 \rightleftharpoons HCO_3^- + H^+$ $H_2CO_3 \rightleftharpoons CO_3^{2-} + 2H^+$	Very weak acid, with a tendency for the carbon dioxide (CO_2) to come out of solution. Most carbonates are insoluble in water (**Table 1.14**)

1.19.1 Water

Water is remarkable as being an ionising solvent i.e., a solvent in which many compounds (acids, bases and salts) dissolve and become dissociated into ions. Water is modelled as a covalently bonded molecule (***Figure 1.30***), but a small fraction is also present as ions.

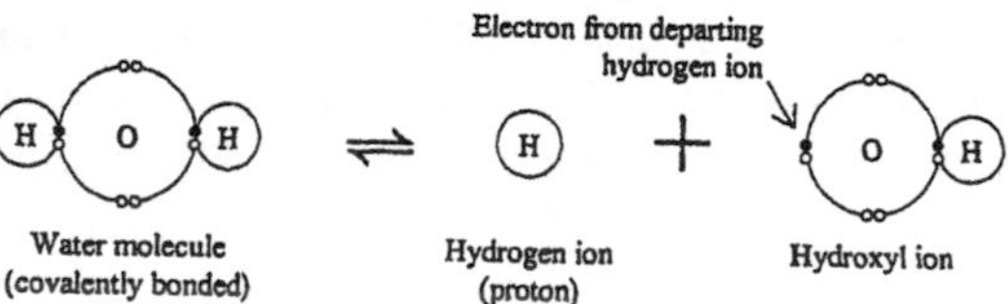

Figure 1.30 *Modelling the dissociation of water*

It is estimated that only one water molecule in every ten million forms ions. Hence water dissociates (ionises) to form hydrogen (H^+) ions and hydroxyl (OH^-) ions, as shown in ***Figure 1.30***. The reaction is

Covalently bonded		Ionically bonded		
H_2O	$\rightleftharpoons$	H^+	+	OH^-
Poor conductor of electricity		Good conductor of electricity		

This reaction is reversible (it can proceed in any direction). We denote this as '$\rightleftharpoons$' in the above equation. However, the reaction is biased towards the left hand side (hence the low concentration of ionically bonded water). Dissociation of an acid in solution results in an excess of hydrogen (H^+) ions, which mop up the hydroxyl (OH^-) ions present in the water to form covalently bonded water. This is shown below, where the bold $\mathbf{OH^-}$

ions are derived from the water and the bold **H⁺** ions are derived from the acid

Water	H_2O	⇌	**OH^-**	+	H^+
Acid	HCl	⇌	Cl^-	+	**H^+**
Products	**OH^-**	+	H^+	⇌	H_2O

The effect of adding an acid to water is therefore to reduce the hydroxyl (OH^-) ion concentration of the water.

The electrical properties of water obey **Ohms Law**, V = i x R, where V is the voltage, i is the current and R is the resistance. The electrical resistance of a conductor is proportional to length (l) and inversely proportional to the cross-sectional area (A), so that R = ρl/A, where ρ is the **resistivity** (having units ohm cm). Thus the resistivity of the conductor is an intrinsic property of the conductor and does not depend upon its dimensions; the higher the resistivity of a material, the greater will be the resistance of a sample of a given size. Resistivities can be quoted for a number of materials (notably pure metals); resistivities of electrolytes cannot be generally quoted because they depend on concentrations (purity) and are therefore variable quantities; **Table 1.16** gives approximate resistivity values for water depending on source.

Table 1.16 Approximate resistivity of water from various sources

Water	Resistivity (ohm cm)
Pure water	20,000,000
Distilled water	500,000
Rain water	20,000
Tap water	1,000 to 5,000
River water	200
Sea water	30

Only very few water molecules are dissociated and able to conduct electricity, so **pure** water will only conduct electricity at high potentials; 1 cm^3 of pure water has the same resistance as approximately 90 million miles of copper of the same cross-sectional area.

1.19.2 Ionic Compounds Containing OH^- Groups

Sodium hydroxide is an **alkali** and is written as **NaOH** (i.e., Na^+OH^-). In solution we would be able to detect both Na^+ and OH^- ions. The excess OH^- ions will combine with any spare H^+ ions to form covalently bonded water. If NaOH was an **acid** it would be written as **HNaO**. There are very few H^+ ions available **(1.19.1)** and so this representation is incorrect.

Magnesium hydroxide is an **alkali** and is written as **$Mg(OH)_2$** (i.e., Mg^{2+} ions and $2OH^-$ are detected). Again the excess OH^- ions will combine with any spare H^+ ions to form covalently bonded water. The dissociation of magnesium hydroxide is written as

$$Mg(OH)_2 \rightleftharpoons Mg^{2+} + 2OH^-$$

As the number of OH^- ions produced from the dissociation is less than the number produced from sodium hydroxide, magnesium hydroxide is a **weaker** alkali. The dissociation reaction indicates that some undissociated $Mg(OH)_2$ are present in the liquid. By writing the formula for magnesium hydroxide as $Mg(OH)_2$ we are indicating that the material is **not soluble. Ionically bonded materials are only soluble if they ionise**, in this case producing Mg^{2+} and $2OH^-$ ions. If magnesium hydroxide was an **acid** it would be written as $\mathbf{H_2MgO_2}$.

Carbon is the start of the non metals in the Periodic table (**Table 1.2**). The properties of carbon are described in section **1.8.4**. The OH^- compound of carbon is $C(OH)_4$ (i.e., C^{4+} and $4OH^-$ ions). This is unstable and loses water H_2O, probably because the cations (C^{4+}) are so small (C^{4+} 0.02 nm; C 0.077 nm, **Table 1.5**). The compound would be much more stable if it were **covalently** bonded (carbon to oxygen). The loss of water is the same as shown for **sulfur** (***Figure 1.28***) and is described by

$C(OH)_4$	→	$CO(OH)_2$	if an **alkali** (which it is not)
	water (H_2O) lost	H_2CO_3	if an **acid** (which it is)

We can detect HCO_3^-, CO_3^{2-} and H^+ ions, but only very few OH^- ions are present. If another water molecule is lost, we get the oxide CO_2 (an acidic gas, **1.11.2**).

Nitrogen continues the properties of the non metals with more acidic properties. In as similar way to carbon, the OH^- compound of **nitrogen** is $N(OH)_5$ (i.e., N^{5+} and $5OH^-$ ions). This is unstable and loses water ($2H_2O$), as described for carbon, as the N^{5+} cations are very small (N^{5+} 0.01-0.02 nm; N 0.071 nm) (**Table 1.5**).

$N(OH)_5$	→	NO_2OH	if an **alkali** (which it is not)
	water ($2H_2O$) lost	HNO_3	if an **acid** (which it is)

The **acid** HNO_3 is formed (***Figure 1.8a***). It is much easier for the scientist to represent the NO_3 unit as **covalently** bonded (nitrogen to oxygen), with the hydrogen ionically bonded to the NO_3^- unit. The NO_3^- unit is called a **radical**. A **radical** is a group of atoms found as a unit in many compounds which obtain the **valency** (**1.5**) shown above on the loss of hydrogen (H^+) ions from the acid. Examples of radicals include CO_3^{2-}, NO_3^-, PO_4^{3-}, SiO_4^{4-}, HCO_3^-, SO_4^{2-}, etc. Structurally, they can be represented by polyhedra (e.g, the silicate and sulfate radicals are represented by a tetrahedron, ***Figures 1.15 and 1.27*** respectively). However, this structural arrangement presents some difficulty, as the positive and negative units greatly differ in size (**Table 1.5**) and yet must be packed together with the minimum of free space. In most cases, packing efficiency is achieved by the incorporation of **water of crystallisation** (**1.18.5**).

In the same way as for **carbon** and **nitrogen**, in the OH^- compound of **sulfur** the cation (S^{6+}) is much more stable when covalently bonded (sulfur to oxygen) (***Figure 1.28***). The OH^- compound of sulfur is $S(OH)_6$ (i.e., S^{6+} and $6OH^-$ ions). This is unstable and loses water ($2H_2O$), as described for carbon and nitrogen, i.e.,

$S(OH)_6$	→	$SO_2(OH)_2$	if an **alkali** (which it is not)
		H_2SO_4	if an **acid** (which it is)
	water ($2H_2O$) lost		

Once again, as there are very few OH^- ions in solution and a very large number of H^+ ions, the OH^- compound of sulfur is an **acid**. The ions found in sulfuric acid are **HSO_4^-**, **SO_4^{2-}**, and **H^+**. The acid dissociates ionically as

$$H_2SO_4 \rightarrow H^+ + HSO_4^- \rightarrow 2H^+ + SO_4^{2-}$$

The SO_4^{2-} radical is all **covalently** bonded (**1.13**). If another water molecule is lost, we get the oxide SO_3 (an acidic gas).

The OH^- compound of **chlorine** is $Cl(OH)_7$. This is unstable and loses water ($3H_2O$)

$Cl(OH)_7$	→	ClO_2OH	if an **alkali** (which it is not)
		$HClO_3$	if an **acid** (which it is)
	water ($3H_2O$) lost		

The acid formed ($HClO_3$) is a **chlorate**; chlorine also forms hydrochloric acid (HCl) and **chlorides**. Similar reactions for **silica** are considered in section **1.13**.

1.19.3 pH

The hydrogen (H^+) ion concentration of a solution can be denoted by means of the **pH** of the solution, which gives a measure of the acidity or alkalinity of a solution. The pH is defined as the logarithm to the base 10 of the reciprocal of the effective concentration of hydrogen ions (in moles per litre).

$$pH = -\log_{10}[H^+]$$

For solutions containing an acid or an alkali, the **product $[H^+] \times [OH^-]$ is always equal to 10^{-14}**. The pH of various solutions can therefore be determined, as shown in **Table 1.17**. Hence if the concentration of hydrogen (H^+) ions exceeds the concentration of hydroxyl (OH^-) ions, the solution is acidic, with a pH from 1 to 7. Conversely, if the concentration of hydroxyl (OH^-) ions exceeds the concentration of hydrogen (H^+) ions, the solution is alkali, with a pH from 8 to 14.

Table 1.17 pH values of acidic and alkali solutions

pH	$[H^+]$	$[OH^-]$	Nature
1	10^{-1}	10^{-13}	Very Acidic
5	10^{-5}	10^{-9}	Fairly acidic
6	10^{-6}	10^{-8}	Acid Rain
8	10^{-8}	10^{-6}	Alkalinity of some brackish waters
10	10^{-10}	10^{-4}	Alkalinity of some oven cleaners
12	10^{-12}	10^{-2}	Strong alkali

1.19.4 Effect of Temperature on pH

Pure water at 25°C contains an equal concentration of hydrogen (H^+) ions and hydroxyl (OH^-) ions, and so the product $[H^+] \times [OH^-] = 1 \times 10^{-14}$. This product is called the

dissociation (or ionisation) constant for water, k_W. Hence the pH of pure water at 25°C is the logarithm of the square root of the dissociation (ionisation) constant for water

$$pH = -\log_{10} k_W^{1/2} = 7$$

At temperatures above 25°C, the dissociation constant for water is greater and so the pH of pure water is temperature dependent, as shown in **Table 1.18**.

Table 1.18 Temperature dependence of pH of pure water

Temperature	$k_w \times 10^{-14}$	pH
0°C	0.115	7.47
10°C	0.29	7.27
25°C	1.008	7.00
40°C	2.916	6.77
60°C	9.614	6.51

1.19.5 Properties of Acids and Alkalis

Figure 1.31 shows the different acidic and alkaline properties of various compounds.

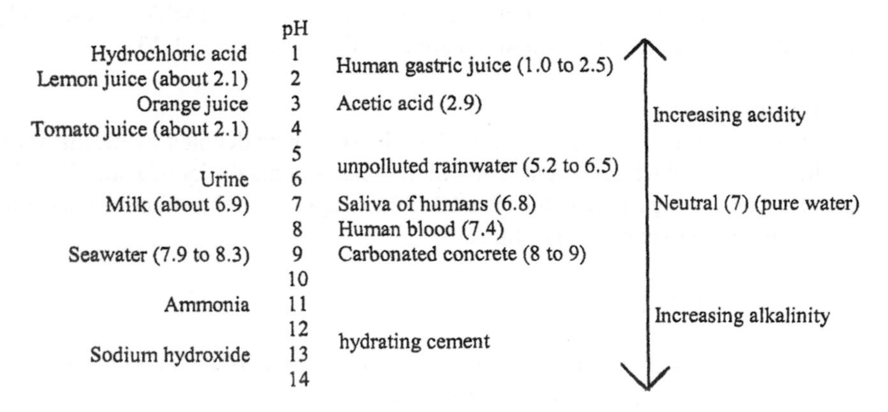

Figure 1.31 *Comparison of the pH of selected acids and alkalis*

1.19.6 Conventions for Chemical Formulae for Acids and Alkalis

Scientists have their own way of writing the formula of any chemical compound (**1.2.1**). For acids and alkalis, the way the chemical elements are ordered in the formula gives information of the acidic or alkali nature of the compound. As an example, consider the possible alternative ways to write the chemical formula of acetic acid (ethanoic acid)

Acetic acid	Notes
CH_3COOH (or CH_3CO_2H)	This is the correct way to write the chemical formula, as it indicates how the acid ionises to produce hydrogen (H^+) ions (giving rise to acidity) and carboxyl (CH_3COO^-) ions
CH_4CO_2	Whilst this is atomically correct, it does not provide information on how the acid ionises and is therefore not used
$C_2H_4O_2$ (or $C_2(H_2O)_2$	This is the accepted method of representing a **carbohydrate**. As acetic acid is not a carbohydrate, this representation is also not used

Acids of carbon (e.g, acetic acid) are called **organic acids**. In organic acids, the hydrogen (H) atoms form the hydrogen (H^+) ions following dissociation of the acid in solution and are responsible for acidity; these are placed at the end of the chemical formula (as in acetic acid, CH_3CO**OH**). Note that the hydrogen atoms at the start of the chemical formula (in acetic acid, **CH_3**COOH) are covalently bonded to the carbon (C) atom, do not dissociate and do not contribute to the acidity of the compound. Note that confusion can arise in the representation of the **organic acids**, since the chemical formula also appears to have an OH group at the end (as in acetic acid CH_3CO**OH**). In the organic acids, the carboxyl (-COOH) group comprises a (-C=O) group and a (-O-H) group. The (-C=O) group attracts electrons away from the (-O-H) group and makes it easy for the hydrogen atom to ionise to produce hydrogen (H^+) ions.

Acids formed from non-carbon elements are termed **inorganic acids**. In the chemical formula of inorganic acids, the hydrogen (H) atoms that dissociate to form hydrogen (H^+) ions are placed at the start of the chemical formula (as in hydrochloric acid **H**Cl, nitric acid **H**NO_3, phosphoric acid **H_3**PO_4, carbonic acid **H_2**CO_3, etc.). Phosphoric acid H_3PO_4 is shown in ***Figure 1.32***.

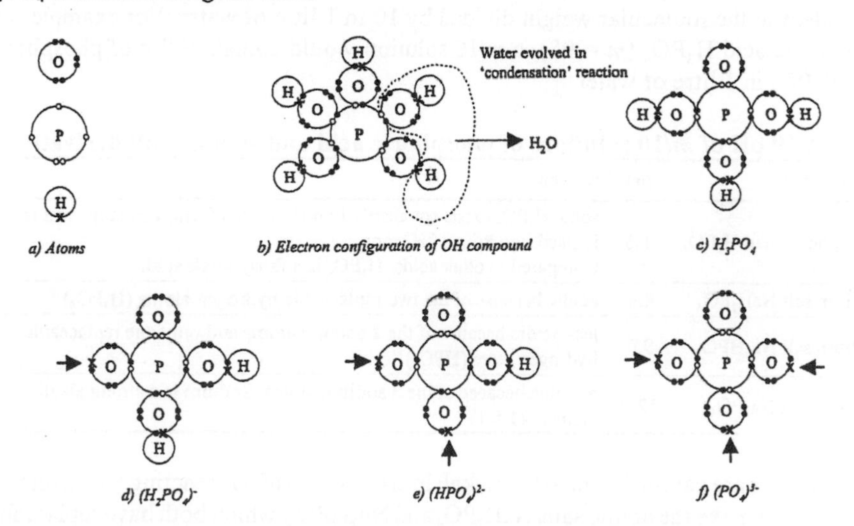

Figure 1.32 Phosphoric acid and the formation of charged radicals

Figure 1.32a shows the outer electron configuration of phosphorous (P), oxygen (O) and hydrogen (H) responsible for chemical activity. ***Figure 1.32b*** shows phosphorous combining with hydroxyl (OH^-) ions in water. This arrangement is unstable and water is lost (in a condensation reaction) to form the acid, phosphoric acid (H_3PO_4). The final electron configuration of phosphoric acid is shown in ***Figure 1.32c***, which illustrates the **covalently** bonded phosphorous (P) and oxygen (O) atoms and the **ionically** bonded hydrogen (H) and oxygen (O) atoms. In solution, the hydrogen can ionise to form hydrogen (H^+) ions, giving the acid character. The hydrogen (H^+) ion is essentially a proton, and the original electron from the hydrogen atom is left attached to the phosphate ion. This leads to the three ionised states of phosphoric acid, shown in

Figures 1.32d to 1.32f, and the electron attached to the phosphate ion as shown by the arrow head (▸). Phosphoric acid can make a number of salts, depending upon the ratio of the amount of acid over the amount of alkali used. Phosphoric acid therefore dissociates in water to form the charged radicals $H_2PO_4^-$, HPO_4^{2-} and PO_4^{3-} (**1.19.7**). Salts formed from these radicals are the basis of 'rust proofing' steel compounds (known as Naval jelly, Kurust etc.) (**11.15.1**). Phosphoric acid (H_3PO_4) is also used as a pretreatment acid to obtain a key or bond between a paint film and steel (**11.12**).

There are no **organic alkalis** as such, the nearest being the **alcohols**, which have the general formula ROH, where R is an alkyl group (C_nH_{2n+1}-). Alcohols from salts of organic acids called **esters**. There are many **inorganic alkalis**, for example sodium hydroxide NaOH, magnesium hydroxide $Mg(OH)_2$, etc. In each case, the OH group is written at the end of the chemical formula; it is these OH groups that form hydroxyl (OH^-) ions following dissociation in solution and are responsible for alkalinity.

Another method of producing an acid is from the salt. **Table 1.19** shows the pH of *m*/10 solutions of various phosphoric acid derivatives. *m*/10 solutions are concentrations equivalent to the molecular weight divided by 10 in 1 litre of water. For example, for phosphoric acid H_3PO_4 ($m = 98$), a *m*/10 solution would contain 9.8 g of phosphoric acid H_3PO_4 in 1 litre of water.

Table 1.19 pH of *m*/10 solutions of phosphoric acid and sodium salt derivatives

Acid or salt	pH	Reason
Phosphoric acid H_3PO_4	1.5	some H_3PO_4 exists covalently bonded in this form, whilst the rest is ionised into $PO_4^- + 3H^+$ ions. Compared to other acids, H_3PO_4 is a fairly weak acid.
sodium salt NaH_2PO_4	4.4	acidic because of the two replaceable hydrogen atoms $(\mathbf{H_2}PO_4)^-$
sodium salt $\mathbf{Na_2}HPO_4$	9.6	less acidic because of the **2** sodium atoms and only one replaceable hydrogen atom $(\mathbf{H}PO_4)^-$
sodium salt $\mathbf{Na_3}PO_4$	12.0	alkaline because of the **3** sodium atoms. Sodium is a strong alkali element (**1.8.1**)

These results are caused by the strong alkali in **Group 1** (NaOH) reacting with a relative weak acid to make the acidic salts NaH_2PO_4 and Na_2HPO_4, which both have replaceable hydrogen ions. As the hydrogen is replaced by an alkali element so the salt becomes less acidic and more alkaline in nature. This is important when we are considering durability of materials in different environments.

1.19.7 Balancing Equations (Acid + Base → Salt + Water)

To illustrate the process for balancing the equations for acid-base reactions, we assume a weak acid (phosphoric acid) and a strong alkali (calcium hydroxide) (refer also to **1.19.2**). Remember that phosphoric acid can form three salts (**Table 1.19**), depending on the ratio of the quantities of acid to base used. As a result of these procedures, we can determine this ratio from the balanced chemical equation. We wish to balance

phosphoric acid + calcium hydroxide → the 3 calcium salts + water

First we examine the periodic table (**Table 1.2**) to determine in which group calcium resides. Using this information, we can determine how calcium is represented in the ionic form (whether it loses or gains electrons and how many are lost or gained). *(we find that calcium is in group 2 of the periodic table and loses 2 electrons from the outside subshell ($4s^2$) to obtain the inert gas electron configuration of argon ($1s^2 2s^2 2p^6 3s^3 3p^6$). Calcium ions have an overall +2 positive charge, denoted Ca^{2+})*

Once we have done this, we know need to determine the chemical formula for calcium hydroxide, remembering that the hydroxide unit is composed of hydroxyl (OH^-) ions (**1.19.2**), which have a -1 charge. We must work out the number of hydroxyl ions required to balance the charge on the calcium ions to obtain electrical neutrality. *(we require $2OH^-$ ions to provide a -2 charge to balance the +2 charge on the calcium ions. Hence the chemical formula for calcium hydroxide is $Ca(OH)_2$)*

We are now in a position to determine the chemical formula of the three salts produced when phosphoric acid reacts with calcium hydroxide (**1.19.6** for how these salts are formed from the three phosphate radicals). The actual salt formed will depend on the ratio of the amount of acid to the amount of base (**Table 1.19**) i.e.,

Ca__ (H_2PO_4)__
Ca__ (HPO_4)__
Ca__ (PO_4)__

We can now write the chemical equation for the reaction of phosphoric acid with calcium hydroxide to form these three salts and water

__ H_3PO_4	+	__ $Ca(OH)_2$	→	Ca__ (H_2PO_4)__	+	__ H_2O
__ H_3PO_4	+	__ $Ca(OH)_2$	→	Ca__ (HPO_4)__	+	__ H_2O
__ H_3PO_4	+	__ $Ca(OH)_2$	→	Ca__ (PO_4)__	+	__ H_2O
Acid	+	Base	→	Salt	+	Water

(the balanced equations are

$2H_3PO_4$	+	$Ca(OH)_2$	→	$Ca(H_2PO_4)_2$	+	$2H_2O$
$2H_3PO_4$	+	$2Ca(OH)_2$	→	$Ca\ (HPO_4)$	+	$4H_2O$
$2H_3PO_4$	+	$3Ca(OH)_2$	→	$Ca_3(PO_4)_2$	+	$6H_2O$
Acid	+	*Base*	→	*Salt*	+	*Water)*

1.19.8 Summary

- Acids dissociate (ionise) to produce an excess of hydrogen (H^+) ions in water. The concentration of (H^+) produced is in excess of that which already exists in pure water. This excess is responsible for the acidity;
- The strength of an acid is expressed as the pH;
- Acids taste sour;

- Acids produce an excess of hydrogen (H^+) ions in water and so often react to produce hydrogen gas with a metal (**11.3.2**);
- Acids change the colour of some dyes and therefore these dyes are used as indicators for acids. The colour change, for particular dyes, are often dependent upon the concentration of hydrogen (H^+) ions in solution;
- Acids react with cementitious products and neutralise the alkalinity of the concrete. They also destroy clothing material with time (usually a slow process, only coming to light when the cloth is subsequently washed);
- Acids react with carbonates (e.g, chalk) to produce a gas (carbon dioxide);
- Acidic properties are destroyed when an acid is mixed with an alkali or a metal; the acid is said to be **neutralised**.

1.20 RADIOACTIVITY

In this Chapter we have considered a variety of chemical reactions in which the atoms that make up the reactants enter into different combinations to form products. In each case, **the nuclei of the atoms remain unchanged**. In certain reactions (**nuclear reactions**), rearrangement of the protons and neutrons in the nucleus of the atoms takes place and new elements can be formed. A number of elements have atoms that are unstable (**1.20.6**) and split up to form smaller atoms. The nucleus splits and the protons and neutrons in it form two new nuclei, and the electrons divide themselves between the two. Sometimes protons, neutrons and electrons are emitted when the original nucleus divides. The process is called **radioactive decay** and the element is said to be **radioactive.** The particles and energy emitted are called **radioactivity**. Isotopes (**1.3.3**), elements of the same type but of different atomic numbers, are usually expressed as $_{\text{atomic number}}\text{Symbol}^{\text{atomic mass}}$ and thus for the two isotopes of chlorine we have

$_{17}Cl^{35}$ 17 protons, (35 -17) = 18 neutrons; and
$_{17}Cl^{37}$ 17 protons, (37 -17) = 20 neutrons.

Radioactive isotopes, sometimes called **radioisotopes**, undergo the same chemical reactions as normal isotopes (**1.3.3**) but decay to produce radiation. Here we represent protons as $_1p^1$, neutrons as $_0n^1$, α-radiation (**1.20.1**) as $_2He^4$ and electrons as $_{-1}e^0$.

1.20.1 Types of Radiation

Three types of radiation are given off by naturally-occurring radioactive elements. All types of radiation cause certain substances to luminesce (e.g, zinc sulphide) and all ionise gases through which they pass. Artificial radioactive decay, produced by bombarding elements with α-particles, protons and neutrons, may result in the emission of other types of radiation. **α (alpha) radiation** is the radiation of α-particles, which are essentially the nucleus of the helium atom, $_2He^4$ (with no orbital electrons). α radiation from an element causes the atomic number of that element to reduce by 2 and the atomic mass to reduce by 4. The chemistry of the decayed element alters as the isotope produced is **two** groups to the **left** of the original element in the Periodic table. α radiation can penetrate no more than about 0.01 m of metal. An example of α radiation is the radioactive decay of uranium (U) to thorium (Th), $_{92}U^{238} \rightarrow {}_{90}Th^{234} + {}_2H^4$. **β (beta) radiation** results from the emission of an electron which causes the atomic number to

reduce by 1 but the atomic mass remains unchanged. The chemistry of the decayed element alters as the isotope produced is **one** group to the **right** of the original element in the Periodic table. β radiation can pass through 0.01 m of metal. An example is the radioactive decay of thorium (Th) to protactinium (Pa), ${}_{90}Th^{234} \rightarrow {}_{91}Pa^{234} + {}_{-1}e^{0}$.
γ (gamma) radiation is the emission of energy from the nucleus of an element as it returns to the ground state from an excited state. The energy radiated is part of the electromagnetic spectrum and is similar to X-rays or photons of light, where the electrons are excited to give characteristic spectral lines of the element. There is no alteration in the mass of the nucleus by emission of γ radiation, and no change in the atomic number or atomic mass. γ radiation has high penetrating power, and can pass through 0.1 m of metal.

1.20.2 Artificial Radiation

Although strictly outside the scope of this text, it is worth noting that some nuclear reactions are not spontaneous, but result from bombarding stable isotopes with particles (such as α radiation or neutrons). Since 1940, a new set of elements with atomic numbers greater than the 92 of uranium (the last naturally-occurring element) have been made. These are called **transuranium elements.** For example, the element neptunium (Np) is made by neutron bombardment of uranium, ${}_{92}U^{238} + {}_{0}n^{1} \rightarrow {}_{92}U^{239}$, followed by radioactive decay of the isotope formed ${}_{92}U^{239} \rightarrow {}_{93}Np^{239} + {}_{-1}e^{0}$.

1.20.3 Balancing Nuclear Equations

In equations for nuclear reactions, the sum of both the atomic mass and the atomic number is the same on both sides of the equation, e.g, nitrogen-16 undergoes β decay to form an isotope of oxygen ${}_{7}N^{16} \rightarrow {}_{b}O^{a} + {}_{-1}e^{0}$; from the atomic mass, $16 = a + 0$, $a = 16$ and from the atomic numbers, $7 = b + (-1)$, $b = 8$. Thus ${}_{7}N^{16} \rightarrow {}_{8}O^{16} + {}_{-1}e^{0}$.

1.20.4 Rates of Radioactive Decay

The rate at which a radioactive isotope decays cannot be speeded up or slowed down. It depends only on the nature of the isotope and the quantity of the isotope present. The nature of radioactive decay is shown in ***Figure 1.33***, and is described as **first order**.

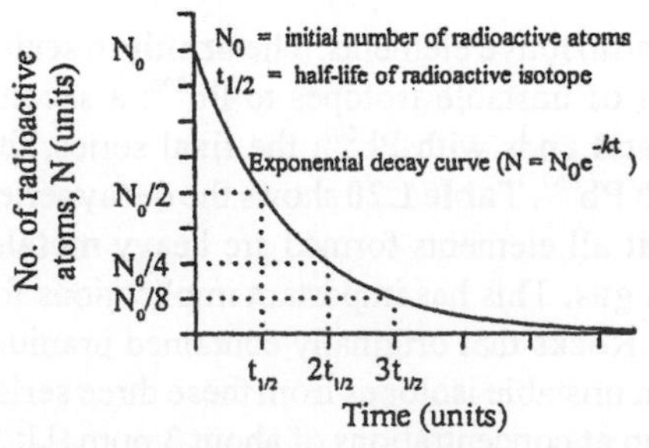

***Figure 1.33** Radioactive decay with time*

We can describe first order equations mathematically. The rate of radioactive decay of the number of atoms, N at time t, i.e., $-dN/dt = kN$, where k is the first order rate constant. If N_0 is the initial number of radioactive atoms, then the integrated form of this equation is $kt = \ln(N_0/N)$, where ln is the natural logarithm (to base e). At $t_{1/2}$, $N = N_0/2$, thus $N_0/N = 2$ and $k\,t_{1/2} = \ln 2 = 0.693$. Therefore the half-life, $t_{1/2}$, of a radioactive isotope is **independent** of the **initial** quantity of that isotope.

The time taken for the number, N_0, of radioactive atoms to decay to $N_0/2$ atoms is called the **half-life,** $t_{1/2}$, of the radioactive isotope. The times taken for $N_0/2$ atoms to decay to $N_0/4$ atoms, and for $N_0/4$ atoms to decay to $N_0/8$ atoms are the same and have the same value, $t_{1/2}$. The rate of decay is thus proportional to the number of atoms present.

1.20.5 Uses of Radioactivity

Radioactive tracers are widely used in medicine to track the path of an element through the body. For example, radioactive iodine is administered to patients with defective thyroids and radioactive barium is used to detect ulcers. As the radioactive half lives of these isotopes is only about 8 to 10 days, the radioactivity soon falls to low levels. In a similar way, air flow rates in buildings have been measured using a radioactive isotope of krypton (Kr), where β particles emitted during decay are counted as pulses of electrical current using a Geiger-Miller tube. Radioactive decay is measured as counts per minute per g of radioactive material (cpm/g) using this apparatus. Radioactive isotopes have several applications in the engineering industry, mainly in relation to radiographic analysis and flaw detection. Examples include

- determination of segregation in concrete;
- determination of the position and diameter of reinforcement bars;
- detecting porosity defects and cracks in welded joints;
- determination of leaks and the corrosion rates of metals used in pipe work;
- smoke detection devices (using unstable radioactive isotopes).
- determination of the thickness of a material from a decrease in intensity I according to the equation $I = I_0\, e^{-Axf(M)}$, where I_0 is incident intensity, A is capture cross section of radiation for the material, x is thickness and f(M) is a function of atomic weight.

The main disadvantage associated with techniques based on radioactive isotopes is that they constitute a radiation hazard. In addition, the detection of the radioactivity in these applications is by photographic techniques, and therefore there is a need to gain access to both sides of the item to be tested. The fact that radiation fogs photographic film is used to detect radiation. Workers exposed to radiation often wear badges containing photographic film, which is examined periodically to ensure that the dose of radiation they receive can be monitored. The advantages result from the fact that the detection tests usually work on density changes due to the relative absorption of air to the material under test. Therefore lower density materials, such as lagging around water pipes, do not have to be removed. Spot checks may be made without too much preparation.

1.20.6 Geological Dating

There are three naturally occurring series of radioactive elements. The **uranium series** starts with U^{238} and decays through as series of unstable isotopes to Pb^{206}; a second series, the **actinum series**, starts with U^{235} and ends with Pb^{207}; the final series, the **thorium series**, starts with Th^{232} and ends with Pb^{207}. **Table 1.20** shows the decay series for the uranium and thorium series. Note that all elements formed are heavy **metals**, with the exception of Radon (Ra), which is a **gas**. This has important implications for the exposure of building occupants (**1.20.7**). Rocks that originally contained uranium or thorium on formation will currently contain unstable isotopes from these three series of radioactive elements. Uranium and thorium at concentrations of about 3 ppm (U^{238}) and 12 ppm (Th^{232}) are still found in geologically younger rocks today. With knowledge of the elements of each series and their respective half lives, the dates when the rocks were formed can be found.

Table 1.20 Unstable isotope half-lives for uranium and thorium series

Uranium series				Thorium series			
Transition			Half-life	Transition			Half-life
U^{238}	→	Ra^{226}	4.51 x 10^9 years	Th^{232}	→	Ra^{224}	1.41 x 10^{10} years
Ra^{226}	→	Rn^{222}	1600 years	Ra^{224}	→	Rn^{220}	3.64 days
Rn^{222}	→	Po^{218}	3.82 days	Rn^{220}	→	Po^{216}	55 seconds
Po^{218}	→	Pb^{214}	3.05 minutes	Po^{216}	→	Pb^{212}	0.15 seconds
Pb^{214}	→	Bi^{214}	26.8 minutes	Pb^{212}	→	Bi^{212}	10.6 hours
Bi^{214}	→	Pb^{210}	19.7 minutes	Bi^{212}	→	Pb^{208}	60.5 months
Pb^{210}	→	Bi^{210}	21 years	Pb^{208}			**Stable**
Bi^{210}	→	Po^{210}	5.0 days				
Po^{210}	→	Pb^{206}	138 days				
Pb^{206}			**Stable**				

Carbon Dating

Carbon-14 ($_6C^{14}$) dating can be used to calculate the age of plant remains. Living plants take in carbon dioxide, which includes a small proportion of the radioactive isotope carbon-14, to form carbohydrates, which are incorporated as the structure of the plants. Since the rate of formation of carbon-14 in the atmosphere is approximately equal to its rate of decay, there is a constant ratio of $_6C^{14}/_6C^{12}$ in organic matter. When a plant dies, it takes in no more carbon of any form, and that which is already present decays. Carbon-14 decays by emitting β radiation to form nitrogen $_6C^{14} \rightarrow {}_7N^{14} + {}_{-1}e^0$. The rate of decay decreases over the years as governed by the first order rate equations, and the activity that does remain can be used to determine the age of the plant material. As an illustration, assume that carbon-14 dating shows that a piece of ancient wood gives 10 cpm/g of carbon, compared to 15 cpm/g of carbon from a modern sample. The half life of $_6C^{14}$ is 5568 years. The age of the ancient wood can then be calculated using the first order rate equations. The first order rate constant

$$k = \frac{0.693}{5568} = 1.24 \times 10^{-4} \text{ /year}$$

$$\text{Since } \frac{N_0}{N} = \frac{_6C^{14} \text{ content in modern wood}}{_6C^{14} \text{ content in ancient wood}} = \frac{15 \text{ cpm/g}}{10 \text{ cpm/g}} = 1.5$$

$$\text{then } t = \frac{\ln 1.5}{1.24 \times 10^{-4}} = 3258 \text{ years (the ancient wood is 3258 years old).}$$

1.20.7 Radon

Radon (Rn) is a radioactive inert gas (atomic number 86) formed when radium (Ra) decays (**1.20.6**). It is the heaviest of the inert gases, has a boiling point of -65°C, a melting point of -71°C and is readily dissolved in water. Radon has 25 isotopes, all of which are radioactive and have short half lives. Examples include thoron (Rn^{220}) and actinon (Rn^{219}). Although the name radon applies strictly to $_{86}Rn^{222}$, the term is often used to refer to all isotopes of radon. Radon is found in rocks that contained uranium or thorium on formation, including many precambrian igneous and metamorphic rocks (granites, gneiss, black shales and phosphate-rich rocks, etc.) common in the UK. Although the majority of radon gas is trapped within the mineral constituents of these rocks, some diffuses to the surface. Ordinarily, the sea emits about 50 atoms of radon per square metre of its surface per second, whilst the earth emits about 7500 atoms of radon per square metre of its surface per second (although higher emission rates occur over precambrian rock formations). Radon is hazardous to human health because it continues to decay radioactively to its short lived isotopes, emitting α-, β- and γ-radiation. This radiation can damage human tissue and increase cancer risk. If inhaled, radon represents an increased risk of lung cancer as solid decay products are deposited in the lungs, where they continue to emit radiation.

The radon concentration in the air is measured in units of Becquerel/m³ (Bq/m³). The unit is named after the scientist Becquerel, who, in 1896, discovered radioactivity. (The discovery was fortuitous. During the course of his experiments on the fluorescence of uranium salts, Becquerel left wrapped photographic plates in a drawer. On developing the plates, he found that they had been exposed. Since no light could enter the wrapping, Becquerel deduced that the plates had been fogged by radiation coming from the uranium salts left in the same drawer. There was at that time no known type of radiation that has this effect). The average radon level in a UK house is 20 Bq/m³. Radon is readily adsorbed by experimental adsorbents (carbon, silica gel, etc.). It is desorbed only at high temperatures (about 350°C). Hence radon can be adsorbed on particulates in the air, and inhalation of particulates will increase radon exposure and consequent health risks. This is illustrated in **Table 1.21**, which compares the incidence of lung cancer in smokers and non-smokers at different radon concentrations.

Table 1.21 Comparison of the radon concentration dependence of the incidence of lung cancer in smokers and non-smokers

Radon concentration (Bq/m³)	Incidence in smokers	Incidence in non-smokers
20	1 in 100	1 in 1000
100	5 in 100	5 in 1000
200	10 in 100	1 in 100
400	20 in 100	2 in 100

Figure 1.34 illustrates sources of the average radiation dose in UK residents.

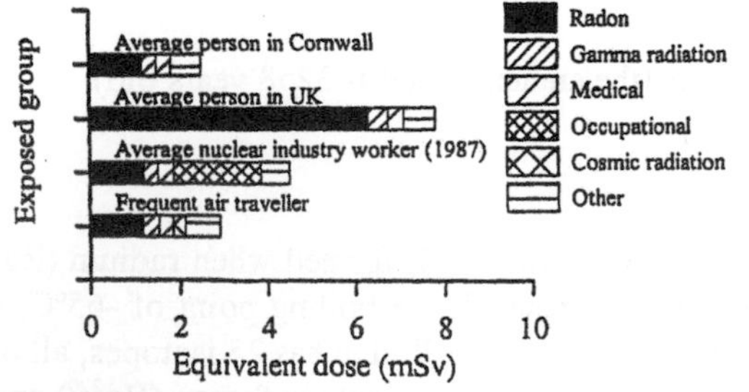

About half the average radiation dose received by an average person in the UK results from exposure to radon. In comparison, a person resident in an area with significant outcrops of precambrian rocks (e.g, Cornwall) will receive approximately three times the UK average radiation dose, due entirely to the increased radon concentration found in these areas.

***Figure 1.34** Sources of the average radiation dose in UK residents*

The incidence of lung cancer of people living beneath power cable is greater than the norm, leading to suggestions that electromagnetic forces beneath overhead power lines concentrate the radon gas in pockets. Radon concentrations in buildings are governed by the rate of internal and external air flow. Recent concerns over energy consumption have resulted in the introduction of a number of energy saving measures in dwellings, some of which (draught exclusion, sealed double glazing, etc.) reduce external air flow and lead to increased internal contaminant concentrations, increasing radon exposure risk. Risks are further enhanced through the use of open fires, since consumption of oxygen in the air will draw radon-containing air from subfloor voids rather than air from outdoors. Extractor fans will have the same effect. Increasing the ventilation rates in buildings will decrease radon concentrations. In addition, as radon is heavier than air, sumps can be used at low level (usually beneath the ground floor) to collect and vent it externally. The relative benefit of these two mitigation measures is shown in **Table 1.22**. The BRE have issued advice on remedial measures that can be adopted to reduce radon concentrations in existing dwellings.

Table 1.22 Relative benefit of two radon mitigation techniques

Measure	Concentration (Bq/m^3)		Reduction factor
	Before	After	(before/after)
500 m^3 sump	1640	140	12
Ventilation	750	160	5

Although uncontrolled exposure to radon gas promotes cancer, controlled exposure is used in the treatment of the disease. Other uses of radon include the prediction of earthquakes (based on recognising abrupt changes in the radon concentrations of seawater and rocks).

1.21 LAND FILL GASES

Landfill gases are a complex cocktail of gases, some of which are flammable (CH_4), asphyxiant (CO_2), or obnoxious (H_2S). Migration of land fill gases into dwellings depends upon the underlying geological structure, particularly permeability (long periods of frost, for example, would seal the pores of the soil). The proportion of gases depends upon the landfill refuse content (organic and acidic content) and whether there is **aerobic** or **anaerobic** decomposition (**1.8.6**). Spontaneous combustion of landfill gases can occur as a result of the temperature rise caused by bacteria living on putrescent matter (see exothermic oxidation reactions, **14.4.1**).

1.21.1 Properties of the Gases Evolved

Methane (CH_4) is a hydrocarbon gas which is colourless, odourless, lighter than air, non toxic (but an asphyxiant). Methane has a boiling point of -164°C and a melting point of -184°C. It is slightly soluble in water (5 cm^3 in 100 cm^3 of water at 20°C). At an atmospheric concentration of 30% v/v, inhalation symptoms are of oxygen starvation (rapid breathing, etc.), and at 75% v/v it is fatal within minutes. It is highly explosive between 5-15% (**Table 14.1**), and has an ignition temperature of 25°C (**Table 14.5**). Methane is produced by decaying vegetation; it is also known as **firedamp** (mines) and **marsh gas** (decaying organic material), e.g,

$$CH_3COOH \text{ (acetic acid)} \xrightarrow{\text{bacterial decay}} CH_4 \text{ (methane)} + CO_2$$

and

$$CH_3COOH \text{ (acetic acid)} \rightarrow C_2H_6 \text{ (ethane)} + CO_2$$

Carbon dioxide (CO_2) is colourless, odourless, heavier than air, and non toxic (but an asphyxiating gas). At 3% v/v, inhalation symptoms are of oxygen starvation (breathing becomes laboured and headaches result), and at 5% v/v these symptoms become severe. At 10% v/v, visual disturbance, body tremors and loss of consciousness may occur; these concentrations are often fatal within a few minutes. This gas is produced by decaying vegetation, as represented above.

Carbon and sulphur are in the same group of the periodic table, and produce similar compounds, H_2O and H_2S, CS_2 and CO_2 (**1.8.6**). **Hydrogen sulphide** (H_2S) is a colourless gas with a characteristic odour detectable at very low concentrations (0.025 ppm). It becomes hazardous at 50 ppm when it dulls the olfactory senses, leading to the presumption that the smell has disappeared. At 20-150 ppm it irritates the eyes, and causes dizziness, headaches or diarrhoea. At 700 ppm it can cause respiratory failure. It would be produced where waste gypsum plaster has been exposed to damp conditions for some time (**1.8.6**). **Carbon bisulphide** (CS_2) is a foul-smelling gas associated with decaying cabbage, formed in refuse containing gypsum plaster in the presence of certain bacteria (**1.8.6**).

2

STATES OF MATTER AND PHYSICAL CONSTANTS

2.1 INTRODUCTION

There are three states of matter, a **gas** or **vapour**, a **liquid** (the most common one being water) and a **solid** (which may be **amorphous** or **crystalline**). These three states (or **phases**) are shown by water as the solid ice, the liquid water and as the vapour steam. In this Chapter we focus on water, as this compound is responsible for a wide range of building defects and deterioration mechanisms (**Chapters 5** and **6**).

2.2 GASES

Gas molecules exhibit constant translocational movement (visualised as billiard balls colliding) and fill the available volume. Gas molecules are modelled as having principally **translocational kinetic energy** and a relatively small amount of **vibrational kinetic energy** (movement which stretches the interatomic bonds, e.g, Cl-Cl, O-O, etc.). An increase in temperature increases the translocational kinetic energy to a greater extent than the vibrational kinetic energy. A decrease in temperature results in the molecules of a gas coming into closer proximity, so that bonding can occur and the gas (water vapour) condenses to form a liquid.

2.2.1 Ideal Gas Laws

If one mole of a gas is confined in volume V at temperature T, the pressure exerted by the gas, p, obeys the relationship pV = RT, where R is the universal gas constant (= 8414 J/kmol.K) and T is absolute temperature (K). For n moles of the gas, pV = nRT, where n = m/*M* and m is the mass of the gas of molecular weight *M*. This equation is the **equation of state** of a perfect gas, also known as the **ideal (perfect) gas law**.

2.2.2 Fundamental Kinetic Theory Equation for a Gas

The kinetic energy possessed by a body by virtue of its motion is called the **kinetic energy** of the body. Suppose a gas molecule of mass m moving with velocity u is bought to rest in a distance s by a constant retarding force F (***Figure 2.1a***). The original kinetic energy of the gas molecule is equal to Fs, and this must therefore be the work done in bringing the molecule to rest. However, since F = ma, where a is acceleration, then (from $v^2 = u^2 + 2as$) we have kinetic energy = Fs = mas = ½mu^2.

a) Kinetic energy of a molecule *b) Gas kinetic theory equation*
***Figure 2.1** Fundamental kinetic theory for a gas*

The kinetic energy of gas molecules results in the gas exerting a pressure, p. Consider a volume of gas, V, enclosed by a cubical box of sides *l* (***Figure 2.1b***). Let the box contain N molecules of the gas each of mass m. Consider a molecule moving in the x direction with velocity u_1 and momentum mu_1. On collision with the face of the cube, the molecule will experience a change in momentum of $[mu_1 - (-mu_1)]$ $2mu_1$. The molecule will then travel back across the box, collide with the opposite face, and thereafter again hit the original face after time t, where $t = 2l/u_1$. The number of impacts on the original face will therefore be $1/t = u_1/2l$. The force exerted on this face due to the collision of one molecule is the rate of change in momentum of the molecule, mu_1^2/l. However, the area of the face is l^2, so the pressure on the face is mu_1^2/l^3. There are N molecules in the box and, if they were all travelling in the x direction, the total pressure on the face would be $m(u_1^2 + u_2^2 + ... + u_n^2)/l^3$. However, on average only one third of the molecules will be travelling in the x direction, so the pressure is $m(u_1^2 + u_2^2 + ... + u_n^2)/3l^3$. If we rewrite $N\bar{c^2} = u_1^2 + u_2^2 + ... + u_n^2$, where $\bar{c^2}$ is the **mean square velocity** of the molecules (i.e. the mean of the sum of the squared velocities of all molecules in the gas), then the pressure exerted by the gas is $mN\bar{c^2}/3l^3$. Since l^3 is the volume of the gas, V, we can write the fundamental gas kinetic theory equation as $pV = ⅓mN\bar{c^2}$. As the total mass of the gas, M = mN, and the density of the gas, ρ = M/V, we can also write $p = ⅓\rho\bar{c^2}$.

2.2.3 Relationships for Gases Based on Fundamental Theory

We can use the fundamental theory (sections **2.2.1** and **2.2.2**) to

- calculate the **root mean square velocity** of molecules in a gas at a given temperature and pressure, $\sqrt{\bar{c^2}} = \sqrt{(3P/\rho)}$, yielding values of 1839 m/s (4400 mph) for hydrogen, 392 m/s (850 mph) for carbon dioxide, 612 m/s (1440 mph) for water vapour (0°C);
- develop a **relationship between the pressure exerted by the gas, p, and the total kinetic energy of the gas molecules, E**. The pressure exerted by a gas, $p = ⅓mN\bar{c^2}$ (**2.2.2**). The total kinetic energy of the molecules is $E = ½mN\bar{c^2}$ (**2.2.2**) and therefore $p = ⅔E$, i.e., the pressure exerted by a gas is ⅔ of the kinetic energy of all molecules in the space;
- develop a **relationship for absolute temperature in the ideal gas scale of temperature.** As $pV = ⅓mN\bar{c^2}$ (**2.2.2**), but for a gram-molecule of a gas, N is given by Avogadro's constant N_0 (= 6.022 x 10^{26} molecules per kg.mole). Thus $pV = ⅓mN_0\bar{c^2} = RT$. Since $p = ⅔E$, then $RT = ½⅔mN_0\bar{c^2}$, giving $½m\bar{c^2} = {}^3/_2T(R/N_0)$, where $½m\bar{c^2}$ is the average translocational kinetic energy of a molecule, $\bar{e}$ (**2.2.2**).

Hence $T = ⅔\,\bar{e}\,(R/N_0)$. We therefore have a physical meaning of absolute temperature T as the average translocational energy multiplied by a constant, $⅔(R/N_0)$ or, as R/N_0 is the Boltzmann constant, k (=1.3806 x 10^{-23} J/K), $\bar{e}$ = ⅔k. W e also have a physical meaning for absolute zero (T = 0K or -273°C), which occurs when the kinetic energy of the gas is zero and the molecules are stationary (with no vibrational or translocational energy).

2.2.4 Intermolecular Distance in Gases

We can actually get a feel for the separation of water molecules in the vapour state knowing that 1 cm^3 (1 cm x 1 cm x 1 cm) of liquid water occupies 1728 cm^3 (12 cm x 12 cm x 12 cm) when changed into steam. Therefore, the molecules in the vapour state are 12 x the distance apart that they are in the liquid state. If we denote the water molecule diameter in the liquid phase as **d**, then the intermolecular distance in the vapour phase is **12d (2.7)**

2.2.5 Dalton's Law of Partial Pressure

Dalton's law of partial pressure states that each gas in a mixture of gases exerts a pressure on the walls of the containing vessel equal to the pressure that the gas would exert if it alone occupied that vessel at that temperature. The pressure exerted by each constituent gas is termed the **partial pressure**, p. Air is a mixture of gases, including oxygen (O_2), nitrogen (N_2) and water vapour (H_2O). The total pressure exerted by air is termed **atmospheric pressure**, P_{atmos}, and is the sum of the partial pressures of all constituent gases, $P_{atmos} = p_{O2} + p_{N2} + p_{H2O}$ = 101.325 kPa (760 mmHg).

There is a maximum partial pressure that any gas can exert, termed the **saturated vapour pressure (SVP)**. The SVP is temperature dependent, but the relationship is not linear, as illustrated in **Table 2.1**.

Table 2.1 Temperature dependence of the SVP of air

Temperature (°C)	0	1	5	10	15	20	25	30	50	100	120
SVP (mmHg)	4.58	4.93	6.54	9.21	12.79	17.53	23.75	31.83	92.59	760	1492
SVP (kPa)	0.611	0.657	0.872	1.227	1.704	2.339	3.166	4.242	12.34	101.33	198.85

Notes: approximate conversion from mmHg to kPa, multiply by = 0.13328

Table 2.1 illustrates that if atmospheric pressure (101.325 kPa) is doubled (e.g, in a pressure cooker), the temperature rises to 120°C. As the temperature rises, water vapour contributes more and more to the atmospheric pressure until, at 100°C, the whole atmosphere is water vapour and the water boils. **The boiling point of a substance is defined as the temperature at which its saturated vapour pressure becomes equal to the external atmospheric pressure**. As the water vapour contributes to the atmospheric pressure with increasing temperatures, the partial pressure of the water vapour must increase; as we have seen (**2.2.2**), $p_{H2O} = ⅓Nm\bar{c}^2$, hence there must be an increase in the number of water vapour molecules, N, and/or their velocity, $\bar{c}^2$.

2.2.6 Relative Humidity

The partial pressure of water vapour is ordinarily expressed as relative humidity. The partial pressure of water vapour is measured by pumping moist air through a volatile liquid contained in a test tube with a highly silvered surface. The latent heat extracted cools the test tube and the temperature is measured when the highly silvered surface mists over; this is the dew point temperature, defined as the temperature at which a specific atmosphere is saturated with water vapour and cooling below which will produce condensation (K or °C). The dew point temperature is directly related to the saturation water vapour pressure. The relative humidity (%RH) of a sample of moist air at temperature T is defined as

$$\text{Relative humidity (\%)} = \frac{\text{partial vapour pressure of water}}{\text{saturation vapour pressure at T}} \times 100\%$$

If the ambient air temperature is 20°C and misting of the silvered surface occurs at 15°C (the dewpoint temperature), then (from **Table 2.1**), the partial pressure of water vapour at the dewpoint is the saturation vapour pressure of 1.704 kPa. The saturation vapour pressure at 20°C is 2.339 kPa. The relative humidity is therefore 1.704/2.339 x 100% = 73%.

2.2.7 Vapour Pressure Stability of Compounds

As we have seen, the partial pressure of water vapour at the boiling point (100°C) is equal to the atmospheric pressure (**2.2.5**). In a similar way, we can model the stability of water soluble salts (**2.9.1**) with the amount of water vapour in the atmosphere. For example, copper sulfate ($CuSO_4$) forms three different compounds containing varied quantities of water of crystallisation (**1.18.5**). The water vapour pressure produced by each compound varies, i.e., at 50°C, $CuSO_4.H_2O$ (0.60 kPa), $CuSO_4.3H_2O$ (4.0 kPa) and $CuSO_4.5H_2O$ (6.26 kPa). If anhydrous copper sulfate ($CuSO_4.0H_2O$) was placed in a closed system at 50°C, the water vapour pressure in the system would initially be 0.60 kPa as the anhydrous copper sulfate ($CuSO_4.0H_2O$) adsorbs water from the air to form the monohydrate ($CuSO_4.H_2O$). Once this reaction is complete, the water vapour pressure would increase to 4.0 kPa as the monohydrate converts to the trihydrate ($CuSO_4.3H_2O$). Finally, the trihydrate converts to the pentahydrate ($CuSO_4.5H_2O$), raising the water vapour pressure in the system to 6.26 kPa. Each salt forms an equilibrium with the water vapour pressure in the system. The salts are **hygroscopic** (**2.9.1**). Other compounds also react in this way; for example, gypsum and calcium hydroxide, $Ca(OH)_2$, which is stable at 100°C, loses 30% of it's weight as water at 400°C and loses all water (forming 'lime', CaO) at 450°C. As these compounds lose water, they are good fire protection materials (**14.7.3**).

Some compounds form equilibriums with gases other than water vapour, e.g, calcium carbonate ($CaCO_3$) is heated to form calcium oxide (CaO); the reaction produces carbon dioxide (CO_2)

$$\underset{\text{(calcium carbonate)}}{CaCO_3} \underset{\text{(Calcination)}}{\overset{\text{Endothermic}}{\rightleftharpoons}} \underset{\text{(calcium oxide)}}{CaO} + \underset{\text{(carbon dioxide)}}{CO_2}$$

The partial pressures of carbon dioxide are 0.01 kPa (500°C), 0.31 kPa (600°C) and 101 kPa (800°C). If these partial pressures are exceeded in the system, the carbon dioxide will recombine with the calcium oxide; hence the carbon dioxide produced must be removed from the system. In practice, to produce lime (CaO) (**7.10.1**), the calcination temperatures are increased (to 1100 to 1200°C) so that the partial pressures are so high that they are never exceeded in the system.

2.3 LIQUIDS

The molecules, within the liquid, are always colliding with each other. At any particular time the arrangement is described as a **chaotic** array of atoms or molecules in three dimensional space. This arrangement is always changing with time, and a liquid molecule can have several different nearest neighbours. Hence a liquid has no definite shape and takes the shape of the containing vessel because the molecules of a liquid are capable of relative movement. Some molecules within a liquid can travel large distances over a period of time as shown by diffusion experiments, eg. a coloured liquid placed into water will eventually be dispersed throughout the water. Therefore the scientist models liquid molecules as having both translocational and vibrational kinetic energy.

2.4 SOLIDS

As liquids cool, they lose their vibrational and translocational energy to such an extent that they form a solid. This energy given up is the latent heat of fusion (**2.5**). There are two solid states which can be formed, the **amorphous state** and the **crystalline state**.

2.4.1 Amorphous State

The amorphous state is modelled as a chaotic array of molecules or atoms fixed in three dimensional space. This arrangement is a "frozen" liquid and therefore there is no latent heat evolved (i.e., there is no horizontal line **RS** in ***Figure 2.2*** as there is in the crystalline state) and no volume abrupt volume decrease (i.e., there is no vertical line **CB** in ***Figure 2.3a***, cf ***Figure 2.3b***). There is *no long range order* (**1.15**) of atoms or molecules in three dimensional space. Examples of materials having an amorphous structure include plastics, bitumens and window glass (which owes its transparency to the amorphous state, **9.1**; glass in the crystalline state would be opaque). There is diffusion of the molecules in the amorphous state, as can be seen by the 'bleeding' of the dark colours of the substrate through paints, etc. Viscosity (**3.5.2**) is the only mechanical property possessed by an amorphous material.

2.4.2 Crystalline State

In the crystalline state, the atoms or molecules of the solid occupy a definite position in a three-dimensional lattice where there *is long range order* (**1.15**). Crystalline solids have a definite volume, shape and resist shearing (but see **creep, 3.5.4**) and diffusion. Crystalline substances are stronger than amorphous solids; numerous solid crystalline materials are utilised in building and their structure and classification is considered in section **1.15**. The transition from liquid to solid is accompanied by a evolution of latent heat (**2.6**). Crystalline solids may be elastic, plastic or brittle

- **elastic** materials have the ability to spring back to their original shape and size following cessation of loading below a loading limit (**3.2.2**). In cases of excessive loading above the loading limit, even elastic materials will not return to their original size and shape i.e., they will suffer **permanent deformation** (a **permanent**

set). Generally, for a material to be elastic, it must have a relatively high resistance to deformation. Steels (**10.9.3**) are a good examples of elastic solids;

- **plastic** materials have little or no elasticity and therefore deform under **most** conditions of loading. Once they have passed the loading point where they will not recover their deformation, plastic materials go through a state of **plastic** flow. Wet clay is an obvious example of a plastic material, but lead can also be included;
- **brittle** materials do not become plastic under heavy loads, but will instead crack suddenly. Cast iron (**10.9.2**) and glass (**9.6.1**) (although no strictly a solid) are examples of brittle materials. The brittleness of a material may increase with use and exposure (e.g, asbestos-cement sheets, plastics, etc.).

2.5 CHANGE OF PHASE AND LATENT HEAT

Transition between the three states of matter for a crystalline material requires **latent heat**. Latent heat is the quantity of heat required to produce a change in phase in unit mass of the substance, **without an accompanying change in temperature of the substance**; it is the energy required to overcome the forces of attraction between the molecules. For water

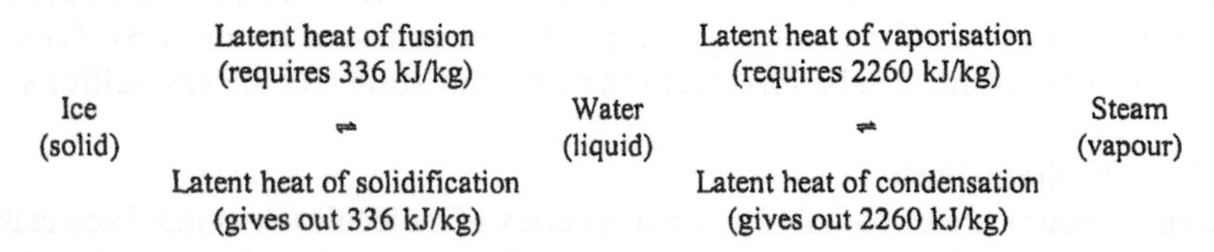

When latent heats are given out (i.e., when the reactions proceed right to left), there is often a problem with the stability of the first nuclei formed, as the latent heat evolved tends to reform the previous phase. In these situations, supercooling has to take place (**2.6.3**).The latent heat of fusion is always less than the latent heat of vaporisation as the phase change solid → liquid does not necessitate breakage of all intermolecular bonds (liquid → vapour phase transitions do require breakage of all intermolecular bonds; in addition, a greater distance is required between vapour molecules than between liquid molecules, **2.7**). The liquid → vapour phase change can also occur when the liquid is not boiling; this is termed the latent heat of evaporation. The latent heat of evaporation is greater than the latent heat of vaporisation; we can illustrate this by considering 1 kg of water at 20°C, assuming **specific heat capacities** (**14.10.2**) of water (4190 J/kg.°C) and water vapour (2016 J/kg.°C). The energy required for evaporation is the sum of

- the energy required to heat 1 kg of water at 20°C to 100°C (80 x 4,200)J;
- the energy required to change 1 kg of water at 100°C into steam (2260)kJ (the latent heat of vaporisation, L);
- the energy **given out** when 1 kg of steam at 100°C is cooled to 20°C (80x2020)J,

i.e., (L + 80(4190-2016)) or 173.92kJ greater than the latent heat of vaporisation, L. Some textbooks refer to the latent heat of vaporisation (2260kJ) as the latent heat of evaporation, calculated here as 2433.92kJ.

No temperature change occurs **during** the change of state; the temperature will only change when all of the substance has changed from one state to another. This can be illustrated in a **temperature-time** graph (***Figure 2.2***).

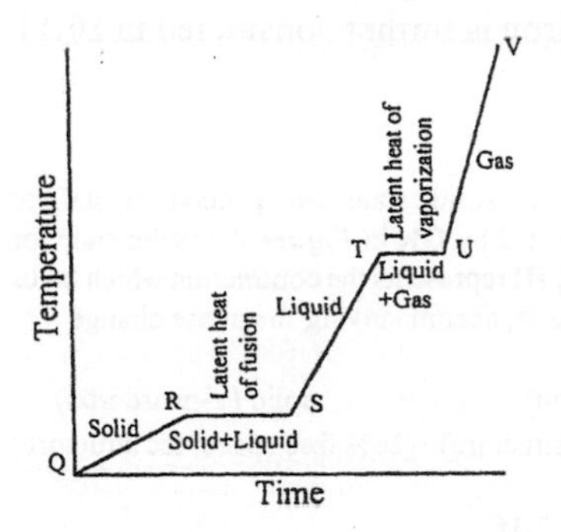

The plateaux **RS** and **TU** arise from the latent heats required to change the state of the crystalline material. Latent heats are a function of crystallinity and bond strengths. Temperature-time graphs give some information on the strength of the bonds, as the plateau **RS** represents the thermal energy required to break down the solid-solid interatomic bonds, whilst the plateau **TU** represents that energy required to break the intermolecular bonds in the liquid. In order to achieve the breakdown of intermolecular bonds in the liquid, the energy required is that to move the molecules 12d away from the adjacent molecules (**2.7**). The slopes **QR** and **ST** will be dependent upon the **specific heat** of the solid and liquid respectively. Note that completely **amorphous** materials do not show any latent heats.

Figure 2.2 *Temperature-time graph for a crystalline material heated at a constant rate*

Changes of state can also be illustrated by comparing changes in **volume** occurring as a result of changes in temperature. In ***Figure 2.3***, the increase in volume associated with an **amorphous** (***Figure 2.3a***) and a **crystalline** material (***Figure 2.3b***) are compared.

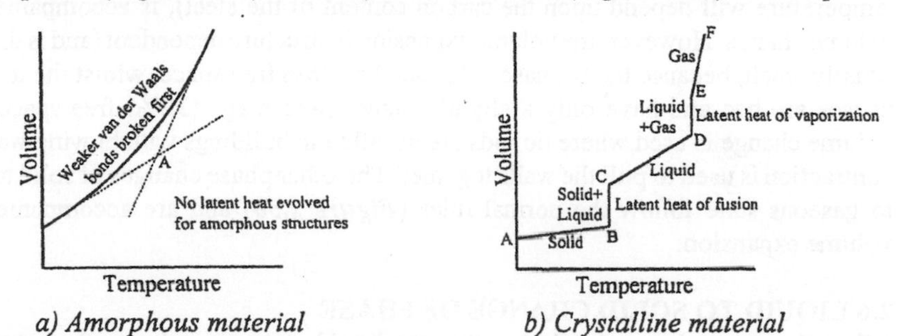

a) Amorphous material *b) Crystalline material*

Figure 2.3 *Increase in volume associated with amorphous and crystalline materials heated at a constant rate*

In an amorphous material (***Figure 2.3a***) there is no evolution of latent heat and thus the volume-temperature relationship is given by a **curve** (with no **abrupt changes**, as for the crystalline material, ***Figure 2.3b***). The weaker Van der Waals bonds (**1.14**) are broken first and the stronger Van der Waals bonds are broken at higher temperatures. The point **A** in ***Figure 2.3a*** (formed at the intersection of the tangents) is equivalent to the **glass transition temperature** (13.9.5). Section **13.9.5** gives further analysis of the **stiffness** of an amorphous (plastic) material (note, however, that a **crystalline** plastic shows a marked melting point). In a crystalline material ***(Figure 2.3b)***, the interatomic distance in the solid state is of the same order of magnitude as that in the liquid state (the expansion **BC** has been exaggerated for diagrammatic purposes). The expansion **BC** takes place at one temperature, called the **melting point**. The temperature does not rise until all the metal has melted (i.e., at point **C**). The gradient of the curve **AB** is determined by the **coefficient of thermal expansion** (2.8.2) and is dependent upon the interbond strength in the solid (**Table 2.5**).

Although not characteristic of all materials, in certain common crystalline solids (e.g., **iron, 10.1**), there is an anomalous expansion due to a phase change that takes place in the **solid state.** In ***Figure 2.4***, we illustrate this by reproducing the solid section **AB** of ***Figure 2.3b*** for iron (note that the phase change for iron is further considered in **10.11** in the **phase diagram** for steel, ***Figure 10.21***).

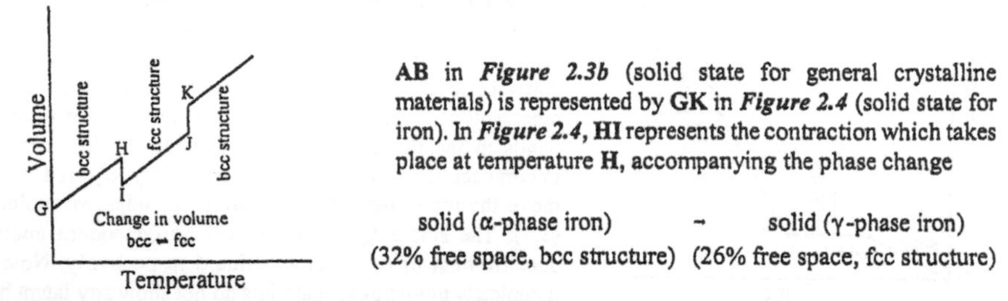

AB in ***Figure 2.3b*** (solid state for general crystalline materials) is represented by **GK** in ***Figure 2.4*** (solid state for iron). In ***Figure 2.4***, **HI** represents the contraction which takes place at temperature **H**, accompanying the phase change

solid (α-phase iron) → solid (γ-phase iron)
(32% free space, bcc structure) (26% free space, fcc structure)

Notes: The crystalline structure of metals is examined in section **1.16**.

Figure 2.4 Increase in volume of iron heated at a constant rate

The α-phase iron solid is only stable over a small temperature range **IJ**. At temperature J, another phase is formed (the δ-phase) which has the same structure as the α-phase (bcc). Hence the phase change (fcc → bcc), which takes place around 723°C (the actual temperature will depend upon the carbon content of the steel), is accompanied by a volume change. However, the volume expansion is structure dependent (and is therefore usually small, because the γ-phase is fcc and has 26% free space, whilst the α- and δ-phases are bcc and have only a slightly higher percentage (32%) free space). This volume change is used where tie rods are installed in buildings with bowing walls; the contraction is used to pull the walls together. The other phase changes of solid to liquid to gaseous state follow the normal rules (***Figure 2.3b***) and are accompanied by a volume expansion.

2.6 LIQUID TO SOLID CHANGE OF PHASE

When the atoms (or molecules) within the liquid state slow down, they can orient themselves into the solid state **provided**

- they are able to form the critical size for the solid phase to be stable (**2.6.3**);
- they lose more thermal energy so there is an overall energy decrease (**2.6.2**);
- there is a nucleus upon which this solidification can take place (e.g, the rough surfaces of the container, dirt or oxide impurities).

We can illustrate these requirements with respect to the liquid → solid change of phase in water. The temperature decline that occurs when water is cooled at a constant rate is shown in ***Figure 2.5a*** and the increase in volume that occurs when ice is heated at a constant rate is shown in ***Figure 2.5b***. The first nucleation of the solid is very difficult because of the latent heat given out (**2.5, 2.6.3**), hence supercooling is usually observed (***Figure 2.5a***). Ice expands uniformly (**MN**) when heated (this is a characteristic of nearly all solids, e.g, **AB** in ***Figure 2.3b***). However, when the ice changes to water (at N), there is a **contraction** in volume of about 9% (see Detail at **NO**). This is not a characteristic of other solid → liquid phase changes (normally substances expand in volume, as shown by **BC** in ***Figure 2.3b***). Water also has exhibits **anomalous**

contraction between 0°C and 4°C (as shown by **NO** in ***Figure 2.5b*** and detail **NO**). This anomaly is accounted for by the differences in the density of ice (920 kg/m^3) and liquid water (1000 kg/m^3). Ice therefore floats on water, whereas, in most other liquids, the solid phase would drop to the bottom of the vessel. ***Figure 2.5b*** indicates that a large increase in volume occurs as the liquid changes to a vapour (**PQ**) (**2.2.4**).

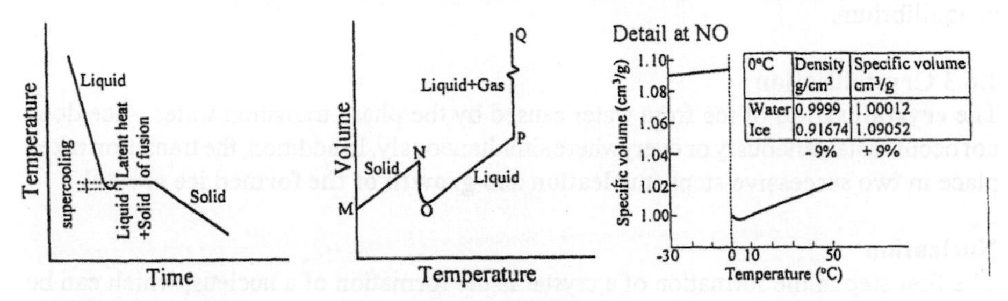

a) temperature-time *b) temperature-volume*

***Figure 2.5** Temperature-time and temperature-volume relationships for water*

2.6.1 Effect of Pressure

Ice will melt with the application of pressure. The degree of pressure required is dependent on temperature of ice below the freezing temperature (***Figure 2.6***).

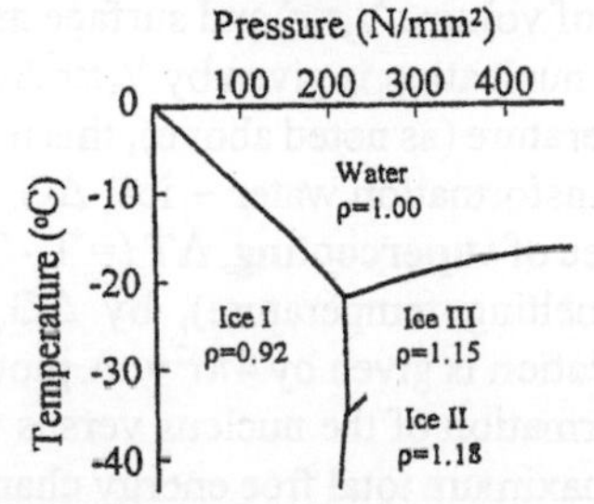

The lower the temperature, the higher the pressure required to transform ice → water. In other words, the application of pressure reduces the freezing point of water. This has important consequences for the freezing of water contained in the capillaries of a porous material, as the pressures generated may be sufficient to liquefy ice (6.6.1).

***Figure 2.6** Relationship between pressure and temperature for water and ice*

2.6.2 Stability of Phases

A phase transformation (e.g, water → ice) involves a complete rearrangement of the atoms or molecules which undergo the transformation. Any system is said to be **in equilibrium** if there are no unbalanced forces either within the system or between the system and its surroundings. In some materials, temperature and/or pressure is rapidly altered such that equilibrium conditions are not immediately obtained and the resultant phase formed is termed a **metastable phase.** The number of atoms within the metastable phase able to surmount the phase change energy barrier is reduced to the extent that the metastable phase appears to be stable. However, true equilibrium has not been obtained and the metastable state will gradually transfer to the stable state (usually over very long time periods). Common examples of the materials in the metastable state include glass (**9.3**), products formed by degradation processes involving water, including the **conversion** of high alumina cement (**7.14.5**) and the formation of ettringite during sulfate attack (**8.19.2**). The relative stability of a system is illustrated using a thermodynamic function called the (Gibbs) free energy, G (**2.6.3**). For a phase transformation (such as water → ice) to occur, the free energy of the final state, G_2, must be **less than** the free energy of the initial state, G_1. Using the water → ice transformation as an example, the change in free energy during the transformation water → ice, $\Delta G_v = G_{ice} - G_{water}$, must be **negative**. If ΔG is positive, the transformation will not occur as the final state has more free energy than the initial state. If ΔG is zero, the two phases are in equilibrium.

2.6.3 Crystallisation

The **crystallisation** of ice from water caused by the phase transition water → ice does not occur instantaneously or everywhere simultaneously. In addition, the transition takes place in two successive steps, **nucleation** and **growth of the formed ice crystal.**

Nucleation

The first step in the formation of a crystal is the formation of a nucleus, which can be regarded as a small cluster of atoms or molecules that have taken on the correct crystalline arrangement. The stability of the nucleus is dependent upon two factors

- the free energy change during the liquid → solid transition, ΔG_v;
- the value of the surface energy of the newly formed nucleus, γ.

The total free energy change, ΔG, associated with formation of the nucleus is the product of these two factors. For a spherical nucleus of volume $\frac{4}{3}\pi r^3$ and surface area $4\pi r^2$, $\Delta G = \frac{4}{3}\pi r^3 \Delta G_v + 4\pi r^2 \gamma$. The **driving force** for nucleation is given by $\frac{4}{3}\pi r^3 \Delta G_v$. Since ΔG_v is always negative below the melting temperature (as noted above), this term is negative. The change in free energy during the transformation water → ice, ΔG_v, is related to the **latent heat of fusion,** ΔH_f and the degree of **supercooling,** ΔT (= $T - T_m$, where T is the bulk temperature and T_m is the melting temperature), by $\Delta G_v = (\Delta H_f \Delta T)/T_m$. The **retarding force** which limits nucleation is given by $4\pi r^2 \gamma$. A plot of the total free energy change, ΔG, associated with formation of the nucleus versus the radius, r, of the nucleus is given in ***Figure 2.7***. The maximum total free energy change associated with the formation of the nucleus, ΔG^*, is reached for $r = r^*$, where r^* is the **critical radius** of the nucleus.

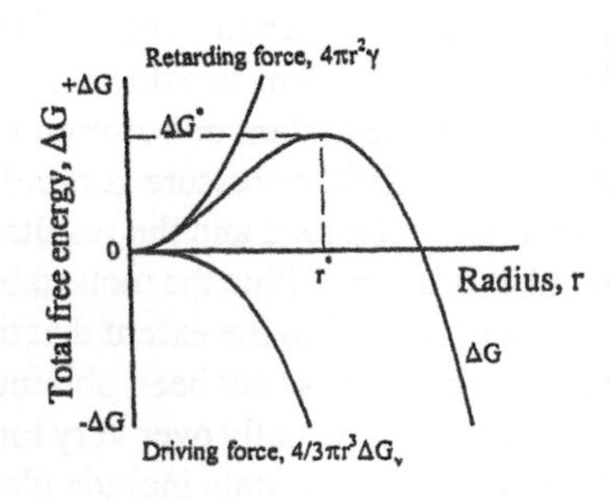

At **r = r*** the free energy decrease during the liquid → solid transition (driving force) just equals the surface free energy of the newly formed nucleus (retarding force). The nucleus is stable. At **r > r*** the total free energy of the system, ΔG, decreases, initiating the formation of a crystal. Further decreases in the total free energy of the system will be followed by continued growth of the crystal. At **r < r*** the surface free energy of the newly formed nucleus (retarding force) is much greater than the free energy change during the liquid → solid transition (driving force) so that the total free energy change associated with the formation of the nucleus, ΔG, will increase. The nucleus is unstable and will dissolve.

Figure 2.7 *Variation of the total free energy, ΔG, associated with the formation of a nucleus with the radius, r, of the nucleus*

These relationships are important in ice formation in porous materials, i.e.,

- there is a **critical radius** required for a stable nucleus to form. The micropores in a porous material may be too small to allow nucleation to take place (***Figure 2.8a***);
- a considerable degree of **supercooling** is required to form a stable nucleus. This is particularly relevant in fine pores, where a considerable temperature reduction is required to allow nucleation (***Figure 2.8a***);
- the formation of a stable nucleus is facilitated by the presence of impurities, surface roughness, dirt particles, etc. This is because the surface free energy term, γ, is **reduced** by these factors so that the degree of supercooling required for formation of a stable nucleus is reduced (see above).

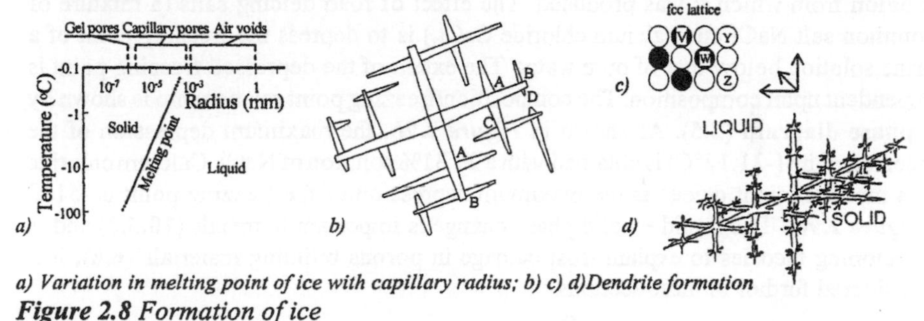

a) Variation in melting point of ice with capillary radius; b) c) d)Dendrite formation

Figure 2.8 *Formation of ice*

Crystal growth

Once a stable nucleus has been formed, it continues to grow to form a crystal. Often growth occurs preferentially in certain directions, giving rise to a **dendrite**. Dendrites are formed because certain directions within the crystal are more efficient at removing the latent heat of fusion, i.e., the structure is anisotropic. This efficiency depends on the crystal lattice structure of the solid. Dendritic growth is shown in ***Figure 2.8b***. Initially, dendritic growth occurs in directions in which atoms or molecules are more closely packed, since in these directions thermal conductivity will be greater ***(Figure 2.8c)*** (compare direction **VWZ** with direction **ZY** in the fcc lattice). However, a considerable amount of latent heat is evolved in this direction, causing the temperature of the adjacent liquid to rise to a temperature that may exceed the freezing temperature. Crystal growth in this direction therefore ceases. In a perpendicular direction, the liquid region will have a lower temperature since there has been no crystallisation and consequently no heat evolution. Therefore the crystal will grow in the new direction (i.e., from point **A** to point **C** in ***Figure 2.8b***) until heat is dissipated away from the initial direction of crystal growth. When the temperature in the initial growth region drops to a sufficient value below the freezing temperature, the crystal will resume its original direction of growth until it reaches point **B**, where again the temperature will rise and further growth of the crystal will stop. Then again growth of the crystals occurs in a direction perpendicular to the original direction of growth (i.e., from point **B** to **D**). This sequence of processes finally leads to the structure characteristic of dendrites (***Figure 2.8d***). In water, these processes lead to the typical dendritic structure of snow-flakes. In metals, cooling produces dendritic growth in the newly formed solid phase (**10.2**). These dendrites grow until they meet other dendritic growths from other nuclei. The junctions between the nuclei are called **grain boundaries (10.2)**. The grain boundary is the last liquid to solidify and marks the orientation difference between the different dendrite nuclei. Often the grain boundaries contain all the insoluble impurities and therefore impart brittleness to cast metal materials.

2.6.4 Effect of Impurities

Dissolved species, whether it is a salt dissolved in water or a metals dissolved in another metal (an alloy), lower the freezing point of the liquid phase, as shown in ***Figure 2.9a***. Sea water is a good example of depressed freezing points; the depression is proportional to the concentration of dissolved substance. The resultant ice produced is purer than the solution from which it was produced. The effect of road deicing salts (a mixture of common salt NaCl and calcium chloride $CaCl_2$) is to depress the freezing point of a brine solution below that of pure water. The extent of the depressed freezing point is dependent upon composition. The composition/freezing point relationship is shown by a **phase diagram** (**2.5**). As shown in ***Figure 2.9b***, the maximum depression of the freezing point (-21.12°C) is obtained with a 23.31% solution of NaCl. Calcium chloride is a more efficient deicer as the maximum depression of the freezing point is -51°C (***Figure 2.9c***). The liquid → solid phase change is important in metals (**10.5.2**) and in developing theories to explain frost damage in porous building materials (**6.6**); it is considered further in these sections.

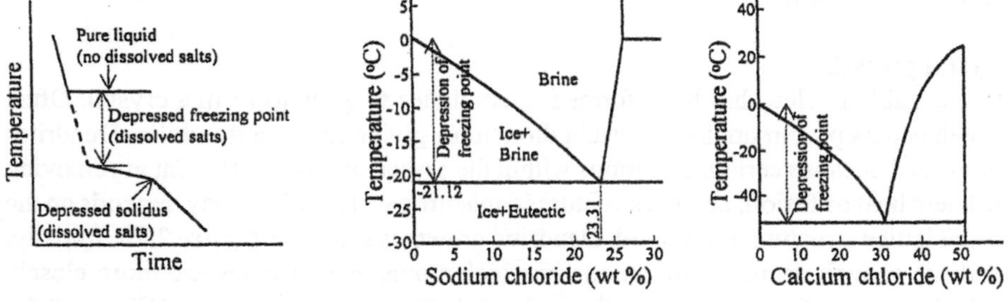

a) Freezing point depression; b) Sodium chloride, NaCl; c) Calcium chloride, $CaCl_2$

***Figure 2.9** Effect of impurities on freezing points*

2.7 LIQUID TO GAS CHANGE OF PHASE

As we have seen (**2.2.2**), the molecules in the vapour state are 12x the distance apart that they are in the liquid state. If we denote the water molecule diameter in the liquid phase as **d**, then the intermolecular distance in the vapour phase is **12d** (***Figure 2.10***).

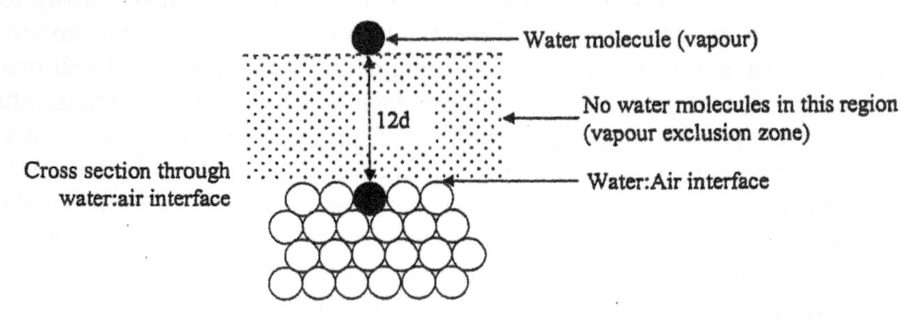

***Figure 2.10** Molecular interpretation of water evaporation and condensation*

Any water vapour molecules coming within 12d of the liquid water surface will be attracted to the liquid surface and **condensation** will result. Thus for **evaporation** to take place, the water loses its most energetic molecules (i.e., those which have sufficient energy to get further than 12d away from the liquid surface). Less energetic molecules are attracted back into the liquid. There is therefore effectively a **vapour exclusion zone** over the surface of the liquid, since water vapour molecules cannot persist within 12d of the liquid surface. When the liquid surface is in a draught, it may lose the less energetic molecules that would otherwise condense, as **any** water molecule with sufficient energy merely to lift out of the surface can potentially be swept away. This phenomenon produces the **wind chill factor**.

2.8 DEFORMATION

External forces applied to a body tend to move the atoms/molecules from their relative positions. This accounts for the elasticity of the body and is dependent upon the bond strength. We can summarise the deformation characteristics of liquids and solids by noting that

- liquids possess **volume elasticity** only;
- solids possess **volume elasticity and rigidity**, which both combine to give **tensile elasticity**. Solids revert to their original shape if not stressed beyond a certain point.

2.8.1 Elastic Constants

When a solid is extended or compressed by an external force P (units: N), the dimensions of the solid change. Conventionally, rather than referring directly to force P, we express the force in terms of the area A (unit: m^2) over which it acts, as **stress**, σ (= P/A) (units: N/m^2). Under simple uniaxial tension (or compression), the material either extends (or shortens). Conventionally, we express this as **strain**, ϵ (= $\Delta L/L$, i.e., the change in length over the original length). Since both the change in length and the original length have the same units, strain is a dimensionless ratio. In many materials, once the force has been removed the material will return to its original dimensions. Such materials are known as **elastic** materials. Experiments show that the extension of an elastic material is directly proportional to the applied force and a graph of force against extension is linear and passes through the origin. However, for most materials, a value of loading can be found beyond which the material will not return to its original dimensions on removal of the force. This point is the limit of proportionality (although other terms are also used, **3.2.2**). Hooke's law expresses these relationships in terms of stress and strain, i.e., the strain produced in an elastic material is directly proportional to the applied stress, provided that the limit of proportionality has not been exceeded

$$E = \frac{\text{Stress } (N/m^2)}{\text{Strain } (m/m)}$$

where E (= σ/ϵ) is a constant known as Young's modulus of elasticity and has units of N/m^2 (i.e., the same as the units of stress) (however, the value of Young's modulus is so large that it is expressed more conveniently in N/mm^2 (or GN/m^2) (1 GN/m^2 = 1 kN/mm^2). This relationship is shown in ***Figure 2.11a***. Young's modulus is a measure of the stress required to produce a given deformation; this property of a material is often referred to as **stiffness**. The Young's modulus of materials is assessed using **stress-strain curves (3.2.2)**; typical values for a variety of common building materials are given in **Table 2.3** (over).

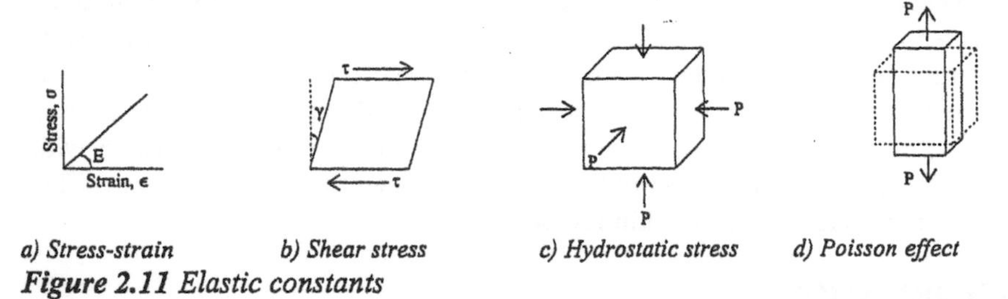

a) Stress-strain *b) Shear stress* *c) Hydrostatic stress* *d) Poisson effect*

Figure 2.11 *Elastic constants*

Similar relationships apply to other forms of stress. Thus in shear (***Figure 2.11b***), the shear stress τ produces a shear strain γ, which is conventionally measured as an angle (and is therefore also dimensionless)

$$G = \frac{\text{Shear stress } (N/m^2)}{\text{Shear strain}}$$

where G (= τ/γ) is a constant known as the **shear modulus** (or modulus of rigidity) and has units of N/m^2 (i.e., the same as the units of stress). The shear modula (and Poisson's ratios, see below) for various materials are illustrated in **Table 2.2.**

Table 2.2 Elastic modula for some common materials

Material	Shear modulus (G) (GN/m^2)	Poisson's ratio (ν)	Material	Shear modulus (G) (GN/m^2)	Poisson's ratio (ν)
Granite	19	0.20	Cast iron	51	0.17
Glass	22	0.23	Mild steel	82	0.26
Concrete	10	0.20	Aluminium	25	0.33
Nylon	1.1	0.40	Copper	44	0.36
Soft rubber		0.49	Lead	6.2	0.40

Notes: $1\ GN/m^2 = 1 \times 10^9\ N/m^2$

Under hydrostatic pressure P (***Figure 2.11c***), the volume changes. Expressing this volume change dimensionally as ΔV/V (i.e., change in volume over original volume), then

$$K = -\frac{P}{\Delta V/V}$$

where K is the **bulk modulus** and the minus sign arises as a consequence of the fact that as pressure increases, volume decreases.

Table 2.3 Modulus of elasticity and coefficients of thermal expansion for materials

Material	Modulus of Elasticity, E	Coefficient of linear thermal expansion, α	Material	Modulus of Elasticity, E	Coefficient of linear thermal expansion, α
	kN/mm^2	$(\times 10^{-6})$		kN/mm^2	$(\times 10^{-6})$
Natural stones			**Metals and alloys**		
Granite	20-60	8-10	Cast iron	80-120	10
Limestone	10-80	3-4	Plain carbon steel	210	12
Marble	35	4-6	Stainless steel		
Sandstone	3-80	7-12	• austenitic	200	18
Slate	10-35	9-11	• ferritic	200	10
Cement based composites			Aluminium and alloys	70	24
Mortar and fine aggregate	20-35	10-13	Copper	95-130	17
Dense aggregate concrete			Bronze	100	20
• gravel aggregate	15-36	12-14	Aluminium bronze	120	18
• crushed rock (except limestone)	15-36	10-13	Brass	100	21
• crushed limestone	20-36	7-8	Zinc		
Steel fibre reinforced concrete	20-41	5-14	• parallel to rolling	140	33
Aerated concrete	1.4-3.2	8	• perpendicular to rolling	220	23
Lightweight aggregate concrete			Lead	14	30
• medium lightweight	8	8-12	**Rubbers, plastics, etc.**		
• ultra lightweight[a]		6-8[d]	Asphalt		30-80
Asbestos cement	14-26	8-12	Pitch fibre		40
Glass reinforced cement	20-34	7-12	Ebonite		65-80
Calcium silicate based composites			Thermoplastics		
Asbestos wall boards and substitutes	8-10	5-12	• PVC, PVC-u, PVC-c	2.1-3.5	50-70
Asbestos insulating boards, substitutes	2.6-3.6	2.5-7.2	• polyethylene (low density)	0.1-0.25	160-200
Gypsum and gypsum based composites			• polyethylene (high density)	0.5-1.0	110-140
Dense plasters; Plasterboard	16	18-21	• polypropylene	0.9-1.6	80-110
Sanded plasters	8.5-16	12-15	• polycarbonate	2.2-2.5	60-70
Lightweight plasters	1.5-4	16-18	• polystyrene	1.7-3.1	60-80
Glass reinforced gypsum	16-20	17-20	Acrylic	2.5-3.3	50-90
Brickwork, blockwork and tiling			Acetal	2.8-3.7	80
Concrete brickwork and blockwork			Polyamide (Nylon 6)	1-2.7	80-130
• dense aggregate	10-25	6-12	ABS	0.9-2.8	60-100
• lightweight aggregate (autoclaved)	4-16	8-12	Thermosets (laminates)		
• aerated (autoclaved)	3-8	8	• phenol and melamine formaldehyde	5.5-8.5	40-55
Calcium silicate brickwork	14-18	8-14	• urea formaldehyde	10	27
Clay, shale brickwork or blockwork	4-26	5-8	Cellular (expanded)		
Clay tiling		4-6	• PVC		35-50
Wood and wood laminates			• phenolic		20-40
Softwoods	5.5-12.5[a]	4-6[e]	• urea formaldehyde		30-90
		30-70[f]	• polyurethane		20-70
Hardwoods	7-21[a]	4-6[e]	• polystyrene		15-45
		30-70[f]	Reinforced		
Plywood	6-12[a]		• GRP	6-12	20-35
Blockboard and laminboard	7-11[a,b]		• carbon fibre oriented		
	5-8[a,c]		parallel to reinforcement	180-220	0-0.05
Woodchip and fibrous materials			perpendicular to reinforcement		30-70
Hardboard	3-6		**Glass**		
Medium board	1.7-3.3		Plain, tinted, opaque	70	9-11
Chipboard	2-2.8		Foamed (cellular)	5-8	8.5
Wood-wool cement slabs	0.6-0.7				

Notes: [a] at 12% moisture content (values reduced at higher moisture contents); [b] with core; [c] across core; [d] based on exfoliated vermiculite and expanded perlite; [e] with grain; [f] across grain.

The general relationships between these three modula are summarised in **Table 2.4**.

Table 2.4 Summary of elastic modula

Name of modula	Nature of stress	Nature of deformation	Examples (GN/m^2) typical metal	water	gas
Bulk modulus (K)	change of pressure	change of size no change of shape	1×10^2	2×10^0	1×10^{-4}
Shear modulus (G)	$\frac{\text{tangential force}}{\text{area}}$	change of shape no change of size	4×10^1	-	-
Young's modulus (E)	$\frac{\text{tensile force}}{\text{area}}$	change of shape change of size	1×10^1	-	-

Finally, when a material is extended by an external force in tension in the longitudinal direction (***Figure 2.11d***), it also contracts laterally in a direction at right angles to the applied tensile force. The ratio of the relative lateral contraction to the relative longitudinal extension is known as **Poisson's ratio, ν**

$$\nu = \frac{\text{relative lateral contraction (m/m)}}{\text{relative longitudinal extension (m/m)}}$$

Poisson's ratio ν for most solids is within the range 0.25 and 0.33 (**Table 2.2**) and this implies a decrease in volume under tension. Note that if $\nu = 0.5$, there is no volume change in tension or compression and this represents complete fluidity. From **Table 2.2** it can be seen that soft rubber comes very close to complete fluidity, but for most other common materials, there is a volume change on deformation.

The elastic constants are related by $G = ½E/(1+\nu)$ and $K = E/3(1-2\nu)$.

2.8.2 Thermally Induced Dimensional Changes

Practically all solids expand when heated and contract when cooled. The expansion affects all dimensions of a body; for example, for an isotropic solid, expansion in length is accompanied by the same expansion in width and thickness. The increase in length per unit degree rise in temperature is called the **coefficient of linear thermal expansion** (α) and is defined as

$$\alpha = \frac{L_{T2} - L_{T1}}{L_{T1} \times (\Delta T)}$$

where L_{T1} is the initial length at any temperature T_1, L_{T2} is the final length at any temperature T_2 and ΔT (= T_2-T_1) is the change in temperature. The thermal expansion of solid arises due to the vibration of the constituent atoms/ions, which increases as the temperature increases. As the atoms/ions move farther away from their average positions, the volume of the solid increases. The vibrational movement of atoms/ions is related to bond strength. If the bond strength is high, then

- the interatomic distance is small;
- the effect of temperature on the amplitude of vibration is small and therefore the coefficient of expansion is small;

- the melting points will higher, because more heat energy is required to melt the material (i.e., thermal energy has to be sufficient to break the bonds and allow the transition of a solid to a liquid).

We can demonstrate these relationships for elementary ionic structures by comparing interatomic distance, coefficient of thermal expansion and melting point (**Table 2.5**).

Table 2.5 Relationship between interatomic distance, coefficient of thermal expansion and melting point for elementary ionic structures

Property	Type of structure				
	AX				AX_2
	NaF	NaCl	NaBr	NaI	CaF_2
Interatomic distance (A-X) (nm)	0.231	0.279	0.294	0.318	0.239
Coefficient of linear thermal expansion (α) (x 10^{-6})	39	40	43	48	19
Melting point (°C)	988	801	740	660	902

←
Smaller interatomic distance
Higher melting point
Smaller coefficient of thermal expansion

Empirical data shows that the total volume change of solids that occurs when they are heated in the temperature range from absolute zero to the melting point is approximately constant. Thus materials with lower melting points (e.g, plastics) have higher coefficients of thermal expansion than materials with higher melting points (e.g, metals).

From the defining equation for the linear thermal expansion coefficient, the change in size, L (= L_{T2}-L_{T1}) of any material can be determined with knowledge of the linear thermal expansion coefficient (α), the initial size of the dimension concerned (L_{T1}) and the temperature change (ΔT). **Table 2.3** lists coefficients of linear thermal expansion for a wide variety of common building materials.

Temperature-induced changes in size occur continuously in all buildings, since no building material is dimensionally unaffected by temperature change. Generally, the dimensional effects of temperature change in materials is reduced below that predicted on the basis of linear thermal expansion due to the presence of restraint offered by mechanical fixings, adhesion and friction. The temperature range experienced by construction materials depends upon

- the environmental temperature change;
- the colour, orientation and exposure of the material;
- the duration of the environmental temperature change in relation to the thermal response time of the material, as determined by the mass and the effect of insulation, both in protecting the material from the temperature change and in limiting the loss of heat gained from the temperature change.

In assessing temperature-induced dimensional movement, direct measurement of surface temperatures will yield values for ΔT. Design temperature change values for various materials used in the external fabric of a building are also available and can be used for design purposes (**Table 2.6**, which includes typical internal surface temperatures likely in occupied and unoccupied buildings).

Table 2.6 Examples of service temperature ranges of materials (valid for UK only)

Building element	Minimum (°C)	Maximum (°C)	Range (°C)
External			
Cladding, walling, roofing			
• heavyweight			
light colour	-20	50	70
dark colour	-20	60	85
• lightweight, over insulation			
light colour	-25	60	85
dark colour	-25	80	105
Glass			
• coloured or solar control	-25	90	115
• clear	-25	40	65
Freestanding or fully exposed structures			
• concrete			
light colour	-20	45	65
dark colour	-20	60	80
• metal			
light colour	-25	50	75
dark colour	-25	65	90
Internal			
Normal use	10	30	20
Empty/out of use	-5	35	40

Notes: The following situations are not included in the Table and may give rise to temperature extremes more severe than those listed: dark surfaces under glass (e.g, solar collectors); materials used in cold rooms or refrigerated stores; materials used for, or in close proximity to, heating, cooking and washing appliances, or flues and heat distribution networks.

Temperature induced changes in size may

- result in stresses in the affected material that may cause it to crack and/or distort. The stress induced in a material is the product of the modulus of elasticity, E, of a material (**Table 2.3**) and the strain (**2.8.1**), which, when temperature induced, is the coefficient of linear thermal expansion (**Table 2.3**), since the units are identical to those for strain. Hence the temperature induced stress can be determined from

  $$\text{Stress} = E \times \alpha \times \Delta T$$

  If the stresses calculated exceed, for example, the ultimate tensile strength of the material, then tensile cracking will occur.

- result in stresses in adjacent materials, particularly where these restrain movement of the affected material. The force (= stress x area) induced in a restraining material for temperature induced stresses is given by

  $$\text{Force} = E \times \alpha \times \Delta T \times A$$

  where A is the area over which the force acts (on the restraining material). As the linear temperature induced dimensional movement of different materials varies

(**Table 2.3**), **differential movement** of materials in juxtaposition may exacerbate problems, particularly when moisture induced dimensional movement (**6.4**) is also be considered.

In practice, the damaging effects of temperature induced dimensional change is commonly limited by allowing for predicted movement, often using jointing techniques. **Table 2.7** summarises recommendations to account for thermal (and moisture) induced dimensional changes in some common materials.

Table 2.7 Recommendations to counteract thermal and moisture movement[1]

Material	Recommendation
Fired clay brickwork[a]	Vertical joints be positioned not more that 15 m apart to avoid cracking due to thermal contraction. The width of joints in millimetres should be about 30% more than the joint spacing in metres. Spacing of joints at intervals closer than 15 m is likely to be needed where comparatively little restraint is placed on the size changes (as in parapet walls, boundary walls, etc.). Account should also be taken at the design stage of irreversible moisture movement, which depends on age of bricks at that time, but can be large (**Table 6.1**), particularly in susceptible bricks fresh from the kiln.
Calcium silicate brickwork[a]	Vertical joints should be located at intervals of between 7.5 m and 9 m, and that the length:height ratio of each panel between joints should not exceed 3:1. As with clay brickwork, additional joints or reinforcement is likely to be required at points of stress concentration. It should be noted that although moisture-induced contraction is the predominant characteristic of calcium silicate bricks (**7.5**); maximum moisture induced movement in calcium silicate bricks can be determined by adding irreversible and reversible movement for a given length. Since irreversible moisture induced movement is shrinkage, the maximum movement calculated assumes shrinkage. Generally, moisture induced dimensional changes should not exceed 0.04%. In addition, their coefficient of thermal expansion is greater than that of fired clay bricks (**Table 2.3**).
Concrete block work walls[a]	Vertical joints should be at intervals of not more than 6 m. The standard also notes that the risk of cracking increases if the length of the panel exceeds twice its height. These recommendations mainly apply to half brick thick walls and, provided suitable precautions are taken, for thicker walls the distances between joints may be increased. As for calcium silicate bricks, moisture induced movement in concrete blocks is normally shrinkage and therefore maximum moisture induced movement (shrinkage) can be determined by adding irreversible and reversible movement.
Ceramic tiles[b,c]	6 mm wide soft joints incorporated at 4.5m intervals. Ceramic tiles expand markedly after manufacture; this is sometimes evidenced by crazing of the glaze. Combined irreversible and reversible moisture induced movement, coupled with temperature induced movement, may impose compressive loads in the plane of the tiling, leading to cracking and/or displacement. The compressibility of the filler material behind the sealant is possibly more critical than that of the sealant itself; a material which allows compression to half its original thickness when subject to a pressure of 0.1 N/mm^2 should be used. Suitable materials include flexible, cellular polythene, cellular polyurethane and foam rubber, whilst hemp, fibreboard and cork should be avoided.
Flat roofs	see below
Cladding	see below
Timber flooring[c]	Provision for a perimeter expansion gap of 10 to 12 mm. For larger areas sufficient space should be provided between successive individual boards or panels.

Notes: [1] Moisture induced dimensional changes are considered in section 6.4; [a] BS 5385. *Code of practice for the design and installation of ceramic floor tiles and mosaics. Part 3:1989. Wall and floor tiling.* [b] BS 5628. *Code of practice for the use of masonry. Part 3:1985. Materials and components, design and workmanship.* [c] BS 8201 (1987) *Code of practice for flooring of timber, timber products and wood based panel products.* BS 8200 (1985). *Code of practice for the design of non-loadbearing external vertical enclosures of buildings* provides data for many aspects of design, including those concerned with temperature induced dimensional changes. The standard recommends that the range of temperatures above and below the installation temperature to which the material may be exposed during service are determined using thermal absorption data included in the standard.

Cladding is generally more responsive to changes in environmental temperature than the relatively more stable building fabric as it is often of low thermal capacity (because of its thin section) and is more directly exposed. The temperature of lightweight cladding can change markedly over a period of a few minutes with corresponding size changes occurring over a short time span. The following points are relevant

- rapid size changes are unlikely to be accommodated by creep in materials;
- shading (e.g, by frame members, such as mullions and transoms) can produce large local temperature differences with correspondingly high local thermal stress;
- thin sheet claddings may relieve rapid stress development by bowing;
- distortions in the mirrored images produced by reflective claddings are indicative of points of local stress in the cladding material.

The most significant aspect with respect to reducing the risk of cracking is achieving joints of adequate thickness. BS 6954: 1988 *Tolerances for building* gives advice on appropriate clearances for a variety of cladding systems, although in practice adequate joint thicknesses are more often related to the degree of supervision available on site.

Flat roof construction experiences large temperature induced dimensional movement, unless the plan dimensions are very small. The following points are relevant

- as flat roofs are horizontal elements, exposure to solar radiation is considerable and, unless shaded (e.g, by adjacent taller buildings), they are potentially exposed throughout the entire day;
- the most common flat roof coverings are black in colour and therefore absorption of radiant solar heat of the roof as a whole is facilitated;
- except for flat roofs of more recent design (warm decked and inverted), thermal insulation within flat roofs effectively raises the temperature experienced in both decks and the roof coverings by reducing downward heat dissipation;
- dark coloured coverings to flat roofs facilitate radiation of heat to the clear night sky in winter so that temperatures of both decks and the coverings can fall well below ambient temperatures;
- thermal insulation within flat roofs impedes of heat lost by radiation to the night sky by heat dissipated from within the building below, reducing still further the temperatures of both decks and coverings;
- materials suitable for flat roof coverings are intolerant of localised strain. For example, a membrane fully bonded over a crack or joint of 0.01 mm width which subsequently opens to (say) 0.05 mm would need to have an extensibility of 500% to remain unbroken; no such roof membrane exists.

These effects ensure that flat roofs of conventional (cold deck) design constitute severely exposed conditions and experience wide ranges of temperature in service both daily and seasonally. As a general rule, intermediate joints in cold deck flat roofs should be introduced when the major dimension exceeds 10 m (or 5 m for roofs with absolute restraint at one end of their longer dimension). The locations of any joints required in the deck to accommodate temperature induced size changes should be selected to take account of support locations, the position of any absolute restraint, the presence of major movement joints in the main structure as a whole and the locations of any changes in roof shape on plan which could have the effect of a change in section (i.e., any departures from simple rectangular plan shape).

2.9 SOLUTIONS

Solutions may be basically of two types, namely **homogeneous** or **heterogeneous**.

2.9.1 Homogeneous Solutions

A homogeneous solution is a mixture of two or more substances whose final composition may vary within very wide limits (unlike molecular compounds, where the composition is in a fixed ratio of elements, **1.2.1**). Homogeneous solutions may be formed by a variety of mixtures, including gases in gases, solid in liquids, liquids in liquids, solids in solids, etc. The most important of these are considered below.

Generally, the term **solvent** is applied to the component in the larger amount and the term **solute(s)** is applied to the other component(s).

Solids in liquids

There are numerous examples of solutions of solids in a liquid. In sea water, for example, common salt (NaCl) dissolves in water. Here the NaCl molecule does not dissolve as NaCl, but as Na^+ and Cl^- ions dissolved in water. A solution that contains as much solute as can be dissolved in the solvent at that temperature in the presence of undissolved solute is called a **saturated solution.** In a saturated solution, a state of dynamic equilibrium exists such that the particles of the solid solute are constantly dissolving and being precipitated. For example, in sea water the undissolved salt remains as NaCl. The quantity of solid solute that dissolves in a given quantity of liquid solvent to form a saturated solution at a given temperature is dependent on the **solubility** of the solid solute. Solubility depends both on the nature of the solid and on temperature. Generally, increased temperature increases the solubility of the solid. There are, however, a few exceptions; for example, calcium hydroxide and sodium sulfate dissolve more freely in cold than hot water. This has important implications in **sulfate attack (8.19.2)** of cementitious products.

Water soluble salts arise from the materials themselves (e.g, in clay bricks, from the clay raw material) or from groundwater, and may include chlorides, sulfates, phosphates, nitrates, etc. Many water soluble salts are also hygroscopic, absorbing water from the air when dry and releasing water to the air when wet. In some cases, the hygroscopic salts can form pools of water, in which case the salt is said to be **deliquescent.** A hygroscopic salt therefore can be considered to artificially increase the relative humidity of its immediate environment. Thus the hygroscopic salt produces an environment which is wetted earlier as the relative humidity rises, and remains wet for a longer period after the relative humidity falls. **Table 2.8** gives the relative humidity of the air in equilibrium with saturated salt solutions at 20°C for certain hygroscopic salts. Note that the road deicing salts NaCl and $CaCl_2$ will not start to "dry out" until the relative humidity falls to 76% and 32% respectively.

Table 2.8 RH% of air in equilibrium with various saturated salt solutions at 20°C

Salt in solution	Relative humidity (RH%)	Salt in solution	Relative humidity (RH%)
$CuSO_4.5H_2O$	98	$NaCl^b$	76
K_2SO_4	98	$CuCl_2.2H_2O^c$	68
$Na_2CO_3.10H_2O$	92	$FeCl_2$	56
$FeSO_4.7H_2O^a$	92	$K_2CO_3.2H_2O$	44
KCl	86	$CaCl_2.6H_2O^b$	32
NH_4SO_4	81	$ZnCl_2.xH_2O$	10

Notes: [a] contaminant on painted steelwork (11.12.1); [b] found in road deicing salts; [c] verdegris on soldering

The addition of a solute to a liquid solvent produces a number of variations in the characteristics of the liquid solvent. For example

- a lowering of the vapour pressure of the liquid solvent, depending on the **concentration** of the solute (not on its identity). The use of saturated salt solutions to obtain known relative humidity environments (**5.11.2**) depend upon this fact;
- an elevation of the boiling temperature of the liquid solvent (resulting from an

increase in the vapour pressure). The boiling temperature elevation of the liquid solvent is proportional to the concentration of the solute and a solvent specific boiling temperature constant;

- a depression in the freezing temperature of the liquid solvent, which also depends on the **concentration** of the solute (not on its identity). This fact is utilised in the application of road de-icing salts, which lower the freezing temperature of water and prevent ice formation (***Figure 2.9***).

Saturated solutions must be formed before a solid solute can be precipitated from solution. A concentrated (unsaturated) solution at a given temperature can become saturated either by loss of the solvent (e.g, by evaporation) to make the remaining solution more concentrated, or by cooling of the concentrated solution. Thus as water passes through a structure, it carries these salts in solution, and when the water reaches the surface of the material and evaporates, the salts are deposited (e.g, as efflorescence on brickwork). Salts in solution account for a number of problems in buildings

- crystallisation of water soluble salts at or near the surface of a porous material results in efflorescence; where crystallisation occurs beneath the surface, this results in cryptoefflorescence (**6.7.1**);
- solutions can concentrate in areas where evaporation of the solvent is restricted, such as crevices and folds, so that the solution can remain for long periods (crevices and surface capillaries are the last areas to dry out) (**5.7**). The presence of solutions in these areas will promote corrosion of metal substrates (**11.7.2**);
- water soluble salts are effective electrolytes (charge carriers) and their presence greatly increases the ability of the water to carry an electrical charge. Thus, for example, a moisture meter may give very high readings if water soluble salts are present at the surface (**5.11.1**), and the presence of water solubles salts will enhance the corrosion of metals (**11.4.7**);
- **sulphate attack** (**8.19.2**), where water soluble sulfates (i.e., the SO_4^{2-} ion), reacts with tri calcium aluminate in the Portland Cement producing a new compound and causing a large expansion. It is important to appreciate that there is **no** reaction if the sulfates are **insoluble**;
- hard waters contain soluble calcium salts (dissolved as the calcium ion Ca^{2+} associated with the bicarbonate ion HCO_3^-). The calcium ions are precipitated by soaps (sodium stearate) as calcium stearate which is **insoluble** and therefore not ionised into calcium and stearate ions. Hence the calcium ions remove all the soluble stearate ions from solution (**4.4.3**).

Gases in liquids

Gases are only partially soluble in liquids. When water is the liquid, the amount of gas that will dissolve in a given amount of water is dependent on

- the nature of the gas and its water solubility;
- the temperature;
- the pressure of the gas in contact with the water.

Unlike most solid solutes, the solubility of gas solutes at constant pressure in liquids diminished with an increase in temperature. Gas in liquid solutions are important in the building context as many acidic gases will dissolve in water to form weak acid solutions, which may contribute to the decay and deterioration of many building materials, notably metals (**11.7.2**) and sedimentary rocks (**4.3.1**). In the latter case, the acid formed is responsible for causing the cementing agent (e.g, calcium carbonate) to dissolve in solution. Alkali solutions may also be formed by alkali gases. Examples of acidic gases include carbon dioxide (CO_2) and, in urban environments, sulfur dioxide (SO_2), which form weak carbonic and sulfuric acid solutions respectively, i.e.,

$$CO_2 + H_2O \rightarrow H_2CO_3$$
$$SO_2 + H_2O \rightarrow H_2SO_3$$

Liquids in liquids
Although pairs of gases are mutually soluble, not all pairs of liquids dissolve. Liquid solutes may be
- **completely immiscible** in a liquid solvent (e.g, oil in water);
- **completely miscible** in a liquid solvent (e.g, alcohol in water);
- **partially miscible** in a liquid solvent (e.g, ether in water).

The constituents of liquid solutions can be separated by centrifuges and flotation methods.

Solids in solids
The solution of a solid in a solid (to give a **solid solution**) usually takes place when the components are in the liquid state. Solid solutions have variable compositions and comprise a mixture of compounds, which are ordinarily represented in the chemical formula. For example, the chemical formula for potash alums (used in fire protection, **14.7.3**) is usually written as $K_2SO_4.(NH_4)_2SO_4.24H_2O$, not $(K_2S_2N_2H_{56}O_{32})$ to reflect the composition of the constituent compounds (K_2SO_4 and $(NH_4)_2SO_4$) (**1.2.2**). Examples of solid solutions include
- glass (main constituents silica, SiO_2 and sodium oxide Na_2O) (**9.1**);
- alloys (**1.8.8**), for example zinc in copper (brass), tin in copper (bronze), carbon in iron (steel). Each new alloy formed is different from the two (or more) individual components (this is very important, for example, in the corrosion process, as can be seen by comparing the **electrochemical** and the **galvanic series (11.4**);
- igneous rocks, e.g, granite, which is a solid solution of quartz, feldspars and mica.

The final constituents of a solid solution are called **phases** and are produced by cooling from the high temperature liquid state. The different phases in a solid solution cannot be separated by mechanical means (unless they are remelted). The separation of the constituents in compounds, solid solutions and mixtures is compared below

		Phase formation		
Compounds	→	Solid solution	←	Mixtures
(not separated)		(separated only in the fused condition)		(separated)

2.9.2 Heterogeneous Solutions

Generally a solution consists of a solvent and a solute. The solute is molecularly dispersed within the solvent (as the unit AX, or even as the ions, **1.4**). A heterogeneous solutions the solute is not molecularly dispersed but exists as aggregates in the solvent (now better called the dispersion medium). The aggregate size is measured in microns (μm). Heterogeneous solutions are therefore two phase systems, and include many substances of overlapping properties, thus classification is complex. **Table 2.9** lists some types of heterogeneous solutions.

Table 2.9 Examples of heterogeneous solutions

Dispersion medium	Disperse phase	Name
	Gas	No heterogeneous solution. Perfect mixture always.
Gas	Liquid	Fog, cloud, mist } aerosols
	Solid	Smoke
	Gas	Foam
Liquid	Liquid	Emulsion } sols
	Solid	Colloid suspension
	Gas	Solid foam (e.g, cavity wall materials), pumice, etc.
Solid	Liquid	Colloidally dispersed crystals (e.g, TiO_2 in paints)
	Solid	Metallic alloys (**10.5.2**), eutectics (**10.6**), igneous rocks, etc.

A **dispersion** is a suspension of two phases. The phase formed by the disperse particles is called the **disperse phase** and the medium in which the particles are suspended is called the **dispersing** (or **continuous**) **phase**. The disperse and continuous phases are analogous to the solute and solvent respectively in homogeneous solutions. Since the disperse and continuous phases may be liquid, solid or gas, a number of dispersion systems are possible and only a few are considered here.

An important characteristic of the disperse phase is that the particles do not settle out under gravity (or only slowly in some cases) and they are not easily removed by filtration. The reason why disperse particles do not coalesce is that their surfaces are charged. For example, disperse phase silica particles in water absorb hydroxide ions on their surface and are negatively charged (**1.11.2**). Repulsion between like charges keeps the particles in suspension. Disperse particles can be precipitated by neutralising the charge on the particles, often by the addition of ions of the opposite charge. This has two interesting applications

- in the initial stages of the purification of drinking water (**4.2.1**), clay and other disperse phase particles must be removed. To achieve this, impure water is treated with aluminium sulfate so that the negative charges on the disperse phase articles are neutralised by the Al^{3+} ions, and the particles coalesce and settle out of solution;
- in the deposition of sedimentary deposits in the sea, the negative surface charge of the disperse phase particles is neutralised by Na^{+} and Mg^{2+} ions in the sea water (**1.11.2**). This process is utilised as a simple method of determining the quantity of 'fines' in aggregate for concrete (called the 'field settling test', **8.6.2**).

Colloids can be classified as **lipophilic** ('solvent loving'; **hydrophilic** if the solvent is water) and **lipophobic** ('solvent fearing'; **hydrophobic** if the solvent is water).

Emulsions are formed from liquids suspended in liquids (e.g, milk, paint, etc.). These liquids are usually immiscible with each other and the emulsions formed are only stable due to the chemical geometry of the molecule. An example of the formation of an emulsion is the action of a detergent on grease and oils (**5.4.2**). The detergent molecule (**5.4.2**) comprises a **hydrophilic** (lipophobic) part and a **hydrophobic** (lipophilic) part. On application to grease and oil, the hydrophobic part of the detergent molecule (a hydrocarbon chain structure, $C_{17}H_{35}$-) orientates itself into the oil droplet whilst the hydrophilic part of the detergent molecule (usually a carboxyl group, -C=O.OH) orientates itself towards the water. This action **emulsifies** the grease and oils. This form of technology is also used in sewage treatment (**4.2.1**) and in the concentration of iron ores from the earth.

A **sol** is a colloidal system in which the continuous phase is a liquid. If the liquid is water they are termed **hydrosols** or **aquasols**, whereas if the liquid s an alcohol they are termed **alcosols**. Sols, like colloids, can be classified as lipophilic and lipophobic. A **gel** is a colloid in a more solid form than a sol. Strong solutions of certain colloids set to a jelly and are called gels. Gels are generally **hygroscopic** and usually expand as they take in water and shrink as the moisture content decreases. This is termed **physical adsorption (5.6)** because the water is not bound chemically (i.e., by covalent, ionic or metallic bonding). Gels are subdivided into

- **Rigid gels** e.g, silica gel, used as a drying agent in packaging and double glazing;
- **Non rigid** (or **elastic**) gels e.g, Gelatin, starch and gum arabic.

Many building materials behave as rigid gels and the uptake of moisture gives rise to dimensional changes (**6.4**). One consequence of this behaviour is that these materials do not go into solution, but have definite saturation levels. For example, the fibre saturation point (**12.6**) of timber is reached when the **cell walls** contain approximately 25-30% of their dry weight as water. Some gels expand **isotropically** (equally in all directions) but some expand **anisotropically**. The moisture content of gels is dependent upon the humidity of the environment.

The most important engineering gel is undoubtedly cement. Powdered cement develops a highly rigid structure when water is added as the individual cement particles take up water of hydration, swell and linkup with each other to give rise to a high-strength but permeable gel of complex calcium silicates (**7.17**). A feature of many gels is their very high **specific surface area**. If the gel is permeable as well as porous, this surface is available for adsorbing large amounts of water vapour (and other airborne compounds) and the gel is an effective drying agent (e.g, silica gel). Adsorption is a reversible process so that, when the gel is saturated, it may be heated to dry off the water and thus can have its 'drying power' restored. Other materials which contain gel-like constituents include

- timber (and other fibrous materials) in which the 'gel' is incorporated into the bulk material;
- clay products (e.g, clay bricks), which are formed from a typically colloidal material, clay. However, the colloidal character is destroyed by firing (**7.3.1**) so that the bricks do not exhibit appreciable reversible moisture movement (**6.4**);

- calcium silicate bricks, which are manufactured by heating mixtures of sand and lime with steam under pressure in a process known as autoclaving (**7.5.1**), exhibit fairly high moisture movement (**6.4**), probably due to the presence of colloidal silicates of calcium (**7.5.2**);
- sandstones show the most pronounced moisture movement of all natural building stones (**6.4**). This may be due to the presence of colloidal minerals, that act as the cementing agent between the silica grains. Movement may be significant when sandstone is used as aggregate in concrete, particularly during initial drying.

2.10 THIXOTROPY AND DILATANCY

If a gel sets by the formation of rather weak links (van der Waals bonds, **1.14**), the linkages may be broken by vigorous stirring so that the gel liquefies again. When the stirring ceases, the bonds will gradually link up and the gel will thicken and return to its original set. Behaviour of this sort, in which an increase in the applied stress causes the material to act in a more fluid manner, is known as **thixotropy**. Examples include

- cement paste only in the early stages of setting (hardened cement paste will not heal itself spontaneously after it has cracked);
- non-drip paints, which liquefy when stirred (and so are easily applied) but set as a gel on completion of brushing (so that dripping or streaks on vertical surfaces are avoided);
- some clay soils, which may cause settlement where applied building loads liquefy the soils.

When an increase in the applied stress causes a viscous material to behave more like a solid, this is termed **dilatancy**. Dilatancy is therefore the reverse effect to thixotropy and is commonly demonstrated with a cornflour-water mixture. The viscous liquid will fracture if stirred vigorously, but the effect is of short duration, since fracture relieves the stress and the fractured surfaces immediately liquefy and run together again. Silicone putty is also a dilatant; it flows very slowly if left alone, but fractures if pulled suddenly, and will bounce like rubber if dropped on a hard surface. Applications of dilatancy in the construction industry are limited.

3

MECHANICAL PROPERTIES AND TESTING

3.1 INTRODUCTION

Building and engineering materials must have adequate strength in service to withstand the loads imposed upon them. To assess mechanical strength properties, a variety of testing techniques are used. Any specific mechanical strength property may be classified

- in terms of the **direction** in which strain is induced in a specimen when a force is applied to it. The four modes of mechanical strength properties are tension, bending (flexure), compression and shear;
- in terms of the **speed** at which the force is applied i.e. **static tests** (which generally involve loading the specimen at speeds ranging from about 5 to 50 mm/minute); **impact tests** (which generally involve loading the specimen at speeds ranging from about 180 to 240 mm/minute); **creep tests** which involve reduced loadings over long periods of time (exceeding 1000 hours); **fatigue tests** which involve cyclic loading over long periods of time (exceeding 1000 hours).

For tensile, compressive, bending and shear strength assessments, materials are tested to destruction in an appropriate and calibrated testing machine. The particular strength properties tested depend on the expected mode of failure of the material. Hence metals and plastics are commonly (though not exclusively) tested in tension whilst concrete, bricks and stone are commonly tested in compression.

3.2 TENSILE TESTING

Tensile testing of metallic materials is governed by BS EN 10002[1] and for plastic materials by BS EN ISO 527[2]. For concrete, BS 1881 Part 117: 1989 deals with tensile splitting strength (cylinders), and Part 119: 1989 for flexural strength (beams). Most modern testing machines are Universal Testing Machines (UTM) and are able to be used to perform several different types of mechanical property tests. A UTM consists basically of two vertical (or horizontal, depending on the machine type) loadbearing columns, on which are mounted both a fixed horizontal crosshead bar and a moveable horizontal crosshead bar. The machines include provision for mounting various kinds of test fixtures at the centres of both the fixed crosshead and moveable crosshead through a universal joint (so that the applied force will pass through the centre of the test specimen). The machine includes a motor for driving the crosshead at a constant speed and a mechanism for setting the speed of the moveable crosshead to any designated constant rate (from about 0.025 mm per minute to about 50 mm per minute). The varying force applied to the test specimen as the crosshead moves is read sequentially by a load cell or other weighing system. A means of reading the linear movement of the crosshead (for flexural, compressive and shear tests) is also included. For tensile tests, the machine incorporates an extensometer, which produces an electrical signal in proportion to the stretching of the specimen. Results are usually recorded on a chart recorder or computer monitor (or, for older machines, plotted directly on a chart recorder). Extensometers are used to measure elongation in test specimens. The extensometer has line contact surfaces ('knife edges') that clamp across the test specimen to establish a gauge length (***Figure 3.1b***). As these points move under the applied force, the change in the distance between them is translated by the extensometer into an electrical signal. For a tensile test, it is this signal that is transmitted to the chart recorder. Extensometers are available with various gauge lengths and with various ranges (up to 1000% in some machines).

To carry out a tensile test, the specimen is gripped at each end by the extensometer and pulled apart. However, as the jaws hold the specimen by exerting a clamping force, they always inflict some damage on the specimen that may cause the specimen to break at or near one of the jaws. In order to minimise this, most tensile specimens are designed to have their centre sections narrower than the ends (***Figure 3.1a***). Careful control over test specimen dimensions is required to ensure consistency in test results. Test specimens for plastics are governed by BS EN ISO 3167[3], BS EN ISO 2818[4] (and BS EN ISO 294[5] for thermoplastics and BS EN ISO 295[6] for thermosetting plastics) and for metals by BS EN 10002: Part 1.

Tensile tests require measurement of the elongation of a specimen. However, as the specimen is not straight sided, it will not stretch uniformly throughout its length and it is therefore important that elongation is measured in the straight-sided section. As shown in ***Figure 3.1b*** for plastic test specimens, this is achieved by establishing, within the centre section of the specimen, initial reference points **A** and **B** that establish a known 'gauge length'. The gauge length does not need to be any specific value but very accurate knowledge of the actual 'gauge length' is imperative. The longer the 'gauge length', the more precise will be the calculated result. The top of the specimen is held stationary and the bottom of the specimen is pulled down. As this occurs, the initial points **A** and **B** both shift, becoming the continuously changing points **A** and **C**. The distance **A** to **C** is continuously measured throughout the test in order to calculate the elongation and subsequently the strain at any point on the load-deformation curve.

$$\text{Elongation} = BC = AC - AB$$

$$\text{Tensile strain} = \frac{\text{elongation}}{\text{gauge length}} = \frac{BC}{AB}$$

As the distance BC is in mm and the distance AB is also in mm, the units of strain are mm/mm. This is, in fact, how strain is expressed in compressive, flexural and shear testing (**2.8.1**). However, in tensile testing, it is more usual to report per cent elongation, rather than strain

$$\% \text{ elongation} = \text{tensile strain} \times 100\% = \frac{BC}{AB} \times 100\%$$

Concrete in tension is extremely brittle, cracking rapidly and without warning. It is difficult to test concrete in tension as the jaws of an extensometer would produce stress concentrations which initiate fracture; in addition, non-axial stresses also cause problems when stressing of concrete in tension. Hence indirect methods of obtaining the tensile fracture stress of concrete are more usually adopted e.g, the tensile splitting stress of concrete cylinders (= $2P/\pi Ld$, N/mm^2, expressed to the nearest 0.05 N/mm^2) and the flexural strength of beams (= PL/bd^2, N/mm^2, expressed to the nearest 0.1 N/mm^2) (**3.3**), or various near-to-surface tests (**3.7.6**).

3.2.1 Load-Deformation Relationships

Results from UTMs are recorded as curved plots representing two components, force (y axis) versus deformation (x axis) i.e., a load-deformation plot, ***Figure 3.1c***.

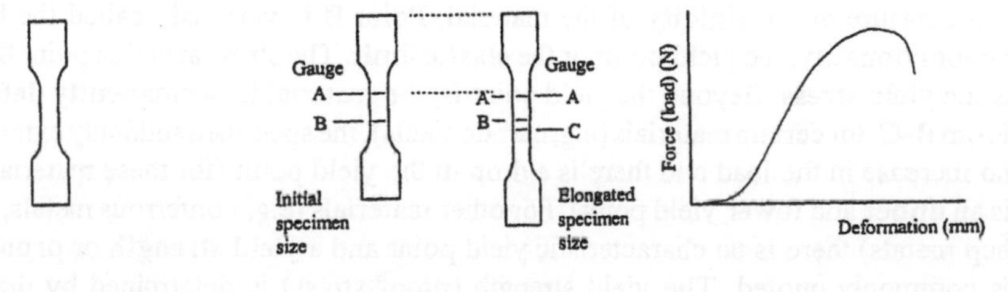

Figure 3.1 Tensile testing

The signal representing the force (load) comes from the load cell in the UTM and is responsible for the y component of the plot, forcing the recorder in the y axis direction proportionately to the force signal. The component representing deformation is dependent on the measurement mechanism in contact with the test specimen, which in turn depends on the type of test. For tensile tests, deformation is recorded from the electrical signal from the extensometer (rather than the crosshead movement). Crosshead movement is used to record deformation in flexural and shear tests. For chart recorders, knowledge of the rate of movement of the crosshead and the rate of movement of the paper (or pen) in the chart recorder, the crosshead movement represented by each division along the x axis of the load-deformation curve can be determined. Note that, in a load-deformation plot, neither the force scale or the deformation scale take into account the size or geometry of the test specimen. Hence it is usually impractical to compare data directly in terms of load and deformation. Ordinarily, therefore, values of load and deformation are converted to values of stress and strain respectively to produce more useful stress-strain curves (**3.2.2**).

3.2.2 Stress-Strain Relationships

A typical stress-strain curve for a ductile material is that for mild steel (***Figure 3.2***).

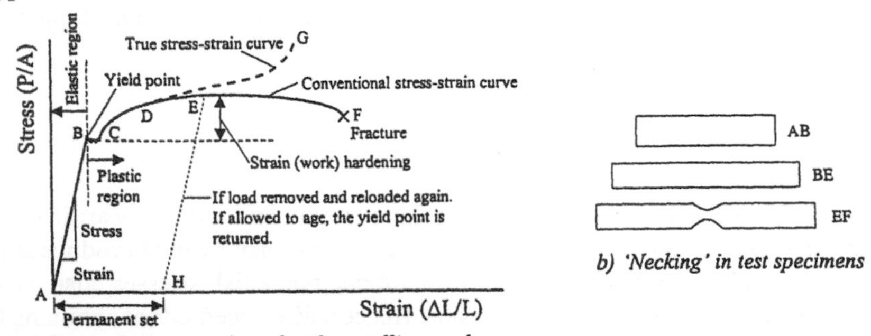

a) Stress-strain curve for a ductile metallic sample

b) 'Necking' in test specimens

***Figure 3.2** Stress-Strain relationships*

Elastic deformation (elasticity) **(A-B)**

In the elastic region, the material is said to show elastic properties, i.e., extension is proportional to the applied load (as described by Hooke's law) **(2.8.1)**. If the load is removed, the specimen returns to its original length. This is often termed **spring back** and has to be allowed for in manufacturing processes **(10.5.3)**. The gradient of **A-B** is **Young's Modulus (E)** (also known as the **elastic modulus**, or the **tensile modulus**) and is a measure of the **rigidity** of the material. Point **B** is variously called the **limit of proportionality**, the **yield point** or the **elastic limit**. The stress at which point **B** occurs is the **yield stress.** Beyond the yield point **B**, the material is **permanently deformed.** From **B-C**, for certain materials (e.g, carbon steels), the specimen suddenly extends with no increase in the load and there is a drop in the yield point (for these materials there is an **upper** and **lower** yield point). For other materials (e.g, nonferrous metals, fcc and hcp metals) there is no characteristic yield point and a **yield strength** or **proof stress** is commonly quoted. The yield strength (proof stress) is determined by drawing a straight line parallel to the initial straight portion of the stress-strain curve in the elastic region and off-setting this curve by (normally) either **0.1%** or **0.2%** extension measured on the strain axis. The point where the newly constructed line intersects the stress-strain curve is the (0.1% or 0.2%) yield strength (or proof stress). Ordinarily, design is at about 50% of the yield stress (or proof stress) to allow for a **factor of safety**.

Plastic deformation (plasticity) (beyond the yield point **B**)

From **C-E**, the extension is no longer proportional to the applied load (i.e., Hooke's law is no longer obeyed). The point **E** is called the ultimate tensile strength (UTS) and measures the **toughness** of the material **(3.2.3)**, provided there is also **ductility** (i.e., the curve continues to fracture at point **F**). If the material fractures within the range **C** to **E**, it would be termed **brittle** (having little **ductility**). If the load is removed, the specimen does **not** return to its original length; this is called a **permanent set** and the material has suffered **plastic deformation**. In strengthening metals, we can make use of permanent set (given by the curve **CDEF**) by a process known as **strain (work) hardening** **(10.5.3)**. On reapplying the load, the stress-strain curve will follow the unloading curve (AE) so that now there is a higher yield point (**E**) and there is no sudden extension (**BC**) (unless the material has been unloaded for a long time, whereupon another yield point will appear but at the much higher stress **E**, a process known as **ageing, 10.13.7**). From **E-F**, the metal appears to be stretching under a reduced **(conventional) stress**. The deformation is actually being concentrated within the gauge length at one particular point (a process called **necking**) (***Figure 3.2a***) and in fact the **true stress** is still increasing (**DG**). Eventually the metal fails at point **F**. ***Figure 3.2*** illustrates that the fracture stress is lower than the ultimate strength of the material. This apparent anomaly occurs because the calculated stress has been based on the **original** cross-sectional area and gauge length of the specimen. The tensile force in the test specimen not only causes an increase in length but also a decrease in the original cross sectional area because the volume of the member must remain constant. Thus the cross sectional area decreases with an increase of strain. If consideration is given to the continuous dimensional changes occurring during plastic deformation, the **true stress** and **true strain** can be determined for the instantaneous area and length of the test specimen under test. True stress and true strain will differ markedly from conventional stress and strain, especially close to the breaking point, where the specimen suddenly decreases in cross-sectional area ('necking'). Conventional stress and strain, rather than true stress and strain, are most often quoted and used since the useful properties of the material all occur well below the ultimate tensile strength (point **E**).

3.2.3 Characterising Materials Using Stress-Strain Curves

The reduction in area of the specimen and the total elongation at fracture are both conventionally used as measures of **ductility**. As yield strength and tensile strength increases, ductility increases. Some materials show little or no plastic deformation before fracture (i.e., they are not ductile, but **brittle**). Typical examples include cast iron, ceramics and concrete. The relationship between the stress-strain curves for a typical brittle material (solid line) and typical ductile material (dotted line) are shown in ***Figure 3.3a***. For the brittle material, (brittle) fracture occurs at a tensile stress S.

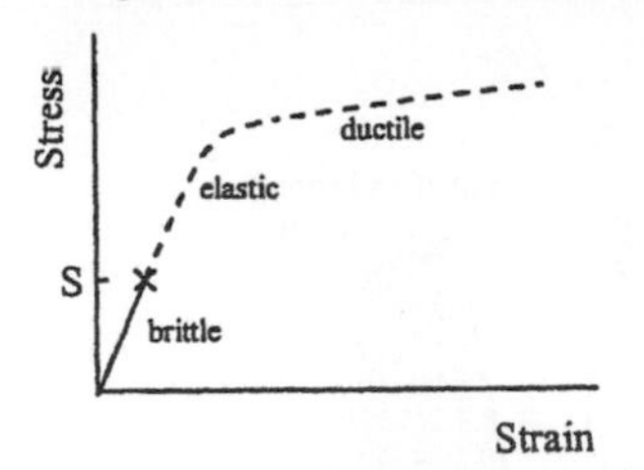

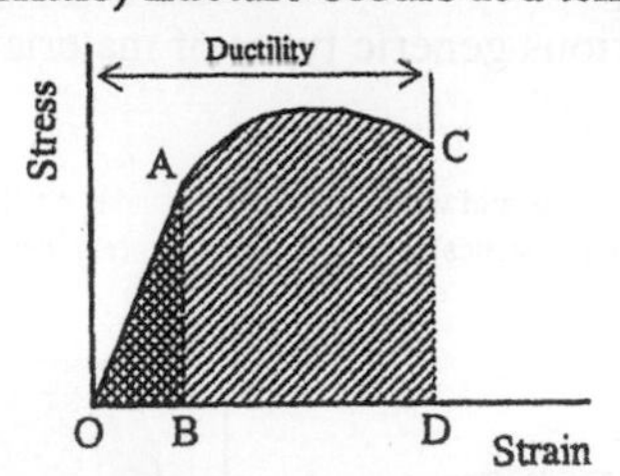

a) a brittle and a ductile material *b) Toughness and Resilience*

Figure 3.3 *Comparison of stress-strain relationships*

Materials may exhibit ductile or brittle modes of fracture. Ductile fracture occurs after a period of plastic flow, as shown by the stress-strain curves in ***Figures 3.2 and 3.3a***. In metals and alloys, plastic flow occurs through the movement of **dislocations (10.4)** under the action of a shear stress. Brittle fracture, however, involves tensile separation with little or no plastic flow. Hence stress systems which involve high ratios of tensile to shear stress are likely to favour brittle fracture. Similarly, any material composition or processing treatment giving a low ratio of tensile strength to shear strength will tend to promote brittle fracture. In construction, inherently brittle materials should be used with caution because the high local tensile stress that can occur in structures at points of stress concentration (such as holes and corners) can lead to instantaneous brittle fracture and consequent catastrophic failure of the component. In contrast, ductile materials exhibit considerable plastic flow before fraction in tension, which acts to relieve stress concentrations. Some materials (for example, most bcc metals, **1.16**) show a temperature dependent transition between brittle and ductile properties (known as brittle transition) (**10.14.1**), although this is generally atypical behaviour. The proposed mechanism by which a brittle material fractures is examined in section **9.6**.

Two related properties are also derived from the stress-strain curve; toughness and resilience (***Figure 3.3b***). **Toughness** is the ability of a material to absorb energy during plastic deformation. The area under the stress-strain curve is a measure of the energy necessary to break the specimen and therefore is a measure of the toughness of the material (area **OACD** in ***Figure 3.3b***). From the shape of the stress-strain curve, it can be seen that materials which have high toughness have high yield strength and high ductility. Brittle materials have low toughness since they show only small plastic deformation before fracture. Toughness is a desirable property for a material subjected to rapid loading, such as shock or impact and is tested using standard impact tests (**3.5.1**). **Resilience** is the ability of a material to absorb energy in the elastic range. Hence the area under the elastic portion of the stress-strain curve is a measure of the resilience of a material (area **OAB** in ***Figure 3.3c***). Resilience is usually measured by determining the rebound of a pendulum or ball after a single impact (see, for example, BS 879[7] for determining the rebound resilience of rubber) and represents the ratio of the energy given up on recovery from deformation to the energy required to produce deformation. For perfect elastic materials, this ratio should be one, but in real materials, there is always a proportion of the deformation energy that is lost (as heat) due to friction (this explains why a rubber ball will not bounce forever !). Resilience is strongly temperature dependent. As shown in ***Figure 3.4***, stress-strain curves can be used to characterise various generic types of materials.

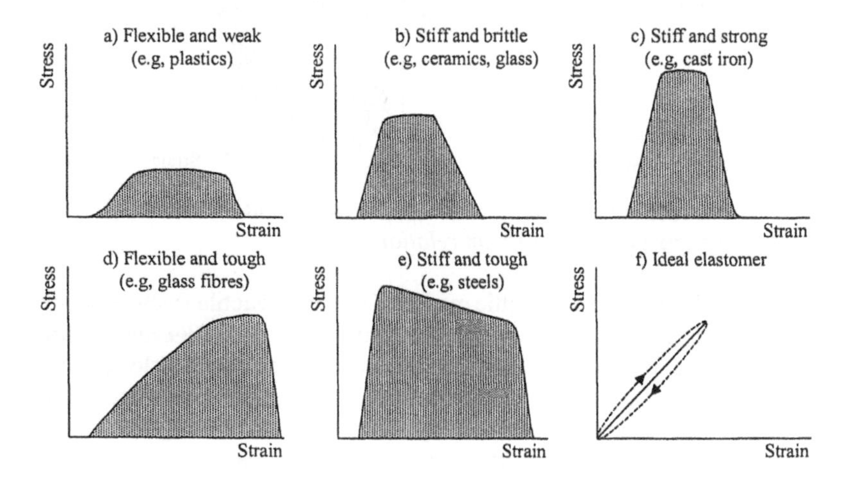

Figure 3.4 *Generic types of materials as classified by stress-strain curves*

3.3 COMPRESSION TESTING

Brittle materials are generally tested in compression since this is the mode of stressing in which they are most frequently used. Requirements for compression testing of plastics are given in BS EN ISO 604[8] and for structural timber in BS EN 408[9]. Compressive tests for concrete are in the process of being harmonised within Europe, as currently incorporated within prEN 12390[10]. Until this standard is published as a British Standard, standard tests for concrete strength are covered by the relevant parts of BS 1881 (Part 116: 1991 for cube compressive strengths). Compression testing for metal is rarely required, but standards are available for specific applications (e.g, BS 7585[11] for bearings and BS EN 24506[12] for hardmetals).

Necking (**3.2.1**) does not occur in compression; instead, the opposite (barrelling) occurs. If it were possible to remove the effects of both friction at the loading surfaces and nonuniform deformation, a true stress-strain curve would be obtained. For optimum results the applied force must be distributed as uniformly as possible against the ends of the specimen. Three fundamental requirements are necessary to achieve this

- opposite ends of the test specimen are as parallel as possible. In addition, where the test specimen is a cube, beam or prism (as is common in most cases), all corners should be as close to 90° as possible;
- the force must be transferred through the specimen by two, rigid, flat metal members (platens) bearing on the ends of the test specimen. In most modern machines, at least one of the flat members is self aligning to ensure that both ends of the specimen will automatically make contact with the flat members.
- the specimen must be placed precisely in the centre of the platen members to achieve axial loading through the specimen.

Concrete compression tests involve placing well mixed concrete test samples in an oiled mould in two layers, ensuring good compaction (by hand tamping or mechanical vibration) of each layer. Once the moulds are full, the surface is trowelled smooth and covered with a plastic sheet to obtain a relative humidity of 95% at 18 °C ± 5 °C. The test specimens are then demoulded after 24 hours and placed in a water tank at 18 °C ± 5 °C until testing. At the required age (normally 28 days), the test specimens are removed from the water tank, surface dried, measured and then placed in the test apparatus so that the trowelled face points sideways (so that the load is applied to a moulded flat face thereby ensuring that stress raisers produced by an uneven surface are eliminated). The load is applied to the specimen without shock at a controlled and even rate until failure. Ordinarily a minimum of five tests are undertaken and the average stress determined. The maximum load (P) and the mode of failure is recorded. If a sudden failure of the specimen occurs (as is common with rigid materials), the specimen will usually exhibit a rupture that is diagonal to the axis of the specimen and the load-deformation curve will drop off sharply. In some cases, unacceptable rupture will result (as a consequence of an unrepresentative test sample or improper loading regimes, etc.) and results from these tests should be discounted. For cube tests, common normal and abnormal failure modes are illustrated in ***Figure 3.5***.

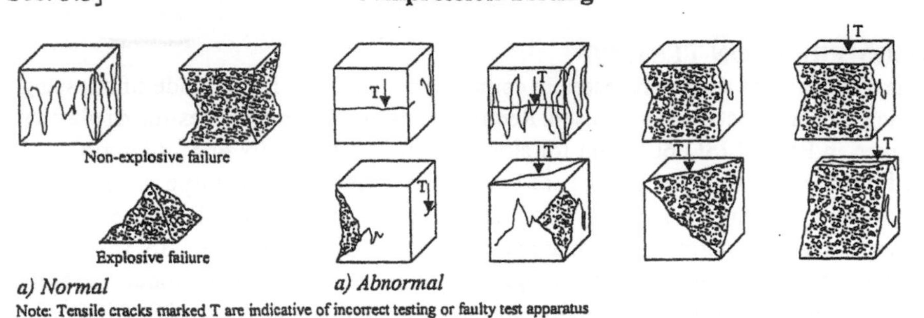

Note: Tensile cracks marked T are indicative of incorrect testing or faulty test apparatus

***Figure 3.5** Characteristic modes of failure in concrete cubes in compression*

For acceptable failure modes, the maximum stress is usually reported as 'compressive strength at failure'. If a sudden failure of the specimen does not occur (as is common in ductile materials), the load-deformation curve will gradually reach a maximum load value which will thereafter decrease. The test should be stopped when this occurs and the maximum stress is reported as 'compressive strength'. Note that if the load is continued, the load-deformation curve would eventually reach a minimum value and then the load would increase. The specimen may be deformed until it becomes a flat disc, but no rupture will occur. (Note that sustained loading in this manner should never be allowed as the force generated may be sufficient to damage the testing machine).

The failing load from these tests are used to calculate the failing stress, $\sigma = P/A$ (N/mm^2, expressed to the nearest 0.5 N/mm^2). Concrete cubes have traditionally been used for compression test of concrete in the UK, due mainly to the fact that cubes are convenient shapes for casting flat sided test specimens and are easy to handle and store. Concrete cube compression tests attempts to load the concrete specimen to produce uniaxial unrestrained stress (likely to be experienced by bulk concrete in buildings). However, compressive test loads induce lateral tensile strains in both the steel platens and the concrete due to the Poisson effect (**2.8.1**). This induces lateral restraint forces in the concrete near the platens as a result of friction between the platen and the concrete. Hence the concrete cube specimen is in fact in a triaxial stress state (as can be seen by the characteristic failure producing pyramidal shapes in ***Figure 3.5***) and therefore exhibits a higher failure stress than the true, unrestrained strength. In an attempt to counter this problem, cylinders with a height to diameter ratio of 2 (normally 300 mm x 150 mm diameter) tested vertically are used in some countries; indeed ENV 206 notes that cylinder testing is the preferred compressive strength test. Generally, concrete cylinder compressive strengths are about 80% of concrete cube strengths, as reflected in the concrete strength classifications of ENV 206 (**8.10.1**); concrete cylinder strength tests do, however, suffer the disadvantage that the tested face is the trowelled face and special provisions are required to produce a flat surface suitable for testing.

The rate at which the load is applied will affect the results obtained (**3.5**) and therefore standards limit the rate of application of the applied load to specific ranges. For example, BS 1881 requires that loads should be applied without shock as a steady rate of 0.2 to 0.4 N/mm^2 to concrete cube specimens (Part 116: 1991), of 0.02 to 0.04 N/mm^2.s for splitting tests of concrete cylinders (Part 117: 1989) and of 0.06 ± 0.04 N/mm^2.s for flexural tests of concrete beams (Part 1881: 1989).

3.4 SURFACE PROPERTIES

Surface properties include hardness, friction and abrasion; special mechanical performance tests have been developed to measure these properties for a variety of different materials. Where these properties are important in terms of the use and performance of a material in the construction context, suitable test methods are described in appropriate standards. In this section, we consider the most commonly measured surface property, hardness.

3.4.1 Hardness

The hardness of a material is most commonly assessed by measuring its resistance to scratching or to indentation. The resistance to scratching can be determined using **Moh's hardness test**, in which the resistance to scratching is of the tested material is matched to the scratch resistance of ten previously classified materials arranged in order of their increasing hardness, as shown in **Table 3.1**.

Table 3.1 Moh's scale of hardness (scratch resistance)

Material	Characteristic hardness	Moh's hardness
Diamond		10
Corrundum (sapphire, emerald, ruby)	only scratched by diamond	9
Topaz	scratched by corundum	8
Quartz	scratched by a file	7
Orthoclase	scratched by quartz	6
Apatite	scratched by steel	5
Fluorite	scratched by glass	4
Calcite	scratched by iron nail	3
Gypsum	scratched by fingernail	2
Talc	crushed by fingernail	1

Moh's scale is arbitrary and the hardness interval between each material is irregular. Hence Moh's scale is only used when quick comparative assessments of the hardness of a material are needed, such as in assessing the hardness of geological materials in the field. For materials used in the engineering context, Moh's scale is an insufficient measure of hardness and more specific and accurate tests have been developed. The most widely used hardness tests involve the determination of the resistance of a material to indentation under strictly standardised conditions. Using a specific load, a hard indenter of standard shape is pressed into the surface of the material under test, causing first elastic then plastic deformation. The resulting area or depth of indentation is measured and assigned a numerical value. Indentation can be achieved by a variety of methods and hence there are a wide variety of hardness test available, including

- **Barcol hardness test** (BS 2782[13]) which is based on a specific commercially available test instrument essentially comprising a spring-loaded point which is forced by hand against the surface of the material. The degree of penetration is read as the Barcol Hardness value from a dial gauge built into the impressor. Barcol tests are suitable for relatively hard plastics, particularly glass reinforced plastics;
- **Durometer hardness test** (BS EN ISO 868[14]) which is similar in principle to the Barcol test and are used predominantly for plastics that are too soft to be measured with the Barcol impressor. The test utilises two instruments, a Type A durometer for plastics in the softer part of the range and the Type D durometer for plastics in the

harder part of the range. Each has its own independent scale of hardness values, referred to as 'Shore A' hardness and 'Shore D' hardness respectively;

- **Rockwell hardness test** (BS EN ISO 2039[15]) which is performed on specialised large scale equipment using one of several possible steel ball or conical (with included angle 120°) indentors. There is a different Rockwell hardness scale for each indentor. The indentor is first forced against the specimen for a short period under a 'minor' load and then under a 'major' load. A hardness reading is then taken from the appropriate scale. The Rockwell hardness test has been traditionally undertaken on metals and to a lesser extent on plastics. Great care is required in performing the test on plastics due to their partially viscous behaviour. Due to the cost of the specialised equipment required to perform the Rockwell hardness test, other hardness test methods are often preferred.
- **Brinell hardness test** (BS EN ISO 6506[16]) which involves forcing a hard steel ball under a standard load for a short period (about 30 seconds) to indent the surface of a material, usually a metal. Various balls and loads are available and the Brinell hardness test should generally be undertaken with the largest indentor and heaviest load consistent with the size and nature of the test specimen. The hardness is expressed as a Brinell Hardness number, which is the quotient of the load in kg divided by the area of the indentation in mm;
- **Vickers hardness test** (BS EN ISO 6507[17]) which is essentially the same in principle as the Brinell hardness test except that the indentor comprises a square diamond pyramid having an apex angle of 136° and the hardness is expressed as a Vickers Hardness number (VHN), which is the quotient of the load in kg divided by the cross-sectional area of the indentation in mm. Portable Vickers hardness test machines are available, permitting tests in situ. Typical values for VHN for heat treated metals are given in **Table 10.5 (10.13.6)**.

For consistent results with the Brinell hardness test, the diameter of the indentation should be between about 0.3 to 0.6 times the diameter of the steel ball. Therefore soft materials have similar Brinell and Vickers hardness numbers since the geometries of the impressions created in the material surface are not too dissimilar. These hardness tests result in plastic deformation of the test material which has been shown to correspond with approximately 8% strain of the material. Thus the measured hardness is a function of the yield stress of the material and, as a rough guide, the Vickers Hardness number is between about 0.2 and 0.3 times the yield stress (MN/m^2) for hard materials and between about 0.06 to 0.09 times the yield stress (MN/m^2) for most metals.

The **setting time (7.16)** of cement paste correlates well to surface hardness, and can be determined using the Vicat apparatus by observing the penetration of a needle into cement paste of standard consistence until it reaches a specified value (BS EN 196-3[18]). 500 g cements and 125 g water is mixed to a consistent paste, which is added to the Vicat mould. The plunger of the Vicat apparatus is released and allowed to penetrate the paste. When penetration has ceased or after 30 seconds, whichever is the earlier, the scale Vicat reading is recorded, which indicates the distance between the bottom face of the plunger and the base plate, together with the water content of the paste expressed as a percentage by mass of the cement. The test is repeated with different water contents until one is found to produce a distance between plunger and base plate of 6±1 mm, at which time the water content of the paste is recorded (to the nearest 0.5%) as the water content for standard consistence. **Initial setting time** for the cement of standard consistence is determined by Vicat plunger operations repeated over suitable times until such time as the plunger needle penetrates 4±1 mm; this is the initial setting time of the cement to the nearest 5 minutes. **Final setting time** for the cement of standard consistence follows the same procedure and is determined when the plunger penetrates only 0.5 mm. Similar methods are used to determine the consistence of fresh mortar (BS EN 1015-4[19]) and the (Vicat) softening temperature of plastics (BS EN ISO 306[20]).

3.5 TIME DEPENDENT MECHANICAL PROPERTIES

The tensile and compressive strength properties described previously are essentially derived from **static tests**, generally involving loading the specimen at speeds ranging from about 5 to 50 mm/minute. However, because the mechanical response of a material also depends on the rate of loading, we often need to measure this response at different speeds. Generally, the very high speeds required for **impact tests** and the very low speeds required for **creep** and **fatigue tests** fall outside the range of operational test speeds in a UTM, so special tests have been developed to assess mechanical performance properties in these cases.

3.5.1 Impact Testing

The stress system to which a material is exposed will affect its behaviour (**3.3**). In addition, two main variables affect the propensity of a material to exhibit brittle fracture

- the rate of straining (***Figure 3.6a***); e.g, bitumen flows slowly at room temperature but will break into pieces if hit by a hammer. In view of the effect of the rate of straining on the mode of fracture, testing standards limit the rate of application of the applied load to specific ranges (see, for example, **3.3**);
- the temperature (***Figure 3.6b***); e.g, glass will flow at elevated temperatures but is brittle at room temperature (**9.5**).

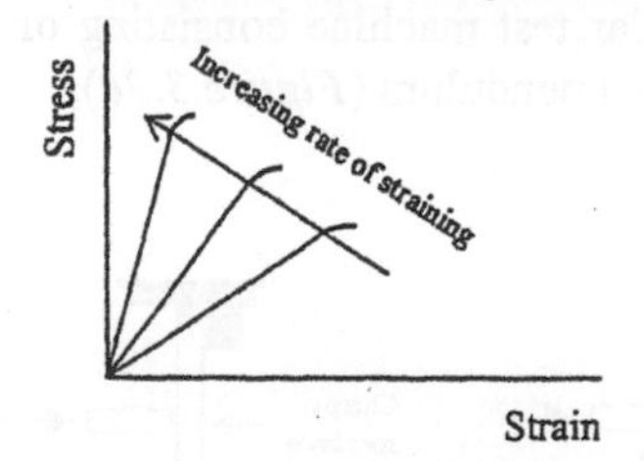

Tensile stress MPa
40
30
20
10
Increasing temperature (°C)
-50°C
-25°C
0°C
30°C
80°C
100 200 300 400 500 600
Elongation %

a) rate of straining (Perspex) *b) temperature (low density polythene)*

Figure 3.6 *Effect of testing variables on the stress-strain curve*

A variety of impact tests have been developed to measure brittle fracture and all measure the energy required to break a specimen. In order to induce brittle fracture in any otherwise ductile material, we can vary either the stress system to which the material is exposed, the rate of straining or the temperature. For most construction materials we are interested in the behaviour of the material at room temperature and therefore in most instances tests are performed at increased rates of straining (i.e., impact tests). In addition, for ductile materials, we commonly vary the stress system to which the material is exposed by introducing notches into test specimens. Effectively, the notches act to induce high local tensile stresses in the vicinity of the notch, so inducing brittle fracture. In essence, we are mirroring the effect of a small crack in the material (**9.6**) when subjected to high (impact) strain. Close production control of test specimens, particularly notch dimensions, is required to ensure consistency in test results (see BS EN ISO 3167 for plastics and BS EN 10002 for metals). The radius of curvature at the base of the notch is very small and the tolerance on this radius is extremely tight. As an example, the British Standard test method for the determination of the Charpy impact strength in plastics (BS EN ISO 179[21]) specifies three acceptable notch types of radii 0.25, 1.0 and 0.1 mm, with tolerances on these radii of 0.05, 0.05 and 0.02 mm respectively. In addition, some materials (e.g, the thermoplastics polycarbonate and polymethyl methacrylate) are very 'notch sensitive', as tiny cracks in the surface of the material make it much more susceptible to rupture than a material having a smoother surface. In these materials, racks may be stopped from spreading by drilling out the front of the crack with a small drill, which effectively blunts the stress raising crack.

In general the mechanism of fracture of any material involves two basic steps

- initiating a crack in the material;
- propagating the crack until a complete break is achieved.

Ideally, an impact test should measure both the energy required to initiate the crack in the test specimen and the energy required to propagate the crack to the point of failure. Impact tests undertaken on notched specimens principally determine only the energy required to propagate the crack (as the initiation energy has already been supplied in forming the notch). Hence unnotched test specimens are also used in impact tests.

The two main impact tests commonly undertaken, the **Izod test** and the **Charpy test**, are described for metals in the various parts of BS 131[22] and for plastics in BS EN ISO 180[23] and BS EN ISO 179. Both tests involve a similar test machine consisting of striking nose mounted at the end of a swinging rigid steel pendulum (***Figure 3.7a***)

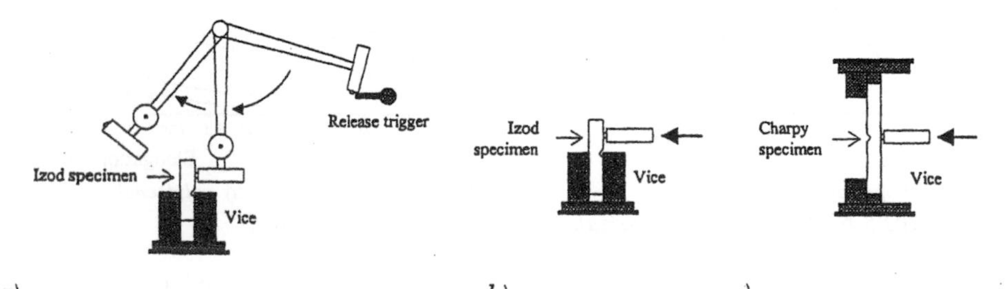

Figure 3.7 *a) Pendulum impact machine for Izod and Charpy impact tests; b) Izod impact test specimen mounting; c) Charpy impact test specimen mounting*

The free end of the pendulum is raised until it becomes automatically held at a preset starting height. This height has been calculated to ensure that, once the pendulum is released from the starting position, the striking nose will reach a defined velocity at the bottom of its swing, just at the instant that it strikes the specimen. The defined velocity is dependent on the test machine type and the pendulum energy (but normally falls within the range 180 to 240 mm/minute). At its starting position, the pendulum possesses an amount of potential energy that is stated for each individual pendulum. When the pendulum is released, this potential energy is gradually converted to kinetic energy as it swings downward until, at the bottom of its swing, all of the potential energy has become kinetic energy. As the pendulum continues to swing passed the vertical in the upward direction, the kinetic energy is converted back into potential energy until the transformation of energy is complete and the pendulum stops swinging upward. The test machines automatically measure the distance of this upward swing and convert it to a scale value of energy lost during the swing. With the specimen clamped firmly in the path of the pendulum (as described below), the difference between the stated pendulum energy and the scale value recorded after the test represents the energy absorbed in breaking the specimen (after allowing for energy losses due to air resistance and friction as described in the relevant standard). The Izod and Charpy impact strength is determined as the scale value (in kJ) divided by the cross-sectional area of the specimen (m^2), expressed in kJ/m^2. For notched specimens, the cross-sectional area is determined from the remaining width at the notch base of the specimen. In all cases, the specimen type (classified according to its dimensions as specified in the relevant standard) should also be reported since dimensions influence the mode of failure of the test specimen, which in turn influences the amount of energy absorbed. For example, the narrower the test specimen, the more it tends to buckle on impact, thus absorbing more energy than is necessary to break it. The only difference between the Izod and the Charpy impact test method is in the way the test specimen is mounted

- in the Izod impact test, the specimen is clamped vertically in a vice that is built into the base of the machine. Where present, the base of the notch in the test specimen must align precisely with the top of the vice. The specimen is struck across a point exactly 22.0 mm ± 0.2-0.5 mm above the top of the vice so that cantilever bending is induced along a precisely controlled moment arm (***Figure 3.7b***);
- in the Charpy impact test, the ends of the specimen rest on two steel supports built in to the base of the machine. The ends of the sample are placed against massive rigid stops so that three point bending is induced when the specimen is struck across the centre of the opposite side (***Figure 3.7c***). Typically, a tough steel absorbs about 130 J of energy in the Charpy test, depending on the specimen size.

3.5.2 Viscosity

A liquid cannot support a shear stress and flows irreversibly and continuously when a stress is applied. This is referred to as viscous flow and the resulting shear strain is a function of both the magnitude of the shear stress and time period over which the shear stress is applied. For an ideal (Newtonian) liquid, the strain is proportional to the shear stress applied and the constant of proportionality is called the **coefficient of viscosity**, or more usually simply **viscosity**, and has units $N/m^2.s$. Thus

$$\text{Viscosity } (\eta) = \frac{\text{Tangential stress}}{\text{Velocity gradient}}$$

Viscous flow occurs in gases and also some solids (including some polymers); in fact viscosity is the only mechanical property possessed by an **amorphous** solid, (e.g, glass). Values of viscosity vary widely for different materials, from about 10^{-3} N/m^2.s for water to about 10^{11} to 10^{19} N/m^2.s for glass (**Table 3.2**). Viscosity is very dependent on temperature; for most materials (except gases, **2.2**) viscosity decreases as the temperature is raised. The viscosity of a liquid is also an important parameter in determining the pressure needed to make that fluid flow through a capillary (**6.6.2**).

Table 3.2 Viscosity of various materials at room temperature

Material	Viscosity (N/m^2.s)
Air	~ 10^{-5}
Water	10^{-3}
Resins	10^{2} to 10^{8}
Asphalt and some plastics	10^{4} to 10^{11}
Glasses	10^{11} to 10^{19}

3.5.3 Viscoelastic Behaviour

A force applied to a viscoelastic material causes three responses

- an **immediate elastic deformation** that is proportional only to the amount of the force. The amount of this purely elastic deformation will remain the same no matter how long the force is applied;
- concurrently, a **slow elastic deformation** that is not only proportional to the amount of the force but depends also on the length of time over which the force is applied;
- concurrently there is a **plastic (viscous) deformation** which is also proportional to the amount of the force and to the length of time over which the force is applied.

However, only two of the above deformations are reversed when the force is removed

- the material will immediately completely recover the initial elastic portion of its total deformation;
- the material will also completely recover the slow elastic portion of its total deformation. However, the recovery will take place gradually over some period of time. The longer the force has been applied, the longer will be the recovery time.

The material will not recover any of the plastic (viscous) deformation and therefore

- there is some degree of permanent deformation induced in every viscoelastic material whenever it is subjected to an external force (in some cases this permanent deformation will be very small and undiscernible to the naked eye);
- permanent deformation is cumulative; if a viscoelastic material is subjected to more than one cycle of loading and unloading, each additional loading cycle will add some more permanent deformation to the material.

Therefore, for a viscoelastic material, at any point along the straight line segment of a load-deformation curve the immediate elastic proportion of the total deformation is very large compared to the two time dependent portions that are also present. The line appears straight because the instrumentation is not sufficiently accurate to measure the time dependent deformations within the short duration of the test.

The behaviour of a viscoelastic material can be illustrated by a simple example. Many car seats contain polyurethane foam and, when we get into the car, we immediately sink down into the seat by an amount proportional to our weight (this is mostly immediate elastic deformation). We then set the rear view mirror to our normal driving position and commence our journey. In the course of a long journey, we may notice that the mirror is no longer exactly centred on the rear window. During the period of our journey we have sunk even lower into the seat to the extent that our eye level is now at a slightly lower level than at the start of the journey (this is mostly slow elastic deformation). We therefore recentre the mirror on the rear window. If we then got out of the car for a short break, we may find on returning to the car that the seat appears to have returned to its original shape (immediate elastic recovery). However, when we got back into the car, the mirror is still properly adjusted (the time span of the unloading our own weight was too short for slow elastic recovery to occur). If we stopped for a day and then resumed our journey, we would notice that the mirror was now no longer adjusted correctly. Our eye level is at a higher level than it was when we finished our journey the previous day. During this longer period of time, the polyurethane foam has exhibited slow elastic recovery. In fact our eye level is now nearly the same as it was when we first sat in the car to start our journey; there is not enough difference in the eye level (and hence in the seat) to be discernable. If we drove the same car for several years, we may notice that we were no longer positioned as high in the seat as we were when we bought the car; the seat has gradually taken on a permanent amount of deformation. This is viscous deformation due to the combined phenomena of **creep** and **fatigue**.

3.5.4 Creep

Under certain combinations of stress and temperature, all materials when subjected to constant stress will exhibit an increase in strain with time. This phenomenon is called creep and leads to the low and progressive deformation of the material over time. The phenomenon of creep occurs in metals, ionic and covalent materials and amorphous materials (such as glass and high polymers e.g, plastics and timber). Metals usually exhibit creep at temperatures $\geq 0.35T_m$, where T_m is the absolute temperature of the melting point of the metal. Amorphous materials, such as plastics and rubbers, are generally highly temperature sensitive to creep. As a general illustration of the phenomenon of creep, consider the various plastic and elastic components of the deformation of timber under constant stress (***Figure 3.8***).

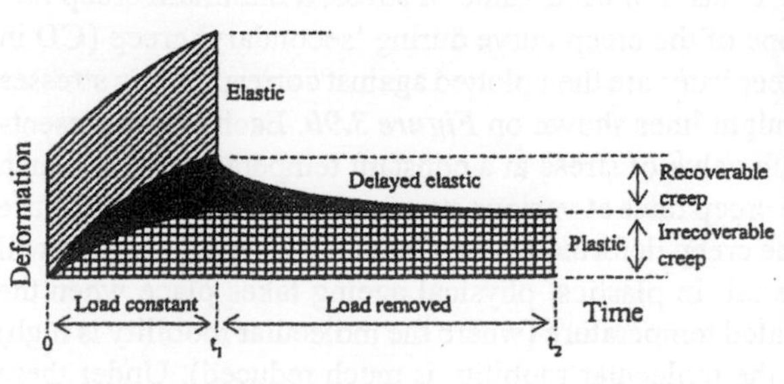

Deformation of timber that occurs when subjected to load comprises **elastic behaviour** (instantaneous and reversible deformation) on application of the load at time zero and **creep behaviour** (increased deformation over time when subjected to a constant load) on maintaining the load to time t_1. The rate of increase in the deformation continually decreases over time. On removal of the load at time t_1, timber exhibits (approximate) **elastic behaviour** (an instantaneous reduction in the deformation approximately equal in magnitude to the initial elastic deformation); **delayed elastic behaviour** (slow recovery over time of that proportion of the deformation due to creep resulting from delayed elastic behaviour); and **plastic behaviour** (no recovery of that proportion of the deformation due to creep resulting from plastic behaviour).

***Figure 3.8** Creep in timber*

Creep tests are all long-term tests which involve applying a constant load without shock to a test specimen and recording deflection measurements over time. Creep tests are described in DD ENV 1156[24] for wood based panel products, BS EN ISO 899[25] for plastics, BS EN 1355[26] for autoclaved aerated concrete and lightweight aggregate concrete and BS 3500[27] for metals. The load applied during these tests depends on the level of stress of interest (that expected for the material under service conditions) and the duration of loading is commonly a minimum of 1000 hours (e.g, for plastics in BS EN ISO 899). The results from creep tests are commonly expressed as strain against time (particularly for metals) to produce a 'creep curve'. A typical creep curve (for lead) is shown in ***Figure 3.9a*** and can conveniently be split into four sections

- section **AB**, the initial instantaneous strain (which is usually elastic);
- section **BC**, a period of decreasing creep rate over time ('primary creep') occurring due to the strain (work) hardening process resulting from the deformation (**10.5.3**); Primary creep is essentially similar in its mechanism to delayed elastic behaviour and as such is recoverable on unloading the specimen;
- section **CD**, a period of dynamic steady state where the creep rate is constant and at a minimum ('secondary' or 'steady state' creep). Secondary creep is believed to be the result of an equilibrium attained between the strain (work) hardening process and the annealing effect (**10.13.3**), and is highly temperature sensitive; and
- section **DE**, a period of accelerating creep culminating in fracture ('tertiary creep'). The tertiary creep occurs at an accelerated rate; it actually represents the process of progressive damage resulting in fracture of the material.

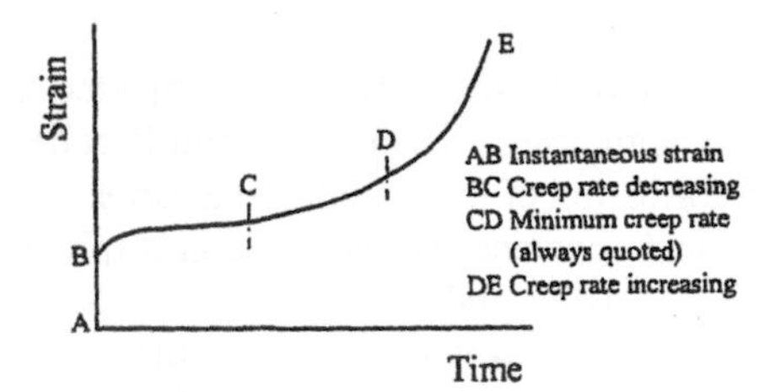

a) Lead at room temperature *b) Minimum creep rates*

***Figure 3.9** Creep curve of strain against time*

Creep tests usually provide data relating to the total creep of a material at a constant temperature for different stress levels. For each value of stress, a minimum creep rate is found by determining the slope of the creep curve during 'secondary' creep (**CD** in ***Figure 3.9a***). The calculated creep rates are then plotted against corresponding stresses on a log-log plot to give the straight lines shown on ***Figure 3.9b***. Each line represents the minimum creep rate for each value of stress at a constant temperature. From such graphs it is possible to estimate creep rates at various stresses and temperatures for use in design. In many materials, the creep deformation produced by an applied stress will depend on the age of the material. In plastics, physical ageing takes place when the polymer is cooled from an elevated temperature (where the molecular mobility is high) to a lower temperature (where the molecular mobility is much reduced). Under these conditions, changes in the structure of the polymer will take place over a long period of time, involving rearrangement in the shape and packing of the molecules as the polymer approaches an equilibrium structural state for the lower temperature. Associated with this ageing process, there is a progressive decrease in the molecular mobility of the polymer, even when the temperature remains constant. Hence creep rates are lower in more highly aged polymers.

3.5.5 Fatigue

Fatigue is the process by which a material fractures when subjected to cyclic stress below the maximum static strength of the material. Fatigue failure is divided into high strain fatigue (low number of cycles to failure) and low strain fatigue (high number of cycles to failure). Fatigue can be illustrated by a simple example; where a tin opener has cut all around the top of a tin can except for a short section where the top remains hinged to the side, we can often remove the top by repeatedly bending it backwards and forwards. The mechanism that slowly allows the complete separation is fatigue. Most materials exhibit fatigue if subjected to repeated cycles of loading and relaxation; it has been estimated that 90% of all mechanical failures are caused by fatigue. Fatigue failures often occur in a catastrophic manner with no visible distortion preceding the failure. The size and location of the cracks formed in a material by the fatigue process often makes fatigue very difficult to detect prior to failure. Fatigue causes the microstructure of a material to suffer some damage on each application of the load, normally at a point of stress concentration. Eventually the damage leads to the formation of a minute crack. The stress at the crack tip will be much greater than anywhere else in the material due to the stress concentrating effect of the tip (**9.6**). The crack therefore grows larger with every cycle of load application. The extent of this propagation stage of the fatigue process depends on the brittleness of the material. In a brittle material, the crack grows to a critical size and then propagates through the structure in a fast brittle manner. In ductile materials, the crack continues to grow until the remaining uncracked area cannot continue to support the load and an almost ductile fracture suddenly occurs.

Various standards are available for testing fatigue, for example for metals BS EN ISO 12737[28], for metals used in air transport engineering prEN 3874[29], for flexural fatigue in rubbers BS 903[30], for thermoplastic valves BS EN 28659[31], etc. A Draft for Public Comment has also been issued in the UK by the BSI for tests for tension crack fatigue in plastics (98/121749 DC[32]). In all fatigue tests it is ordinarily desirable to apply forces to the test specimen which result in a finite life. The testing machine is supplied with the maximum and minimum load limits for the cycled loading regime. The maximum limit must be such that the specimen will not be stretched to its static breaking strength, otherwise the specimen will break immediately. The selection of these limits is based on past experience of the ratio of the fatigue strength (see below) and the tensile strength of the material. For example, the ratio limit (see below) in rotating bending to the ultimate tensile strength is typically about 0.5 for steels, with 70% of all steels having a value in the range 0.4 to 0.55. For cast irons, slightly lower values (in the range 0.35 to 0.5) are usual. Once the maximum and minimum load have been established, the number of cycles of loading is selected from the relevant standard. For example, the application of 5 x 10^6 cycles is normally sufficient to establish a fatigue limit in ferrous metals. The frequency of cycling is also specified in the relevant standard; metals are often tested at frequencies of up to 600 Hz (cycles per second), whereas for plastic frequency is usually limited to about 3 Hz (in order to prevent premature fa resulting from increased friction temperatures).

In most fatigue tests, a series of specimens of the material are tested to failure by c loading at different stresses and each test gives one point on a plot of stress again logarithm of the number of cycles to failure. These plots are used to determin fatigue strength of a material, which is quoted relative to the number of specified c (and depends on the material and type of cyclic test undertaken, as noted above). S materials (e.g, most steels) exhibit a fatigue curve which has a definite stress b which a fatigue failure will never occur (called the fatigue limit) (***Figure 3.10***). M factors affect the data obtained from fatigue tests, particularly surface finish, s concentrations, welds, temperature and corrosive environments (***Figure 3.10***).

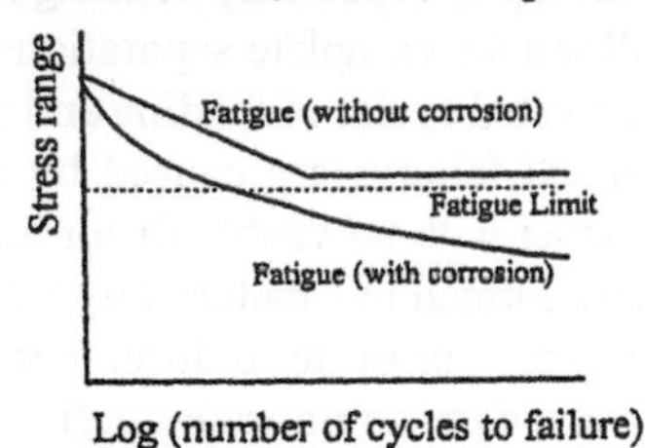

The effect of corrosive environments is shown in ***Figur*** which indicates that corrosion greatly reduces the strength of a metal and results in a loss of the fatigue Hence corrosive environments represent particularly haz conditions for a metal subjected to cyclic loading.

Figure 3.10 *Fatigue results for typical metals in normal and corrosive environ*

3.6 MECHANISMS RESULTING IN CRACKING IN MATERIALS

There are many mechanisms which may give rise to cracks in materials. In the prec sections, we have described some of these mechanisms (tension, compression, fle bending, creep, fatigue, etc.). Other examples include expansive disruption (by ph or chemical means), rotation, expansive distortion etc. In all cases the crack mech is a physical process by which changes of size in a material are translated i displacement or stress greater than the material can withstand. In practice, materia not normally used in isolation in a building and therefore the deformation and forces developed within a material may be transmitted to neighbouring materials paths taken by these forces depend upon the manner in which the materials are lc relative to each other, and on the location and effectiveness of restraints which movement (such as mechanical fixings, adhesion and friction). Cracks within mat used within a building are unavoidable, though only some may impair the service of the building (or do so if they widen further). It is only for these cracks that rep some other remedial action is required.

In subsequent Chapters, the properties and uses of common building materia examined and the processes responsible for their degradation explained. In **Tab** the degradation mechanisms that may induce cracking in materials are summarise their relevant Chapter locations highlighted. **Table 3.3** serves as a useful ai overview to readers interested predominantly in degradation mechanisms and proc

Table 3.3 Degradation mechanisms and processes in building materials and structures

Action	Effect and duration	Materials affected	Components affected
Physical changes			
Temperature changes (2.8.2)	Expansion and contraction. Continuous daily and annually	All materials except special ceramics and alloys	Walls and roofs, especially well-insulated dark cladding and south-facing roofs
Initial drying (irreversible shrinkage) (6.4)	Shrinkage over weeks or years	Mortar, concrete, aerated concrete, sandlime units, timber	Large, low aspect ratio masonry walls. Timber frames and floors. Large concrete frames.
Wetting and drying (reversible) (6.4)	Expansion and contraction. Seasonal and weather related. Intermittent for life of material	Most materials, especially timber, porous concrete products, shrinkable clay soils (especially if affected by large trees)	Timber and concrete directly exposed to weathering. Shallow foundations on shrinkable clay soils.
Loss of volatiles (13.12)	Shrinkage over hours to years. Irreversible.	Solvent-based (paints, mastics, plastics).	Finishes. Movement joints. Sealants.
Freezing and thawing of absorbed water (6.6)	Expansion. Internal damage. Spalling. Intermittent related to weather.	Porous materials, especially fired clay products, natural stones, weaker concrete materials, soils.	Walls and roofs exposed to high levels of driving rain and associated freeze/thaw action. Services and very shallow foundations in soil subject to freezing.
Crypto-efflorescence (6.7.1) 'Lime staining' (4.3.1)	As above, with associated staining	Porous materials, especially fired clay bricks and tiles, natural stones which contain salts or are subject to contamination from the environment	Walls. Roofs. Floors subject to wetting and drying cycles. Structures in contact with contaminants (from groundwater, sea water, effluents, acid rain).
Chemical changes			
Sulfate attack (8.19.2)	Permanent expansion over months to years	Portland cement and hydraulic lime mortars. Concrete. Concrete blocks.	Structures that remain wet for long periods (e.g. exposed walls, slabs).
Corrosion (Chapter 11)	Permanent expansion over months to years	Metals, especially mild and high tensile steels	Fixings, especially in exposed structure. Reinforcement in carbonated or contaminated concrete.
Moisture expansion of ceramics (6.4)	Permanent expansion over decades. Worse in 'young' structures.	Fired clay bricks and tiles. Due to chemical reaction with the atmosphere.	Slender brick walls (especially cladding and parapets). Tiled floors and walls.
Carbonation (8.21.4)	Permanent shrinkage over one to 50 years (depending on porosity)	Porous concrete and calcium silicate products	Walls, especially where they are weakened by openings
Alkali aggregate reaction (AAR) (8.20.1) Alkali silica reaction (ASR) (8.20.2)	Irreversible expansion over many years	Concrete containing reactive aggregates and sufficient alkali	Concrete that remains wet for long periods, e.g. slabs, bridges, dams, earth retaining structures
Hydration of oxides (7.3.4) and unstable slag aggregates (8.5)	Permanent expansion over months to a few years	Usually occurs as particles of unhydrated lime or magnesia in fired clay products. May occur in slag aggregates not complying with relevant British Standard.	Most common as aesthetic spalling of fair-faced brickwork. May be a contributory cause of volumetric moisture expansion in some bricks and concrete blocks. Heave of floor slabs above unstable fills.

Table 3.3 Degradation mechanisms and processes in building materials and structures (contd.)

Action	Effect and duration	Materials affected	Components affected
Other effects			
Imposed load effects			
Dead and imposed loading on structure within design limits	Elastic (instantaneous) and creep deflection over years	All materials in the load path	Seldom causes degradation
Structural loading	Elastic (instantaneous) and creep deflection over years	All materials in load path	Walls subject to concentrated loads (e.g. where joist and joist hangers bear).
Loading of ground/foundations	Consolidation. Settlement over months to years.	Especially silty or peaty soils and any made ground	Mainly walls of brittle material where differential settlement occurs
Differential soil movements			
Settlement, mining, subsidence, swallow holes, land slips, soil creep, earthquakes	Differential settlement. Can occur at any time in the life of the building.	Foundation strata	Walls and floors. Usually affects both leaves of a cavity wall but degradation is different due to different loading and restraints
Vibration			
Traffic, machinery, sonic booms, mining, explosions	Oscillating strain fields	Brittle walling materials	Probable cause of damage to bridges and buildings near transport facilities, but difficult to prove.

3.7 IN SERVICE INSPECTION AND TESTING

Many mechanisms may give rise to incipient problems in service and investigations are required away from the laboratory. This section describes some of these techniques. Many in service tests are non-destructive or relatively non-destructive; they can be used to detect the onset of deterioration prior to failure or physical, chemical or mechanical properties of materials. Many in service tests are available; Annex A, DD ENV 1504-9: 1997 (**8.25**) lists 99 tests currently under development for testing products and systems for the repair of defects in concrete. In this section, we concentrate on the most widely used non-destructive tests, those for concrete.

3.7.1 Inspection Aids

Examination of large inaccessible voids can be undertaken using a borescope (***Figure 3.11a***), comprising a thin industrial rigid telescope which introduces light and permits visual examination in dark cavities. Fibrescopes (***Figure 3.11b***) are also used; these have the added advantage of high flexibility.

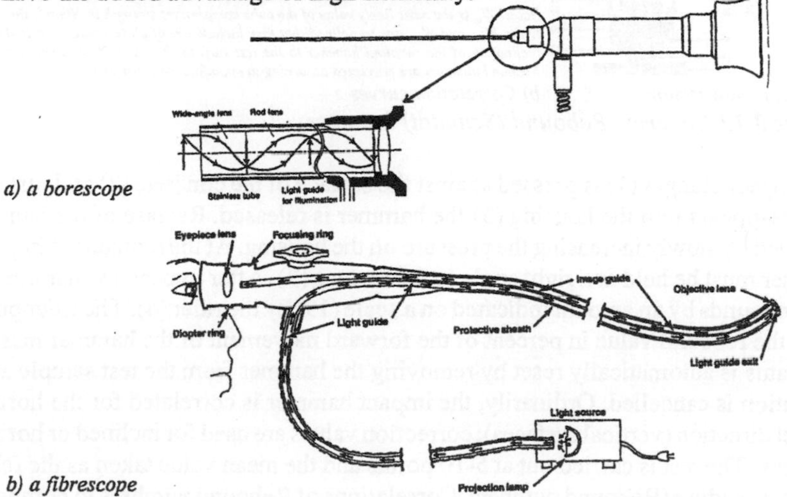

a) a borescope

b) a fibrescope

***Figure 3.11** Equipment for inspection of internal cavities and voids*

3.7.2 Surface Hardness (BS 1881:Part 202:1986)

The surface hardness test is based on the principle that the rebound of the elastic mass depends on the hardness of the surface which it strikes. A number of hand held spring loaded steel rebound hammers are available to suit various concrete types. ***Figure 3.12a*** shows a Schmidt rebound hammer. Results are expressed in terms of a rebound number, which is affected by near surface conditions. Testing of concrete by hardness methods is not a substitute for other well-established methods, but rather a useful preliminary or complementary method; it is particularly suitable for comparative surveys but results may also be correlated with other concrete properties. Hardness measurements provide information on the quality of the surface layer (about 30 mm deep) only; relationships to other properties are empirical only. Correlation curves enable prediction of concrete strength from hardness measurements (***Figure 3.12b***), but estimates of *in situ* concrete strength are unlikely to be better than ± 25% under ideal conditions.

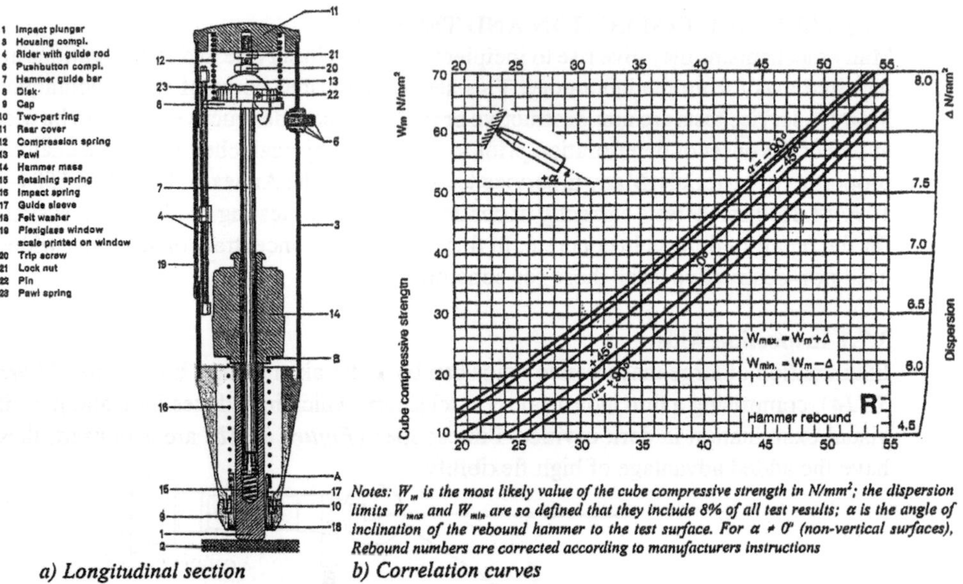

Notes: W_m is the most likely value of the cube compressive strength in N/mm²; the dispersion limits W_{max} and W_{min} are so defined that they include 8% of all test results; α is the angle of inclination of the rebound hammer to the test surface. For α ≠ 0° (non-vertical surfaces), Rebound numbers are corrected according to manufacturers instructions

a) Longitudinal section *b) Correlation curves*

***Figure 3.12** Concrete Rebound (Schmidt) hammer*

The impact plunger (1) is pressed against the surface of the concrete (2) and just before it is disappears into the housing (3) the hammer is released. Release of the hammer is achieved by slowly increasing the pressure on the housing. At the moment of impact the hammer must be held at a right angle to the surface (2). After impact, the hammer mass (14) rebounds by an amount indicated on a scale (19) by the rider (4). The rider position gives the rebound value in percent of the forward movement of the hammer mass. The apparatus is automatically reset by removing the hammer from the test sample and the indication is cancelled. Ordinarily, the impact hammer is correlated for the horizontal impact direction (vertical surfaces); correction values are used for inclined or horizontal surfaces. The test is carried out at 5-10 points and the mean value taken as the rebound hammer reading (Rebound number). Correlations of Rebound numbers to compressive strength are given by manufacturers for a variety of situations (***Figure 3.12b***). **Table 3.4** outlines the factors affecting Rebound hammer readings.

Table 3.4 Factors affecting Rebound hammer readings

Factor	Notes
Concrete strength	As affected by: type of cement (compared to Portland cement concrete, HAC can give strengths 100% higher, and supersulfated cement can give strength 50% lower), fineness of cement (± 10% variations in strength), cement content (high cement contents give lower rebound hammer readings and typical variations of ± 10% in strength are likely), type of aggregate (lightweight aggregates required special correlations), type of curing, concrete age and compaction (rebound hammers are not suitable for detecting different degrees of compaction);
Type of surface	Only smooth surfaces should be tested. Trowelled surfaces are generally harder than those cast against formwork and should give more reliable results. Tests on moulded surfaces are preferred;
Type of concrete	Rebound hammer tests are suitable for open textured concrete but unsuitable for open textured masonry blocks, honeycombed or no-fines concrete;
Surface moisture condition	A wet surface gives a lower rebound hammer reading than a dry surface. A reduction in the rebound number of about 20% is typical for structural concrete;
Carbonation	Carbonation increases the surface hardness of the concrete (8.21.4), which then ceases to be representative of the whole concrete; the effect is not significant in concrete less than 3 months old;
Movement	No movement of the test specimen should be allowed during the test;
Direction of the test	Surfaces at any angle can be tested, although the apparatus is usually calibrated for testing vertical surfaces and manufacturers supply correlations for other test directions.

3.7.3 Ultrasonic Pulses (BS 1881:Part 203:1986)

A pulse of longitudinal vibrations is produced by an electro-acoustical transducer which is held in contact with one surface of the concrete under test. The transducer is commonly called a PUNDIT™ meter (Portable Ultrasonic Nondestructive Digital Indicating Tester). Suitable pulse frequencies range from about 20 to 250 kHz, with 50 kHz usually being adopted for concrete. These frequencies correspond to wavelengths in the range 200 mm (for the lower frequency range) to about 16 mm (at the higher frequency range). After traversing a known path length in the concrete, the pulse of vibrations is converted into an electrical signal by a second transducer. Electronic timing circuits enable the transit time of the pulse to be measured. A pulse of vibrations of ultrasonic rather than sonic frequency in used to give the pulse a sharp leading edge and to generate maximum energy in the direction of propagation of the pulse. When the pulse is coupled into concrete from a transducer, it undergoes multiple reflection at the boundaries of different material phases within the concrete. A complex system of stress waves within the concrete is developed and their form can be measured at the surface. Measurements can be interpreted to identify

- the uniformity of concrete in or between members;
- the presence and approximate extent of cracks, voids and other defects;
- changes occurring within time in the properties of the concrete;
- the correlation of the pulse velocity and strength as a measure of concrete quality;
- the modulus of elasticity and dynamic Poisson's ratio of the concrete.

The receiving transducer detects the arrival of that component of the pulse which arrives earliest (generally the leading edge of the longitudinal vibration). Although the direction in which the maximum energy is propagated is at right angles to the face of the transmitting transducer, it is possible to detect pulses which have travelled through the concrete in some other direction. It is possible, therefore, to make measurements of the pulse velocity by placing the transducer on either opposite faces (direct transmission), adjacent faces (semi-direct transmission) or the same face (indirect or surface transmission), as shown in ***Figure 3.13***. Therefore the method can be used in situations where access to both sides of the concrete under test is limited.

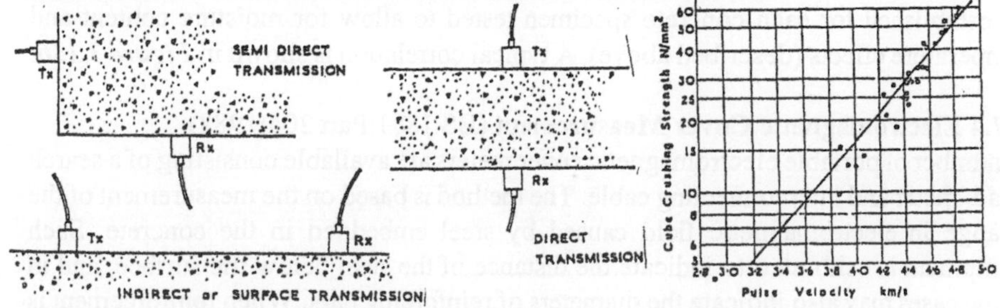

a) Methods of propagating and receiving ultrasonic pulses *b) Typical correlation curve*

***Figure 3.13** Ultrasonic pulse (PUNDIT) measurement and correlations*

To ensure that the ultrasonic pulses pass through the concrete and are detected by the receiving transducer, it is essential that there adequate acoustic coupling between the concrete and the face of each transducer. In many instances the concrete surfaces are sufficiently smooth for good contact to be achieved by the use of a coupling medium (such as petroleum jelly) and by pushing the transducers into close contact with the surface. Uneven surfaces require smoothing and levelling or the application of a thin layer of levelling medium (quick-setting epoxy resin or plaster) with good contact to the concrete surface. **Table 3.5** outlines the factors affecting pulse velocities in concrete.

Table 3.5 Factors affecting pulse velocities in concrete

Factor	Notes
Moisture content	Differences in pulse velocities obtained using a properly cured standard test cube and in situ structural concrete will occur. These differences are due to the effect of different curing regimes on the hydration of cement and on the presence of free water in the pore structure of concrete. It is therefore important that these effects are correlated, as described below;
Temperature	Variations in concrete temperature between 10°C and 30°C have been found to cause no significant change in strength predictions. Outside this range, BS 1881: Part 203: Table 1 gives corrections for pulse velocity measurements.
Path length	The method is suitable for measurement of pulse transit times over path lengths of up to 3 m in concrete to an accuracy of ± 1%. Minimum path lengths are controlled by the nominal size of the aggregate; for nominal ≤ 20 mm aggregate, a minimum path length of 100 mm is permitted whilst for nominal 20 to 40 mm aggregate, a minimum path length of 150 mm is permitted;
Specimen dimensions	The velocity of short pulses of vibrations is independent of the size and shape of the specimen in which they travel, provided the least lateral dimension is greater than a certain minimum value. Minimum lateral dimensions for test specimens for different transducer frequencies are given in BS 1881: Part 203: Table 2. For a typical transducer frequency suitable for concrete, these minimum lateral dimensions range from 65 to 83 mm for pulse velocities 3.5 to 4.5 km/s.

It can be shown that the velocity (V) of a pulse of longitudinal ultrasonic vibrations travelling through an elastic solid is given by

$$V = \sqrt{\frac{E(L-\nu)}{\rho(L+\nu)(L-2\nu)}}$$

where L is path length, E is Young's modulus of elasticity (**2.8.1**), ρ is the density of the elastic solid and ν is Poisson's ratio (**2.8.1**). BS 1881: Part 203: 1986 gives advice on performing the tests and interpreting the results for the variety of different purposes outlined above. The pulse velocity is not effected by the frequency of the pulse so that the wavelength of the pulse vibrations is inversely proportional to this frequency. Thus the pulse velocity will generally depend only on the properties of the materials, i.e., $V \propto \sqrt{(E/\rho)}$ and measurement of this velocity enables an assessment to be made of the condition of the material. The correlation of concrete strength and pulse velocity should be established for each concrete specimen tested to allow for moisture content and temperature effects (described above). A typical correlation is shown in ***Figure 3.13b***.

3.7.4 Electromagnetic Cover Measurement (BS 1881:Part 204:1988)

A number of portable electromagnetic cover meters are available consisting of a search head, meter and interconnecting cable. The method is based on the measurement of the change in electromagnetic field caused by steel embedded in the concrete. Such equipment is calibrated to indicate the distance of the steel below the surface, and in some cases may also indicate the diameters of reinforcing bars. When reinforcement is parallel to the surface, rotation of the search head will enable the alignment of the reinforcement to be identified. To use the device, the concrete surface is scanned with the search head kept in contact with it while the meter indicates by analogue or digital means the proximity of the reinforcement. The search head may consist of a single of multiple coil system and the physical principle involved can either be one utilising eddy current effects or magnetic induction effects

- with covermeters utilising eddy current effects, alternating currents in the search coil set up eddy currents in the reinforcement which in turn cause a change in the measured impedance of the search coil. Instruments working on this principle operate at frequencies above 1 kHz and are thus very sensitive to the presence of any conducting metal in the vicinity of the search head;
- with covermeters utilising magnetic induction effects, a multicoil search head is used with a lower operating frequency than the eddy current type of device (typically below 90 Hz). The principle used is similar to that of a transformer in that one or two coils carry the driving current while one or two further coils pick up the voltage transferred via the magnetic circuit formed by the search head and the embedded steel reinforcement. Such instruments are less sensitive to non-magnetic materials than those using the eddy current principle.

The search head of a typical covermeter is shown in ***Figure 3.14***. Since the covermeter (like a transformer) cannot operate from DC supplies, a circuit oscillator is included to turn the DC supply into AC form.

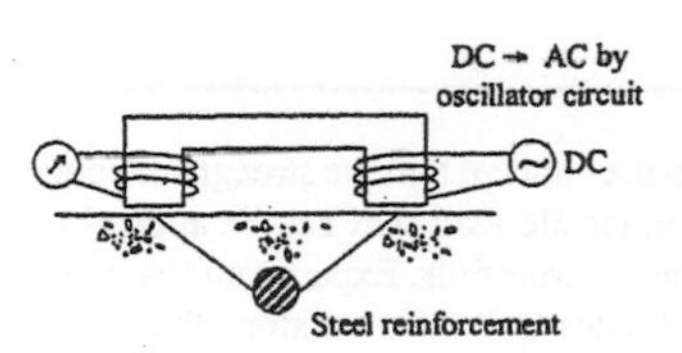

Concrete is examined by systematically traversing the search head over the concrete and, where the reinforcement is located, rotating the search head until a position of maximum disturbance of the electromagnetic field is indicated by the meter. In such a position, under ideal conditions, the indicated cover to the nearest piece of reinforcement may be read if the reinforcement bar size is known. If the cover depth is known, the reinforcement bar size can be estimated. In addition, the axis of the reinforcement will then lie in the plane containing the centre line through the poles of the search head. Where the reinforcement is not too congested, it is possible to map out all the reinforcement bars within the area under examination which lie sufficiently close to the surface. Note, however, that the covermeter cannot detect corrosion of the reinforcement.

Figure 3.14 *Schematic diagram of the search head of a typical covermeter*

The equipment needs careful calibration (away from external ferromagnetic materials such as scaffolding). When correctly calibrated, covermeters are usually accurate within ± 5% of ± 2 mm, whichever is the greater. Potential loss in accuracy (which can normally be minimised with experience) is due to the many factors affecting the magnetic field within the range of the covermeter and to other physical influences

- the type of reinforcement, including the steel type, the reinforcement cross-section, shape and orientation, the presence of multiple reinforcement and tie wire;
- the type of concrete, including the type of aggregate (particularly if aggregates have magnetic properties), the type of matrix and the surface finish;
- the temperature (some search heads are sensitive to changes in temperature caused by the operators hand);
- extraneous effects, particularly the presence of other magnetic materials (embedded screws, wire, etc.) and where corrosion of the steel reinforcement has occurred (scaling and the presence of solutions of corrosion products may influence readings).

3.7.5 Radiography (BS 1881:Part 205:1990)

Radiography provides a method of obtaining a photographic image of the interior of a member from which variations in density may be identified. The image is produced on a suitable film held against the rear face of the member, while a beam of gamma rays (up to 500 mm thicknesses) or high energy X-rays (up to 1.6 m thicknesses) is directed at the front face. The presence of high density materials (such as reinforcement) or low density areas (such as voids) will produce light and dark areas on the film. The principal applications of radiography are to determine locations of areas of variable compaction or voids in concrete (particularly for these areas in the grouting of post-tensioned steel reinforcement) and for the location and sizing of steel reinforcement. Radiography requires highly specialised equipment and experienced operators.

3.7.6 Near-to-Surface Tests (relatively non-destructive) (BS 1881:Part 207:1992)
Typical near-to-surface tests are shown in ***Figure 3.15***. Strengths obtained by these tests strictly relate only to the concrete quality in the immediate vicinity of the test point. The mean result from a number of tests is needed to estimate the average quality of the surface zone. All near-to-surface tests produce a localised zone of surface damage.

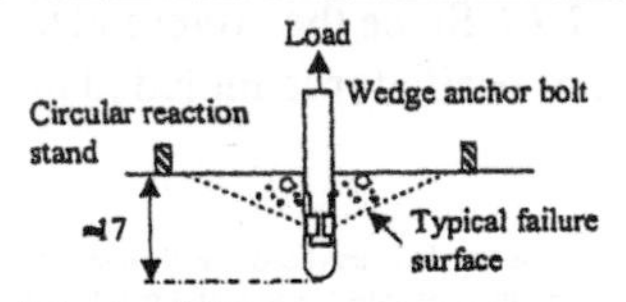

a) Internal fracture tests

Internal fracture tests are based on the concept that a measurement of the tensile force required for a wedge anchor bolt to cause failure of the concrete can be related to the concrete compressive strength.

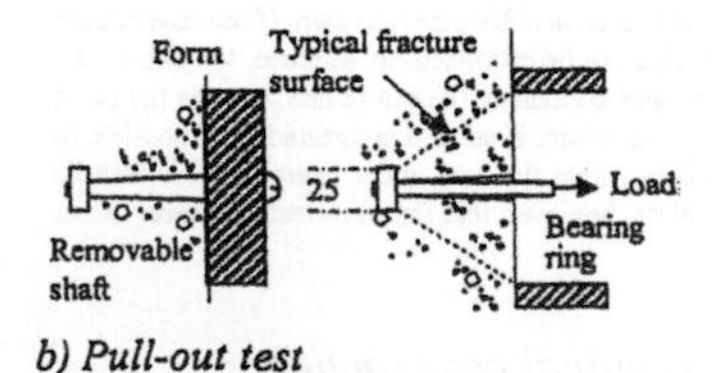

b) Pull-out test

Pull-out tests are based on the concept that the strength of concrete is related to the maximum tensile load that can be applied to an embedded insert before the concrete fails. Experiments have shown that this test gives more reliable results at lower strengths than other near-to-surface tests and so is useful for the estimation of early age strength.

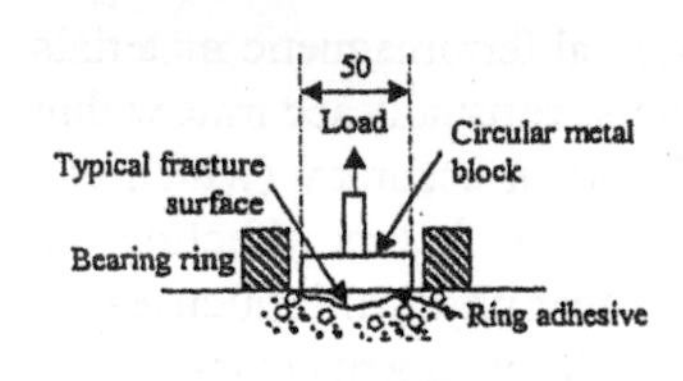

c) Pull-off test

Pull-off tests are based on the concept that the force required to pull a metal block, together with a layer of mortar or concrete, from the surface to which it has been attached is related to the strength of the concrete. If the block is attached directly to the surface, then the stressed volume of concrete lies close to the face of the block and the results may be less related to the body of the concrete than with other near-to-surface tests. If, however, the test is carried out by partially coring the concrete and bonding a block of the same nominal diameter to the top of the of the cylinder thus created, the fracture surface will occur deeper into the concrete.

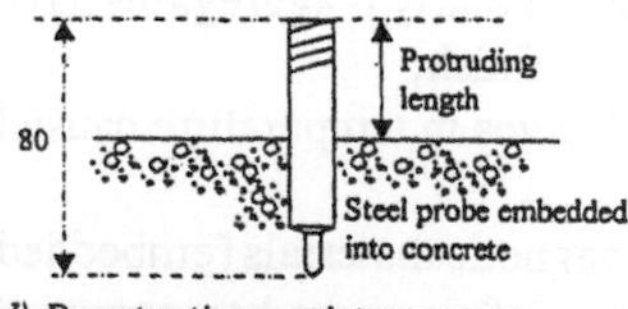

d) Penetration resistance test

Penetration resistance tests are based on the concept that the depth of penetration of a metal probe fired into a concrete surface (normally by a powder charge) depends on the strength of the concrete.

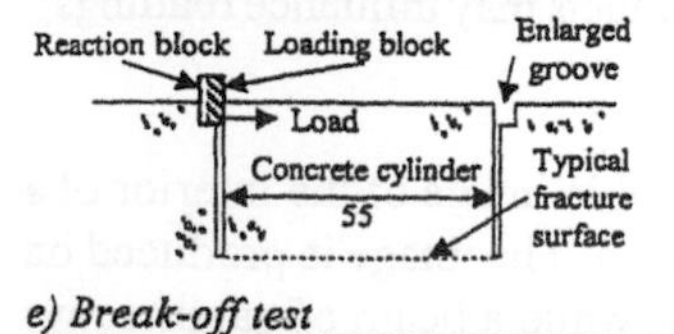

e) Break-off test

Break-off tests determine directly the flexural tensile strength in a plane parallel to the concrete surface at a predetermined distance below the surface. A transverse force is applied to the top of a cylinder of concrete (made either by partial coring of the hardened concrete or formed in fresh concrete during casting). The force required to break this cylinder of concrete at its base from the parent material (known as the break off force) is measured.

***Figure 3.15** Near-to-surface tests using commercially available test apparatus*

3.7.7 Initial Surface Absorption Test (ISAT) (BS 1881:Part 208:1996)

The ISAT test involves measurement of the rate of flow of water per unit area into the concrete surface subjected to a constant applied head. The equipment consists of a cap clamped and sealed to the concrete surface, with an inlet connected to a reservoir and an outlet connected to horizontal calibrated capillary tube and scale (***Figure 3.16***).

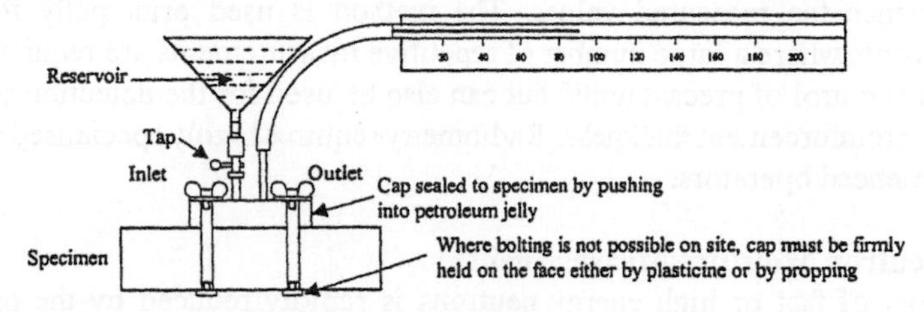

Figure 3.16 Initial surface absorption test system

Measurements are made of the movement of the water in the capillary tube over a fixed period of time following closure of a tap between the cap and the reservoir. The absorption of the water by a dry surface is initially high but decreases as the water-filled length of the capillaries increases. Therefore measurements should be taken at shorter time intervals (specified in the standard) at early times. Results are affected by the moisture content of the concrete prior to commencement of the test and, for in situ concrete, the method is best used for comparative measurements. Due to the low pressure head adopted in the test (200 mm), the ISAT is most relevant to assessment of surface water exposure (rain).

3.7.8 Resonant Frequency (BS 1881:Part 209:1990)

Measurement of the resonant frequency of a prismatic specimen and determination of its density may be used to yield a value for the dynamic elastic modulus of the concrete of which the specimen is formed. The procedure is quick and reliable. However, the test has to be performed on a standard specimen of hardened concrete and is usually carried out on laboratory cast samples; it is not usually suitable for site use without adaptation.

3.7.9 Subsurface Radar Surveys[33]

High frequency (1-4 GHz) electromagnetic impulses are transmitted from an antenna traversed over the surface under investigation. Reflections are received from internal features and discrete boundaries between materials having different electrical properties. The resolution and penetration achieved depends upon the frequency of radiation employed, which is controlled by the antenna being used. Moisture and the presence of chlorides affects behaviour and performance. In favourable circumstances, radar can also be used to provide some broad indications on a number of other parameters (e.g, density, delamination or the presence of chlorides and moisture. Experienced radar operators are required to implement subsurface radar surveys.

3.7.10 Radiometry

A narrow beam of gamma rays is directed into the concrete and the intensity of radiation emerging is measured by means of a Geiger counter. Measurements may be made either of the radiation passing through a body of concrete or of radiation reflected back to the same surface by collision with electrons within the concrete. In either case, the mass per unit area of the concrete is the property which has the greatest influence on the attenuation of the beam of gamma rays and hence the measured value of radiation. Steel reinforcement has about three times the effect of normal concrete and its presence will thus influence the measured values. The method is used principally for density measurements where a large number of repetitive measurements are required (e.g, for the quality control of precast units) but can also be used for the detection of concrete member or reinforcement thickness. Radiometry requires highly specialised equipment and experienced operators.

3.7.11 Neutron Moisture Measurement

The energy of fast or high energy neutrons is rapidly reduced by the presence of elements of low atomic weight. The resulting slow or low energy neutrons may be counted by a detector designed for this purpose. Few elements of low atomic mass are found in concrete other than hydrogen contained in water, and the counts may be used to provide an indication of the moisture content. Measurements may either be made of the scattered neutrons reflected back to the same surface as the source or a direct transmission value may be obtained by lowering the source into a predrilled hole. The best results are obtained by the direct method when the moisture content is high. Neutron moisture measurement devices which incorporate a microprocessor to compute results are commercially available. The main limitation of the method is that results will only relate to a surface zone of the concrete a few millimetres deep when using the backscatter method (direct measurements are limited to about 300 mm deep). In addition, moisture measurements may be influence by moisture gradients near to the surface and the presence of other neutron absorbers. The accuracy of the method is relatively poor for concrete of low moisture content.

3.7.12 Surface Permeability

Several permeability tests are under development which permit an assessment of the permeability of the concrete in the surface zone to water, air, carbon dioxide or other gases under pressure. Typical examples are shown for water in ***Figure 3.17***.

Figure 3.17 *Near surface water permeability tests for in situ concrete*

Measurements of this type may provide valuable information on the durability of the concrete. However, the results are strictly only valid for the concrete under test and measurements at repeated test points over the face of the concrete are required. Although the use of tests of this type is increasing, experience in the interpretation of test results is still limited.

3.8 REFERENCES

1. BS EN 10002. Tensile testing of metallic materials. Part 1:1990. Method of test at ambient temperature. Part 3:1995. Calibration of force proving instruments used for the verification of uniaxial testing machines. Part 4:1995. Verification of extensometers used in uniaxial testing. Part 5:1992. Method of test at elevated temperatures.
2. BS EN ISO 527. Plastics. Determination of tensile properties. Part :1996. General principles. Part 2:1996. Test conditions for moulding and extrusion plastics. Part 3:1996. Test conditions for films and sheets. Part 4:1997. Test conditions for isotropic and orthotropic fibre-reinforced composites. Part 5:1997. Test conditions for unidirectional fibre-reinforced plastic composites.
3. BS EN ISO 3167:1997. Plastics. Multipurpose test specimens.
4. BS EN ISO 2818:1997. Plastics. Preparation of test specimens by machining.
5. BS EN ISO 294. Plastics. Injection moulding of test specimens of thermoplastic materials.
6. BS EN ISO 295:1999. Plastics. Compression moulding of test specimens of thermosetting materials.
7. BS 903. Physical testing of rubber. Section A8:1990. Method for determination of rebound resilience.
8. BS EN ISO 604:1997. Plastics. Determination of compressive properties.
9. BS EN 408:1995. Timber structures. Structural timber and glued laminated timber. Determination of some physical and mechanical properties.
10. prEN 12390:1996. Testing concrete. Determination of compressive strength. Specification for compression testing machines.
11. BS 7585:Part 2:1992. Metallic multilayer plain bearings. Method for destructive testing of bond for bearing metal layer thickness 2 mm.
12. BS EN 24506:1993. Specification for hardmetals. Compression test.
13. BS 2782. Methods of testing plastics. Part 10. Glass reinforced plastics. Method 1001:1977. Measurement of hardness by means of a Barcol impressor. (= EN 59:1977).
14. BS EN ISO 868:1998. Plastics and ebonite. Determination of indentation hardness by means of a durometer (Shore hardness).
15. BS EN ISO 2039. Plastics. Determination of hardness. Part 1:1997. Ball indentation method. Part 2:2000. Rockwell hardness.
16. BS EN ISO 6506. Metallic materials. Brinell hardness test. Part 1:1999. Test method. Part 2:1999. Verification and calibration of Brinell hardness testing machines. Part 3:1999. Calibration.
17. BS EN ISO 6507. Metallic materials. Vickers hardness test. Part 1:1998. Test method. Part 2:1998. Verification of testing machines. Part 3:1998. Calibration of reference blocks.
18. BS EN 196. Methods of testing cement. Part 3: 1995. Determination of setting time and soundness.
19. BS EN 1015. Method of test for mortar for masonry. Part 4: 1999. Determination of consistence of fresh mortar (by plunger penetration).
20. BS EN ISO 306: 1997. Thermoplastic materials. Determination of VICAT softening temperature (VST).
21. BS EN ISO 179. Plastics. Part 1:1997. Determination of Charpy impact strength. Part 2:1999. Instrumented impact strength.
22. BS 131. Notched bar samples. Part 1:1961. Izod impact test of metals. Part 4:1972. Calibration of pendulum impact testing machines for metals. Part 5:1965. Determination of crystallinity. Part 6:1998. Method of precision of Charpy impact energies for metals. Part 7:1998. Specification for verification of the test machine used for precision determination of Charpy V-notch impact energies for metals.
23. BS EN ISO 180:1997. Plastics. Determination of Izod impact strength.
24. DD ENV 1156:1999. Wood-based panels. Determination of duration of load and creep factors.
25. BS EN ISO 899. Plastics. Determination of creep behaviour. Part 1:1997. Tensile creep. Part 2:1997. Flexural creep by three-point loading.
26. BS EN 1355:1997. Determination of creep strains under compression of autoclaved aerated concrete or lightweight aggregate concrete with open structure.
27. BS 3500. Methods for creep and rupture testing of metals. Part 3:1969. Tensile creep testing.
28. BS EN ISO 12737:1999. Metallic materials. Determination of plane-strain fracture toughness.
29. prEN 3874:1996. Published by the BSI in the UK as a Draft for Public Comment 96/705057 DC. Test methods for metallic materials. Constant amplitude force-controlled low cycle fatigue testing.
30. BS 903. Physical testing of rubber. Part A-49:1984. Determination of temperature rise and resistance to fatigue in flexometer testing (basic principles).
31. BS EN 28659:1992. Thermoplastic valves. Fatigue strength. Test method.
32. 98/121749 DC. Plastics. Test methods for tension. Tension fatigue crack propagation.
33. BRE (1998) Application of subsurface radar as an investigative technique. Report 340. Garston, BRE.

4

WATER

4.1 INTRODUCTION

Water in the environment is constantly being re-used and is changing from one state to another by a variety of complex processes. The principal mechanisms of transportation of water are evaporation of liquid water, subsequent condensation and final precipitation as liquid water. This process is usefully summarised by the **water cycle** (***Figure 4.1a***).

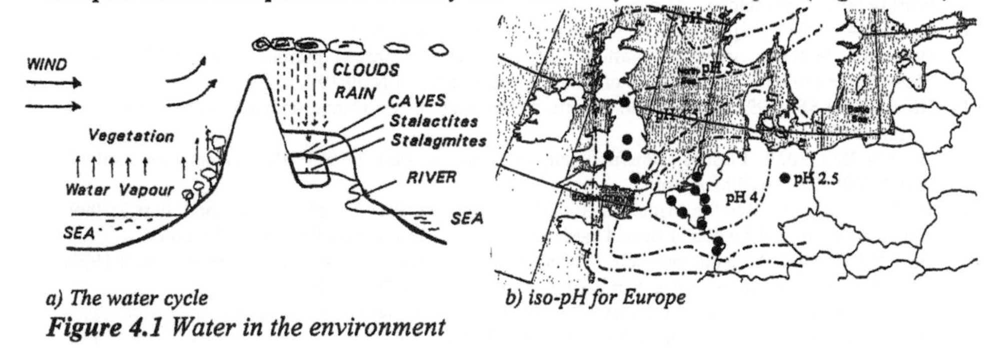

a) The water cycle *b) iso-pH for Europe*

***Figure 4.1** Water in the environment*

The presence of water in the atmosphere is expressed as the **relative humidity (2.2.6)**. Water vapour enters the atmosphere from the large expanses of water on the surface of the earth (glaciers, lakes and seas) (***Figure 4.1a***); it is free from **all contaminants**, which are left behind in the source (glaciers, lakes and seas). Seas which became cut off from the oceans in geological times past evaporate away to provide materials used for building (e.g, the evaporite gypsum, **14.7.3**). A large amount of water also enters the atmosphere from green vegetation when chlorophyll, in the presence of sunlight, produces the sugars which are necessary for growth (**12.4.1**). During the process, carbon dioxide is consumed and oxygen is evolved together with the loss of some water vapour by a process called **transpiration**. On a hot day, this can be as much as 80 gallons of water per tree. Finally, most combustion processes produce water, derived from the action of oxygen in the atmosphere with hydrogen contained within the structure of the material (**14.3**). Atmospheric water vapour is often condensed into rain or clouds by the rapid adiabatic expansion of the moist air mass as it is forced over mountain ranges, which produces a drop in temperature. The air becomes saturated with moisture vapour, which is deposited as very small (micro-) droplets of water to form clouds. These micro-droplets of water are too small to drop to earth as rain. However, they may coalesce to form larger droplets, for example on the lea side of the mountain range. The rain water is practically pure water, free from **all** solid contaminants (with the exception of the nuclei required for droplet growth, **2.6**). Typically, nuclei are particulates such as dirt particles, soot, etc; rain water in the UK can contain particulates of Sahara dust storms. Due to absence of minerals, rain water is tasteless and would not produce good drinking water because minerals are necessary for good health and taste. However, rain water does absorb gases, for example oxygen, carbon dioxide and sulphur dioxide. Absorption of acidic gases (e.g, the carbon dioxide, CO_2 and sulfur dioxide, SO_2 and trioxide, SO_3) forms **acid rain** (**4.3.1**), which is capable of dissolving other materials, notably calcium and magnesium ions, thereby increasing the mineral content of the water. As the quantity of acidic gases in the atmosphere is greater in industrialised regions compared to rural areas, the atmospheric pH will vary both locally and globally. ***Figure 4.1b*** illustrates the pH variation (degree of acidity) of the atmosphere of Europe. The dotted lines are iso-pH lines (connecting equal pH values), indicating increased levels of atmospheric acidity concentrated around the industrialised area of the Ruhr valley. This increased acidity will, for example, accelerate the corrosion of metals exposed to the atmosphere (**11.4.7**).

Rain falls to the ground and runs down streams, which can erode away geological strata over long time periods. Some water penetrates the soil to enter the water table, which acts as an underground reservoir for water. Other streams may cut into the strata (e.g, limestone strata) to produce underground caverns where, on evaporation of the water, dissolved limestone can be deposited as stalagmites and stalactites (**4.3.1**). Some water lies in lakes which contain a large amount of organic matter. The organic matter can remove oxygen from the water when it decays, rendering the water unsuitable for other forms of life. Organic matter in reservoirs is removed by the water board by filtering through beds of 'sharp sand'. Water which flows over certain soils can contain agricultural fertilisers, industrial pollutants and bacteria, and may pick up minerals as a direct result of the acidity of the water. For example, aluminium salts are dissolved from certain soils, and the aluminium can then be precipitated as the hydroxide (**1.8.3**) when the streams reach larger lakes. As aluminium hydroxide is a gel (**2.9.2**), it can block up the gills of fish. Most of these pollutants are removed by the water board. Hence there is a big difference between the water which falls as rain and the water which comes out of a tap or exists in one of the underground reservoirs.

4.2 WATER SUPPLIES

Water is the universal solvent and is required for industry (for cooling, washing and carrying out most chemical reactions). Water is also required domestically; consumption of water per person in third world countries is about 12 L/day, compared to 150-250 L/day in the Western countries. Clearly, water is a scarce resource which must be recycled to meet these demands. Domestic water comes from a constantly circulating national resource. It is estimated that the river Thames water is drunk 11 times from its upper reaches to the point where it reaches the sea (at which point the salinity increases to such a degree that the water becomes unpalatable). Dissolved oxygen in the water is essential for aquatic life, including the bacteria responsible for the removal of natural organic matter from the water. For example, fish require a minimum of about 6 mg of oxygen per litre of water for survival. Water can dissolve 14.2 mg of oxygen per litre of water at 1°C and about 9.2 mg of oxygen per litre of water at 20°C. Rivers and lakes can be polluted by the following mechanisms

- increases in the temperature of the water (e.g, from industrial cooling processes);
- organic wastes, particularly untreated sewage, which produce nitrates, carbon dioxide and generally lower the oxygen content of the water;
- accelerated growth of aquatic weeds, which make excessive demands upon the available dissolved oxygen.

Large expanses of water and streams have traditionally been used as dumping grounds for waste. Sewage was disposed of by making use of natural purification methods (dilution by rain, oxidation by air, putrefaction and filtration by the soil). In 1849, London's sewage was discarded into the Thames without treatment, causing the death of the river fish, which could not survive in the oxygen denuded waters. In 1899, the first sewage plant was built, but this improvement was short lived due to the rapid population expansion in London (from 2 million in 1900 to 6 million by 1945). New sewage works have since been built and the benefits in terms of cleaner water are evidenced by the presence of fish found further up the river, due to the return of the microorganisms responsible for the removal of the natural organic matter from the water. In view of the large human population densities today, natural purification methods cannot be relied upon and water must be treated artificially at sewage stations.

4.2.1 Water Board Sewage Treatment

The water board classifies contaminants by the **biological oxygen demand (BOD)** made upon the water. Sewage is treated so that the water may be returned to the rivers with less than 30 parts per million (ppm) of suspended solids and will not require more than 20 ppm of dissolved oxygen in five days (see equation (4.2)). This is undertaken by the water board using screening, sedimentation, thorough oxygenation, filtration and sterilisation processes; ***Figure 4.2*** illustrates a typical treatment procedure.

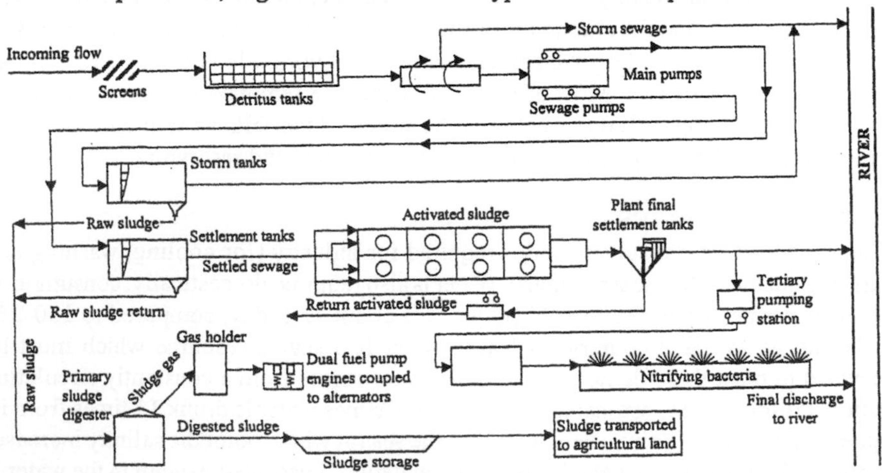

Figure 4.2 Schematic flow diagram of typical sewage treatment operations

The inflowing sewage is **screened** to remove insoluble solid materials (such as grit, paper, rags, etc.), which are burnt to destroy the bacteria which they contain. The resulting ash is added to the fertiliser and the heat generated by this process is utilised at the sewerage works. **Sedimentation** involves removal of the small particles of organic matter and silt by adding a **flocculent** (normally ferrous sulfate alum, $FeSO_4.Al_2(SO_4)_3$ and lime slurry, $Ca(OH)_2$). The flocculent causes the small particles to aggregate together, whilst the lime slurry will **soften (4.3.1)** the water

$$\underset{}{Ca(HCO_3)_2} + \underset{\text{hydrated lime}}{Ca(OH)_2} \rightarrow \underset{\text{sludge}}{CaCO_3\downarrow} + H_2O \qquad ...(4.1)$$

The sedimented material and precipitated $CaCO_3$ are treated separately from the liquid sewage. Organic material is removed by **digestion** by aerobic microorganisms, which produces ammonia, methane and carbon dioxide as waste products

$$\text{sludge} + \text{Air}(O_2) \xrightarrow[\text{(aerobic micro-organisms)}]{} NH_3 + CO_2 + CH_4 \qquad ...(4.2)$$

The methane produced is used on the site to heat the digestion tanks

$$2NH_3 + \text{Air}(O_2) \xrightarrow[\text{(aerobic micro-organisms)}]{} 2NO_3^- + 3H_2O \qquad ...(4.3)$$

The ammonia (NH_3) and nitrate (NO_3) form a fertiliser, ammonium nitrate, and are combined with the ashes from previous processes and sold for agricultural use. The microorganisms responsible for the degradation process at the sewage works and within the rivers also require oxygen. It is therefore important to extract the ammonium nitrate before it reaches the rivers as this will only make the river weeds grow much quicker, putting a greater oxygen demand upon the river water. Small particulates are removed by a process of **filtration** by allowing the water to flow through beds (100 m x 40 m) of 'sharp sand', where harmless bacteria decompose any remaining soluble organic material. The unpleasant smells and taste are removed by passing the water over **activated carbon**. The high carbon dioxide content in the resulting sewage would lead to problems in pipe work (causing the corrosion of some metals) and therefore it is reduced by **oxygenation**. This artificial oxygenation aids the natural purification by aerobic bacteria which thrive on dissolved oxygen. Once oxygenated, the water can be returned to the rivers. For **domestic supplies**, the water is **sterilised** by the addition of 0.001% chlorine to remove all harmful microorganisms. The sterilised water is dechlorinated by a sulphur dioxide treatment. The water is then pumped into the mains supply where it can be stored for future use.

Prior to 1984, surface water from building roofs was discharged into the sewage system. This meant that, following storms, a larger volume of sewage than normal had to be treated and the excess was accommodated in storm tanks. Currently, BS EN 752[1] and BS 6367[2] require that rain water discharge from roofs should be to a drain (separate from the sewerage) or to a soakaway. Part of the domestic charge levied by the water board is to deal with the water from roofs (the charge is waived if the rain water goes direct to a soakaway or river).

4.3 HARD AND SOFT WATER

Domestic water supplies can be **soft** or **hard**, depending upon the strata above the water table, but is never **brackish** (salty, perhaps containing organic material). A **soft water** is a water which lathers readily with soap, whilst a **hard water** is a water which is difficult to lather with soap. The ability of water to lather with soap depends simply on the mineral content (as ions) present in the water. A hard water contains ions which form an insoluble compound with the soap, preventing the soap from reducing the surface tension of the water (**5.4.2**). The insoluble compound is called **scum** (a *trivial* name for calcium stearate). Any ion which forms an insoluble compound with the soap will form hard water, for example ions of **magnesium** and **calcium**. The insoluble compound formed by calcium ions is calcium stearate (used industrially as a **water proofing agent, 5.8.1**, and is **completely insoluble in water**). Insoluble compounds formed, when compacted as a thin film on bathroom sanitary ware (as scum), are very difficult to remove (although they are, however, soluble in alcohol). Hard water may exhibit **temporary** or **permanent hardness.**

4.3.1 Temporary Hardness

Temporary hardness is caused by any metallic cation (e.g, a positive ion like Ca^{2+}, Mg^{2+}, etc.) which forms a soluble **bicarbonate** $(HCO_3)^-$ in water and an insoluble compound (salt) with soap. The soluble bicarbonate is **unstable when heated** (stability depends upon the carbon dioxide content of the water, **2.2.7**), breaking down into the **carbonate** which is **insoluble in water**. If these parameters are satisfied, then heating the water will reduce the number of cations in solution (because the carbonate is insoluble in water), thus giving rise to **temporary hardness**. The solubility of the cation is temporary because it is dependent upon the temperature-dependent solubility of carbon dioxide and the temperature stability of the bicarbonate.

As rain water falls through the atmosphere, it dissolves both oxygen and carbon dioxide. In industrial regions, where sulphur dioxide is present, this gas is also dissolved. The overall solution is therefore weakly acidic (***Figure 4.1b***)

CO_2	+	H_2O	→	H_2CO_3	...(4.4)
carbon dioxide (acidic gas)		Water		carbonic acid 'acid rain'	

Although only a weak acid, carbonic acid is capable of dissolving some forms of strata and certain building materials, especially when based upon a cementitious matrix of calcium or magnesium carbonate, $CaCO_3$ or $MgCO_3$, e.g, limestone or chalk ($CaCO_3$) or dolomite (a mixture of magnesium and calcium carbonates, $CaCO_3.MgCO_3$). This leads to the formation of large gullies, erosion of substrata causing subsidence beneath buildings, underground caves in limestone rock, etc. The reaction can be shown as follows ($CaCO_3$ rather than the mixed carbonates are used for illustrative purposes)

H_2CO_3	+	$CaCO_3$	→	$Ca(HCO_3)_2$	...(4.5)
		Calcium carbonate (limestone) (insoluble)		Calcium bicarbonate (soluble)	

This reaction is capable of going in the reverse direction at higher temperatures (e.g, in electric kettles, central heating pipes, etc.) or when carbon dioxide is lost from water, for example in caves to produce **stalactites** (this process takes some time, and occurs as the water drips from the cavern ceiling). When the water droplet passes through the air, it loses more carbon dioxide, and so when it drips onto the floor, more calcium carbonate must be deposited (as **stalagmites**). The reaction is shown below, where the '$\rightleftharpoons$' indicates that the reaction can proceed in either direction

(solubility direction)

$$Ca(HCO_3)_2 \rightleftharpoons CaCO_3\downarrow + H_2O + CO_2\uparrow \quad ...(4.6)$$

(precipitating direction)

Calcium bicarbonate	Calcium carbonate
(soluble)	(insoluble)

Note that a similar reaction occurs where carbonic acid ('acid rain') falls continuously over cementitious products containing 'free lime' (calcium hydroxide, $Ca(OH)_2$, **7.16**), a process producing 'lime' washings.

$$H_2CO_3 + Ca(OH)_2 \rightarrow CaCO_3\downarrow + 2H_2O \quad ...(4.7)$$

$Ca(OH)_2$	$CaCO_3$
Calcium hydroxide	Calcium carbonate
(cement products)	('lime' washing)
(slightly soluble)	(insoluble)

'Lime' staining of adjacent building materials can occur and, since calcium carbonate is insoluble, lime staining is difficult to remove (unlike other salt deposition defects, such as efflorescence, **6.7.1**). Note also that the solubility of most compounds increases with an increase of temperature (**2.9.1**), but the deposits of calcium carbonate are **not** dissolved by further boiling (i.e., it is **insoluble**). However, in accordance with equation (4.6), **temporary water hardness can be removed by heating**, i.e.,

$$Mg(HCO_3)_2 \text{ or } Ca(HCO_3)_2 \xrightarrow{70 \text{ to } 80^\circ C} MgCO_3 \text{ or } CaCO_3 + H_2O + CO_2 \quad ...(4.8)$$

$Mg(HCO_3)_2$ or $Ca(HCO_3)_2$	$MgCO_3$ or $CaCO_3$	CO_2
Magnesium bicarbonate	'fur'	gas evolved when temporarily
Calcium bicarbonate	'limescale'	hard water is heated
(soluble)	(insoluble)	

In effect, the water has been 'softened' because the cations responsible for the water hardness have been removed as insoluble calcium carbonate $CaCO_3$ ('limescale'). This process can be made more efficient by incorporation of larger surface areas over which the reaction can proceed. For example, stainless steel mesh is sometimes placed in electric kettles. The function of this mesh is to expose a larger surface area for scale deposition. These meshes do not **stop** the scale building up on the heating element. They work because the same amount of scale must be deposited and, for deposition, the scale requires a nucleus upon which to deposit. The mesh acts as a nucleus and so the thickness deposited upon the heating element is reduced by the ratio of the surface area of the heating element to the total surface area (element + mesh) available.

4.3.2 Permanent Hardness

Permanent hardness occurs where the stability of the compound and its solubility is **not** affected by heat. Several compounds are present in domestic water, either by deliberate additions (e.g, calcium fluoride, CaF_2, to improve the resistance of teeth to decay) or as impurities (e.g, calcium nitrates, $CaNO_3$).

4.3.3 Total Hardness

The total hardness of water is the combined hardness due to temporary and permanent hardness. About 60% of the households in Britain are supplied with hard water.

4.3.4 Advantages and Disadvantages of Hard Water

The disadvantages of hard water include

- The build up of scale around electric immersion heaters, electric kettle elements, heat exchanger coils, etc. reduces their efficiency by reducing heat transfer. In the case of electric elements, this reduction in heat transfer can cause the element to overheat and burn out, or increase the operational temperature, giving a shorter life. Often, however, the scale deposit is shed from the electric element when the thickness increases to such an extent that thermal stresses build up within the electric element sufficient to crack the scale deposit. Scale can also reduce the sensing elements of safety devices;
- When the scale is deposited over a period of time, the layers can restrict the flow of water in pipes. Where these pipes are used for venting and/or safety processes, this restriction can lead to explosions caused by excess pressure. The recommendations of manufacturers of water heater elements is not to store hard water above 60°C. Practical requirements, however, may often require higher temperatures (e.g, where the length of pipe and draw off volumes are high);
- Both dyes and soaps can combine with the soluble calcium and magnesium ions and form an insoluble compound, making both ineffective. Soaps will not be able to perform their function of reducing the surface tension of water to allow formation of bubbles (a lather) (**5.4.2**);
- The insoluble compound formed with soaps (calcium or magnesium stearate, 'scum') spoils the appearance of highly coloured bathroom suites;
- The insoluble compound (calcium carbonate, $CaCO_3$) formed when the hard water dries out is responsible for the harsh, rigid feel of fabrics (e.g, towels).

The advantages of hard water include

- Calcium, as calcium phosphate, is important for bone formation. The human body obtains calcium ions from hard water in just the same way that shellfish obtain calcium ions for shell formation from the calcium bicarbonate in the sea. There is some evidence to suggest that hard water is better for the heart than soft water;
- Calcium fluoride prevents the decay of the enamel on teeth and as such is often added to the water supplies under "mass medication" routines;
- Certain soluble ions give water its "palatable taste";
- Hard water is required for beer brewing (e.g, at Burton on Trent);
- Lead pipes were used extensively for plumbing in construction. Great flexibility, excellent corrosion resistance and ease of soldering were advantageous, but lead is

dissolved in certain waters (e.g, peaty waters containing organic acids and rain water). Although dissolution would not cause perforation of the pipe, lead is a cumulative poison. The protective carbonate film formed on the lead pipe in hard water areas reduces the rate of dissolution of lead.

4.4 WATER SOFTENING

Water softening is a process that removes **both** temporary and permanent hardness at the same time by chemical treatment (except Method 1 below). This is achieved by turning the soluble calcium and magnesium compounds into other compounds which have low solubility. The compounds of calcium and magnesium are precipitated by forming a sludge **(1.2.2)** so that they cannot affect the action of soap. Sludge is produced wherever calcium carbonate or calcium stearate is deposited. The methods by which water is softened are considered below.

4.4.1 Method 1. Water Board Methods

The soluble compounds responsible for water hardness are made insoluble by the addition of lime, $Ca(OH)_2$ (equation (4.1)). This process is generally undertaken by the water board, and purifies the water by producing better settlement of the fine materials. The method has two advantages

- the addition of calcium hydroxide reduces the acidity, producing a pH of about 8.5, which reduces the corrosion of the old cast iron pipes;
- the excess calcium is beneficial for the human body.

Note that if an excess of calcium hydroxide is added, then the hardness of the water is not decreased because the hardness can be due to any soluble calcium ion; the method can therefore harden water if incorrectly used and so the exact amount of calcium hydroxide must be added. This addition, together with sulphuric acid (which is a necessary pretreatment for the water to ensure that the alum flocculation process operates efficiently) reduces the acidity of the water. Note that this method **reduces temporary hardness only**. For **permanent hardness**, the addition of calcium hydroxide will **not** reduce the calcium ion concentration, i.e.,

$$CaF_2 + Ca(OH)_2 \rightarrow CaF_2 + Ca(OH)_2 \quad ...(4.9)$$

(no reduction in soluble calcium ions)

4.4.2 Method 2. Base Exchange Process (domestic system)

The base exchange process is usually known as the **Permutit**™ process. It is a domestic method and is usually installed in the incoming water main. The Permutit process removes both permanent and temporary hardness at the same time. The soluble calcium compounds responsible for water hardness are removed from water by replacement with sodium ions from an active mineral. The active mineral is a **heulandite** or **zeolite**; a zeolite is an alumino silicate of general formula $(Na,Ca)_{4\,to\,6}.\,Al_6(Al,Si)_4Si_{26}O_{72}.\,24H_2O$ **(1.18.5)**. Zeolites occur naturally in the environment. The zeolite structure is an open ring which contains the cations, either sodium (Na^+), magnesium (Mg^{2+}) or calcium (Ca^{2+}) (***Figure 1.15***). The ions in the zeolite structure are capable of interchange in the solid structure within an aqueous medium, one calcium ion site being occupied by two sodium ions within the zeolite anion. For the purposes of this text, we represent this complex zeolite formula as **'Na zeolite'**. The **ionic substitution (1.18.6)** does **not** alter the structure of the zeolite material. In the base exchange system, the zeolite is first charged with common salt solution (brine) by replacing all the calcium and magnesium ions with sodium ions, producing Na zeolite. Domestic water, passing through the Na zeolite structure and containing the soluble bicarbonates of magnesium or calcium, will substitute the calcium (in the water) for two of the sodium (in the zeolite structure).

$$Ca(HCO_3)_2 \quad + \quad \text{Na zeolite} \quad \rightarrow \quad NaHCO_3 \quad + \quad \text{Ca zeolite}$$

Eventually, the Na zeolite will transform to a Ca zeolite by ion exchange. The Ca zeolite can be regenerated by running a concentrated solution of sodium chloride through the water softener and the resulting calcium chloride is run to waste, i.e.,

$$\text{Ca Zeolite} \quad + \quad NaCl \quad \rightarrow \quad CaCl_2 \quad + \quad \text{Na Zeolite}$$

The zeolite is regenerated by the base exchange, one Ca^{2+} ion being exchanged for two Na^+ ions. Note that sodium bicarbonate, $NaHCO_3$, produced by the process does not cause temporary hardness because the sodium salt of stearic acid is soluble (in fact soap is sodium stearate, **4.4.3**). There is no sludge produced by this method of softening. The capital expense is high as expensive plumbing is needed and a regeneration unit (to recharge the active Na zeolite agent when this becomes exhausted) is required. The system does, however, give the cheapest running costs and, once installed, is simple to operate. As regeneration is achieved by the passage of a concentrated solution of sodium chloride, care should be taken to ensure that this does not cause corrosion of the steel pipe work. In addition, this system does waste water on regeneration because the excess salt must be washed away (due to its unpalatable taste).

4.4.3 Method 3. Addition of Soap (domestic system)

Soluble calcium compounds responsible for water hardness are made insoluble by the addition soap to form scum (calcium stearate), i.e.,

$$Ca(HCO_3)_2 \quad + \quad \underset{C_{17}H_{35}\text{-}CO.O^- \, 2Na^+}{\text{2 Na stearate}} \quad \rightarrow \quad \underset{(C_{17}H_{35}\text{-}CO.O^-)_2Ca^+}{\text{Ca stearate} \downarrow} \quad + \quad 2\,NaHCO_3$$

The process is a waste of soap (since the soap must remove all soluble calcium ions before it can act to reduce the surface tension of water, **5.4.2**) and produces an insoluble scum of calcium stearate, $Ca(C_{17}H_{35}\text{-}CO.O)_2$. Note that the ordinary domestic soap molecule, $C_{17}H_{35}$-CO.ONa, possesses both a hydrophobic part and a hydrophilic part. Mixing of softened water with unsoftened water causes a problem if the calcium salt is insoluble. This arises from the fact that all the soap will go to make an insoluble compound and there will be no soap left to suspend the dirt and oil in the water:oil emulsion. By law, all domestic detergents must be **biodegradable**. Washing up liquid does not give a scum like domestic soap because the calcium compound of the acid from which a detergent is made is soluble in water. The detergent molecule contains a hydrophilic part (usually a $-CO.O^-$ group for soap and a $-SO_3^-$ group for the detergent, **5.4.2**) and a lipophilic part, usually a long hydrocarbon chain (e.g, a $C_{17}H_{35}$- group). Detergents can be considered as the sodium salt of $C_{15\text{-}17}H_{31\text{-}35}\text{-}SO_3.ONa$ (as a first approximation), and the calcium salt $Ca(C_{17}H_{35}\text{-}SO_3)_2$ is soluble in water. Detergents have the advantage of being biodegradable and more soluble in water than the stearate soaps. In addition, they do not form insoluble compounds with calcium or magnesium and therefore do not produce a scum.

4.4.4 Method 4. Addition of Washing Soda Crystals (domestic system)

Soluble calcium compounds responsible for water hardness are removed from the water by replacement with sodium ions using washing soda crystals, $Na_2CO_3.10H_2O$. This is a cheap domestic alternative to water softening and only softens the water required for use. The method produces a slightly alkaline solution

$$Ca(HCO_3)_2 \quad + \quad Na_2CO_3 \quad \rightarrow \quad NaHCO_3 \quad + \quad CaCO_3 \downarrow$$

4.4.5 Method 5. Garage Forecourt System (commercial system)

The positive metallic ions, e.g, Ca^{2+}, Mg^{2+}, etc. are replaced by H^+ ions, whilst the negative ions e.g, $(HCO_3)^-$; $(SO_4)^{2-}$; Cl^-, etc. are replaced by OH^- ions. This form of ion exchange is very expensive and the water produced is used as an alternative to distilled water. The reliability of this method is dependent upon the frequency at which the ion exchange cartridges are changed. It is the most expensive method of softening water.

4.4.6 Method 6. Calgon^tm (domestic system)

Softeners, such as polyphosphate or hexametaphosphate, which are a polymers with the formula $(NaPO_3)_n$, where n is about 150, are often added to the water in doses not considered harmful to health (about 1.5 ppm). Although the exact method of operation of these chemicals is not fully understood, they must remove the calcium and magnesium salts by adsorption in the polyphosphate colloid. These softening chemicals dry to produce very fine crystals (which do not have the stiffening effect that hard water has on towels, etc.). However, these chemicals are relatively expensive

4.4.7 Other Water Softening Methods

More recently, a number of non-chemical methods, sold as inhibiting systems, have become available. Such systems are relatively cheap compared to most domestic systems. Examples include capacitance systems, electronic discharge systems and permanent magnets. These items are not water softeners, as the Ca^{2+} ions are not removed from the water. In all cases they are attached to copper pipes; copper is a paramagnetic material which does not allow electric/magnetic fields to pass through it (hence the efficiency of these systems is questionable). These systems are best described as **conditioners** as they **do not inhibit the scale from forming**, but are reputed to alter the form of the precipitated molecules so that instead of forming a hard scale, a soft sludge is produced. In most domestic heating systems, the flow of water will be sufficient to carry the sludge through the pipes into the waste system.

4.5 REFERENCES

1. BS EN 752. Drain and sewer systems. Outside buildings.
2. BS 6367:1983. British standard code of practice for drainage of roofs and paved areas.

5

MOISTURE EFFECTS IN BUILDINGS

5.1 INTRODUCTION

Moisture in buildings may result in cold (or hot), damp, uncomfortable internal environments; excessive heating, cooling and maintenance costs; deterioration of internal finishes (including mould growth); unsightly damp areas on wall and ceiling surfaces; and increased concentrations of harmful pollutants affecting occupant health. The exclusion of moisture from buildings is therefore desirable. Direct signs of moisture ingress include visible moisture films, drops of water, puddles and flooding. Indirect signs, which may by visible internally and externally, include damp patches, peeling and blistering of wall and ceiling finishes, deposition of water soluble salts (efflorescence), mildew and mould growth, metallic corrosion, timber deformation and decay, etc.

5.2 MOISTURE SOURCES

There are a wide variety of moisture sources which may produce dampness in buildings. For the purposes of this text, we group moisture sources into three areas as follows

- **construction moisture sources**, which can be substantial but are generally important only during the first two to three years in the life of a new building;
- **interior moisture sources** resulting from the activities inside the building, such as cooking, showering, respiration of the occupants, etc.;
- **exterior moisture sources** that result in the entry of moisture into the building by processes such as air infiltration, vapour diffusion, capillarity, etc. (**Table 5.1**).

5.2.1 Construction Moisture Sources

Newly constructed buildings give off significant quantities of moisture during their early life as a result of evaporation from moisture trapped within the materials use. This water may be present in the material prior to incorporation into the building fabric, may be water added during a construction process (e.g, concrete, mortar, etc.) or may be absorbed from the environment into a material supplied to the construction site dry but inadequately stored on site. A typical three bedroomed house of traditional masonry construction contains an estimated 7000 litres of construction water, most of which is dispersed in the first heating season[1]. To appreciate the quantities of water released from construction sources, it has been estimated that

- an 'average' house containing timber at 19% moisture content can release a total of 200 litres of water from the timber as it dries to equilibrium conditions[2];
- an estimated 90 litres of water per m^3 of poured concrete is released over the first two years following construction;
- for new houses, the total moisture input internally from construction sources may average 10 or more litres/day during winter in the first year and about 5 litres/day in the second year[3].

5.2.2 Interior Moisture Sources

Major interior moisture sources include

- people, who generate moisture by respiration and perspiration. Generation rates will be dependent upon the level of activity (generation rates of 0.03 to 0.06 litres/hour for light activity, 0.12 to 0.2 for medium activity and 0.2 to 0.3 litres/day for hard work have been recorded at typical indoor temperatures[4]) and temperature[5];
- combustion processes, for example complete combustion of 1 m^3/h of methane (e.g, for cooking, heating, etc.) results in a moisture vapour production rate of 0.0026 litres/hour[5], whilst the use of a paraffin heater during the evening results in a moisture vapour production rate of 1 to 2 litres/day[1];
- clothes washing and drying operations, for example it is estimated that 0.5 to 1.0 litres/day of water vapour is produced by washing clothes and about 3.0 to 7.5 litres/day by drying clothes in unvented tumble driers[1];
- moisture generated in bathrooms, for example, from baths (0.7 litres/hour[4]) and showers (0.22 litres assuming a 5 minute shower[4]);
- moisture generated in kitchens, for example, for a family of four, the daily average moisture load from gas cooking is about 2.4 litres/day[4] (cooking with electricity will reduce this load) and from dishwashing about 0.5 to 1.0 litre/day[1];
- moisture from plants (only about 0.2% of the water used to water plants is used for growth and the vast majority enters the air). Moisture generation rates have been estimated[4] for medium sized household ferns (between 0.17 and 0.37 litres/day), rubber plants (up to 0.5 litres/day) and larger trees such as beech (2 to 3 m tall) (50 to 100 litres/day).

Additionally, wet surfaces (from cleaning operations, spillages, leakages, etc.) may contribute significantly to building moisture loads. A major variable in moisture generation rates in domestic environments is the number of household members. For example, the BRE quotes values of 1.0 to 2.0 litres/day for four persons asleep for 8 hours and 1.5 to 3.0 litres/day for two persons active for 16 hours[1], whilst ASHRAE states that a typical family of four may produce as much as 11.4 litres/day of water vapour, or more automatic washing machines and clothes driers are used[6]. Corrections for the number of household members can be obtained by assuming each additional member contributes an additional 2.1 litres/day[7]. Most interior moisture sources increase the humidity of the internal air which may lead to problems of condensation (**5.12**).

5.2.3 Exterior Moisture Sources

One of the primary functions of a building envelope is that of **moisture exclusion** (**5.8**). However, if the external envelope fails in this function, exterior moisture sources are likely to exceed all interior moisture sources discussed above. Exterior moisture sources include both liquid water and water vapour. Although we distinguish liquid and vapour moisture sources, in actual buildings the two types are often interrelated. For example, liquid rainwater main leak into buildings or building components and then evaporate, only to condense again at some later time (in possibly quite a remote location).Conversely, water vapour can condense, move as liquid water and reappear at a different location (where it may appear to be rainwater).

Liquid water sources include rain, fog, dew, blown snow, etc. Above ground level, the driving rain index (**5.9.2**) can be used to determine the degree of exposure to rain. Theoretically, exterior moisture sources in the form of rain water can result in internal moisture loads which vary from zero (in a well constructed building) to the average annual rainfall multiplied by the roof area. Below ground level, the ground itself provides a reservoir of moisture which depends upon local climate, topography, etc. and the water table level. One of the largest sources of moisture in building enclosures is the migration of moisture from the surrounding soil into foundations, basements, etc. and subsequently into living spaces. A common example is rising damp (**5.10**).

Vapour water sources may arise from the infiltration of humid air from outside. In cooling climates, the outside air is warm and humid relative to the inside air and the greater the degree of air exchange of interior air with exterior air, the greater the rate of inward moisture migration. This moisture source depends on the air leakage rate and/or the controlled ventilation rate. High internal humidity may also result in adsorption of moisture by building materials and furnishings, particularly in summer. Water adsorbed can subsequently be desorbed later when the ambient humidity level starts to drop. These effect operate daily and seasonally. For example, 3 to 8 litres/day may be released from furnishings during autumn. Additionally, moist air does leak into ground floor spaces from around ground floor perimeter wall joints, through cracks and around drains. This air may contain a significant amount of moisture and the moisture input rate would be at a maximum during the coldest part of winter due to the stack effect.

In well constructed and maintained residential buildings, the largest single source of moisture is from people (as respiration and perspiration). For a family of four, this is around 5 to 6 litres/day. Other activities, such as cooking, cleaning, etc. approximately double this rate to around 10 litre/day. Construction sources may double the total residential moisture load in the first year to about 20 litres/day. All other major potential moisture sources should be controlled in a well constructed and maintained building.

5.3 MOISTURE TRANSPORT AND STORAGE

Water, like any other pure substance, can exist in three states (**2.1**): solid (ice), liquid (water) and gas (vapour). These three states of moisture can exist in buildings. In addition, various building materials can capture water molecules from the surrounding air and localise them on their surfaces. Localised moisture in this state is said to be **adsorbed** (**5.6**). Subject to changes in temperature and vapour pressure, it is possible for moisture in any state to undergo a change of state (or phase transition) (**2.5**). Moisture can be transported and stored in a porous building material in all four states.

To describe these transport and storage processes, four mechanisms can be considered
- vapour transport and storage;
- liquid transport and storage;
- solid transport and storage;
- adsorbate transport and storage.

In the vapour and liquid states, water molecules are more mobile than in the solid (**2.4**) or adsorbed state and the former transport processes may therefore significantly dominate the latter transport processes in most cases. The mobility of the vapour state and the limited occurrence of the solid state indicates that storage of moisture in porous building materials is most significant in the liquid and adsorbate states. Any transport process is brought about by a driving force or potential and is limited by a resistance. The most well-known examples are heat transport due to a temperature gradient across two locations within a material (Fourier's law) and electron transport due to a voltage difference (Ohm's law); in each case transport is restricted by the thermal or electrical properties of the material under examination. Moisture transport also operates in this way; however, it has not been possible to identify an experimentally realisable single potential that causes moisture transport. Thus, for practical reasons, the driving potential for moisture transport has been considered to be the result of experimentally realisable driving potentials, leading to a variety of moisture transport processes (**Table 5.1**).

Table 5.1 Moisture transport processes

Process	Moisture state	Driving force
	DIFFUSION	
Adsorbate	Adsorbate	Concentration gradient
Gas	Vapour	Vapour pressure gradient
Liquid	Liquid	Concentration gradient
Thermal	Vapour and liquid	Temperature gradient
	FLOW	
Capillary	Liquid	Capillarity
Convection	Vapour	Air pressure gradient
Gravitational	Liquid	Height
Poiseuille	Liquid	Liquid pressure gradient

Many of these processes are in themselves complex and, due to the dynamic nature of moisture transport, interact with one another in complex ways. For the purposes of this text, we therefore consider only the following effects in porous building materials
- absorption (**5.6**) and liquid transport and storage by capillarity (**5.5**);
- the effect of adsorption on moisture storage (**5.7**).

Both absorption of water vapour and absorption of liquid water give rise to moisture movement in building materials (**6.4**) and the phenomenon of bulking (**6.5**).

5.4 CAPILLARITY (LIQUID WATER)

Capillarity is a process by which the surface of a fluid (water) in contact with a solid is raised proportional to surface wetting. When placed in contact with a liquid, a fine capillary will draw up water against gravity to a definite height dependent upon

- the internal radius of the capillary;
- the surface tension of the liquid; and
- the contact angle within the liquid at the liquid:capillary interface.

This is the process of **capillarity** and accounts for the ability of blotting paper to soak up ink, the rise of water through the capillaries in the stems of plants (in this case, osmotic pressure accounts for a large part of the rise), the ingress of liquid water into porous building materials, etc.

5.4.1 Capillary Radius

To illustrate the effect of capillary radius, consider the crevice formed by two glass plates (***Figure 5.1a***) and the capillaries of different radii (***Figure 5.1b***) placed in water.

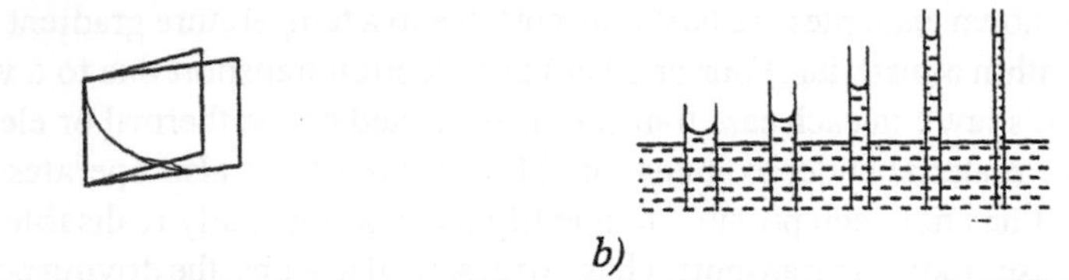

Figure 5.1 *The effect of crevice width and capillary radius on capillarity*

Figure 5.1a illustrates that the capillary rise, h, is **inversely proportional to the crevice width,** w, i.e., $h \propto 1/w$ and ***Figure 5.1b*** illustrates that the capillary rise, h, is **inversely proportional to the capillary radius,** r, i.e., $h \propto 1/r$. Note that these effects are not instantaneous, but depend on the viscosity of the liquid. In addition, the height rise, h, depends on the **surface tension** of the water and the **contact angle** formed by the water and a particular porous substrate. The wetting action of liquid water therefore depends on the nature of the surface in contact with it (effecting the contact angle) and the presence of any additives in the water (effecting the surface tension of the water).

5.4.2 Surface Tension

The free surface of a liquid can exhibit many interesting properties due to a phenomenon known as surface tension. Surface tension explains why liquid drops are spherical (in the absence of a gravitational field), why insects can walk on the surface of ponds, why water proof materials will not let in the rain, etc. Surface tension is known to be due to the intermolecular attractions in the liquid surface that produce a skin effect on the surface. Although it is rather difficult to represent the three-dimensional structure of water in the liquid state, we illustrate the skin effect by considering the water molecules to be hard spheres, as illustrated in ***Figure 5.2***.

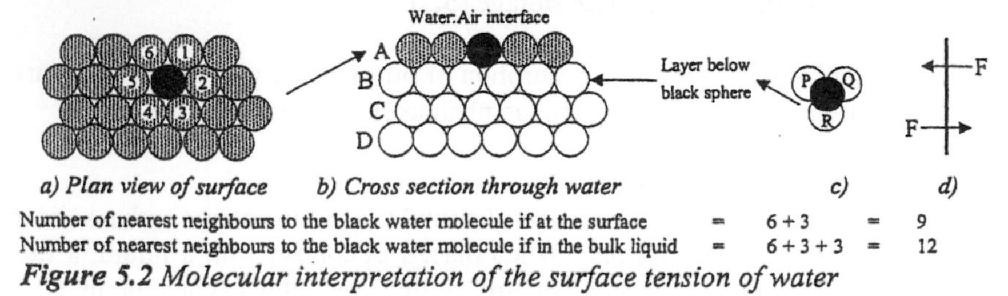

Figure 5.2 *Molecular interpretation of the surface tension of water*

Figure 5.2a depicts the surface molecules (in the layer **A**) as seen by a fish. The black sphere has 6 nearest neighbours. ***Figure 5.2b*** illustrates the molecular layers below the surface layer **A** (denoted layer **B, C, D**). ***Figure 5.2c*** depicts the three molecules **P, Q** and **R** in layer **B** which **are nearest** to the black water molecules in the layer **A** (in ***Figures 5.2a and 5.2b***); hence there are **3** nearest neighbours in the layer below the black surface molecule. There are therefore **9** nearest neighbours to the black **surface** molecule in the liquid water. By inspection, it can be seen that if the black water molecule was **inside** the liquid (in layer **C**) it would have another **3** neighbours, making **12** nearest neighbours for a molecule **inside** the liquid. The molecules in the liquid surface are attracted only to the molecules at their sides and below (9 nearest neighbours), while the molecules within the liquid are attracted to liquid molecules all around (12 nearest neighbours). There is therefore a net attraction of the surface molecules back into the bulk of the liquid and it is this force which accounts for surface tension. The surface tension force is modelled as a line of unit length drawn in the surface of the liquid ***(Figure 5.2d)***. Clearly, forces act in all directions in the surface, but we will consider only those forces acting at right angles to the line. The force on the whole line is the sum of all the forces on the individual molecules (note that any given molecule is in equilibrium due to the equal and opposite forces acting upon it). We can therefore define the surface tension force, T, of a liquid as the force acting in the surface of a the liquid and at right angles to a line of unit length drawn in the surface of the liquid, having units N/m. Surface tension can be measured using a tensiometer, which involves the determination of the force required to detach a ring or loop of wire from the surface of a liquid. The tensiometer is calibrated to account for the weight of the loop itself and allows the surface tension of the liquid to be read off directly.

Table 5.2 gives the surface tensions of some common liquids. As the surface tension of a liquid decrease with increasing temperature (and vanishes at the critical temperature), all values given in **Table 5.2** are for temperatures of 20°C.

Table 5.2 Surface tension values for some common liquids

Liquid	Surface tension, T (x 10^{-3}) (N/m)
Water	72.7
Paraffin	25
Benzene	28.9
Turpentine	27
Alcohol	23
Mercury	472

A useful illustration of the implications of the surface tension of water concerns the action of soaps. In order to clean clothes, dirt must be removed from the fabric of a material. This can be achieved by creating air bubbles (a froth) and forcing them through the fabric. The air bubbles encapsulate the dirt particles and carry them away from the fabric to the surface of the liquid within the froth (***Figure 5.3***). Note that the soap itself does not "dissolve" the dirt. This process is also used to separate metals from poor quality ores.

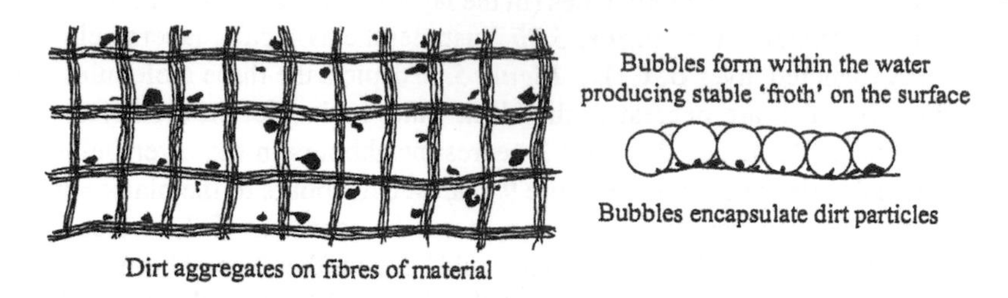

***Figure 5.3** Removal of dirt from fabric materials*

The fact that air has to be blown into a drop of soap solution to make a bubble suggests that the pressure within the bubble is greater than that outside. This is in fact the case; the excess pressure creates a force that is just balanced by the inward pull of the soap film of the bubble. We can determine the excess pressure within the bubble by considering the forces acting across an imaginary section through the centre of the bubble shown in ***Figure 5.4***.

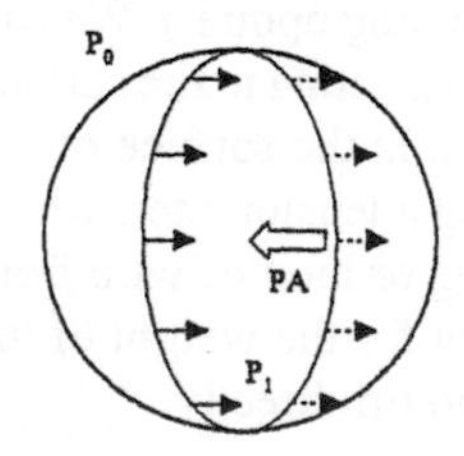

Let the external pressure be P_0 and the internal pressure be P_1, so that the excess pressure within the bubble, $P = P_1 - P_0$. The area of the section through centre of bubble, $A = \pi r^2$.
Consider the left hand side of the bubble. The force acting from right to left due to the internal excess pressure (which operates over the cross-sectional area A) is PA ($= P\pi r^2$). Since the bubble is in equilibrium (i.e., it has not burst), this force is balanced by a force due to the surface tension of the soap film acting from left to right (which operates over the perimeter of the circle formed by cross-sectional area A, i.e., $2\pi r$). As the bubble is in air, there are two soap film:air interfaces (one on the outside and one on the inside) and thus the surface tension force is $2 \times 2\pi r T$. At equilibrium we can equate these two forces to give

$$P = \frac{4T}{r}$$ (for a bubble in air with two soap film:air interfaces)

$$P = \frac{2T}{r}$$ (for a bubble within a liquid with only one soap film:air interface)

***Figure 5.4** Excess pressure within a soap bubble*

Both excess pressure formula indicate that the excess pressure, P, within a small bubble is greater than that within a large bubble. Bubbles cannot be formed with water alone, as the excess pressure is too great to allow the bubble to remain. Soaps lower the surface tension of the water, so that the excess pressure required to maintain the bubble is lowered, allowing the bubble to remain. The effect on the surface tension of water of adding soap can be demonstrated using the tensiometer; the surface tension of water can be reduced from 0.073 N/m to about 0.033 N/m by addition of a dilute soap solution.

When dirty soapy water meets water with no soap, the soap solution is diluted, increasing the surface tension and thus the excess pressure, causing the bubbles to burst and the dirt to be deposited. Waste water pipes should therefore be designed with an adequate fall to allow water to run with sufficient velocity to remove deposited dirt.

Soaps are commonly based on sodium stearates (***Figure 5.5***), which comprise a lipophilic, hydrophobic part (a long hydrocarbon chain) and a hydrophilic part (usually a $-CO.O^-$ group).

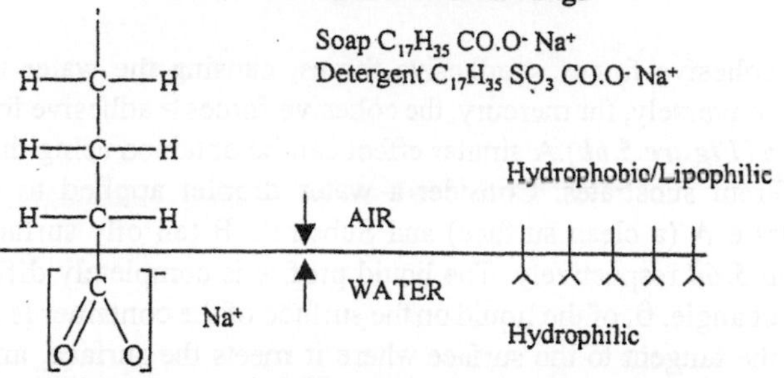

Soap $C_{17}H_{35}$ $CO.O^{-}$ Na^{+}
Detergent $C_{17}H_{35}$ SO_3 $CO.O^{-}$ Na^{+}

Hydrophobic/Lipophilic

AIR

WATER

Na^{+}

Hydrophilic

Figure 5.5 Chemical structure of sodium stearates used in soaps

Oil and grease is removed by forming a hydrophilic 'skin' around the oil droplet. The lipophilic hydrocarbon part of the soap molecule is miscible with the oil so that the soap molecule orientates itself to present the hydrophilic part of the molecule ($C{=}O.O^{-}$) to the water. Emulsion paints are oil-based and utilise similar technology to make the paint water soluble. The two-part composition of stearates also makes them useful water-repellents for porous building products, where the hydrophilic end is bound to the substrate and the hydrophobic hydrocarbon end serves to repel water (**5.8.1**). The effect of soaps in hard water areas is considered in section **4.4.3**.

5.4.3 Contact Angle

When a liquid surface meets a solid surface, the angle formed between the two surfaces is important. This angle is known as the contact angle θ and its size determines whether a liquid will spread over the surface or whether it will form droplets on it. For example, a droplet of water placed on a glass surface will spread over the surface, whereas a droplet of mercury gathers itself into globules (***Figure 5.6***).

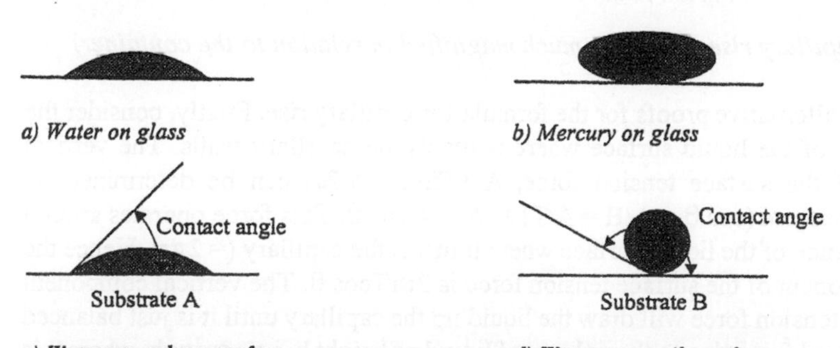

Figure 5.6 Contact angle effects of different liquids and substrates

This behaviour can be explained in terms of

- **adhesive forces** (i.e., attractive forces between the molecules of different substances, for example, a liquid and glass in ***Figure 5.6***); and
- **cohesive forces** (i.e., attraction between the molecules of a particular substance, for example the water or the mercury in ***Figure 5.6***).

For water, the cohesive forces < adhesive forces, causing the water to spread out (***Figure 5.6a***). Conversely, for mercury, the cohesive forces > adhesive forces, causing globules to form (***Figure 5.6b***).A similar effect can be obtained using the same liquid applied to different substrates. Consider a water droplet applied to two different surfaces, Substrate **A** (a clean surface) and Substrate **B** (an oily surface), shown in ***Figure 5.6c and 5.6d*** respectively. The liquid profile is completely different in each case. The contact angle, θ, of the liquid on the surface of the container is defined as the angle between the tangent to the surface where it meets the surface, and the surface itself, **measured through the liquid** (***Figure 5.6***). Liquid:surface interactions that produce a contact angle $\theta < 90°$ cause liquids to spread on surfaces whilst those which produce a contact angle $\theta > 90°$ cause liquids to form globules.

5.4.4 Mathematical Relationship for Capillarity

Water of density ρ with surface tension, T, is placed in a container and a narrow capillary of radius, r, is added, as shown in ***Figure 5.7***. Water is drawn up the capillary to height, h, and forms a contact angle θ with the walls of the capillary, producing a complete liquid surface which is concave in shape. For a circular capillary, the meniscus can be considered to be a portion of a sphere of almost constant radius. The radius of this sphere depends on the contact angle, θ.

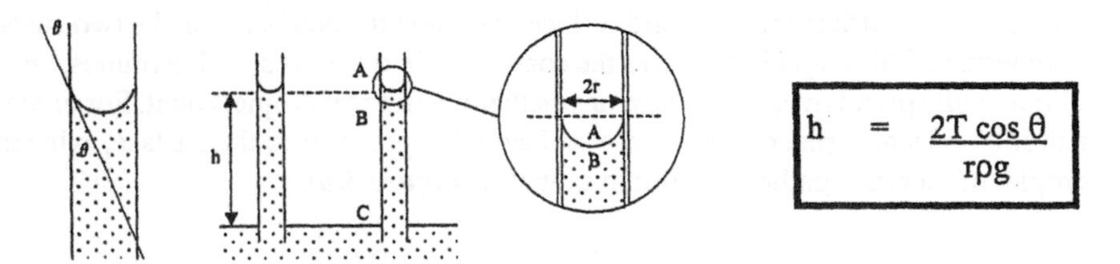

***Figure 5.7** Capillary rise (capillary much magnified in relation to the container)*

There are two alternative proofs for the formula for capillary rise. Firstly, consider the circumference of the liquid surface where is meets the capillary walls. The vertical component of the surface tension force, A (***Figure 5.7a***) can be determined by Pythagoras' theorem ($\cos\theta = A/H = A/T$) as $A = T\cos\theta$. This force operates around the circumference of the liquid surface where it meets the capillary ($= 2\pi r$). Hence the vertical component of the surface tension force is $2\pi r T\cos\theta$. The vertical component of the surface tension force will draw the liquid up the capillary until it is just balanced by the downward force due to the column of liquid of height h, i.e., $\pi r^2\rho g h$, where g is the acceleration due to gravity. Therefore at equilibrium $2\pi r T\cos\theta = \pi r^2\rho g h$, yielding $h = (2T\cos\theta)/r\rho g$. For a clean water:glass interface, $\theta = 0$ and therefore $h = 2T/r\rho g$.

The second proof utilises the concept that the excess pressure within an air bubble with one air:liquid interface is 2T/r (***Figure 5.4***). The pressure at **A** (***Figure 5.7b***) must be atmospheric pressure, P_{atmos}, but since **A** is within a hemispherical surface, the pressure at **B** must be less than the pressure at **A** by an amount 2T/r, i.e., pressure at **B** = (P_{atmos} - 2T/r). The pressure at **C** is also atmospheric, but is greater than the pressure at **B** by the hydrostatic pressure hρg, i.e., pressure at **C** = (pressure at **B** + hρg). Thus the pressure at B = (P_{atmos} - $h\rho g$). Therefore, at equilibrium, (P_{atmos} - $2T/r$) = (P_{atmos} - $h\rho g$), yielding $h = (2T/r\rho g)$.

Both proofs illustrate that the capillary rise, h is greater in small radius capillaries and for zero contact angles. Indeed the relationships indicate that the contact angle, θ, is the key parameter in determining whether a liquid rises up a capillary or is depressed

- if $\theta < 90°$, cos θ is **positive**, and the porous surfaces 'absorbs' the liquid;
- if $\theta > 90°$, cos θ is **negative**, and the porous surfaces 'repels' the liquid.

Hydrocarbon chains are hydrophobic since they form contact angles, $\theta > 90°$ with water, and chemicals incorporating these structures form the basis of water-repellent chemicals for various building materials (**5.8.1**).

5.5 ABSORPTION (LIQUID WATER)

Absorption is the process by which a porous material extracts a fluid (water) from an environment. When measured, the results are normally represented as the percentage of the dry weight by weight or volume, i.e.

$$\text{Absorption (\%)} = \frac{\text{Weight of water absorbed}}{\text{Weight of material}} \times 100\%$$

A porous material in contact with liquid water **absorb** water by capillarity (**5.4**) into the porous structure. The magnitude of liquid water absorbed will be far in excess of water derived from **adsorption** of water vapour from the air (**5.6**) and therefore moisture induced dimensional movement (**6.4**) will also be greater. In addition to those factors already described (**5.4**), the rate of absorption will depend predominantly on the length of time the material remains in contact with the liquid. For materials submerged in water, a plot of the moisture content of the material against time characteristically shows a rapid absorption rate at early times, reducing to constant weight thereafter (***Figure 5.8***). The moisture content plateau obtained after longer submersion times represents the saturated moisture content.

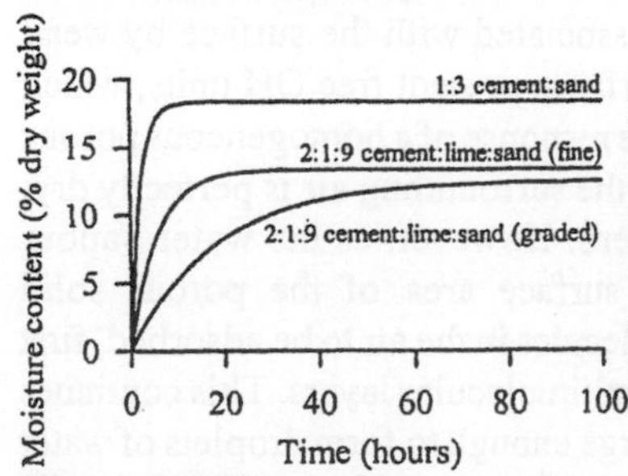

In addition to the expected absorption characteristics, the curves illustrate a definite correlation between the cement content of the mix and water absorption. Contrary to a widely held belief, these curves imply that cement:sand renderings, when mixed in proportions desirable from the point of view of limiting drying shrinkage (6.4), i.e., in the range 1:4½ to 1:6 by volume cement:sand, are far from impermeable. They merely behave as a thick layer of absorbing material on the outside of a building capable of holding a fairly large quantity of water in their porous structure. Water is absorbed during wet conditions and lost during drying conditions. For example, when applied at a thickness of 12 mm, the curves indicate that renders are capable of holding about 2.5 litres/m². On further wetting, no more moisture can be absorbed by the material and will either run down the outside face of the material or be passed through the material to the adjacent materials.

Figure 5.8 Water absorption for cement:sand mixes

The rate of absorption of a fluid (water) into a porous material is often a balance between the rate of absorption and the rate of evaporation (this balance is often of more importance than the total quantity of the fluid absorbed) (**5.10**). Factors affecting the quantity and rate of absorption of water are

- the number of pores actually in contact with the water;
- the size of the interconnecting pores and capillaries, and their radii (**5.4.1**);
- the contact angle of the water with respect to the substrate (**5.4.3**);
- the chemical nature of the surface on the microscopic scale (e.g, surface electron configuration, **1.11.2**);
- the degree of micro-porosity and the distribution of pores.

5.6 ADSORPTION (WATER VAPOUR)

Adsorption is the process by which fluid (water) molecules are concentrated on the surfaces of a porous material by chemical or physical forces, or by a combination of both. In the absence of other materials, the equilibrium between solid, liquid and gas is well defined. At any given temperature, there is a well defined maximum vapour pressure that moisture can establish (the saturation vapour pressure, **2.2.5**). There is only one temperature and saturation vapour pressure at which all three states coexist (known as the triple point of water). At any other temperature, approximate relationships are available to determine the saturation vapour pressure. The unique relationship between the saturation vapour pressure and temperature is the basis for the various psychrometric calculations in building applications. However, within **porous building materials**, the unique relationships between saturation vapour pressure and temperature outlined above does not exist. If a porous building material is homogeneous and isotropic, it may have its own unique relation for the dependence of temperature and maximum vapour pressure. However, such relationships are virtually unknown because most building materials are nonhomogeneous and anisotropic, therefore, **sorption isotherms** are used to indirectly approximate the temperature and moisture vapour pressure behaviour of porous building materials.

The maximum amount of moisture adsorbed by a given amount of a solid depends on the temperature, the partial pressure of water vapour and the surface area available for adsorption. Furthermore, each material has its own characteristic affinity for water, commonly referred to as **hygroscopicity**, which results from the nature of the chemical bonds at the surface; for example, the surfaces of most covalently bonded materials are electron deficient (**1.11.2**), allowing water to be associated with the surface by weak chemical and/or physical bonds, whereas some surfaces present free OH units, which are hydrophilic (e.g, cellulose, **12.4.1**). Consider the response of a homogeneous porous material to water vapour at a fixed temperature. If the surrounding air is perfectly dry, the amount of water adsorbed by the materialis zero. However, as the water vapour pressure is progressively increased, the whole surface area of the porous solid participates in providing a surface for the water molecules in the air to be adsorbed, first in the form of a monomolecular layer and then in multimolecular layers. This continues until the surface layers at various locations grow large enough to form droplets of water (**5.7**). From the absolute dry state to the point where droplets of water begin to form, the material is said to be in its **hygroscopic range**. Note that the hygroscopic range of moisture contents will be much lower than the moisture contents obtained when the material is in contact with liquid water (**5.5**). In the hygroscopic range, the maximum amount of adsorbed moisture is restricted by the hygroscopicity of the material. Once the vapour pressure is above the hygroscopic range, larger amounts of moisture begin to deposit in the pores until the structure is filled with condensed moisture. The maximum amount of moisture that can be accommodated by a porous material is limited only by its porosity (6.2). In principle, this behaviour appears well-defined, but in practice each specimen of a building material will have its own individual response to water vapour, not least because of variations in the proportion, distribution and structure of the pore space.

The relationship between the amount of water adsorbed and the vapour pressure of moisture at a given temperature is termed the **adsorption isotherm.** Obviously, this relationship is temperature dependent; however, for inorganic materials the relationship is virtually independent of temperature and for organic materials the effect of temperature is small. Therefore, if the adsorption isotherm is expressed in terms of the **relative humidity** (ϕ) of the surrounding air, all isotherms for a given material tend to merge into a single relationship. The merged relationship is termed a **sorption isotherm.** Even though no unique sorption isotherm can be obtained for any building material, researchers have determined representative sorption isotherms for many common building materials[8] by conditioning samples at a range of relative humidities at temperatures normally encountered in buildings. The reverse process of adsorption, **desorption**, presents further complexity. If a porous building material is saturated with water and allowed to dry in air at different relative humidities, it does not retrace the adsorption isotherm. Usually, the material retains more moisture during desorption than it can adsorb at any given relative humidity. This phenomenon is referred to as **hysteresis.** These principles are illustrated in ***Figure 5.9a***, which shows the adsorption and desorption isotherms for gypsum board and the hysteresis exhibited. Alternatively, these curves can also be represented by regression equations, with values of the parameters of the equation recorded for different building materials.

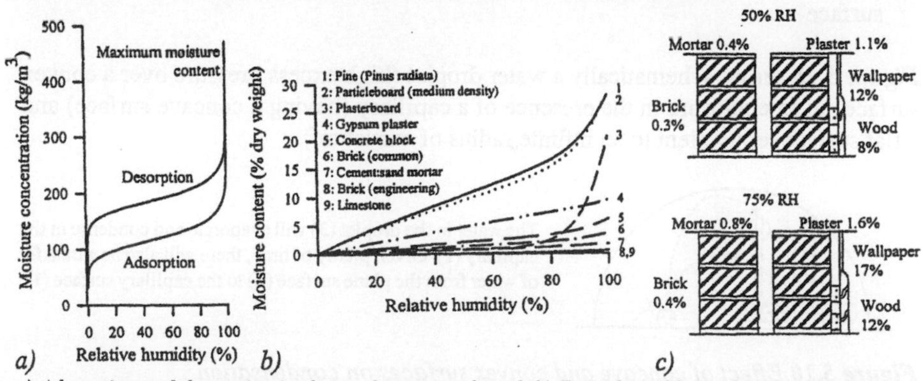

a) Adsorption and desorption isotherms for gypsum board; b) Typical equilibrium hygroscopic moisture contents of building materials at different relative humidities ($MC = a\phi^b + c\phi^d$); c) Equilibrium hygroscopic moisture content of materials in a wall at 50% and 75% RH

***Figure 5.9** Adsorption effects*

Adsorption isotherms may be used to show the equilibrium hygroscopic moisture contents of materials at various relative humidities (RH) (***Figure 5.9b***). Note that the curves relate only to the hygroscopic moisture of the materials. The relationship between hygroscopic moisture content and relative humidity changes if water is present in the liquid form. The curves in ***Figure 5.9b*** are produced by conditioning samples at the range of relative humidities illustrated. Similar curvès can be produced for different materials, notably different timber species. Since building materials of the same class vary widely (in terms of the proportion and structures of pores) curves are merely typical. In addition, they do not take account of the effect of capillarity on liquid water transfer between materials of different pore size and structure and so simplify the situation in real wall assemblies. Nevertheless, if steady state conditions are assumed, ***Figure 5.9b*** can be used to illustrate the equilibrium hygroscopic moisture content of materials comprising a wall in contact with various environmental relative humidities (***Figure 5.9c***).

5.7 CAPILLARY CONDENSATION

Atmospheric pressure, P_{atmos}, is the sum of the partial pressures of all constituent gases, $P_{atmos} = p_{O2} + p_{N2} + p_{H2O}$ **(2.2.5)**. If the atmospheric pressure is to be reduced, then there is only **one** partial pressure which can reduce under normal circumstances and that is the partial pressure of water vapour, P_{H2O}. (Unusual circumstances may result in the removal of oxygen or nitrogen, such as combustion in a fire.). The partial pressure of water vapour can only be reduced if the water condenses as a liquid, and for water vapour to condense, the partial pressure of water vapour must equal the saturated vapour pressure of water at that temperature. Therefore, the saturation vapour pressure of water over the concave surface is very small compared to that over a convex surface, as shown in Proof 2 **(5.4.4)**. In addition, the saturation vapour pressure of water above a concave surface is less than that across a flat surface. This results in two important effects

- liquid water will **condense** more easily on a concave surface than a flat or convex surface;
- liquid water will **evaporate** more easily from a convex surface than a flat or concave surface

Figure 5.10 shows schematically a water droplet (high excess pressure over a convex surface) in an enclosure in the presence of a capillary (having a concave surface) and a flat surface (equivalent to an infinite radius of curvature).

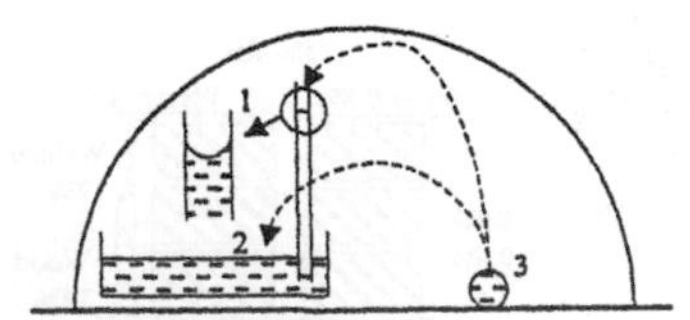

The water in the droplet (**3**) will evaporate and condense in the capillary (**1**). Given plenty of time, there will also be a transfer of water from the plane surface (**2**) to the capillary surface (**1**).

Figure 5.10 *Effect of concave and convex surfaces on condensation*

Once capillaries contain liquid water, it is very difficult to dry them out as their surface profile is always concave. This phenomenon is used in nature where plants have to survive in very dry climates, or where the rainfall is very low. Their leaves are a filled with microscopic pores and so rely upon condensed (liquid) water from the atmosphere for their water intake. As the saturation vapour pressure of water above a concave surface is less than that across a flat surface, then, under equilibrium conditions, atmospheric water will condense into capillaries or crevices and thus **produce a liquid**

environment where there should be dryness for that relative humidity. Table 5.3 gives the relative humidity for condensation for particular capillaries of different radii.

Table 5.3 Capillary radii for condensation at various relative humidities (%RH)

Capillary radius nm (10^{-9} m)	%RH for condensation
36	98
10.4	90
4.7	80
3.0	70
2.1	60
1.5	50

Thus, as the capillary radius or crevice width decreases, so condensation can be obtained within the capillary or crevice with drier air. The converse is also true, so that the capillary or crevice will remain wetter for long periods and, in some locations, narrow capillaries will never dry out. This effect produces the following consequences

- porous materials act as humidity 'sinks' due their water uptake;
- porous materials control the internal relative humidity in domestic dwellings;
- crevice corrosion of metals (**11.6.3**) and the ingress of carbon dioxide into reinforced concrete (**8.21.4**);
- When oxide films (such as rust) are over-painted, the porous nature of the oxide can trap water, even in very dry atmospheres (**11.12.3**).

5.8 MOISTURE CONTROL IN NEW BUILDINGS

The design, construction and operation of an efficient building depends on having sufficient numbers, types and combinations of materials available to provide a protective function against the effects of moisture and moisture movement from each of these sources. Note that it is not necessary for the internal environment to be moisture free; in fact large amounts of moisture are produced in some buildings, whilst, for human comfort, a certain amount of moisture in the environment is necessary. The prime objective is therefore one of moisture control. Normally, moisture control is concerned primarily with the control of moisture from the outside, principally the exclusion of rainwater and ground moisture. However, in adopting exclusion methods, regard must be given to control of moisture from the inside, principally to limit the effects of condensation and interstitial condensation.

The three basic conditions that must exist before water penetration can occur are illustrated in ***Figure 5.11***.

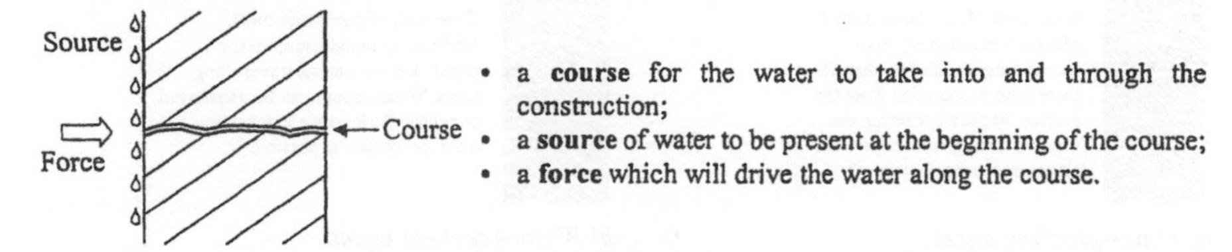

***Figure 5.11** The three preconditions for moisture penetration*

Remove any one of these preconditions and moisture exclusion is achieved. In practice, one of three basic methods are usually adopted to provide exclusion

- the use of **completely impermeable** materials and joints (the 'raincoat' principle);
- the use of porous or permeable materials which allow **controlled penetration** of water into the fabric, which is thick enough to prevent complete penetration (the 'overcoat' principle);
- the use of materials arranged in **multiple layers** in a construction so that water is drained away ('water-shedding' or 'rainscreen' construction) and capillary paths are broken, either by continuous cavities or by discontinuities between components.

To illustrate these principles, common exclusion methods applied to major building elements are shown in ***Figure 5.12***.

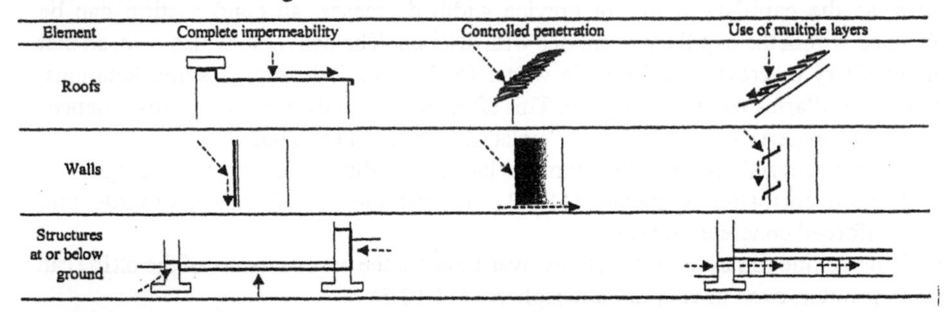

***Figure 5.12** Three methods of moisture exclusion applied to major building elements*

All elements and assemblies within the building fabric function to exclude moisture by one of these three methods, and numerous examples can be used to illustrate the principles involved[9]. For the purposes of this text, we consider the types and mode of operation of surface treatments (water-proofing and water-repellent agents), and illustrate the moisture exclusion principles with reference to pitched roofs.

5.8.1 Surface Treatments

Surface Treatments are grouped into two distinct classes, **water proofing agents** and **water repellent agents** depending upon how the movement of water is controlled. Water-repellent treatments for masonry surfaces are governed by BS 6477[10]. The two types of surface treatment are shown in ***Figure 5.13***. Both types of treatment provide a high water repellency by producing a high contact angle (**5.4.3**) with water. Since cementitious products are alkaline (**7.16**), surface treatments should be alkali-resistant.

External surface

Open ends of capillaries sealed. Moisture as liquid or vapour cannot enter capillaries, therefore there is no evaporation from the surface. Moisture entering the material through cracks or from elsewhere becomes trapped.

a) Water-proofing agents

External surface

Open ends of capillaries lined. Moisture as liquid cannot enter capillaries, but cannot travel along them. Water vapour can be transferred along capillaries, thus evaporation from the surface is possible.

b) Water-repellent agents

***Figure 5.13** Application of thin coatings to make porous materials impermeable*

Water-proofing agents are an ideal way of sealing against moisture ingress in both the **vapour** and **liquid** states by forming a thin impervious film at or near the surface of the substrate. This form of treatment blocks the pores of the substrate thus stopping water from going **either** way (***Figure 5.13a***). Hence if the wall is wet when treated, it cannot dry out by evaporation to the external environment. They are based on organic polymers (e.g, acrylic resins) dissolved in a solvent. The solvent type depends upon the choice of active component and the required viscosity. It is often found that larger quantities of solvents give better coatings as the solvent can evaporate away too quickly, leaving brittle coatings. This is particularly important for treatments carried out in the summer. If an increase in viscosity is required to stop the material running down vertical surfaces, a high molecular weight material is preferable. These high molecular weight materials also give good mechanical properties and better durability. **Water-repellent coatings** are usually based on alkoxysilanes or metallic stearates. **Alkoxysilanes** are monomeric organo-silicon compounds that contain oxygen, hydrogen, carbon and silicon atoms. Most of these compounds are colourless, low viscosity liquids that react with water and cure to form a three-dimensional polymer network. The molecular weight of alkoxysilanes allows deep penetration (up to 80 mm) in porous materials. Subsequent curing is a two stage process of hydrolysis and condensation (**13.5.2**) to form a silica lattice that, in the correct conditions, forms a molecular bond to the substrate. If a large number of alkoxysilanes are joined together, the resulting material is known as a **siloxane**. When several siloxanes are joined together chemically, the resulting material is known as a **silicone**. These larger molecules can only penetrate porous materials superficially and therefore form ideal water-repellent surface coatings. **Siloxane** production and products are shown below.

```
      H                        OH
      |                        |
H — Si — R  +  MOH  →  HO — Si — OM
      |                        |
      Cl                       R
                  Siloxane (MRH2SiO3) ←
```

R is the active water repellent, usually a hydrocarbon chain (alkyl group, CH_3, C_4H_9, C_4H_{17})

M is a metal, usually Na or K

The siloxane material is dissolved in a solvent and applied.

Polymerisation reactions produces **silicone** complexes

```
      R        R        R        R        R        R   ← active water
      |        |        |        |        |        |     repellent
HO — Si —O— Si —O— Si —O— Si —O— Si —O— Si —
      |        |        |        |        |        |
      O        O        O        O        O        O
      |        |        |        |        |        |
HO — Si —O— Si —O— Si —O— Si —O— Si —O— Si —
      |        |        |        |        |        |
      R        R        R        R        R        R   ← active water
                                                         repellent
```

The structure of **stearates** is considered above (**5.4.2**); they consists of $C_{17}H_{35}$-CO.OM, where M is sodium, potassium or aluminium, e.g, polyoxoaluminium stearate.

Polyoxoaluminium stearates, Al.O.(O.stearate), are readily broken down by carbon dioxide from the atmosphere to produce a gelatinous precipitate of aluminium hydroxide and the alkali carbonate. The aluminium hydroxide, $Al(OH)_3$, is amphoteric (**1.8.3**) i.e., it can behave as an alkali, AlO.OH (less one molecule of water) or as an acid, H_3AlO_3. The precipitate can subsequently form an insoluble calcium stearate when it reacts with alkalis, derived from leachates from cement products, e.g, $Ca(OH)_2$.

Water-repellent compounds are also used for injection dpcs. These compounds leave the treated substrate **permeable to water vapour** whilst restricting the passage of water in the liquid phase. This is achieved by lining the pores (without blocking them) with the compound (***Figure 5.13b***), whereupon various curing and solvent evaporation processes occur resulting in polymerisation reactions and weak bonding of the hydrophilic part of the compound to the material substrate. Water repellent agents require good penetration into brickwork (favoured by a small molecule and therefore low molecular weight) **and** a high degree of water repellency, breathability and resistance to alkaline materials (e.g, cement hydration products in the mortar bed). The penetration of silane-based products (based upon the silicon compound SiH_4, whose molecular diameter is 0.5 to 1.4 nm) into clay bricks is about 30 mm, whilst the larger siloxane-based products will only penetrate about 6 mm into brickwork. The size of the water molecule is 0.3 nm, the surface tension of water is 0.07 N/m and of common solvents (turpentine) is 0.027 N/m (**Table 5.2**). This means that the ability of a water repellant material suspended in a turpentine derivative to displace water from the capillaries is limited. Thus water proofing should only be undertaken when the water table is low and when the capillaries within the masonry are dry.

Note that the difference between water-proof and water-repellent coatings is not always evident from trade names and product information. This is important as pore blocking agents should not be used where there is a risk of interstitial condensation (**5.12.4**). Natural stones, which contains water-soluble salts, should not be treated with any water-proofing or water-repellant agent. If water is trapped beneath the surface, crypto-efflorescence (**6.7.1**) will occur, which may make the outer layers spall off.

5.8.2 Pitched Roofs

Roofs are exposed to moisture in the form of downward flowing water and internally produced water vapour. The pitch of the roof is critical in determining which of the three exclusion methods is feasible (***Figure 5.14a***).

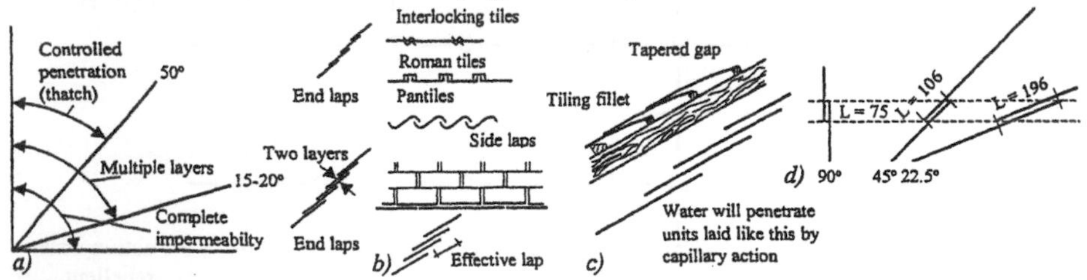

a) Relationship between roof pitch and exclusion method; b) Overlapping roof units; c) Importance of tiling fillets in slated roofs; d) Relationship of pitch, lap and head

Figure 5.14 *Pitched roof techniques and requirements*

Traditional methods of roofing comprise overlapping small units (tiles, slates, shingles, etc.) which exclude rainwater by the use of multiple layers. Effective exclusion depends on using the correct number and size of laps for given pitches and exposure to ensure that water finding its way to the edge of the unit is drained to the outside. **Single lap** tiles include traditional pantiles and Spanish tiles, and more modern interlocking concrete tiles that can be used down to pitches as low as 17.5° (***Figure 5.14a***). Slates and plain tiles require **double lap** (***Figure 5.14b***). In section, the effective lap is the portion where there are three layers, and this lap must be maintained at ridge and eaves. The lap controls the penetration of water driven by the direct effect of gravity and by wind up the roof. Water travelling between the units by capillarity is prevented by tiles having a camber in two directions and by slates being set off from a tiling fillet so that they only touch at the bottom edge (***Figure 5.14c***). Shingles made from Oak or Cedar are **triple lapped** because the irregularity of the units makes the size of the bond harder to control. The lap size in relation to the pitch required to prevent rain being driven under the units is established empirically for the different materials, depending on their exposure, as illustrated in ***Figure 5.14d***. Further protection has traditionally involved bedding the heads of the units on mortar and 'torching' (pointing the back of the tiles or slates). Neither practice is recommended, as movement of the units works the mortar loose, and it may drop out or retain moisture. Currently, a second line of defence in the form of sarking felt is used, which drains any water which gets past the laps to the gutter. The felt should be vapour permeable.

5.9 PENETRATING DAMPNESS (LIQUID WATER)

The risk of penetrating dampness depends on the condition of the external envelope and the degree of exposure to the weather.

5.9.1 Condition of the External Envelope

Factors which affect the performance of the various elements of the external envelope with respect to penetrating dampness are summarised in **Table 5.4**.

Table 5.4 Factors affecting penetrating damp performance of external envelope

Element	Factors
Walls	Adequacy of detailing (vertical dpcs, cavity trays, weatherings, throatings, flashings, etc.)
	Compatibility of bricks and mortar (with respect to porosity)
	Thickness (if solid construction) (minimum provision 215 mm)
	Lack of bridging (if cavity construction)
	State of repair (eroded mortar, pointing, etc)
	Insulation fill to cavities can cause problems in exposed locations
	Defective external finishes (e.g, cracked render, broken/missing slate/tile hangings, etc.)
	Defects at openings (e.g, rotten cills and frames, poor threshold details, etc.); Leaking rainwater goods
Pitched roofs	Adequacy of coverings (e.g, defective/missing slates, etc.) and pointing (at ridge and eaves, etc)
	State of repair of flashings (at parapet walls, chimneys, dormer windows, pipe penetrations, etc.)
	State of repair of copings (cracked copings, defective joints between copings, absence of dpc)
	Defects at chimneys (defective flaunching, missing/defective caps to redundant chimney pots)
	State of repair of sarking felt (missing/defective, etc.)
	Sate of repair and maintenance of gutters (leaking, overflowing, blocked)
	State of repair of valley and parapet gutter linings (splits, perforations, etc.)
Flat roofs	State of repair of waterproof membrane (cracks, splits, blisters, effect of UV light, etc.)
	Adequacy of falls (inadequate falls encourage vegetation, premature failure of membrane, etc.)
	Adequacy of flashings (particularly at abutments and upstands)
	Effectiveness of rainwater outlets (blockages, etc.)
	State of repair of parapets

5.9.2 Degree of Exposure to the Weather

Assessment of the degree of exposure of a building can be undertaken by determining the **Driving Rain Index**. BS 8104[11] can be used to predict the quantity of rain falling on a vertical wall (in litres of water/m^2 wall area, l/m^2). The method is based on the premise that the quantity of rain falling on a vertical surface is proportional to the rate falling on a horizontal surface and to the local wind speed. By summing the hourly product of the wind speed and rainfall during a 33 year period, the Meteorological Office have produced maps of the ('airfield') annual wind-driven rain index. These maps show variations in wind-driven rain over distances greater than 1 km. Variations in wind-driven rain according to the orientation of the wall are included as spell and annual rose values for various geographical areas; together these maps form the basis of the assessment method.

5.9.3 Visual Diagnosis of Penetrating Dampness

An experienced building professional will often be able to identify penetrating damp merely by a visual inspection by identifying features characteristic of penetrating damp

- uniform dampness on **solid** external walls from driving rain (internal walls will remain dry);
- damp patches at the top of ceilings and/or walls (roof and/or gutter defects);
- damp patches on the internal leaf of **cavity** walls (bridging by mortar snots, incorrectly installed wall ties, insulation foam or batts, etc.);
- dampness around window/door openings in cavity walls (absent or defective dpcs);
- penetration associated with windborne rain.

5.10 RISING DAMP (LIQUID WATER)

Rising damp is the ingress of moisture in to the structure from ground water; the **course** of the water is through the complex pore structure of the material components of the wall (brick, mortar. plaster, wood, etc.), the **source** of the water is the ground in contact with the base of the wall, and the **force** is provided by capillarity. Under real, dynamic conditions in a building, rising damp in a wall is a sensitive equilibrium (***Figure 5.15***).

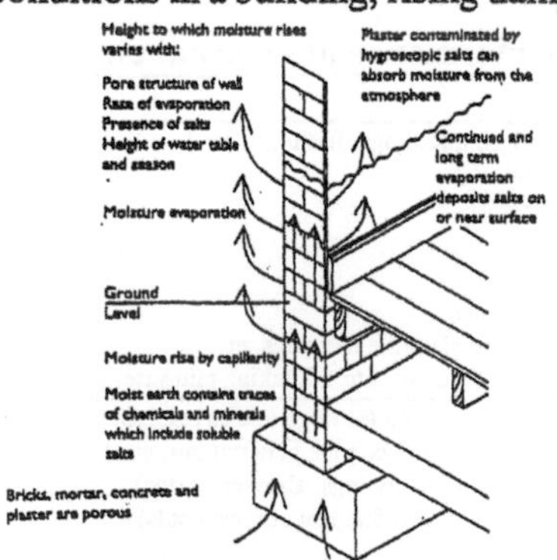

Evaporation occurs from the surface of the wall, and the height that the water may rise will be influenced by the complex pore structure of the materials used. In wet weather, the groundwater table may rise and rates of evaporation will decrease, with consequent increases in the severity and height of the rising damp. Conversely, in dry weather (or with the installation of cental heating), evaporation will increase. The frequent presence of water soluble salts, derived from the soil or from the bricks or mortar, further complicates the situation. Not only may they increase the surface tension of the water (increasing the upward 'pull') but, as evaporation occurs, the soluble salt solution becomes more concentrated at the evaporating surface, and finally salts will crystallise out (efflorescence, 6.7.1). These salts tend to block the pores, reduce evaporation and consequently raise the apparent height of dampness on the wall. In these cases, water movement may be facilitated by osmosis (6.6.4). Often the water soluble salts are **hygroscopic** and will absorb water from the air. The surface therefore appears wet in wet weather, but the dampness seems to disappear when the air becomes drier again. The presence of hygroscopic salts (2.9.1) tends to keep the wall more moist than it would otherwise have been.

***Figure 5.15** Rising damp in a wall*

5.10.1 Visual Diagnosis of Rising Damp

An experienced building professional will often be able to identify rising damp merely by a visual inspection by identifying certain features characteristic of rising damp

- semi-permanent dampness at the bases on internal and external walls, rising to a height of about 1 m;
- a distinct line of dampness at about 1 m height, approximately horizontal and often higher at the edges;
- the presence of salts at the surface, where present, commonly associated with mortar joints and plaster;
- noticeable signs of moisture ingress at skirting level (blistered paintwork, peeling wallpaper, timber decay).

5.11 DAMP DETECTION EQUIPMENT

Moisture meters and Calcium Carbide meters (Speedy™) are commonly used to support identification and diagnosis of dampness problems in buildings

5.11.1 Moisture Meters

Ordinarily, diagnosis of dampness is supported using non-destructive **moisture meter** tests, which measure changes in the electrical properties of moist materials. Such examination may be quickly undertaken if part of a mortgage valuation survey or in support of a loan application. If rising damp is suspected, the usual procedure is for the mortgage lender to recommend a more detailed inspection, commonly undertaken by a damp-proof contractor. The contractor's surveyor commonly use the same instruments as the mortgage lender's surveyor, and therefore reach the same conclusion, confirming the need for their services and products! However, it is clearly important to identify the source of the moisture so that expensive remedial treatments can be fully justified.

Liquid moisture ingress may result in conditions of dampness in building materials which are visible to the eye (for example, blistering paint and wallpaper finishes, visible dampness, efflorescence, etc.) (**5.9.3, 5.10.1**). However, the exchange of moisture between building materials and the atmosphere is generally invisible to the eye. Therefore, to aid diagnosis, moisture meters are often used to identify unacceptable levels of dampness. Moisture meters are normally used by pressing two probes against the surface. The electrical properties of the material are calibrated to give an equivalent moisture reading directly on a scale or series of scales of different materials. The moisture meter operates on the 'Wheatstone Bridge' principle (***Figure 5.16***).

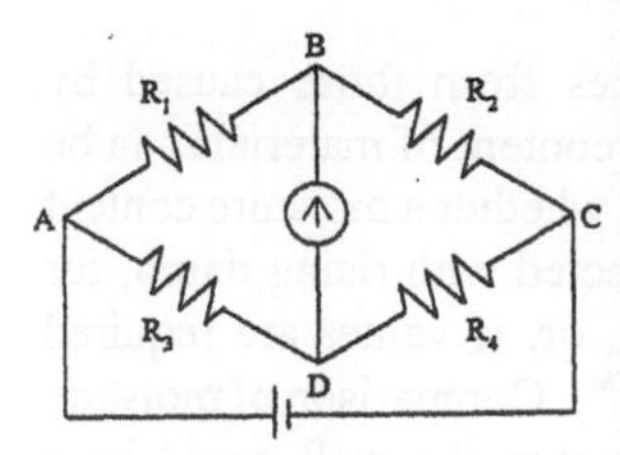

AB, BC, CD and DA are four conductors of resistances R_1, R_2, R_3 and R_4 ohms respectively. A battery is connected between A and C, and a galvanometer (which measures the strength of the current) between B and D. If B and D are at the same potential, no current flows in the galvanometer. In this arrangement, it can be shown that the current in AB must be must be equal to the current in BC (= x amps, say). Similarly, the currents in AD = DC (= y amps, say). If B and D are at the same potential, the potential difference between A and B is equal to the potential difference between A and D, i.e., $xR_1 = yR_3$. In addition, the potential difference between B and C is equal to the potential difference between D and C, i.e., $xR_2 = yR_4$. Thus $R_1/R_2 = R_3/R_4$. In the moisture meter, the value of one resistance and the ratio of two more are predetermined constants (based on dry wood), so that the value of the fourth resistance (produced by the wet building material, measured between the two sharp prongs of the meter) can be determined. The 'out of balance' current is calibrated to give the moisture content.

Figure 5.16 The operation of a moisture meter using the 'Wheatstone Bridge' principle

Moisture meters are manufactured by a number of companies, the most well known is probably Protimeter. Typically, moisture meters (e.g, the Protimeter Mini III) is

- calibrated to give **direct** moisture content readings in **wood only**. A calibration chart is included for determination of the moisture content of different timber species;
- colour coded for use on plaster, brick, etc. on a **relative** basis, where high readings indicate a high degree of moisture and vice versa. A reference scale (0 to 100) is also often included.

Some moisture meters include a 'wood moisture equivalent scale' which records the moisture level in any building material as if it were in close contact, and in moisture equilibrium, with timber and is based on the equilibrium relationships explained earlier (**5.6**). Where the material under consideration is not wood, relative moisture levels are recorded. A valuable clue to the source of an area of dampness can therefore be obtained by outlining its limits and marking out ('pinpointing') relative degrees of dampness within these limits. This is best achieved by taking readings at regular intervals all over the suspect area and recording the readings accurately on a scale diagram of the area. By joining up areas of equal reading, contours of equal dampness are established, which will help to identify the area responsible.

The electrical properties of the moist building material are affected not only by the quantity of water present, but also by

- the presence of water soluble salts (**2.9.1**);
- the presence of metals in the structure (e.g, aluminium backed plasterboard, nails);
- the presence of carbon (sometimes present in breeze blocks).

In addition, pinpointing degrees of dampness with a moisture meter cannot distinguish a liquid source of water (e.g, rising damp) from a vapour source of water (i.e., condensation) (although pinpointing exercises could help to distinguish condensation). To account for the problem of condensation, moisture meters are commonly sold with insulated deep wall probes to ensure that the moisture actually present in the building material is measured. In addition, surface thermometers are available which allow prediction of the likelihood of surface condensation (**5.12.3**). All these factors may result in erroneous moisture meter readings being obtained.

5.11.2 Calcium Carbide Meter (Speedy™)

To distinguish dampness caused by liquid water sources from those caused by condensation, the moisture content and hygroscopic water content of materials can be measured[12]. Moisture content readings are used to establish whether a moisture content gradient exists over the height of a wall (as would be expected with rising damp, for example). These can be obtained by laboratory weighing, or, if values are required immediately, by a portable calcium carbide meter (Speedy™). Comparison of moisture content with hygroscopic moisture content indicates whether the wall could have absorbed from the atmosphere the quantity of water found in the samples. Briefly, the method involves obtaining drilled mortar samples from successive heights up the internal face of a wall. The outer 10 mm or so of the sample is rejected to eliminate the possibility of contamination with hygroscopic salts (which commonly accumulate at material surfaces). Subsequently the sample is stored in a screw topped glass specimen bottle (of unlidded **weight W_o**); the unlidded bottle and sample (as found) are weighed (**Weight W_w**); the unlidded bottle and sample are conditioned in an air tight enclosure at 20°C and 75% relative humidity (characteristic of the upper range of humidities found in residential buildings) for 1 week and reweighed (**Weight W_{75}**); the unlidded bottle and sample are conditioned in an oven at about 110°C for about 1 hour and reweighed (**Weight W_d**). The relative humidity conditions required by the method can be obtained using saturated salt solutions (**2.9.1**), where the salt selected is that which will produce the required relative humidity (NaCl in this case, **Table 2.8**). From these measurements, the moisture content and hygroscopic moisture content of the samples can be obtained

$$\text{Moisture content} = \frac{W_w - W_d}{W_w - W_o} \times 100\%$$

$$\text{Hygroscopic moisture content} = \frac{W_{75} - W_d}{W_{75} - W_o} \times 100\%$$

The calcium carbide (Speedy™) meter provides a quick method to obtain the moisture content of granular material. It consists of a portable air tight vessel into which measured quantities of the sample and calcium carbide are added. The instrument includes two steel ball-bearings to crush the sample and ensure maximum exposure of incorporated moisture to the calcium carbide. The calcium carbide reacts with any moisture present

$$\underset{\text{(calcium carbide)}}{CaC_2} + \underset{\text{(moisture)}}{2H_2O} \rightarrow \underset{\text{(acetylene gas)}}{C_2H_2\uparrow} + \underset{\text{(calcium hydroxide)}}{Ca(OH)_2}$$

The production of acetylene gas produces an increased pressure within the meter which is recorded on a pressure gauge at the bottom of the meter, calibrated directly to the moisture content of the sample. If the samples are obtained at a reasonable number of heights, the results can be plotted as sown in ***Figure 5.17***, which illustrates two typical cases of rising damp (moisture content profiles decreasing with height).

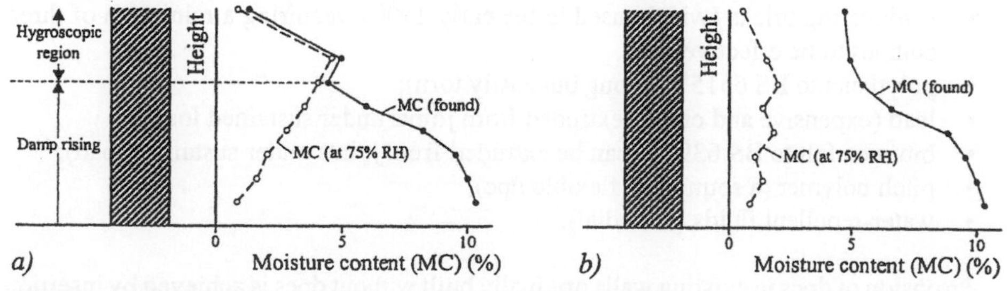

Figure 5.17 *Moisture content and hygroscopic moisture content results from two typical cases of rising damp*

Figure 5.17a illustrates a situation where the moisture content in the lower part of the wall is dominated by rising damp, above which there is a region in which the moisture content is controlled by the hygroscopicity of the wall. ***Figure 5.17b*** illustrates a situation where the conditions at the base of the wall are similar to ***Figure 5.17a***, but where the upper region of the wall is gaining water rom other sources (e.g, driving rain). The moisture content of the upper portion is above that due to any hygroscopic salts present and no significant moisture gradient exists. A wide variety of moisture content gradients are obtainable in this way and will aid diagnosis of the source of any dampness. As a general guide to interpretation[12]

- moisture contents of greater than 5% most probably indicate that some form of remedial action is required (although some building materials possess a hygroscopic moisture content in excess of 5% **without** the introduction of salts from external sources, the proposed 5% threshold does provide a reasonable guide as to whether remedial treatment is necessary);
- if only rising damp is present, there should be a definite moisture gradient from the base of the wall, tapering off to a point where the hygroscopic moisture content becomes equal to the moisture content. Other sources of moisture in the wall (e.g, rain penetration) will alter this gradient so that precise diagnosis of rising damp cannot be made. It is therefore important to remedy other causes and allow time for remedial measures to take effect;
- rising damp is a seasonal effect, increasing in winter with rising water tables and falling in summer. This seasonal effect must be taken into account in any diagnosis since the problem could disappear in the summer months and return in winter.

5.11.3 Exclusion Methods for Rising Damp

Exclusion methods historically relied on the principles of **controlled penetration**. The Public Health Act 1875 made the installation of **impermeable membranes** in the form of damp-proof courses (dpcs), mandatory in new dwellings. Many older buildings therefore do not have dpcs and many dpcs in older buildings may be inadequate or have become defective or bridged over the years. Damp-proof courses comprise a **continuous** barrier across the width of the wall to prevent capillarity (BS 8215[13]) and have been composed of

- slate (widely used from the early 1900s, but prone to crack);
- tar (used for the 1880s to the early 1900s, but prone to crack);
- engineering bricks (widely used in the early 1900s, requiring a minimum of three courses to be effective);
- polythene to BS 6515[14] (strong but easily torn);
- lead (expensive and can be extruded from joints under sustained load);
- bitumen felt to BS 6398[15] (can be extruded from joints under sustained load);
- pitch polymer (a sound and flexible dpc);
- water-repellent fluids (remedial).

Provision of dpcs in existing walls originally built without dpcs is achieved by insertion (by cutting out short lengths of brickwork and inserting impermeable materials) or, more usually, by injection. For injection dpcs, a number of holes are drilled at regular intervals along the base of the wall, into which water-repellent compounds (**5.8.1**) are injected under pressure or gravity. To avoid problems with penetration (**5.8.1**), injection dpc fluids often include water-displacing fluids to counter the existing dampness of the wall, but the effectiveness of injection dpcs in wet walls remains contentious.

5.12 CONDENSATION (WATER VAPOUR)

Psychrometrics is the study of the thermodynamic properties of moist air. Moist air is a mixture of dry air and water vapour and therefore the pressure of moist air is the sum of the pressure of water vapour and that of dry air (**2.2.5**). The temperature of the moist air is an important physical quantity in all psychrometric calculations. A temperature attributed to a sample of moist air implies that the water vapour and the dry air have the same attributed temperature. This temperature then puts a limit on the maximum possible partial pressure for water vapour as the partial pressure of the water vapour cannot exceed the saturation vapour pressure at that temperature. If the partial pressure is equal to the saturation pressure, the sample of moist air is saturated. Hence the amount of water vapour in moist air varies from zero for dry air to a maximum, which depends on the temperature and pressure, at which point the air is saturated with water vapour. At all other partial pressure, the air is unsaturated. Whether the air is saturated or unsaturated, it is always assumed in psychrometric calculations that water vapour follows the ideal gas law (**2.2.1**). Thus the partial pressure and temperature define the state of water vapour in a sample of moist air. If the total pressure of the moist air is also known, the state of the moist air is completely defined.

5.12.1 Heating Moist Air

The air in buildings undergoes heating and cooling for various reasons and therefore psychrometric calculations can be used in a number of application to quantify the changes. When a sample of moist air is heated, one major change occurs, **the relative humidity of the sample decreases**, as illustrated by the following examples. Firstly, consider a sample of moist air at 5°C with a partial water vapour pressure of 600 Pa which is heated to 20°C at constant volume. From **Table 2.1**, the saturation vapour pressure is 872.5 Pa at 5°C and 2338.9 Pa at 20°C. For a given amount of a gas, the ideal gas law (**2.2.1**) relates any two states as $(p_1V_1)/T_1 = (p_2V_2)/T_2$, where subscript 1 and 2 relate to state 1 and 2 respectively. For heating at constant volume, this relationship becomes $p_1/T_1 = p_2/T_2$. In this example, $p_1 = 600$ Pa, $T_1 = 278.15$ K and $T_2 = 293.15$ K, hence $p_2 = 632.4$ Pa. Using the saturation vapour pressures given above, the relative humidity of the moist air initially at 5°C is 69% and that after heating to 20°C is 27%. This is a substantial decrease in relative humidity.

Secondly, consider two samples of 1 m^3 of saturated air, one at 5°C and one at 20°C. Rearranging an appropriate form of the ideal gas law (**2.2.1**) to give $m = pM/RT$, the mass of water vapour in each sample can be found, given that the molar mass of water is 0.018016 kg/mol. Thus saturated air at 5°C contains only 0.0068 kg of water vapour whilst saturated air at 20°C contains 0.0173 kg. This indicates a substantial increase in the capacity of air to accommodate water vapour at higher temperatures.

5.12.2 Cooling Moist Air

Theoretically, cooling moist air reverses the effect of heating moist air i.e., the relative humidity of the air increases and the capacity of the air to accommodate water vapour decreases. However, the relative humidity of the air cannot exceed 100%. Therefore, as moist air is cooled, the air reaches 100% relative humidity at some temperature and, if cooling is continued below this temperature, the capacity of the air to accommodate water vapour will be less than whit is available. The moist air must then discard the excess moisture and this is where **condensation** occurs. The temperature at which condensation occurs is the **dew point temperature**. If the dew point temperature is above 0°C, the condensed water appears as liquid water whereas if the temperature is below 0°C, frost results. The dew point temperature is 100°C when the partial pressure of water vapour is equal to atmospheric pressure (i.e., when water boils). Two types of condensation can be considered

- surface condensation, when water vapour condenses **upon the surface** of a material;
- interstitial condensation, when water vapour condenses **within** a structural element.

Condensation may lead to mould growth, deterioration of materials and finishes, reduction in the effectiveness of insulation materials and occasionally structural damage.

5.12.3 Surface Condensation

As shown above, water vapour condenses on a surface whenever the temperature of the surface is at or below the dew point temperature. The velocity of the water molecules in the gas (vapour) state are reduced at the dew point temperature and no longer have sufficient energy to remain as a vapour. They therefore condense as a liquid. Surface condensation risk may be predicted by comparison of the surface temperature and the dew point temperature; typical results are shown in ***Figure 5.18***.

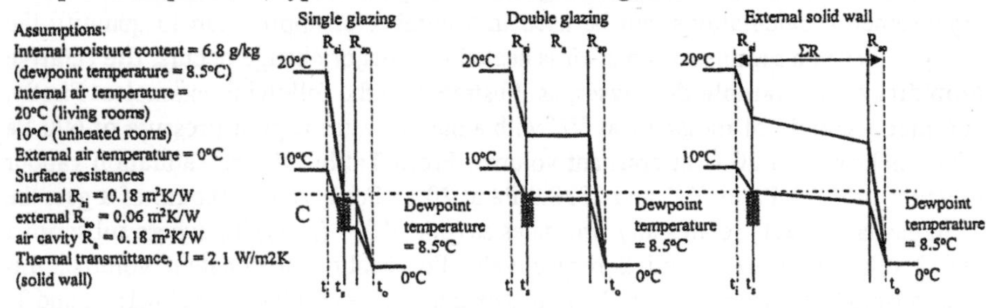

***Figure 5.18** Surface condensation (surface condensation risk zone shaded)*

Surface condensation will occur on the coldest available surface, commonly mains water intake pipes and the water reservoir of the WC toilet system. Mains water pipes usually rise from ground level and, where they are not insulated, the surface condensation draining down the pipe can give the appearance that the pipe is leaking. In areas in which there is high water vapour production (kitchens, bathrooms, etc.), surface condensation is commonly seen on windows. Double glazing reduces considerably the formation of surface condensation on the windows themselves, but the water will still condense on the next coldest surface below the dew point temperature. Commonly, this is the perimeter glazing seals around the double glazed unit where it is bonded to the outer pane of glass (i.e., a **thermal bridge**). The surface condensation may lead to mould growth. If there is a lot of condensation and mould growth, the internal surface of the double glazed units may get very dirty (hazy or milky) (which will be more noticeable in bright sunshine). Note that this is not double glazed unit breaking down, merely the presence of mould. Mould may be removed effectively using household bleach (1 part bleach to 4 parts water, washed off after treatment), but this provides no residual protection. Proprietary fungicidal treatments with some residual effects are also available, e.g, Nuodex 87 (1 to 5% solution in water).

Water vapour requires a nucleus upon which to condense (2.6.3). Dirt particles and dust provide a nucleus for condensation (this can be seen in the formation of snow, where minute dust particles provide the nucleus for the formation of ice crystals, which are left behind when the snow melts to produce a dirty 'slush'). Hence cleanliness (removal of dust and dirt particles) will help limit surface condensation.

5.12.4 Interstitial Condensation

Generally (and particularly in winter), interior moist air temperature and vapour pressure is higher than that outside, causing transfer of heat and moisture outward. Transfer occurs both through cracks in the external building envelope and through the fabric; in the latter case the total thermal and vapour resistance of the elements of the structure determine the rates of flow. Prediction of interstitial condensation can be achieved by

- by comparison of the structural and dew point temperature profiles (as for surface condensation prediction); or
- by comparison of the vapour pressure and the saturation vapour pressure at the predicted structural temperatures[16].

The former method is more often adopted due to its simplicity although the latter method allows both the amount and rate of condensation at each condensation plane to be determined. Both methods are commonly represented graphically, enabling the areas of potential risk to be readily identified. Typical results are illustrated in ***Figure 5.19***.

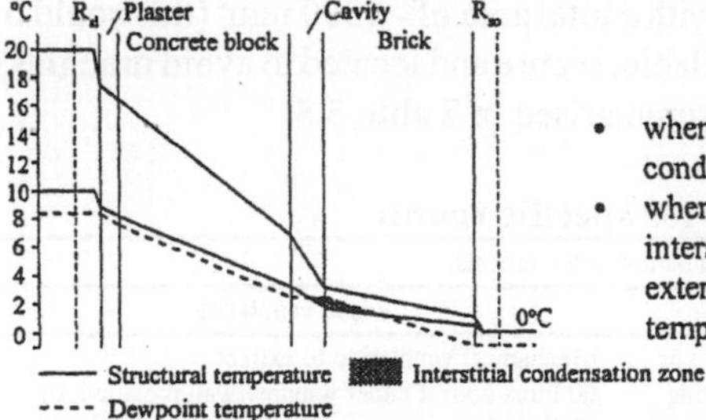

- when the internal temperature is maintained at 20°C, condensation does not occur;
- when the internal environment is maintained at 10°C, interstitial condensation occurs at the inner face of the external (brick) leaf of the cavity wall as the structural temperature has fallen below the dew point temperature.

Figure 5.19 Interstitial condensation predictions for a simple cavity wall

5.12.5 Limiting Condensation in Buildings

A number of inter-related factors contribute to condensation risk, principally moisture production, ventilation, heating, insulation and thermal response. In addition, the absorbency of the surfaces, the vapour and thermal resistance of the materials, constructional details and workmanship are also important. Local weather conditions govern the amount of water vapour in the external air at any time.

Moisture production by internal occupant activities (**5.2.2**). Without the production of internal moisture production, there would be few condensation problems in buildings. Moisture vapour production gives rise to high internal relative humidities and may result in the internal environment feeling cool. This arises because materials absorb water and latent heat is required to vaporise this water again. The latent heat must come from the materials and so the bodies of the occupants are always radiating heat to cold, heat absorbing material surfaces.

Ventilation. Ventilation acts to replace warm, moist internal air with external air of low moisture content. In the past, ventilation of buildings happened fortuitously by the combined use of suspended wooden floors, badly fitting doors and windows and the large volume of air drawn up chimneys. With a coal fire, to remove 1 m^3 of burnt gases, 3 to 4 m^3 of the room air had to go up the chimney to prevent back draughts (which would result in smoke reentering the house). With a typical room size of 40 m^3, this would mean that the room air was changed about 10 times per hour. With central heating replacing coal, and better fitting doors and windows, current natural ventilation rates are below 2 air changes per hour. Background ventilation ought to be 0.5 to 1.5 air changes per hour. In today's house, the air changes averaged over a 24 hour day are minimal, with the greatest air change often occurring when the front door is opened to arrive or depart from the home. To combat the trend of reduced ventilation rates, the government has regulated ventilation requirements, notably in the Building Regulations (Part F1 Means of Ventilation). These regulations specify minimum ventilation rates to ensure human health and help reduce condensation risk. The spirit of these regulations is to ensure some **background** ventilation **at all times** and **rapid** ventilation when odours and moisture are produced (e.g, kitchens and bathrooms). Generally, for **rapid ventilation** opening windows with a total area of 1/20th of the room's floor area are suggested. This facility allows the occupant to ventilate the room by allowing large air changes for short periods of time and to cool the room in hot weather. For **background ventilation**, openings should be provided with a total area of >4000 mm^2 (this could be a trickle ventilator, which should be controllable, secure and located to avoid draughts). Recommendations for specific rooms are summarised in **Table 5.5**.

Table 5.5 Ventilation recommendations for specific rooms

Room	Ventilation requirements	
	Background ventilation	Rapid ventilation
Kitchens	• openings with a total area of >4000 mm^2; or • mechanical ventilation capable of operating continuously at 1 air change per hour	Mechanical ventilation to extract 60 litres/hour if either window/wall mounted; or 30 litres/hour if incorporated into a cooker hood
Bathrooms	Mechanical ventilation should be provided and be capable of operating intermittently and extracting at least 15 litres/hour	
Sanitary accommodation	Mechanical ventilation should be provided capable of extracting 3 air changes per hour with 15 minute over run or natural ventilation by opening windows with a total area of 1/20th of the room floor area.	

It is important that ventilation openings are not inadvertently obstructed by the building occupants. Dead end areas, such as passages beneath the stairs, built in clothing cupboards, etc. usually resist the flow of heat more readily than the flow of water vapour. These spaces therefore usually act in a manner analogous to an insulating material (high thermal resistance and a low vapour resistance). These types of areas can therefore reduce the temperature of an external wall while having little effect on the vapour pressure, and the risk of condensation is therefore increased. In addition, these areas often have limited ventilation, which exacerbates the problem.

Heating. Heating helps to combat condensation by warming internal surfaces. In addition, heating internal air increases the capacity of the air to hold more water vapour, so preventing the air from becoming saturated. Effective limitation of condensation risk requires good heat distribution and that heating appliance have flues so that the products of combustion (waste gases and water vapour) can be vented outdoors.

Structural insulation. Insulation of the structure effects both the thermal response of the structure and may increase the risk of interstitial condensation. In heavy weight buildings which are intermittently heated, surface condensation can occur before the mass of the structure has had time to warm up. If insulation substantially reduces the temperature within the construction and traps water vapour within the fabric, interstitial condensation may result (***Figure 5.20a***). Note that the high thermal resistance of the insulant accounts for the steep structural temperature gradients across the filled cavity. Although the vapour resistance of the insulant is quite high compared to the overall vapour resistance of the wall, it is not sufficient to reduce the dew point temperature below the structural temperature. The results indicate the condensation is now predicted within the insulant and at the internal face of the external leaf of the cavity wall when the internal temperature is 20°C and 14°C, so the incorporation of wall insulant has **increased** the likliehood of interstitial condensation. In addition, the effect of condensation may reduce the thermal resistance of the insulant material.

Absorbent surfaces. Condensation can take place on absorbent surfaces by capillary condensation (**5.7**). This may usually pass unnoticed by the occupants of the house. A reservoir of water is formed at high internal humidities and may by given up to the atmosphere when the humidity is low. These fluctuations occur between heating cycles and are influenced by daily and seasonal variations in temperature and humidity. House hold fabrics start to disintegrate at relative humidities above 70% (often without the occupants being aware that the fabric is wet).

Vapour resistance of materials. Interstitial condensation risk may be eliminated if moisture laden air inside is prevented from diffusing through the fabric. Vapour barriers (through which water vapour cannot pass) may therefore be incorporated into the structural elements of the building, as illustrated in ***Figure 5.20b*** (plaster painted with a gloss paint and plaster replaced with aluminium foil backed plasterboard).

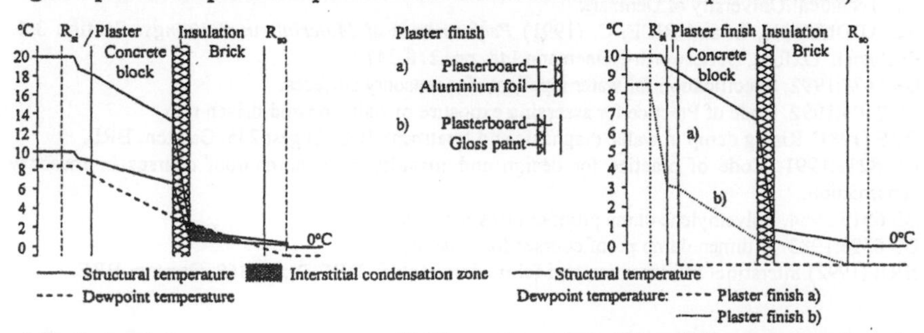

a) Cavity insulation *b) Gloss paint and aluminium foil backed plasterboard*

***Figure 5.20** Interstitial condensation prediction. Cavity insulation and vapour barriers.*

Note that the thermal resistances of the gloss paint and the aluminium foil are negligible so that the structural temperatures remain largely unchanged. The results indicate that, whilst the gloss paint does not provide the very high vapour resistance of the aluminium foil, it affords some resistance to moisture vapour transfer from the interior. Both vapour barrier techniques reduce the risk of interstitial condensation within the wall so that the full insulation value of the insulant material is retained; aluminium foil backed plasterboard performs better than gloss paint.

Generally, to reduce condensation risk

- the vapour resistance of materials in a construction should **decrease** through the construction from inside to outside (i.e., high vapour resistance inside, low vapour resistance outside);
- the thermal resistance of materials in a construction should **increase** through the construction from inside to outside (i.e., low thermal resistance inside, high thermal resistance outside).

5.12.6 Summary

The factors which can be considered in reducing condensation risks are summarised in **Table 5.6** (over).

5.13 REFERENCES

1. BRE (1985) Surface condensation and mould growth in traditionally-built dwelling. BRE Digest 297. Garston, BRE.
2. ROUSSEAU, M.Z. (1984) Sources of moisture and its migration through the building enclosure. *ASTM Standardisation News, November 1984*. pp. 35-37.
3. QUIROUTTE, R.L. (1984) Moisture sources in houses. *Condensation and ventilation in houses,* Document No. NRCC 23293, Ottawa, National Research Council of Canada.
4. ERHORN, H. and GERTIS, K. (1986) Minimal thermal insulation and minimal ventilation. *Gesundheits Ingenieur, Vol. 107*. pp. 12-14.
5. HARRIMAN, L.G. (1990) *The Dehumidification Handbook, 2nd ed.* Massachusetts, Munters Cargocaire.
6. ASHRAE (1989) *ASHRAE Handbook of Fundamentals 1989.* Atlanta, USA, ASHRAE.
7. STUM, K.R. (1992) Winter steady-state relative humidity and moisture load prediction in dwelling. *ASHRAE Transactions, Vol. 98, Part 1.* pp. 262-272.
8. HANSEN, K. K. (1986) *Sorption Isotherms. A catalogue.* Technical Report 162/86, Lyngby, Denmark, The Technical University of Denmark.
9. see ADDLESON, L. and RICE, C. (1991) *Performance of Materials in Buildings.* Section 3.4. Exclusion. Oxford, Butterworth-Heinemann Ltd. pp. 278-347.
10. BS 6477:1992. Specification for water repellents for masonry surfaces.
11. BS 8104:1992. Code of Practice for assessing exposure of walls to wind-driven rain.
12. BRE (1981) Rising damp in walls: diagnosis and treatment. BRE Digest 245. Garston, BRE.
13. BS 8215:1991. Code of practice for design and installation of dampproof courses in masonry construction.
14. BS 6515:1984. Polyethylene damp proof courses for masonry.
15. BS 6398:1983. Bitumen damp proof courses for masonry.
16. BRE (1992) Interstitial condensation and fabric degradation. BRE Digest 369. Garston, BRE.

Table 5.6 Summary of factors to be considered in reducing condensation risk

Factor		Means of reducing condensation risk
Moisture input	From occupants and activities	Moisture input from occupants cannot be controlled. Moisture input from occupant activities could be controlled by • change process to generate less moisture (e.g. electric cooking instead of gas); • reduce wetted area surrounding wet activities; • select apparatus with built in exhaust • extract water laden air near point of moisture input.
Internal temperature	Of air and environment	Although relative humidity varies with air temperature, it is absolute humidity (i.e. moisture content of the air) which governs condensation, and this is not affected by temperature changes. Internal temperatures, however, govern the structural temperatures in walls, floors and roofs. Increases in air temperatures are not normally possible since temperatures are governed by thermal comfort considerations. If heating is intermittent, however, more continuous heating will raise structural temperatures and thereby reduce condensation risk.
Ventilation	Of rooms	Increased ventilation results in reduced levels of moisture in the air. This is a very effective way of reducing condensation risk.
	Of cavities in walls and roofs	Unventilated cavities present a barrier to heat flow but not to vapour flow; they therefore increase condensation risk in the outer leaf. The effects of ventilation on the incidence of interstitial condensation depend on the relative positions of the condensation plane and the ventilated cavity, i.e. • if the cavity is between the internal (warm) side and the condensation plane, ventilation to the outside air will reduce the condensation risk and reduce the rate of condensation if it does occur, but may reduce the thermal performance of the structure; • if the cavity is between the condensation plane and the external (cold) side, ventilation will increase the risk of condensation as the condensation plane will be cooled with little reduction in vapour pressure. However, if some water vapour is transferred through this plane, ventilation may be required to reduce subsequent water deposition. • if condensation occurs on the cold face of a cavity, the effect of ventilation will depend markedly on the outside conditions. Sufficient ventilation under favourable conditions will eliminate condensation but levels needed may markedly reduce the thermal performance of the structure.
Structural temperature	Of walls, floors and roofs	Increases in structural temperature reduce condensation risk. This can be achieved by • increasing the internal air temperature; • making heat more continuous; or • providing insulation towards the outer face of the wall (note that insulation on the inner face, if not inherently vapour resisting or associated with a vapour barrier, increases condensation risk). Wall surface temperatures in external corners can be as much as 3°C less than those in the centre of the wall and cold bridges in construction may also create areas where structural temperatures are very much lower than on the wall generally. Cupboards, fitted furniture, dead-end corridors, etc. can also reduce wall temperatures while not affecting vapour pressure.
Vapour pressure	On walls, floors and roofs	Reducing vapour pressure will reduce condensation. Increased ventilation achieves this. Vapour barriers towards the inside surface can limit the diffusion of water vapour. Note that problems arising from inadequate jointing or puncturing of sheet type vapour barriers are frequent. Flat roofs, which have an effective vapour barrier on the cold side, present a critical condensation problem since the internal vapour pressure will be maintained all the way through the construction and condensation on the cold underside is very common. the most effective remedies are • provide thermal insulation outside the barrier; or • provide ventilation of the roof cavity.
Pipe, etc. temperatures		In winter, water from the mains supply will often enter buildings at temperatures of about 5°C. Condensation on cold pipes is likely in many places, particularly kitchens and bathrooms. Cold pipes may therefore be at temperatures below dew point temperature for considerable periods and will inevitably lead to condensation. Vapour barriers on the outside of insulation will limit the rate at which condensation will take place but the effect cannot be totally avoided.

6

MOISTURE EFFECTS IN MATERIALS

6.1 INTRODUCTION

Water is directly or indirectly responsible for a range of physical and chemical degradation processes (**Table 3.3**), including

- **physical changes**, including weathering (**6.3**), moisture induced dimensional changes (**6.4**), bulking (**6.5**), freezing and thawing of absorbed water (**6.6**) and action of water soluble salts (**6.7**); and/or
- **chemical changes**, including sulfate attack (**8.19.2**), corrosion of metals and alloys (**Chapter 11**), moisture expansion of fired clay products (**6.4**), alkali-silica and alkali-aggregate reactions (**8.20**) and the hydration of oxides (**7.3.4**) and unstable slag aggregates (**8.5**).

In this text we deal with material-specific degradation processes in the relevant Chapters (noted above). In this Chapter we examine those water-induced degradation factors that may affect any building material.

In general terms, the type and extent of water-induced degradation depends on the **porosity**. Porosity is a measure of all the pores present in a material. Some materials are fundamentally **non-porous** and have mechanical properties which are independent of their environment. Other materials, in contrast, are fundamentally **porous** and have mechanical properties which are environmentally dependent (properties which, for example, are quite different when the material is immersed in a fluid compared to when surrounded by air). Generally, non-porous materials are predominantly affected by chemical processes, whereas porous materials are affected by both chemical and physical processes. Characteristically **non-porous materials** are formed from the liquid or molten state. They are usually characterised by solids which have a **non-directional bond** (e.g, metals, ionic structures, etc. **1.9**). Where the structure is large and bonded by covalent bonds, a non-porous structure may be formed by slow cooling under pressure (e.g, rocks and high density plastics). Examples of non-porous materials include metals, some igneous rocks, metamorphic rocks (such as marble and quartzite), glass and polymers (such as epoxy resins, polyesters, perspex and PVC). **Porous materials**, on the other hand, are mainly formed from raw materials which contained water (e.g, clays used in fired clay products, wood) or from processes in which water is added (e.g, concrete, aerated concrete blocks, etc.). The water present or formed by chemical reactions is evolved as a liquid or a vapour during the production process to produce porosity in the finished material. Examples include **concrete**, which exhibits limited porosity when made with a w/c ratio below about 0.28 to 0.35 because below this w/c ratio the cement remains unhydrated. Concrete porosity is affected by

- the w/c ratio. It is essential to have a low w/c ratio for severe exposure. Most lightweight concrete must be protected by other means;
- the degree of compaction;
- the use of poorly graded aggregates;
- the use of porous aggregates;
- use of air entraining agents.

In some cases, other gases formed produce porosity. A good example is the evolution of hydrogen gas in the manufacture of **autoclaved aerated concrete blocks (AAC)** by the reaction of aluminium with the alkali produced from the hydrating cement **(1.8.3)**

$$\underset{\text{(cement)}}{2(3CaO.SiO_2)} + \underset{\text{(mix water)}}{6H_2O} \rightarrow \underset{\text{(alkali)}}{3Ca(OH)_2} + \underset{\text{(hydrated cement)}}{3CaO.2SiO_2.3H_2O} \quad \text{(+ Heat)}$$

and the reaction of alkali lime (calcium hydroxide) from cementitious products on aluminium metal (e.g, windows) **(1.8.3)**

$$\underset{\text{(lime)}}{Ca(OH)_2} + 2Al + 2H_2O \rightarrow CaO.Al_2O_3 + 3H_2 \quad \text{(+ Heat)}$$

The porosity of a material affects both strength and durability, although the strength parameters (compression, tension, bending and shear) are not always related. The strength parameters will usually be higher when the porous material is tested wet compared to dry; is tested at a faster rate of loading compared to a slower rate of loading; specimen is small; is unrestrained compared to when restrained. These parameters may be in part explained using **Griffith's crack theory (9.6)**.

6.2 STRUCTURE OF POROUS MATERIALS

The pores within a porous material may be open or closed. In an **open pored structure**, pores are generally interconnected with each other to form channels or capillaries (e.g, by pits in timber, **12.3**, or by gel and capillary pores in cementitious materials, **7.17**), thereby making the material **permeable** (the permeability of a porous body is the property that permits liquids and/or gases to flow through the pores under a pressure gradient). In a **close pored structure**, pores may be enclosed within individual particles or may form isolated spaces within the matrix, thereby making the material **impermeable** to liquids and/or gases. For example, in aerated concrete blocks, the pore structure is of uniform size without any connection. The close pored structure arises from the evolution of hydrogen surrounded by calcium silicate hydrates in the cementitious matrix **(6.1)**. Note that the pore structure of hydrating cementitious materials is different to that of aerated concrete blocks, as in the former the unreacted water evaporates to leave continuous capillary pores **(7.17)** through which rainwater can permeate. Typically, porous material density increases as the volume fraction of voids decrease and mechanical strength increases as the volume fraction of voids decreases.

Two types of porosity can be distinguished. **Apparent** (or **effective**) **porosity,** expressed as the percentage of the volume of the open pores with respect to the exterior volume of the material under consideration, i.e.,

$$\text{Apparent porosity (\%)} = \frac{W - D}{W - S} \times 100\%$$

where D is the dry weight of the solid, S is the weight of the suspended solid in water (after having been soaked in water so that all open pores in the solid are completely filled with water) and W is the weight of the soaked solid determined by weighing the soaked solid from which all excess surface water has been removed. Hence (W - D) determines the weight of the water that filled all the open pores within the solid and therefore represents the volume of the pores within the solid. (W - S) determines the weight of the water in the volume of the solid and therefore represents the whole volume of the body, including the pores.

True porosity represents the total volume of both the open and closed pores in the volume of the solid, i.e.,

$$\text{True porosity (\%)} = \frac{\rho_s - \rho_b}{\rho_s} \times 100\%$$

where ρ_s is the true density (specific gravity) of the solid (which refers only to the solid matter in the solid and does not include any pores) and ρ_b is the **bulk density** of the solid (i.e., the weight per unit volume of the solid, which includes any pore space present). Bulk density can be determined from

$$\text{Bulk density, } \rho_b = \frac{D}{W - S} = \frac{D}{V}$$

where V is the volume of the solid including the pores (= W - S), and D and S are defined above. The true density (specific gravity) is an inherent property of a material whereas the bulk density is affected by the way the material has been produced.

The **permeability** of a porous body depends not only on the pore size, structure and distribution but also on the properties of the fluid (for water, for example, the surface tension, **5.4.2** and contact angle, **5.4.3**). Permeability is a key parameter in determining the quantity of a fluid present in the porous structure of the solid; fluid retention arises from both **absorption (5.5)** (where the fluid exists in the same state within the porous structure of the material) and **adsorption (5.6)** (where the fluid combines with the porous solid substrate chemically and/or physically such that its form and structure is different from that prior to adsorption, i.e., the fluid exists in the **adsorbed** state). One important consequence of the change in the moisture content of a porous material is **dimensional movement** of the solid (**6.4**). Dimensional instability arises as there is a contraction when the water content is reduced and an expansion when the water content increases. Particulate materials also exhibit large volume expansions depending on the quantity of moisture associated with the particles; a good example of this is the **bulking** of clays and aggregates (**6.5**).

6.3 WEATHERING

In geological terms, weathering processes are continually operating on the solid environment, albeit slowly, and give rise to the gradual degradation of materials in the environment. Although building materials are also weathered, significant degradation will only be caused in exposed conditions; correct specification and detailing should limit effects to a permissible level. In particular, weathering causing the degradation of stone should be carefully considered (**4.3.1**). Weathering processes include

- **erosion** i.e., the degradation of materials by running water, ice, wind, waves, etc. and the subsequent removal of the resulting debris by the same degradation agent;
- **abrasion** i.e., the combined work of the transporting agent and the applied load which acts as the abrasive in the wearing process;
- **attrition** i.e., the wearing away of the abrasive agent or load itself.

Commonly, these processes are influenced by the chemical nature of the water, which may not be pure and is often acidic (**4.1**). Weathering processes are thus affected by

- highly polluted atmospheres (giving rise to weak carbonic acid or sulphuric acid solutions, **4.3.1**);
- acid fumes or sewage discharged in the locality;
- in ground waters, contamination by industrial wastes;
- contact with water supplies which are very pure (e.g, mountain water in sandstones or igneous rock areas) or are acid in character (e.g, water from peat bogs containing organic acids from decaying vegetation);
- conditions of exposure to spillage or splashes of acid or certain other liquids (e.g, vegetable oils, fruit acids, milk products, beer and acid preparations for metals, etc.).

Where these arise, precautions should be taken at the design stage by using special quality materials and/or methods of protection e.g, asphalt coatings, ceramic tiles etc.

6.4 MOISTURE INDUCED DIMENSIONAL CHANGES

Whereas all building materials are affected by temperature change (**2.8.2**), there are several building materials in common use that are dimensionally unaffected (or insignificantly affected) by moisture content changes, including gypsum plasters and plasterboards; metals and alloys; glass; plastics (although some may shrink irreversibly due to loss of volatile constituents, **13.12**). Unlike temperature induced size changes, size changes in response to changing moisture conditions may either be **reversible** (where the material expands when wetter and contracts when drier) or **irreversible**. With the exception of fired clay products (**7.3**), **all** irreversible size changes due to moisture movement are **negative** (i.e., building materials experience irreversible **shrinkage**). Typically, products and materials which require water during their manufacture are associated with irreversible shrinkage due to moisture movement arising from evaporation of the water during drying. All materials that experience irreversible size change also experience reversible size change. The reversible size change is additional to the irreversible size change.

Moisture induced changes in size may result in stresses in the affected material that may cause it to crack and/or distort. The stress induced in a material is the product of the modulus of elasticity, E, of a material and the strain (**2.8.1**), which, when moisture induced, is the percentage moisture induced change in size (**Table 6.1**). Hence the moisture induced stress is

$$\text{Stress} = \frac{\text{E x percentage moisture movement}}{100}$$

If the stresses calculated exceed, for example, the ultimate tensile strength of the material, then tensile cracking will occur. In addition, moisture induced changes in size may result in stresses in adjacent materials, particularly where these restrain movement of the affected material. The force (= stress x area) induced in a restraining material for moisture induced stresses is

$$\text{Force} = \frac{\text{E x percentage moisture movement x A}}{100}$$

where A is the area over which the force acts (on the restraining material). As the linear moisture induced dimensional movement of different materials varies (**Table 6.1**), **differential movement** of materials in juxtaposition may exacerbate problems, particularly when temperature induced movement (**2.8.2**) is also be considered.

Irreversible size changes are largely independent of changes in ambient relative humidity within normal service ranges and are due to chemical and/or physical adsorption (**5.6**) rather than absorption (**5.5**). In materials bonded by a gel like matrix (**2.9.2**), such as cementitious products, the adsorption of water molecules onto the surface increases the volume of the product without accompanying chemical reaction in the matrix itself. On reacting with water, the gel like matrix acts as a semipermeable membrane (**7.17**). Fired clay materials (**7.3**) may also act in this way to produce anomalous dimensional changes in response to movement (this may be complicated by the possible presence of free alkalis, such as calcium oxides, CaO, or magnesium oxide, MgO, which may not have been removed if under-fired, **6.7.3**). Typical irreversible moisture movement in clay bricks is shown diagrammatically in ***Figure 6.1a***. Fired clay products should not normally be used in construction until about 127 days following manufacture, to allow the product to stabilise. Note that irreversible moisture movement in fired clay products **cannot** be accelerated by immersion in water.

Colloidal (**2.9.2**) calcium silicate is formed when steamed at low temperatures in the manufacture of **calcium silicate bricks** (**7.5**). The crystalline network ordinarily formed by silicates (**1.13**), which is stable with water, will not form readily and therefore calcium silicate bricks are unstable in moist environments. Naturally formed **sandstones** are held together with cementitious gels between the sand grains and will undergo changes in size due to the reaction of water with uncombined alumina (Al_2O_3) and aluminium silicates. Although irreversible moisture induced dimensional changes may continue for very long periods, the rate of change diminishes with time and for practical purposes can be assumed negligible about 6 months after manufacture (***Figure 6.1a***). As with thermal responses, moisture induced size changes in materials and can be modified by restraints, which influence the extent and distribution of the size changes and may possibly produce distortion.

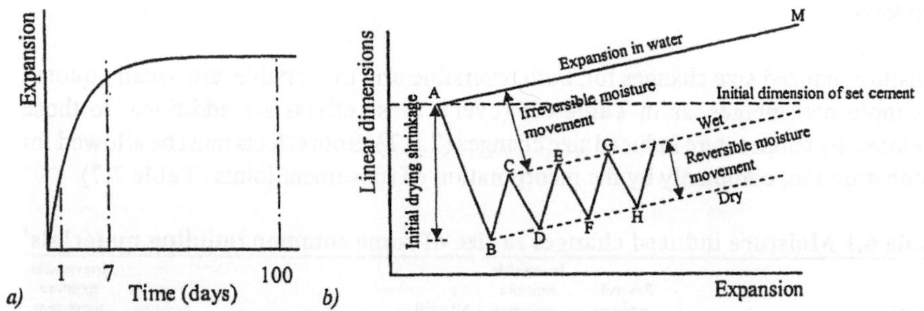

a) *b)*

a) Irreversible expansion of fired clay bricks; b) Moisture movement in concrete

Figure 6.1 *Moisture induced dimensional movement*

Cementitious products show reversible moisture induced size changes, which are superimposed on initial (irreversible) drying shrinkage (***Figure 6.1b***). Most exhibit comparatively large size changes due to moisture content reduction on drying. Movement is shrinkage, of which most is irreversible (this is often restrained to some extent by other external influences, e.g, reinforcement). Cementitious products increase in both dimensions and weight when stored in water, as shown in ***Figure 6.1b*** by the line **AM**, but decrease in weight and length if stored in air at 50% humidity, as shown by the line **AB**. If the dried cementitious product is then immersed in water, it will expand, as shown by the line **BC**. Subsequent drying and wetting will produce the lines **CD, DE** and **EF**. The range of these cycles represents reversible moisture induced size changes and the loci of the points **C** and **E** from the initial state **A** are irreversible moisture induced size changes (shrinkage). Shrinkage occurs because the water used in mixing forms a gel with the cement; a w/c ratio of only about 0.28 is required to fully hydrate the cement (**7.17**). The excess water modifies the gel such that the greater the water cement ratio, the greater the shrinkage. As the quantity of the mix water increases, so does the moisture induced size change, as the excess water over that required for full hydration will evaporate, leaving capillary pores (**7.17**), which provide a large surface area for adsorption. In general, the chemically inert aggregate restrains moisture induced size changes in cementitious products. Angular, rough textured aggregates require more water for a given workability and therefore indirectly increase the moisture induced size change by evaporation. Crushed rock aggregate, such as sandstone, may expand and contract due to water absorption. Aggregates of the dolerite family give large moisture induced size changes in cementitious products (about 4 times that of products made from other rocks). Other aggregates may react adversely with the high concentration of alkali in the cement to form a silicate jelly which is often exuded from the surface (alkali-silica attack, **8.20.2**). Portland cements show up to a 50% variation in moisture induced size changes. The amount of initial shrinkage in cementitious products varies with cement content. Generally, consequential moisture induced cracking will be reduced if the leanest mixes consistent with adequate strength are used. Continuous moist conditions maintained during the hydration reactions (curing) produces expansion and therefore the longer the cementitious product is kept moist, the lower the shrinkage. Steam curing (autoclaving) (used to produce aerated concrete products, e.g, Thermalite™ and Celcon™ blocks, and calcium silicate bricks, **7.5**) reduces initial shrinkage.

Moisture induced size changes for, both reversible and irreversible, are usually quoted as simple percentages, as in **Table 6.1** (over); these effects are additional to those produced by temperature induced size changes (**2.8.2**). Both effects must be allowed for in construction, commonly by the incorporation of movement joints (**Table 2.7**).

Table 6.1 Moisture induced changes in size of some common building materials[1]

Material	Reversible moisture movement (%)	Irreversible moisture movement (+)expansion (-)contraction (%)
Natural stones		
Granite		
Limestone	0.01	
Marble		
Sandstone	0.07	
Slate		
Cement based composites		
Mortar and fine aggregate	0.02-0.06	0.04-0.10 (-)
Dense aggregate concrete		
• gravel aggregate	0.02-0.06	0.03-0.08 (-)
• crushed rock (except limestone)	0.03-0.10	0.03-0.08 (-)
• crushed limestone	0.02-0.03	0.03-0.04 (-)
Steel fibre reinforced concrete	0.02-0.06	0.03-0.06 (-)
Aerated concrete	0.02-0.03	0.07-0.09 (-)
Lightweight aggregate concrete		
• medium lightweight	0.03-0.06	0.03-0.09 (-)
• ultra lightweight[a]	0.10-0.20	0.20-0.40 (-)
Asbestos cement	0.10-0.25	0.08 (-)
Glass reinforced cement	0.15-0.25	0.07 (-)
Brickwork, blockwork and tiling		
Concrete brickwork and blockwork		
• dense aggregate	0.02-0.04	0.02-0.06 (-)
• lightweight aggregate (autoclaved)	0.03-0.06	0.02-0.06 (-)
• aerated (autoclaved)	0.02-0.03	0.05-0.09 (-)
Calcium silicate brickwork	0.01-0.05	0.01-0.04 (-)
Clay, shale brickwork or blockwork	0.02	0.02-0.10 (+)
Clay tiling		
Wood and wood laminates		
Softwoods[b]	0.6-2.2[c,d] 0.45-2.0[c,e]	
Hardwoods[b]	0.8-4.0[c,d] 0.5-2.5[c,e]	
Plywood	0.15-0.20[f,g] 0.20-0.30[f,h]	
Blockboard and laminboard	0.05-0.07[f,i] 0.15-0.35[f,j]	

Material	Reversible moisture movement (%)	Irreversible moisture movement (+)expansion (-)contraction (%)
Woodchip and fibrous materials[k]		
Hardboard	0.30-0.35[f]	
Medium board	0.30-0.40[f]	
Softboard	0.40[f]	
Chipboard	0.35[f]	
Wood-wool cement slabs	0.15-0.30[l] 0.25-0.40[m]	
Rubbers, plastics, etc.		
Asphalt		
Pitch fibre	0.2-0.3	
Ebonite		
Thermoplastics		
• PVC, PVC-u, PVC-c		
• polyethylene (low density)		
• polyethylene (high density)	Not subjected to moisture effects (although some plastics are liable to progressive irreversible contraction due to loss of volatiles (**13.12**)	
• polypropylene		
• polycarbonate		
• polystyrene		
Acrylic		
Acetal		
Polyamide		
ABS		
Thermosets (laminates)		
• phenol and melamine formaldehyde		
• urea formaldehyde		
Cellular (expanded)		
• PVC		
• phenolic		
• urea formaldehyde		
• polyurethane		
• polystyrene		
Reinforced		
• GRP		
• carbon fibre oriented		
Calcium silicate based composites		
Asbestos wall boards and substitutes	0.14-0.27	
Asbestos insulating boards, substitutes	0.16-0.25	

Notes: [a] Exfoliated vermiculite and expanded perlite. Values are for plain concrete; moisture movements may be partially restrained by appropriately placed reinforcement. [b] Negligible with the grain; [c] based on 60% and 90% relative humidities; [d] tangential; [e] radial; [f] based on 33% and 90% relative humidities; [g] with grain; [h] across grain; [i] with core; [j] across core; [k] on length or width (values for thickness may be up to 30 times greater; [l] on length; [m] on width.

6.5 BULKING

When excavated, soils and aggregates do not readily settle to a close packed state and therefore occupy a greater volume than before they were disturbed. This phenomenon is termed **bulking**, and will have implications for provision of extra carrying capacity in transport Typical volume increases are 10% for gravel, 12.5% for sand, 20% and above for clay, 25% for garden earth, 33% for chalk and 50% for rock. These values depend upon the size of particles and the degree of wetness. For example, sand exhibits a volume distribution dependent on water content (***Figure 6.2a***).

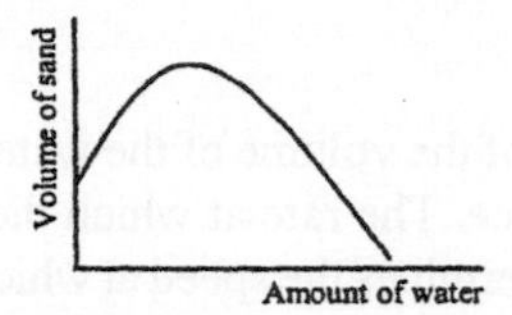

a) Variation in volume with moisture content

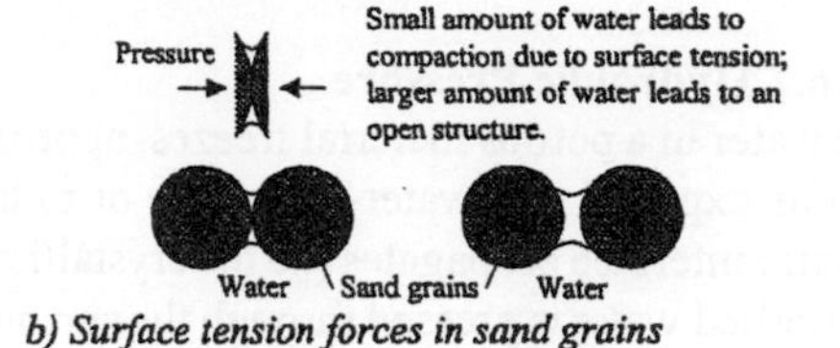

b) Surface tension forces in sand grains

Figure 6.2 Bulking in aggregates

As the amount of water between each sand particle increases, the minimum surface area for the associated water can only be obtained by pushing the particles further apart (***Figure 6.2b***). As more water is added, the surface tension forces act to pull the sand particles together and the volume decreases. There is an approximate 30% increase in volume of sand with 8% moisture content compared to the same volume of dry sand. Bulking is more pronounced in finer particle sizes.

The **bulk density** of a material is the weight of the material in a given volume (kg/m^3). It depends upon the density of packing of the aggregate particles, which in turn depends upon the particle shape and size distribution. Two simple methods can be used to determine bulk density using a container having dimensions appropriate to the size of aggregate and known empty weight (see also BS EN 1097-3[2]). **Compacted bulk density** can be determined by filling a container of known volume with aggregate at ⅓ increments using a scoop discharging from a height not greater than 50 mm above the top of the container. The aggregate is tamped a suitable number of times (varying with the size of the aggregate). The container is filled to overflowing, tamped off to a level surface and then weighed. The compacted bulk density is determined as the weight of the sample divided by the volume of the container. **Uncompacted bulk density** can be determined in a similar way, except that compaction is omitted.

6.6 FROST DAMAGE

A variety of mechanisms have been forwarded to explain the damage caused in porous materials resulting from the phase transformation water → ice.

6.6.1 Volume Expansion Water → Ice

The conventional theory is based on the increase in volume that accompanies the formation of the ice (**2.6**). During cooling, the volume of water obtains a minimum at + 4°C, and then increases rapidly (by about 9%) at the freezing point (***Figure 2.5***). Water is one of the only liquids that **expands** when it transforms to the solid (ice) phase; most liquids show a contraction. In a confined space, such as the small capillaries of a porous building material, this volume increase can amount to a pressure increase of around 200 N/mm^2 (***Figure 2.6***), much higher than the normal tensile strength of bricks and blocks, for example. It is the expansion associated with the water → ice transformation that is used to account for frost damage in porous materials. The advantage of this simple theory is that it is well known. However, it does not fully explain the process as the theory cannot account for the breakdown of porous materials containing liquids which **decrease** in volume on freezing (observed in the laboratory using benzene in cement paste). Other mechanisms must therefore operate.

6.6.2 Hydraulic Pressure

If water in a porous material freezes, approximately 9% of the volume of the water has to be expelled into water-free pores or to the outer surface. The rate at which the ice-water interface propagates (i.e the crystallisation rate) determines the speed at which the expelled water is pressed through the surrounding pores. The greater the crystallisation rate, the greater will be the pressure exerted by the process. In very fine pores, the critical size required for nucleation of the solid ice phase (**2.6.3**) may be too large to allow ice formation (***Figure 2.8a***). Hence micro-capillaries feed water into larger pores where the critical size may be achieved. The pressure generated by this movement of water contributes to pressures obtained in porous materials. The crystallisation rate depends primarily on the cooling (or supercooling) rate. The phase diagram for water (***Figure 2.6***) shows the highest pressures theoretically possible during the freezing of water in capillaries. Pressures of this order of magnitude would crack most porous materials. The fact that porous materials can survive freezing in practice indicates that the crystallisation rate (and therefore the pressures) actually obtained are limited. In addition, internal pressures can be relieved by the plastic flow of ice (which can move under pressure in a similar way to liquid water).

Poiseuille showed experimentally that the flow of liquids through narrow capillaries obeyed the relationship

$$\frac{V}{t} = \frac{\pi P r^4}{8 \eta L}$$

where V/t is the volume of liquid flowing in time t (m^3/sec), P is the pressure in the capillary (N/m^2), r is the radius of the capillary (m), L is the length of the capillary (m) and η is the viscosity of the liquid (kg/m.s). This relationship can be rearranged to illustrate the effect of capillary radius on the pressure inside the capillary, i.e.,

$$P = \frac{8 V \eta L}{\pi r^4 t}$$

This indicates that the pressure in a capillary is inversely proportional to the reciprocal of the fourth power of the radius of the capillary (i.e., to $1/r^{-4}$). Thus the pressure generated by the flow of water through narrow capillaries is very high (for a 10 fold decrease in radii, there is a 10^4 increase in pressure). This relationship indicates, for example, that the pressure required to force a damp proofing liquid through the narrow capillaries of a brick and to displace any water already filling those capillaries is very high. In addition, the pressure generated when water freezes, expands and forces unfrozen water through the narrow capillaries is very high. This means that very high bursting pressures can be experienced in very narrow capillaries. Note that this pressure is generated whether the liquid is being forced through capillaries by either an **expansion** (water → ice) or a **contraction** (ice → water).

6.6.3 Capillary Effects

The capillary effect occurs in porous materials having different sized pores in which two phases (e.g, water and ice) coexist. If a material is cooled until the water freezes in the larger capillaries, the water in the smaller pores remains as a liquid due to the smaller capillary size (***Figure 2.8a***). This is shown in ***Figure 6.3*** for water and ice in concrete.

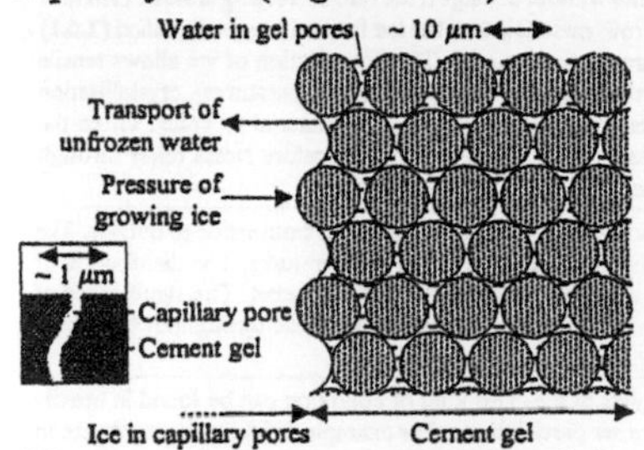

In this scenario, a thermodynamic disequilibrium will arise between the unfrozen water and the newly formed ice. In order to balance the disequilibrium, unfrozen water will move from the gel pores (which consequently will shrink) to the ice in the capillary pores, where it can freeze immediately. Movement of water in the gel pores to form ice in the capillary pores steadily increases the volume of ice, hence the pressure on the surrounding solid barrier increases

***Figure 6.3** Pressure generated (in concrete) by capillary effect*

This mechanism accounts for the breakdown of porous materials containing liquids which freeze to produce a **decrease** in volume. Note that only the solid phase produces the increased pressure, in contrast to the hydraulic effect (**6.6.2**), where movement of the liquid phase is responsible for pressure increases. The capillary effect accounts for deterioration of parapet walls and other highly exposed areas (***Figure 6.4***).

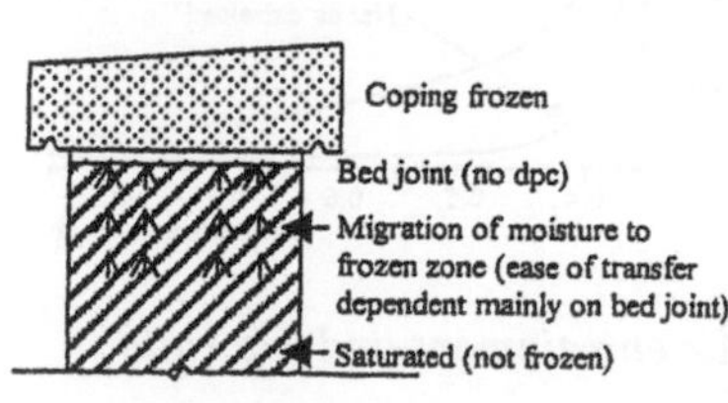

In the absence of a dpc under the coping stone to a parapet wall, saturation of a considerable part of the parapet may occur. Initial freezing of the water in the coping may not result in any deleterious effect in the coping stone itself. However, water can migrate from unfrozen to frozen parts of the parapet through the capillary effect. Water drawn from underlying masonry may then contribute to the overall volume of ice formed, enhancing deterioration. Water movement will depend on the nature of the bed joint. Dpcs should be inserted under coping stones to increase durability of the coping stone.

***Figure 6.4** Capillary effect. A deterioration mechanism in exposed parapet walls*

6.6.4 Osmosis

Osmosis is the movement of water from a low to a high solution concentration through a semi-permeable membrane. Water in the capillaries of a porous material is not pure, but contains a variety of dissolved salts. On freezing, pure water only transfers to the solid phase, and the concentration of dissolved salts in the remaining liquid increases. This leads to concentration gradients within the material and osmotic pressures, which may increase the flow of liquid to the freezing front.

6.6.5 Factors Influencing Freeze-thaw Deterioration

Whether or not deterioration occurs as a result of one or more, or a combination, of the mechanisms outlined above depends on a variety of factors (**Table 6.2**). Relationships between these factors is extremely complex, and only an overview of each is given here.

Table 6.2 Factors influencing freeze-thaw deterioration

Factor	Effect
Moisture content	The severity of mechanical damage is directly proportional to the water content of the porous material. In a fully saturated state, few (if any) systems can endure even a single freeze-thaw cycle without some degree of damage.
Moisture gradient	Moisture gradients occur between the outer and inner part of a material of construction. In exposed external walls, for example, there is generally a high moisture content near the (colder) exposed face and a lower moisture content at the (warmer) internal face. In general, subject to the pore structure (see below), a moisture gradient is desirable as there is room within the material for water squeezed out by freezing at the surface to expand into.
Rate of freezing	If the rate of freezing is slow, crystallisation rates are slow and the pressure exerted is gradual. Even the most vulnerable porous materials can be taken through repeated freeze-thaw cycles without damage if the rate of freezing is slow. Provided the pressures generated are below the tensile strength of the porous material, then the ice formed may be liquefied (**2.6.1**) and the surplus water extruded to the surface or to empty pores in the material. The liquefication of ice allows tensile stresses to be relieved. Conversely, if the rate of freezing is fast (e.g, due to low night-time temperatures), crystallisation rates are rapid and the pressure is exerted rapidly. This in itself may cause the porous material to crack. Often the magnitude of stresses produced are far in excess of the tensile strength of the material, therefore stress relief through liquefication of the formed ice is not possible and damage usually results.
Temperature gradient	In most structures, there is a drop in temperature across the thickness of the external elements from inside to outside. The extent of this temperature difference is dependent upon the internal and external air temperatures. The distribution of temperatures is dependent upon the thermal properties of the materials which make up the element. The significance of temperature gradient is that the zone of freezing is generally restricted and can only take place throughout the whole thickness of the element in highly exposed conditions.
Freeze-thaw cycles	Severe damage may occur if water is trapped between two layers of ice. This kind of condition can be found in heavily saturated walls subjected to severe frosts, between which there are partial thaws. For example, brickwork may freeze to a depth of about 75 to 100 mm under conditions of severe frost during one night. The following day, the sun may thaw the surface and moisture from melted snow may trickle over it. There may then be a severe frost the following night, trapping water between two layers of ice at different depths.
Pore structure	The frost resistance of porous materials depends greatly on the size and distribution of pores and other voids. Pore structure influences both the moisture content and the moisture gradient (above). Most porous materials contain a distribution of pore diameters rather than a number of pores of the same diameter. In most porous materials, it is the size distribution of the pores rather than the total porosity that determines frost resistance (***Figure 6.5a***).

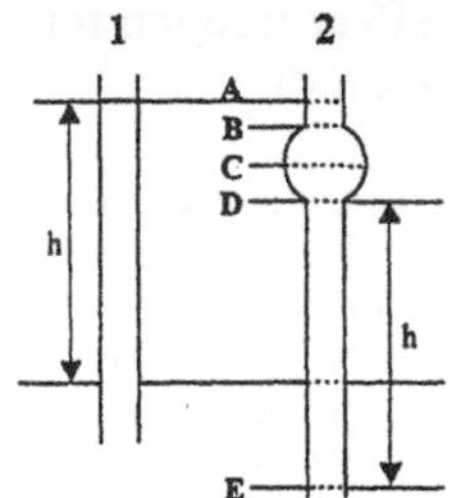

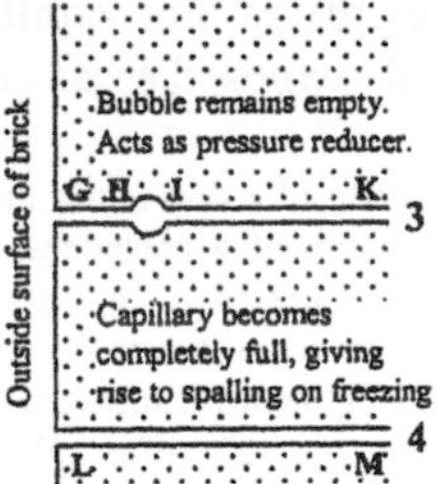

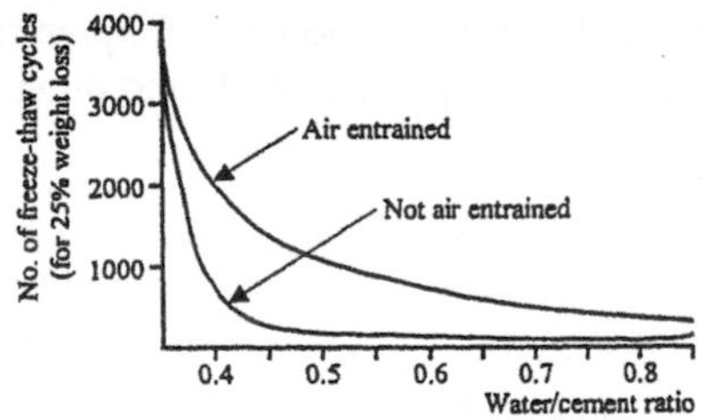

a) Variation in pore size *b) Air entrainment and w/c ratio*

***Figure 6.5** Effect of pore size on frost resistance*

In ***Figure 6.5a***, four capillaries are denoted **1, 2, 3** and **4**. Capillaries **1** and **2** are of equal radii, except in capillary **2** there is a bulge. If water moves up capillary **1** to a height of **h**, it will only reach height **D** in capillary **2**. If the water table is raised to **A**, the bulge will be filled with water, but on lowering the water table, the water level in capillary **2** will remain at **B** until the water table is significantly reduced (below **E**). Consider capillary **3**, with a temperature gradient from **G** to **K**. From the discussions above, section **GH** will be full of water and section **HJK** will be empty. The ice will nucleate in the cavity **GH**, and the bulb **HJ** will fill with any unfrozen water. The pressure generated will be small because the length of the water filled capillary is small.

This pressure will be relieved by the water having room to go into the reservoir **HJ**, thus reducing the back pressure associated with the volume flowing per second. On the other hand, if the capillary is of uniform radius, as shown with capillary **4**, the associated expansion will force the water through the capillaries **LM** and will cause a very high pressure gradient within the walls of the capillary. These principles are incorporated into many porous building materials; e.g,, by entraining 4 to 5% of minute discontinuous and uniformly distributed air bubbles into concrete, the resistance to frost action is improved by allowing capillary reservoirs to form which alleviate pressure. Air entrainment also increases the workability of the concrete (allowing a lower w/c ratio to be used), but the reduction in density causes loss of strength (up to 15% in richer mixes). The combined effect of w/c ratio and air entrainment on the number of freeze-thaw cycles to which concrete can be exposed before failure is illustrated in ***Figure 6.5b***.

6.6.6 Degradation Tests for Frost Resistance

For many years, attempts were made to subject samples to freezing conditions in a laboratory, but it was found that these tests did not simulate natural conditions with accuracy. One reason for this was that samples of the materials only were used, rather than samples of the constructional form in which they occur in building (e.g, samples of brick rather than masonry wall samples). Currently, there are no standard laboratory freezing tests as a means of assessing durability. Reliance is placed in the tray tests for bricks and stones subjected to the environment for many years. Useful properties of porous materials that give some indication of frost resistance are the **saturation coefficient** of the material, defined as the ratio of the volume of water absorbed to the total volume of voids in a material, i.e.,

$$\text{Saturation coefficient} = \frac{\text{Volume of water absorbed}}{\text{Total volume of voids}}$$

Thus when all voids are filled with water the saturation coefficient equals 1.0. A brick with a saturation coefficient of 0.6 would probably be more durable under exposed conditions than a brick with a saturation coefficient of 0.8. However, it should be noted that the relationship between water absorption and frost resistance cannot be universally applied because there is a wide variation in pore structure in materials of the same kind. Simple charts based on the relationship between adsorption and saturation coefficient have been suggested, as shown in ***Figure 6.6***.

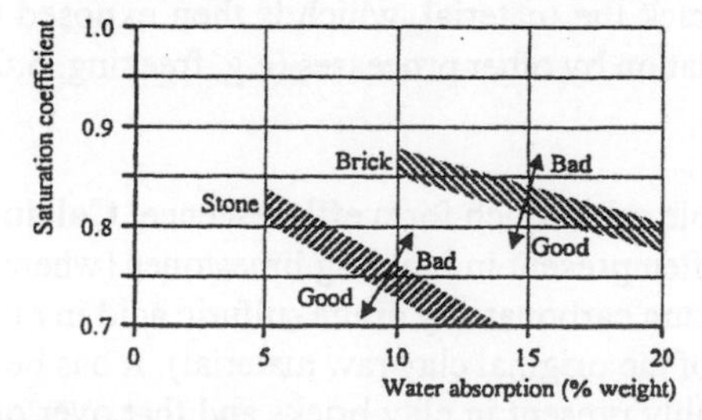

For example, a brick with an absorption of 12% and a saturation coefficient of 0.85 is likely to be as durable as one with and absorption of 18% and a saturation coefficient of 0.80.

***Figure 6.6** Different relationships between frost resistance and water absorption and saturation coefficients of bricks and certain limestones*

In general, we can summarise the affects of frost on porous building materials as follows

Class of material	How affected
Natural stone	Variable. Best stones unaffected. Some stones with pronounced cleavage along the bedding planes are unsuitable for cornices and copings.
Clay products	Variable. Best bricks and tiles are unaffected. Some products, sufficiently fired or with flaws of structure originating in the machine, may deteriorate, especially bricks in copings and tiles on flat pitched roofs
Cast stone, concrete, asbestos cement	Material of good quality is rarely affected

6.7 WATER SOLUBLE SALTS

Water soluble salts are responsible for the degradation of a wide range of building materials by a variety of different processes. These effects and processes include

- efflorescence and crypto-efflorescence (**6.7.1**);
- lime staining (**4.3.1**);
- sulfate attack of cementitious products (**8.19.2**);
- corrosion of metals (**Chapter 11**).

In this section, we outline the basic processes involved in the formation of efflorescence (other effects of water soluble salts are considered elsewhere, as indicated above). More specific information on the effect of water soluble salts on particular materials is given in appropriate Chapters.

6.7.1 Efflorescence and Crypto-efflorescence

Efflorescence is the name given to the deposits of soluble salts at or near the surface of a porous material as a result of evaporation of the water in which the salts are dissolved. On evaporation, the concentration of the salts in solution increases until they finally crystallise out. As evaporation of water from materials usually takes place at the surface, efflorescence is a surface phenomena and, in general, is not harmful to the material. However, efflorescence on internal finishes (which, due to the light colours, e.g, plaster, may not be noticeable) may break down the adhesion of paints and distempering, and bleach out colours from wallpaper coverings. A generally more damaging phenomenon is where evaporation of pore water occurs within the material. Here solubles salts may be deposited within the material, a phenomenon known as **crypto-efflorescence**. Under these circumstances, crystallisation of the solid salt occurs within the material and the volume expansion may be sufficient to crack the material, which is then exposed to further water ingress and enhanced degradation by other processes (e.g, freezing, **6.6**).

6.7.2 Types of Water Soluble Salts

Sulfates are the most common water soluble salts which form efflorescence. **Calcium sulfate** ($CaSO_4$) (gypsum or selenite) is often present in building limestones (where it is derived from the decomposition of calcium carbonate by dilute sulfuric acid in rain) and in clay bricks (as a gypsum impurity of the original clay raw material). It has been estimated that calcium sulfate is unavoidably present in clay bricks and that over one third of all bricks produced in the UK contain $\geq 3.0\%$ calcium sulfate[3]. However, calcium sulfate does not normally cause efflorescence because it is relatively insoluble in water (**Table 1.14**). Solubility is increased by the presence of other soluble materials and facilitated by prolonged damp conditions. If these either of these conditions exists, calcium sulfate can cause sulfate attack in cementitious materials (**8.19.2**) and, if present in limestones, calcium sulfate can form a hard glassy skin which may lead to surface disruption of the material. Potassium sulfate combines with calcium sulfate to produce a double salt which is more soluble and, where present together, these salts are present in large quantities in efflorescence. **Potassium sulfate** (K_2SO_4), which produces a hard, glassy deposit, and **sodium sulfate** (Na_2SO_4) (Glauber's salt), which produces a fluffy deposit, produce decay where the salts concerned can exist in more than one state of hydration. For example, sodium sulfate crystallises with 10 molecules of water of crystallisation (**2.2.7**) below 32.5°C (as $Na_2SO_4.10H_2O$) but reverts to the anhydrous state above this temperature (**2.2.7**). Cyclical temperature changes can therefore result in transformation of the salt between these two states and, since $Na_2SO_4.10H_2O$ occupies 4 times the volume of Na_2SO_4, this may lead to disruption of the porous material. **Magnesium sulfate** ($MgSO_4$) (Epsom salt) is probably responsible for most failures by soluble salts. It may be present in magnesium limestones (where it is derived from the decomposition of magnesium carbonate by dilute sulfuric acid in rain) and in clay bricks (< 0.5%). Magnesium sulfate is relatively soluble and therefore surface deposits may be removed by rain, etc. Lime plasters are particularly susceptible, where crystallisation at the interface of the substrate material and the plaster may exert sufficient pressure to separate the plaster (and sometimes the surface of the brick as well). **Ferrous sulfate** gives rise to the rust coloured stains common in efflorescence, particularly at mortar joints. Ferrous sulfate arises when iron salts from the corrosion of metals are changed to ferric oxide in the presence of oxygen and lime, which is subsequently converted to ferrous sulfate in the presence of rainwater contaminated with sulfur dioxide.

Other water soluble salts include carbonates (Mg, Na, K) (although these are most commonly converted to sulfates in the presence of acid rainwater), nitrates and chlorides (from groundwater). The metallic hydroxides (Na, K) left in solution when Portland cement is mixed with water do not normally form efflorescence because they are **hygroscopic (2.9.1)** and are unlikely to be dried sufficiently by ambient temperatures to a level at which a saturated solution would be formed.

6.7.3 Sources of Water Soluble Salts

Water soluble salts may be

- **originally present in the raw material.** A good example is fired clay products (**7.3**), where the clay raw material may contain traces of gypsum ($CaSO_4$), iron pyrites (iron sulfide, which reacts with oxygen during firing to yield gaseous oxides of sulfur, which react further with metal alkalis to form sulfates). The firing temperature of clay products is an important parameter; for example, firing temperatures of around 1000 to 1500°C will decompose magnesium sulfate and sodium chloride (but not calcium sulfate). Firing temperatures may vary throughout the kiln, producing under-fired clay products which may contain a larger quantity of water soluble salts. In addition, under-fired clay products are susceptible to increased moisture movement and may contain free alkali oxides, which may then react with water in the atmosphere to produce localised disruption of the clay product (this is sometimes seen in clay products used in exposed conditions, e.g, clay roof tiles). Finally, the fuel used to fire clay products may yield acidic oxides of sulfur, which may then be available to react with bases to form sulfates.

- **derived by decomposition of a material.** Acidic gases in the atmosphere (carbon dioxide, sulfur dioxide, sulfur trioxide, etc.) react with rainwater to produce weak acids (acid rain), which subsequently react with calcareous materials present in cementitious products to produce water soluble salts (**4.3.1**). This process is called **carbonation**, an important degradation process in reinforced concrete (**8.21.4**);
- **derived from external sources.** A large variety of external sources may lead to the ingress of water soluble salts into porous building materials, notably soils and groundwater, cleaning fluids and adjacent building materials (in this respect, porous sedimentary stones fixed to concrete constitute a particularly hazardous situation).

6.7.4 Formation of Efflorescence

The formation of efflorescence depends on a number of physical factors involving the salts themselves, and also on the transfer of water through the porous material. Many of these factors have been consider in previous sections of the text, as noted below.

Solubility

A solution of a soluble salt in water is an example of a **homogeneous solution** of a solid in a liquid (**2.9.1**). For deposition of the solid salt from the liquid to occur, the solution **must** be saturated. Saturation of the solution occurs when no more salt can be dissolved in a given volume of water, and this depends on the temperature and on the properties of the salt. Ordinarily, warm water will dissolve more salt than cold water; for example, 100 parts water will dissolve 9 parts sodium sulfate (Na_2SO_4) at 10°C, but 19.4 parts sodium sulfate (Na_2SO_4) at 20°C. Once a saturated solution has been formed, water soluble salts crystallise out as a solid. The requirement for a saturated solution can be illustrated by the following example. If a brick containing 1% sodium sulfate (Na_2SO_4) is exposed to rain and absorbs 25% water, the relationship between the quantity of salt and the quantity of water available is insufficient to produce a saturated solution. However, if the brick absorbs only 2% water, the resulting solution is nearer the solubility limit and efflorescence is more likely to appear on subsequent drying.

Unsaturated solutions may become saturated by

- cooling of the solution. This explains why efflorescence is more noticeable in summer than in winter. Compared to winter, high summer temperatures result in a relatively large quantity of water soluble salts in solution. Whilst solutions may not be saturated, diurnal temperature changes may be sufficient to result in the formation of saturated solutions. In winter, the quantity of water soluble salt capable of dissolving initially is severely restricted by lower temperatures, so that any subsequent temperature reduction produces only limited salt deposition.
- loss of water through evaporation (at constant temperature). As only pure water is lost by evaporation, the remaining solution may reach the solubility limits for the particular salt dissolved.

Pore structure

The pore structure of a porous building material influences both the rate of evaporation of water and the position at which the water soluble salts are deposited. Although the pore structure of the material as a whole will influence the rate of water transfer (**6.6.2**), the **pore structure at the surface** is particularly significant with regards to where the salts are deposited. To illustrate this, consider two fundamental surface pore structures, a fine pored structure (exemplified by a close-textured brick) and a coarse pored structure (exemplified by an open textured brick). In the fine pored structure, capillarity (**5.4**) will tend to draw the solution from the depth of the material so that the free water surfaces in the pores are near the face exposed to evaporation. On evaporation, crystallisation of salts **at the surface** is most likely (i.e., efflorescence). In a coarse pored structure, the larger dimensions of the pores cause the free water surfaces to fall below the surface of the face exposed to evaporation. Evaporation of pore water may still occur, but crystallisation of the salts **below the surface** is most likely (i.e., **crypto-efflorescence**). This process helps to explain

- why efflorescence is normally found on denser bricks (e.g, facings) than on commons (denser bricks have a finer pore structure than common bricks);
- why efflorescence is sometimes found only in certain areas of materials (which most probably have a finer pore structure than the remainder of the material).

In addition, the process outlined provides a theoretical justification for the common advice adopted for jointing brickwork, namely that the mortar should match the porosity of the jointed units and that pointing should not be denser than the mortar. If either the mortar or, more importantly (as it represents a drying surface), the pointing is dense, drying out at the mortar joints will be prevented or restricted. Drying out of the wall therefore takes place mostly through the bricks, and therefore soluble salt deposition is concentrated on the bricks and degradation of the facing may result. Conversely, if the bricks are denser than the mortar or pointing, soluble salt deposition is concentrated in the mortar or pointing, which may lead the degradation of the joints thorough sulfate attack (**8.19.2**).

6.8 REFERENCES

1. BRE (1979) *Estimation of thermal and moisture movements and stresses. Part 2*. BRE Digest 228. Garston, BRE.
2. BS EN 1097. Tests for mechanical and physical properties of aggregates. Part 3: 1998. Determination of loose bulk density and voids.
3. ADDLESON, L. and RICE, C. (1991) *Performance of materials in building*. London, Butterworth-Heinemann. p. 355.

7

MASONRY MATERIALS

7.1 INTRODUCTION

Masonry comprises units of stone, brick, concrete or calcium silicate packed and bonded together in mortar. In **Chapter 7**, common masonry units (bricks and blocks), and cementitious materials and products (cement, mortar and render) are examined (the cementitious product concrete is examined in **Chapter 8**). The strength of masonry construction is derived not only from the strength of the individual masonry units and mortar, but also from the form of the construction, the bond patterns and strengths, and the orientation of the applied load. For this reason, tests on masonry construction are used to provide information on performance more related to that actually occurring in practice. For example, BS EN 1052[1] covers tests methods for masonry construction, which comprises

- Part 1. Compressive strength of masonry construction perpendicular to bed joints, determined from the strength of small masonry specimens tested to destruction;
- Part 2. Flexural strength of masonry construction for the two principal axes of loading, determined from the strength of small masonry specimens tested to destruction under four point loading. The maximum load achieved is recorded and the characteristic value, calculated from the maximum stresses achieved by the samples, is considered to be the flexural strength of the masonry.

In simple axial loading of squat masonry construction, experimental evidence indicates that overall masonry strength is broadly proportional to (masonry unit strength)$^{0.7\text{-}0.8}$ and (mortar strength)$^{0.2\text{-}0.3}$ i.e., broadly related to the volume proportion of each. However, in most masonry construction, these simple relationships are less valid since masonry units are far stronger than the mortar and the three-dimensional confining restraint increases the effective strength of the mortar beds; in addition, simple axial loading rarely occurs.

The most common masonry units, clay bricks, belong to a group of related materials known as **ceramics**, which comprise inorganic, nonmetallic materials that are processed or used at high temperatures. Ceramics include a broad range of silicates, metallic oxides and combinations thereof and can be broadly grouped in to categories according to their common characteristic features, i.e., clay products (**7.3**), refractories (**7.4**) and glasses (**Chapter 9**). Closely related to ceramics in chemical composition are inorganic cements used as binding materials to produce cementitious materials, including mortar (**7.19**) and render (**7.20**), and concrete (**Chapter 8**). Both ceramics and cements are **solid solutions (2.9.1)** of variable compounds, principally calcium oxide (CaO), silica (SiO_2), alumina (Al_2O_3) and iron oxide (Fe_2O_3). We can illustrate these close compositional relationships in a (ternary) **phase diagram (2.5)**, as in ***Figure 7.1***.

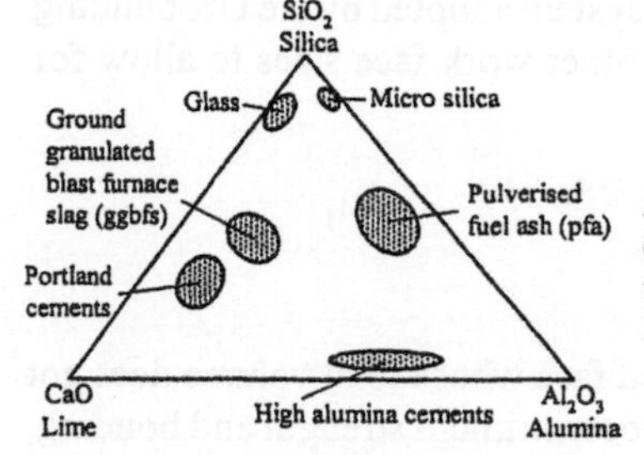

Figure 7.1 illustrates that, for example, High Alumina Cement (HAC) (**7.14.5**) is based principally on solid solutions within the $CaO.Al_2O_3$ system, whilst glass is predominantly SiO_2.

***Figure 7.1** Ternary phase diagram ($CaO.SiO_2.Al_2O_3$). Ceramics and related materials*

Ceramic products generally have

- high strength and thermal shock resistance, which arises from the fine crystallite sizes induced by the alumina, which also produces a white opaque medium that is quite unlike glass (thus ceramics can be translucent or coloured like glass);
- extremely low coefficients of thermal expansion, enabling ceramics to be used for heat and shock resistance kitchen and tableware utensils and the new, easy cleaned tops to electric cookers.
- variable degrees of porosity (**6.2**), primarily dependent on composition and firing temperature.

7.2 BRICKS AND BLOCKS

Currently, standard specifications for brick and other masonry units are controlled by relevant British Standard specifications, although a unified preEuropean Standard (prEN 771[2]) has been published by the CEN for all masonry units.

A **brick** is a walling unit composed of a hard, durable inorganic material of a size easily handled in one hand. It is usually rectangular in form and of such proportions that the length exceeds twice the width plus one mortar joint, whilst the depth is less than the width. By far the most common single standard metric brick of standard actual size 215 mm x 102.5 mm x 65 mm, weighing between 2 kg and 4 kg. The modular coordination system adopted by the UK building industry (BS 6750[3]) is based on 300 mm multiples. Four courses of 65 mm height brickwork with 10 mm joints gives a vertical height of 300 mm whilst four stretchers with 10 mm joints gives 900 mm. Ideally, masonry designs and specifications should therefore be kept to the modular coordination format so that bricks do not have to be cut to fit between openings (e.g, windows, doors, etc.).

Blocks, principally **concrete blocks** in the UK, are available in a range of densities, offering a wide range of loadbearing, sound and thermal insulation properties. Concrete blocks weigh about 4 kg/m^3 and can be lifted with one hand. Hence blockwork offers the considerable advantage over brickwork in respect of speed of work due to their larger face dimensions. Concrete blocks are available in a range of work face dimensions and thicknesses and are manufactured to produce a wide range of loadbearing and thermal insulation characteristics. Concrete blocks are covered by BS 6073[4] which specifies a standard work face size of 440 mm x 215 mm, which coordinates to three courses of metric brickwork (including 10 mm mortar joints), thus ensuring conformity with the modular coordination system adopted by the UK building industry (BS 6750[3]). In addition, BS 6073 allows other work face sizes to allow for special structural or aesthetic considerations.

7.3 CLAY BRICKS

Clay bricks (***Figure** 7.2*) may be (BS 3921[5])

- **solid** (free from holes, cavities or depressions);
- **frogged** (having depressions in one or more bed face whose total volume does not exceed 20% of the gross volume of the brick. For maximum strength and bonding, bricks with a single frog should be laid frog-up so the frog is filled with mortar);
- **perforated** (having holes not exceeding 25% of the gross volume of the brick and so disposed that the aggregate thickness of the solid material measured horizontally across the width of the unit at right angles to the face is nowhere less than 30% of the overall width of the brick. The area of any one hole should not exceed 10% of the gross area of the brick. Structurally, perforated bricks generally behave as if they were solid);
- **cellular** (having frogs or cavities exceeding 20% of the gross volume of the brick).

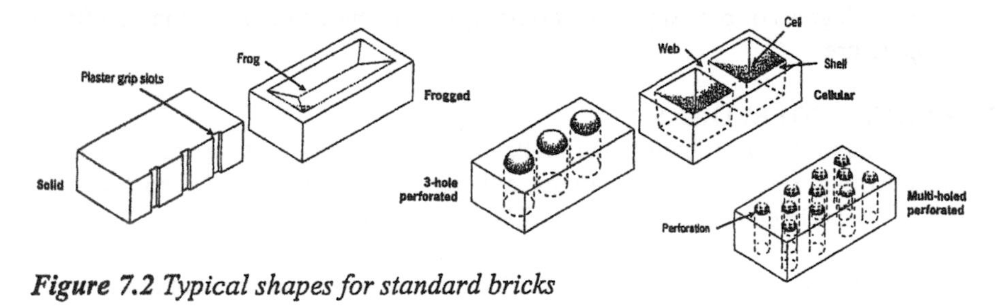

***Figure** 7.2 Typical shapes for standard bricks*

7.3.1 Manufacture

Clays, the raw material for clay bricks, contain a variety of minerals which comprise predominantly silica (SiO_2) and alumina (Al_2O_3). Additionally, clays contain a variety of impurities (e.g, alkali oxides, organic matter, etc.), depending on the source of the clay deposit. During manufacture, clay is **tempered** with water to produce the required plasticity for **moulding**. Water forms a film around the flaky clay particles so that they become orientated parallel to one another. This facilitates moulding and compaction required during **forming** by allowing the particles to 'slip' across one another. Forming processes for clay bricks include moulding (by hand or 'soft mud' processes), pressing (e.g, Fletton bricks), casting and extrusion followed by wire cutting to size. Formed clays are **dried** from the plastic state using temperatures up to 110°C. Drying causes water to be lost from between the clay particles, which results in an overall reduction in volume (typically about 10%). In addition, the clay particles come closer together so that mutual attraction between particles occurs, resulting in a much higher strength compared to the plastic state. At very low water contents, the clay volume approaches a constant value. The minimum volume that clay particles can occupy occurs at a water content that will just fill all the microvoids in the clay particles when the surface is completely dry (sometimes known as the **shrinkage limit**). This definition is analogous to **saturated surface dry (SSD)** conditions in aggregates for concrete (**8.6.5**).

The formed, dried clay bricks are converted to a permanent product possessing the required characteristics (e.g, of strength, porosity, appearance, etc.) by **firing**. Drying the formed clay from the plastic state is required before firing to limit excessive shrinkage, although often in practice the two processes form part of a continuous manufacturing operation. Firing temperatures range from 900°C to 1400°C, depending upon the type (chemical constituents) of the clay and the desired properties of the product (**Table 7.1**). During the firing process

- **hygroscopic** water is completely removed (110°C to 260°C). At this stage, the clay is hard and brittle, though can still absorb water and return to the plastic state;
- **chemically-bound** water is removed (425°C to 650°C), i.e.,
  $$Al_2O_3.2SiO_2.2H_2O \rightarrow Al_2O_3 + 2SiO_2 + 2H_2O$$
  At this stage, the clay cannot return to the plastic state by the addition of water as the OH groups become chemically joined;
- adjacent particles melt at points of contact, resulting in the transfer of ions between particles to form a larger agglomeration, a process known as **sintering** (800°C to 1000°C). Sintering is a complex process affected by the compounds present and involving diffusion and partial melting of the chemical constituents. However, sintering produces the interconnected pore structure common to most bricks;
- further heating (> 900°C) causes the partial formation of a liquid from the solid clay particles which fills up the pore space, a process known as **vitrification.**

The addition of water, followed by a high temperature cycle, is sometimes used to accelerate the processes of sintering and vitrification. Sintering and vitrification processes result in a decrease in porosity and an increase in strength. The degree of sintering and/or vitrification therefore controls the main properties of the brick (strength, porosity, density, etc.). Commonly, clay bricks are fired in the range 850°C to 1300°C and are sintered/vitrified to varying degrees. Clay bricks are available with compressive strengths ranging from about 7 N/mm^2 (for under fired soft mud brick) to 200 N/mm^2 (for a solid engineering brick). Strengths for common clay bricks are included in **Table 7.2**. Note that increased firing temperature produces highly vitrified products (e.g, porcelain, which is nonporous and virtually translucent in thin sections). Glass (**Chapter 9**) is another example of a highly vitrified product (here the raw material is principally silica, SiO_2).

The mineral constituents of the original clay are important as they affect the sintering/vitrification temperature and also the properties of the brick. For example, the presence of alkali oxides reduces the melting point of silica (about 1700°C) to about 800°C (depending upon the alkali content) (as shown in ***Figure 9.1***). In the presence of alkalis, a high firing temperature results in a highly sintered/vitrified ('glass-like') brick. Clays containing large quantities of alumina (e.g, fireclays) are used to produce **refractory bricks (7.4)**, capable of withstanding very high temperatures without loss of strength, used in kilns, furnaces, etc.

7.3.2 Properties

BS 3921 specifies a minimum compressive strength of 5 N/mm^2 for clay bricks. This is sufficient for low loading in houses and ensures that the bricks have sufficient strength to be handled. High compressive strength is not necessarily indicative of high **durability** (for example, clay bricks of 50 N/mm^2 compressive strength may still decay in conditions of severe exposure). The strength and porosity of the clay product will depend upon the degree of melting or vitrification which has taken place. This will depend upon the firing temperature, because the higher firing temperatures favour the melting of more complex oxides. The general relationships between firing temperature, porosity and compressive strength are given in **Table 7.1**.

Table 7.1 Generalised relationship between firing temperature, porosity and compressive strength for clay bricks

	Firing temperature (°C)	Density (kg/m^3)	Porosity (%)	Compressive strength (N/mm^2)
Common	950 to 1150	1200 to 1600	55 to 45	10 to 35
Facing	1000 to 1250	1800 to 2000	35 to 26	35 to 45
Engineering B	1200	2500	7.0	48.5
Engineering A	1300	2600	4.5	69.0

Table 7.2 gives typical properties for commonly available clay bricks.

Table 7.2 Properties of some common clay bricks[6]

Brick type	Compressive strength (N/mm^2)	Water absorption (weight %)	Water porosity (volume %)	Bulk density (kg/m^3)	Flexural strength (N/mm^2)
Handmade facing	10-60	9-28	19-42	-	-
London Stock	5-20	22-37	36-50	1390	1.6
Gault wirecut	15-20	22-28	38-44	1720	-
Keuper marl wirecut	30-45	12-21	24-37	2030	-
Coal measure shale	35-100	1-16	2-30	2070	-
Fletton	15-30	17-25	30-40	1630	2.8
Perforated wirecut	72.4	3.3	5.8	-	7.0
Solid wirecut (high density)	109.9	4.2	10.0	2370	6.5
Solid wirecut	55.5	8.9	17.5	2110	-
Solid wirecut (low density)	21.3	21.2	35.2	1710	-

The thermal conductivity (k) of clay masonry is moisture content and density dependent; for masonry of density 2000 kg/m^3, k = 0.96 W/mK at 1% moisture content by volume (sheltered conditions) and about k = 0.65-1.95 W/mK at 5% moisture content by volume (exposed conditions). Clay brick masonry offers excellent fire resistance by retaining its loadbearing capacity, integrity and insulating properties. BS 5268: Part 3 indicates that 100 mm and 200 mm thickness clay brick masonry will provide 2 hours and 6 hours fire resistance respectively. Clay brickwork also provides an effective barrier to sound transmission, with typical average sound insulation values over the normal frequency range (100 to 3150 Hz) of about 42 dB for a half brick wall and about 50 dB for a single brick wall plastered on both sides.

7.3.3 Classification

Traditionally, clay bricks have been referred to by a number of classifications

- **Place of Origin** (e.g, Leicester Red, Staffordshire Blue, London stock, etc.);
- **Raw Material** (e.g, marl, Gault clay, shale, etc.);
- **Manufacture** (e.g, handmade, wire cut, stiff plastic-pressed, semi-dry pressed, etc.);
- **Uses** (e.g, facings, engineering, common, paviour, squint, plinth, bullnose, etc.);
- **Colour** (often associated with the origin of the clay) (e.g, yellow, red, buff, multicolour, brindle, etc.);
- **Surface Texture** (e.g, smooth, sand faced, glazed, etc.).

The use classification has traditionally been adopted in the UK and bricks are commonly referred to as

- **Common bricks**, suitable for general building work but are not designed for good finished appearance or high strength. These are the cheapest clay bricks available;
- **Facing bricks**, specially made or selected to give good finished appearance;
- **Engineering bricks**, designed to be a dense and strong **semi-vitreous (7.3.1)** body characterised by limited water absorption and high strength.

The term **Flettons** is widely applied in the UK to bricks made from Lower Oxford Clay (which occurs in economically accessible geological deposits in Peterborough, Bedford and Buckinghamshire). Flettons comprise a predominant proportion of the UK clay brick output. The uniquely high carbonaceous content of the Lower Oxford clays reduces the cost of firing Fletton bricks. On many occasions, reliance on the traditional classifications for clay bricks can be misleading. For example, **'stock' bricks** were originally made by hand on a 'stock' (a piece of wood which located the mould), although nowadays the term is loosely applied to the usual brick manufactured in a particular area and is commonly associated with the yellow London stock bricks.

Currently, clay bricks are more formally classified by British Standard BS 3921: 1985; there is also a draft European Standard, prEN 771-1. These standards specify those properties of clay bricks which are most critical in use (with particular regard to durability), including

- **frost resistance**, i.e., *Frost resistant (F) (BS 3921); Severe exposure (F2) (prEN 771-1)*. Bricks durable in all building situations, including those where they are in a saturated condition and subjected to repeated freezing and thawing;. *Moderately frost resistant (M) (BS 3921); Moderate exposure (F1) (prEN 771-1)*. Bricks durable except in a saturated condition and subjected to repeated freezing and thawing. *Not frost resistant (O) (BS 3921); Passive exposure (F0) (prEN 771-1)*. Bricks liable to be damaged by freezing and thawing if not protected (according to BS 5628[7]) during construction and afterwards (e.g, by an impermeable cladding). These bricks are suitable for internal use.
- **soluble salt content**, i.e., *Low (L) (BS 3921)*. Maximum levels (by mass) of

magnesium 0.03%, potassium 0.03%, sodium 0.03%, sulfate 0.5%. *Normal (N) (BS 3921)*. Maximum levels (by mass) of magnesium+potassium+sodium 0.25%, sulfate 1.6%. *(S2) (prEN 771-1)*. Maximum levels (by mass) of potassium+sodium 0.06%, magnesium 0.03%. *(S1) (prEN 771-1)*. Maximum levels (by mass) of potassium+sodium 0.17%, magnesium 0.08%. *(S0) (prEN 771-1)*. No requirements.

- **compressive strength** *(BS 3921)*, i.e., not less than the strength for the appropriate class of brick given in **Table 7.3** (not specified in prEN 771-1).
- **water absorption** *(BS 3921)*, i.e., not greater than the water absorption for the appropriate class of brick given in **Table 7.3** (prEN 771-1 requirements are the same, but limits are extended to include other clay bricks).

Efflorescence properties are no longer specified, although the first version of BS 3921 classified efflorescence according to the sample showing the greatest amount of efflorescence, i.e., *Nil*. No perceptible deposit of salts; *Slight*. Up to 10% of the area of the face covered with a deposit of salts, but unaccompanied by powdering or flaking of the surface; *Moderate*. More than 10% but not more than 50% of the area of the face covered with a deposit of salts, but unaccompanied by powdering or flaking of the surface; *Heavy*. More than 50% of the area of the face covered with a deposit of salts and/or by powdering or flaking of the surface.

The requirements of BS 3921 in respect of durability, compressive strength and water absorption are summarised in **Table 7.3**.

Table 7.3 Performance requirements for clay bricks to BS 3921

Durability designation

Designation	Frost resistance	Soluble salt content
FL	Frost resistant (F)	Low (L)
FN	Frost resistant (F)	Normal (N)
ML	Moderately frost resistant (M)	Low (L)
MN	Moderately frost resistant (M)	Normal (N)
OL	Not frost resistant (O)	Low (L)
ON	Not frost resistant (O)	Normal (N)

Compressive strength and water absorption designation

Class	Compressive strength (N/mm^2)	Water absorption (% dry mass)
Engineering A	≥ 70	≤ 4.5
Engineering B	≥ 50	≤ 7.0
Damp-proof course 1 (buildings)	≥ 5	≤ 4.5
Damp-proof course 2 (external works)	≥ 5	≤ 7.0
All others	≥ 5	no limits

7.3.4 Durability

Durability is most likely to be the key performance criteria for clay bricks since, in most situations, the strength of clay bricks exceeds that required for structural stability. It is important to point out that strength is not necessarily an index of the durability of the brick. The loss of water during the manufacturing process produces a **porous** product (**6.1**) which can take in both liquid water (**5.5**) and water vapour from the air (**5.6**). Almost all durability problems in clay bricks are associated with the ingress of moisture into the porous structure and it is therefore important that adequate design and workmanship standards are adopted to limit this. For example, it is important to cover new brickwork during wet weather to avoid over saturation of the bricks and the leaching of salts from the mortar bed into the facing (excessive water within the brickwork can lead to efflorescence during the drying out period) (**6.7.1**). For similar reasons, over soaking of the bricks before laying should be avoided. Since by far the largest source of moisture ingress in brickwork is from external liquid water (**5.2**), design for durability commonly involves establishing brick types required for aesthetic or structural reasons, selecting bricks types appropriate for the most exposed position occupied by the brick types required and using selected units in all remaining areas.

The principle causes and mechanisms of degradation of clay bricks are
- moisture induced dimensional movement (**6.4**);
- temperature induced dimensional movement (**2.8.2**);
- water → ice phase transformation of pore water (**6.6**);
- water soluble salt (sulfate) content (**6.7**);

These factors are affected by the **sintering temperature** and the **quantity and type of alkali oxide impurities** present in the raw materials. If the sintering temperature is too high the shape of the brick will distort and, at very high temperatures, the raw materials may fuse together and become "glass like" (**vitrification**). Often more critical is the result of a low sintering temperature, which may leave free alkalis (e.g, calcium oxide, CaO) present in the finished brick which will subsequently react with water in service to produce the alkali hydroxide (e.g, $Ca(OH)_2$, causing localised expansion and disruption of the brick. Spalling of the surface brick will undoubtedly occur in severe exposure conditions. The correct sintering temperatures will combine the alkali oxides with the silica and the presence of alkalis act to reduce the sintering temperature. In addition, correct firing temperatures will decompose any persisting salt impurities. If the temperature of the kiln varies widely, there may be large variations in the salt content of bricks; however, the quantity of salts present in a brick should not necessarily condemn a brick (unless $MgSO_4$ is present in quantity).

In general, the best clay bricks and variable resistance clay tiles are unaffected by freeze-thaw cycles. However, products which are under-fired or have flaws in their structure due to strong pressing or extrusion may be affected. A particular risk of frost damage occurs in those parts of building where materials may be saturated with water when exposed to frost (e.g, parapets, work below dpc, retaining walls, horizontal surfaces, copings, cornices, string courses, etc.). These will require special care in the choice of materials. Besides the actual incidence of frost itself, is a sharp frost immediately following rain or mist represents a particularly hazardous exposure condition. There are currently no freeze-thaw tests which result in a 'number rating' for clay bricks; frost resistance is commonly a question of experience and test exposures carried out by manufacturers (who will often supply evidence of frost resistance in severe conditions of exposure for a particular brick type).

Efflorescence (**6.7.1**) on new brickwork is unsightly but not damaging and will disappear in time. Efflorescence is a solid and, on wetting by rain, the salts will be dissolved into solution and be either washed off or passed back into the pores of the brickwork (only to reappear again usually in greater quantity as the water evaporates). Much of the efflorescence on new brickwork will disappear as the building dries out due to the combined action of wind and rain. However, if removal is required, it is not recommended that chemicals (such as hydrochloric acid) are used to dissolve the salts deposits but rather that the efflorescence is removed mechanically (e.g, by brushing).

7.4 REFRACTORIES

The tertiary phase diagram (***Figure 7.1***) can be used to examine the properties of the $SiO_2.Al_2O_3$ system, which give a series of products known as **refractories** (which have excellent high temperature properties), as shown in ***Figure 7.3***.

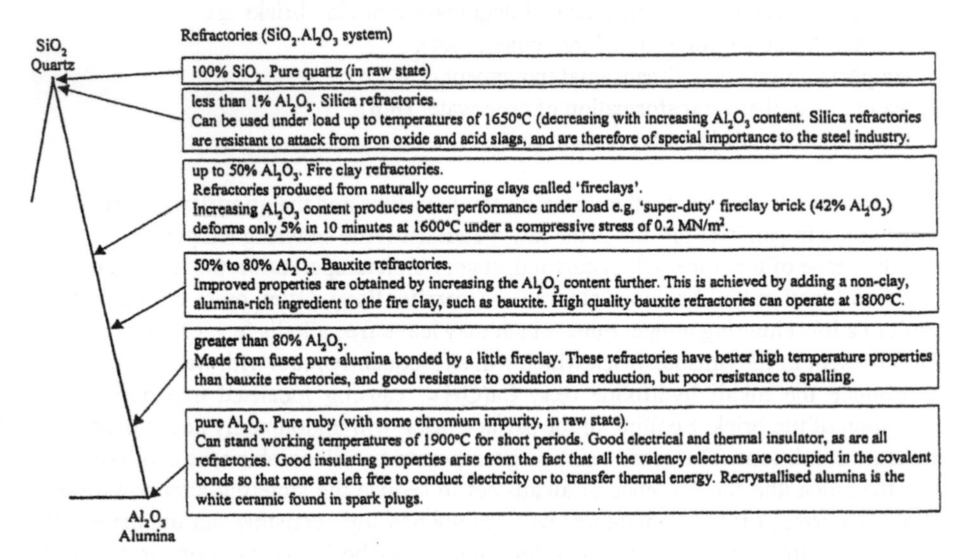

Figure 7.3 *Ternary phase diagram ($CaO.SiO_2.Al_2O_3$ system). Refractories.*

The higher strength and thermal shock resistance of refractories (and certain ceramic products) arise from the fine crystallite sizes induced by the alumina, which also produces a white opaque medium which is quite unlike glass (thus ceramics may be translucent or coloured like glass). The extremely low coefficient of thermal expansion of certain ceramics allows their use for heat and shock resistance kitchen and tableware utensils and the new, easy cleaned tops to electric cookers. Ceramics were originally developed for missile nose cones; in the building industry, they are now also used for fire resistant windows for integrity but not for insulation (due to the high radiant heat transfer through the ceramic).

7.5 CALCIUM SILICATE BRICKS

Calcium silicate bricks are manufactured from mixes of sand (or uncrushed or crushed siliceous gravel or rock, or a combination of these) and lime in the approximate ratio 10:1 sand:lime. Additionally, admixtures (such as colorants) may be added to produce a wide range of colours, commonly, whites, blacks, buffs and grey-blues. Calcium silicate bricks include **sandlime bricks** (in which only natural sand is used with lime) and **flintlime bricks** (which contain a substantial proportion of crushed flint).

Since sand (or siliceous materials) is the main constituent of calcium silicate bricks, it is common for manufacturing facilities to be located in close approximation to appropriate sand deposits. The suitability of sand depends upon the grading, the nature of the surface of the grains and any coatings on the surface of the grains. The properties of the aggregates have an important influence on the quality of the bricks. Either quicklime or hydrated lime (**7.10.2**) is used, but hydration must be complete before the bricks are pressed, to avoid expansion during subsequent steam treatment.

7.5.1 Manufacture

The process by which the constituents are mixed depends upon whether ground quicklime or hydrated lime is used. In the **silo** (or **reactor**) process, ground quicklime and sand are mixed with excess water above that required to hydrate the lime. The mix is then stored in the silo for 3-24 hours to fully hydrate the lime before being remixed and passed to the press. In the **drum hydration** process, ground quicklime and sand are mixed with a small excess of water over that required to hydrate the lime in a large revolving drum in the presence of low pressure steam (which increases the temperature and so accelerates the hydration). Mixing is completed with additional water required for pressing. In processes using dry hydrated lime, the sand and lime are mixed directly in the mixer with the necessary water for pressing. The mixture of sand and lime is then compressed to high pressures in a static press, ejected then autoclaved in high pressure steam for 4-15 hours (depending on the steam pressure used).

Under the autoclaving part of the manufacturing process, the silica (present in sand) reacts slowly with the hydrated lime (calcium hydroxide) and water to form calcium hydrosilicates whose exact composition is dependent upon the chemical constituents of the sand. Ideally, the autoclaving procedure results in the calcium hydrosilicates completely enveloping each sand grain to form strong, fused bridges between them. In this respect the process is very similar to embedding aggregate in a matrix of cement to form concrete and, as with concrete, an excess of binder would lower the strength.

7.5.2 Properties

The strength of dry calcium silicate bricks is approximately 30%-50% higher than when saturated. Prolonged exposure to acidic atmospheric gases (**4.3.1**) in moist air may degrade calcium silicate bricks. Sulfur dioxide decomposes the hydrated calcium silicate cementing agent, forming a skin of calcium sulfate (gypsum) and hydrated silica. Carbon dioxide ingress marginally increases strength but causes slight shrinkage. Absorption (**5.5**) of liquids (about 7-16% by weight) is similar to some clay bricks (except engineering bricks). Persistent rain will penetrate the bricks when saturation values are reached. The thermal conductivity (k) of calcium silicate masonry is moisture content and density dependent; for masonry of density 2000 kg/m^3, k = 0.92 W/mK at 1% moisture content by volume (sheltered conditions) and k = 1.24 W/mK at 5% moisture content by volume (exposed conditions). Ordinarily there are no water soluble salts present in calcium silicate bricks; however, repeated crystallisation of sea salt may cause surface deterioration; they are not recommended in these locations. The resistance of calcium silicate bricks to frost damage is mainly related to their mechanical strength and they are not recommended for use in highly exposed conditions (e.g, pavings).

7.5.3 Classification

BS EN 771-2 permits classification by national standards. BS 187[8] classifies calcium silicate bricks according to their compressive strength (**Table 7.4**). The class numbers have been retained from classifications based on imperial units (where the compressive strength = class number x 1000 lbf/in^2 approximately).

Table 7.4 Compressive strength class and requirements for calcium silicate bricks

Designation	Class	Compressive strength[a] (N/mm^2)	Colour marking (if applied)	Uses
	7	48.5	Green	
	6	41.5	Blue	
Loadbearing or Facing	5	34.5	Yellow	Copings and retaining walls
	4	27.5	Red	
	3	20.5	Black	Free-standing external; below dpc
Facing or common	2	-	-	Protected external applications

Notes: BS 187 also specifies limits on the minimum predicted lower limits of compressive strength ranges for 10 bricks; [a] mean for 10 bricks not less than value stated (wet compressive strengths).

7.6 CONCRETE BRICKS

Concrete bricks can be accurately made to close tolerances and, by addition of appropriate metallic oxides or coloured aggregates, are available as a variety of colour stable products. Concrete bricks are manufactured from blended dense aggregates and cement under high pressure in steel moulds.

Concrete bricks are controlled by BS 6073[4]; a European Standard also exists (prEN 771-3). A standard concrete brick weighs 3.2 kg and typically has a compressive strength in the range 20 to 40 N/mm^2. Engineering quality concrete bricks (compressive strength 40 N/mm^2) are also available made using a sulfate-resisting Portland cement (or equivalent) (**7.14.1**) with minimum cement content 350 kg/m^3. These are suitable for situations of high exposure to aggressive conditions (e.g, retaining walls, below dpc level, inspection chambers). Water absorption in concrete bricks is typically about 8%, but engineering quality bricks average less than 7% after 24 hours cold immersion. Moisture and temperature induced dimensional movement is greater than that for clay and calcium silicate bricks. Drying shrinkages of up to 0.06% are allowed in BS 6073. In consequence, movement joints at approximately 5 to 6 m intervals are required in concrete brick masonry. The thermal and sound transmission, and fire resistance properties of concrete bricks is broadly compatible with that of clay and calcium silicate bricks of similar densities. Thermal conductivities range from about 1.4 to 1.8 W/mK.

7.7 CONCRETE BLOCKS

Concrete blocks may be

- **solid** (but may have grooves or holes not exceeding 25% of the gross volume of the block to reduce weight or facilitate handling);
- **hollow**, having voids passing right through the block, which can be filled e.g, with concrete (to improve strength and/or sound insulation);
- **cellular**, in which the cavities are closed at one end, where the solid end is normally laid upwards to make it easier to produce an effective bed joint.

7.7.1 Manufacture

Autoclaved aerated concrete (AAC) blocks are formed from the addition of aluminium powder to a fine mix of sand, lime, fly ash and Portland cement (**6.1**). Aggregates for dense concrete blocks are principally natural aggregates (including crushed granite, limestone and gravel) whilst those for medium and lightweight concrete blocks come from a wide range of sources, including expanded clay, expanded blast furnace slag, sintered ash and pumice. The resultant concrete is cast into moulds, thoroughly vibrated and cured (the process is accelerated by pressured steam curing in an autoclave).

7.7.2 Properties

Concrete block densities range from about 420 to 2200 kg/m^3 corresponding to a range of compressive strengths of about 2.8 to 35 N/mm^2 and a range of thermal conductivities of about 0.10 to 1.5 W/mK at 3% moisture content. The requirements of the UK Building Regulations with respect to thermal insulation of external walls and sound transmission across party walls is most readily obtained using concrete blocks in conjunction with brickwork and cavity insulation; guidance is given in Approved Document L (thermal insulation) and Approved Document E (sound insulation).

Standards for concrete blocks are BS EN 771-4, which will replace BS 6073 in the near future. BS 6073 requires minimum compressive strengths for all concrete blocks of

- **≥ 75 mm thickness**. Average compressive strength of 10 blocks $\nless$ 2.8 N/mm^2 (with no individual block to be less than 80% of this value);
- **< 75 mm thickness**. Average transverse strength of 5 blocks $\nless$ 0.65 N/mm^2.

and maximum drying shrinkages for all concrete blocks of

- $\ngtr$ 0.06% (except autoclaved aerated blocks);
- $\ngtr$ 0.09% (autoclaved aerated blocks).

BS 6073 provides no guidance on the durability of concrete blocks, although the principles outlined in **8.13** for concrete will apply. Dense concrete blocks and certain aerated lightweight concrete blocks are more frost resistant than would be expected from consideration of their strength alone, probably as a consequence of the type and distribution of the voids present and the fact that saturation is rare. However, lightweight concrete blocks (with compressive strengths below about 7 N/mm^2) should not be used under conditions of severe exposure or excessive pollution. Concrete block construction provides good fire protection; blockwork comprising solid unplastered 90 mm blocks can give up to 1 hour fire protection and most 215 mm blocks (and certain 150 mm blocks) can achieve 6 hours fire protection.

7.8 CLAY BLOCKS

Clay blocks have been used in the UK for floor construction and comprise extruded hollow units, where the voids formed in the product reduce firing times and weight but increase handling ability and thermal insulation. However, clay blocks for floor construction have been superseded in the UK by the use of reinforced concrete inverted T-beams with concrete infill blocks. Currently, clay blocks are not commercially available in the UK but are still used extensively in Europe.

7.9 CEMENTS

Cements are inorganic (**1.2.2**) materials that exhibit characteristic properties of setting and hardening when mixed to a paste with water. This makes cements capable of joining rigid solid masses into coherent structures. Ordinarily, the function of cement is as a binder which binds together mixtures (e.g, of sands, aggregates, fillers, plasticisers, pigments and other additions) to make a solid mass or unit (e.g, mortars, concretes, grouts, renders, etc). Cements must be finely divided to be able to penetrate the spaces between the particles of the mixture and react in some way to give the change from a formable plastic material to a hard adhesive. As such, hardened cements are inherently more chemically reactive than the other components of the mixture and it is this reactivity which is the major weakness, as they are often used in environments containing reactive chemicals which can result in their deterioration. Inorganic cements can be divided according to the way they set and harden into

- **non hydraulic cements** (e.g, lime) which harden only in air and will not harden under water;
- **hydraulic cements** (e.g, Portland cement), which when mixed with water, form a paste which sets and hardens by means of hydration reactions and processes and which, after hardening, retains its strength and stability even under water.

7.10 NON-HYDRAULIC CEMENTS. LIME.

Lime is the most important non-hydraulic cement. Until the development of Portland cement (**7.15**), lime was the main cementing material for both building mortar and plasters. It was subsequently found that Portland cement hardened more quickly and produced a higher strength than lime. As a result, there was a tendency to replace lime with cement (**7.12**). More recently, mortars based on both lime and Portland cement have been developed which combine the advantage of lime (e.g, workability during mixing and the ability to accommodate some movement before cracking) and cement (e.g, quicker hardening and increased strength). Currently, products based solely on lime are preferred for restoration work, where a principal requirement is that the replacement material matches the properties and performance of the original material.

Raw materials for UK lime are principally derived from white chalk, carboniferous and oolitic limestones (which produce a non-hydraulic lime); grey chalk, siliceous and argillaceous limestones (which produce a semi-hydraulic lime); and Lias limestone and chalkmarl (which produce hydraulic limes). Dolomitic limes are produced when raw limestone contains magnesium as well as calcium carbonate. These limes are also suitable for use in mortar. Hydraulic lime results when an argillaceous limestone is used as the raw material feed to the kiln. It has the property of setting and hardening when in contact with water, although the rate and ultimate strength is inferior to that of Portland cement. As with non-hydraulic (air lime) mortars, hydraulic lime mortars also gain additional strength from carbonation in air. Some standards classify hydraulic limes by the constituent amount of active clay materials responsible for hydraulic properties (e.g, BS 8221, **7.19.4**) as shown in **Table 7.5**. Feebly hydraulic limes set more slowly and have lower compressive strength than eminently hydraulic limes. Some eminently hydraulic limes are pregauged with cement to ensure consistency of hydraulicity, which leads to variability in colour and salt content.

Table 7.5 Classification of hydraulic limes

Classification	Active clay materials	Typical colours
Feebly hydraulic	< 12%	off-white or pale grey
Moderately hydraulic	12% to 18%	pale grey or pale buff
Eminently hydraulic	18% to 25%	grey or brown

7.10.1 Manufacture

The raw materials are heated in kilns to about 1100 to 1200°C (a process known a calcination) to produce calcium oxide (CaO) (for the purposes of this section, it is assumed that the raw material is limestone, $CaCO_3$; however, reactions are equally valid for dolomitic limes). The reaction is endothermic and reversible (dependent upon the partial pressure of carbon dioxide in the kiln atmosphere, **2.2.7**). Where the raw materials are pure calcium carbonate, the products will be pure calcium oxide (CaO) (commonly referred to as 'quicklime'). 'Quicklime' was widely used in the past for mortars for stonework and is currently used in the preparation of ready-mix mortars. Where the raw materials contain impurities, the products contain a proportion of hydraulically active compounds (normally calcium silicates and/or calcium aluminates).

$$\underset{\text{(raw material)}}{CaCO_3} \underset{\text{Endothermic}}{\overset{\text{Exothermic (calcination)}}{\rightleftharpoons}} \underset{\text{(calcium oxide) ('quicklime')}}{CaO} + \underset{\text{(carbon dioxide)}}{CO_2}$$

'Quicklime' has no hydraulic properties and can therefore be kept for weeks provided it is covered and drying out is prevented. Before it can be used, 'quicklime' is hydrated (a process known as slaking) to produce calcium hydroxide, $Ca(OH)_2$ (known as hydrated lime). This reaction is exothermic and results in considerable volume expansion. The quantity of water is carefully controlled so that excess is driven off as steam by the heat of the reaction. In the past, hydrated lime was produced on site by adding water to 'quicklime'. Excess water (1:1 'quicklime':water) was always added to produce a 'lime putty'. Common practice is for slaking to be undertaken in the factory.

$$\underset{\text{calcium oxide ('quicklime')}}{CaO} \overset{\text{Slaking}}{+} \underset{\text{(water)}}{H_2O} \overset{\text{Exothermic}}{\rightarrow} \underset{\text{(calcium hydroxide) (hydrated lime)}}{Ca(OH)_2}$$

Non-hydraulic hydrated lime, as a finely ground powder, is available in the UK either in bulk or bagged and is used both for ready-mix mortars and for mixing on site. It has no setting (i.e., hydraulic) properties when mixed with water.

7.10.2 Setting and Hardening

Non-hydraulic lime mortars set when excessive water evaporates off. These mortars harden only in contact with carbon dioxide from the atmosphere, which combines with the calcium hydroxide to form a carbonate (i.e., by a process of carbonation, **8.21.4**). The reaction is exothermic.

$$\underset{\text{(calcium hydroxide)}\\\text{(hydrated lime)}}{Ca(OH)_2} \overset{\text{carbonation}}{+} \underset{\text{(carbon dioxide)}}{CO_2} \xrightarrow{\text{Exothermic}} \underset{\text{(calcium carbonate)}}{CaCO_3} + \underset{\text{(water)}}{H_2O}$$

Lime with this ability and which is composed mainly of calcium hydroxide is described as **air lime** in CEN Standards.

Carbonation of lime products commences at the surface, forming a hard skin of calcium carbonate which results in increased strength but retards further carbonation of the interior. In addition, evaporation of water results in shrinkage of the product, which can result in cracking. Therefore, lime plasters are applied in several coats up to the required thickness. Traditionally, three coats were used

- backing coat (3:1 sand:lime putty, sometimes with the addition of fibrous materials) to a thickness of about 6 to 9 mm which was allowed to carbonate for several weeks;
- floating coat (3:1 sand:lime putty) which, due to a suction effect from the backing coat, stiffens more rapidly;
- finishing coat (1:1 fine washed sand:lime putty).

7.10.3 Standards

Specification of building limes is covered by BS 890[9] and test methods by BS 6463[10]. However, these standards are set to be superseded on full publication of BS EN 459. CEN specification of building limes is covered by DD ENV 459-1[11] and test methods by BS EN 459-2[12]. This latter standard has already replaced most parts of BS 6463. Building limes to DD ENV 459-1 are called CEN limes. CEN building limes to ENV 459-1 are calcium limes (CL), dolomitic limes (DL) or hydraulic limes (HL), subclassified according to their (CaO + MgO) content, or in the case of hydraulic limes, their compressive strength, as summarised in **Table 7.6**. The standard includes requirements for additional physical properties (reactivity, water demand, water retention, bulk density). Hydraulicity (**Table 7.5**) is defined as HL 2 (Feebly hydraulic), HL 3.5 (Moderately hydraulic) or HL 5 (Eminently hydraulic).

Table 7.6 Chemical and physical requirements for CEN building limes

Type	Chemical requirements					Physical requirements	
						Compressive strength (N/mm²)	
	CaO+MgO	MgO	CO_2	SO_3	available lime	7 days	28 days
CL 90	≥ 90	≤ 5[a,c]	≤ 4	≤ 2	-	-	-
CL 80	≥ 80	≤ 5[a]	≤ 7	≤ 2	-	-	-
CL 70	≥ 70	≤ 5	≤ 12	≤ 2	-	-	-
DL 85	≥ 85	≥ 30	≤ 7	≤ 2	-	-	-
DL 80	≥ 80	> 5[a]	≤ 7	≤ 2	-	-	-
HL 2	-	-	-	≤ 3[b]	≥ 8	-	2 to 5
HL 3.5	-	-	-	≤ 3[b]	≥ 6	≥ 1.5	3.5 to 10
HL 5	-	-	-	≤ 3[b]	≥ 3	≥ 2	5 to 15[d]

Notes: Values given in percent by mass; these values are applicable for all kinds of lime. For quicklime, these values correspond to the 'as delivered' condition; for all other kinds of lime (hydrated lime, lime putty and hydraulic lime) the values are based on the water free and bound water free product. [a] for soil stabilisation ≤ 10%; [b] SO_3 content of more than 3% and up to 7% is permissible if soundness is demonstrated at 28 days of water curing; [c] MgO content up to 7% is acceptable if the soundness test given in EN 459-2 is passed; [d] HL 5 with a bulk density < 0.9 kg/dm³ is allowed to have a strength of not more than 20 N/mm².

7.10.4 Use

A wet mortar mix of hydrated lime and sand is commonly referred to as 'coarse stuff'. Ready-mixed sand:lime is available commercially in which hydrated lime is premixed with damp sand in proportions required for a particular site. Over several years, some binding together of sand particles occurs in mortars exposed to air, as the calcium hydroxide (of which calcium limes are composed) reacts with the carbon dioxide in the air to form calcium carbonate (**8.21.4**). Note that mortar that carbonates in this way is never as hard or durable as properly specified hydraulic cement mortars (**7.19.3**). Non- and semi-hydraulic limes can be used with gypsum and cement, whereas hydraulic lime can be used in mortars mixed with sand, but should not be used with gypsum or cement.

Modern lime plasters contain a proportion of quicker setting material to impart higher early strengths, which may be additions or may derive from impurities in the initial raw materials. The presence of these materials produce 'hydraulic limes' (i.e., are able to set and harden under water), and they interfere with the setting mechanism so that strength is a product of the crosslinking of chemical bonds produced by the hydraulic constituents rather than by carbonation. This is the same mechanism responsible for the strength production in cementitious materials.

Various degrees of hydraulicity (e.g, semi-hydraulic limes, moderately hydraulic, eminently hydraulic) are possible, depending on the nature of the materials present. However, all hydraulic limes set to various degrees by combination with water rather than by carbonation. Most of the hydraulic cements (**7.11**) may be blended with pure hydrated lime in various proportions to make hybrid cements which give mortars with a lower strength and rigidity but still retain the plasticity of the 1:3 binder:sand ratio. This leads to mortars which are more tolerant of movement and more economical.

7.11 HYDRAULIC CEMENTS

Hydraulic cements examined are Portland cement (**7.14.1**), Portland replacement cements (**7.14.4**), high alumina cement (HAC) (**7.14.5**) and supersulfated cement (**7.14.6**). As Portland cement is the most important commercial cement, we consider this cement in some detail (**7.15 to 7.18**). Most hydraulic cements comprise a wide variety of complex chemical compounds and, to simplify the representation of these compounds, we use an abbreviation system known as cement chemists notation.

7.11.1 Cement Chemists Notation

Portland cements are based on the CaO-Al_2O_3-Fe_2O_3-SiO_2 system and contain an extremely complex array of compounds. To simplify these compounds, an abbreviation system known as **cement chemists notation** is adopted in which complex compounds are separated in to their mixed alkali and acidic oxides (and water), and each oxide is referred to by a letter code (**Table 7.7**). As an example, tri-calcium silicate (Ca_3SiO_5) (a compound present in Portland cement, **7.15.2**) is split into the mixed oxides 3CaO and SiO_2), which are substituted for the letter codes from **Table 7.7** to give C_3S, the formula for tri-calcium silicate using cement chemists notation. We will refer to cement compounds using cement chemists notation throughout this chapter.

Table 7.7 Cement chemists notation

Alkali or acidic oxide	Chemical formula	Cement chemists notation
Alumina	Al_2O_3	A
Calcium oxide	CaO	C
Carbon dioxide	CO_2	$\bar{C}$
Magnesium oxide	MgO	M
Sodium oxide	Na_2O	N
Iron oxide	Fe_2O_3	F
Silicon dioxide	SiO_2	S
Sulfur trioxide	SO_3	$\bar{S}$
Water	H_2O	H
Calcium silicate hydrate fibres	Complex	CSH

7.12 HISTORICAL DEVELOPMENT OF HYDRAULIC CEMENTS

The requirement to bind stones to form a solid, formed mass was appreciated by early civilisations. The Assyrians and Babylonians used clay for this purpose; the Egyptians advanced the process by the incorporation of lime and gypsum in to a mortar used, for example, in the Egyptian pyramids. These processes were further improved by the Greeks and finally the Romans, who developed the durable cement incorporated in structures that remain today. The Roman Baths, built in about 27 BC, the Colosseum and the Basilica of Constantine are examples of early Roman architecture in which cement mortar was used. Most of the foundations of the Roman Forum were constructed in a form of concrete, placed at some locations to a depth of 3½ m. Roman cement was created by mixing slaked lime with a natural pozzolana (**7.14.4**), consisting of volcanic ash from Mount Vesuvius, to produce a hydraulic cement. Significant scientific advancement into the properties of cements was not undertaken until the 1800s. Repeated structural failure of the Eddystone Lighthouse in Cornwall led a British engineer, John Smeaton, to conduct experiments with mortars in both fresh and salt water. In 1756, these tests led to the discovery that cement made from limestone containing considerable proportions of clay was hydraulic. Making use of this discovery, Eddystone lighthouse was rebuilt in 1759 and it stood for 126 years before replacement was necessary. Prior to the discovery of Portland cement, large quantities of natural cement were used, produced by burning a naturally occurring mixture of lime and clay. However, little was known about the effect of these constituents on the final product, and consequently the properties of the final product varied widely depending on the initial raw material used. In the period 1756 to 1830, further advances in the scientific understanding of cements were made by Vicat and Lesage in France, and Parker and Frost in England, particularly with regard to producing a standard product with known and reproducible properties. In 1824, Aspdin, a bricklayer and mason from Leeds, took out a patent on a hydraulic cement he called Portland cement, because its colour resembled the limestone quarried at the Isle of Portland. Aspdin's method involved the careful proportioning of limestone and clay, subsequent crushing and burning to form a clinker, which was then ground to a finished cement with known and reproducible properties. Aspdin established a Portland cement factory in Wakefield that produced Portland cement used in the construction of the Thames River tunnel in 1828. Subsequently, J.D. White and Sons set up a factory in Kent which led to the greatest period of expansion in cement manufacture, both in the UK and in Germany and Belgium. Portland cement was used to build the London sewage system in 1859-1867.

7.13 CLASSIFICATION OF HYDRAULIC CEMENTS

Over the years, several types of Portland cement have been developed, including Ordinary Portland Cement (OPC), rapid-hardening Portland cement (RHPC), sulfate resisting Portland cement (SRPC), white Portland cement and low heat Portland cement (LHPC). In addition, by incorporating other materials during manufacture, further cement types have been developed, for example Portland blastfurnace cement, coloured, water-repellent and hydrophobic cements. Standard specifications for most of these cements are governed by British Standards, which uses a classification system for cements based on

- strength classes (based on 28 day compressive strength of mortar prisms, made and tested in accordance with EN 196[13]) i.e., 32.5, 42.5, 52.5 and 62.5;
- rate of development of early strength, i.e., rapid (R), normal (N) and low (L).

Currently, Portland cements standards are being harmonised within Europe and a new standard is set to govern the specification of cements (ENV 197-1[14]). However, the current draft of ENV 197 excludes the special properties of low heat Portland cement conforming to BS 1370 or of sulfate resisting Portland cement conforming to BS 4027. In addition, other types of cement whose hardening is not primarily due to the hydration of calcium silicates, notably high alumina cement (HAC) to conforming to BS 915[15] and supersulfated cement conforming to BS 4248[16], are also currently excluded. It is intended that future sections of ENV 197 will incorporate these cements and it is anticipated that, in these publications, HAC will be known as Calcium Aluminate Cement (CAC).

Portland cements to ENV 197-1 are referred to as **CEM cements**. The nomenclature for CEM cements is more elaborate than in British Standards because of the need to accommodate the wide variety of European cement types and their constituents, strength class and rate of strength development. Despite this complexity, there are essentially identifiable equivalents to traditional British Standard cements (**Table 7.13**). In CEM cements conforming to ENV 197-1, the sum of the proportions of reactive calcium oxide (CaO) and reactive silica (SiO_2) should be at least 50% by mass. ENV 197-1 classifies cements into one of five types based on the main constituent

CEM I Portland cement
CEM II Portland composite cement
CEM III Blastfurnace cement
CEMIV Pozzolanic cement
CEM V Composite cement

Within each of these five types, subdivisions indicate

- the proportion of cement clinker i.e., high (A), medium (B) and low (C);
- the second main constituent (denoted by letter code, see **Table 7.12**);
- the standard strength class, i.e., 32.5, 42.5 and 52.5 (strength class 62.5 in the British Standards will be withdrawn in the final version of EN 197-1);
- the rate of early strength gain. i.e., rapid (R) or left blank if normal.

Compressive strengths are determined in accordance with EN 196-1 at 28 days to determine the standard strength class, and at 2 and 7 days to determine early strength class. Prisms of mortar are used instead of cubes, giving slightly higher compressive strength values (**Table 7.8**).

Table 7.8 Comparison of compressive strengths of prisms (ENV 196-1) and cubes

Age at test (days)	2	3	7	28
Mortar prism compressive strength (N/mm^2) / Cube compressive strength (N/mm^2)	1.48	1.41	1.34	1.30

The compressive strength for all CEM cement types should conform with the requirements of **Table 7.9**.

Table 7.9 Mechanical and physical requirements for CEM cement to ENV 197-1

Class	Compressive strength (N/mm^2)				Initial setting time (min)	Expansion (mm)
	Early strength		Standard strength			
	2 days	7 days	28 days			
32.5	-	≥ 16	≥ 32.5	≤ 52.5	≥ 60	≤ 10
32.5R	≥ 10	-				
42.5	≥ 10	-	≥ 42.5	≥ 62.5		
42.5R	≥ 20	-				
52.5	≥ 20	-	≥ 52.5	-	≥ 45	
52.5R	≥ 30	-				

Table 7.9 is illustrative rather than definitive. Noted that if the R (higher early strength) classification is reached, the lower classification is not claimed in the Table but a manufacturer might choose to claim the lower early strength classification; e.g, a coarsely ground cement Type CEM I with a 28 day compressive strength > 32.5 N/mm^2 and ≤ 52.5 N/mm^2 (Class 32.5) is likely to substantially exceed the 2 day strength requirement of 10 N/mm^2 for Class 32.5R. However, the cement manufacturer may choose to market the cement claiming to meet the less onerous strength requirements of Class 32.5 only (16 N/mm^2 at 7 days is far more easy to attain than 10 N/mm^2 at 2 days). Each class is represented by a 'conformity band' of 20 N/mm^2 (**Table 7.10**), with the strength class of the cement designated as the lower characteristic value.

Table 7.10 Standard strength of cements EN 196-1

Strength class (N/mm^2)	Conformity band
32.5	32.5 to 52.5
42.5	42.5 to 62.5
52.5	≥ 52.5

In addition to the mechanical and physical requirements for CEM cements, ENV 197-1 specifies certain chemical requirements, summarised in **Table 7.11**. The lack of loss on ignition and insoluble residue requirements for CEM II, IV and V follows from the possibility that materials exhibiting high loss on ignition (e.g, limestone) or high insoluble residues (e.g, fly ashes and pozzolana) might be in present these cements.

Table 7.11 Chemical requirements for CEM cement to ENV 197-1

Property	Test reference	Cement type	Strength class	Requirements (by mass)
Loss on ignition	EN 192-2	CEM I CEM III	all classes	≤ 5%
Insoluble residue	EN 196-2	CEM I CEM III	all classes	≤ 5%
Sulfate	EN 196-2	CEM I CEM II[a] CEM IV CEM V	32,5 32.5R 42.5	≤ 3.5%
			42.5R 53.5 52.5R	≤ 4%
		CEM III[b]	all classes	
Chloride	EN 196-21	all types[c]	all classes	≤ 0.1%
Pozzolanicity	EN 196-5	CEM IV	all classes	satisfies the test

Notes: [a] This indication covers cement types CEM II/A and CEM II/B, including Portland cement composite cements containing only one other main constituent (e.g, II/A-S or II/B-V), except type CEM II/B-T, which may contain up to 4.5% SO_3 for all strength classes; [b] Cement type CEM III/C may contain up to 4.5% SO_3; [c] Cement type CEM III may contain more than 0.1% chloride (actual chloride content declared by manufacturer)

As an example, the CEM designation for Ordinary Portland cement (OPC) to BS 12

PC	42.5	N	**PC42.5N**
(type of cement)	(strength class)	(normal strength development)	(BS 12 classification)

would be

CEM I	42.5	**CEM I 42.5**
(type of cement)	(strength class)	(ENV 197-1 classification)

In a similar way, a Portland slag cement with a high proportion of Portland cement clinker with rapid hardening characteristics would be designated under ENV 197-1 as

CEM II/A	S	42.5	R	**CEM II/A-S42.5R**
(cement type/ proportion of Portland cement clinker)	(second main constituent)	(strength class)	(rapid strength development)	(ENV 197-1 classification)

Permitted cement constituents corresponding to the various cement types listed above are Portland cement clinker; calcium sulfate (set regulator); additional main constituents (e.g, fly ash, slag, limestone, etc.); minor additional constituents (e.g, fly ash, slag, limestone, etc., to optimise properties) and/or pigments; additives (e.g, grinding aids, air entraining improvers, etc). Permitted constituents are outlined in ENV 197-1 (**Table 7.12**) and may be the 'additional main constituent' or a 'minor constituent'.

ENV 197-1 identifies 25 CEM cements types depending on the constituents described above and the wide range of national practices existing within Europe. These CEM cement types are summarised in **Table 7.13**, which also indicates the relationship between the British Standard (where available) and the CEM cement type.

Table 7.12 Summary of permitted constituents for CEM cements (ENV 197-1)

Constituent	Designation	Specification requirement (by mass)	Notes
Portland cement clinker	K	$C_3S+C_2S \geq 66.7\%$ $C/S \geq 2.0\%$ $MgO \leq 5.0\%$	Portland cement clinker is a hydraulic material made by burning, at least to sintering, a homogeneous mixture of raw materials containing CaO, SiO_2, Al_2O_3 (with minor quantities of other materials)
Granulated blastfurnace slag	S	'glassy slag' $\geq 66.7\%$ $C+S+MgO \geq 66.7\%$ $(C+MgO)/S > 1.0\%$	Granulated blastfurnace slag is a latent hydraulic material, i.e. it possesses hydraulic properties when suitably activated. Granulated blastfurnace slag is made by the rapid cooling of slag melt of suitable composition as obtained from smelting iron ore in a blastfurnace.
Natural pozzolana	P	-	Pozzolanic materials are natural or industrial pozzolanas, siliceous or silico-aluminous, or a combination thereof. They do not harden in themselves when mixed with water but, when finely ground and in the presence of water, they react with dissolved calcium hydroxide, $Ca(OH)_2$, at normal ambient temperatures to form strength developing calcium silicate and calcium aluminate compounds. These compounds are similar to those formed in the hardening of hydraulic materials. Pozzolanas consist essentially of reactive silica (SiO_2) and aluminium oxide (Al_2O_3) with minor proportions of Fe_2O_3 and other oxides (the proportion of reactive CaO is negligible). **Natural pozzolana** are usually substances of volcanic origin or sedimentary rocks with suitable chemical and mineralogical composition. **Industrial pozzolana** are thermally treated and activated clays and shales, air cooled slags from lead, copper, zinc and other products from the ferroalloys industry.
Industrial pozzolana[a]	Q	-	
Siliceous fly ash	V	loss on ignition $\leq 5\%$ 'reactive' $C < 5\%$ 'reactive $S > 1.0\%$	Fly ash may be silico-aluminous or silico-calcareous in nature, both of which have pozzolanic properties and the latter having, in addition, hydraulic properties. Fly ash is obtained by electrostatic or mechanical precipitation of dust-like particles from the flue gases from furnaces fired with pulverised coal. **Siliceous fly ash** is a fine powder of mainly spherical grains comprising predominantly reactive silica (SiO_2) and aluminium oxide (Al_2O_3) with some other oxides (e.g. Fe_2O_3). **Calcareous fly ash** comprises predominantly reactive calcium oxide (CaO), silica (SiO_2) and aluminium oxide (Al_2O_3) with some other oxides (e.g. Fe_2O_3).
Calcareous fly ash	W	loss on ignition $\leq 5\%$ Expansion < 10 mm For 'reactive' C 5-15%, 'reactive' $S \geq 25\%$ For 'reactive' C >15%, strength $\geq 10\ N/mm^2$	
Burnt oil shale	T	28 day compressive strength $\geq 25\ N/mm^2$ Expansion < 10 mm	Burnt oil shale is produced in a special kiln at temperatures of approximately 800°C. Owing to the composition of the natural material and the production process, burnt oil shale contains clinker phases (mainly dicalcium silicate and monocalcium aluminate), pozzolanically reacting oxides (especially silica), and small amounts of free calcium oxide (CaO) and calcium sulfate ($CaSO_4$). As a consequence, finely ground burnt oil shale shows pronounced hydraulic properties like Portland cement and, in addition, pozzolanic properties.
Limestone	L	$CaCO_3 \geq 75\%$ Clay content ≤ 1.2 g/100 g Organic material content $\leq 0.2\%$ Organic material content $\leq 0.5\%$[b]	
Silica fume	D	amorphous $S \geq 85\%$ loss on ignition $\leq 4\%$ specific surface $\geq 15\ m^2/g$	Silica fume comprises very fine spherical particles with a high content of amorphous silica, is sometimes used to provide reactive silica. Silica fume originates from the reduction of high purity quartz with coal in electric arc furnaces in the production of silicon and ferrosilicon alloys.
Filler[a]	F	-	Fillers are specially selected natural or artificial inorganic mineral materials which, after appropriate preparation, improve the physical properties of the cement (such as workability or water retention) as a consequence of the particle size distribution. Fillers can be inert or have slightly hydraulic, latent hydraulic or pozzolanic properties.
Calcium sulfate	-	-	Small quantities of calcium sulfate (gypsum, 14.7.3, as dihydrate, anhydrate or combinations thereof) are added to the other constituents of cement during manufacture to control setting. Gypsum and anhydrite are found naturally and are available as by-products of certain industrial processes.
Additives[a]	-	Total quantity $\leq 1\%$	Additives are added to improve the manufacture or properties of cement (e.g. grinding acids).

Notes: [a] Industrial pozzolanas, fillers and additives should selected so as not to increase water demand of the cement, impair resistance of the product to deterioration or reduce corrosion protection of reinforcement.; [b] national option

Table 7.13 Cements to ENV 197-1 corresponding to British Standards

CEM cement type	Cement type (ENV 197-1)	CEM notation	Portland cement clinker content (K) (%)	Additional main constituent[a] (%)	Type	Cement Group[b]	Corresponding British Standard
Portland cement	I	I	95-100	0		1	BS 12
Portland slag cement	II	II/A-S	80-94	6-20	S	1	BS 146[17]
		II/B-S	65-79	21-35	S	1	BS 146
Portland silica fume cement	II	II/A-D	90-94	6-10	D		None
Portland pozzolana cement	II	II/A-P	80-94	6-20	P		None
		II/B-P	65-79	21-35	P		None
		II/A-Q	80-94	6-20	Q		None
		II/B-Q	65-79	21-35	Q		None
Portland fly ash cement	II	II/A-V	80-94	6-20	V	1	BS 6588[18]
		II/B-V	65-79	21-35	V	1,2	BS 6588
		II/A-W	80-94	6-20	W		None
		II/B-W	65-79	21-35	W		None
Portland burnt shale cement	II	II/A-T	80-94	6-20	T		None
		II/B-T	65-79	21-35	T		None
Portland limestone cement	II	II/A-L	80-94	6-20	L	1	BS 7583[19]
		II/B-L	65-79	21-35	L		None
Portland composite cement	II	II/A-M	80-94	6-20	M		None
		II/B-M	65-79	21-35	M		None
Blastfurnace cement	III	III/A	35-64	36-65	S	1	BS 146[c]
		III/B	20-34	66-80	S	1	BS 4246[20,c]
		III/C	5-19	81-95	S	1,2	
Pozzolanic cement	IV	IV/A	65-89	11-35	M		None
		IV/B	45-64	36-55	M	1	BS 6610[21]
Composite cement	V	V/A	40-64	36-60	S+M		None
		V/B	20-39	61-80	S+M		None

Notes: the 'additional main constituent' component additionally comprise up to 5% filler (F). For Portland composite cement, the designation M ('mixed') refers to several different constituents. It should be noted that the 100% total is considered as the 'cement nucleus' and does not include calcium sulfate or additions. ENV 197-1 places no limit on calcium sulfate addition but the amount is controlled by the cement SO_3 limits specified in the Standard. Note that the chemical requirements of ENV 197-1 are not exactly duplicated in the British Standards and an important exception is for sulfate (as SO_3), where the 4.0% SO_3 limit (Table 7.11) is not permitted in the British Standards (which retain an effective limit of 3.5% SO_3 for all cements). [a] Code letters used in ENV 197-1 for the additional main constituent are given in Table 7.12. [b] for use with concrete specification in aggressive ground conditions, 8.19.3; [c] BS 4246 covers a blastfurnace slag content of 50% to 85%.

7.14 TYPES OF HYDRAULIC CEMENT

7.14.1 Portland Cements

Ordinary Portland Cement (OPC) develops strength sufficiently rapidly for most purposes; the rate of hardening is accelerated by increased temperature. Resistance to aggressive environments (acids and sulfates) is generally low.

Rapid hardening Portland cement (RHPC) does not set any faster than OPC but, after setting, hardens and develops strength more rapidly (**Table 7.19**). This occurs as a result of the finer particle size of RHPC (**7.16.4**) and the higher proportion of calcium silicates (***Figure 7.4***)

Sulfate resisting Portland cement (SRPC) is more resistant to water soluble sulfates. Aggressive sulfate ions in solution will react with the hydration products of tri-calcium aluminate in the hardened cement paste, i.e., the hydrated calcium aluminate crystals, to produce metastable ettringite, in a manner identical to that noted for gypsum in the curing of Portland cement (**7.16**). In addition, it is known that alkali metal water soluble sulfates react with free calcium hydroxide to form more gypsum, which is then available to react further with the hydration products of C_3A (**7.16**) to form ettringite. The problem of **sulfate attack** of cementitious products arises because the volume expansion associated with the formation of ettringite **occurs in the hardened cement paste**, disrupting the matrix. Note that in the curing of cement, the volume expansion can both be accommodated within the plastic paste and is of short duration, due to the secondary transformation of ettringite to the more stable monosulfate form ($C_3A.C\bar{S}.H_{16}$); hence the formation of ettringite during curing is non-disruptive. Sulfate attack is examined in more detail in section **8.19.2**.

The property of resistance to aggressive sulfates exhibited by SRPC is achieved by

- increasing the proportion of iron oxide so that the aluminates are taken up as tetra-calcium aluminoferrites (C_4AF) rather than as tri-calcium aluminate (C_3A), thereby reducing the quantity of hydrated calcium aluminate available. Thus, compared to other Portland cements, the proportion of tri-calcium aluminate (C_3A) is reduced and the content of tetracalcium aluminoferrite (C_4AF) increased. The iron content, as C_4AF, produces a darker colour and therefore SRPC is normally darker in colour than other Portland cements.
- reducing the quantity of free calcium hydroxide by reducing the quantity of the calcium silicates C_3S and C_2S (responsible for the formation of free calcium hydroxide on hydration). Since tri-calcium silicate (C_3S) produces more free calcium hydroxide on hydration than di-calcium silicate (C_2S) (as evidence by the simplified hydration reactions in section **7.16**), the proportion of tri-calcium silicate is limited in SRPC (***Figure 7.4***).

SRPC is used for concrete foundations laid in clays (such as the Keuper marl of the Midlands) and in mortars for bricks made from the sulfate containing London clays. It is useful in severely exposed areas such as copings, flues, retaining walls, etc. This cement is also used for the manufacture of concrete sewer pipes where the sulfur/ sulfate content as high. However, SRPC is similar to other Portland cements in that it is not resistant to acids, nor immune to the effects of some other dissolved salts, such as magnesium compounds (**8.19.2**), which may occur in natural waters or effluents, although low alkali sulfate resisting Portland cement (LASRPC) is available.

Low heat Portland cement (LHPC) (BS 1370[22]) produces cementitious products which gain strength and evolve heat more slowly during hydration than products of similar composition made with other Portland cements, although ultimately the strength and heat evolved are virtually the same. LHPC has been developed for use in large masses where the rapid evolution of heat would produce high temperatures resulting in stresses which may lead to cracking (e.g, concrete dams and large mass foundations, etc). However, the strength development in LHPC is slower than that in OPC, but does offer better resistance to water soluble sulfates. LHPC is not normally available except to special order and is only usually considered for very large structures. For smaller works where excessive heat evolution may be a problem SRPC is commonly used, as this has a lower rate of heat evolution than OPC. Note that the total heat evolved by most Portland cements is similar, but in LHPC the rate of heat evolution is slower.

White Portland cement is distinguished by its low iron compound content which impart the grey-green colour to ordinary Portland cements. White Portland cement is made using white china clay and limestone as the raw materials. Gypsum is added to control setting (**7.16.1**) and special care is taken during manufacturing stages to ensure that coloured contaminants are not introduced. White Portland cement is used for decorative white cementitious products (notably concrete) and also for some coloured products, where a suitable pigment is added to the mix.

7.14.2 Compositional Relationships of Portland Cements

As previously noted, the four main constituents of cement react differently with water during hydration and therefore cements with different properties can be produced by varying the quantities of these compounds, as indicted in ***Figure 7.4***.

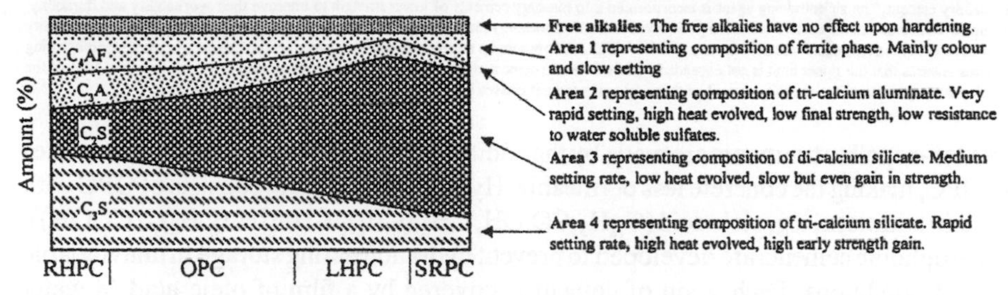

Figure 7.4 *Relative composition of various cements and the function of constituents*

7.14.3 Cement Additives

Cement additives comprise a wide range of materials incorporated to enhance various properties of the hardened cement paste. In essence, these additives are similar to the admixtures applied to concrete (**8.7**) and give rise to the following cement types

Masonry cement consists of OPC with certain additions at the manufacturing stage to increase workability (including water-retaining mineral fibres, usually ground limestone and air-entraining agents). Masonry cements have been developed because OPC and sand mixes are often too strong for rendering, brickwork and blockwork mortars and the addition of lime to produce a more suitable mortar often involves unpleasant working conditions. Masonry cements should not normally be blended with further admixtures on site other than sand. It is not suitable for concrete work. On mixing with sand to give between 1:4 and 1:6½ mixes (depending on the degree of exposure), masonry cement produces a mortar suitable for most brickwork. Masonry cement has be controlled in the UK by BS 5224[23]. More recently, ENV 413[24] has been published which covers a wider range of types and compositions than BS 5224. ENV 413 introduces significant changes in the manner in which masonry cements are specified i.e., on the basis of 28 day compressive strength. The requirements for masonry cement to ENV 413 are summarised in **Table 7.14**.

Masonry cement that conformed to BS 5224 is equivalent to Class MC 12.5 of ENV 413-1 with air entraining agent. There is no UK experience of lower strength Class MC 5 with air entraining agent or Classes MC 12.5X and MC 22.5X without air entraining agent. Class MC 5 would be considered unsuitable for external masonry in the UK and non air entrained Classes MC 12.5X and MC 22.5X would be considered more acceptable to the action of freezing and thawing in fresh and hardened mortar.

Table 7.14 Requirements for masonry cement to ENV 413

Type	Strength class	Air-entraining agent	Content[a] (%)		Fresh cement properties		Compressive strength (N/mm^2)			Chemical requirements	
			Portland cement clinker	Organic material	Air content[b] (% by volume)	Water retention[c] (% by mass)	7 day	28 day		SO_3	Cl
MC 5	5	required	≥ 25	≤ 1	≥ 8 ≤ 20	≥ 80 ≤ 95	-	≥ 5	≤ 15	≤ 2	≤ 0.1
MC 12.5	12.5	required	≥ 40	≤ 1	≥ 8 ≤ 20	≥ 80 ≤ 95	≥ 7	≥ 12.5	≤ 32.5	≤ 3[e]	≤ 0.1
MC 12.5X	12.5	not permitted	≥ 40	≤ 1	≤ 6[d]	≥ 80 ≤ 95	≥ 7	≥ 12.5	≤ 32.5	≤ 3[e]	≤ 0.1
MC 22.5X	22.5	not permitted	≥ 40	≤ 1	≤ 6[d]	≥ 80 ≤ 95	≥ 10	≥ 22.5	≤ 42.5	≤ 3[e]	≤ 0.1

Notes: [a] constituents should not promote corrosion of embedded metal or impair the properties (including fire behaviour) of mortar made from a masonry cement; [b] an air entraining agent is incorporated into masonry cements of lower strength to improve their workability and durability. An upper limit is set for air content to ensure good bond strength to masonry units; [c] water retention limits are specified for all classes of masonry cement to provide performance suitable for use with both high and low suction masonry units; [d] the control of the masonry cement manufacturing process ensures that this upper limit is not exceeded; [e] if the Portland cement clinker content can be demonstrated to be not less than 55% of the whole masonry cement then the SO_3 content of these masonry cement classes is permitted to be not more than 3.5% by mass.

Water-repellent cements are made by the addition of a metallic soap (calcium stearate) to OPC, making the concrete less permeable. Hydrophobic cement consists of OPC with certain additions, e.g, oleic acid ($C_{18}H_{33}CO.OH$, $CH_3.(CH_2)_7.CH{=}CH.(CH_2)_7.(COOH)$). Hydrophobic cements are developed to prevent hydration during storage in unavoidable humid conditions. Each grain of cement is covered by a film of oleic acid (a water repellent film) at the manufacturing stage to increase its shelf life. The additions of oleic acid prevent hydration by coating the particles with this protective film. The protective film is removed by abrasion when it is mixed in with aggregates, and thereafter hydration takes place normally. The cement is manufactured only to special order.

Coloured cements are white Portland cement (**7.14.1**) with 5-10% pigment added in the form of metal earth oxides (**Table 7.15**). Greater than 10% pigment addition reduces cement strength. Pigments have to be finely ground as they reduce the effective surface area available for hydration, reducing the cohesive nature of the product.

Table 7.15 Colouring for cement by mineral oxide additions

Colours	Mineral oxide (BS 1014[25])
Brown	Manganese oxide (Mn_2O_3)
Black	Iron oxide (Fe_3O_4), carbon black
Green	Chromium oxide (Cr_2O_3)
Red	Red iron oxide (Fe_2O_3)
Blue	Cobalt oxide (Co_2O_3)
Yellow	Synthetic material

7.14.4 Portland Cement Replacement Materials

Materials other than Portland cement may contain hydraulic properties and can be used in the manufacture of cement, usually by partial replacement of Portland cement to produce **blended Portland cements**. In many cases there are considerable cost savings, since many replacement materials are produced as by-products from other processes (e.g, fuel ash, blastfurnace slag, burnt shale, etc.) Suitable materials are usually **pozzolanic** to some degree; those currently used within Europe and incorporated in European Standards are discussed in section **7.13**.

Pozzolanas. Naturally occurring pozzolanas were used in early cements (**7.12**) and are named after Pozzuoli in Italy where natural pozzolana is found. A pozzolanic material is one which contains 'reactive' silica (SiO_2) and is not cementitious in itself, but will, in a finely divided form and in the presence of moisture, chemically react with calcium hydroxide at ordinary temperatures to form secondary hydrated calcium silicates (the cementitious compounds). When blended with Portland cement, the calcium hydroxide is supplied from the hydration reactions of Portland cement (**7.16**)

$$\underset{\text{(cement clinker)}}{(CaO.SiO_2)} + \underset{\text{(water)}}{6H_2O} \rightarrow \underset{\text{(hydrated cement)}}{3CaO.2SiO_2.3H_2O} + \underset{\text{(free alkali)}}{3Ca(OH)_2}$$

which yields an alkali, $Ca(OH)_2$, which reacts with the pozzolan to form hydrated calcium silicates which possesses cementitious properties

$$\underset{\text{(Pozzolana)}}{2SiO_2} + \underset{\text{(water)}}{3H_2O} + \underset{\text{(free alkali)}}{3Ca(OH)_2} \rightarrow \underset{\text{(calcium silicate)}}{(CaO.SiO_2)} + \underset{\text{(water)}}{6H_2O}$$

The formation of secondary hydrated calcium silicates cannot be distinguished from the primary calcium silicates formed during the hydration of di- and tri-calcium silicates from Portland cement (**7.16**). The pozzolana has to be finely ground because, as with Portland cement, hydration is a surface area phenomena. In addition, the structure of the silica within the pozzolana is important, e.g, sand (SiO_2) will not react in this way as the structure is uniformly crystalline and chemically unreactive (**1.8.4**). The required structure must be glassy or amorphous with a disordered structure, as formed, for example, in the rapid cooling of volcanic magma to form pumice (a natural pozzolana).

Pozzolans are added to Portland cement for the following reasons

- to economise on cement;
- to improve workability and reduce porosity;
- to reduce the evolution of heat during the curing process (because the amount of Portland cement required can be reduced);
- to lower the specific gravity of the cement (many pozzolanas have significantly lower specific gravities than Portland cement, and therefore substitution of a proportion of the cement on a weight for weight basis will result in a greater volume of the paste).

Examples of naturally occurring pozzolanas include diatomaceous earth, pumice, opaline, chert, tuffs, and pumicite. In addition, many waste products from various manufacturing processes possess pozzolanic qualities which are together referred to as industrial pozzolanas.

Portland fly ash and granulated blastfurnace slag. Portland fly ash (formally known in the UK as pulverised-fuel ash, pfa) and/or granulated blastfurnace slag (formally known in the UK as ground granulated blastfurnace slag, ggbs) may be intimately mixed with Portland cement clinker during manufacture, although usually these constituents are mixed with Portland cement on site. Cements incorporating Portland fly ash and/or granulated blastfurnace slag produce cementitious products which have lower permeability than those based on Portland cement alone. This therefore enhances resistance to attack from water soluble sulfates and weak acids (**8.15**) and to the ingress of chlorides (**8.21.5**). Typically, a granulated blastfurnace slag content of about 70% provides adequate resistance to water soluble sulfates in groundwater of levels up to 6.0 g/l (Class 3 and 4 soils). Cementitious products made with a mixture of 25% to 40% by weight Portland fly ash and Portland cement have good sulfate resisting properties although in the presence of high magnesium concentrations (> 1.0 g/l), SRPC or granulated blastfurnace slag cements should be used (**8.19.3**).

The rate of strength development of Portland fly ash and/or granulated blastfurnace slag cements is much slower than that for Portland cement over the first 28 days, although the ultimate strength of the hardened cement paste is comparable to that of an equivalent Portland cement. However, due to the slower rate of hydration, the rate of heat evolution during the hydration of Portland fly ash and/or granulated blastfurnace slag is lower than that of Portland cement alone and hence additions (of up to 70% blastfurnace slag and up to 25% Portland fly ash) can be used to reduce the high heats of hydration evolved in mass concrete. In addition, the total quantity of heat evolved during curing of Portland fly ash and/or granulated blastfurnace slag cements is lower than comparable Portland cements.

Portland limestone cement. The addition of up to 5% limestone filler to Portland cement has little effect on the properties of the cement. The addition of up to 25% limestone gives comparable performance to Portland cement, but this is achieved with proportionally lower cementitious content and therefore, if comparable durability is to be obtained, cement contents must be increased.

Silica fume cement. Silica fume can be added as a filler (up to 5%) or between 6% to 10% to produce silica fume cement. Silica fume has a high surface area:mass ratio and, when blended as a minor constituent with Portland cement, acts to increase the rate of the hydration reactions by providing a readily accessible source of 'reactive' silica (SiO_2). The resulting cement exhibits characteristically high early strength development and reduced permeability, which enhances resistance to chemical attack and abrasion.

Burnt shale cement. Burnt (oil) shale can be added as a filler (up to 5%) or between 6% and 35% to produce burnt shale cement. Burnt shale cement is not currently produced in the UK and there are currently no British Standards for this material.

7.14.5 High Alumina Cement (HAC)

This cement, known in the UK as High Alumina Cement (HAC), is referred to as Calcium Aluminate Cement (CAC) within Europe. HAC was developed by the Frenchman Jules Bied as a solution to the problem of sulfate attack of Portland cements (**8.19.2**). Bied realised that the calcium silicates in Portland cement were responsible for producing free calcium hydroxide on hydration, which in turn reacted with alkali metal water soluble sulfates to produce gypsum, enhancing sulfate attack (**7.14.1**) and that tri-calcium aluminate hydration products were closely associated with the severity of sulfate attack (**8.19.2**). Hence a product which contained a high proportion of calcium aluminates in place of the less resistant calcium silicates, together with no tri-calcium aluminate, would provide a high degree of resistance to sulfate attack. HAC was first patented in France in 1908 and subsequently registered in the UK in 1909. During the development of HAC prior to the First World War, it was realised that HAC has the added advantage of developing very high early strength. HAC is also known by the trade names Lightning and Cement Fondu, the former relating to the rapid development of early strength and the latter relating to the process of solidification of a completely fused mass in the manufacturing process (unlike the clinkered mass of Portland cement).

HAC is fundamentally different in chemical composition, manufacture and characteristics to Portland cements. However, like Portland cement, it is a hydraulic cement. HAC is made from limestone (calcium carbonate, $CaCO_3$) and bauxite (alumina, Al_2O_3), which are progressively dehydrated and decarbonated are increasing temperatures until calcium and aluminium oxides are formed. Heating of these oxides is continued to the point of fusion (about 1600°C, much higher than that required to produce Portland cement clinker and hence the process is expensive). The molten material ('*cement fondu*') flows continuously to a cooling conveyor where it is cooled at a controlled rate to form flat slabs. The flat slabs ('pigs') are broken to form clinker, which is subsequently further cooled and ground to produce HAC. No further additions are made to the HAC clinker (unlike for Portland cements, where gypsum is added to retard the setting time, **7.16.1**). The setting time of HAC is controlled by its composition. The final product has a specific surface in the range 2250 cm^2/g to 3000 cm^2/g (minimum 2250 cm^2/g to BS 915) somewhat lower than that of Portland cement. As a result of the high (fusion) temperatures involved, HAC forms a ceramic bond and is hence widely used in refractory products (**7.4**), such as flue and furnace linings.

The main compounds in HAC are substantially different to those present in Portland cement (**Table 7.17**); they are calcium aluminates of low basicity, monocalcium aluminate $CaO.Al_2O_3$ (CA) and $5CaO.3Al_2O_3$ (C_5A_3). These compounds are responsible for the high compressive strengths achieved by HAC. On hydration

- CA results in the formation of monocalcium aluminate decahydrate $CaO.Al_2O_3.10H_2O$ (CAH_{10}), a small quantity of dicalcium aluminate octahydrate $2CaO.Al_2O_3.8H_2O$ (CAH_8) and alumina gel (Al_2O_3.aq). The hydrated calcium aluminate CAH_{10} is unstable both at normal and higher temperatures and becomes transformed to crystals of $3CaO.Al_2O_3.6H_2)$ (C_3AH_6), a process known as **conversion** (discussed below);
- C_5A_3 is believed to result in the formation of dicalcium aluminate octahydrate $2CaO.Al_2O_3.8H_2O$ (CAH_8).

The quantity of water that combines chemically with the anhydrous cement during hydration is about 50% of the weight of the cement. This is approximately twice as much as that required for the hydration of Portland cement (w/c = 0.28) (**7.16**). One effect of this is that, for the same mix proportions, HAC results in a lower porosity and therefore higher impermeability than would be the case for Portland cement.

HAC to BS 915 should have a total alumina content of not less than 32% by weight of the whole and the ratio by weight Al_2O_3:CaO should be not less than 0.85 and not more than 1.3. The minimum compressive strength of HAC is specified in BS 915 is 42 N/mm^2 at 1 day and 49 N/mm^2 at 3 days, with the proviso that the 3 day compressive strength must be higher than the 24 hour compressive strength. The initial setting time specified in BS 915 not less than 2 hours and the final set should occur not more than 8 hours after the initial set. HAC is somewhat slower in setting than Portland cements, but the subsequent development of strength is extremely rapid (much more rapid the RHPC, for example). In 12 hours it will effectively carry traffic and within about 24 hours it has gained approximately 90% of its ultimate maximum strength. This does, however, result in a much greater rate of heat output during hydration and the heat output may need to be checked by artificial cooling if the temperature reaches 30°C. The ultimate strength developed by HAC exceeds that for Portland cement, typically by about 25 to 30%. Other properties of HAC products (coefficient of thermal expansion, modulus of elasticity, moisture movement) are all of the same order as those of comparable Portland cement products.

The calcium aluminates in HAC, unlike the calcium silicates in Portland cement, do not liberate free calcium hydroxide during hydration. Since it is this constituent in the hardened cement paste that is attacked by a variety of substances, the absence of this constituent is the reason why HAC is resistant to many forms of chemical attack. Unlike Portland cement (which contains the alkali calcium hydroxide), HAC does not attack aluminium and zinc, so that these metals can be used in HAC concrete. However, the aluminium hydroxide gel (Al_2O_3.aq) produced during hydration is attacked by caustic alkalis (such as sodium and potassium hydroxide) and HAC is therefore not resistant to such agents. In addition to the good performance of HAC exposed to aggressive water soluble sulfates, HAC can also resist many weak acids (e.g, sulfuric, lactic, humic, carbonic). The over-riding problem with HAC is that of **conversion**. In addition to the deterioration effects of the process itself, conversion also decreases the resistance of HAC to sulfate attack (especially by magnesium sulfate), increases porosity (and hence the permeability of HAC to aggressive species) and reduces the resistance to frost damage (as a consequence of increased porosity).

Conversion

HAC has a very important and serious disadvantages in that it suffers from a degradation process known as **conversion**. This process was at least partly responsible for major structural collapses in the UK at a school in Camden, a swimming pool at Stepney and at the University of Leicester Bennett Building. From compressive strength measurements of HAC concrete taken from the Bennett Building, the least strength measured was about 13% of the design strength specified and the degree of conversion ranged from 85% to 95%. As a direct result of these failures, HAC has not been recommended for structural use in the UK since the mid-1970s. The chemical and mineralogical changes that take pace during conversion have been the subject of extensive study over many years, leading to a fairly good understanding of the process. The important reaction during the initial setting of HAC cement is the formation on hydration of monocalcium aluminate decahydrate (CAH_{10}), dicalcium aluminate octahydrate (CAH_8) and alumina gel (Al_2O_3.aq). These aluminates give the high strength to hardened HAC but they are only **metastable** and convert gradually to more stable calcium aluminate hydrates. Although the actual reactions and products are complex, conversion is often illustrated with respect to the formation of the stable compounds tricalcium aluminate hexahydrate (C_3AH_6) and gibbsite (AH_3), as shown below

$3(CaO.Al_2O_3.10H_2O)$	→	$3CaO.Al_2O_3.6H_2O$	+	$2(Al_2O_3.3H_2O)$	+	$18H_2O$
$3CAH_{10}$		C_3AH_6		$2AH_3$		$18H$

Experimental evidence suggests that this reaction is the most important conversion reaction. It was initially thought that these reactions were associated with high temperatures and humidities (as might occur in a swimming pool), though it is know appreciated that these reactions will take place at ambient conditions. The change in the composition is accompanied by a loss of strength and a change in the crystal form from hexagonal to cubic and rhombohedral (**Table 7.16**) with the release of water, both of which results in increase porosity of the HAC product. A number of mechanisms by which conversion reduces the strength of HAC have been proposed, although the most commonly accepted principal reason is due to increases in the density of the hydrate calcium aluminates. This can be appreciated by consideration of the densities of the compounds involved (**Table 7.16**).

Table 7.16 HAC hydration and conversion products

Compound	Occurrence	Crystal system[a]	Density (g/ml)
$CaO.Al_2O_3.10H_2O$ (CAH_{10})	HAC hydration product	Hexagonal	1.72 to 1.78
$2CaO.Al_2O_3.8H_2O$ (C_2AH_8)	HAC hydration product	Hexagonal	1.95
$3CaO.Al_2O_3.6H_2O$ (C_3AH_6)	HAC conversion product	Rhombohedral	2.52 to 2.53
$Al_2O_3.3H_2O$ (AH_3)	HAC conversion product	Cubic	about 2.4

Notes: [a] see section **1.15.3**

It can therefore be seen that, even though the conversion products occupy some space, a change from the hexagonal form to the cubic form under conditions such that the overall dimensions of the HAC product are constant (as is the case in the hardened HAC paste) results in an increase in the porosity of the paste. The volumes occupied by the compounds can be found from their densities, as shown for the most important conversion reaction below

Conversion reaction	$3(CaO.Al_2O_3.10H_2O)$	→	$3CaO.Al_2O_3.6H_2O$	+	$2(Al_2O_3.3H_2O)$	+	$18H_2O$
Molecular weight	1014		378		312		324
÷ Density	÷ 1.72		÷ 2.52		÷ 2.4		÷ 1
= Molecular volume	589		150		130		324
Ratio	1	:	0.255	:	0.220	:	0.550

Thus 1 ml of solid HAC hydration products yields only 0.475 ml of solid conversion products and 0.550 ml of water (or voids) (this, of course, would only apply if the starting point were a solid mass of calcium aluminate decahydrate and, in the case of HAC products such as concrete, the volume of voids would be proportionally lower). Experiments indicate that the porosity of converted HAC is about 20% and that of HAC concrete about 10%. Note that all concretes are porous (**6.1**) and approximately a 7% porosity results in a 50% reduction in strength, The effect of conversion of HAC is to increase the porosity by an additional amount, resulting in a considerable decrease in strength over and above that which would ordinarily be expected. Other reasons forwarded for the reduction in strength of HAC on conversion include

- the dissociation of the hydrates, generally resulting in a decrease in strength (whether due to the hydration process itself or to the formation of the reaction products);

- the degree of grain perfection and size resulting from conversion;
- the aging of microcrystalline alumina gel (Al_2O_3.aq) to macrocrystalline alumina gel (Al_2O_3.aq);
- the rate of conversion (the faster the rate, the greater the loss in strength).

It is likely that these mechanisms all contribute to the strength reduction of HAC on conversion but the relationship between them is highly complex and not fully elucidated. The precise way in which the changes on conversion take place depends on

- **temperature**, which affects the rate of conversion (the higher the temperature, the faster the rate of conversion). At temperatures lower than 55°C, the relationship between mineralogy and strength is that initially the formation of monocalcium aluminate decahydrate leads to a rise in strength to a maximum, which then falls to a minimum due to conversion. Subsequently, a small rise in strength has been noted as result of the gradual formation of Stratlings compound (C_2ASH_8). The minimum strength is reached quicker at higher temperatures and the amount of conversion at minimum strength is also affected by temperature.
- **w/c ratio**, the higher the w/c ratio the greater the rate of conversion. Examination of the crystal size of the conversion product tricalcium aluminate hexahydrate (C_3AH_6) shows that it increases with w/c ratio and with the rate of conversion (the crystal size, and hence the number of contacts between grains, has a significant effect on strength, smaller crystals leading to greater strength). Significant conversion has not been noted at w/c ratios of less than 0.5 and it is important to appreciate that original design advice specified maximum w/c ratios of 0.4 to prevent conversion.
- **chemical environment**, particularly the availability of alkali soluble sodium and potassium, present in some aggregates (e.g, those used at Stepney), are known to increase the rate of conversion.

In the past, the rate of conversion was also thought to related to the applied stress and the presence of moisture. Since the products of the conversion reaction have a higher density, stress should increase the rate of conversion but there is no clear evidence that the stresses normally induced in concrete by prestressing (**8.9.3**) or under load are of sufficient magnitude to produce a significant effect. Although water is required for conversion to proceed (conversion will not occur, for example, in desiccated HAC), the conversion process itself releases water and it is only in exceptionally dry conditions that conversion is likely to be delayed. In some cases, the conversion reaction has been observed to convert HAC from dark grey to a browner colour. This is mainly due to the oxidation of ferrous compounds which can take place in the absence of conversion and therefore colour changes are not conclusive evidence that conversion has taken place. The relationship between colour change and conversion is probably due to the fact the increased porosity caused by conversion allows greater exposure and therefore oxidation of ferrous compounds. The majority of HAC used in the UK for structural purposes prior to the mid-1970s was in the manufacture of precast, prestressed concrete I- and X-beams, where the rapid hardening characteristics made it an ideal product for prefabricated components. As a result of the effective withdrawal of HAC in the UK for structural purposes, HAC in concrete in buildings is now over 25 years old and the main degradation problems are associated with carbonation of concrete cover to steel reinforcement (**8.21.4**) and, to a lesser extent, chemical attack (**8.19**). Advice on the recognition of buildings which may contain HAC and on monitoring of such buildings has been published by the BRE[26,27].

7.14.6 Supersulfated Cement (BS 4248)

Supersulfated cement comprises a mixture of finely ground granulated blastfurnace slag and calcium sulfate together with Portland cement, Portland cement clinker or other source of lime. Under BS 4248, the granulated blastfurnace slag component should not be less than 75% by weight of the total quantity. Ethylene glycol is sometimes added to the cement clinker before grinding to reduce the energy required. It is driven off by the heat generated during grinding and only traces remain. However, since this compound will reduce performance properties, the proportion of ethylene glycol in the final product should not exceed 0.1%. Supersulfated cement contains no free lime and no tri-calcium aluminate (C_3A) and therefore confers a high chemical resistance to aggressive conditions; in particular, cementitious products made with supersulfated cement have high resistance to water soluble sulfates (**8.19**). It has been found in practice that dense concretes made with supersulfated cement and a w/c ratio of 0.45 or less have given an acceptable life in contact with weak acidic solutions (pH 3.5 upwards). Supersulfated cement is finely ground (specific surface $\not> 4000\ cm^2/g$) and has intrinsically low heat of hydration. Supersulfated cement is not currently produced in the UK.

7.15 PORTLAND CEMENTS

Portland cement is the most important hydraulic cement and accounts for approximately 90% of all cement production in the UK. Portland cement is used extensively in construction to produce cementitious products, including mortars (**7.19**), renders (**7.20**) and concrete (**Chapter 8**).

Portland cement is obtained by heating an intimate mixture of mainly calcareous materials (principally limestones, chalk, etc.) which provide the reactive calcium oxide, CaO and argillaceous materials (principally clays, shales, etc.) which provide the reactive silica, SiO_2, and other silica, aluminium and iron oxide bearing materials at a temperature of about 1400°C to produce a partially sintered material (called **cement clinker**). A small quantity of calcium sulfate (about 2 to 4%) is added and the mixture is then ground to a very fine powder (Portland cement). The final chemical composition of Portland cement is extremely complex, but comprises principally calcium and magnesium oxide (CaO, MgO), silica (SiO_2), alumina (Al_2O_3) and iron oxide (Fe_2O_3), with minor quantities of other metallic oxides. Hydraulic hardening of Portland cement is primarily due to the hydration of calcium silicates, but other constituents may also participate in the hardening process (e.g, aluminates).

7.15.1 Manufacturing Process

Portland cement manufacture is essentially the same as Aspdin's process. The raw materials are mixed very roughly in the proportion of 2-4 parts of calcareous material to 1 part argillaceous material. Limestones and shales have first to be crushed, ordinarily be a two stage process; first crushing reduced the rocks to a maximum size of about 150 mm and secondary crushers or hammer mills reduce particles to about 75 mm or smaller. They may then be ground in ball mills in a dry state (the 'dry' process) or mixed in a wet state (the 'wet' process). In the UK, the 'dry' process is most often used as this reduces the quantity of heat absorbed by the slurry water during the manufacture. The dry powder is initially passed through a kiln which uses recovered kiln gases to preheat it to about 750°C, then with added fuel to precalcine at 900°C prior to passage to a rotary kiln for clinkering at clinkering temperatures of about 1400 to 1500°C (i.e., below the point of complete fusion). In the dry process, the slurry is fired directly at the clinkering temperature. In both processes, an intimately mixed feedstock to the kiln is required for maintaining the correct proportions. Progression of the dry powder through successively higher temperatures produces a number of effects

- initially water is driven off;
- at about 600°C, calcium carbonate ($CaCO_3$) decomposes to calcium oxide (CaO) with the evolution of gaseous carbon dioxide (CO_2);
- finally, fusion reactions start at about 1200°C and calcium silicates, calcium aluminates and smaller amounts of other compounds are formed. The oxides of iron, aluminium and magnesium present in the argillaceous raw material assist in this process by acting as a flux to enable the calcium silicates to be formed at a lower temperature than would otherwise be possible.

The material emerges from the kiln as a **cement clinker** of particle size of the order of a few millimetres. The cement clinker is first cooled with heat recovery, then passed to ball mills where gypsum (and, in some countries in Western Europe, up to 5% inert filler, such as limestone dust) is added. It is ground to the requisite fineness according to the class of product (ordinarily 2 - 80 µm, equivalent to a specific surface area of about 300 m^2/g). The high surface area:mass ratio facilitates hydration reactions (**7.16**).

7.15.2 Composition of Cement Clinker

There are four main compounds (or, more correctly, phases) formed during clinkering and these exist in a complex and impure semi-crystalline form (**Table 7.17**). Strictly, tetracalcium aluminoferrite (C_4AF) is not a true compound, but represents the average composition of a solid solution. The impure forms of C_3S and C_2S in cement are commonly called alite and belite respectively. Also shown in **Table 7.17** are the strength gain characteristics of each compound. It has been shown that C_3S is the main contributor to strength development, with C_2S contributing more to late strength. The C_3A and C_4AF are important from the point of view of sulfate resistance (**7.14.1**), and, although they do not contribute significantly to cement strength, they play an essential role as fluxes in the sintering reactions that take place in the kiln.

Table 7.17 Compounds present in the cement clinker

Compound (phase)		Mixed oxides	Chemical formula	Strength gain characteristics
C_3S	Tri-calcium silicate	3CaO. SiO_2	Ca_3SiO_5	Rapid. High heat evolved
C_2S	Di-calcium silicate	2CaO. SiO_2	Ca_2SiO_4	Medium. Low heat evolved
C_3A	Tri-calcium aluminate	3CaO. Al_2O_3	$Ca_3Al_2O_6$	Slow. Reacts with sulfate
C_4AF	Tetra-calcium alumino ferrite	4CaO. Al_2O_3. Fe_2O_3	$Ca_4Al_2Fe_2O_{10}$	Very slow. Dark colour

Each cement grain will contain and intimate mixture of these compounds but the exact composition will depend on the precise proportions of the raw materials used, the kiln temperatures and the impurities present. In addition, these compounds are indeterminate by direct chemical analysis, with assessment ordinarily undertaken by calculation based on the mixed oxide proportions present. Typical oxide compositions of some common cements (**7.13**) are shown in **Table 7.18**.

Table 7.18 Typical oxide compositions of selected cements

Cement	Oxide							
	SiO_2	Al_2O_3	Fe_2O_3	MgO	CaO	Na_2O	K_2O	TiO_2
Portland cement	20	5	4	1	64	0.2	0.5	-
Portland fly ash cement								
• low CaO	48	27	9	2	3	1	4	-
• high CaO	40	18	8	4	20	-	-	-
Blastfurnace slag cement	36	9	1	11	40	-	-	-
Silica fume cement	97	2	0.1	0.1	-	-	-	-
High alumina cement (HAC)	5	39	11	0.5	37	-	-	2

Table 7.18 illustrates the close chemical composition relationships of different cements (see also ***Figure 7.1***). 'Reactive' calcium oxide (CaO) and silica (SiO_2) are the principal compounds present in Portland cement, typically in the ratio 3:1 CaO:SiO_2 by weight. Variations in the exact composition produce variations in the properties of the resultant cement (**7.14.2**). Note that, relative to other cements, HAC has a decreased proportion of SiO_2 and an increased proportion of Al_2O_3 due to the different raw materials used in the manufacture of this product (**7.14.5**).

7.16 HYDRATION REACTIONS FOR PORTLAND CEMENT

The setting and hardening of hydraulic cements is the result of hydration reactions occurring between the four cement compounds (**Table 7.17**) and water. When the cement is mixed with water to form a paste, hydration reactions begin, resulting in the formation of gel and crystalline products capable of binding inert particles (such as aggregate, **8.3**) into a coherent mass. **Setting** can be defined as the stiffening of the originally plastic paste of cement and water such that no significant indentation is obtained when subjected to certain standardised pressures in accordance with BS EN 196-3 (**3.4.1**). **Hardening** follows setting and is the result of further hydration reactions advancing gradually into the interior of the cement particle. The strength development of cement depends on the amount of the gel formed and the degree of crystallisation. The hydration reactions proceed through a series of stages (***Figure 7.5***), as follows

- a **dormant period** immediately after mixing, where the plasticity of the paste of cement and water remains relatively constant, and any loss of plasticity can be recovered on remixing;
- the **initial set** when the mix starts to stiffen at a much faster rate (normally commencing between two to four hours after mixing at normal temperatures). As this time the mix has little or no strength;
- the **final set** when the mix begins to harden (and gain strength). This occurs a few hours after the initial set and proceeds rapidly for a period of one or two days, and thereafter for a period of at least a few months and a steadily decreasing rate.

The hydration reactions involving the four cement compounds are

$2C_3S$	+	$6H$	$\rightarrow$	$C_3S_2H_3$	+	$3CH$	+	$2X$ joules
(tri-calcium silicate)		(water)		(CSH fibres) (gel) (strength)		(calcium hydroxide) (crystals) (alkali)		(502 J/g) (Heat)

$2C_2S$	+	$4H$	$\rightarrow$	$C_3S_2H_3$	+	CH	+	X joules
(di-calcium silicate)		(water)		(CSH fibres) (gel) (strength)		(calcium hydroxide) (crystals) (alkali)		(260 J/g) (Heat)

C_3A	+	$6H$	$\rightarrow$	C_3AH_6	+	$3X$ joules
(tri-calcium aluminate)		(water)		(calcium aluminates) (crystals) (absorbs Cl^- ions) (reacts with SO_4^{2-} ions)		(867 J/g) (Heat)

complex reactions

$$C_4AF + xH \rightarrow C_3(A.F).3C\bar{S}.H_{32} \text{ (high sulfate form)}$$

$$\rightarrow C_3(A.F).C\bar{S}.H_{18} \text{ (low sulfate form)}$$

(tetra-calcium alumino ferrite) (water) (calcium sulphoaluminates) (419 J/g) (Heat)

7.16.1 Hydration Product Development Over Time

The cumulative quantity of individual products formed are illustrated in ***Figure 7.5***.

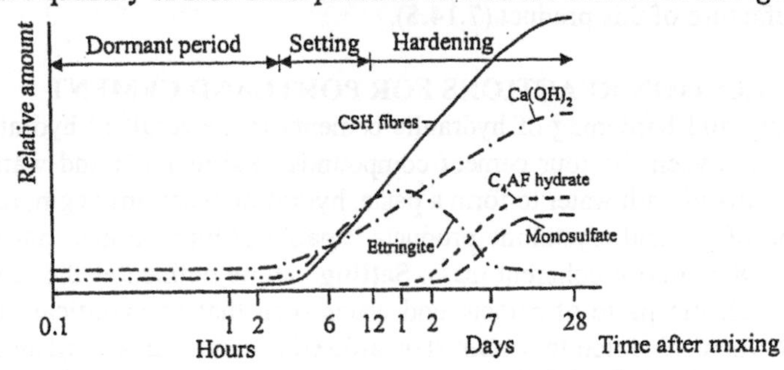

Figure 7.5 *Typical hydration product development in Portland cement paste*

Initially, hydrates are formed from the corresponding anhydrous products that passed into the solution. The hydrates have lower solubility that their corresponding anhydrous products and begin to crystallise from solution when it becomes saturated with respect to the anhydrous products. Hydration of C_3A occurs very rapidly with the formation of calcium aluminate hydrate (C_3AH_6) crystals, which are important from the point of view of sulfate attack (**8.19.2**). The reaction is fast enough to result in **'flash-setting'** in a few minutes. The hydrate crystals form a film over the silicate particles (**7.17**), inhibiting their further hydration, so that subsequent development of strength is slow and incomplete. In addition, when this rapid evolution of heat stops, the subsequent cooling of the cementitious product could be severe and may cause thermally induced cracking (**2.8.2**). For these reasons, gypsum (**14.7.3**) is added at the grinding stage to retard the reaction by retarding the hydration of C_3A . This is achieved by the reaction of gypsum with C_3A to produce a metastable calcium sulfoaluminate (called **ettringite**)

C_3A	+	$3C\bar{S}H_2$	+	$26H$	→	$C_3A.3C\bar{S}.H_{32}$
(tricalcium aluminate)		(gypsum)		(water)		(calcium sulphoaluminate) (ettringite) (crystals)

The ettringite is insoluble and therefore the high concentration of aluminates in solution is prevented, retarding the initial set of the cement. Ettringite crystallises out of solution causing a volume expansion. However, unlike in sulfate attack of the hardened cement paste (**8.19.2**), this volume expansion is not disruptive as secondary transformation of the metastable ettringite into a more stable monosulfate form ($C_3A.C\bar{S}.H_{16}$) occurs. Over a similar time frame, the tetra-calcium aluminoferrite (C_4AF) reacts with water (via an intermediate compound with gypsum) to form products of variable composition (approximated by a high and low sulfate form in the above equations and similar to the C_3A products). The products of these reactions (sulphoaluminates and aluminoferrites) are important from the point of view of **absorbing aggressive chlorine (Cl^-) ions**, which may otherwise promote the corrosion of steel in reinforced concrete (**8.21.5**).

Both C_2S and C_3S (or, more correctly, belite and alite) hydrate at a slower rate than C_3A and are responsible for most of the significant strength and reactivity characteristics of the hardened cement. C_3S hydrates more rapidly than C_2S but C_2S contributes little heat. However, both constituents yield an amorphous mass of tricalcium disilicate hydrate (a gel) and crystals of calcium hydroxide, $Ca(OH)_2$. The calcium silicate hydrate gel (commonly referred to as CSH fibres) is responsible for the **strength** of the hardened cement paste (**7.17**) and the calcium hydroxide formed is an **alkali** and is very important in protecting steel reinforcement from corrosion (**8.21.3**).

7.16.2 Rate of Strength Gain Over Time

Each constituents present in the cement adhesive hydrates and hardens at a different rate. The relationship between the rate of development of compressive strength of the four principal compounds in cement is shown in ***Figure 7.6***, which indicate that

- the ordered rate of strength development is $C_3S > C_2S > C_3A > C_4AF$;
- C_2S and, particularly C_3S, contribute most strength to the hardened cement paste. A cement containing a high proportion of C_3S will promote rapid strength gain in the hardened cement paste, as in rapid hardening Portland cement (**7.14.1**);
- hydration reactions proceed **provided** the mixture remains wet (if insufficient water to allow hydration reactions to produce the CSH gel which fills the available space, the hydration reaction ceases, a process known as **self-desiccation**. For this reason, **curing** is an important parameter in determining concrete quality (**8.13.1**). Once the available water has been consumed, the hydration reaction will cease and the process cannot be restarted by the addition of further water;
- hydration reactions proceed over long time periods and can in fact never be considered as complete (the extent of the completeness is termed the **degree of hydration**, an important parameter in determining the porosity of the hardened cement paste, **7.17**).

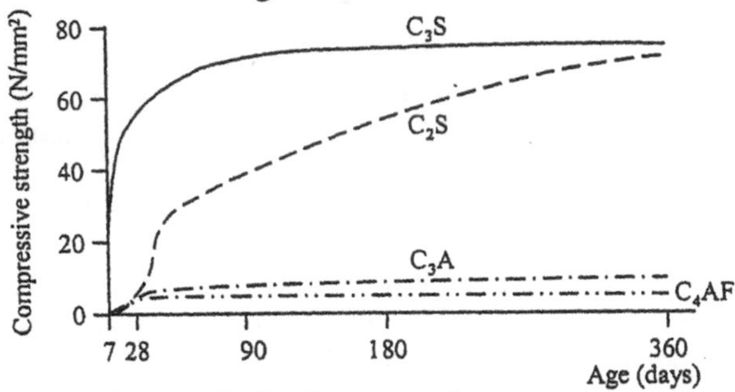

Figure 7.6 *Comparison of strength development of pure cement compounds from Portland cement*

As a result of this increase in strength with time, the scientist considers the compressive strength of cementitious products to be that at 28 days, as at this time the strength developed is a large proportion of the ultimate strength. To illustrate this, **Table 7.19** shows the approximate compressive strength of OPC (CEM I 42.5) and RHPC (CEM I 42.5R) at various times expressed as a percentage of the 28 day compressive strength.

Table 7.19 Strength development of cement with respect to time

Age (days)	Fraction of 28 day compressive strength	
	OPC (CEM I 42.5)	RHPC (CEM I 42.5R)
3	0.40	0.55
7	0.65	0.75
28	1.00	1.00
90	1.20	1.15
360	1.35	1.20

The CSH fibre hydration product is a gel (**2.9.2**), which has an amorphous or noncrystalline structure (and therefore little strength) and is responsible for the **setting** process. The setting process is accompanied by a change in flow characteristics (or **stiffness**) characteristics. To improve strength, the gel must form a compound in the **crystalline** state (this is the **hardening** process). The ions responsible for the promotion of the crystalline state are the alkali ions Ca^{2+}, Na^{+}, etc. (which are also responsible for alkali silica attack under certain conditions, **8.20**). The covalently bonded silicates (**1.13**) and aluminates do not favour the formation of the crystalline state. It is not until some hours after the initial set that the rate of hardening and strength gain increases rapidly (final set). This proceeds rapidly for about one or two days, and thereafter the cement continues to harden and gain strength but at a decreasing rate (***Figure 7.6***). The cementitious matrix is very strong in compression, but weak in tension. In many cementitious products, metals (predominantly steels) are embedded in the matrix to improve tensile properties (**8.9.3**).

7.16.3 Heat Evolved Over Time

The hydration reactions are exothermic and the rate of heat output from Portland cement involve all four main compounds (**Table 7.20**) and include the heat output from the reactions of hydration products with one another.

Table 7.20 Heats evolved from the cement constituents

Constituent	Heat evolution (J/g) at the following ages					
	3 days	7 days	28 days	90 days	360 days	6½ years
C_3S	243	222	377	436	490	490
C_2S	50	42	105	176	226	222
C_3A	888	1559	1378	1303	1169	1374
C_4AF	289	494	494	410	410	465

The heat output associated with the hydration of the four compounds in cement are in the order $C_3A > C_4AF > C_3S > C_2S$. In oversimplified form, the heat evolution is noted in the chemical reactions above (**7.16**). Measurement of the total rate of heat output of Portland cement at constant temperature gives the form of the curve in ***Figure 7.7.***

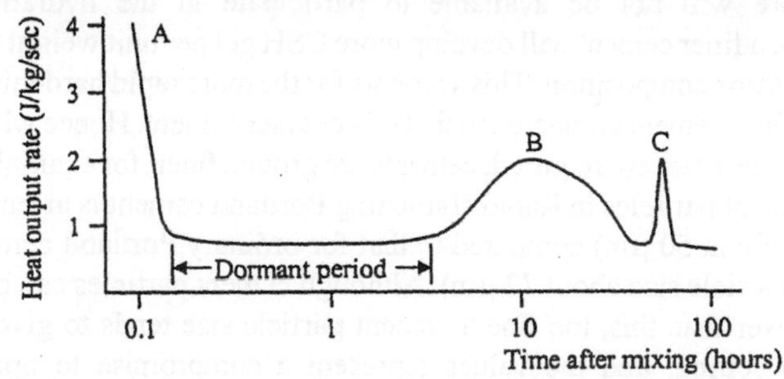

***Figure 7.7** Typical rate of heat output from Portland cement during hydration at constant temperature*[28]

Although not fully understood, the main contributions to the rate of heat output are thought to derive from the following reactions

- the first peak **A** occurs on mixing and is very high but of short duration (lasting only a few minutes). Peak **A** is thought to derive from the rehydration of calcium sulfate hemihydrate ($CaSO_4.½H_2O$) which arises from the thermal decomposition of gypsum in the grinding stage of the manufacturing process. The rehydration is exothermic and gypsum is reformed (**14.7.3**)

$$C\bar{S}.½H \quad + \quad 3H \quad \rightarrow \quad C\bar{S}H_2$$

(calcium sulfate hemihydrate) (water) (gypsum)

  Additional contributions to peak **A** come from the hydration of free calcium oxide (CaO), the heat of wetting, the heat of solution and the initial reactions of the aluminate compounds (as described above).
- thereafter, the rate of heat output declines to a low constant value associated with a dormant period when the cement is relatively inactive, lasting for up to 2 to 3 hours;
- thereafter, at a time roughly corresponding to the initial set, the rate of heat output starts to increase, reaching the broad peak **B** sometime after the time of the final set. Peak B is though to derive from the hydration of the tri-calcium silicate (C_3S). Di-calcium silicate (C_2S) contributes little to the heat output. A cement containing a high proportion of C_2S will promote a low heat output in the hardening cement paste, as in low heat Portland cement (**7.14.1**).

- in cements with a substantial C_3A content (greater than about 12%), a shape peak **C** is often observed toward the end of the heat output curve. This is associated with the secondary transformation of the metastable ettringite to the stable monosulfate form ($C_3A.C\overline{S}.H_{16}$) (**7.16.1**).

7.16.4 Effect of Fineness and Temperature

The fineness of cement greatly affects that setting time and the strength of the hardened cement paste because the chemical activity of the solid is directly proportional to its surface area, which greatly increases with the increased fineness of the particles. As hydration proceeds from the outer to inner core of the cement particle, the smaller the particle, the greater the probability that nearly the whole core will be converted to CSH gel and calcium hydroxide crystals. For coarser particles, a considerable proportion of the inner core will not be available to participate in the hydration reactions. Consequently, a finer cement will develop more CSH gel per unit weight than a coarser cement of the same composition. This accounts for the more rapid hardening and greater strength of a finer cement compared to that of a coarser cement. Hence where high early strength characteristics are required, cements are ground finer; for example, the specific surface of cement particles in Rapid Hardening Portland cement is around 3250 cm^2/g (particle size about 50 μm) compared to that for ordinary Portland cement of around 2250 cm^2/g (particle size about 72 μm). Although cement particles can be produced at sizes even lower than this, too fine a cement particle size tends to give considerable shrinkage on setting, and the values represent a compromise to optimise overall properties. For many years there has been a demand by specialist users in the UK for a cement which makes it easier to remove excess water from concrete during compaction. In some applications, the fineness of the cement is more critical than its compressive strength and, as a result, 'controlled fineness Portland cement' was introduced in BS 12[29]. 'Controlled fineness Portland cement' is a Portland cement having a specific surface controlled within a small agreed range that usually lies below that for ordinary Portland cement. In common with other chemical reactions, increasing temperature increases the rate of the hydration reactions and conversely decreasing temperature will slow the rate of hydration. Provided a significant quantity of the mix water has been incorporated in the hydration reactions so that insufficient is free to dramatically disrupt the immature matrix on freezing, hydration reactions will proceed below 0°C, although hydration will stop completely at about -10°C. The effects of variable curing temperatures are, in fact, more complex than this relationship would suggest. Whilst high temperatures during the early curing of concrete increases the rate of strength gain, at later stages of the curing process higher strengths are obtained at lower temperatures. It is thought that, although the CSH gel will be formed more rapidly at increased temperatures, its structure is more coarse and less uniform than that formed at low temperatures. Since the strength of the hardened cement paste is largely a surface effect (**7.17**), coarser structures have reduced specific surface area compared to finer structures and therefore reduced strength. Although the optimum temperature for maximum long-term strength gain will vary for different cements, on average it is around 13°C. Steam curing of concrete under pressure (autoclaving) allows the use of temperatures of greater than 100°C without causing drying out. However, autoclave treatment may have a detrimental affect on strength by producing CSH gel of less uniform structure (as described above). In addition, autoclaving may produce different hydration products; for example, the formation of dicalcium silicate hydrates (C_2SH), which have lower strength and increased porosity compared to the calcium silicate hydrates ($C_3S_2H_3$) normally produced (**7.16.1**). Despite these disadvantages, autoclaving is widely used for the manufacture of precast units (**7.7.1**), such as autoclaved aerated (AAC) concrete blocks, where the benefits of high early hardening (enabling formwork to be struck early, increasing productivity) may outweigh the reductions in strength.

7.17 STRUCTURE OF HARDENED PORTLAND CEMENT PASTE

The sequence of events responsible for the formation of the hardened cement past is schematically represented in ***Figure 7.8***.

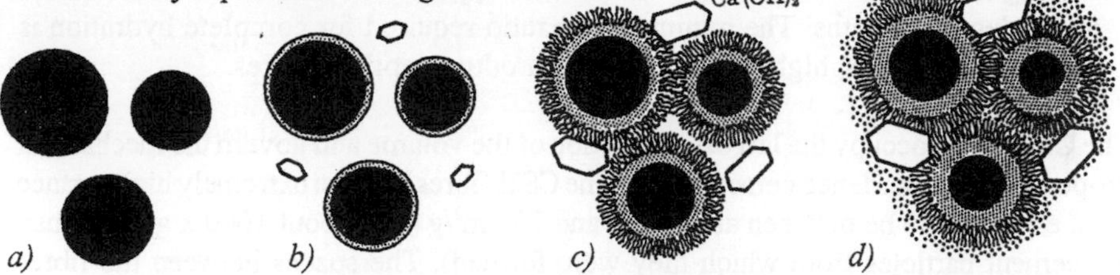

***Figure 7.8** Schematic representation of the sequence of hydration of cement*

In ***Figure 7.8a***, the cement grains are shown, initially surrounded by water. Hydration reactions produce a precipitation of CSH coatings around the cement grains (***Figure 7.8b***) which provides a barrier that inhibits the access of water to the cement surface. The initial formation of CSH coatings therefore retards hydration, and this is associated with the reduced activity of the cement during the dormant period. The end of the dormant period occurs when the CSH coatings are disrupted, thus leading to accelerated hydration and the growth of secondary CSH gel hydration products (***Figure 7.8c***) and calcium hydroxide, $Ca(OH)_2$. This is associated with the increase in the rate of heat output giving rise to peak **B** (***Figure** 7.7*). The period of accelerated hydration is followed by a gradually decreasing rate of reaction attributed to the later infilling and accretion by hydration products and calcium hydroxide (***Figure 7.8d***). Effectively, as the layers of hydration products thicken by growth around the cement grains and as free water in the microstructure is consumed, longer range diffusional processes dominate and progressively slow down hydration.

A number of theories have been forwarded to explain the disruption of the initial CSH coating (***Figure 7.8c***). The most commonly accepted theory is that the coatings are ruptured by **osmotic pressure** (**6.6.4**) effects due to the selectively permeable character of the colloidal CSH gel. The CSH gel coating acts as a semi-permeable membrane that allows the inward diffusion of water to the cement grains and the outward diffusion of calcium and hydroxyl ions released by hydrolysis. The membrane is, however, effectively impermeable to the outward diffusion of larger species, such as silicate ions. Preferential diffusion leads to a rising osmotic pressure within the CSH gel coating eventually sufficient (at the end of the dormant period) to cause rupture. The hydrous silicates, originally retained within the membrane formed by the CSH gel coating, is extruded into the outer calcium oxide rich bulk solution, where it is immediately precipitated as secondary CSH gel products growing from the surrounding faces of the original coating.

The final (hardened) cement paste therefore consists of

- a residue of unhydrated cement at the centre of the original cement grains;
- the products of the hydration reactions, principally CSH fibres;
- the alkali calcium hydroxide, $Ca(OH)_2$;
- unfilled spaces between cement grains (**capillary pores**), resulting from evaporation of unreacted water. The proportion of the hardened cement paste that is capillary pores is dependent on the **w/c ratio** (**8.13.2**); e.g, a w/c ratio of 0.4 produces 30% porosity whereas a w/c ratio of 0.7 produces 50% porosity. Capillary pores increase diffusion of hostile species (chlorides, acidic gases, etc.) through the hardened paste, leading to reduced durability (**8.22**). In addition, porous pastes have lower densities and reduced strengths. The minimum w/c ratio required for complete hydration is approximately 0.28; higher w/c ratios will produce capillary pores.

The CSH fibres occupy the largest proportion of the volume and govern the mechanical properties of the hardened cement paste. The CSH fibres have an extremely high surface area, estimated to be between about 100 and 700 m^2/g (i.e., about 1000 x greater than the cement particles from which they were formed). The spaces between the fibres (known as **gel pores**) (about 0.5 to 5 nm wide) are about two orders of magnitude smaller than the capillary pores. The strength of the hardened cement paste is derived from the van der Waals type forces between adjacent CSH fibres. Although individually weak, the integrated effect over the enormous surface area is considerable, with the result that the concrete matrix is very strong in compression (but weak in tension).

7.18 DEGRADATION OF HARDENED PORTLAND CEMENT

Cements are usually used in combination with aggregates to produce various cementitious materials, the most important of which is concrete. A variety of mechanisms may be responsible for the degradation of concrete, including interactions with the hardened cement paste. We therefore examine all mechanisms responsible for the degradation of cementitious products in **Chapter 8**.

7.19 MORTAR

Mortar consists of sand, a binder (such as cement, **7.11**) and/or lime (**7.10**) and water. Additional admixtures (**8.7**) and/or additions (**8.8**), including plasticisers and colorants, may be used. Mortar is required to perform one or more of the following functions

- to take up the tolerances between the building units (bricks, blocks, stone, etc.) and provide a bearing for adjacent units;
- to permit movement allowing the transmission of shear or tensile forces (any movement that occurs within a masonry construction should take place in the form of microcracks within the mortar rather than within the building units themselves). Thus the general working rule for mortar use is that the mortar should be weaker than the individual building units bonded;
- to be durable and develop compressive strength quickly. A compressive strength of fully cured mortar for low rise construction of about 2 to 5 N/mm^2 is required;
- to act as a sealant between the building units to resist the penetration of water;
- to contribute aesthetically to the appearance of the wall.

Mortars should be workable yet cohesive in the fresh state (i.e., they should hang on the trowel and not slip off to easily, and should spread easily), and not readily lose water (e.g, they should not stiffen in contact with absorptive bricks) and so reduce workability. Mortars should remain sufficiently plastic for the building units to be adjusted to the operatives building lines. Strong mortars (high cement:sand ratio) are not necessarily the most durable as they tend to form large cracks when differential movement takes place. Weaker mortars (low cement:sand ratio) help to accommodate dimensional movement and any microcracks formed are more visually acceptable; this is particularly important in building units which undergo large temperature induced dimensional movement (**2.8.2**), e.g, clay bricks, calcium silicate bricks, concrete bricks and blocks.

7.19.1 Lime Mortar

Traditionally, mortars based on lime (**7.10**) were used in the UK, commonly based on one volume of lime for every three volumes of sand. The proportion of lime in 1:3 lime:sand mixes is the optimum to ensure that the binder fills the voids between the sand particles, thereby increasing workability. Lime mortars are generally more flexible and exhibit less shrinkage than cement mortars but do not attain the same strength. This is a direct result of the curing process; lime mortars do not 'set', but stiffen as they lose water by absorption to the porous building units and by evaporation. In inclement weather, evaporation rates are slow, limiting work to dry weather conditions. Further strength gain is achieved by carbonation (**7.10.3**) in contact with carbon dioxide in the surrounding air. Hence lime mortars cannot be used below ground level. The carbonation reaction is very slow and is limited only to the surface of the lime mortar and therefore the rate of strength gain is slow. For this reason, lime mortars are rarely used today, except for restoration work. The binder in modern mortars is more usually based on Portland cement (**7.14.1**) which provide increased strength, quicker rates of strength gain and can be used in all weathers and in all locations.

7.19.2 Cement Mortar

For adequate strength in fully-hardened mortar and suitable workability, about one volume of Portland cement is required for three volumes of sand. 1:3 cement:sand mixes are suitable for work subjected to high stress and made with appropriate building units of suitable strength; they are relatively impermeable (therefore resisting the effects of frost damage and the deposition of water soluble salts). Although these properties make 1:3 cement:sand mixes suitable for below ground work (foundations, below dpcs, etc.), they are too strong for most other purposes. Hence, in many mortars

- part of the Portland cement in a 1:3 cement:sand mix is replaced by an equal volume of lime to give **Portland cement:lime:sand ('compo') mortars**. In this way the advantages of both lime and cement are obtained; the resulting mortar has good workability, water retention, bonding properties and early strength gain. However, there is a small reduction in strength; e.g, a 1:½:4 cement:lime:sand mix is about 50% weaker (a 1:2:9 mix about 10% weaker) than a 1:3 cement:sand mix (**Table 7.22**);
- the mortar is air-entrained using a mortar plasticiser (complying with BS 4887[30]) to give **air-entrained Portland cement:sand mortars**. Air-entrainment enhances workability and frost resistance whilst reducing the quantity of mix water required. Although not highly critical, too much air entrainment leads to poor durability, low compressive and bond strength, whilst too little air entrainment reduces workability and frost resistance. The air content should be between about 10% and 18%.

Admixtures for mortar are governed by the recently published BS EN 934[31]. The special properties of mortar needed for laying masonry have been found to vary widely depending on the type of building unit, the grading of the available sand and the laying technique. This has led to the development of **ready-mix mortar** (BS 4721[32]) and to a variety of cements designed specifically for use in mortars, known as masonry cements (**7.14.3**). Masonry cements have a lower Portland cement clinker content than OPCs, leading to a requirement for higher cement contents in masonry cement:sand mortars (as shown by comparison of equivalent strengths of cement:lime:sand mortars in **Table 7.22**). Specifications for masonry mortar are governed by prEN 998[33].

7.19.3 Mortar Mixes

In the UK, five levels of replacement of Portland cement with lime are prescribed (mortar designations i. to v.) based on the type of construction for which they are suitable (**Table 7.22**, which shows the compressive strength associated with each designation). Mortar designation (v) is currently little used except for repairs of historic buildings. Under prEN 998[34], the designation of mortar is performance based and includes information on mortar type, and fresh and hardened properties. Classification is based on compressive strength (N/mm^2), prefixed with 'M' (for mortar). Mortar mixes of equivalent compressive strength (proportions by **volume**) are shown in **Table 7.22** (over). Variations mortar properties can be achieved by additives or by special treatment. For example, the strength of the bond formed by air entrained mortars can be improved by the addition of water retaining additives (e.g, cellulose esters); early damage by freezing can be reduced by providing expansion cavities within the joint. Where mixes contain hydrated lime (lime putty), this is best made 1-2 days before use. The hydrate should be preferably soaked overnight. **Table 7.21** illustrates suitable mortar mixes from within any one mortar designation based on strength and resistance to frost and rain penetration. Generally, mortar should contain no more cement than is necessary to give adequate strength in the whole masonry unit, unless there is a good reason for choosing a richer mix (for example, for winter construction, where richer mixes will provide increased protection from frost due to more rapid strength gain, or where enhanced resistance to sulfate attack is required).

Table 7.21 Selection of mortar designations (to BS 5628) for construction

	Clay bricks		Concrete and calcium silicate bricks	
Early frost hazard[a]	No	Yes	No	Yes
Internal walls	(v)	(iii) or (iv)[b]	(iv)	(iii)
Inner leaf of cavity walls	(v)	(iii) or (iv)[b]	(iv)	(iii)
Backing to external solid walls	(iv)	(iii) or (iv)[b]	(iv)	(iii)
Outer leaf of external cavity walls				
• above dpc	(iv)[d]	(iii)[d]	(iv)	(iii)
• below dpc	(iii)[e]	(iii)[b,e]	(iii)[i,e] or (ii)[e,k]	(iii)[i,e] or (ii)[e,k]
Parapet walls. Domestic chimneys				
• rendered	(iii)[f,g]	(iii)[f,g]	(iii)	(iii)
• not rendered	(ii)[h] or (iii)	(i)	(iii)	(iii)
External free-standing walls	(iii)	(iii)[b]	(iii)	(iii)
Cills, Copings	(i)	(i)	(ii)	(ii)
Earth-retaining walls	(i)	(i)	(ii)[e]	(ii)[e]

Notes: [a] during construction, before mortar has hardened (say 7 days after laying) or before wall is completed and protected against the entry of rain at the top; [b] special precautions and/or air-entrainment may be required in cold weather; [c] if not plastered, use designation (iv); [d] if rendered, use designation (iii) made with SRPC; [e] if sulfates are present, use SRPC; [f] parapet walls of clay units should not be rendered on both sides (if unavoidable, select mortar as though not rendered); [g] use SRPC; [h] with 'special' quality bricks, or with bricks that contain appreciable quantities of water soluble sulfates; [i] but $\ngtr$ 150 mm above finished ground level; [k] within 150 mm of finished ground level or below ground.

Table 7.22 Comparison of mortar mixes of equivalent compressive strength (proportions by volume)

Mortar designation		Proportions by volume													
BS 5628	prEN 998	Non-hydraulic lime:sand[a,b]		Hydraulic lime:sand[a]		Portland cement:sand		Portland cement:lime:sand[c]			Air-entrained Portland cement:sand		Masonry cement[d]:sand		
		Lime	Sand	Lime	Sand	Cement	Sand	Cement	Lime	Sand	Cement	Sand	Cement	Sand	
(i)	M10	1	12	-	-	1	3	1	0-¼	3	-	-	1	-	Increasing strength but decreasing ability to accommodate movement caused by settlement shrinkage, etc. ↑
(ii)	M5	1	8-9	-	-	1	4-4½	1	½	4-4½	1[e]	3-4	1	2½-3½	
(iii)	M2.5	1	5-6	-	-	1	5-6	1	1	5-6	1[e]	5-6	1	4-5	
(iv)	M1	1	4-4½	1	2-3	1	8-9	1	2	8-9	1	7-8	1	5½-6½	
(v)	-	1	3½-4	-	1	1	10-12	1	3	10-12	-	-	1	-	

← Equivalent strengths in each mortar designation →

Increasing frost resistance →

← Improving bond strength (and, in consequence, increasing resistance to rain penetration)

Notes: Where a range of sand contents is given, the larger quantity should be used for sand that is well graded and the smaller for coarse or uniformly fine sand. [a] In the UK, the only significant use of lime as a binder for building mortar is in restoration work; [b] Proportions given are for lime-putty. If hydrated lime is batched dry, the volume may be increased by up to 50% to obtain adequate workability; [c] An air-entraining admixture may be incorporated into these mixes; [d] BS 5224[35] or ENV 413 MC12.5 cements only; [e] Sulfate-resisting cement can be used if required.

7.19.4 Mortar Replacement, Repair and Repointing

Repair or replacement mortar should match, as far as possible, the unweathered interior of the original mortar in composition, strength colour and texture. BS 8221: Part 2[35] provides advice on appropriate mixes for typical applications and exposure conditions, summarised in **Table 7.23** (over).

7.20 RENDER

Renders to external walls are used to provide a relatively impermeable coating to more absorbent substrates. This function is historically important since substrates (such as rammed earth and wattle and daub) were highly absorptive. Modern renders additionally provide an aesthetically pleasing appearance and may increase thermal and sound insulation properties of the underlying substrate. In this respect, special renders incorporating insulation granules or applied to metal lathes over insulation may be particularly beneficial. Specifications for mortars for renders are given in prEN 998 and BS 5262[36]; the latter standard includes design advice for renders together with specifications for materials and details appropriate for various exposure conditions. Modern renders are basically Portland cement:lime:sand mortars (**9.17.3**) (with appropriate additions to achieve e.g, enhanced insulation properties) and mixes should be selected to give strong adhesion properties, reasonable impermeability, freedom from cracks and adequate durability.

As with mortars, rich mixes (high cement content) produce strong, relatively impervious renders with high susceptibility to cracking. Rich mixes should be restricted to strong substrates, highly exposed areas or for application during winter. Generally, permeable renders are more durable than dense, impermeable renders. Cracking in strong renders may lead to localised water penetration. Where render is applied to clay bricks, water penetration produces severe exposure conditions, since evaporation of water trapped between the render and the clay brickwork is reduced, increasing the risk of sulfate attack (**8.19.2**) of the mortar joints and frost damage (both effects promote loss of adhesion of the render to the substrate, promoting further degradation). Sands should be carefully selected; coarse sands lead to renders with poor adhesion, whilst fine sands (or those containing fine particles, e.g, clay) require more mix water, which may lead to shrinkage. Sand should comprise angular particles (sharp sand) rather than rounded particles (soft sand) and conform to Type A or Type B gradings in BS 1199[37] (BS 1199 will be withdrawn in December 2003 and replaced with BS EN 13193[38]; see **8.6.3**).

Renders are applied in a series of coats; each successive coat should be weaker (i.e., less rich in cement) than the coat to which it is applied. In addition, sufficient time should be given to allow each coat to cure (about 3 days) and dry (a further 4 to 5 days, to produce a fine network of shrinkage cracks) before application of the next coat. The background substrate should be neither too wet (as this reduces adhesion) or too dry (as this increases water absorption from the render) prior to application of the render. Where the substrate is smooth e.g, common bricks, mortar joints should be raked to improve adhesion.

Table 7.23 Replacement mortar mixes for typical applications and exposure conditions

Typical applications and exposure conditions	Mortar mix group	Standard mortar mixes								Non-hydraulic mortar mixes			Lime mortar mixes with pozzolanic additives			
		Binder[a]						Aggregate[a]		Binder[a]		Aggregate[a]	Binder[a]	Aggregate[a]		
		White cement /OPC[b]	Masonry cement[c]	Eminently hydraulic lime[d]	Moderately hydraulic lime[d]	Feebly hydraulic lime[d]	Non-hydraulic lime[e]	Brick dust pozzolan[f]	Sand (or other fine aggregate)[g]	White cement /OPC[b]	Non-hydraulic lime[e]	Sand (or other fine aggregate)[g]	Non-hydraulic lime[e]	Pozzolanic additives	Sand (or other fine aggregate)[g]	
Use with dense, impermeable, durable materials (e.g. granite, basalt, flint or well-vitrified brick), especially where there is severe exposure (sea and river walls, retaining walls, paving, plinths and copings).	A1[h]	1					½		4 to 4½	1	2	4 to 4½				Decrease in strength and resistance to frost and salt damage ↓
	A2[h]		1						2½ to 3½							
	A3[h]		1						4 to 5							
	A4	1					1		6	1	1	6				
	A5			1				½	2½							
	A6			1					3							
Use with durable, moderately permeable limestones and sandstones, semi-vitrified brick for all exposures and high demands (e.g. mortar fillets). Suitable for group A materials in less severe exposures.	B1	1					2		9	1	2	9				
	B2			1					4							
	B3				1			½	2½							
	B4				1				3							
Use with weathered materials from groups A and B that are tending to scale and powder, in all exposures and locations.	C1	1					3		12	1	3	1½				
	C2					1		½	2							
	C3					1		½	2½							
	C4					1			3							↑
Use with less durable material (e.g. some calcareous or argillaceous sandstones, limestones or soft gauged bricks). Group D mortar mixes may also be used with materials in Group B and C in sheltered environments (other than pavings, etc.).	D1						1	½	2½				1	½	2½	Decrease in workability and ability to accommodate movement
	D2						1	½	3½				1	½	3½	
	D3						1		3		1	3				
	D4						1		4		1	4				
	D5[j]						1		1		1	1				

Notes: [a] proportions are by volume; [b] either cement can be used. Cement should be slurried to a thin cream before blending with lime putty and aggregates, especially if small amounts of cement are used; [c] Masonry cement should conform to strength class MC 12.5 of ENV 413 (Table 7.14); [d] hydraulic limes are natural hydraulic limes classified in DD ENV 459-1: 1995; [e] Non-hydraulic lime is high calcium or dolomitic lime conforming to BS 890: 1995. It is available as putty or as hydrated lime powder; [f] Brick dust is fired, ground clay brick in particle size range 38 μm to 75 μm. It is used primarily as a pozzolanic material, reacting with lime and water to form a hydraulic set. Most reactive brick is fired at temperatures of 900°C to 1000°C) (7.3.1); [g] in general, sand and other fine aggregate should conform to BS 1199 and BS 1200; [h] A1, A2 and A3 mortar mixes should not be used on historic masonry; [j] D5 mortar mix should be used for very fine joints in ashlar or gauged brick, using soft (rounded particle) sand or stone dust as aggregate.

The first coat applied is about 15 mm thick and functions to level the substrate and to even out water absorption. Where excessive, surface treatments e.g, PVA (**Table 13.3**) emulsions may be applied prior to the first coat to reduce absorption to acceptable levels. After drying, the first coat is scratched to produce a suitable key to increase adhesion of the second coat. The second coat, commonly the finishing coat, is thinner than the first (usually about 3 mm thickness) and should cover any shrinkage cracks that have occurred. Shrinkage in this coat is retrained by the fully shrunk undercoat so that any crack that are formed should be very fine. External coats may be

- **Smooth,** commonly 1:2:9 cement:lime:sand, often textured by working with a float;
- **Roughcast,** commonly 1:½:3 cement:lime:sand with 1½ parts of 5 to 15 mm shingle or crushed stone added, applied to walls by throwing from a hand scoop;
- **Pebble dashed,** commonly 1:1:5 cement:lime:sand, applied to walls and, whilst still wet, dashed with 3 to 10 mm shingle, pebbles or crushed rock and tamped with a float. Pebble dashed finishes help to obscure shrinkage cracks.

Following application of successive coats, render should be protected from the weather until curing and shrinkage is complete. In hot weather, this may involve damping down newly applied coats to avoid excessive evaporation and shrinkage. The durability of render is directly related to the adequacy of the detailing used to protect the render from rainwater ingress, in particular at the top (through provision of flashings, copings, etc.) and at openings (through provision of cills to ensure water is shed away from the render). Render should stop a minimum of 150 mm above dpc level (to prevent bridging of the dpc) and the lower edges should be formed in to a drip with an appropriate detail (usually by finishing with a bellcast or external render stop).

7.21 REFERENCES

1. BS 1052. Methods of test for masonry. Part 1: 1999. Determination of compressive strength. Part 2: 1999. Determination of flexural strength.
2. prEN 771. Specification for masonry units. Part 1: 1992. Clay masonry units. Part 2: 1992. Calcium silicate masonry units. Part 3: 1992. Aggregate concrete masonry units (dense and lightweight aggregates). Part 4: 1992. Autoclaved aerated concrete masonry units. Part 5: 1992. Manufactured stone masonry units. Part 6: Natural stone masonry units.
3. BS 6750: 1986. British standard specification for modular coordination in building.
4. BS 6073. Precast concrete masonry units. Part 1: 1981. Specification for precast concrete masonry units. Part 2: 1981. Method for specifying precast concrete masonry units.
5. BS 3921: 1985. British Standard specification for clay bricks, incorporating amendment 1 1995.
6. data from DE VEKEY, R.C. (1994) Part 5. Brickwork and blockwork. In: ILLSTON, J.M. (ed.) *Construction Materials. Their nature and behaviour.* London, E&FN Spon. pp. 272.
7. BS 5628. Code of practice for the use of masonry. Part 3:1985. Materials and components, design and workmanship.
8. BS 187: 1978. Calcium silicate (sandlime and flintlime) bricks.
9. BS 890: 1995. Specification for building limes.
10. BS 6463. Quicklime, hydrated lime and natural calcium carbonate. Part 2:1984. Methods of chemical analysis. Part 101: 1996. Methods of preparing samples for testing. Part 103: 1999. Methods for physical testing.
11. DD ENV 459. Building lime. Part 1: 1995. Definitions, specifications and conformity criteria.
12. BS EN 459. Building lime. Part 2: 1995. Test methods.
13. BS EN 196. Methods of testing cement. Part 1:1995. Determination of strength.
14. ENV 197-1:1995. Cement. Composition, specifications and conformity criteria. Part 1. Common cements.

15. BS 915. Specification for high alumina cement. Part 2: 1995. Metric units.
16. BS 4248: 1974. Specification for supersulfated cement.
17. BS 146: 1991. British Standard specification for Portland blastfurnace cements.
18. BS 6588: 1991. British Standard specification for Portland pulverized-fuel ash cement.
19. BS 7583: 1992. British Standard specification for Portland limestone cement.
20. BS 4246: 1991. British standard specification for high slag blastfurnace cement.
21. BS 6610: 1991. British Standard specification for pozzolanic pulverized-fuel ash cement.
22. BS 1370: 1979. Specification for low heat Portland cement.
23. BS 5224: 1995. Specification for masonry cement.
24. ENV 413. Masonry cement. Part 1: 1995. Specification. Part 2: 1997. Test methods.
25. BS 1014: 1975. Specification for pigments for Portland cement and Portland cement products.
26. BRE (1994) *Assessment of existing high alumina cement concrete in the UK*. BRE Digest 392. Garston, BRE.
27. BRE (2000) *Durability of pre-cast HAC concrete in buildings*. BRE Information Paper IP 8/00. Garston, BRE.
28. FORRESTER, J.A. (1970) A conduction calorimeter for the study of cement hydration. *Cement Technology, May/June*. pp. 95-99.
29. BS 12: 1991. British Standard specification for Portland cement.
30. BS 4887. Mortar admixtures. Part 1: 1986. Specification for air-entraining (plasticising) admixtures. Part 2: 1987. Specification for set retarding admixtures.
31. BS EN 934. Admixtures for concrete, mortar and grout.
32. BS 4721: 1981. Ready-mixed building mortars.
33. prEN 998-2 Specification for mortar for masonry. Part 2. Masonry mortar.
34. prEN 998. Specification for mortar for masonry. Part 1: 1993. Rendering and plastering mortar with inorganic binding agents. Part 2: 1993. Masonry mortar.
35. BS 5224:1995. Specification for masonry cement.
36. BS 8221. Code of practice for cleaning and surface repair of buildings. Part 2: 2000. Surface repair of natural stones, brick and terracotta.
37. BS 5262: 1991. Code of practice for external renderings.
38. BS 1199: 1976. Sands for external renderings, internal plastering with lime and Portland cement, and floor screeds.
39. BS EN 13193. Aggregates for mortar.

8

CONCRETE

8.1 INTRODUCTION

Concrete is a material formed by mixing cement (**7.11**), coarse and fine (sand) aggregate (**8.3**) and water (**8.2**) in defined proportions so that a plastic and workable mass results that can be moulded into the desired shape. In addition to these materials, concrete can contain admixtures (**8.7**) and/or additions (**8.8**) to enhance certain properties. In normal concrete, the aggregate constituents comprise the dominant proportion (about 70 to 80% by volume), the remainder comprising cement paste (about 30 to 50% cement by volume) and water.

Fresh concrete should have properties which facilitate ease of handling and placing of the material on site (workability, consistence) whilst hardened concrete should have adequate performance properties (compressive strength, durability). Generally, higher workability is obtained by increasing the quantity of mix water in relation to the quantity of cement (i.e., increasing the w/c ratio), but this is achieved at the expense of reduced performance. Hence a balance between acceptable workability and performance must be found for each particular application. On site, the standard of workmanship and level of supervision will have a large bearing on the strength and durability of the concrete product. In addition, variations in concrete properties will occur throughout a job as a consequence of

- variations in the quality of the materials used;
- variations in the mix proportions due to the batching process;
- variations due to sampling and testing.

8.2 WATER

Water should be of appropriate quality to ensure that the hydration reactions occurring with cement are not adversely affected by any contaminants within the water, including

- contamination by organic acids derived from organic vegetation (notably peat). The presence of organic acid contaminants can reduce the rate of hydration by increasing the pH value of the wet concrete;

- contamination by algae (which may turn water a green or blue-green colour). Algae liberate oxygen by photosynthesis (**12.4**), which may cause localised air entrainment and consequent strength reduction;
- contamination by water soluble salts (**2.9.1**). The presence of sulfates can cause disruption of the hardened cement paste (**8.19.2**) and the presence of chlorides may promote degradation of reinforcement (**8.21.5**). Relevant standards limit the quantity of water soluble salts derived from aggregates (**Table 8.13**) and BS 3148[1] includes advice on procedures to be adopted to ensure water is of suitable quality.

Generally, clear water of drinking quality to BS 3148 is suitable (following harmonisation of European standards, mix water for concrete will be governed by prEN 1008[2]).

8.3 AGGREGATES

Aggregates are added to cement to produce many composite materials, notably cementitious products such as mortar (**7.19**), render (**7.20**) and concrete. The aggregate becomes dispersed within the cement matrix to become part of the cohesive mass. Aggregates are added to cementitious products for the following reasons

- **economy**. Aggregates are cheaper than cement and are used as fillers or extenders. Large lumps of consolidated rocks (called 'plums') are added to large scale mass concrete (e.g, dam building) to control heat emitted during cement hydration (**7.16.3**);
- **durability**. Aggregates increase the resistance to weathering and abrasion);
- **volume stability**. Aggregates resist the initial shrinkage of concrete on drying (**6.4**). Expansion on setting is reduced by larger additions of aggregate for the same w/c ratio;
- **density adjustment**. Barytes ($BaSO_4$) is added for high specific gravity concrete (used for nuclear reactors). Lightweight concretes (**8.9.1**) are made by the addition of aerated blast furnace slags (slags cooled in water to promote gas bubbles).

Aggregates can be grouped according to their specific gravity into the following classes

- **normal density aggregates**. Aggregates with specific gravities within a limited range (approximately 2.55 to 2.75) producing concretes of similar densities (in the range 2250 to 2450 kg/m^3). Obtained from natural sources;
- **lightweight aggregates**. Produce concrete densities ranging from approximately 320 to 1900 kg/m^3. Important for reducing the self-weight of structures and improved thermal insulation (thermal conductivity 0.09 W/m°C compared to 1.4 to 3.6 W/m°C in normal density aggregate concrete). However, increased porosity lowers compressive strength. Lightweight aggregates include **sintered pulverised fuel ash** (formed by heating pelletised ash from pulverised coal used in power stations until partial fusion and binding occurs); **expanded clay or shale** (formed by heating suitable sources of clay or shale until gas given off is trapped in the semi-molten mass); **foamed slag** (formed by jets of water, steam and compressed air directed on to molten slag from blast furnaces);
- **heavyweight aggregates**. Densities of about 3500 to 4500 kg/m^3 can be obtained using barytes, and of about 7000 kg/m^3 using steel shot. Used for radiation shielding, etc.

Aggregates are currently controlled by BS 882[3] (natural sources), BS 1047[4] (air-cooled blastfurnace slag aggregate) and BS 3797[5] (lightweight aggregates). These standards will be withdrawn on 1 June 2004 and replaced with European Standards, BS EN 12620[6] (concrete), BS EN 13055[7] (lightweight aggregate) and BS EN 13139[8] (mortar).

8.4 NATURAL AGGREGATES

Many different natural materials have been used for making concrete, including clastic (broken or fragmented) materials, such as sands and gravels, igneous rocks (granite, basalt, etc.) and stronger sedimentary rocks (limestones, sandstones). Gravels from suitable sources (rivers, shallow coastal waters) have particles sizes suitable for direct use in concrete, and therefore only require washing and grading before use. Bulk rock sources require crushing to produce a suitable particle size. Particles are therefore sharp and angular (compared to naturally rounded particles in gravels). Particle shape has a significant effect on fresh and hardened concrete properties. In geological terms, the particle size of aggregates tends to be graded coarse to fine away from the source area. Clastic materials can be classified according to their particle size (**Table 8.1**).

Table 8.1 Classification of clastic aggregates by particle size

Sediment	Consolidated rock	Average particle diameter (mm)
Gravel (pebbles, cobbles, etc.)	Conglomerate	> 2
Sand	Sandstone	2 to 0.06
Silt	Siltstone	0.06 to 0.004
Clay	Shale	< 0.004

In concrete technology, we distinguish two types of aggregate; **coarse aggregate** (> 5 mm particle size) and **sand (fine aggregate)** (5 mm to 300 μm). Two of the characteristics of aggregate particles that affect the properties of concrete are particle shape (which affects workability) and particle surface texture (which mainly affects the bond between the matrix and the aggregate particles, and thus the strength of the concrete).

8.4.1 Particle Shape

Particles can be rounded or angular, depending upon the distance they were transported from the parent rock (by rivers, etc.). Five categories can be defined, as shown in **Table 8.2**.

Table 8.2 Classification of aggregates by particle shape

Classification	Description	Examples
Rounded	Fully water worn or completely shaped by attrition	River, seashore gravels, wind blown sands
Irregular	Naturally irregular or partly shaped by attrition and having rounded edges	Other gravel, land or dug flint
Flaky	Material of which the thickness is small relative to the other two dimensions	Laminated rock
Angular	Possessing well-defined edges formed at the intersection of roughly planar surfaces	Crushed rocks of all types
Elongated	Material, usually angular, in which the length is considerably larger than in the other two dimensions	Crushed rock of all types

8.4.2 Particle Surface Texture

Aggregates exhibit a variety of surface textures which will influence the strength of the cementitious material formed (**8.14.2**). Classification of aggregates by particle surface texture is shown in **Table 8.3**.

Table 8.3 Classification of aggregates by particle surface texture

Surface texture	Characteristics	Examples
Glassy	Conchoidal fracture	Black flint, vitreous slag
Smooth	Water worn or smooth due to fracture of laminated of fine grained rock	Gravels, chert, slate, marble, some rhyolites
Granular	Fracture showing more or less uniform rounded grains	Sandstone, oolitic limestone
Rough	Rough fracture of fine or medium grained rock containing no easily visible crystalline material	Basalt, felsite, porphyry, limestone
Crystalline	Containing easily visible crystalline constituents	Granite, gabbro, gneiss

8.5 AGGREGATE REQUIREMENTS FOR USE IN CONCRETE

For use in concrete, aggregates must be

- **correctly graded** in order to form a compact, pore free matrix;
- **clean** in order to provide the cement with a coherent, adherent bond to the particle substrate (i.e.,no dust or clayey materials);
- **free from organic material** which may weather adversely when exposed (e.g, coal). Organic materials may adversely affect the hydraulic activity of the cement, swell when exposed to high humidity or react adversely with the hydroxyl bonding of the cement. Aggregates containing high silica content should be avoided (**8.20**).

The whole concrete material must also be free from chlorides, sulfides and sulfates which may affect the corrosion of steel in reinforced and prestressed concrete structures. Iron pyrites (FeS_2) is a constituent of silts and clays. It readily oxides in air to form the sulphate and gives rise to iron staining on concrete. Dissolved salts such as the soluble sulfates of Ca, Mg and Na give rise to efflorescence. **Flaky** or **elongated** particles are detrimental to the workability of concrete, whilst **laminated** materials (such as shale and mica) give rise to poor durability.

To produce concrete of required strength and durability, free (void) space must be minimised. It has been traditional practice in the UK to batch concrete by volume, particularly for general concrete work. However this process offers little control over variations (particularly in the quantity of water already present in the aggregate) and so is not generally acceptable for structural concrete. Volume batching does, however, help to illustrate the scientific basis of mix design, as follows. **Coarse aggregates** have (approximately) a void space of 48% ($\approx$ 50%). Thus 4 volumes of coarse aggregate which has 2 volumes of void space. This 2 volumes of void space is filled with **Fine aggregate**, which has a void space of 30% (33%). Thus 2 volumes of fine aggregate which has 2/3 volumes of void space. This 2/3 volumes of void space is filled with **Cement**. Note that some of the fine aggregate will 'jack up' some of the coarse aggregate, and so more cement is added to give the common volume mix design of 4:2:1 (coarse:fine:cement). Also, there is void space between the cement particles. If the these are assumed to be perfect spheres, the void space can be calculated as 26% (**Table 1.12**). This void space is filled with water, giving a w/c ratio of 0.26. Water is needed for hydration (a rough minimum figure is a w/c ratio of 0.28, **7.16**), making a total w/c ratio of 0.54 for our volume mix. However, some of the water filling the voids may be used for the hydration process and therefore the void space is filled with the CSH fibres on hydration (**7.17**). A w/c ratio of 0.5 is usually adopted as a balance between workability and strength. The problem for concrete technology is therefore to specify a mix in which the free space is minimised, bearing in mind that the aggregate is not spherical and contains foreign materials. Hence certain tests are required

- Particle shape and texture;
- Sieve analysis;
- Determination of clay fine silt and fine dust;
- Flakiness and elongation index;
- Specific gravity density voids and absorption;
- Organic impurities;
- Mechanical properties e.g, crushing value;
- Aggregate abrasion value.

8.6 TESTS FOR AGGREGATES

Standard tests for aggregates have recently been harmonised within Europe by publication of the following standards which supersede the current British Standard tests for aggregates (in the various parts of BS 812[9]). These tests are used to indicate compliance with relevant British and European Standards. Tests most relevant to concrete technology are considered in following subsections.

BS EN 932[10]	Tests for general properties of aggregates
BS EN 933[11]	Test for geometrical properties of aggregates
BS EN 1097[12]	Tests for mechanical and physical properties of aggregates
BS EN 1367[13]	Tests for thermal and weathering properties of aggregates
BS EN 1744[14]	Tests for chemical properties of aggregates
prEN 13179[15]	Tests for filler aggregate used in bituminous mixtures

8.6.1 Sampling

It is **very important** that the test sample is representative of the supply (i.e.,taken from various parts of the stock pile, preferably whilst damp). The aggregate should be stored on a self draining hard based stock pile. The size of stock pile should be as large as possible to ensure uniformity of particle size and moisture content. Samples of the aggregate at the base of the stockpile should be avoided since dirt and washed debris from above accumulates in this region. Stockpiles are often radially arranged around the mixer unity. In practice, quantities sampled depend on the nominal size of material tested, ranging from 50 kg (coarse aggregate, 63 mm nominal aggregate size) to 0.1 kg (fine aggregate, < 5 mm nominal aggregate size) (BS EN 932-1). The sample is further reduced in quantity appropriate for particular test by **quartering** or by use of a **riffle box.** The sample may also be damp, air dry or oven dry. **Quartering** (***Figure 8.1a***) involves dividing a pile of aggregate on a large tray is divided into 4 as shown. Two diametrically opposite quarters 2 & 3 are discarded. The process is repeated by mixing together the remaining quarters (1 & 4) and again quartering until the requisite amount for the test is reached. The **riffle box** (***Figure 8.1b***) is another method of achieving the same result. The riffle box consists of 3 boxes. The riffle box is designed to separate an aggregate sample by bisecting the contents poured therein. The apparatus is set up as shown and the sample poured in. The contents of box 1 is discarded. The apparatus is reset, and the contents retained in box 2 are used to partially fill boxes 1 and 3. This sequence is repeated until the required mass of aggregate is obtained for the test to be carried out, box 1 always being discarded.

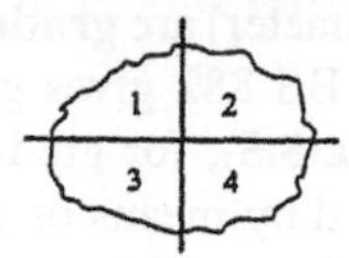

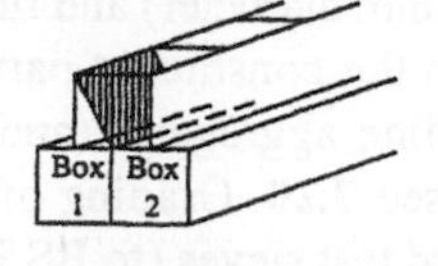

a) quartering *b) Riffle box*
Figure 8.1 Aggregate sampling techniques

8.6.2 Cleanliness

The surface area per unit mass of material is termed the **specific surface** and is a higher number for finer particles. The specific surface of cement particles of 50 to 70 μm in diameter is of the order 2500 cm²/g, whilst the specific surface of the aggregate is around 50 cm²/g. Fineness of **cement** has two effects (**7.16.4**), firstly on the rate of hydration of the cement particle and secondly on the amount of cement required in a mix containing a larger quantity of fines. The more finely ground cements have a higher specific surface and this increases the amount of surface available for hydration and therefore the rate of hardening of the cement will alter. For example, the specific surface of OPC is approximately 2250 cm²/g whilst that for RHPC is approximately 3250 cm²/g (**7.16.4**). These cements contain similar compounds but are ground to a different particle sizes (< 50 μm for RHPC against 72 μm for OPC). Fineness of **aggregate** is important since very fine material will cover the surface of larger aggregate particles, reducing the effectiveness of the cement bond (**8.14.1**). For this reason, for use in concrete, BS 882 limits the amount (% mass) of material passing a 75 μm sieve to a maximum of 2% for uncrushed, partially crushed or crushed gravel and 4% for crushed rock (for coarse aggregate) and 4% for uncrushed, partially crushed or crushed gravel, 16% for crushed rock (9% for use in heavy duty floor finishes), 3% for all-in aggregate and 11% for all-in crushed rock (for fine aggregate).

A measure of the cleanliness of the aggregate can be obtained in the laboratory using the sand equivalent test (BS EN 933-8) or methylene blue test (BS EN 933-9), or on site using the **Field Settling test**. If there is an excessive amount of dust (termed 'fines') in the aggregate it will prevent the particles of aggregate being coated with cement. The fines are attracted electrostatically to the freshly fractured surface. This electrostatic attraction is not broken down by tap water but is by saline solutions. Hence in the Field Settling Test sodium chloride is used to make the fine clayey/silt particles settle rapidly. Alternatively, the **Silt test** can be used, which is a more precise method of determining the quantity of silt and requires a laboratory to undertake weighings. For this test, about 100 g of **oven dry** sand (**weight A**) is placed onto the surface of a 1.18 mm sieve with the 75 μm sieve below it. The sand is washed beneath a running tap, allowing the residue to go to waste, dried (24 hrs) and the quantity remaining on both sieves (75 μm and 1.18 mm) is weighed. The combined weight is calculated (**weight B**). Then % silt passing the 75 μm sieve = (A - B)/B x 100%.

8.6.3 Grading (BS EN 933-1)

Coarse aggregate (> 5 mm diameter) and fine aggregate (< 5 mm diameter) are **graded** into zones depending upon the constituent particle size. For concrete, BS 882 gives grading limits for coarse and fine aggregates (and all-in aggregate) (**Table 8.5**); for grading for mortars and renders, see **7.20**. Grading of aggregates is determined by means of a **sieve analysis** using standard test sieves (to BS EN 933-2) of 50 mm, 37.5 mm, 20 mm, 14 mm, 10 mm and 5 mm (for coarse aggregate) and 5 mm, 2.36 mm, 1.18 mm, 600 μm, 300 μm, 150 μm and 75 μm (for sand) (1 μm = 10^{-6} m).

The sample (**air dry** aggregate must be used) is quartered until 400 to 500 g are available for sieving and the weight of the sample is recorded. The sample is passed through the successively diminishing sieves stacked one on top of the other. The sieving action should be an elliptical motion in the horizontal plane. When a negligible amount is passing each sieve, the **fraction retained** in each sieve is weighed and recorded (after lightly brushing any material free from the sieves). The amount passing the 150 μm sieve is also recorded. The **% retained** on each sieve is recorded to the nearest 0.1%. The **cumulative % passing** each sieve is then calculated. For example, the % passing the 150 μm sieve is first recorded; the amount passing the 300 μm sieve then is equal to the amount retained on the 150 μm sieve plus the amount passing the 150 μm sieve, and so on (**Table 8.4**).

Table 8.4 Example of results from an aggregate grading exercise

Sieve size	mass retained on sieve (g)	% retained in each sieve	cumulative % passing each sieve
10 mm	0	0	100
5 mm	18	3.6	96.4
2.36 mm	118	23.7	72.7
1.18 mm	96	19.3	53.4
600 μm	62	12.5	40.9 (25.6 + 13.5 + 1.8)
300 μm	128	25.6	15.3 (13.5 + 1.8)
150 μm	67	13.5	1.8
passing 150 μm	9	1.8	

Graded aggregate is compared to the grading requirements for concrete (**Table 8.5**).

Table 8.5 Grading requirements for coarse aggregate and sand for concrete (BS 882)

Coarse aggregate			Sand				
Sieve size	% passing sieve for nominal sizes		Sieve size	% passing sieve for nominal sizes			
	Graded aggregates			Overall limits	Additional limits for grading		
	40 to 5 mm	20 to 5 mm			C	M	F
50 mm	100	-	10 mm	100	-	-	-
37.5 mm	90 to 100	100	5 mm	89 to 100	-	-	-
20 mm	35 to 70	90 to 100	2.36 mm	60 to 100	60 to 100	65 to 100	80 to 100
14 mm	25 to 55	40 to 80	1.18 mm	30 to 100	30 to 90	45 to 100	70 to 100
10 mm	10 to 40	30 to 60	600 µm	15 to 100	15 to 54	25 to 80	55 to 100
5 mm	0 to 5	0 to 10	300 µm	5 to 70	5 to 40	5 to 48	5 to 70
			150 µm	0 to 15[a]	-	-	-

Notes: Individual sands may comply with the requirements of more than one grading. Alternatively, some sands may satisfy the overall limits, but may not fall within one of the additional limits C, M or F. In this case and where sands do not comply with the grading limits given in the Table, an agreed grading envelope may be used provided the supplier can satisfy the purchaser that such materials can produce concrete of the required quality. [a] Increased to 20% for crushed stone sands containing rock fines, except where they are used for heavy duty floors.

Commonly, sieve test results are expressed graphically as a plot of the BS grading requirements (known as a **grading envelope**) relative to the grading determined for the test aggregate. Grading envelopes for BS 882 are shown in ***Figure 8.2***.

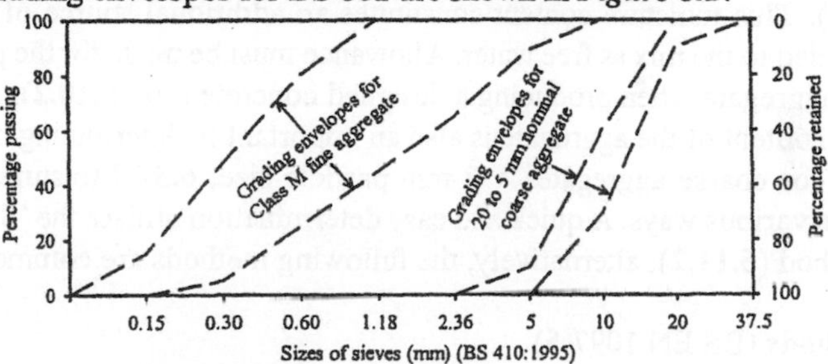

Figure 8.2 Grading envelopes for grading Class M fine aggregate and 20 mm graded coarse aggregate (to BS 882)

8.6.4 Bulk Density

When excavated, all soils or aggregates do not readily settle to a close packed state and therefore occupy a greater volume than before they were disturbed. This is phenomenon is termed **bulking (6.5)**. The bulk density of a material is the weight of the material in a given volume (kg/m^3). It depends upon the density of packing of the aggregate particles which in turn depends upon the grading and particle shape. A quick test that can be undertaken to determine percentage bulking is to pour damp sand into a glass cylinder until it is about ⅔ full and the measure the depth of the sand (**height H**). The sand is then removed into a metal tray. The glass cylinder is half filled with water and the sand carefully replaced (a tamping rod can be used to ensure that no air bubbles are trapped). The depth of the saturated sand is measured (**height h**). The percentage bulking is given by (H - h)/h x 100%. Two simple methods can be used for the determination of bulk density using a container having dimensions appropriate to the size of aggregate (see also BS EN 1097-3) and known empty weight. **Compacted bulk density** can be determined by filling a container of known volume with aggregate at ⅓ increments using a scoop discharging from a height not greater than 50 mm above the top of the container. The aggregate is tamped a suitable number of times (varying with the size of the aggregate). The container is filled to overflowing, tamped off to a level surface and then weighed. The compacted bulk density is determined as the weight of the sample divided by the volume of the container. **Uncompacted bulk density** can be determined in a similar way, except that compaction with the tamping rod is omitted.

8.6.5 Moisture Content

The hydration of cement is an exothermic reaction with water (**7.16.3**). The strength of the resultant mass is dependant upon the both the age of concrete and the w/c ratio at the point of fabrication. It is necessary to know the water content of the aggregates as well as the free water content added to the mix. Due to the problem of bulking, concrete mixes should be determined by weighing and not by volume. Aggregate is a porous material (some pores can be seen by eye whilst most require a microscope). The smaller the radius of pore the more water will be absorbed due to the low vapour pressure required for condensation at curved surfaces (**5.7**). This moisture content constitutes an additional source of water over and above that added to the mix as free water. Allowance must be made for the presence of pore water in the aggregate when producing a designed concrete mix (**8.10.2**). As noted above, the moisture content of the aggregate is also an important in determining bulking (bulking is negligible for coarse aggregates > 5 mm particle size, **6.5**). Moisture content can be determined in various ways. A quick and easy determination utilises the 'Speedy' (calcium carbide) method (**5.11.2**); alternatively, the following methods are commonly used.

Drying methods (BS EN 1097-5)

About 200 g of damp sand is weighed (**weight A**) then spread on a tray and air dried for about 24 hours (or carefully dried with a hair dryer) until the sand is free running and reweighed (**weight B**). The (free) moisture content (% SSD weight) = (A - B)/B x 100%. Alternatively, the (total) moisture content found by drying the sand for 24 hours in the oven at 110°C ± 5°C and reweighing (**weight B**), where the total water content (% dry weight) = = (A - B)/B x 100%.

Displacement Methods

In the displacement method, the moisture content of any sample of aggregate (which may have varying moisture content) is compared with the moisture content of a saturated surface dry (SSD) sample of the same aggregate. In the saturated surface dry condition, aggregate of a given type will have (effectively) a constant water content. The moisture content in the saturated surface dry condition therefore acts as a reference moisture content against which variable moisture contents can be measured. A 1000 ml cylinder is filled with exactly 300 ml of water and aggregate (in the saturated surface dry condition) is slowly added up to the 600 ml mark. Full compaction is required (and can be obtained by rotating the measuring cylinder backwards and forwards between the hands with a jerky motion). When enough aggregate has been added to reach the 600 ml mark and has settled to a flat surface at this level, the volume of the aggregate will be about 600 ml (**volume V_s**) and the level of the water is recorded (**volume V_d**). This is a reference mark for future tests as all the pores are full of water (saturated surface dry aggregate was used). The aggregate is then discarded and the cylinder washed out and carefully dried. The test is then repeated with wet or damp sand and the level of water is again recorded (**volume V_w**). Theoretically,

Mass of 'free' water between the particles = $\rho_w \times (V_w - V_d)$
Mass of sand in SSD condition = $\rho_s \times (V_d - 300)$

and therefore the % moisture content = (Mass of 'free' water in aggregate)/(Mass of sand) = $(\rho_w[V_w - V_d])/(\rho_s[V_d - 300]) \times 100\%$. Since ρ_s/ρ_w is the specific gravity of the aggregate (ordinarily averaging 2.6 for sand and uncrushed aggregate and 2.7 for crushed aggregate), then the % moisture content = $(V_w - V_d)/2.6(V_d - 300) \times 100\%$.

8.6.6 Specific Gravity

The density (or specific gravity) of a material is the ratio of the mass of a given volume of the material to the mass of an equal volume of water. The presence of pores in aggregates makes it necessary to be very careful in defining density (or specific gravity) (**6.2**). To obtain aggregate densities (or specific gravities), the following procedure can be used. Initially, the aggregate is kept in water for 24 hours and the weighed in water (**weight A**). This is achieved by weighing the aggregate contained in wire basket in water. The aggregate is then dried with a cloth and left in the atmosphere until SSD conditions are obtained, whereupon the aggregate is weighed in air (**weight B**). The aggregate is then oven dried for 24 hours at 110 ± 5°C and weighed again in air (**weight C**). Then bulk density = C/(C - A), true density (on an oven dried basis) = C/(B - A) and density (on a SSD basis) = B/(B - A). This test procedure also allows the water absorption of the aggregate to be determined, i.e., water absorption (% dry weight) = (B - C)/C x 100%.

8.6.7 Fragmentation Resistance

Fragmentation resistance is measured by the Los Angeles test method and the impact test method (BS EN 1092-2). Generally, more experience has been gained in the UK with the impact test (since this test method was contained in BS 812: Part 112: 1990). BS 882: 1992 gives requirements for maximum impact values for aggregates used in concrete of 25% for heavy duty concrete floor finishes, 30% for pavement wearing surfaces and 45% for general uses. Although now superseded by BS EN 1092, another fragmentation test commonly undertaken in the UK was the Aggregate Crushing Value (ACV) test (BS 812: Part 110: 1990), which gives a relative measure of the resistance of an aggregate to crushing under gradually applied compressive load. In this test, the aggregate sample was passed through 14 and 10 mm sieves until sufficient is retained on the 10 mm sieves to the fill a specified cylinder. The weight of the sample (**weight A**) is then taken. The specified cylinder apparatus is then filled in 3 layers, each layer tamped 25 times. A plunger is inserted and positioned in the cylinder and a uniform load of 406.5 kN is then applied uniformly over 10 minutes. After release of the load, the sample is passed through a 2.36 mm sieve and the fraction passing weighed (**weight B**). This is the weight of the fines formed. The aggregate crushing value, ACV = B/A x 100. A mean of at least two determinations is required. For wearing surfaces, not more than 25% should pass a 2.36 mm sieve whilst for normal concrete not more than 30% should pass the 2.36 mm sieve. If the ACV is more than 30, the result may be anomalous, in which case the test is repeated.

8.6.8 Other Tests

In addition to the requirements above, requirements for lightweight aggregates are given in BS 3797; BS 882 sets the following requirements for natural aggregates for concrete

- **flakiness index** of combined coarse aggregate not exceeding 50 for uncrushed gravel and 40 for crushed rock or gravel;
- **shell content** not exceeding 20% for fractions of 10 mm single size, or of graded or all-in aggregate that is finer than 10 mm and coarser than 5 mm, and not exceeding 8% for fractions of single sizes, or of graded or all-in aggregate that is coarser than 10 mm. (No requirements are given for aggregates finer than 5 mm);
- guidance on limits for **chloride content** (**8.21.5**).

8.7 ADMIXTURES

The performance requirements for admixtures for cementitious materials (concrete, mortars and grouts) is controlled by BS EN 934[16] and tests for admixtures are controlled by BS EN 480[17]. Together these standards currently supersede the requirements of BS 5075[18] (admixtures for concrete) and BS 4887[19] (admixtures for mortar) for single function admixtures and will eventually supersede these standards for dual function admixtures (such as retarding water-reducing admixtures and retarding superplasticising admixtures) by publication of further parts of BS EN 934 and 480. Admixtures are relatively common additions to cementitious products; for example, in 1989 it has been estimated that 40% of all concrete produced in the UK contained an admixture. Although an extremely large range of admixtures are currently available in the UK, most admixtures can be broadly placed into four single function categories, namely retarders, accelerators, plasticisers and air entraining agents (although these can be combined to produce dual function admixtures).

8.7.1 Retarders

Set retarders extend the time to commencement of the transition of the mix from the plastic to the rigid state by decreasing the rate of hydration of the cement (**7.16.3**). An example of a set retarder is gypsum, added during the manufacturing of Portland cement to prevent 'flash-setting' by altering the hydration products of tri-calcium aluminate (C_3A) (**7.16.1**).

Other examples of set retarders include

- sucrose (0.1% by mass of cement). Sucrose is a highly effective set retarder (about 0.5-1.0% sucrose will completely suppress the hydrating mechanism) but the effects are difficult to control;
- the soluble calcium salt derived from casein in milk (precipitated when milk goes sour);
- sodium bicarbonate $Na(HCO_3)$, sodium hexa-metaphosphate, starch, etc.

Set retarders are often counteract the accelerated setting effect of hot weather, particularly hot climates, or where the concrete has to be transported over large distances. However, set retarder may decrease in compressive strengths and may lead to bleeding and shrinkage (**8.18**). Compared to a control test mix of equal consistence, the set retarded concrete mix is required by BS EN 934-2 to have an initial set occurring ≥ +90 minutes and a final set occurring ≤ +360 minutes of the control; a 7 day compressive strength ≥ 80% and a 28 days compressive strength ≥ 90% of the control; and an air content in fresh concrete ≤ 2% by volume above the control.

Surface retarders are used to produce 'natural aggregate textures' to the faces of concrete in contact with formwork, where the formwork is treated with a surface retarder. The action of the surface retarder is to reduce the rate of hydration of the cement in contact with the surface retarder so that a surface layer of fine aggregate and cement may easily be washed on striking the formwork so as to expose the coarse aggregate beneath. Note that surface retarders are not mixed with the concrete; they act as an interface retarder between the formwork and concrete surface.

8.7.2 Accelerators

Set accelerators decrease the time to commencement of the transition of the mix from the plastic to the rigid state and hardening accelerators increase the rate of development of early strength in the product, with or without affecting the setting time. In practice, both types are generally referred to as set accelerators and act to increase the rate of hydration of the cement (**7.16**). This may be particularly desirable to counter the effects of increased curing time for concrete placed in cold weather or to allow early removal of formwork. However, in comparison to a control mix of equal proportion, the high early strength benefits obtained by a set accelerator are always obtained at the expense of lower final strength. Examples of set accelerators include calcium chloride ($CaCl_2$), sodium chloride (sea water), sodium sulfate Na_2SO_4, calcium formate $Ca(HCOO)_2$, sodium and potassium hydroxides NaOH and KOH. In the past, calcium chloride (1% to 2% by mass of cement) was used extensively to counteract slow setting times in cold weather. However, this practice is not now encouraged due to problems with increased corrosion rates in reinforcement, and strict limits on chloride contents in cementitious products are now required in relevant standards (**8.21.5**). Compared to a control test mix of equal consistence, the set accelerated concrete mix is required by BS EN 934-2 to have an initial set occurring ≥ 30 minutes faster than control at 20°C and ≤ 60% of the control set time at 5°C; a 28 day compressive strength ≥ 80% of the control mix and a 90 day compressive strength ≥ 28 day compressive strength of the control mix; and an air content in fresh concrete ≤ 2% by volume above the control. A hardened accelerated concrete mix is required by BS EN 934-2 to have at 20°C, a 24 hour compressive strength ≥ 120% and a 28 day compressive strength ≥ 90% of the control mix and at 5°C a 48 hour compressive strength ≥ 130% of the control mix; and an air content in fresh concrete ≤ 2% by volume above the control.

8.7.3 Plasticisers

Workability can be considered to be the amount of vibration or work which has to be done on the concrete during placement to obtain full compaction (i.e., to eliminate entrapped air which will would otherwise produce very large voids). Full compaction is dependent upon the size and shape of aggregate forming the mix, crushed rock aggregate being more difficult to compact than rounded aggregates. In addition, full compaction is easier to achieve if the concrete contains more water, but this unreacted water evaporates to leave capillary voids which may reduce strength and durability. Plasticisers (also known as workability aids or water reducing agents) are admixtures which, without affecting the consistence, permit a reduction in the water content of a given concrete mix, or which, without affecting the water content, act to increase the workability (slump/flow), or produce both effects simultaneously. Plasticisers with water reducing properties typically reduce the water content in the range 5 to 15%.

BS EN 934-2 distinguishes two types of plasticisers

- normal plasticisers (water reducing agents), normally based on lignosulfonates or hydrocarboxylic acids;
- superplasticisers (high range water reducing agents), normally based on modified lignosulfonates or sulfonated melamine or napthelene formaldehyde condensates.

The plasticising action of all plasticisers arises due to the absorption of the molecules of the plasticiser onto the surface of the cement grains and early hydration products in such a way that the ionic (charged) part of the molecule points outwards. Adjacent particles are therefore repelled, so agglomeration of particles is avoided with the net effect of greater lubrication and increased fluidity. This action is analogous to the action of soap on oil and grease (**5.4.2**). However, as a result of this process, plasticisers may act as a barrier to hydration and therefore may retard setting and hardening. In addition, some plasticisers have a tendency to entrain air which may increase porosity and reduce strength, although will increase workability (**8.7.4**). Whilst superplasticisers have enhanced properties, their effective plasticising action is limited to about 30 to 60 minutes. Compared to a control test mix of equal consistence, the plasticised concrete mix is required by BS EN 934-2 to have water reduction ≥ 5% by volume compared to control; a compressive strength ≥ 110% of the control at 7 and 28 days; and an air content in fresh concrete ≤ 2% by volume above the control. A superplasticised concrete mix is required to have water reduction ≥ 12% by volume compared to control; a 1 day compressive strength ≥ 140% of the control and a 28 day compressive strength ≥ 115% of the control; and an air content in fresh concrete ≤ 2% by volume above the control. When used as admixtures for concrete, normal plasticisers (water reducing agents) can typically reduce the water content of concrete by about 8% to 12%, whilst mid-range plasticisers (water reducing agents) can reduce the water content by 10% to 15%. Superplasticisers (high range water reducing agents) can reduce water contents by as much as 30%, but are not commonly used in the UK. Further information on water reducing agents has been summarised by the BRE[20].

Certain clays (e.g, Bentonite, a valuable clay formed from the decomposition of volcanic ash) and diatomaceous earths (a soft, fine grained, whitish rock consisting of the siliceous remains of diatoms deposited in small ponds or lakes; a diatom is any microscopic, unicellular alga occurring in water and is composed of two halves impregnated with silica) have water reducing properties and swell on exposure to water. Such materials have been used as water absorbers in civil engineering constructions (e.g, tunnelling, etc).

8.7.4 Air Entraining Admixtures

Air entraining admixtures allow a controlled quantity of small, uniformly distributed air bubbles to be incorporated during mixing with remain after hardening. The bubble diameters produced are much smaller (generally less than 0.1 mm) than air voids left by poor compaction, excess water (capillary pores), etc. The main reason for entraining air in to cementitious products is to provide freeze-thaw resistance (**6.6.5**), where the air entrained bubbles act as 'relief reservoir chambers' for water under pressure during ice formation during freeze thaw cycles (***Figure 6.5***). Entrained air volumes of only about 4 to 7% by volume of the cementitious product are required to provide effective freeze-thaw resistance. Air entrainment has a number of important secondary effects

- increased workability, where the bubbles act as small ball-bearings to reduce friction during placement;
- reduced compressive strength as a result of increased porosity, by a factor of about 6% for each 1% of air by volume. However, improvements in workability means that compressive strength reductions can be partially offset by reducing the w/c ratio;
- increased resistance to segregation and bleeding (**8.18**).

Air entraining admixtures are usually based on organic materials, including natural wood resins, fatty acids e.g, oleic acid, $C_{18}H_{34}O_2$, or $CH_3.(CH_2)_7.CH{=}CH.(CH_2)_7.(COOH)$ and sulfonated compounds e.g, $C_6H_4CH_3.SO_3H$ (a cyclic carbon compound formed by a condensation reaction (**13.5.2**) with sulphuric acid). All these compounds act to reduce the surface tension of the mix water, so allowing bubbles to form and persist throughout the hydration process (**5.4.2**). More recently, more effective organic compounds have been introduced, notably vinsol resins, which are based upon calcium lignosulfonate complexes derived from wood pulp. This compound acts both as a foaming agent and to reduce the attractive forces between cement particles, thereby increasing the workability of wet concrete and retarding the setting. Compared to a control test mix of equal consistence, the air entrained concrete mix is required by BS EN 934-2 to have an air content in fresh concrete ≥ 2.5% by volume above the control and a total air content of 4% to 6% by volume; a spacing factor between air voids in the hardened concrete of ≤ 0.2 mm; and a 28 day compressive strength ≥ 75% of the control.

8.7.5 Permeability Reducing Agents

Concrete is often used in intimate contact with water and there are several agents which may be added to concrete to prevent water uptake by absorption in the capillaries. Some materials (water repellents) reduce water absorption by modifying the contact angle between the water and the hardened concrete (**5.8.1**). However, they do not provide an effective barrier to water if the water is under a pressure (e.g, hydrostatic pressure at the base of a dam). Examples of permeability reducing agents include calcium or aluminium stearate (**5.8.1**), some resins and vegetable oils. Masonry cements (**7.14.3**) commonly contain permeability reducing agents added during manufacture.

8.8 ADDITIONS

ENV 206 classifies additions into two categories: **Type I additions**, which comprise nearly inert additions such as pigments (**7.14.3**) and fillers (**7.13**); and **Type II additions**, which comprise pozzolanic (**7.14.4**) or latently hydraulic (**7.10**) additions

8.9 TYPES OF CONCRETE

8.9.1 Lightweight Concrete

Lightweight concrete with densities ranging from about 500 kg/m^3 to 2000 kg/m^3 can be produced for a variety of purposes, including enhanced thermal insulation, increased high-frequency sound absorption (although sound insulation is reduced), enhanced fire resistance (as the aggregate in most dense aggregate concrete, e.g, granite, expands, causing spalling of the concrete, **14.6.3**), reduced self-weight (as a consequence of lower density, therefore reducing structural support requirements compared to denser concretes) and is easy to finish (i.e., site operations, such as cutting, nailing, etc., are easier compared to denser concretes and, as surfaces are often rough, they provide a better substrate for the application of surface finishes, such as renders, plasters, etc.). However, because lightweight concretes are less dense than normal concretes, their compressive strengths are generally lower and therefore their main application is where one of the above properties is more important in a particular location than high compressive strength (typically, where high thermal insulation properties are required). The three main methods of producing a low density, lightweight concrete are by using lightweight aggregates (to produce lightweight aggregate concrete), aeration (to produce aerated concrete) and using no fine aggregate (to produce no-fines concrete).

Lightweight aggregate concrete. To produce lightweight aggregate concrete, low density aggregates are selected with inherently porous structures. Care should be taken that specified w/c ratios are actually obtained, as the porous structure of lightweight aggregates can absorb large quantities of mix water (up to 15% by weight). Typical examples (**8.3**) include pulverised fuel ash, foamed blastfurnace slag and expanded clay and shale. Additions of similar materials to the mix may also increase thermal insulation, including

- **expanded perlite**, a naturally occurring glassy volcanic rock which, when sintered, evolves steam to produce a cellular material of low density (**14.6.3**) and good thermal insulation properties but low compressive strength and high drying shrinkage;
- **exfoliated vermiculite**, a naturally occurring mineral composed of thin layers like mica. When heated very rapidly, the layers separate such that the material can expand up to 30 times its original volume (**14.7.3**). Exfoliated vermiculite has very good thermal insulation properties but low compressive strength and very high drying shrinkage;
- **expanded polystyrene**, introduced as beads, offers the highest level of thermal insulation at the expense of very low compressive strength (in consequence, polystyrene bead aggregate cement is usually used in conjunction with structural materials, e.g, as a coring insulation material within precast concrete units).

Aerated concrete. Aerated concrete is produced by the addition of air-entraining agents (**8.7.4**) or by the addition of aluminium powder (**1.8.3**) to produce densities in the range 400-800 kg/m^3 corresponding to compressive strengths in the range 2-5 N/mm^2 and thermal conductivities as low as 0.1 W/m°C. Conventionally cured aerated concretes have high drying shrinkage (**6.4**); high temperature steam curing (autoclaving) is used to reduce this in aerated concrete blocks. Precast autoclaved aerated concrete (AAC) blocks (**7.7**) ordinarily include pozzolanic materials (**7.14.4**) to enhance compressive strengths.

No-fines concrete. No-fines concrete contains no fine aggregate (sand) and is manufactured from cement, water and normal, high- or low density coarse aggregate ranging from 10 to 20 mm nominal size. The absence of fine aggregate and the increase in pore space packing reduces the density of no-fines concrete, increasing thermal insulation performance. Care is required during placing to ensure that the cement paste remains in contact with the coarse aggregate and does not settle into bands (e.g, under gravity) (known as **bleeding, 8.18**); compaction by vibration is not usually used. The inclusion of coarse aggregate with high resistance to moisture induced movement results in a product of inherently low shrinkage; any initial shrinkage occurs quickly due to the high air permeability of the product. No-fines concrete has been used as an *in situ* walling material for low rise dwellings and is also manufactured as lightweight precast blocks, where the controlled production process ensures homogeneous distribution of the aggregate throughout the concrete.

8.9.2 Heavyweight Concrete

Heavyweight concretes are produced to obtain high compressive strengths in the range 60 to 100 N/mm^2 at 28 days and have the following advantages

- reduced construction times as a consequence of enhanced rates of hardening;
- reductions in space occupancy as a consequence of higher compressive strength, which enables a smaller section size to carry a given load;

- enhanced mechanical performance (stiffness, hardness, compressive strength), chemical resistance and durability, although it should be noted that very strong concretes do not necessarily have proportionately higher stiffness (elastic modulus).

Heavyweight concretes with enhanced compressive strength properties can be obtained by
- using as w/c ratio of 0.3 or less;
- increasing the cement content;
- using vacuum de-watering followed by power trowelling (e.g, for floors);
- autoclaving (**7.16.4**) mixes containing pozzolanic materials (**7.14.4**) (for precast units).

Reduced w/c ratios and increased cement contents will lead to reductions in workability, although this can be partially offset by the used of plasticising admixtures (**8.7.3**).

8.9.3 Reinforced Concrete

Concrete is very strong in compression, showing good engineering properties. However, concrete is very weak in tension, due to the many stress raisers which are present in the surface of the concrete matrix. Unlike metals (**10.4**), concrete does not possess dislocations and therefore the crack blunting mechanism (provided by the aggregates) is less effective. Most concrete structures are therefore designed so for use in compression only or with reinforcement to improve the tensile properties. The reinforcement may be alkali resisting glass, a polymer or steel wires/rods. The usual method of reinforcement is by steel tendons, which carries the **total** tensile stresses. Steel tendons can be used in several ways. **Reinforced concrete** (***Figure 8.3a***) is produced by casting concrete around the steel reinforcement (primary) or mesh (secondary). The concrete sets and hardens without being placed in compression; **Pre-tensioned concrete** (***Figure 8.3b***) is produced by casting concrete around steel tendons which are stressed in tension. When the concrete hardens, the tensile stresses are released. The reduction in stress allows the steel to expand thereby increasing the friction between steel and hardened concrete, placing the concrete in compression. The steel outside the concrete is then cut and each individual casting is separated. Note the steel is flush with the external face and is not covered with concrete; **Post-tensioned concrete** (***Figure 8.3c***) is produced by casting concrete around the sheathing ducts which will house the steel tendons. The concrete sets and hardens without being placed in stress. After hardening, steel tendons are threaded into the sheathing ducts, anchored, stressed and then grouted in order to keep the environment out of the sheathing.

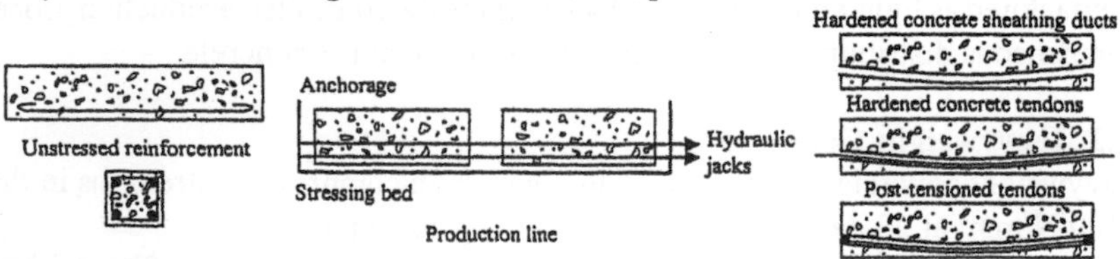

a) Reinforced concrete *b) Pre-tensioned reinforced concrete* *c) Post-tensioned reinforced concrete*

Figure 8.3 *Types of steel reinforcement techniques for concrete*

8.10 CONCRETE STANDARDS

Currently, concrete standards are being harmonised within Europe with the publication of Eurocode 2 and associated standards. Once completed, British Standard codes of practice for the specification (BS 5328[21]), structural use (BS 8110[22]) and testing (BS 1881) of concrete will be withdrawn. The European Standard for the performance, production, placing and compliance criteria for concrete is ENV 206[23]. Whereas Eurocode 2 is the European standard for the design of concrete structures, ENV 206 is the reference document for concrete as a material. During the transition period, either Eurocode 2 or the relevant British standard code of practice (BS 8110) can be used as a basis of design for concrete structures. For the purposes of this text we concentrate on concrete material requirements within harmonised European standard system, as embodied in ENV 206, although reference to British standards will be made where relevant (e.g, where only preEuropean Standards, prEN, are currently available). European Standards for concrete testing including testing fresh concrete (BS EN 12350) (**8.12**); testing hardened concrete (prEN 12390) (**8.15**); and testing concrete in structures (prEN 12504[24]) (see, e.g, **3.7.6**).

8.10.1 Concrete to ENV 206

The basic requirements for concrete to ENV 206 are shown in **Table 8.6.**

Table 8.6 Basic requirements for concrete to ENV 206

Requirement	Notes
Structure of concrete	Following compaction, normal concrete (without entrained air and excluding aggregate pores) should have a closed structure such that the air content by volume should not exceed the 3% for nominal aggregate size ≥ 16 mm and 4% for nominal aggregate size < 16 mm.
Cement content and w/c ratio	For concrete made with aggregates of nominal maximum size ≤32 mm, the minimum cement content and maximum w/c ratio depend on the environmental conditions (**Table 8.18**) and on the required properties of the concrete cover to reinforcement (**8.22.1**). Suitable concrete mixes for a variety of conditions are given in ENV 206 (**Table 8.19**). For concrete with aggregate sizes >32 mm (e.g, mass concrete), lower values for cement content than those given in **Table 8.19** may be acceptable.
Particle size of aggregates	The nominal maximum size of aggregates should not exceed one quarter of the smallest dimension of a structural member, the distance between reinforcing bars less 5 mm (unless special provision, such as grouping the reinforcing bars, is made), or 1.3 times the thickness of the concrete cover (not required for exposure class 1 in **Table 8.19**).
Chloride content	see **8.21.5**.
Consistence	To ensure proper compaction of concrete cast in situ and the consistence of the concrete at the time of placing should be equal to slump class S3 (BS EN 12350. Part 2) (**8.12.1**) or flow class F3 (BS EN 12350. Part 5) (**8.12.5**)
Resistance to AAR	see **8.20.1**.
Admixtures	Where included, the total quantity of admixtures should not exceed 50 g/kg cement and should not be less than 2 g/kg cement.
Additions	Where included, additions should not impair the durability of concrete or reinforced concrete.
Temperature	The temperature of fresh concrete in the time between mixing and placing should not exceed 30°C and should not be less than 5°C.
Curing	The required curing time depends on the rate at which a certain impermeability (resistance to penetration of liquids and gases) is reached in the surface zone (cover to reinforcement) of the concrete. Ordinarily, this is dependent on local conditions, but ENV 206 gives advice on minimum curing times.
Resistance to abrasion	To produce a concrete with high resistance to abrasion the concrete strength class should not be less than C30; well graded and hard aggregates with a rough surface texture and a high proportion of coarse particles should be used; normal curing times should be doubled; for severe abrasion conditions, a special wearing surface should be provided.
Resistance to water penetration	A mix is considered suitable for impermeable concrete if the resistance to water penetration when tested according to ISO 7031[25] (at present at draft stage) results in maximum values of penetration of less than 50 mm and the mean values of penetration of less than 20 mm. The w/c ratio should not exceed 0.55.

Concrete to ENV 206 is classified according to its **compressive strength** at 28 days (**7.13**), as derived from compression tests using 150/300 mm cylinder or 150 mm cube moulds (**8.16**). Under ENV 206, compressive strength is expressed in terms of the characteristic strength, i.e., that value of strength below which 5% of the population of all possible strength measurements of the specified concrete are expected to fall; and its **oven dry density**, as **normal weight concrete** (density greater than 2000 kg/m^3 but not exceeding 2800 kg/m^3) denoted with the prefix 'C', **lightweight concrete** (density not exceeding 2000 kg/m^3) produced by the incorporation of lightweight aggregates and denoted with the prefix 'LC' and **heavyweight concrete** (density greater than 2800 kg/m^3) denoted with the prefix 'HC'. Lightweight concrete is further subdivided into density classes based on density ranges of 200 kg/m^3, where the density class is derived from the upper limit of the density range divided by 1000. Hence a C50 concrete is a normal weight concrete (density 2000 to 2800 kg/m^3) with a compressive strength of 50N/mm^2 at 28 days.

8.10.2 Specification of Concrete

To specify concrete to meet strength, durability or any other special requirements, it is necessary to select the characteristic strength, together with any limits required on the mix proportions, the requirements of fresh concrete and the type of materials that may or may not be used.

Details of the concrete mix needed to meet the end-use requirements are specified in BS 5328 and ENV 206 in the form of one of the following types of mix

- a **designed** mix specified by its required performance in terms of a grade, subject to any special requirements for materials, minimum or maximum cement content, free w/c ratio or any other properties. Assessment of conformity is by strength testing;
- a **prescribed** mix specified by its constituent materials and the properties of qualities of those constituents to produce a concrete with the required performance. Assessment of conformity is by assessment of the mix proportions;
- a **standard** mix selected from a restricted range (given in BS 5328; see **Table 8.7**). Assessment of conformity is by assessment of the mix proportions. Standard mixes have been developed to produce concrete of the required characteristic strength. They are applicable for concrete for housing and similar construction and should be specified only where the scale of the work or economy does not justify mix design procedures;
- a **designated** mix specified by identifying the application for which the concrete is to be used and citing the corresponding designation, as given by BS 5328. It is the purchasers responsibility to specify the appropriate designated mix.

Under ENV 206, requirements for designed and prescribed mixes only are specified. However, the standard does allow specification of concrete based on information obtained from the long experience with comparable concrete; this provision allows the continued use of standard and designated mixes to BS 5328.

Section 4 of BS 5328: Part 2:1997 gives the weight of materials (except water) to produce certain **standard mixes** (ST1, ST2, ST3, ST4 and ST5), as shown in **Table 8.7**.

Table 8.7 Mix proportions for standard mixes and related strengths[21]

Standard mix	Compressive strength^a (N/mm²)	Constituent	Nominal maximum size of aggregate: 40 mm, Slump 75 mm	40 mm, Slump 125 mm	20 mm, Slump 75 mm	20 mm, Slump 125 mm
ST1	7.5	Cement (kg)	180	200	210	230
		Total aggregate (kg)	2010	1950	1940	1880
ST2	10.0	Cement (kg)	210	230	240	260
		Total aggregate (kg)	1980	1920	1920	1860
ST3	15.0	Cement (kg)	240	260	270	300
		Total aggregate (kg)	1950	1900	1890	1820
ST4	20.0	Cement (kg)	280	300	300	330
		Total aggregate (kg)	1920	1860	1860	1800
ST5	25.0	Cement (kg)	320	340	340	370
		Total aggregate (kg)	1890	1830	1830	1770
ST1 ST2 ST3	7.5 10.0 15.0	Fine aggregate (% by mass of total aggregate)	30 to 45	30 to 45	35 to 50	35 to 50
ST4 ST5	20.0 25.0	Fine aggregate (% by mass of total aggregate)				
		Grading limits C^b	30 to 40		35 to 45	
		Grading limits M^b	25 to 35		30 to 40	
		Grading limits F^b	25 to 30		25 to 35	

Notes: The mix proportions are based on the use of cements of standard classes 42.5 or higher, and will normally provide concrete having the characteristic strengths given. The mass of cement shall be increased by 10% when cements of standard strength classes 32.5 and 37.5 are used. The cement contents, together with the total masses of SSD aggregates and added water, will produce approximately 1 m³ of concrete. ^a Characteristic compressive strength at 28 days assumed for structural design (N/mm²); ^b Grading limits to BS 882.

8.11 PROPERTIES OF FRESH CONCRETE

Fresh concrete should be sufficiently **workable** and of homogeneous **consistence** to allow handling, placing and compacting. Workability is required to ensure sufficient fluidity to be handled and to flow into the available space (around reinforcement, formwork, etc.) and be compactable to remove air entrapped during mixing. Consistence is required to ensure concrete does not segregate into a noncohesive mass (e.g, by bleeding, **8.18**), which results in reduced performance. **Table 8.8** lists factors affecting workability and consistence.

Table 8.8 Factors affecting the workability and consistence of fresh concrete

Factor	Notes
Water content	Water content is the major factor affecting workability. The higher the water content the higher the workability.
Aggregate type	Generally, uncrushed aggregates require a lower water content than crushed aggregates to produce equal workability.
Aggregate particle size	Generally, the smaller the maximum aggregate size, the higher the water content required to maintain equal workability. Workability can be maintained at the same water content when using a fine sand by reducing the proportion of 'fines'.
Nature of the particles	Concrete is effectively a suspension of cement and aggregate particles. Compared to gravitational forces, surface attractive forces are high between cement particles but insignificant between aggregate particles. Hence consistence is governed predominantly by the fluidity of the mixture and the gravitational forces acting on the aggregate.
Aggregate particle shape	Coarse aggregate particle shape affects the fluidity of the mix by influencing the ease with which particles pass one another. Angular particles reduce workability by increasing the friction resistance to flow compared to rounded particles.
Grading of aggregates	Aggregate grading affects the water requirements of a concrete mix needed to produce equivalent workability. In this respect, grading of coarse aggregate has little effect on the water requirement, whereas grading of fine aggregate (sand) has a pronounced effect; e.g, changing the grading of fine aggregate from a coarse sand (20% passing a 600 μm sieve) to a fine sand (90% passing a 600 μm sieve) can increase the water content by 25 kg/m³ to maintain equivalent workability.
Cement type	Different cement types and strength classes of have different water requirements to produce a cement paste of standard consistence (although, compared to the factors above, such differences generally have little effect on concrete workability).

8.12 TESTS FOR FRESH CONCRETE

The measurement of the workability of fresh concrete is of importance in assessing the practicability of compacting the mix and also in maintaining consistency throughout a job. Workability tests are often used as an indirect check on the water content and, therefore, on the w/c ratio. Standard tests for fresh concrete are described in BS EN 12350[26] (which will eventually supersede the relevant parts of BS 1881[27]), including the slump test (Part 2); the Vebe test (Part 3); a compactability test (Part 4); and the flow table test (Part 5). All methods are 'single point' tests, providing only a single value which is strictly related only to the specific sample and test conditions. Although these tests provide only limited information on the true workability of concrete, they are convenient and cheap to perform and are widely adopted in practice.

8.12.1 Slump Test

The workability of concrete is most often determined by the Slump test, which is used on site as an easy method of quality control. For a particular mix and job specification, the Slump should be a constant. If the Slump changes then this may reflect the fact that more water is being added to the mix or there is some other change in mix proportion.

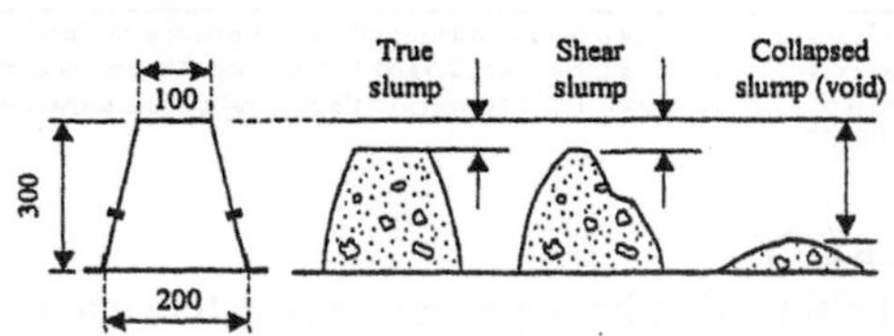

The Slump test is carried out using a slump cone (made to standard dimensions). The cone is placed on a flat surface with the small end uppermost and filled with concrete in 4 layers. Each layer is tamped 25 times (evenly distributed over the surface). The time to fill the cone should not be any longer than 2 minutes and the layer below the added layer should be penetrated with the tamping bar. The surface is smoothed off with a float and the cone is then lifted vertically leaving the concrete unsupported. The decrease in height of the cone (in mm) as it slumps is measured and recorded, as is the type of slump. The test should be repeated if the slump is not true, that is the concrete has not slumped evenly but has either slid down an inclined plane (shear slump) or even collapsed completely.

Figure 8.4 Single point workability tests for concrete: Slump test

8.12.2 Vebe Test

The Vebe test (***Figure 8.5***) is useful for low workability and air-entrained concrete.

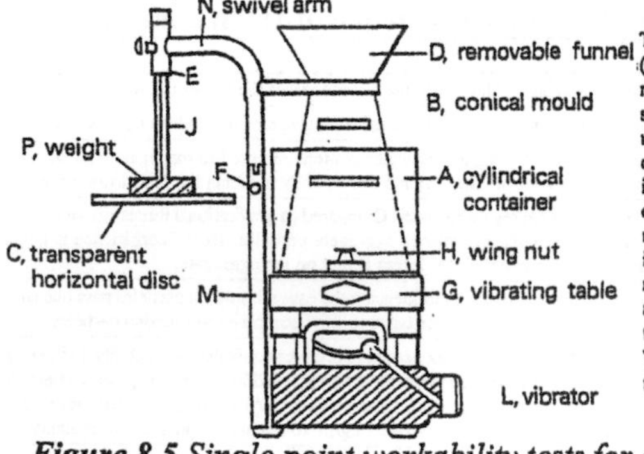

The main parts of the Vebe test apparatus are a metal cylindrical container (A) which can be clamped to the top of a vibrating table (G), a conical mould, a funnel (D) and a transparent horizontal disc (C) mounted on a swivel arm (N). The container (A) is first securely fastened to the table (G) using the wing nuts (H). The mould (B) is placed concentrically in the container (A). The screws at (F) are tightened and the funnel (D) is swung into position. The mould (B) is filled with concrete in four layers, tamping each 25 times. The funnel is swung away and the mould is struck level. The mould is then removed using a vertical lifting action. The disc (C) is swung into the container and carefully lowered until it touches the concrete. The slump is measured using scale (J) and recorded. The screw at (F) is tightened and the vibrating table and a stopwatch are started simultaneously. The time taken for the transparent disc to become covered with cement grout is noted. This is the time for full compaction. The workability of the concrete is recorded as 'X' V-B seconds (to the nearest 0.5 V-B's).

Figure 8.5 Single point workability tests for concrete: Vebe test

8.12.3 Compaction Factor Test

The compacting factor test measures the degree of compaction of concrete which has been allowed to fall through a standard height. The compacting factor is defined as the ratio of the weight of partially compacted concrete to the weight of fully compacted concrete (*Figure 8.6*).

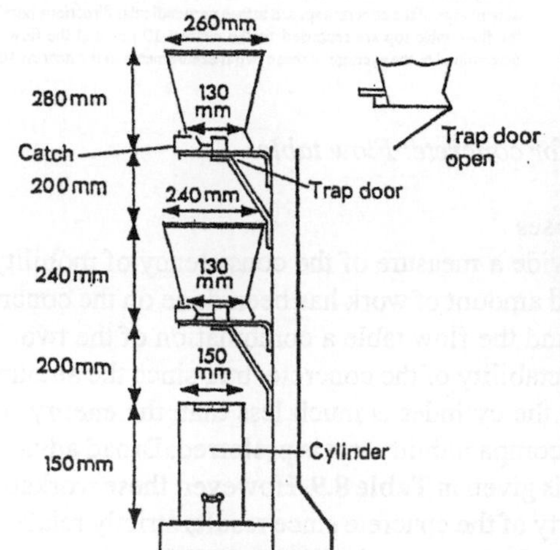

The concrete to be tested is mixed to a uniform consistency. All trapdoors in the apparatus are closed and secured and the cylinder covered. The concrete is gently placed in the upper hopper using a hand scoop, and the surface of the concrete is made level with the brim of the hopper. The trapdoor to the upper hopper is then opened so that the concrete falls into the lower hopper. Immediately the concrete has come to rest, the cylinder is uncovered and the trapdoor to the lower hopper opened. The excess concrete remaining above the top of the cylinder is removed by holding a trowel in each hand, with the plane of the blades horizontal, and moving them simultaneously (one from each side) across the top of the cylinder. The trowels must be kept pressed to the top edge of the cylinder during this operation. The concrete in the cylinder is then weighed to give 'the weight of partially compacted concrete'. The cylinder is then refilled from the same sample of concrete in layers approximately 50 mm deep, each heavily tamped to produce full compaction. The excess concrete is struck off level with the top of the cylinder. The concrete in the cylinder is then weighed to give 'the weight of fully compacted concrete'. Then

$$\text{Compacting factor} = \frac{\text{weight of partially compacted concrete}}{\text{weight of fully compacted concrete}}$$

Figure 8.6 Single point workability tests for concrete: Compaction Factor test

8.12.4 Compactability Test

The compactability test involves careful filling of a container (200 ± 2 mm diameter and 400 ± 2 mm height) of height (h_1) with fresh concrete, avoiding any compaction. When the container is full, the top surface is struck off level with the top of the container. The concrete is then compacted by vibration (vibrating rod or table) and the distance from the surface of the compacted concrete to the upper edge of the container is measured. The test is repeated four times and the mean distance computed (h_2). The degree of compactability is then expressed as $h_1/(h_1 - h_2)$, to two decimal places. The BS EN compactability test is a simplified version of the compacting factor test (BS 1881:Part 103[28]). Both tests are more sensitive methods of measuring the workability of concrete than the slump test and enable values to be obtained for mixes which are too dry to give a reasonable slump. If strictly comparable results are to be obtained, each test should be carried out at a constant time interval after mixing is completed (about 2 minutes is optimum). This is because the sensitivity of the tests is sufficiently high to detect differences in workability arising from the initial processes in the hydration of the cement.

8.12.5 Flow Table Test

The flow table test measures the consistency of fresh concrete by measuring the spread of concrete on a flat plate which is subjected to jolting (*Figure 8.7*).

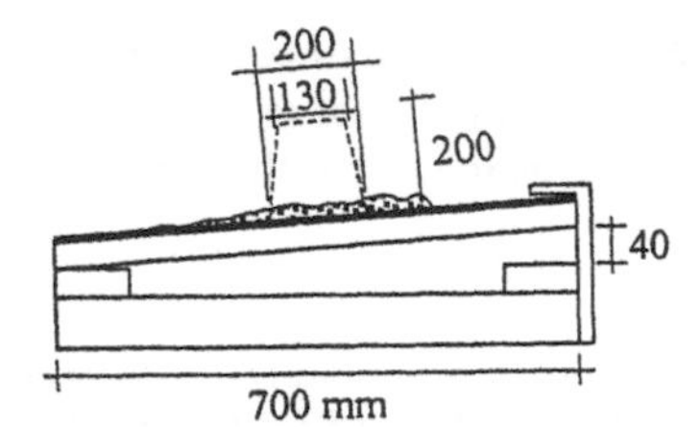

A cone (of slightfy smaller dimensions than that used in the slump test, 8.12.1) is filled with concrete in two equal layers, each lightly tamped 10 times. Excess concrete is struck off level with the base of the cone, and the cone is then inverted over the flow table. The flow table top is then raised to the upper catch position and allowed to drop freely to the lower catch position. This cycle is repeated too give a total of 15 drops, each cycle taking not more than 2 seconds or more than 5 seconds. The maximum dimensions of the concrete spread in two perpendicular directions parallel to the flow table top are recorded to the nearest 10 mm and the flow value determined as the average of these two measurements to the nearest 10 mm.

Figure 8.7 Single point workability tests for concrete: Flow table

8.12.6 Workability For Different Purposes

The slump, Vebe and flow table tests provide a measure of the consistency of mobility of the concrete, the slump test after a standard amount of work has been done on the concrete, the Vebe during a standard energy input and the flow table a combination of the two. The compacting factor test assesses the compactability of the concrete, but, since the amount of work done on the concrete in falling into the cylinder is much less than the energy input from mechanical compaction devices, the compactability test is preferred. Broad advice for suitable workabilities based on these tests is given in **Table 8.9**. However, these workability tests do not measure a fundamental property of the concrete since results strictly relate only to the specific sample and test conditions. This means that the correlations between the results of different workability tests, used as a basis for the advice given in **Table 8.9**, are relatively broad and therefore **Table 8.9** is indicative only.

Table 8.9 Suitable workabilities for concrete based on single point workability tests

Purposes	Compacting factor	Slump (mm)	Vebe degrees
Very high strength concrete for prestressed concrete sections compacted by heavy vibration.	0.70 to 0.78	0	over 20
High strength concrete sections, paving and mass concrete compacted by vibration.	0.78 to 0.85	0 to 25	7 to 20
Normally reinforced concrete sections compacted by vibration. Hand compacted mass concrete.	0.85 to 0.92	25 to 50	3 to 7
Heavily reinforced concrete sections compacted by vibration. Hand compacted concrete in normally reinforced slabs, beams, columns and walls.	0.92 to 0.95	50 to 100	1 to 3
Heavily reinforced concrete sections compacted without vibration. Work where compaction is particularly difficult. Cast in situ piling.	over 0.95	100 to 150	0 to 1

8.13 PROPERTIES OF HARDENED CONCRETE

Hardened concrete should have sufficient **compressive strength** to carry imposed loads and should have adequate tensile and flexural strength to resist deformation. The hardened concrete should be **durable** so that it can continue to perform its function with respect to serviceability, strength and stability for the life of the structure without significant loss of utility or excessive unforseen maintenance. Whilst strength and durability are generally the most important properties of hardened concrete, other properties, such as thermal, acoustic or fire performance or visual appearance may influence the production of concrete. The strength and durability properties of hardened concrete depend largely on the **quality** of concrete, as determined by the following general factors: curing conditions; cement type and content; degree of compaction; w/c ratio.

The quality of concrete produced on site is dependent of the practicality of the design; it must be easy to specify and achieve all of the necessary requirements on site. Generally, it should be remembered that, in contrast to concrete test specimens, it is relatively difficult to achieve uniform compaction and curing on site (in a cast beam, for example) and this will lead to zones of poorer quality concrete within a single structural member (***Figure 8.8***)

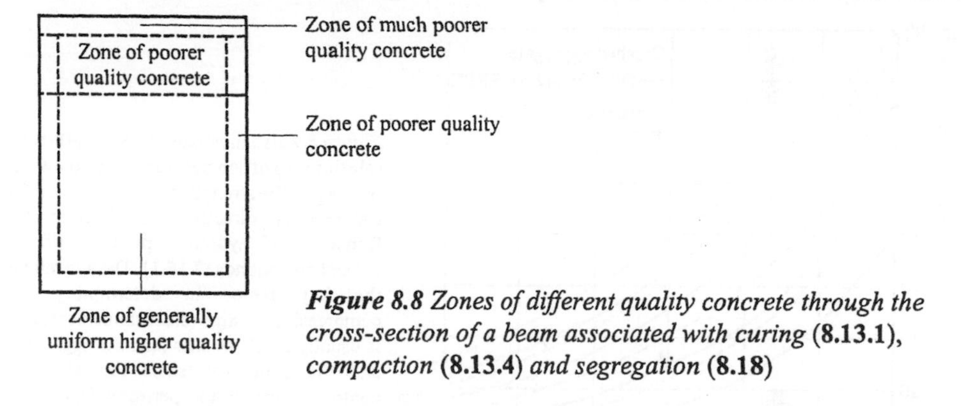

Figure 8.8 *Zones of different quality concrete through the cross-section of a beam associated with curing* (**8.13.1**), *compaction* (**8.13.4**) *and segregation* (**8.18**)

8.13.1 Curing Conditions

The compressive strength developed by concrete increases for many months under favourable conditions due to the continued hydration of cement (**7.16.2**). Strength development depends on the temperature and humidity conditions during curing. Generally higher temperatures increase the speed of the hydration reactions and thus the rate of strength gain, although the effect of temperature throughout the curing process is quite complex (**7.16.4**). Hydration reactions will proceed provided water is present (**7.16.2**), and a variety curing methods are used, e.g, spraying or ponding the surface of the concrete with water; protecting exposed surfaces from excessive evaporation (by the wind and sun); covering surfaces (e.g, with wet Hessian or polythene), which limits evaporation and may also provide a warm moist environment and/or water as condensate derived from the heat of hydration; applying proprietary curing membranes (usually spray-applied seals), which weathers away within a few weeks; storage under water (BS 1881: Part 111[29] requires that concrete test specimens for strength tests are stored under water at a constant temperature of 20°C). Flooding of the surface of the concrete for 7-14 days produces an extremely impermeable skin on the surface, which will subsequently reduce the penetration of gaseous species by about 10,000 times that for concrete cured in air, even when protected with polythene sheets. Curing membranes may also enhance penetration of gaseous species, although the performance in service of these membranes has not been fully investigated. Concretes containing cement replacement materials (**7.14.4**) may require extended periods of wet curing to ensure adequate durability compared to similar OPC concretes.

8.13.2 W/C Ratio

The weight of water to the weight of cement (the w/c ratio) affects both the strength and durability of the hardened concrete. A w/c ratio of approximately 0.28 produces complete hydration of the cement and any excess water will evaporate to produce capillary pores (**7.17**), increasing porosity and permeability and reducing compressive strength. Permeability to aggressive chemicals reduces the durability of concrete (**8.22**). The relationship between w/c ratio and compressive strength is illustrated in ***Figure 8.9***.

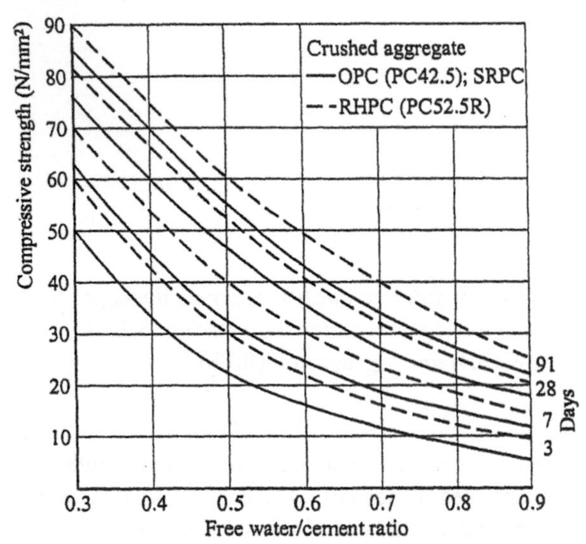

Figure 8.9 The relationship between free w/c ratio and compressive strength

Figure 8.9 also illustrates the **strength-time relationship** of concrete (a process known as **'ageing'**). The strength-time relationship for concrete is directly derived from the formation of hydration products of the cement compounds (**7.16.1**). The w/c ratio is the key factor in determining the compressive strength (since the cement paste is usually much weaker than the aggregate) and the durability (since the w/c ratio controls porosity and permeability) of the hardened concrete.

8.13.3 Cement Type and Content

Different types and strength classes of cement produce concretes having different rates of strength development (**7.16.2**). Increasing the cement content to produce richer mixes produces concretes which initially gain strength more rapidly, whereas reducing the cement content to produce leaner mixes has the converse effect. The strength increases as a result of the increase in cement content is demonstrated for the **standard mixes** (**8.10.2**) in **Table 8.7**. In addition, there is inevitably some variability in the strength of concrete that results from the inherent variability of any particular cement supplied from different works and from a single works over a period of time (this is allowed for in the concrete mix design process (**8.10.2**) by specifying the quality of concrete in relation to a 'characteristic strength' below which a specified proportion of test results may be expected to fail).

8.13.4 Degree of Compaction

Compaction reduces the quantity of air entrapped in the fresh concrete on placing and the degree of compaction therefore has a direct influence on the compressive strength of the hardened concrete. As a general rule, hand compaction is likely to produce a lower strength concrete than compaction by mechanical means and therefore poker vibrators are commonly used on site to ensure adequate compaction. Although the compressive strength of concrete is not directly affected by the type of mixer used, hand-mixing is likely to produce a lower strength concrete than machine-mixed concrete of similar proportions.

8.14 STRENGTH OF HARDENED CONCRETE

Factors affecting the strength of the hardened cement paste have been considered previously (7.16). These factors are equally valid in determining the strength of hardened concrete. However, the introduction of aggregate into the cement paste to form concrete introduces other factors which affect the strength of hardened concrete arising from the properties of the aggregate itself and the nature and strength of the bond between the cement paste and the aggregate. The influence of these additional factors largely depends on whether the aggregate is stronger or weaker than the cement paste. For normal concrete, aggregates are much stronger than the hardened cement matrix and failure occurs predominantly in the cement paste, commonly in the region of the cement paste:aggregate interface (since this is weaker than the bulk cement paste and commonly contains microcracks prior to loading produced, for example, by differential shrinkage between the cement paste and the aggregate). Aggregates particles stronger than the cement paste have the beneficial effect of acting as crack arresters, preventing the propagation of cracks. For a crack to progress into the aggregate, an energy barrier must be exceeded and the magnitude of this energy barrier for strong aggregates far exceeds that required either to propagate the crack through the hardened cement paste (***Figure 8.10***) or propagate other, smaller cracks (at other cement paste:aggregate interfaces or within the hardened cement paste matrix).

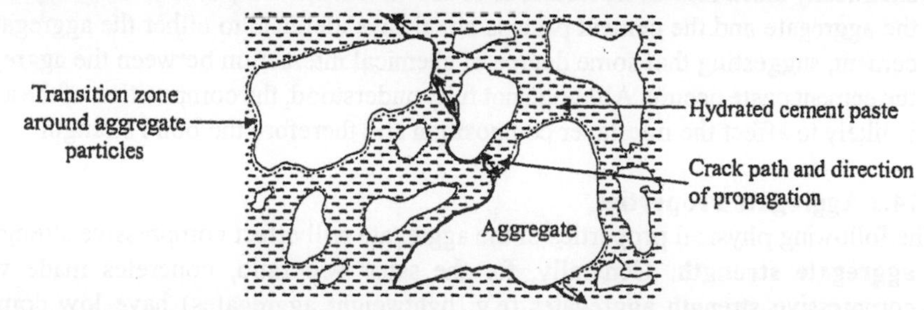

Figure 8.10 Typical crack path propagation through normal concrete

Hence the presence of aggregates stronger than the hardened cement paste delays fracture of the concrete, thereby increasing its strength. For lightweight aggregates, the compressive strength of the aggregate is ordinarily less than that of the surrounding hardened cement paste, and therefore cracking can proceed through the aggregate. Under these circumstances, the overall strength of the concrete for a given w/c ratio will be less than that of normal concrete. As with other cementitious products, the w/c ratio is the single most important factor in determining concrete strength (**8.13.2**). In addition to the other factors affecting the strength of the hardened cement paste (**8.13**), the bond at the cement paste:aggregate interface and the properties of the aggregate are critical.

8.14.1 Bond at the Cement Paste:Aggregate Interface

The bond between the cement paste and the aggregate is due to mechanical and physical effects and, in some cases, to the chemical reaction between the aggregate and the cement paste. The influence of bond strength on the compressive demonstrated by a number of researchers; for example, the compressive strength of concrete made with aggregate comprising glass marbles was found to be approximately 2.5 times that of the same concrete made with identical, but lubricated, glass marble aggregate[30]. The strength of the bond depends upon the properties of the paste and of the aggregate and is affected by

- the **w/c ratio**. Decreases in the w/c ratio, in addition to increasing the compressive strength of the cement paste (**8.13.2**), also increases the strength of the cement paste:aggregate bond;
- the **surface roughness** of the aggregate. Generally, the rougher the aggregate surface the greater the compressive strength of the hardened concrete. The effect of surface roughness is important for strong concretes (w/c around 0.4), but the effect decreases at increasing w/c ratios, becoming negligible at w/c ratios of greater than about 0.65. Generally, for strong concretes, the effect of surface roughness is more important than the effect of other aggregate properties (e.g, particle shape), whereas for weaker concretes the effect of the w/c ratio in reducing strength renders the effect of surface roughness insignificant;
- the **chemical composition of the aggregate**. Generally, and with the exception of aggregates that react with alkali hydration products, common concrete aggregates are chemically inert. However, studies have shown that the composition of the layer between the aggregate and the cement paste is not always identical to either the aggregate or the cement, suggesting that some degree of chemical interaction between the aggregate and the cement paste occurs. Although not fully understood, the composition of the aggregate is likely to affect the interlayer composition and therefore the bond strength.

8.14.2 Aggregate Properties

The following physical properties of the aggregate will effect compressive strengths

- **aggregate strength**. Generally, for the same w/c ratio, concretes made with low compressive strength aggregates (e.g, lightweight aggregates) have low compressive strengths due to the formation of cracks within the aggregate (**8.14**). In comparison, the strength of the aggregate, the effect of the type of fine aggregate, the maximum size of aggregate, overall grading and the particle surface roughness is relatively insignificant.
- **aggregate concentration** in the concrete. Generally, for the same w/c ratio, an increase in the concentration of aggregates in the concrete increases the compressive strength. This effect occurs because the stress concentrations surrounding adjacent aggregate particles

interact so that the average stress concentration induced by a high aggregate concentration is reduced compared to a similar concrete with a low aggregate concentration. This effect will occur provided that the reduced cement paste content remains high enough to form a matrix around all aggregate particles (otherwise air voids will be formed and the concrete will be substantially weaker). Clearly, there will be an optimum aggregate concentration at which strength is maximised; this has been reported for cement mortars as 30% for compressive strength and 36% for flexural strength[31].

- **sand:cement ratio**. Laboratory tests have shown that the sand:cement ratio affects the strength of medium and high strength concretes (≥35 N/mm^2), although this is only a secondary factor in the strength of concrete. In theory, a leaner mix (higher sand:cement ratio) should lead to a higher compressive strength. This is associated with the absorption of water by the aggregate; a larger amount of aggregate absorbs a greater quantity of water thus reducing the effective w/c ratio. Also, the total water content per cubic metre of concrete is lower in a leaner mix compared to a richer mix. Thus the voids in a leaner mix form a smaller fraction of the total volume of the concrete. As it is the voids which have a detrimental effect on the compressive strength of concrete, leaner mixes tend to have higher compressive strengths compared to richer mixes.
- **aggregate particle size**. Generally, an decrease in the aggregate particle size results in an increase in the compressive strength of the hardened concrete. Since the presence of aggregate in the hardened concrete induces stress concentrations at the cement :aggregate interface, a reduction in the particle size reduces the volume of the stress concentration associated with the aggregate particle, thereby increasing compressive strength.

8.15 TESTS FOR HARDENED CONCRETE: CHEMICAL COMPOSITION

Chemical analysis of hardened concrete is governed by BS 1881: Part 124: 1988[32], which describes the sampling procedures, treatment of samples and analytical methods to be used to determine cement content, aggregate content, aggregate grading, original water content, type of cement, type of aggregate, chloride content, sulfate content and alkali content. Methods are highly complex and beyond the scope of this text. The methods of analysis are generally very time consuming (some taking about 3 days to complete) and consequently very expensive. The complexity is illustrated by the following (oversimplified) examples. Methods to determine the type of cement used are based on determination of the quantity of soluble silica (SiO_2) (which is associated with the calcium silicate hydrates formed on hydration of the cement, **7.16**). However, aggregates also contain soluble silica and so for a truly accurate picture an analysis of the aggregates used is also required. Similar requirements exist for the determination of the quantity of calcium oxide (CaO). As a consequence of these complexities, cement content analysis of hardened concrete is usually only undertaken as a last resort. In addition, simplified methods for determining the quantity of aggressive compounds present in hardened concrete, which are sufficiently accurate for most purposes, are generally available, for example for determining the quantity of water soluble sulfates (**8.19.2**) in concrete, carbonation depths (**8.21.4**) and chloride content (**8.21.5**) in reinforced concrete.

8.16 TESTS FOR HARDENED CONCRETE: STRENGTH TESTS

Compressive, flexural and tensile splitting strength tests are ordinarily undertaken on moulded concrete samples which are tested to failure using appropriate testing machines. Standard tests for hardened concrete are in the process of being harmonised within Europe, as currently incorporated within prEN 12390[33]. Until this standard is published as a British Standard, standard tests for concrete strength are covered by the relevant parts of BS 1881

- methods of making test specimens (Part 108: 1999 cubes, Part 109: 1989 beams, Part 110: 1989 cylinders);
- methods of curing (Part 111: 1997 normal curing, Part 112: 1983 accelerated curing);
- methods for no-fines concrete (Part 113: 1989);
- specification for testing machines (Part 115: 1990 compression testing machines);
- methods for determining strength (Part 116: 1991 compressive strength (cubes) (**3.3**), Part 117: 1989 tensile splitting strength (cylinders), Part 119: 1989 flexural strength (beams) (**3.2**))

8.17 DEGRADATION OF CONCRETE

Figure 8.11 summarises the most important degradation mechanisms in concrete.

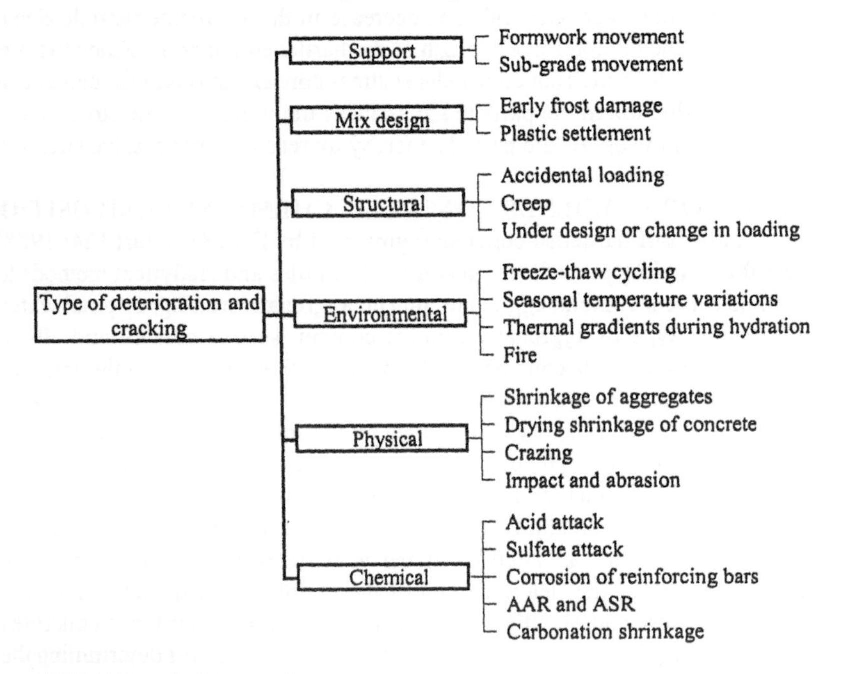

Figure 8.11 *Degradation mechanisms in concrete*

In following sections, we concentrate on chemical deterioration and degradation mechanisms, specifically

- defects in fresh concrete (**8.18**);
- mechanisms causing degradation of the hardened cement paste (acid attack and sulfate attack) in any cementitious product. Degradation (conversion) of hardened HAC paste has been considered in section **7.14.5**;
- mechanisms causing degradation of the hardened concrete matrix (alkali-silica attack). This mechanism may cause deterioration of a cementitious matrix containing specific alkali and silica constituents;
- mechanisms causing degradation of reinforced concrete (carbonation and chloride ingress). These mechanisms promote the corrosion of steel reinforcement.

8.18 DEFECTS IN FRESH CONCRETE

During the period between placing and final set, the placed fresh concrete is in a plastic, semi fluid state and the constituents are relatively free to move. Generally, denser, heavier materials tend to settle under gravitational forces whilst water tends to rise to the concrete surface. This can lead to

- **Bleeding**. Bleeding is a term describing the upward movement of water prior to the final set. Water at the surface can evaporate or be reabsorbed into the concrete to contribute to hydration reactions, but in either case will lead to a net reduction in the original concrete volume. Since evaporation rates are usually greater than the rate at which water can migrate through the fresh concrete, plastic shrinkage of the surface may occur, leading to characteristic map cracking ('crazing'). Hydration at the water rich surface will produce an inherently weak structure (a problem, for example, in floor finishes required to have a hard wearing surface) and any upwardly migrating water may become trapped under larger coarse aggregate particles, reducing the bond strength formed by the cement paste to the aggregate. Bleeding is phenomenon commonly encountered using poorly graded aggregate mixes, particularly where there is a low content of fine material (< 300 μm), or where high workability mixes are required. In either case, bleeding can be reduced by ensuring adequate proportions of fine material (due to its high specific surface, silica fume, **7.14.4**, is a particularly effective bleed control agent), by using air-entrainment and by ensuring adequate curing.
- **Plastic settlement**. Plastic settlement cracking may result where the movement occurring during the overall settlement of concrete after placing is restrained (for example, by reinforcement), articularly at the concrete surface.

8.19 DEGRADATION OF THE HARDENED CEMENT PASTE

8.19.1 Acid Attack

Portland cement is a 'calcium alkali adhesive' and will react with any chemical which reacts with calcium hydroxide, $Ca(OH)_2$ or calcium carbonate, $CaCO_3$. Hence any cementitious product containing Portland cement is theoretically at risk from deterioration by aggressive acids. Most acids attack the hardened cement paste by converting its constituents into readily soluble salts which can recrystallise on or within the surface of the cementitious material

(**1.19.2**). However, satisfactory durability to acid attack is achieved by good quality concrete with good compaction and curing. Good quality concrete is ordinarily at risk only if the pH of the aggressive acid is less than about 6.5 to 6. Reactions with acidic gases from the environment which form acidic solutions have been examined previously (**4.3.1**) and are particularly important in the deterioration of reinforced concrete (**8.21.4**). Additional examples of aggressive agents include fats and some oils. Recommendations for concrete in acid environments is given in section **8.19.3**.

8.19.2 Sulfate Attack

Concrete exposed to water soluble sulfates may deteriorate due to the reactions of the sulfates on the cementitious matrix. Any cementitious product may be affected, although products used underground (where a constant supply of water soluble sulfates, e.g, from soil or mobile groundwater) are more susceptible. Sulfates of various types found in soil or groundwater may be derived from their presence in many geological deposits. Additionally, sulfates may be found in brick rubble (particularly where gypsum plaster, **14.7.3**, adheres to the surfaces), poor quality blast furnace slag (**7.13**) and industrial waste, particularly from the mining or combustion of coal. Burning fossil fuels, such as oil, coal and gas, may also liberate sulfates. Use of these fuels commonly leads to sulfate attack of mortar joints, as evidenced by leaning chimney stacks in many domestic properties, particularly in coal mining regions of the UK. These are formed by the expansive reaction of water soluble sulfates from the domestic fuel with the mortar, causing the chimney to lean. As the wind controls the evaporation of water, more evaporation occurs on the windward side of the chimney compared to the leeward side. Hence water soluble sulfates are continuously dissolving and being deposited within the mortar joints on the windward face, Forcing the chimney to lean in the windward direction.

Most sulfates are very aggressive to the hardened cement paste and the severity of attack depends to a large extent on the solubility of the sulfate in water. Insoluble sulfates (such as barium sulfate) are not aggressive, whilst most other sulfates are water soluble to a greater or lesser degree, and are therefore aggressive. Water soluble sulfates react with the following hydration products of cement (note that the reactions shown are simplified; in practice the reaction compounds more complex.)

- the free calcium hydroxide, $Ca(OH)_2$ from the hydration of calcium silicates (**7.16**) e.g,

$$\underset{\text{(calcium hydroxide)}}{Ca(OH)_2} + \underset{\text{(sodium sulfate)}}{Na_2SO_4} + \underset{\text{(water)}}{2H_2O} \rightarrow \underset{\text{(gypsum)}}{CaSO_4.2H_2O} + \underset{\text{(sodium hydroxide)}}{2NaOH}$$

- the hydration products of tricalcium aluminate, C_3A, particularly the reaction with the hydrated monosulfate aluminate, $C_3A.C\overline{S}.H_{16}$ (**7.16.1**) e.g,

$3CaO.Al_2O_3.C\underline{a}SO_4.18H_2O$	+	$2Ca(OH)_2$	+	$2S\underline{O}_4$	+	$12H_2O$	→	$3CaO.Al_2O_3.3C\underline{a}SO_3.32H_2O$
$C_3A.CS.H_{16}$		2CH		2S		12H		$C_3A.3CS.H_{32}$
(hydrated monosulfate aluminate)		(calcium hydroxide)		(sulfate)		(water)		(ettringite)
hydration product of tricalcium aluminate, C_3A (7.16.1)		hydration product of calcium silicates, C_2S and C_3S (7.16)		water soluble sulfate				

The reaction of water soluble alkali metal sulfates (shown in the first set of reactions for sodium sulfate, Na_2SO_4, but applying equally to other alkali metal sulfates, such as potassium sulfate, K_2SO_4) forms gypsum, which may cause a loss of strength and stiffness of the hardened cement paste, promoting degradation and deterioration. In addition, the gypsum formed is then available to react further with the hydration products of tricalcium aluminates, as shown in the second set of reactions. The reaction of water soluble sulfates with the hydrated monosulfate aluminate, $C_3A.CS.H_{16}$, to form **ettringite** is by far the most important reaction. Ettringite is a sulfate and, in common with other sulfates (e.g, gypsum, **14.7.3**), contains water of crystallisation (**1.18.5**). The incorporation of water of crystallisation in ettringite results in this reaction product having a larger solid volume than the compounds from which it was derived. Hence the formation of ettringite leads to volume expansion (of the order of 0.2-2%), sufficient to crack the hardened cement paste.

Whilst the formation of ettringite is the most widely accepted mechanism producing sulfate attack of the hardened cement paste, more recent evidence suggests that the formation of **thaumasite** ($CaSiO_3.CaCO_3.CaSO_4.15H_2O$) may also contribute to deterioration. Whilst the formation of natural thaumasite is rare, it is most often found associated with basic igneous rocks and, more importantly, in limestones (the principal raw material for Portland cement). Thaumasite is formed by the action of water soluble sulfates on the calcium silicates in the presence of reactive alumina (Al_2O_3). Most commonly, the reactive alumina is present in the form of ettringite and hence the presence of ettringite is an important initiator in the formation of thaumasite. Hence ettringite and thaumasite are commonly found together in cementitious products which have been affected by water soluble sulfates. Although the reactions involving the formation of thaumasite are complex and not fully understood, the degradation mechanism is similar to that occurring with the formation of ettringite, i.e., a volume expansion associated with the incorporation of water of crystallisation. Experimental evidence suggests that the highest risk of thaumasite formation occurs under conditions of constantly high humidity; at temperatures around 4°C; in the presence of reactive alumina (Al_2O_3) contents of between 0.4 and 1.0% (commonly as ettringite); in the presence of an adequate supply of sulfate and carbonate anions.

When all factors are present, thaumasite has been observed to form within a couple of weeks[34]. In some cases where concrete is in persistently cold and very wet conditions, concrete with finely divided internal calcium carbonate has developed sulfate attack through thaumasite formation even though sulfate-resisting Portland cement has been used.

Magnesium sulfate is particularly aggressive to the hardened cement paste. Whereas the reactions of most of the water soluble alkali metal sulfates with the calcium silicate hydrates (CSH fibres) is insignificant, magnesium sulfates do react with, and deteriorate, these products. These reactions occur because, although magnesium sulfate reacts with calcium hydroxide, $Ca(OH)_2$ in a similar way to the alkali metal sulfates (according to the first set of reactions above), the product, magnesium hydroxide, $Mg(OH)_2$, has low solubility in water (**1.19**) and produces saturated solutions with a pH of about 10.5. Thus magnesium hydroxide forms more acidic saturated solutions than the other alkali metal hydroxides e.g, sodium hydroxide (pH about 14.0) or calcium hydroxide (pH about 12.4) and, in addition, forms saturated solutions at lower concentrations. The pH of magnesium hydroxide is lower than that required to stabilise the calcium silicate hydrates (CSH fibres), which therefore liberate free calcium hydroxide to maintain the pH equilibrium. Calcium hydroxide liberated is then available to react further with magnesium hydroxide, as before. This cycle of reactions can cause complete disintegration of the calcium silicate hydrates in the hardened cement paste. For dicalcium silicate

$3CaO.2SiO_2.2H_2O$	+	$3MgSO_4$	+	$8H_2O$	→	$3(CaSO_4.2H_2O)$	+	$3Mg(OH)_2$	+	$2SiO_2.H_2O$
(dicalcium silicate)		(magnesium sulfate)		(water)		(gypsum)		(magnesium hydroxide)		(silica)

The calcium silicate hydrates are predominantly responsible for hardened cement paste strength (**7.16.2**), hence decomposition of these compounds leads to rapid strength loss.

As a consequence of these degradation mechanisms, four essential requirements for sulfate attack to occur can be defined

- presence of hydrated monosulfate aluminates derived from hydration of the cementitious constituent tri-calcium aluminate (C_3A) (**7.16.1**);
- water;
- water soluble sulfates;
- sufficient time for the reactions to occur (generally, cracks caused by the ettringite form of sulfate attack take many years to form; most other crack producing mechanisms occur relatively rapidly after construction is completed, e.g, moisture induced cracking, **6.4**).

Degradation of cementitious products can be arrested by the absence of any one factor. For example, SRPC has a reduced amount of the tricalcium aluminate and calcium silicates relative to other Portland cements (**7.14.2**), so that the hydration products of these cement compounds required for sulfate attack are limited.

The **rate** of sulfate attack is affected by

- the water table and mobility of the groundwater (mobile groundwater containing water soluble sulfates represents a more hazardous environment since it continually replenishes the sulfates consumed by the reactions, thereby allowing the process to continue);
- the quality of the concrete and particularly permeability (as affected by compaction, cement type and content, type of aggregate, w/c ratio and curing);

- the form of construction (concrete in massive forms of construction will deteriorate less quickly than similar quality concrete in thin sections);
- the concentration and type of aggressive sulfate (magnesium > sodium > calcium);
- the pH of the water.

Although concrete below ground is unlikely to be affected by frost, the combination of sulfate attack and exposure to frost represents particularly severe conditions.

8.19.3 Design of Concrete in Aggressive Ground

Detailed design advice exists for concrete exposed to water soluble sulfates and acidic conditions[21,35]. Firstly, the site is classified on the basis of the sulfate level (Design Sulfate (DS) Class) and aggressive chemical level (Aggressive Chemical Environment for Concrete (ACEC) Class) of the site in accordance with **Table 8.10.**

Table 8.10 Aggressive Chemical Environment for Concrete (ACEC) site classification

DS Class	Sulfate and Magnesium: 2:1 water/soil extract		Groundwater		Total potential sulfate[b]	Natural Soil: Static water	Mobile water	Brownfield[c]: Static water	Mobile water	ACEC Class[e]
	SO₄ (g/l)	Mg (g/l)	SO₄ (g/l)	Mg (g/l)	SO₄ (%)	pH	pH	pH[d]	pH[d]	
DS-1	<1.2		<0.4		<0.24	all pH values		all pH values		AC-1s
							>5.5		>6.5	AC-1
							≤5.5		5.5-6.5	AC-2z
									4.5-5.5	AC-3z
									<4.5	AC-4z
DS-2	1.2-2.3		0.4-1.4		0.24-0.6	>3.5		>5.5		AC-1s
							>5.5		>6.5	AC-2
						≤3.5		≤5.5		AC-2s
							≤5.5		5.5-6.5	AC-3z
									4.5-5.5	AC-4z
									<4.5	AC-5z
DS-3	2.4-3.7		1.5-3.0		0.7-1.2	>3.5		>5.5		AC-2s
							>5.5		>6.5	AC-3
						≤3.5		≤5.5		AC-3s
							≤5.5		5.5-6.5	AC-4
									<5.5	AC-5
DS-4	3.8-6.7	≤1.2	3.1-6.0	≤1.0	1.3-2.4	>3.5		>5.5		AC-3s
							>5.5		>6.5	AC-4
						≤3.5		≤5.5		AC-4s
							≤5.5		≤6.5	AC-5
DS-4m	3.8-6.7	>1.2[a]	3.1-6.0	>1.0[a]	1.3-2.4	>3.5		>5.5		AC-3s
							>5.5		>6.5	AC-4m
						≤3.5		≤5.5		AC-4ms
							≤5.5		≤6.5	AC-5m
DS-5	>6.7	≤1.2	>6.0	≤1.0	>2.4	>3.5	all pH values	>5.5	all pH values	AC-4s
						≤3.5		≤5.5		AC-5
DS-5m	>6.7	>1.2[a]	>6.0	>1.0[a]	>2.4	>3.5	all pH values	>5.5	all pH values	AC-4ms
						≤3.5		≤5.5		AC-5m

Notes: [a] The limit on water soluble magnesium does not apply to brackish groundwater (chloride content between 12g/l and 18g/l); this allows **m** to be omitted from the relevant ACEC classification; [b] Applies only to sites where concrete will be exposed to sulfate ions (SO_4) which may result from the oxidation of sulfides, such as pyrite, following ground disturbance; [c] 'Brownfield' is defined as sites which may contain chemical wasters remaining from previous industrial use or from imported wastes; [d] An additional account is taken of hydrochloric and nitric acids by adjustment to sulfate content. [e] Suffix **s** indicates that, as water has been classified as static, no Additional Protective Measures (APMs) are generally necessary. Concrete placed in ACEC Classes which include the suffix **z** have primarily to resist acid conditions and may be made with cements from any of the Cement Groups (**Table 7.12**), subject to the restrictions concerning Portland limestone cement. Suffix **m** relates to the higher levels of magnesium in Sulfate Classes 4 and 5.

Secondly, the Structural Performance Level (SPL) of the concrete is specified according to the following criteria: **Low** (service life of the structure less than 30 years; unreinforced concrete; non-critical structural details; temporary structures; long service life structures, but with associated low stress levels (e.g, unreinforced house foundations); **Medium** (intermediate service life (30 to 100 years); not falling in either **low** or **high** categories); **High** (Long service life structures (greater than 100 years); vulnerable critical details, e.g, slender structural elements, joints, etc.; structures retaining hazardous materials). Thirdly, the suitable types of cement and aggregate, and the type and quality of the concrete for use below ground level in the aggressive chemical environment identified. Concrete quality is specified in terms of a Design Chemical Class (DC Class) corresponding to each ACEC Class and use of concrete below ground (**Table 8.11**).

Table 8.11 Concrete quality and number of Additional Protective Measures (APMs) recommended for the general use of *in situ* and precast concrete

ACEC Class'	Design Chemical (DC) Class/Number of Additional Protection Measures (APMs)[a,b]							
	Low structural performance level			Normal structural performance level			High structural performance level[c]	
section thickness →	<140 mm	140-450 mm	>450 mm[d]	<140 mm	140-450 mm	>450 mm[d]	140-450 mm	>450 mm[d]
AC-1	DC-2/0	DC-1/0	DC-1/0	DC-2/0	DC-1/0	DC-1/0	DC-1/0	DC-1/0
AC-1s	DC-2/0	DC-1/0	DC-1/0	DC-2/0	DC-1/0	DC-1/0	DC-1/0	DC-1/0
AC-2	DC-3/0	DC-2/0	DC-1/0	DC-3/0	DC-2/0	DC-1/0	DC-2/0	DC-1/0
AC-2s	DC-3/0	DC-2/0	DC-1/0	DC-3/0	DC-2/0	DC-1/0	DC-2/0	DC-1/0
AC-2z	DC-3z/0	DC-2z/0	DC-1/0	DC-3z/0	DC-2z/0	DC-1	DC-2z/0	DC-1/0
AC-3	DC-3/2[e]	DC-3/1[e]	DC-2/1	DC-3/3[e]	DC-3/2[e]	DC-2/2	DC-3/3[e]	DC-2/3
AC-3s	DC-4/0	DC-3/0	DC-2/0	DC-4/0	DC-3/0	DC-2/0	DC-3/0	DC-2/0
AC-3z	DC-4z/0	DC-3z/0	DC-2z/0	DC-4z/0	DC-3z/0	DC-2z/0	DC-3z/0[e]	DC-2z/0[e]
AC-4	DC-4/2[e,f]	DC-4/1[e]	DC-3/1[e]	DC-4/3[e]	DC-4/2[e]	DC-3/2[e]	DC-4/3	DC-3/3
AC-4s	DC-4/0	DC-4/0	DC-3/0	DC-4/0	DC-4/0	DC-3/0	DC-4/0	DC-3/0
AC-4z	DC-4z/1[g]	DC-4z/0	DC-3z/0	DC-4z/1[g]	DC-4z/0	DC-3z/0	DC-4z/0	DC-3z/0
AC-4m	DC-4m/2[e,f]	DC-4m/1[e]	DC-4m/0	DC-4m/3[e]	DC-4m/2[e]	DC-4m/1[e]	DC-4m/3[e]	DC-4m/2[e]
AC-4ms	DC-4m/0	DC-4m/0	DC-3/0	DC-4m/0	DC-4m/0	DC-3/0	DC-4m/0	DC-3/0
AC-5	DC-4/1[g]	DC-4/1[g]	DC-3/1[g]	DC-4/3[e,h]	DC-4/2[e,h]	DC-3/2[e,h]	DC-4**/1[g]	DC-3**/1[g]
AC-5m	DC-4m/1[g]	DC-4m/1[g]	DC-4m/1[g]	DC-4m/3[e,h]	DC-4m/2[e,h]	DC-4m/1[g]	DC-4m**/1[g]	DC-4m**/1[g]
AC-5z	DC-4z/1[g]	DC-4z/1[g]	DC-3z/1[g]	DC-4z/1[g]	DC-4z/1[g]	DC-3z/1[g]	DC-4z/1[g]	DC-4z/1[g]

Notes: [a] where carbonation of the concrete prior to exposure can be assured, the recommended DC Class/APM (other than those carrying a z suffix) may be relaxed by 1 DC Class or 1 APM, provided that for AC-5 and AC-5m conditions the resulting package includes APM3; [b] where the hydrostatic head of groundwater is greater than 5 times the section thickness, one APM over and above the number indicated in the Table should be applied (preferably APM5 but, if not practicable, APM1 or APM3). [c] For High SPL, a section thickness of <140 mm is generally not recommended in aggressive ground conditions (although specific precast units are an exception). [d] Use of this column is inappropriate where the surface of the concrete is required to remain unaffected by aggressive ground (e.g, friction piles). [e] where DC-3, DC-4 or DC-4m is given, the number of APM may be reduced (provided the reduction does not override the recommendation to use APM3) by either one provided DC-3*, DC-4* or DC-4m* concrete is specified; or by two provided DC-3**, DC-4** or DC-4m** concrete is specified. [f] If APM3 is selected for Low SPL concrete, no further APM are necessary. [g] Only APM3 is recommended (not applicable to bored piles). [h] To include APM3, where practicable, as one of the recommended number of APM.

Each DC Class defines concrete quality in terms of minimum cement content and maximum free w/c ratio for each cement group (**Table 8.12**), and allows for any aggregate carbonate content by additionally specifying concrete in terms of Aggregate Carbonate Range (ACR). The ACR is expressed as calcium carbonate equivalent as a percentage of the total aggregate mass; for carbonate in both the coarse and fine aggregate fraction, **Range A** is 10-30% to 100%, **Range B** is 2-12% to 10-30%, **Range C** is 0% to 2-12%. The ACR has been introduced to combat the thaumasite form of sulfate attack (**8.19.2**). Relaxation or increases in the concrete quality requirements, as incorporated in the DC Class, are allowed according to the element thickness and the exposure conditions.

Table 8.12 Concrete quality to resist chemical attack

Design Chemical (DC) Class	Aggregate Carbonate range (ACR)	CEM Cement or combination group[a]	Dense fully compacted concrete made with aggregate conforming to BS 882 or BS 1047	
			Minimum cement content (kg/m³)	Maximum free w/c ratio
DC-1	No restriction	1,2,3	-	
DC-2	A[d],B,C	1[b]	340	0.50
	A[d],B,C	2,3	300	0.55
DC-2z[c]	No restriction	1[b],2,3	300	0.55
DC-3	A	2a	400	0.40
	A	2b,3	380	0.45
	B,C	2,3	340	0.50
DC-3*[f]	B	2,3	380	0.45
DC-3**[g]	C	2,3	380	0.45
DC-3z[c]	No restriction	1[b],2,3	340	0.50
DC-4	A	2a	400	0.35
	A	2b,3	400	0.40
	B,C	2,3	380	0.45
DC-4*[f]	B	2,3	400	0.40
DC-4**[g]	C	2,3	400	0.40
DC-4z[c]	No restriction	1[b],2,3	380	0.45
DC-4m	A	2b,3	400	0.40
	B,C	3	380	0.45
DC-4m*[f]	B	3	400	0.40
DC-4m**[g]	C	3	400	0.40

Notes: [a] for cement or combination groups, see **Table 7.12.** [b] Portland limestone cement may be used only in concrete specified as DC-2, DC-2z, DC-3z, DC-4z where the DS class (**Table 8.10**) of the site does not exceed DS-1. [c] Classes DC-2z, DC-3z and DC-4z apply where chemical resistance is recommended primarily to resist acid attack. [d] In addition to the requirement for a minimum cement content and maximum free w/c ratio, a minimum concrete grade of C35 is recommended when using Range A aggregate combination. [e] Use of APM1 (enhanced concrete quality) can be satisfied by using the recommendations of the next numerically higher DC Class (where the starting point is a starred DC Class, the step is to a similarly starred higher DC Class (e.g, DC-3* plus a 1 step enhancement in concrete quality leads to a specification of DC-4**). [f] Using a single-starred DC Class may permit the recommended number of APMs to be reduced by 1 (**Table 8.11**). [g] Using a double-starred DC Class may permit the recommended number of APMs to be reduced by 2 (**Table 8.11**).

In some circumstances, Additional Protective Measures (APMs) are recommended to further protect concrete (**Table 8.11**). These include **APM1** (enhanced concrete quality); **APM2** (use of controlled permeability formwork); **APM3** (provide surface protection); **APM4** (provide sacrificial layer); **APM5** (address drainage of the site).

Most standard test procedures[36,37] for the determination of the water or acid soluble sulfate content of samples (soils, bricks, groundwater, etc.) involve a gravimetric method in which barium chloride is added to the water or acid extract and the resulting precipitate of barium sulfate is dried and weighed. The sulfate content of the sample is then calculated from the mass of the material used in the analysis and the mass of the barium sulfate precipitated.

8.20 DEGRADATION OF THE HARDENED CONCRETE MATRIX

The aggregates present in concrete are usually considered to be inert. However in a strongly alkaline medium, they can chemically react in a manner disruptive to concrete. There are two types of reactions, alkali aggregate (AAR) and alkali silica (ASR) reactions, sometimes referred to as '**concrete cancer**'.

8.20.1 Alkali aggregate reaction (AAR)

Under moist conditions, aggregates containing carbonates will react with the alkaline pore water in the cement paste disruptively and cause degradation within the concrete matrix. The reaction was first reported in Ontario in 1955 and was found to occur with certain dolomitic limestones which contain intergranular clays. It is the reaction with this clay which leads to expansion. Most studies have indicated that there are three interactions with carbonates

- reactions which improve the aggregate cement bond, and do not further react with water. Most carbonates fall into this category;
- reactions which involve a chemical interaction around the rim of the carbonate aggregate. This rim will dissolve in hydrochloric acid at the same rate as the innermost part of the aggregate which has not been 'attacked' by the alkali. These reactions do not produce a disruptive expansion under moist conditions;
- reactions with fine grained dolomitic limestone containing interstitial clay or calcite. This reaction produces a rim around the aggregate which does not dissolve in hydrochloric acid due to an enrichment of silicon within the rim.

A possible reaction is

$Ca.Mg(CO_3)_2$	+	$2MOH$	→	$Mg(OH)_2$	+	$CaCO_3$	+	M_2CO_3
Calcium magnesium carbonate Dolomite		M hydroxide cement alkali		Magnesium hydroxide		Calcium carbonate		M carbonate

followed by

M_2CO_3	+	$Ca(OH)_2$	→	MOH	+	$CaCO_3$
M carbonate		Calcium hydroxide cement alkali		M hydroxide alkali		Calcium carbonate

where **M** is either sodium (Na), potassium (K) or lithium (Li). These are group 1 elements and therefore their alkalis are much stronger than those formed by the group 2 elements magnesium (Mg) and calcium (Ca) (**1.7**). The reaction is also complicated by the release of sulfate ions from the calcium sulphoaluminate hydrates, which react with the '**M** carbonate'. The strong group 1 element hydroxides are regenerated by this mechanism and these diffuse back into the dolomitic aggregate. The mechanism by which the silica ring is formed is through the magnesium hydroxide reacting with silicate ions $(SiO_4)^{4-}$.

8.20.2 Alkali Silica Reaction (ASR)

Some aggregates react with the alkali constituent of the cement and form a gel (***Figure 8.12a***) which swells by absorbing water through osmosis (**6.6.4**). The expansive forces cause cracking and disruption of the concrete. ASR first came to prominence in California in 1940, where the alkali pore water reacted with certain silica containing aggregates to form an alkali-silica gel. This gel is hygroscopic and has the property of absorbing water to form an internal expansive product which cracks the concrete and its' aggregates. This cracking reduces the degree of protection from the environment which the physical concrete cover can offer the embedded steel reinforcement. The alkalis may be derived from the cement, admixtures, pigments or from sodium chloride.

$$\underset{\text{(alkali)}}{OH^-} + \underset{\text{water}}{H_2O} + \underset{\text{(aggregate)}}{\text{reactive silica}} \rightarrow \text{gel}$$

The gel can react further with free water producing expansion and consequent disruption. The gel has the following properties

- hygroscopicity, attracting water and swelling;
- it readily reacts with carbon dioxide (hence carbonation stops the expansion);
- the expansive reaction can take place in the absence of liquid water provided the relative humidity RH > 75%.

Three conditions are necessary before damage can be caused by ASR in concrete

- sufficient moisture (since most concretes are exposed to frequent wetting, it is not possible to remove this contributory element);
- free alkali, usually sodium (Na_2O) and/or potassium (K_2O) oxides;
- a critical amount of **reactive silica**.

The reactivity of the aggregate is often dependent upon minor constituents that may be non-uniformly distributed within the aggregate source (examined by petrographical methods). The most potentially reactive forms of silica have been found to be glassy, amorphous or poorly crystalline or disordered forms of silica, such as opal, whilst the least reactive is quartz. Potentially reactive forms of silica are chert, flint, chalcedony, recrystallised quartz and volcanic glasses; hence an aggregate sample containing 5% to 60% flint should be considered potentially liable to ASR. Many aggregates contain small percentages of reactive materials. Different w/c ratios can adversely affect ASR; i.e., low w/c ratios will give a higher incidence of ASR at a high flint content, whereas a high w/c ratio will give no expansion at all. The expansion of the gel results in a series of cracks which follow the highly stressed regions, causing a map pattern (***Figure8.12b***)

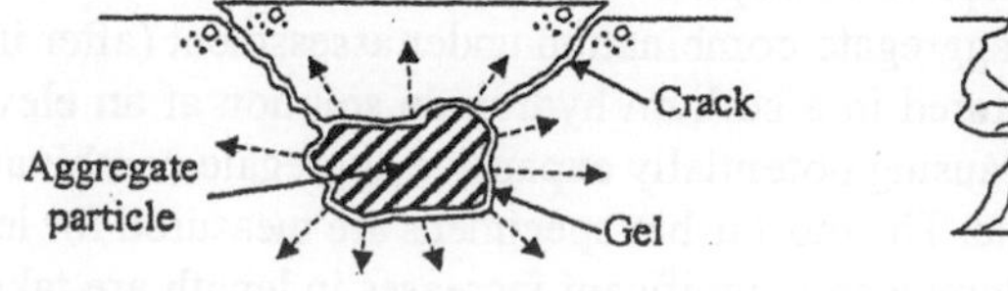

a) Pop-outs

b) Surface map cracking

***Figure 8.12** Typical cracking patterns resulting from ASR*

However, this map pattern damage also exists in concrete affected by other deterioration mechanisms, such as frost attack, sulfate attack, corrosion of reinforced steel and drying shrinkage and care must be taken to distinguish their effects. ASR is often characterised by a **pinkish discolouration** along the cracks, which is associated with the exudation or efflorescence of calcium salts. The surface of the concrete is also characterised by sweaty areas which are not a characteristic of these other defects. Therefore it is best to examine for ASR on a drying day after rainfall and in good light. Note should be made of sources of water, for example water table level, areas of poor detailing, the likelihood of condensation, location of leaking joints and the prevailing wind (for rain) direction. As a result of the similarity in the crack patterns caused by different defects, laboratory tests must include

- the examination of thin films made from the concrete, as these are the only unequivocal way of determining the existence or otherwise of the ASR reactive gel. The concrete sample, usually obtained by coring, should reach the petrographic laboratory in the wet condition. It should be wrapped in cling film, and placed in sealed polythene bags to ensure that the reactive gel does not dry out. If the gel dries out, it shrinks and the petrographic analysis is void;
- determination of the alkali content of the concrete;
- expansion testing of the concrete in a humid environment.

BS 7943[38] gives guidance on the interpretation of petrographical sections of concrete for alkali-silica reactivity and includes a summary of rock types associated with alkali-silica damage in the UK (as at 1997) and the suggested potential reactivity of these rocks when used as aggregates in concrete. In addition, there are currently two British Standard tests for determining alkali-silica reactivity in concrete

- the **concrete prism method** (BS 812: Part 123[39]) measures the expansion of concrete produced by alkali-silica reactions involving specific combinations of aggregates. The method involves preparing concrete test prisms from the aggregate combinations under test and storing them for a period of 52 weeks in conditions known to promote alkali-silica attack. During this time, measurements are made at intervals to determine whether any expansion has occurred. The test is accelerated by formulating the concrete to possess higher than average cement and alkali content and by storing the prisms at high humidity and temperature (38°C) to increase the rate of the reaction.
- the **potential accelerated mortar-bar method** (DD 249[40]) is a rapid procedure for predicting the expansion potential of concrete by alkali-silica reactivity of aggregates and aggregate combinations over a period of 16 days. Test specimens are cast from a mortar mix made using the aggregate or aggregate combination under assessment (after initial conditioning in hot water), and stored in a sodium hydroxide solution at an elevated temperature for at least 14 days, causing potentially expansive aggregate combinations to react at a greatly accelerated rate. The mortar-bar specimens are measured for length change after various periods of storage and significant increases in length are taken to indicate a potential for ASR or AAR expansion. The expansions in the test are used to classify the potential expansivity of the aggregate or aggregate combination.

Both methods are designed as a accelerated laboratory tests for aggregates or combinations of aggregates and are not therefore considered to be performance tests for concrete. Further advice on appropriate test methods is given by the British Cement Association[41].

With respect to limiting the potential for alkali-silica reactions, ENV 206 states that the appropriate national standard requirement should be adopted which, in the UK, is BS 5328. The control of ASR requires knowledge of the chemical and petrographic nature of the aggregate used in the concrete. BS 5328: Part 1:1997 gives the following guidance on limiting the potential for alkali-silica reaction

- the concrete should be kept dry after curing;
- the maximum reactive alkali content of the concrete mix is 3.0 kg/m^3 (expressed as Na_2O equivalent). Where the reactive alkali content exceeds 0.2 kg/m^3 from sources other than the cementitious materials, the reactive alkali content should be taken into account. The 3.0 kg/m^3 limit may be achieved by limiting the cement content, selecting an appropriate Portland cement (or Portland replacement cement) (e.g, a concrete comprising Portland cement with an certified average alkali content of 0.6% or less with a maximum cement content of 500 kg/m^3 will conform to this limit) or using at least 25% blastfurnace slag or fuel ash in a composite Portland cement;
- where the reactive alkali content of the concrete from other sources does not exceed 0.2%, the reactive alkali content of the cement is limited by using a cement with an alkali content of less than 0.6% or using by using a blastfurnace slag composite cement with a minimum proportion of 50% blastfurnace slag;
- a combination of aggregates likely to be unreactive is used.

8.21 DEGRADATION OF REINFORCED CONCRETE

There are three essential conditions that must exist before corrosion of reinforcement in concrete can occur

- a low concrete electrical resistivity is required to permit electrochemical corrosion;
- ordinarily, oxygen from the atmosphere must be present at the steel surface before corrosion can proceed;
- the chemical passivity conferred on the steel reinforcement by the alkaline nature of the cement must be destroyed.

8.21.1 Electrical Resistivity

The electrical resistivity of concrete is very important in many corrosion processes since, for corrosion to proceed, concrete must be able to transfer charged ions through its porous structure. Electrical resistivity (**1.10.1**) values in the range 5000 to 10000 ohm cm appear to be critical; above this range the corrosion process is significantly reduced. In practice, the electrical resistivity of concrete is dependent on porosity, chloride content and moisture content and the general relationship of these last two factors is illustrated in ***Figure 8.13***. ***Figure 8.13a*** shows how, with moisture contents of over about 4%, the electrical resistivity is below the critical value and concrete in this condition would allow corrosion reactions to proceed. In ***Figure 8.13b*** shows that wetting and drying cycles can reduce the electrical resistivity of concrete through absorption of water. In practice, the electrical resistivity is rarely a limiting factor in the corrosion of steel reinforcement in most concrete, except in very dry conditions (for example, the Middle East, where high concrete electrical resistivities in excess of 100000 ohm cm are common).

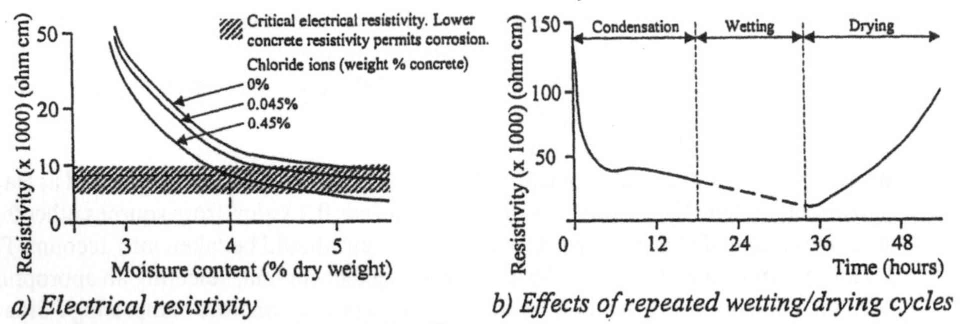

a) Electrical resistivity *b) Effects of repeated wetting/drying cycles*

Figure 8.13 *The electrical resistivity of concrete cover*

8.21.2 Presence of Oxygen

Oxygen is required at the surface of the reinforcement steel to form the passive oxide film (**11.4**). In addition, the presence of oxygen is required in the corrosion process to allow the cathode reaction

$$2H_2O \quad + \quad 4OH^- \quad + \quad 4e \quad \rightarrow \quad 4OH^-$$

to occur. Unless the concrete is saturated with water, the oxygen can ordinarily reach the reinforcement steel through the permeable concrete cover in a matter of hours. Under certain circumstances (e.g, concrete submerged under water or buried under ground) the supply of oxygen may be restricted so that the passive oxide film may not be formed on the steel. Under these circumstances, the cathodic reaction shown above cannot take place (due to the absence of oxygen) but corrosion can still theoretically occur through another cathodic reaction, the reduction of water to hydrogen (**11.3.2**). However, the kinetics of this process are extremely slow and it is unlikely that corrosion of steel reinforcement by this process will be significant, except when the kinetics of this process are artificially speeded up (using, for example, cathodic protection by impressed current, **11.13.3**), sufficient hydrogen may be evolved to cause problems of hydrogen embrittlement (**8.25.5**).

8.21.3 Chemical Passivity

The presence of calcium hydroxide $Ca(OH)_2$ is important because it provides an alkaline environment, which confers a **chemical passivity** upon embedded steel. The term chemical passivity means that a very adherent and coherent oxide film is formed which prevents the metal from corroding (rather like the aluminium oxide film on the metal aluminium). The chemical protection conferred on steel is through a passive protective iron oxide film (of

γ-Fe_2O_3), which forms on steel in an environment at or above a pH of 9-10. The steel reinforcement is prevented from corroding as long as this passive oxide film is maintained.

A Scientist called **Pourbaix** first examined the relationship of pH and standard electrode potential for a variety of metals. For the iron (Fe)-H_2O system, the Pourbaix diagram is shown in ***Figure 8.14***.

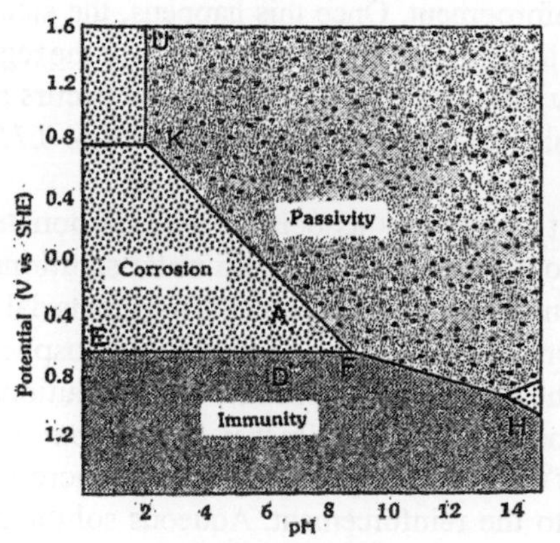

The Pourbaix diagram indicates the potentials and pH values which lead to corrosion, passivity and immunity. The figure also shows how the hydrogen potential varies with respect to the pH of the environment (line **H**) and the cathodic reaction equilibrium between O_2 and OH^- (line **O**). In the area of **passivity**, a stable oxide film (of γ-Fe_2O_3) will form. Thus steel embedded in concrete will not corrode as long as the pH remains sufficiently high in relation to the potential of the metal. In the area of **immunity**, steel cannot corrode. Immunity occurs through maintaining a supply of energy of a magnitude that is equal to or greater than the energy liberated when the reaction occurs spontaneously. Hence if the potential of the steel is depressed sufficiently to prevent it oxidising, the steel cannot corrode. Modern repair systems for reinforced concrete (known as **cathodic protection** systems) are based on this principle and operate using sacrificial anodic metals or, more usually, impressed current (**8.21.1**).

Figure 8.14 *Pourbaix diagram for Fe-H_2O system*

There are two processes by which the stable oxide film formed on embedded steel can be destroyed; **carbonation** and **chloride ion ingress**.

8.21.4 Carbonation

Consider, for example, embedded steel initially surrounded by concrete at a pH of, say, 9 (point B in ***Figure 8.14***). Under these circumstances, the steel occupies the passive region of the Pourbaix diagram, forming a stable oxide film. If, however, the concrete is then exposed to an acidic environment that reduces the pH of the surrounding concrete to, say, 7 (point A in ***Figure 8.14***), the embedded steel then occupies a region of corrosion. Within this region, the passive oxide film is no longer stable and deteriorates, allowing oxidation (corrosion) of the steel. The process by which the alkalinity of the cementitious products is reduced by the action of acidic environments is termed **carbonation** (also responsible for the hardening of building limes, **7.10.2**). In reinforced concrete, however, carbonation is a potentially damaging phenomenon which may lead to structural degradation.

As we have already seen (**7.16**), two compounds present in Portland cement (C_3S and C_2S) hydrate to produce calcium hydroxide, $Ca(OH)_2$. Calcium hydroxide is an alkali but is not very soluble (0.0019%) and produces pore water solutions in cementitious materials with a pH normally exceeding 12.6. The presence of other alkali salts (e.g, sodium and potassium, derived from the aggregate) can also increase the pH further, as these salts are more soluble (**Table 1.14**). In this pore water environment, the steel acquires passivity with great ease because only a small amount of oxygen is required to exceed the critical current. However, concrete is porous and permeable to aggressive species present in the surrounding environment (**8.13**). The diffusion of acidic gases from the environment through the porous cementitious material gradually lowers the alkalinity so that eventually the depth of penetration may reach the embedded steel reinforcement. Once this happens, the steel is depassivated allowing the corrosion mechanism to commence. The progression of the region of reduced alkalinity (increased pH) caused by the ingress of carbon dioxide occurs as a front (called the carbonation front) from atmospherically exposed surfaces (***Figure 8.15a***).

The acidic gas carbon dioxide (CO_2) is most commonly associated with carbonation, although sulfur dioxide (SO_2) is sometimes also present. Aqueous pore water solutions of sulfur dioxide will tend to react with the alkaline surface of the cementitious material but usually presents only a superficial problem, even in highly polluted industrial atmospheres. Carbon dioxide will, however, dissolves into the pore water to produce weak solutions of carbonic acid (H_2CO_3). Carbon dioxide can penetrate more readily than sulfur dioxide and the rate of penetration will depend the quality of the concrete, placement of the concrete and the nature of thickness of the concrete cover to the reinforcement. Aqueous solutions of carbon dioxide react with the calcium hydroxide in the cementitious product to produce a neutral salt, calcium carbonate ($CaCO_3$)

$$Ca(OH)_2 \rightarrow Ca^{2+} + 2OH^-$$

$$CO_2 + 2OH^- \rightarrow CO_3^- + H_2O$$

$$Ca^{2+} + CO_3^- \rightarrow CaCO_3$$

From these reactions it can be seen that the process of carbonation effectively removes alkali hydroxyl (OH^-) ions from the pore water solution thereby reducing the pH of the concrete. These reactions are ordinarily abbreviated to

$$\underset{\text{alkaline}}{Ca(OH)_2} + \underset{\text{acidic}}{H_2CO_3} \rightarrow \underset{\text{neutral}}{CaCO_3} + \underset{\text{water}}{2H_2O}$$

Similar sets of reactions can be written for the other hydration products. These reactions illustrate that water is released. In addition, there is an increase in the weight of the cement paste; shrinkage (broadly equivalent to the amount of drying shrinkage, **6.4**); and the paste also increase in strength and decreases in permeability. This can be explained by the fact that calcium hydroxide is dissolved from the more highly stressed regions (resulting in shrinkage) and the calcium carbonate crystallises out in the pores on the concrete, reducing permeability and increasing strength. These processes are not necessarily detrimental to the concrete itself.

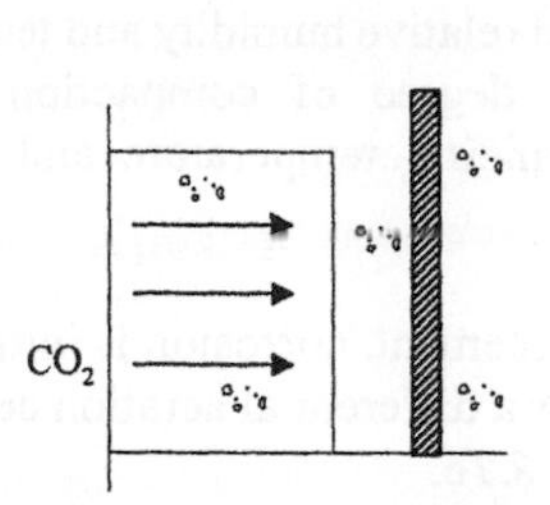

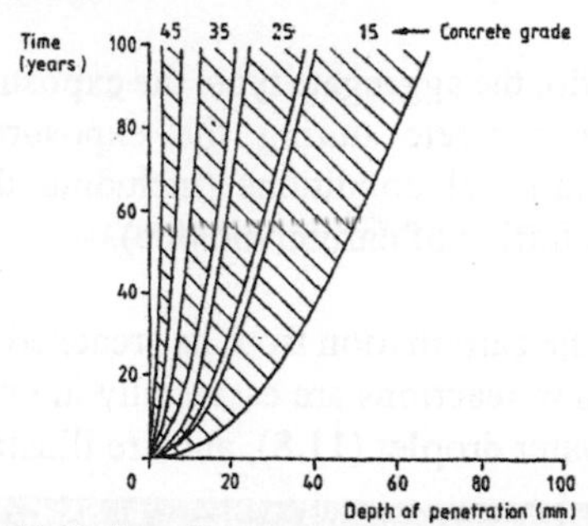

a) Progression of the carbonation front *b) Carbonation penetration curves*

***Figure 8.15** Carbonation of reinforced concrete*

The carbonation front is only significant when it reaches the depth of the steel reinforcement. The rate of penetration of the carbonation front depends primarily upon

- the permeability of the concrete to carbon dioxide, which strongly influenced by the w/c ratio (**8.13.2**), the cement type and content (**8.13.3**) and the aggregate porosity (**8.14.2**);
- the moisture content of the concrete;
- the total alkali content of the concrete.

The greatest rate of carbonation occurs when the relative humidity of the concrete is in the range 60 to 70%. This fact explains why damage does not always occur in reinforced concrete where the carbonation zone has reached the embedded steel. For moisture contents lower than this, there is insufficient water available for the carbonation reactions to take place, whilst at higher relative humidities the liquid water present in the pores reduces the rate of diffusion. Dense, well compacted concrete with adequate cover will be highly resistant to the ingress of aggressive gaseous species and the carbonation front may only penetrate a few centimetres in many years. However, in low quality, porous concrete with low cover or containing cracks (from other degradation process, such as settlement) have low resistance. Reinforced concrete is expected to exhibit some cracking (***Figure 8.11***) and design codes of practice seek to limit the width of cracks caused by these mechanisms depending on the exposure conditions of the structure. Under normal service conditions, crack widths of up to about 0.3 mm are anticipated in reinforced concrete and cracks may lead to the more rapid penetration of chlorides or carbonation and may cause very severe local corrosion. However, this effect is partially offset because the corrosion products can block the crack and protect the underlying steel in the localised vicinity of the crack. Generally, crack widths appear to be less important to the risk of corrosion than crack frequency, cover depth and concrete quality.

The relationship between concrete quality and depth of penetration of the carbonation front is shown in ***Figure 8.15b***. Commonly, carbonation is assumed to follow a parabolic ingress rate, $d = k\,t^n$, where d is the depth of carbonation, t is time, n an exponent lower than 1 (commonly taken as 0.5, hence the carbonation ingress rate is commonly referred to as the 'square root law') and k is a constant which will depend on the cement type and content, the w/c ratio, the aggregate type, the exposure duration (and relative humidity and temperature of the concrete during the exposure period), the degree of compaction and the environmental conditions (including the relative humidity, temperature and the local concentration of carbon dioxide).

Once the carbonation front has reached the steel reinforcement, corrosion is initiated. The corrosion reactions are essentially identical to those for a differential aeration cell formed by a water droplet (**11.8**), and are illustrated in ***Figure 8.16***.

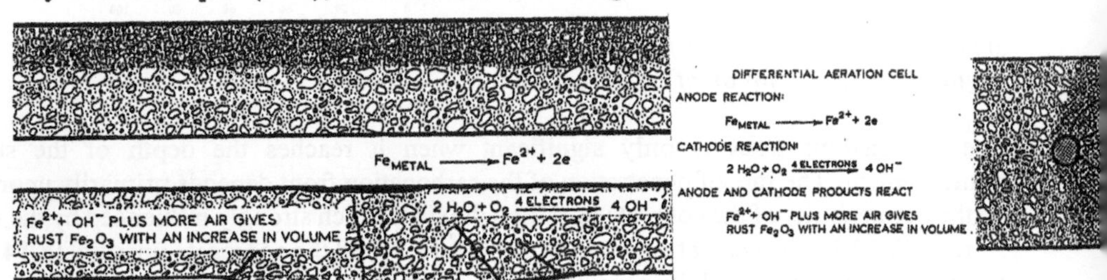

***Figure 8.16** Mechanism of corrosion of steel reinforcement in concrete*

Subsequent deterioration of the concrete occurs because the corrosion products have a volume 3-4 times that of the steel from which they were derived. The increased volume produces tensile stresses which, if they exceed the tensile strength of the concrete (which is in any case relatively low), will produce cracks. Only a relatively small build up of corrosion products is needed to produce cracking (only about 0.01 mm in many cases). The variety of cracks formed (***Figure 8.17***) differ according to the spacing and size of the reinforcement and the type and thickness of the concrete cover.

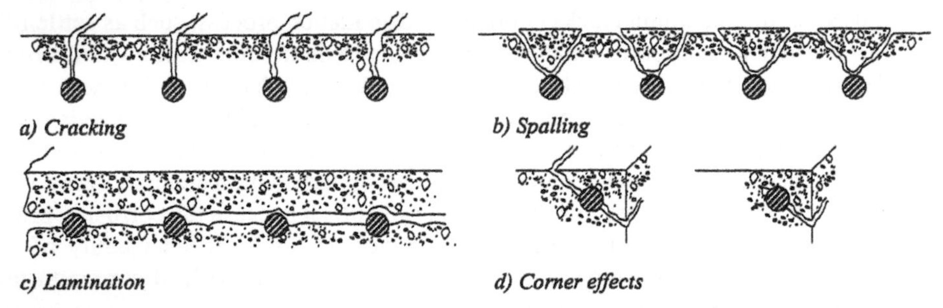

***Figure 8.17** Cracking in concrete associated with corrosion reinforcement*

The depth of carbonation in a concrete can be readily assessed by applying a suitable pH indicator to a freshly broken surface of the concrete taken perpendicular to the surface. Indicators change colour at specific pH levels and include phenolphthalein at pH ≈ 10, thymolphthalein at pH ≈ 11.5, titan yellow at pH ≈ 12.5, indigo carmine at pH ≈ 13 and universal indicator over a range of pHs. For Portland cement concrete[42], phenolphthalein is commonly used and this will turn from a clear solution to pink at or above pH ≈ 10, whilst for HAC concrete[43], bromocresol purple (pH range 5.2-6.8) or m-cresol purple (pH range 7.6-9.2) can be used. Where indicators are unclear or unreliable (e.g, where the pink colour change with phenolphthalein indicator does not occur within 30 seconds or where the development of pink colouration spreads with time), petrographic examination can be used.

8.21.5 Chlorides (Cl^- ions)

It should be appreciated that the Pourbaix diagram (***Figure 8.14***) is shown for the Fe-H_2O system; relationships will change in the presence of dissolved ions, the most important of which is the **chloride (Cl^-) ion**. Chlorides present in the pore water of concrete are effective charge carriers, enhancing the corrosion process. In addition, chlorides cause localised weaknesses in the protective oxide film to the steel reinforcement. Hydroxyl ions counter this process to some extent by repairing the damaged film but, at a critical chloride:hydroxyl ion ratio, the breakdown of the protective is irreversible and pitting corrosion (**11.8**) occurs.

The effect of presence of chloride ions in the concrete surrounding the steel is to raise the potential of the steel (relative to the standard hydrogen electrode), effectively shifting the line **UKFH** in ***Figure 8.14*** to the right, changing the regions of passivity, immunity and corrosion. Consider, for example, embedded steel initially at, say, point **B** in ***Figure 8.14*** in the passive region of the Pourbaix diagram. As the presence of chlorides will shift the line UKFH to the right, this may change the steel from a region of immunity to a region of corrosion. The presence of chlorides may therefore cause corrosion of the steel reinforcement **irrespective of the alkalinity of the concrete.** In addition, the presence of chlorides increases the rate at which the corrosion process takes place to a degree dependent upon the pH of the medium, as shown in ***Figure 8.18.***

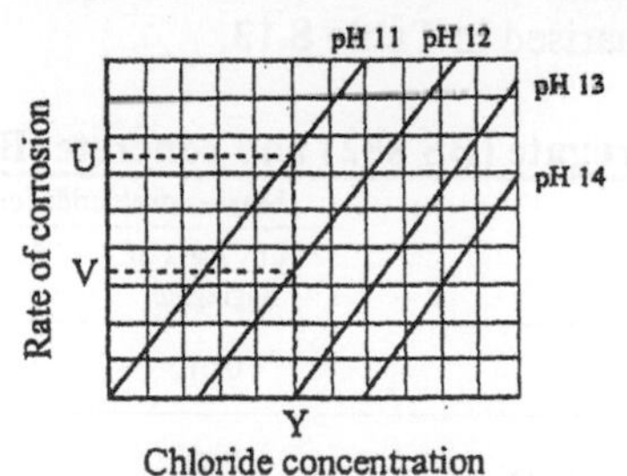

Figure 8.18 **illustrates that for a concentration (Y) of chlorides, there will be a corrosion rate of V in a concrete of pH 12 but a corrosion rate of U in a concrete of pH 11.**

Figure 8.18 Corrosion rate and chloride concentration at various pH

Chlorides can be derived either

- from the materials incorporated in the mix e.g, from admixtures of calcium chloride used as accelerators (**8.7.2**), from aggregates or from cement. Note that the use of calcium chloride or chloride based admixtures as accelerators in reinforced concrete has been banned in the UK since 1977 (ENV 206 maintains this exclusion); or

- from the external environment e.g, sea water (for coastal areas and marine applications) and road de-icing salts (particularly for road bridge decks).

Chlorides in hardened concrete are generally considered either
- fixed (chemically/physically bound to cement particles and hydration products); or
- free (present in the pore water of the concrete).

A significant proportion of the total chloride content present in the concrete derived from the mix materials is fixed by the hydration products of the cement compounds, principally tricalcium aluminate (C_3A) and tetracalcium aluminoferrite (C_4AF), as chloride containing phases (e.g, Friedel's salt, $3CaO.Al_2O_3.CaCl_2.10H_2O$). In this state, the chlorides are immobilised. The remainder are present in the pore water in the free state and are available to facilitate corrosion. The amount of bound chloride present is influenced by the presence of both water soluble sulfates and by carbonation. The products formed by the action of water soluble sulfates on the cement hydration products, and by the action of carbonation, are formed preferentially to the chloride containing phases and therefore these two processes may **act to liberate free chloride ions**. Generally, an equilibrium exists between free and bound chloride in the concrete; any removal of chloride ions from the pore water solution will tend to be replenished from the bound chloride in order to maintain equilibrium.

The chloride content of reinforced concrete should be limited to a level where there is minimal free chloride. With the current limits on chloride contaminants within concrete mix constituents, particularly admixtures and aggregates, the risk of corrosion from free chlorides cast into current concrete construction is very low. Maximum values of chloride content of aggregates are given in Appendix C of BS 882. ENV 206 states that the chloride ion content of concrete should not exceed the values laid down in national standards (BS 5328 in the UK). These requirements are summarised in **Table 8.13**.

Table 8.13 Limits of chloride content in aggregate (BS 882) and concrete (BS 5328)

Type or use of concrete	Maximum chloride ion content	
	% by mass of aggregate	% by mass of cement
Prestressed concrete. Heat-cured concrete containing embedded metal.	0.01	0.10[a]
Concrete containing embedded metal made with SRPC (7.14.1) Concrete made with supersulfated cement (7.14.6) with or without embedded metal.	0.03	0.20
Concrete containing embedded metal and made with Portland cement (7.14.1), Portland blastfurnace cement (7.13), high slag blastfurnace cement (7.13), Portland fuel ash cement (7.14.4), Pozzolanic cement (7.14.4), Portland limestone cement (7.14.4) or combinations (7.13).	0.05	0.40
Other concrete	no limit	no limit[b]

Notes: Where no national advice is given, ENV 206 gives similar limits, with the following exceptions: [a] ENV 206 specifies a maximum chloride content (Cl^- by mass) of 0.2% for prestressed concrete; [b] ENV 206 specifies a maximum chloride content (Cl^- by mass) of 1% for plain concrete.

Due to the porous nature of concrete, chlorides from the environment can also enter hardened concrete and, as these will be largely contained as free chlorides within the pore water, they are particularly aggressive to embedded steel reinforcement. The rate of penetration of the chloride ions from an external source depends primarily upon

- the permeability of the concrete (where the concrete is exposed to a liquid containing chloride ions, e.g, seawater, solutions of road de-icing salts, etc);
- the diffusivity of the concrete (where the chlorides are deposited onto saturated concrete);
- the capillary suction of the concrete (where chlorides are deposited onto a partially saturated surface).

These factors depend largely on the quality of the concrete and the general relationship between concrete quality and chloride penetration is shown in ***Figure 8.19a***.

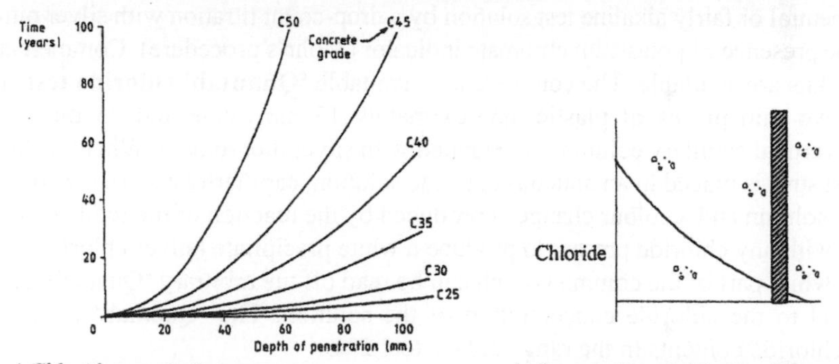

a) Chloride penetration curves *b) Progression of chlorides in concrete*

***Figure 8.19** Chlorides in reinforced concrete*

In concrete exposed to environmental sources of chloride, special protection measures are usually required. The chloride content in the vicinity of the steel reinforcement primarily controls the risk of corrosion. This content (and hence the risk of corrosion) will increase as the chloride ions ingress over time (***Figure 8.19b***) and is influenced by the quality of the concrete (***Figure 8.19a***) and the nature of the environment (***Figure 8.18***). Although there will be a minimum total chloride content of concrete required for corrosion to occur (the threshold value), this value will depend not only on the above factors, but also upon the w/c ratio, the cement type and content, the pH of the pore solution, the capacity of the cement matrix to bind chloride ions, the proportion of the total chloride content present as free (rather than bound) chloride ions and the exposure conditions. It is therefore very difficult to represent the threshold value is a single value and most design advice is based on estimated risk categories in relation to total chloride ion concentration. For example, the BRE[44] gives advice for situations where the chlorides were originally present in the mix materials and for ingressed chlorides by relating the chloride ion concentration (% by weight of cement) in the concrete to a risk category (representing the likelihood of corrosion occurring at the chloride ion concentrations noted). Generally, this advice indicates that

- chloride ion concentrations of 1-2% by weight of cement represent high corrosion risk;
- damp environments are more hazardous situations than equivalent dry environments;

- steel in concrete at a pH < 10 (carbonated) is at greater risk of corrosion in the presence of chlorides than steel in equivalent concrete at a pH > 10 (not carbonated).

Once the threshold value of total chloride content for the initiation of corrosion is reached in the concrete adjacent to the steel reinforcement, corrosion of the steel in the presence of water and oxygen will proceed, as described above (***Figure 8.16***).

Simplified methods to determine the chloride content of hardened concrete are have been outlined by the BRE[45] based on 25 g drilled powdered samples of the concrete. The tests involve treatment of the test sample with an excess of nitric acid solution and then determination of the amount of chloride in the solution. The methods give the total (free plus bound) chloride content of the sample. The **'Hach' chloride test** determines the chloride content of a neutral or fairly alkaline test solution by a drop-count titration with silver nitrate solution in the presence of potassium chromate indicator (Mohr's procedure). Commercially available test kits are available. The commercially available **'Quantab' chloride test** strip consists of two thin pieces of plastic, approximately 15 mm wide and 75 mm long, containing a vertical capillary column impregnated with silver dichromate. When the lower end of the test strip is placed in an aqueous chloride solution, capillarity causes the solution to rise up the column and a colour change is produced by the reaction of the reddish brown silver nitrate with any chloride present to produce a white precipitate (silver chloride). The length of the white part of the column (which can be read off the arbitrary 'Quantab' scale) is proportional to the chloride concentration of the solution. The 'Quantab' method is suitable for chloride contents in the range 0.03% to 1.2%.

8.22 REQUIREMENTS FOR DURABLE CONCRETE

Concrete structures will start to deteriorate from the time they are constructed. Any requirement for durable concrete implies that the structure will not deteriorate to an unacceptable level within the design life. This can be visualised graphically where the level of deterioration over time is plotted over the life of the structure (***Figure 8.20a***).

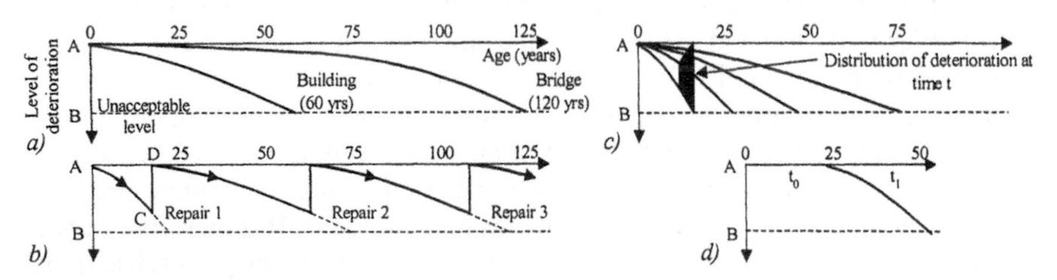

Figure 8.20 *The deterioration process in hardened concrete*

Although the shape of the deterioration curve may be influenced by many factors, for the purposes of this illustration it is assumed that all deterioration factors lead to an overall observable level of deterioration. In a (hypothetical) ideal structure, no deterioration would excluded, there is a need to implement measures to return the deterioration curves to line **A**. This can be achieved by planned maintenance (periodic inspection and sampling) so that the progress of the level of deterioration can be monitored and remedial action taken before an unacceptable level (line **B**) is reached (***Figure 8.20b***). ***Figure 8.20b*** shows that remedial actions at point **C** can restore the structure to line **A** again at point **D**, with (hopefully) a lower rate of deterioration and repair frequency thereafter. Little information exists on satisfactory repair of concrete structures beyond 15-20 years; thus at present 3 to 5 repairs may be necessary to meet most design life demands, although future improvements in repair systems may reduce this frequency. However, the client and designer must be aware that deterioration could occur and could well evaluate the cost of extra protection against future remedial costs. In practice, the rate of deterioration of a particular concrete component will vary over the building (***Figure 8.20c***), depending typically on variations in concrete quality, cover and on the levels of exposure at the point of use. Most durability related problems arise from the environment penetrating into the concrete; deterioration occurs from the inability of the concrete to exclude the environment. It may not be necessary to exclude the environment totally, as the consequences of penetration of a few millimetres may not be significant. All deterioration processes occur in two stages (***Figure 8.20d***)

- the time taken for the environment to penetrate into the concrete to a level where attack and deterioration starts (t_0);
- the time taken for deterioration to reach a level of concern (t_1).

Thus durability and deterioration are generally dependent solely on the performance of the outer layers rather than the body of the concrete. As it is difficult to predict the corrosion rate once the steel reinforcement has been depassivated, the primary objective of design for durable concrete is to extend the time taken for the environment to penetrate the concrete (i.e., extend the time t_0 in ***Figure 8.20d***). This is generally achieved by designing a dense, impermeable outer layer for the concrete using the methods given in **Table 8.14**.

Table 8.14 Design methods for dense, impermeable outer layers for durable concrete

- low w/c ratio (≤ 0.4) to minimise the number of interconnecting pores in the hardened concrete;
- carefully selected concrete mix with reasonable compaction (low volume), no deficiencies in the aggregate grading and high quality materials (e.g, no porous aggregates);
- high cement contents to ensure that the concrete will contain few porous voids caused by evaporation of the mix water. High cement contents also help to mop up more carbon dioxide (CO_2) (8.21.4). However, high cement contents may lead to other problems (e.g, excessive heats of hydration leading to high early age cracking) which may outweigh the benefits obtained;
- high strength grade concretes (≥35-40 N/mm²) can indirectly ensure low w/c ratios and high cement contents in the mix to achieve low levels of environmental penetration. However, this can be an expensive design solution if only low strength concretes are required for structural purposes;
- high tri-calcium aluminate (C_3A) contents in cement can mop up penetrating chlorides (8.21.5) but may enhance sulfate attack (8.19.2);
- slags and fuel ash cement replacement materials (7.14.4) can produce dense concrete without the use of high cement contents;
- workability admixtures such as plasticisers (8.7.3) allow low w/c ratios to be produced without detriment to workability;
- cohesive mixes can be obtained using sand contents which take account of the placing method to limit honeycombing and holes in the concrete cover;
- minimum working of the top surface of placed concrete will prevent sinking of the aggregate, sand and cement (bleeding, 8.18);
- adequate curing (8.13.1), which may be enhanced by using controlled penetration formwork which allows the drainage of mix water from the surface zone while the concrete hardens. This greatly reduces the absorption properties of the surface zone and the associated risk of chloride and carbonation ingress in uncracked concrete;
- high levels of workmanship, particularly to ensure that the specified cover to reinforcement is maintained everywhere throughout the construction (particularly at corners, where penetration may be almost double that for a flat surface) and that thermal, moisture or load induced cracking of the concrete is not beyond acceptable levels.

Many of these principles are included in relevant national standards (**8.23**). Other methods used to limit the risk of corrosion of the steel reinforcement are given in **Table 8.15**.

Table 8.15 Methods to limit the risk of corrosion of the steel reinforcement[46]

- concrete surface treatments that can be used to inhibit the ingress of moisture and depassivating agents (carbon dioxide, chlorides, etc). Coatings can provide a barrier to liquid water and, hence, chlorides. Surface penetrating treatments such as silanes and siloxanes (**5.8.1**) line the capillary pores close to the surface of the concrete to produce a hydrophobic finish (**5.4.4**);
- protective coatings to reinforcement, principally epoxy coatings, which provide a barrier between the embedded steel and the aggressive environment, and galvanised coatings, where the zinc acts sacrificially to protect the steel (**11.13.2**);
- corrosion resistant reinforcement, such as austenitic stainless steels (**10.10.7**) (which are very corrosion resistant in chloride bearing concrete) or fibre reinforced plastics (**9.13**);
- corrosion inhibitor (**11.16**) admixtures, which are both organic and inorganic inhibitive compounds. Inorganic compounds most commonly used in the UK are based on calcium nitrite, $Ca(NO_2)_2$ (**11.16.4**). More recently, organic compounds such as amino alcohols have become available;
- cathodic prevention and protection (**8.25.4**), which is being used increasingly to achieve long service in environments containing high levels of chlorides (e.g, road bridges);
- overcladding and enclosure, which have been used to physically isolate parts of a structure from an aggressive environment.

8.22.1 Concrete Cover

In protecting embedded steel from these effects, the concrete cover acts both as

- a **chemical barrier** (conferring chemical passivity on the steel due to its alkalinity and ability to absorb aggressive chloride ions); and
- a **physical barrier** (as the aggressive chlorides and acidic gases must permeate through the depth of the cover to reach the steel). The permeability of the concrete is greatly influenced by the quality of the concrete.

As a physical barrier to the ingress of oxygen and water vapour, concrete is very poor, due to the micro-voids produced by the hydration mechanism (**7.17**). The chemical protection of the steel by passivation is very efficient as it can only be destroyed by the ingress of carbon dioxide (which reduces the alkalinity of the concrete) or chlorides. Chemical passivation thus relies upon the physical nature of the concrete barrier. Breakdown occurs where the barrier thickness is insufficient or the quality of the concrete cover is poor. The greater the depth of concrete cover to the steel reinforcement, the longer the alkaline environment adjacent to the steel survives (**Table 8.16**).

Table 8.16 Carbonation time (years) and the cover depth for various w/c ratios

Water/cement ratio	28 day strength (N/mm²) average of 40 results	Depth of cover					
		5 mm	10 mm	15 mm	20 mm	25 mm	30 mm
0.45	55	19 yrs	75 yrs	100+ yrs	100+ yrs	100+ yrs	100+ yrs
0.50	47	6 yrs	25 yrs	56 yrs	99 yrs	100+ yrs	100+ yrs
0.55	40	3 yrs	12 yrs	27 yrs	49 yrs	76 yrs	100+ yrs
0.60	35	1.8 yrs	7 yrs	16 yrs	29 yrs	45 yrs	65 yrs
0.65	31	1.5 yrs	6 yrs	13 yrs	23 yrs	36 yrs	52 yrs
0.70	28	1.2 yrs	5 yrs	11 yrs	19 yrs	30 yrs	43 yrs

As the use of reinforced concrete has expanded and the demand for low maintenance buildings has increased, there have been dramatic increases in the cover requirements in relevant codes of practice (**Table 8.17**). Current cover requirements for concrete to ENV 206 are given in relation to a variety of exposure classes in **Table 8.20 (8.23)**.

Table 8.17 Changes in UK Codes of Practice for concrete cover over time

Code of Practice (Year)	Cover specified (mm)	Life span expected with respect to 1982 specification
1948	25	15 years
1957	37	30 years
1969	40	37 years
1972	50	60 years
1982	60	100 years

In spite of the requirements for minimum cover specified in standards, it should be appreciated that the cover actually achieved on site is variable. Observations of the actual cover achieved in a wide variety of reinforced concrete structures[47] found best practice produced a range of ±5 mm of the specified mean cover (though seldom in cast *in situ* concrete); good practice achieved a range of ±10 mm; other cases (e.g, with little site supervision), the range may be ±20 mm. It is therefore apparent that adequate cover will only be achieved by good reinforcement detailing and rigorous quality control.

8.23 STANDARDS FOR DURABLE CONCRETE

Compared to the durability requirements of current British Standards, the approach adopted in Europe is considerably more sophisticated and many more exposure situations are recognised, as summarised in **Table 8.18**.

Table 8.18 Exposure classes related to environmental conditions (ENV 206)

Exposure class		Examples of environmental conditions
1. Dry environment		Interior of dwellings or offices[a].
2. Humid environment	a. Without frost	Interior of buildings where humidity is high (e.g, laundries); Exterior components; Components in non-aggressive soil and/or water.
	b. With frost	Exterior components exposed to frost; Components in non-aggressive soil and/or water and exposed to frost; Interior components where humidity is high and exposed to frost.
3. Humid environment with frost and de-icing agents		Interior and exterior components exposed to frost and de-icing salts.
4. Seawater environment	a. Without frost	Components completely or partially submerged in seawater or in splash zone; Components in saturated salt air (coastal areas).
	b. With frost	Components partially submerged in seawater or in splash zone and exposed to frost; Components in saturated salt air and exposed to frost.
5. Aggressive chemical environment[b]	a.	Slightly aggressive chemical environment (gas, liquid or solid); Aggressive industrial atmosphere.
	b.	Moderately aggressive chemical environment (gas, liquid or solid).
	c.	Highly aggressive chemical atmosphere (gas, liquid or solid).

Notes: [a] This exposure class is valid as long as, during construction, the structure (or some of its components) is not exposed to more severe conditions over a prolonged period of time. [b] Conditions within this exposure class may occur alone or in combination with preceding classes.

The associated recommended minimum compressive strength classes and concrete mix requirements for durability in the specified exposure condition are summarised in **Table 8.19**, which includes UK national minimum strength grades derived from the UK National Application Document to Eurocode 2 (DD ENV 1992-1-1: 1992).

Table 8.19 Durability requirements related to environmental exposure (ENV 206)

	Exposure class (see **Table 8.18**)								
	1	2a	2b	3	4a	4b	5a	5b	5c[a]
Minimum grade									
Plain concrete	C16/20	C20/25[b,c]	C35/45[d]	C35/45	C35/45	C35/45	British Standards do not give		
Reinforced concrete	C25/30	C30/37	C35/45[d]	C35/45	C35/45	C35/45	minimum strength grades for		
Prestressed concrete	C30/37	C30/37	C35/45[d]	C35/45	C35/45	C35/45	aggressive chemical environments		
Maximum water:cement (w/c) ratio[e]									
Plain concrete	-	0.70	0.55	0.50	0.55	0.50	0.55	0.50	0.45
Reinforced concrete	0.65	0.60	0.55	0.50	0.55	0.50	0.55	0.50	0.45
Prestressed concrete	0.60	0.60	0.55	0.50	0.55	0.50	0.55	0.50	0.45
Minimum cement content[e] (kg/m³)									
Plain concrete	150	200	300	300	300	300	200	300	300
Reinforced concrete	260	280	280	300	300	300	280	300	300
Prestressed concrete	300	300	300	300	300	300	300	300	300
Minimum air content of fresh concrete (%) for nominal aggregate size[f]									
32 mm	-	-	4[g]	4[g]	-	4[g]	-	-	-
16 mm	-	-	5[g]	5[g]	-	5[g]	-	-	-
8 mm	-	-	6[g]	6[g]	-	6[g]	-	-	-
Frost resistance of aggregates required[j]									
	-	-	Yes	Yes	-	Yes	-	-	-
Impermeable concrete required									
	-	-	Yes	Yes	Yes	Yes	Yes	Yes	Yes
Types of cement for plain and reinforced concrete									
	CEM	CEM	CEM	CEM	CEM	CEM	CEM[h]	CEM[h]	CEM[h]

Notes: [a] in addition, the concrete should be protected from direct contact with the aggressive media by coatings, [b] the value recommended in Table 6.2 of BS 8110: Part 1 for similar exposure conditions corresponds to C25/30; [c] may be reduced by one grade for foundations to low rise structures in non-aggressive soil conditions provided that the requirements for cement content and w/c ratio are met; [d] may be reduced one grade if air entrained; [e] for minimum cement content and maximum w/c ratio only CEM cement should be used; [f] with a spacing factor of the entrained air void system < 0.2 mm measured in the hardened cement paste; [g] where the degree of saturation is high for long time periods; [h] SRPC for sulfate content > 500 mg/kg in water or > 300 mg/kg in soli. The sulfate resistance of the cement should be judged on the basis of national standards; [j] assessed against national standards.

Cover requirements for normal weight concrete for reinforced concrete and prestressed reinforced concrete to ENV 206 are given in the UK National Application Document to Eurocode 2 (DD ENV 1992-1-1: 1992) and are summarised in **Table 8.20**.

Table 8.20 Cover requirements for normal concrete to ENV 206

		Exposure class (see **Table 8.18**)								
		1	2a	2b	3	4a	4b	5a	5b	5c
Reinforcement	Nominal[a]	20	35	35	40	40	40	35	35	45
	Minimum	(15)	(30)	(30)	(35)	(35)	(35)	(30)	(30)	(40)
Prestressing	Nominal[a]	25	40	40	45	45	45	40	40	50
	Minimum	(20)	(35)	(35)	(40)	(40)	(40)	(35)	(35)	(45)

Notes: To satisfy ENV 206, cover values should be associated with the particular concrete qualities given in **Table 8.19**. A reduction in cover of 5 mm may be made where concrete of strength class C40/50 or above is used for reinforced or prestressed reinforced concrete in exposure classes 2a to 5b. For slab elements, a further reduction of 5 mm may be made for exposure classes 2 to 5. For exposure class 5c, a protective barrier should be provided to prevent direct contact with aggressive media. [a] nominal cover values have been obtained from minimum values by allowing a negative tolerance of 5 mm.

8.24 TESTS FOR REINFORCED CONCRETE

8.24.1 Concrete Resistivity Measurement

The electrolytic resistivity (**8.21.1**) of concrete is an important parameter in controlling the corrosion of reinforcement. If the resistivity of the concrete is low it provides little resistance to the transport of the ionic current through the concrete. Hence concrete of low electrolytic resistivity provides little protection for the reinforcement. Electrolytic resistivity is known to be influenced by many factors, including moisture and salt content and temperature, as well as mix proportions and the w/c ratio. Reinforcement may also influence resistivity. In situ measurements may be made using a Wenner four-probe technique[48] in which four electrodes are placed in a straight line on or just below the concrete surface at equal spacings. An electrical current is passed through the outer electrodes while the voltage drop between the inner electrodes is measured. The apparent resistivity of the concrete may be calculated from a knowledge of the current, voltage drop and electrode spacing. For practical purposes, the depth of the zone of concrete affecting the measurement may be taken as equal to the concrete spacing.

8.24.2 Half-cell Potential Measurements

A characteristic feature for the corrosion of steel in concrete is the coexistence of passive and corroding areas on the same reinforcement bar forming a short-circuited galvanic cell, with the corroding area as the anode and the passive surface as the cathode. The voltage of such a cell can reach as high as 0.5V or more, especially where chloride ions are present. The resulting current flow (which is directly proportional to the mass lost by the steel) is determined by the electrical resistance of the concrete and the anodic and cathodic reaction resistance. The current flow in the concrete is accompanied by an electrical field which can be measured at the concrete surface, resulting in equipotential lines that allow the location of the most corroding zones at the most negative values. This is the basis of **potential mapping**, the principal electrochemical technique applied to the routine inspection of reinforced concrete structures.

The method of half cell potential measurements normally involves measuring the potential of an embedded reinforcement bar relative to a reference half-cell placed on the concrete surface. The reference half cell is usually a copper/copper sulfate or silver/silver chloride cells but other combinations are used. The concrete functions as an electrolyte and the risk of corrosion in the reinforcement in the immediate region of the test location may be related empirically to the measured potential difference, as outlined below. The equipment is simple and enables a non-destructive survey to produce an isopotential contour map of the surface of the concrete member. Zones of varying degrees of corrosion risk can be identified from these maps. Commonly, the reference half cell is incorporated in a wheel so allow coverage of very large surface areas. The stages in understanding potential mapping are outlined in ***Figure 8.21*** and the reader is referred to **Chapter 11** for the general fundamental principles involved.

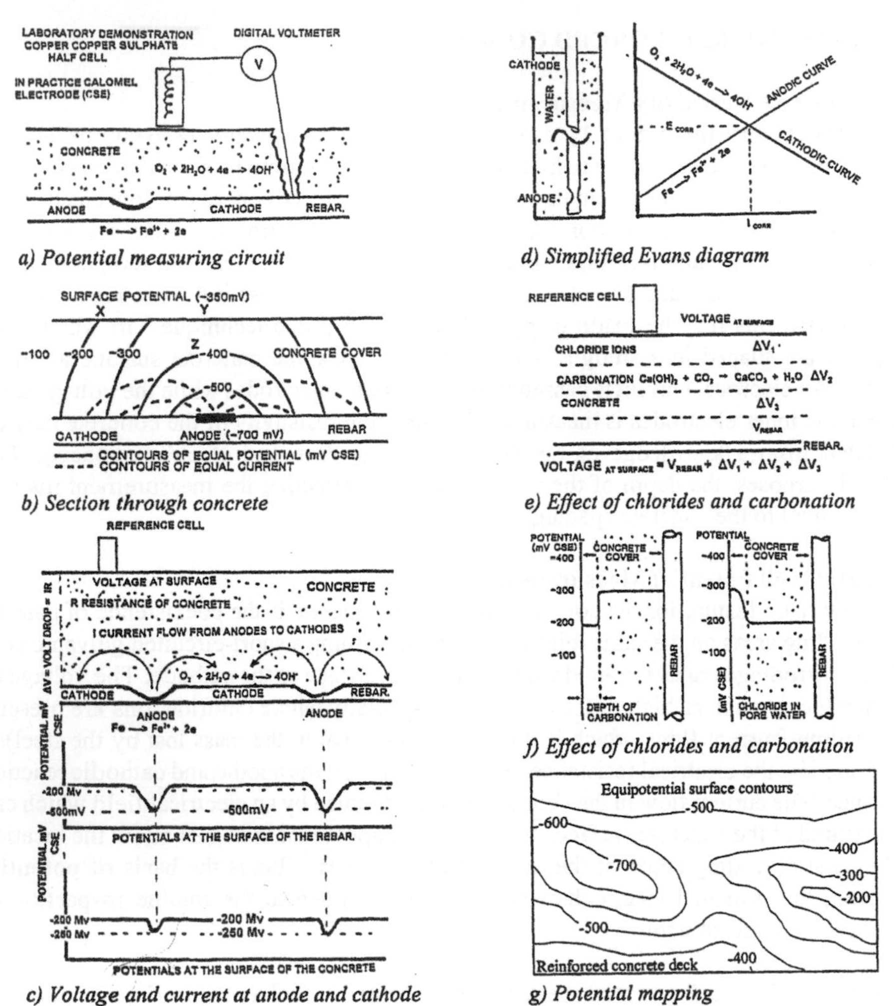

a) Potential measuring circuit

b) Section through concrete

c) Voltage and current at anode and cathode

d) Simplified Evans diagram

e) Effect of chlorides and carbonation

f) Effect of chlorides and carbonation

g) Potential mapping

Figure 8.21 *Non-destructive examination for corrosion of steel reinforcement*

The concrete is broken away to expose the reinforcement and an electrical connection is made to the steel reinforcement (***Figure 8.21a***) via a high impedance digital millivoltmeter, often backed up with a data logging device. The other connection to the millivoltmeter is taken to an external half cell, which has a porous connection at one end that can be touched to the concrete surface, whereupon the corrosion potential of the steel reinforcement nearest the point of contact will be registered (***Figures 8.21b*** and ***8.21c***). By measuring and plotting results on a regular grid, an equipotential contour map is drawn and areas of corroding steel may readily be seen (***Figure 8.21g***). More recently, 3D mapping techniques have become available. The theory of potential mapping can be seen with reference to ***Figure 8.21d***. If a steel rod is immersed in water, and an imaginary cut is made through the middle to produce two halves, the upper half will produce an open circuit voltage with respect to the hydrogen electrode of about +1.40 V (**cathode**) and the lower half will produce an open circuit voltage with respect to the hydrogen electrode of -0.440 V (**anode**) (**Table 11.2**). If the imaginary cut is now removed, the top and bottom of the steel rod must be at the same potential with respect to the hydrogen electrode (= E_{corr}), as shown in the simplified **Evans diagram** (***Figure 8.21b***). Effectively, the anode and cathode are short circuited. The short circuit of the anode and cathode will generate a corrosion current (= I_{corr}). The function of the potential mapping is to measure E_{corr} with respect to the hydrogen electrode. ***Figure 8.21c*** represents the flow of current (dotted lines) and electro-potential (as full lines). If the half cell is located at **X**, the potential will read -200 mV, at **Y** -350mV and at **Z** -400mV. Traversing the reference cell over the concrete surface will produce a drop in potential difference between the reinforcement and the reference electrode (E_{corr} - E_{ref}) (***Figure 8.21c***). If all the sites giving the same (E_{corr} - E_{ref}) readings are joined, the resulting lines produce contours of equal potential (***Figure 8.21f***). Areas producing the largest negative reading are anodes. These areas can be broken out and treated. Note that carbonation and chloride ingress will effect the actual potential difference (E_{corr} - E_{ref}) readings obtained (***Figures 8.21e*** and ***8.21f***). The potential difference (= IR) drop is represented for different concentrations of chloride ions as ΔV_1, ΔV_2, ΔV_3 etc. Similar arguments apply to carbonation at differing depths, producing different (E_{corr} - E_{ref}) profiles. Allowance must be made for both carbonation depth and chloride ingress in interpreting results from these tests.

8.25 ELECTROCHEMICAL PROTECTION FOR REINFORCED CONCRETE

A new European Standard, DD ENV 1504[49], deals with repair and remediation methods for concrete structures, including electrochemical remediation techniques for reinforced concrete. This standard currently has Draft for Development status (only Parts 1 and 9 have been published) though is set to become mandatory for specifying all concrete repair works. The standard groups concrete repair and remediation methods according the principle by which the method operates. **Table 8.21** summarises this information and gives examples of methods which comply with each principle. General repair strategies for defects in concrete are outside the scope of this textbook and specialised texts deal with the subject in more detail[50]. In particular, recent guidance on the protection and remediation of reinforced concrete in relation to the DD ENV 1504 has been produced by the BRE[51]. Here we concentrate on electrochemical protection/remediation for reinforced concrete.

A general overview of cathodic protection is given in section **8.25.1**. The requirements for successful cathodic protection are the same as those for corrosion to proceed (**11.3**), i.e.,

- the **electrons** formed at the **anode** must be used up by a **cathodic reaction**; and
- electrons can **only move in a metallic conductor** and therefore the anode and cathode regions must be joined by an electrical conduction path; and
- **both** the anode and the steel must be in the **same liquid environment**. In reinforced concrete, this is the permeable concrete cover (**8.22.1**).

Table 8.21 Principles and methods for remediation of concrete defects

Principle	Principle definition	Methods based on principle
Principles and methods relating to defects in concrete		
Principle 1. Protection against ingress (PI)	Reducing or preventing the ingress of adverse agents e.g. water, other liquids, vapour, gas, chemicals and biological agents	1.1 Impregnation 1.2 Surface coating with and without crack bridging ability 1.3 Locally bandaged cracks[a] 1.4 Filling cracks 1.5 Transferring cracks into joints[a] 1.6 Erecting external panels[a,b] 1.7 Applying membranes[a]
Principle 2. Moisture control (MC)	Adjusting and maintaining the moisture content in the concrete within a specified range of values	2.1 Hydrophobic impregnation 2.2 Surface coating 2.3 Sheltering or overcladding[a,b] 2.4 Electrochemical treatment[a,b]
Principle 3. Concrete restoration (CR)	Restoring the original concrete of an element of the structure to the originally specified shape and function Restoring the concrete structure by replacing part of it	3.1 Applying mortar by hand 3.2 Recasting with concrete 3.3 Spraying concrete or mortar 3.5 Replacing elements
Principle 4. Structural strengthening (SS)	Increasing or restoring the structural loadbearing capacity of an element of the concrete structure	4.1 Adding or replacing embedded or external steel reinforcement 4.2 Installing bonded steel reinforcement in preformed or drilled holes in the concrete 4.3 Plate bonding 4.4 Adding mortar or concrete 4.5 Injecting cracks, voids or interstices 4.6 Filling cracks, voids or interstices 4.7 prestressing or post-tensioning[a]
Principle 5. Physical resistance (PR)	Increasing resistance to physical or mechanical attack	5.1 Overlays or coatings 5.2 Impregnation
Principle 6. Resistance to chemicals	Increasing resistance of the concrete surface to deteriorations by chemical attack	6.1 Overlays or coatings 6.2 Impregnation
Principles and methods relating to reinforcement corrosion		
Principle 7. Preserving or restoring passivity (RP)	Creating chemical conditions in which the surface of the reinforcement is maintained in, or is returned to, a passive condition	7.1 Increasing cover to reinforcement with additional cementitious mortar or concrete 7.2 Replacing contaminated or carbonated concrete 7.3 Electrochemical realkalisation of carbonated concrete[a] 7.4 Realkalisation of carbonated concrete by diffusion 7.5 Electrochemical chloride extraction[a]
Principle 8. Increasing resistivity (IR)	Increasing the electrical resistivity of the concrete	8.1 Limiting moisture content by surface treatments, coatings or sheltering
Principle 9. Cathodic control (CC)	Creating conditions in which potentially cathodic areas of reinforcement are unable to drive an anodic reaction	9.1 Limiting oxygen content (at the cathode) by saturation or surface coating[b,c] 9.2 Applying cathodic inhibitors to concrete[a,b,c]
Principle 10. Cathodic protection (CP)	Polarising the steel reinforcement cathodically so as to reduce the rate of anodic reactions	10.1 Impressed current[a] 10.2 Sacrificial anode systems[a]
Principle 11. Control of anodic areas (CA)	Creating conditions in which potentially anodic areas of the reinforcement are unable to take part in the corrosion reaction	11.1 Painting reinforcement with coatings containing active pigments[c] 11.2 Painting reinforcement with barrier coatings[b,c] 11.3 Applying inhibitors to the concrete[a,b,c]

Notes: [a] These methods may make use of products and system not covered by the EN 1504 series of standards; [b] Inclusion of methods in this table does not imply EU approval or confirmation of their effectiveness;
[c] Existence of supporting evidence that the method achieves protection by the appropriate principle confirmed by BRE[51]

8.25.1 General Principles of Electrochemical Remediation/Protection

Options for electrochemical remediation of reinforced concrete include

- electrochemical realkalisation of carbonated concrete (Method 7.3);
- electrochemical chloride extraction (Method 7.5);
- cathodic protection of prevention (CP) (Principle 10).

Although the same concept applies for all electrochemical remediation methods (**11.13.3**), CP methods are grouped together under a separate Principle since they involve long term permanent installations compared to the short term treatments (usually less than three months) of other electrochemical remediation methods. Cathodic protection relies upon there being a good electrolytic path between the anodic and cathodic sites. In long term cathodic protection systems (Principle 10), this is provided by pore water within the concrete. In short term (Principle 7) systems, the electrolyte is enhanced by introducing an external electrolyte, as described for each system below. Although all methods are based on the same fundamental theory, there are some differences in the impressed current densities used and the duration of the polarisation period, as summarised in **Table 8.22**.

Table 8.22 Typical current densities for electrochemical remediation techniques

Technique	Current density (per m^2 of steel)	Typical period of polarisation
Realkalisation	0.5-1.0 A	3-10 days
Chloride extraction	1.0-2.0 A	2-6 weeks
Cathodic protection	5.0-20.0 mA	Permanent
Cathodic prevention	0.4-2.0 mA	Permanent

A new standard, prEN 12696[52] will in future govern cathodic protection systems for steel in concrete.

8.25.2 Electrochemical Realkalisation of Carbonated Concrete (Method 7.3)

This method is applied to carbonated reinforced concrete (**8.21.4**) and involves fixing a temporary anode material (commonly a metal cage or mesh, activated by titanium or steel) to the surface of the concrete (***Figure 8.22***). An external electrolyte, normally an alkali carbonate such as sodium carbonate (Na_2CO_3) or potassium carbonate (K_2CO_3), is introduced to the surface (sometimes within in a cellulose mulch sprayed over the surface of the concrete). The electrolyte penetrates the concrete by absorption and/or diffusion to a depth dependent on the permeability of the concrete and its moisture content (surface drying is sometimes used to enhance penetration). The alkalinity around the steel reinforcement as raised by

- production of hydroxyl ions by cathodic reactions induced on the reinforcement (**11.13.3**);
- the ingress of the electrolyte (particularly in the cover concrete).

The introduction of alkali carbonate electrolytes may enhance alkali-aggregate or alkali-silica reactions (**8.20**) and lithium based solutions (which are less alkaline, **1.8.1**) have been suggested as possible electrolytes to circumvent this problem. Once the polarisation period is ended (after 3-10 days), the steel reinforcement is likely to remain polarised to a degree for only several weeks or months and the long term benefits of this system have yet to be fully investigated.

8.25.3 Electrochemical Chloride Extraction (Method 7.5)

This method, sometimes referred to as **desalination**, is very similar to realkalisation (**8.25.2**), although the electrolyte is commonly calcium hydroxide, $Ca(OH)_2$ (even tap water has been used). The method aims to remove free chlorides from the reinforced concrete (**8.21.5**) as chlorine gas according to the following reaction at the artificial anode

$$\underset{\text{chlorine ions in solution}}{2\,Cl^-} \rightarrow \underset{\text{chlorine gas}}{Cl_2} + \underset{\text{electrons}}{2e}$$

The chloride ions are repelled from the cathodic steel reinforcement under the impressed current towards the anode (***Figure 8.22***), which is covered in a paper mulch saturated with the electrolyte over the surface of the concrete. The acidic chlorine gas reacts with the alkali electrolyte and the resulting salt present in the paper mulch is periodically removed, thereby removing the chlorine ions. As a consequence of equilibrium considerations, only a proportion of the free chlorine ions in the pore water solution can be removed in practice during a single treatment. The rate of chloride extraction also reduces with decreasing chloride ion concentrations in the pore water, reducing the cost effectiveness of the process with successive reapplications. In addition, chloride ions present behind the reinforcement outside the field of the current could diffuse towards the steel over time. As with realkalisation, once the polarisation period is ended (after 2-6 weeks), the steel reinforcement is likely to remain polarised to a degree for only short periods; the long term benefits of this system have yet to be fully investigated.

8.25.4 Cathodic Protection or Prevention (CP) (Principle 10)

These are permanent systems where permanent anodes are introduced. Typical anodes include organic conductive coatings, activated titanium mesh, thermally sprayed zinc, etc. which are painted, fixed or sprayed over the whole surface of the concrete. Zinc coatings require little maintenance but the sacrificial nature of the zinc means that the coating must be periodically replaced. Conductive coating systems usually employ one or two anodes. For high rise buildings, the anode is usually graphite in a resin binder (paint) applied to the surface of the concrete in a thickness of about 250-400 μm. For bridge decks, primary anodes are thin wires of titanium fixed to the surface beneath a slurry of super plasticised mortar or concrete slurry ('**Gunite**') combined with secondary (graphite) anodes contained in the road finish itself. ***Figure 8.22*** illustrates a typical system, where the conductive coating (anode) is applied to the entire concrete surface so that all the concrete matrix (liquid electrolyte environment) is in contact with the cathode (steel reinforcement).

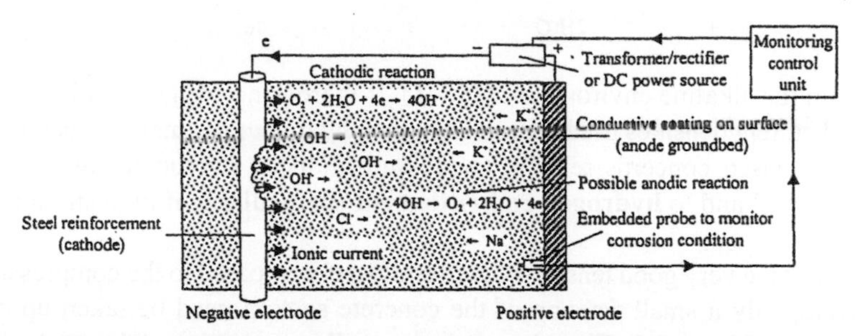

Figure 8.22 Typical conductive coating cathodic protection system

8.25.5 Problems Associated with Electrochemical Protection

The disadvantage of these systems arises from the acid produced at the anode, which in some cases can lead to separation of the concrete surface from the main body of the structure and may corrode any surface fixings (e.g, fixings to rain water pipes, external fire escapes, TV aerials etc.) even though these items are electrically insulated from the anode of the cathodic protection. The acidic environment at the anode can arise through a number of mechanisms, e.g, in the presence of chlorides

$$2Cl^- \rightarrow Cl_2 + 2e$$

and subsequently

$$\underset{}{Cl_2} + H_2O \rightarrow \underset{\text{Acid}}{HCl} + \underset{\text{(bleach)}}{HClO}$$

or by the production of nascent oxygen, where hydroxyl ions present in water

$$\underset{\text{Water}}{H_2O} \rightarrow \underset{\text{hydrogen ions}}{H^+} + \underset{\text{hydroxyl ions}}{OH^-}$$

are continuously removed according to

$$\underset{\text{hydroxyl ions}}{2OH^-} \rightarrow \underset{\text{water}}{H_2O} + \underset{\text{oxygen}}{O_2} + \underset{\text{electrons}}{2e}$$

leaving the **H^+ ions** in excess thus rendering the solution acidic **(1.19)**. Acidic environments may enhance the rate of corrosion of the anode **(11.7.2)**.

In addition, it is important to appreciate that oxygen is required at the surface of the steel reinforcement to ensure that the cathode reaction that will actually take place on the surface of the steel is

$$O_2 \quad + \quad 2H_2O \quad + \quad 4e \quad \rightarrow \quad 4OH^-$$

to maintain the alkaline environment (**11.7.1**). If there is no oxygen available or when the impressed current densities used are too high, then hydrogen gas may be liberated (**11.13.3**), which has caused concerns relating to pressure induced microcracking in the concrete around the steel and to **hydrogen embrittlement** (particularly of high strength steels).

As a result of the very good tensile properties of steel compared to the compression strength of concrete, only a small fraction of the concrete section need be taken up by the steel reinforcement (***Figure 8.3***). Thus the relatively small cross section of the steel reinforcement means that the steel must not suffer any **localised** form of corrosion (**pitting**) which would increase the stress in one particular area. Pitting is an unfortunate characteristic of steel when subjected to localised oxygen concentration cells (**11.8**) (note that, in the hydrating and setting state, concrete produces the alkali calcium hydroxide, $Ca(OH)_2$ (**7.16**) that causes pitting in glass fibres (**9.12**), but this same alkali produces chemical protection upon steel). However, the liberation of hydrogen gas at the anode may produce localised pitting corrosion, lead to loss of cross section and in some cases to **brittleness**. The effect of hydrogen on steel reduces the strength of high strength steels and reduces the strain at fracture. The combination of these two effects is to reduce the energy absorbed by the steel and the effect is known as **hydrogen embrittlement** of the steel.

The application of cathodic protection must therefore be carefully controlled, not only to prevent hydrogen evolution, but also to ensure that the protection system adopted prevents corrosion in the reinforcement, to monitor the level of chlorides (particularly where applied to roads and bridges), to prevent alkali-aggregate or alkali-silica reaction (**8.20**) and to make due allowance for the acidic environment produced at the anodic sites.

8.26 REFERENCES

1. BS 3148: 1980. Methods of test for water for making concrete (including notes on the suitability of the water).
2. prEN 1008: 1993. Mixing water for concrete. Specification and tests.
3. BS 882: 1992. Specification for Aggregates from natural sources for concrete.
4. BS 1047: 1983. Specification for air-cooled blastfurnace slag aggregate for use in construction.
5. BS 3797: 1990. Lightweight aggregates for masonry units and structural concrete.
6. BS EN 12620. Aggregates for concrete.
7. BS EN 13055. Lightweight aggregates.
8. BS EN 13139. Aggregates for mortar.
9. BS 812. Testing aggregates. Part 130: 1985. Methods for determination of particle size distribution.
10. BS EN 932. Tests for general properties of aggregates. Part 1: 1997. Methods for sampling. Part 2: 1999. Methods for reducing laboratory samples. Part 3: 1997. Procedure and terminology for simplified petrographic description. prEN 932. Part 4: Quantitative and qualitative system for description and petrography. Part 5: 2000. Common equipment and calibration. Part 6: 1999. Definitions of repeatability and reproducibility.
11. BS EN 933. Test for geometrical properties of aggregates. Part 1: 1997. Determination of particle size distribution. Sieving method. Part 2: 1996. Determination of particle size distribution. Test sieves, nominal size of apertures. Part 3: 1997. Determination of particle shape. Flakiness index. Part 4: 2000. Determination of

of particle shape. Shape index. Part 5: 1998. Determination of percentage of crushed and broken surfaces in coarse aggregate particles. prEN 933. Part 6: 1999. Assessment of surface characteristics. Flow coefficient of aggregates. Part 7: 1998. Determination of shell content. Percentage of shells in coarse aggregates. Part 8: 1999. Assessment of fines. Sand Equivalent test. Part 9: 1999. Assessment of fines. Methylene blue test. prEN 933. Part 10: 1999. Assessment of fines. Grading of fillers (air jet sieving).

12. BS EN 1097. Tests for mechanical and physical properties of aggregates. Part 1: 1996. Determination of resistance to wear (Micro-Deval). Part 2: 1998. Methods of determination of resistance to fragmentation. Part 3: 1998. Determination of loose bulk density and voids. Part 4: 1999. Determination of the voids of dry compacted filler. Part 5: 1999. Determination of the water content by drying in a ventilated oven. prEN 1097. Part 6: 1997. Determination of particle density and water absorption. Part 7: 1999. Determination of the particle density of filler. Pyknometer method. Part 8: 2000. Determination of the polished stone value. Part 9: 1998. Determination of the resistance to wear by abrasion from studded tyres. Nordic test.
13. BS EN 1367. Tests for thermal and weathering properties of aggregates. Part 1. 2000. Determination of resistance to freezing and thawing. Part 2: 1997. Magnesium sulfate test. prEN 1367. Part 3: 1998. Boiling test for 'Sonnenbrand basalt' and disintegration of steel slag. Part 4: 1997. Determination of drying shrinkage. prEN 1367. Part 5: 1998. Determination of resistance to thermal shock.
14. BS EN 1744. Tests for chemical properties of aggregates. Part 1: 1998. Chemical analysis. prEN 1744. Part 3: 2000. Preparation of eluates by leaching of aggregates.
15. prEN 13179. Tests for filler aggregate used in bituminous mixtures. Part 1: 1998. Delta ring and ball test. Part 2: 1998. Bitumen number.
16. prEN 934. Admixtures for concrete, mortar and grout. Part 1: (not yet published). General definitions and general requirements for all the types of admixtures. BS EN 934: Part 2: 1997. Concrete admixtures. Definitions, specifications and conformity criteria. Part 3: 1998. Admixtures for masonry mortar. Definitions, requirements and conformity criteria. Part 4: 1995. Admixtures for grout. Definitions, specifications and conformity criteria. Part 5: 1998. Admixtures for sprayed concrete. Definitions, specifications and conformity criteria. Part 6: 2000. Sampling, quality control, evaluation of conformity and marking and labelling.
17. BS EN 480. Admixtures for concrete, mortar and grout. Test methods. Part 1: 1998. Reference concrete and reference mortar for testing. Part 2: 1997. Determination of setting time. prEN 480. Part 3: 1991. Determination of shrinkage and expansion. Part 4: 1997. Determination of bleeding of concrete. Part 5: 1997. Determination of capillary absorption. Part 6: 1997. Infrared analysis. prEN 480. Part 7: 1991. Determination of the density of liquid admixtures. Part 8: 1997. Determination of the conventional dry material content. prEN 480. Part 9: 1991. Determination of the pH value. Part 10: 1997. Determination of water soluble chloride content. Part 11: 1999. Determination of air void characteristics in hardened concrete. Part 12: 1998. Determination of the alkali content of admixtures. prEN 480. Part 13: 1998. Reference masonry mortar for testing mortar admixtures.
18. BS 5075. Concrete admixtures. Part 1: 1982. Specification for accelerating admixtures, retarding admixtures and water-reducing admixtures. Part 2: 1982. Specification for air-entraining admixtures. Part 3: 1985. Specification superplasticising admixtures.
19. BS 4887. Mortar admixtures. Part 1: 1986. Specification for air-entraining (plasticising) admixtures. Part 2: 1987. Specification for set-retarding admixtures.
20. BRE (2000) Water reducing admixtures in concrete. Information Paper IP 15/00. Garston, BRE.
21. BS 5328. Concrete. Part 1: 2002. Guide to specifying concrete. Part 2: 1997. Methods for specifying concrete. Part 3: 1997. Specification for the procedures to be used in producing and transporting concrete. Part 4: 1997. Specification for the procedures to be used in sampling, testing and assessing compliance of concrete.
22. BS 8110. Structural use of concrete. Part 1: 1998. Code of practice for design and construction. Part 2: 1985. Code of practice for special circumstances. Part 3: 1985. Design charts for singly reinforced beams, doubly reinforced beams and rectangular columns.
23. ENV 206: 1992. Concrete. Performance, production, placing and compliance criteria.
24. prEN 12540. Testing concrete. Cored specimens. Taking, examining and testing in compression.
25. ISO 7031 (draft stage). Concrete, hardened. Determination of the depth of penetration of water under pressure.
26. BS EN 12350. Testing fresh concrete. Part 1: 2000. Sampling. Part 2: 2000. Slump test. Part 3: 2000. Vebe test. Part 4: 2000. degree of compactability. Part 5: 2000. Flow table test. Part 6: 2000. Density. prEN 12350.

Part 7: 2000. Air content. Pressure methods.
27. BS 1881. Methods of testing concrete.
28. BS 1881: Part 103: 1993. Method of determination of compacting factor.
29. BS 1881: Part 111: 1997. Method of normal curing of test specimens (20°C method).
30. NEPPER-CHRISTENSEN, P. and NIELSEN, T.P.H. (1969) Modal determination of the effect of bond between coarse aggregate and mortar on the compressive strength of concrete. *American Concrete Institute Journal, Vol. 66(7) January*. pp. 69-72.
31. SOROKA, I. (1979) *Portland cement paste and concrete*. London, MacMillan Press. pp. 205.
32. BS 1881: Part 124: 1988. Methods of analysis of hardened concrete.
33. prEN 12390: 1996. Testing concrete. Determination of compressive strength. Specification for compression testing machines.
34. CRAMMOND, N.J (1985) Thaumasite in failed cement mortars and renders from exposed brickwork. *Cement and Concrete Research, Vol. 15*. pp. 1039-1050.
35. BRE (2001) BRE Special Digest 1. Concrete in aggressive ground. Part 1: Assessing the aggressive chemical environment. Part 2: Specifying concrete and additional protective measures. Part 3. Design guides for common applications. Part 4. Design guides for specific precast products. Garston, BRE.
36. BS 7755. Soil quality. Part 3. Chemical methods. Section 3.10: 1995. Determination of water soluble and acid soluble sulfate.
37. BS 1377. Test methods for soils for civil engineering purposes. Part 3:1990. Chemical and electro-chemical tests.
38. BS 7943: 1999. Guide to the interpretation of petrographical examinations for alkali-silica reactivity.
39. BS 812. Testing aggregates. Part 123. Method of determination of alkali-silica reactivity. Concrete prism method.
40. BS DD 249: 1999. Testing aggregates. Method for the assessment of alkali-silica reactivity. Potential accelerated mortar-bar method.
41. BCA (1992) *The diagnosis of alkali-silica reaction. Report of a working party, 2nd ed.* Telford, BCA.
42. BRE (1995) Carbonation of concrete and its effect on durability. BRE Digest 405. Garston, BRE.
43. BRE (1998) Assessing carbonation depth in ageing high alumina cement concrete. BRE Information Paper IP 11/98. Garston, BRE.
44. BRE (2000). Corrosion of steel in concrete. Part 2. Investigation and assessment. BRE Digest 444 Part 2. Garston, BRE.
45. BRE (1986) Determination of the chloride and cement contents of hardened concrete. Information Paper IP 21/86. Garston, BRE.
46. BRE (2000). Corrosion of steel in concrete. Part 1. Durability of reinforced concrete structures. BRE Digest 444 Part 1. Garston, BRE. p.9.
47. CLARK, L.A; SHAMMAS-TOMA, M.G.K.; SEYMOUR, D.E.; PALLETT, P.F. and MARSH, B.K. (1997) How can we get the cover we need? *The Structural Engineer, Vol. 75(17) (2 September 1997)*. pp. 289-296.
48. McCARTER, W.J.; FORDE, M.C and WHITTINGTON, H.W. (1982) Resistivity characteristics of concrete. *Proceedings of the Institution of Civil Engineers, Part 2, Vol 71*. pp. 107-117.
49. DD ENV 1504. Products and systems for the protection and repair of concrete. Part 1: 1998. General scope and definitions. Part 2. Surface protection systems. Part 3. Structural and non structural repair. Part 4. Structural bonding. Part 5. Concrete injection. Part 6. Grouting to anchor reinforcement of to fill external voids. Part 7. Reinforcement corrosion prevention. Part 8. Quality control and evaluation of conformity. Part 9: 1997. General principles for the use of products and systems. Part 10. Site application of products and systems and quality control of the works.
50. MAYS, G. (1992) *Durability of concrete structures. Investigation, repair, protection*. London, E&FN Spon.
51. BRE (2000). Corrosion of steel in concrete. Part 3. Protection and remediation. BRE Digest 444 Part 3. Garston, BRE.
52. prEN 12696. Cathodic protection of steel in concrete. Part 1: 1999. Atmospherically exposed concrete.

9

GLASS AND FIBRE COMPOSITES

9.1 INTRODUCTION

Glass is a hard, brittle, transparent or translucent non-crystalline substance made from silica (sand) (**1.17**), soda (sodium carbonate) and lime (calcium oxide and carbonates) from limestone, i.e., a 'soda lime silicate' (average composition $Na_2SiO_3.CaSiO_3.4SiO_2$). Glassy materials occur naturally as by-products of volcanoes, where high temperatures and pressures cause silicate rocks to melt and, on ejection from the volcano, they cool to form glassy materials (e.g, obsidian). Glass belongs to a group of related materials known as **ceramics**, which comprise inorganic, nonmetallic materials that are processed or used at high temperatures. Ceramics include a broad range of silicates, metallic oxides and combinations thereof and can be broadly grouped in to categories according to their common characteristic features, i.e., clay products (**7.3**), refractories (**7.4**) and glasses. The close compositional relationships between these materials are examined in section **7.1**.

The most remarkable property of glass is that, when heated, it does not melt at a specific temperature, but becomes more and more **plastic**. It can therefore be moulded when hot by blowing, casting, rolling, pulling and extrusion. Moreover, when glass is cooled, it does not crystallise or become a solid at any particular temperature. Glass, to the scientist, is not really a **solid** but rather is a **super-cooled liquid**. Glass becomes a solid simply because its **viscosity** has increased sufficiently (with a fall in temperature) to become a rigid substance, loosely termed a solid. Silicates, tars and bitumens are the only materials to show this large viscosity variation with temperature.

Glass has good optical properties, and hence is used as glazing (windows). In order to be transparent to light, the glass must be **amorphous** (non-crystalline) (**2.4.1**). If the molten liquid were to crystallise, the resulting glass would be opaque to light; this is just the same phenomena that results in water in the liquid state being 'transparent' to light, whilst ice (crystallised water) being opaque. The opacity of crystallised ice and glass results from the fact that the light is reflected at the **grain boundaries** (**2.6.3**) between each crystal. Molten glass must be cooled sufficiently quickly for it **not** to form a crystalline solid, but slowly enough for it not to induce too high a thermal stress.

9.2 COMPOSITION

On heating the raw materials, the carbonates decompose at about 1400 to 1550°C to evolve carbon dioxide (CO_2) gas and the respective alkali oxide. The gas evolved (CO_2) provides the stirring action within the molten mass, whilst the alkali oxides provide the flux to lower the melting point of the sand (silica) SiO_2 (***Figure 9.1***). This action is important as it means that glass can be manufactured at temperatures very much lower than the melting point of silica (the main constituent), with consequent energy savings. The effect on the melting point of silica (SiO_2) of the addition of one alkali oxide (sodium oxide, Na_2O, obtained by the thermal decomposition of soda, sodium carbonate, Na_2CO_3) is shown on a binary ($SiO_2.Na_2O$) phase diagram (***Figure 9.1***).

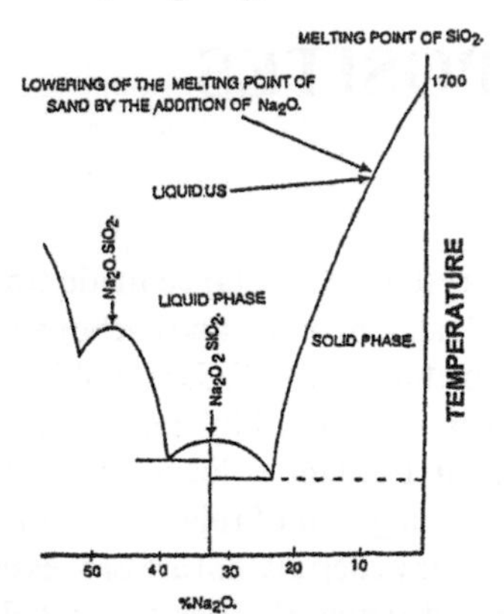

Addition of alkali metal oxides (sodium and calcium) to silica sand at elevated temperatures depresses the melting point of sand (from 1700°C for pure silica to 800°C with the addition of about 22% sodium oxide). The manufacture of glass utilises this depression of the melting point by sodium oxide as the lower melting point makes the moulding of glass artefacts cheaper. However, there is a limit to the amount of sodium oxide that can be added because the product (sodium silicate, known as **water glass**) becomes soluble in water. Water glass is a glass like substance which is soluble in water (and was used in the past to pickle eggs, as the soluble water glass does not allow oxygen into the egg shell). To prevent the formation of water glass, calcium oxide (CaO, produced by the thermal decomposition of $CaCO_3$) is added to form a durable product.

Figure 9.1 *Binary phase diagram for the $SiO_2.Na_2O$ system in glass manufacture*

The manufacture of glass is very energy intensive as the decomposition of both Na_2CO_3 and $CaCO_3$ to provide the alkali oxides Na_2O and CaO respectively, and to melt silica sand, are endothermic reactions. The manufacture of glass ceramics is the same as for glass except the melt has a higher alumina (Al_2O_3) content, which promotes crystallisation upon cooling. By varying the proportions of the oxide additions, glass suitable for a wide variety of end uses can be produced, as summarised in **Table 9.1**.

Table 9.1 Typical compositions of various glass types

Glass material	Composition			
	% SiO_2	% Na_2O	% CaO	Other
Fused silica[a]	100	-	-	-
Soda lime silicate glass[b]	69-74	12-16	5-12	0-6% MgO, 0-3% Al_2O_3
Window glass	72	13	10	1.1% Al_2O_3
Container glass	72	14.4	10.5	-
Fluorescent tubes	71	15	4.6	3.9% MgO
TV screens	66	8	-	12.7% BaO, 4.8% Al_2O_3
Lead crystal glass	57	4	-	29% PbO
Borosilicate glass oven ware[c]	80	4	-	12.7% B_2O_3

Notes: [a] Sheaths for heating elements, laboratory equipment; [b] Defined by BS EN 572-1: 1994[1]; [c] Trade name Pyrex

Metallic oxide additions can also be added as modifiers (**9.3.1**) to remove unwanted impurities that affect glass colour (**Table 9.2**).

Table 9.2 Colour control by the addition of metallic oxides

Colorant	Glass colour	Colorant	Glass colour
Iron	Blue, brown, green	Selenium	Pink, red
Manganese	Purple	Vanadium	Green, blue, grey
Copper	Blue, green, red	Cobalt	Blue, green, pink
Chromium	Green, yellow, pink	Nickel	Yellow, purple

9.3 GLASS AS A SUPERCOOLED LIQUID

Silica sand (SiO_2) is the main component of glass. Molten silica has a very high melting point (1700°C) and is a very viscous liquid in the molten state. If silica is melted and cooled very slowly, it will crystallise at a particular temperature T_m, the freezing or melting temperature, in exactly the same way as a metal. A graph of specific volume against temperature for temperatures around the melting temperature is shown in ***Figure 9.2***, where the solid curve indicates a liquid slowly cooled through the melting temperature, and the dashed line indicates a liquid rapidly cooled.

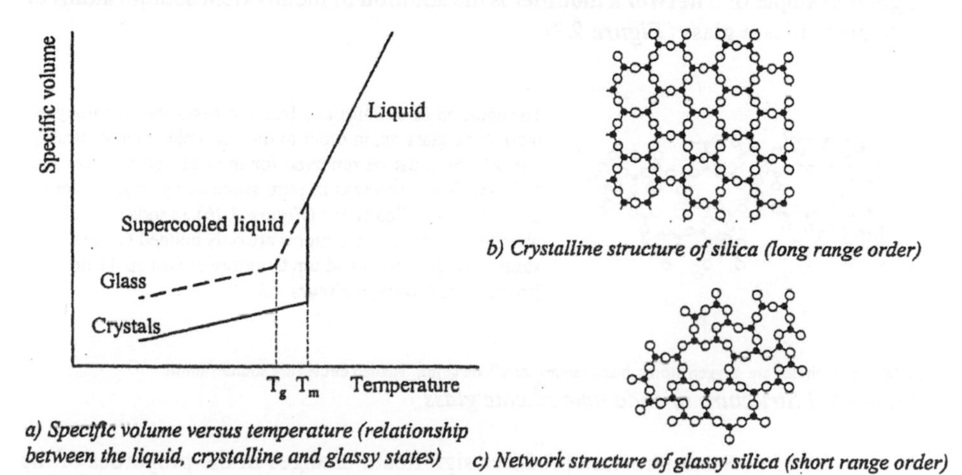

a) Specific volume versus temperature (relationship between the liquid, crystalline and glassy states)

b) Crystalline structure of silica (long range order)

c) Network structure of glassy silica (short range order)

Notes: Open circles are oxygen atoms, black circles are silicon atoms

Figure 9.2 Relationship between temperature and structure for glass

On cooling, the liquid (shown by the solid line in ***Figure 9.2a***) exhibits a discontinuity in the melting point, as the liquid changes phase to a solid with the evolution of latent heat. Silica can crystallise in a number of forms, all of which can be regarded as a network of oxygen atoms (forming a cubic or tetragonal lattice), with silica atoms in the tetrahedral spaces between them (***Figure 9.2b***). If the silica is cooled more rapidly from the molten state, it is unable to attain the **long range order (1.15)** of the crystalline state and the temperature dependence of the specific volume is given by the dashed curve in ***Figure 9.2a***. The temperature T_g is the **glass transition temperature (13.9.5)**. The slope of the curve between T_g and T_m is the same as that above T_m, indicating that there is no change in structure at T_m, i.e., between T_g and T_m, the material is a **supercooled liquid**. Unlike the slowly cooled material, there is no evolution of latent heat at T_g. ***Figure 9.2a*** indicates that, below T_g, the material is in a glassy state very closely related to the liquid state. The structure of this glassy, supercooled liquid is shown in ***Figure***

9.2c, i.e., it exhibits **short range order** (**1.15**). This is a **metastable** structure (**2.6.2**), and will very slowly tend to change to a crystalline form (a process aided by temperature and the application of stress). Glass is termed a **vitreous** solid (a term used to describe the conversion of a material into a glassy substance by fusion due to heat). The process of **de-vitrification** is used to describe the change from a metastable structure to a more crystalline structure. Roman glass can show this change quite clearly, where a fine network of lines indicates the formation of cracks initiating from crystallisation. In this state, the glass becomes quite fragile.

9.3.1 Network Modifiers and Glass Formers

Glass manufactured in this way has a very open network structure (***Figure 9.2b***) which can easily accommodate atoms of different species, such as sodium, potassium, calcium and boron atoms. These atoms can act as **network modifiers**, disrupting the continuity of the network, or as **glass formers**, which contribute to the formation of the network. A good example of a **network modifier** is the addition of monovalent sodium atoms to soda lime silicate glass (***Figure 9.3***).

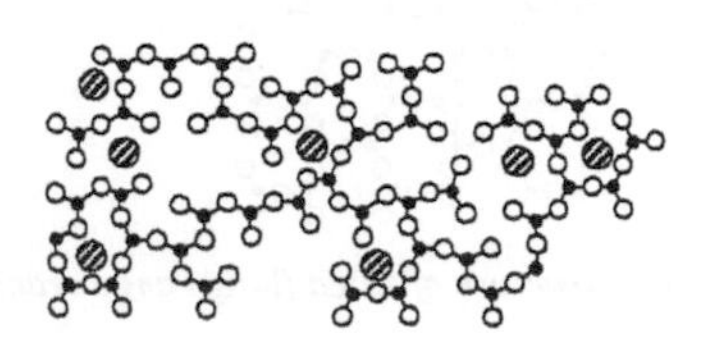

The addition of sodium to silica decreases the silica/oxygen ratio of the glass as, in order to maintain electrical neutrality, one Si^{4+} ion must be removed for the addition of every four Na^{+} ions. Thus, whereas in pure silica every oxygen atom is bonded to two silica atoms (***Figure 9.2b***), in soda lime silicate glass some of the oxygen atoms are only bonded to one silicon atom. The addition of sodium therefore breaks up the network structure, as shown in ***Figure 9.3***.

Notes: Open circles are oxygen atoms, black circles are silicon atoms, shaded circles are sodium atoms

***Figure 9.3** Structure of soda lime silicate glass*

The modified network structure produces significant changes in the properties of the glass. For example, at high temperatures the viscosity of soda lime silicate glass is much less than that of pure silica and is therefore easier to fabricate and process (**9.5**). A good example of a **glass former** is B_2O_3, added to silica to form **borosilicate glass** (Pyrex™). The characteristics of Pyrex™ glass (i.e., high viscosity, resistance to chemical attack and low coefficient of thermal expansion) arise from the network being undisrupted.

9.4 MANUFACTURING PROCESSES

Man-made glass artifacts (in the form of glazed coatings for beads) dating from around 4000 BC have been recovered in Egypt. Hollow glass vessels dating to about 1500 BC have been found through Syria, Italy and along the Rhine and Rhone valleys. These vessels were made by covering a sand core with a layer of molten glass. It was not until about the first century BC that glass vessels as we know them today were made, formed by blowing. In the first century AD (Roman Empire) colourless glass was introduced which could be intentionally coloured by the addition of various materials. At least one Roman glass works has been identified in England (in Lancashire). Medieval glass, produced in Surrey-Sussex area, was used in Westminster Abbey (in about 1240 AD).

Wood was the main fuel for the glass-making process at this time but, due to denudation of hunting forests, in 1615 King James I forbade the use of wood in glass (and iron) manufacture. In consequence, glass-making moved to the coal fields of Newcastle-upon-Tyne, Lancashire, Yorkshire and around Birmingham. In the absence of any scientific knowledge of glass-making (and metal smelting), the selection and proportions of the raw materials added was by trial and error.

Early processes for the production of window glass were the **Cylinder process** and the **Crown glass process**. In the **Cylinder process**, a bubble of glass was blown (by mouth) and elongated by swinging into a cylinder closed at both ends. The ends were cut off and the cylinder was cut lengthwise ('developed'), reheated and opened out into a flat sheet. The sheet was then slowly cooled in a Lehr. The process produced a surface which was uneven, resulting in considerable viewing distortion. However, from around 1615, polishing techniques were developed to reduce this. An improvement to the cylinder process was introduced in 1832 in which the cylinder of glass was drawn from a double sided pot to a length of about 12 m in length and 0.75 m in diameter. The cylinder was then cut and separated by a hot wire into sections, and flattened before annealing in the Lehr. About 930000 m^2 of glass was supplied by this method for the 1851 Crystal Palace, designed by Joseph Paxton. The **Crown glass process** was introduced from Normandy, and involved gathering a glob of glass onto a blow pipe to produce a large bubble of glass by blowing. The bubble was then spun rapidly whilst the glass was still hot to produce a disk of glass fixed at the end of the blowpipe. The blow pipe was then removed and the glass annealed and cut to size. Obviously, the size of the panes was limited, leading to the small window sizes characteristic of old houses. Where the blowpipe was attached, there was an attractive 'bulls-eye' left behind. This feature is now expensively reproduced to provide feature windows.

Currently, two techniques for producing flat glass are used in the UK, the **Rolled glass process** and the **Float glass process**. The **Rolled glass process** (***Figure 9.4a***) is used for the manufacture of patterned and wired glass. Here a continuous stream of molten glass is poured between water-cooled rollers as a controlled ribbon, passed in to an annealing Lehr and then cut to size. The process readily allows wire to be introduced between two continuous streams of molten glass to produce wired glass. The **Float glass process** (***Figure 9.4b***), developed in 1959 by Pilkington Brothers plc, is the main process worldwide for the formation of flat glass for window glazing. Here a continuous ribbon of molten glass up to 3.3 m wide moves out of a melting furnace (at 1500°C) and floats along the surface of a bath of molten tin. The glass is held in a chemically controlled atmosphere at a high enough temperature (1000°C) for sufficient time to allow irregularities to melt out and for the surfaces to become flat and parallel. The ribbon is cooled while still advancing along the molten tin, until the surfaces are hard enough (600°C) to be lifted on to conveyor rollers without marking the bottom surface. The ribbon passes through the annealing lehr and is cut to size. Float glass has a uniform thickness and bright fire-polished surface without the need for grinding and polishing. The molten tin gives the float glass an optically flat surface, such that the glass appears polished. Float glass is used for all glazing (industrial, domestic and motor vehicles).

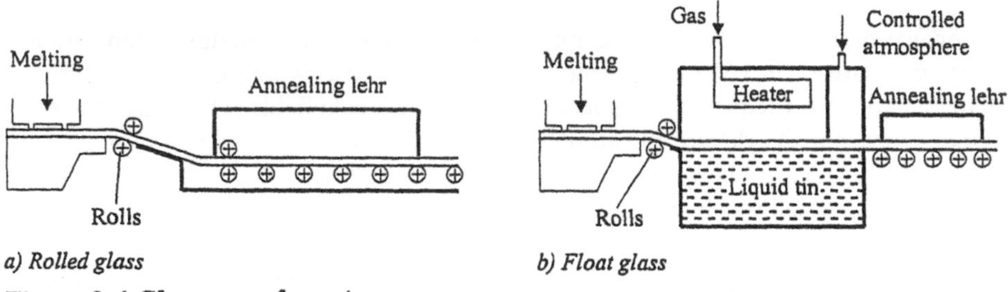

Figure 9.4 *Glass manufacturing processes*

In each process, the purpose of annealing is to remove the inherent stresses formed during cooling. Annealed glass can be further processed to produce **thermally toughened glass (9.7.2)**, and for decorating and engraving. If the quantity of metallic oxides in the glass are kept low, the glass is insoluble in water and forms a very good corrosion resistant material; indeed glass is used to store many aggressive acids. Glass is also used to package food and drink because it does not impart a 'taste' or emit any foreign substance (toxic or nontoxic). Glass is chemically attacked by acidic fluorides; these acids are used to engrave glass (sand blasting is also used for security engraving, for example, where number plates are abraded onto vehicle windows). Glass does not discolour and has very good dimensional stability under all humidity conditions. **Cullet** (old broken glass) can be added to the mix, but in general the quantity of cullet added is restricted because the metallic oxides present **(Table 9.2)** may adversely affect the colour of the article to be manufactured.

9.5 PROCESSING OF GLASS

During the manufacture of glass, the molten liquid glass is cooled to temperatures at which the working plasticity of the glass is suitable for forming glass products. The working plasticity of the glass is dependent on viscosity. The viscosity of glass is affected both by composition and by temperature, as shown in ***Figure 9.5***.

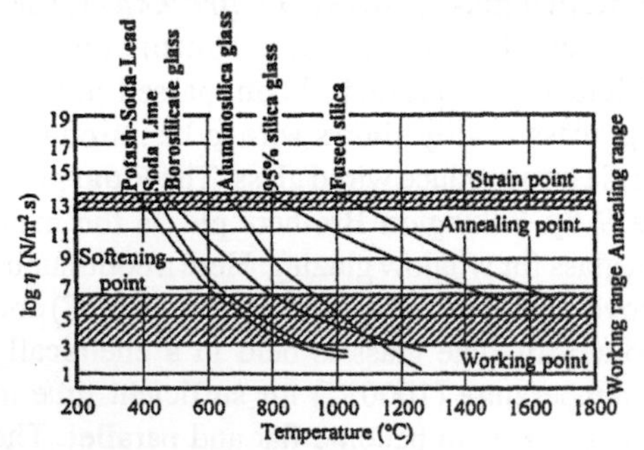

The viscosity of glass is both temperature and composition dependent. For example, at about 1400°C, the viscosities of silica and silica with a 20% Na_2O addition are 10^{11} $N/m^2.s$ and 10 $N/m^2.s$ respectively. Addition of alkali metallic oxides lowers the softening point and allows the glass to be worked at a lower temperature. In addition, the amount of alkali oxide combined with the silica is important as it governs the type of 'glass' produced (**Table 9.1**).

Figure 9.5 *Viscosity-temperature curves for glasses*

From the melt (about 1500°C), glass is cooled to the working temperature (about 1200°C) (viscosity about 10^3 to 10^6 $N/m^2.s$, **Table 9.3**). For working, the temperature is maintained above the softening point (about 1000°C) as below this temperature the glass is too viscous (about 10^7 to 10^8 $N/m^2.s$) to be worked. When the glass is annealed **(9.4)** to remove all the thermal stresses, it has a viscosity of about 10^{13} $N/m^2.s$.

Table 9.3 Viscosity at various temperatures within the manufacturing process

Process	Temperature (°C)	Viscosity ($N/m^2.s$)
Melting	1400	100
Working	1200	$10^3 - 10^6$
Softening	1000	$10^7 - 10^8$
Annealing	500	10^{13}

9.6 FRACTURE OF GLASS

One of the most important deficiencies that an engineering material can have is a lack of **toughness (3.2.3)** as this implies that the material is unable to stop or blunt cracks. There are two principle ways of stopping a crack from propagating

- **placing a surface in compression**. For example, the surface of **whole structures** can be placed in compression (e.g, the arches of railway bridges, aqueducts, etc. and Cathedral domes and buttresses, etc.). Alternatively, the surface of the **material** can be placed in compression (e.g, the surface treatment of toughened glass, **9.7.2**, and gypsum plaster, **14.7.3**). Generally, it is good practice to **design for the structure and/or component to be in compression;**
- **grain boundaries** in crystalline materials (e.g, metals) provide an effective crack blunting mechanism, as they prevent movement of the line defects (dislocations, **10.4**) responsible for deformation. However, it is very easy to break a metal by subjecting it to very large stress cycles, as these stresses tend to sharpen up the blunted crack by repeatedly placing it in the compression cycle of the alternate bending programme. This phenomenon is called **high strain fatigue (3.5.5)**. Fatigue is damaging to metals and exceedingly damaging to composites.

To understand the mechanism of brittle fracture, it is best to describe the original work by Griffiths (on glass). Griffiths realised that the theoretical strength required to physically separate the constituent atoms of a solid material was some several orders of magnitude larger than the measured strength of the material. Griffiths postulated that the presence of **flaws** would act as sources of weakness by concentrating the stress at their tips (i.e., they act as **stress raisers**). Griffith's concept was that the stress σ_t at the tip of a crack of length l with radius of curvature r (***Figure 9.6***) magnifies the applied stress σ_a according to the formula

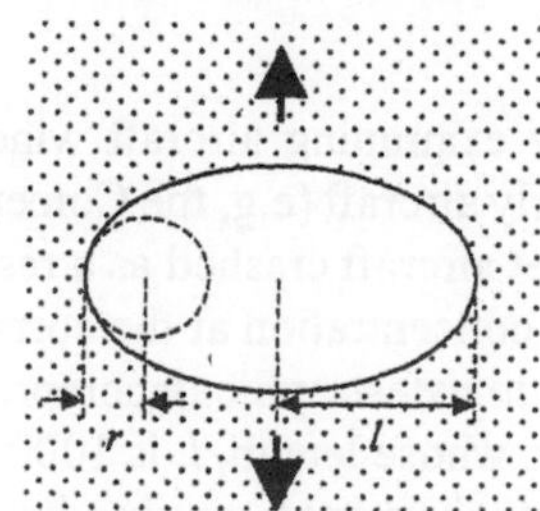

$$\sigma_t = \sigma_a \left(\frac{l}{r}\right)^{1/2}$$

where $\left(\frac{l}{r}\right)^{1/2}$ is the **stress concentration factor**.

Figure 9.6 Stress raisers and the Griffiths crack hypothesis

Figure 9.6 illustrates an elliptical stress raiser with a radius of curvature, r, for which the crack length on the major axis is l. If the stress σ_t exists within a distance of approximately r of the tip and if it exceeds the strength of the inter-molecular bonds in the material, then the crack will propagate through the material. Griffiths also modelled the fracture process by correlating the relationship between the fracture stress (σ_f) required to form a crack in non-crystalline solids which behave elastically and stretch up to their breaking point. At fracture, the inter-atomic bonds are broken and a new surface is created. This new surface requires energy (γ) to form. Griffiths postulated that the crack would propagate when the released strain energy is just sufficient to provide the surface energy required for the formation of the new surface, i.e.,

$$\text{Stress required to form a crack} = \left(\frac{\text{Surface Energy } (\gamma) \text{ x Young's Modulus (E)}}{\text{Crack length}} \right)^{1/2}$$

Note that

- the stress required to form a brittle fracture is inversely proportional to the crack length. Hence the fracture stress of a brittle material is determined by the largest crack existing before loading;
- Substituting typical values of surface energy and Young's modulus for glass fibres show that a critical crack length of 2.7 x 10^{-7} m is required to initiate fracture. This is approximately **1000 times the interatomic distance**;
- once a crack begins to spread, the stress required for propagation falls (as the crack length is increasing), hence the crack propagates through the material very quickly.

For toughened glass (**9.7.2**), surface imperfections in the glass are 'locked in' by ensuring that the surface is in compression. However, the internal area of the glass is in tension, and so toughened glass has high elastic stored energy. Therefore, a high stress is required initially to propagate a crack, as the built in compressive forces must be initially exceeded to form the crack. Thereafter, once fracture starts, the elastic stored energy is so high that failure is often explosive (a small impact with a stone on a toughened glass windscreen would rapidly fracture the whole windscreen into small cubes because there is insufficient energy absorbed by the creation of the new (fracture) surface). As the crack propagates, it rapidly gains kinetic energy and propagates very quickly (almost instantaneously in a car windscreen hit by a small stone). The way to overcome this problem is to both to attempt the crack initiating and thereafter to prevent the crack from propagating.

A good example of **crack initiators** can be seen by examining aircraft windows. Modern aircraft windows are oval in shape, whereas early aircraft (e.g, the Comet) had square windows (***Figure 9.7***). A large number of Comet aircraft crashed as a result of fatigue failures which were initiated by the high stress concentration at the corners of the windows, producing cracks in the fuselage leading to catastrophic decompression in high altitude flight. Any small surface defect or flaw, whose length, l, is 1000 times the interatomic bond distance and which has been sharpened up by being in compression so that the radius of curvature, r, is of the order of the interatomic bond, produces a **stress concentration factor** of some 300 times. These regions are therefore high risk areas and, to prevent these high stress concentrations, aircraft windows were redesigned to produce a curved profile to ensure a high radius of curvature.

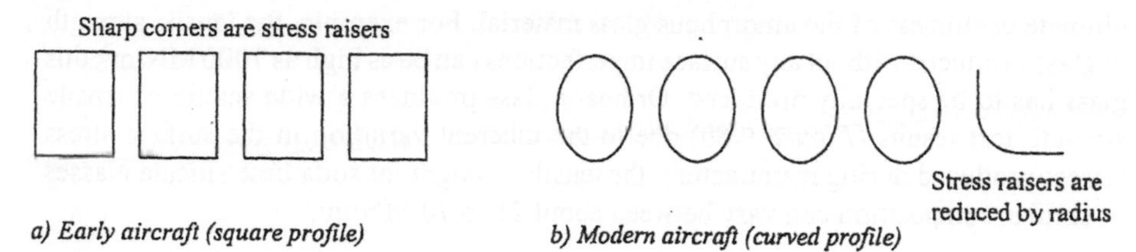

Figure 9.7 Stress concentrations in windows

Other examples of the development of high stress concentration factors in brittle materials and their prevention by design include

- in tensile tests of brittle materials, the mere action of clamping the material in the jaws of the testing apparatus produces a high stress concentration factor at the point of clamping arising from the "non-slip serrations" on the jaws. To counteract this problem, most tensile test specifications include requirements for the gripped shoulders to have a specified radii to remove this stress concentration (for example, BS EN 10002[2] for metals and BS EN ISO 3167: 1997[3] for plastics) (**3.2**);
- in many engineering design functions, where sharp 90° angles are replaced by fillets with large radii of curvature (***Figure 9.7b***);
- cracks may be stopped from spreading in e.g, Perspex (**13.6.1**) by drilling out the front of the crack with a small drill, effectively blunting the crack as a stress raiser;
- for certain materials, the problem of the development of stress concentration factors can be addressed by testing in compression (e.g, concrete cubes are commonly tested in compression, 3.3) and by eliminating stress concentrators by providing large radius holes at crack fronts or a right angle bends.

9.6.1 Stress-Strain Behaviour

The stress-strain curve for glass is compared to steel in ***Figure 9.8a***. Glass undergoes elastic deformation only up to the point of fracture (Point 4). The fracture stress and strain to fracture are low (Point 1), and the elastic (Young's) modulus (Point 2) of glass is lower than steel. The low strain to fracture and elastic modulus mean that glass has a low energy of fracture (Area 3), obtained from the area under the stress- strain curve.

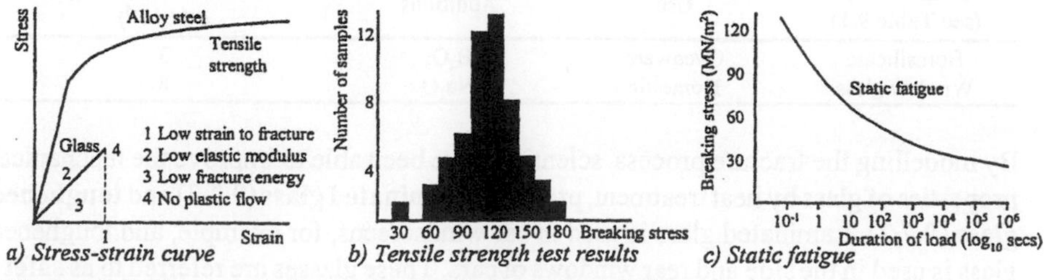

Figure 9.8 Characteristic mechanical properties of glass

The shape of the stress strain curve for glass is characteristic of an amorphous supercooled liquid. Surface micro cracks and scratches act as **stress raisers**, sites that concentrate stresses sufficiently to initiate fracture (**9.6**). Glass is therefore susceptible to brittle fracture. Surface imperfections and micro cracks reduce the strength and ultimate usefulness of the amorphous glass material. For example, the tensile strength of glass produced without any surface imperfections can be as high as 7000 MN/m^2; this glass has to be specially produced. Ordinary glass produces a wide scatter of tensile strength test results (*Figure 9.8b*) due to the inherent variation in the surface stress raisers produced during manufacture; the tensile strength for soda lime silicate glasses of similar composition can vary between about 25 to 70 MN/m^2.

Glass has poor thermal conductivity. Unequal cooling rates during manufacture will set up thermal gradients within the glass, and the resulting forces developed within the glass by differential expansion can exceed the mechanical strength of the glass and cause breakage. The ability of glass to withstand thermal gradients is dependent upon its thickness. The energy absorbed at fracture (toughness, the area under the stress strain curve, **3.2.3**) is very low as a result of the amorphous structure and the absence of dislocations. The absence of dislocations means that the glass lacks ductility and cannot be permanently deformed by an applied load. All permanent deformation must be carried out at an elevated temperature and the resultant glass product annealed, otherwise the differences in cooling rates between the centre and external regions will freeze in stresses which will cause the glass to shatter unexpectedly.

The strength of glass is very much dependent upon the rate of deformation and the length of time that the glass is under stress (often referred to as **static fatigue**). If glass is loaded rapidly, it will withstand a much higher load than if it were loaded slowly (***Figure 9.8c***). Glass has a history of failing under static fatigue, where sudden unexpected failure arises at stresses which the component has withstood before. Glass gives no warning as to when it is about to break and does not withstand thermal or mechanical shocks, unless certain additions are made to the melt process which confer upon the glass a much lower thermal expansion coefficient (**Table 9.4**). Borosilicate glass is very resistant to thermal shock whilst window glass (soda lime silicate glass) is not resistant to thermal shock.

Table 9.4 Thermal conductivities of two different glasses

Glass material (see **Table 9.1**)	Use	Additions	Coefficient of thermal expansion α (x 10^{-6})
Borosilicate	Ovenware	B_2O_3	3
Window glass	Domestic	Na_2O	8

By modelling the fracture process, scientists have been able to improve the mechanical properties of glass by heat treatment, producing **laminated glass (9.7.3)** and **toughened glass (9.7.2)**. Laminated glass is used in car windscreens, for example, and toughened glass is used in the side and rear windows of cars. These glasses are referred to as **safety glass** (defined by BS 6206: 1995[4] and BS 6262: Part 4: 1994[5] as glass which, when fractured, is less likely to cause severe cuts or serious physical injury than ordinary glass). Ordinary glass fractures to give long razor sharp 'splines' that are not restrained (**9.7.1**). The Road Traffic Act 1930 made compulsory the fitting of safety glass in cars from 1932. The manufacture of laminated and toughened glass are completely different, but both manufacturing processes start with **Float glass** (**9.4**) (ordinary domestic glass).

9.7 REINFORCED AND STRENGTHENED GLASS

9.7.1 Wired Glass

Contrary to popular belief, wired glass is **not** reinforced glass. The wire is sandwiched between two sheets of molten glass (at 1050°C) and rolled when hot to produce a uniform composite material (**9.4**). The function of the wire is to hold any broken pieces together to prevent the glass from falling from roofs or skylights and causing injury to pedestrian traffic. The wired glass also acts to spread heat evenly through the glass in the event of a fire. It is used, for example, in doors where **laminated glass** (**9.7.3**) would be too expensive. Note that if wired glass is used for roofing, the exposed ends of the wire should be passivated and protected with zinc chromate (**Table 11.7**) to prevent the wire from corroding and the corrosion product initiating a crack within the sandwich.

9.7.2 Toughened Glass

The weakness of glass lies in the defects at the air-glass interface. Small scratches, which the eye cannot resolve, act as stress raisers (**9.6**). In order to visualise the effect of stress raisers, imagine the glass is cut with a diamond scribe. This will produce a deep scribe line and, in addition, **much finer cracks at right angles to the main scribe line**. These finer cracks act as stress raisers and, for the glass to break along the scribe line, the stress concentration has to exceed the stress concentrations of the stress raisers formed as a consequence of the normal inherent defects at the air-glass interface (**9.4**). If the stress raiser caused by the inherent defects is greater than the stress raiser of the scribe line, the glass fracture will run off the scribe line.

In toughened glass, the air-glass interface is placed into **compression** so that the inherent flaws and scratches do not act as stress raisers until this compressive stress has been overcome by any externally applied tensile forces. The external forces in the outer layers of glass are therefore balanced by tensile forces within the central region of the glass (which has no air-glass interface and so cannot initiate fracture). In soda lime silicate glasses, the surface compressive forces can be some 2 to 2.5 times the central tensile forces, rising to 3.0 to 4.0 times the central tensile forces in toughened glass.

Toughened glass can be produced by **thermal** processes; or **chemical** processes. **Thermal processes** are utilised where the glass thickness is uniform, whereas **chemical processes** can be used where the thickness is not uniform (e.g, milk bottles). Chemical toughening involves the substitution of a sodium or lithium ion by the much bulkier potassium and/or tin ion in the surface layers. The toughening arises because the potassium and tin ions (0.133 nm and 0.140 nm respectively) are larger than both the sodium (0.098 nm) and lithium (0.078 nm) ions (**Table 1.5**). For most applications, the stress profile for thermally toughened glass is a parabolic distribution, whilst chemical toughening produces a stress profile with a more flattened distribution (***Figure 9.9a***). In ***Figure 9.9a***, ordinary toughened glass with the compressive (100 MN/m^2) and tensile (50 MN/m^2) stresses is shown.

Thermally induced compressive forces must be produced after all the cutting, forming and bending processes have been undertaken. The preformed unit is heated to a temperature of about 600 to 670°C, where the viscosity of the glass is just sufficient to relax most of the manufacturing stresses, but not high enough to allow any shape deformation. The glass is then cooled quickly by cold air jets which are played onto **both** surfaces of the glass. This cools the surface layers more than the central region, which remains hot as a direct result of the low thermal conductivity of glass (0.8 to 1.3 W/m°C). The areas of the glass which have been thermally toughened with air jets can be seen by wearing sunglasses correctly orientated (e.g, car windscreens). After the external surfaces have cooled and stiffened, the central hot region cools to room temperature, putting the outer pre-cooled layers in compression. This makes the toughened glass very strong (some three times stronger than annealed glass of the same thickness). In order to break the glass by propagating a small surface flaw, the outer surface compressive stresses have to be exceeded, in conjunction with the bulk internal central stress distribution which is free of surface stress raising defects. When the toughened glass does break, the tensile (locked-in) stresses within the central region initiate multiple cracks. Since the fracture surfaces are perpendicular to the surface, the razor sharp dagger like splines are not produced; **toughened glass** breaks into fairly small cubes as the complex locked-in stresses are relieved.

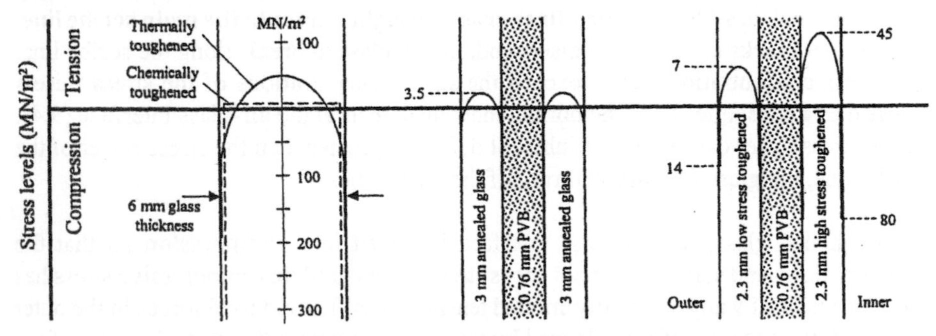

a) Thermally and chemically toughened b) Conventional laminate c) Modern laminate (windscreen)

***Figure 9.9** Stress distribution in various types of glass*

Toughened glass was developed initially for the motor industry. When car windscreens were first glazed they were flat and nearly upright and therefore the cracking of toughened glass windscreens into small cubes did not present a great problem in obscuring the drivers' vision of the road ahead. As cars became more aerodynamic, the front windscreen became curved and raked backwards. This feature causes vision distortion problems, as the crack interfaces reflect light so that the driver's vision is obscured. This led to the development of the **laminated** windscreens and, thereafter, laminated glass for use in building (high security areas, patio doors, etc.).

9.7.3 Laminated Glass

Laminated glass is produced by sandwiching a 0.76 mm thick transparent plastic sheet (of plasticised **polyvinyl butyryl, PVB**) between two sheets of glass and firing the resultant matrix at 150°C to for an adherent sheet. At this very low temperature, there is no formability left in the two sheets of glass and therefore the two sheets of glass have to be made as a pair. In addition, any forming processes required must be applied to the two sheets as a pair (an infusible dust layer is usually incorporated to keep the two sheets apart). Note that, like toughened glass, laminated glass components must be made to size (they cannot be cut to size after manufacture).

The manufacturing process allows the two sheets to have different stresses induced (***Figure 9.9b*** and ***9.9c***), depending on the end use application. ***Figure 9.9b*** illustrates a conventional laminate between annealed float glass (used, for example, in patio doors), where the laminate material is a polyvinyl butyryl (PVB) inter-layer. ***Figure 9.9c*** illustrates a modern laminated windscreen, made from low stress toughened glass on the outside and high stress toughened glass on the inside (towards the occupants). The low stress toughened glass fractures producing single cracks, while the inner higher stressed glass fractures to produce small 'cubes'. Hence either (or both) of the two sheets of glass can break without producing complete disintegration of the windscreen. As outer surfaces are not toughened to the same extent as toughened glass, there is no shattering into small pieces (for example, on impact with road stones) as there would be with toughened glass screens. Hence only limited cracking is obtained within the outer sheet and the driver's vision is not impaired. The inner sheet of the pair is toughened to a higher degree so that it will shatter into small pieces. This is to reduce the amount of cutting and head injury that would occur in a car crash when the occupants head is thrown forward. The strength of laminated car windscreens is often demonstrated by manufacturers advertisements which show the laminated glass being attacked with hammers to no detriment.

The middle sheet of plastic (PVB) can undergo elongations of about 200 to 350% before fracture. The outer sheets of glass, being good corrosion resistant coating, prevent the degradation of the plastic and ensure complete transparency over the service life of the laminated glass. The function of the PVB sheet is to hold together any glass that fractures, preventing the formation of dangerous glass shards.

9.8 SURFACE COATINGS

Glass is transparent to solar radiation. The degree of transmission can be modified by coating the glass surface with oxides of dielectric materials, altering both the reflection and scattering properties at the surface. Glass is not transparent to some thermal wavelengths associated with heat (***Figure 9.10a***) (transparency to heat is called a **diathermic** process). Transmission and reflection characteristics for solar radiation of toughened and heat absorbing glass are shown in ***Figure 9.10b*** and ***9.10c*** respectively.

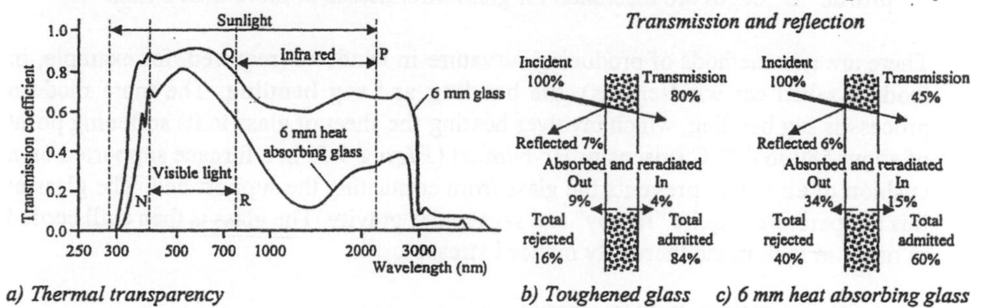

a) Thermal transparency *b) Toughened glass* *c) 6 mm heat absorbing glass*

Figure 9.10 *Transparency, transmission and reflection characteristics of glass*

As shown in ***Figure 9.10a***, the energy produced by the sun is of many wavelengths, but only a small fraction of this radiation (**N-R**) produces the sensation of vision. Much of the radiation is the infrared radiation (**Q-P**). Only a small fraction of the total radiation is transmitted by glass. A heat-absorbing glass absorbs the infrared (**Q-P**) and the red end of the visible spectrum and therefore appears with a blue or grey colour. This absorption heats up the glass and therefore the glass re-radiates more heat than conventional toughened glass (***Figure 9.10b*** and ***9.10c***). Toughened glass reflects 16% and admits 84% of the incident radiation falling upon it. Heat-absorbing glass rejects 40% and admits 60% of the incident radiation falling upon it so that, for example, the inside of a car remains cool. It is also important to appreciate that the reflected light may be wavelength specific; for example, light reflected off sand makes some glasses appear red. Demisting of aircraft windows and modern car windscreens is achieved by thin metallic coatings to the surface of the glass.

9.9 SURFACE FINISHES

Glass cutting is usually undertaken by making a diamond scribe line on the surface. This method becomes more difficult as the thickness of glass increases (maximum glass thickness about 38 mm; above this thickness, diamond sawing is used). Various methods are available for decorating glass, for example

- **machining** of glass by grinding, either in cast iron mills fed with loose abrasives (e.g, silicon carbide, SiC and water) or by alumina wheels. Polishing is carried out by various rouges on cork or felt pads, pumice or Al_2O_3 powder on willow wood. Rubber wheels are used for edge polishing;
- **drilling** is undertaken using ultrasonic techniques or with a triangular drill in a carpenter's brace, with paraffin or turpentine as a lubricant;
- **sandblasting** (sand and compressed air). The area not required to be sand-blasted is protected by an abrasion resistant surface. Sandblasting is commonly used for lettering, pictorial decoration, labels for containers, etc.;
- **acid etching** using hydrofluoric acid, which etches glass. The area not to be etched must be protected by wax;
- **patterned glass** is made by rolling the glass with a flat roll on one side and a patterned roll to give the textured surface.
- **arris edges** are obtained by removing the sharp edges from sheet glass. Standard profile "A" edges can be machined to shape for thicknesses up to 6 mm. Standard profile "B" edges are machined for glass thicknesses of more than 6 mm.

There are two methods of producing curvature in windows (required, for example, in modern raked car windscreens), **die bending** and **sag bending**. The more modern process is sag bending, which involves heating the sheet of glass to its softening point of about 580 to 670°C (viscosity 10^6 N/m^2.s) (***Figure 9.5***) in a furnace supported on a cushion of air which prevents the glass from contacting the support bed. The glass at this temperature is quite "floppy" and sags under gravity. The glass is then chill cooled in order to lock in the thermally induced stresses.

9.10 GLASS FIBRE INSULATION

Glass wool is made by rapidly ejecting a stream of molten glass through spinners into a blast of hot gases. The resultant effect of centrifugal force and the rapid gas stream is to produce very fine fibres. These fibres are coated with a bonding agent, which is cured in ovens and cut to size. This mat can be further processed into many products for sound and heat insulation, including rigid pipe insulation.

9.11 FIBRE COMPOSITES

The dispersal of a fibre of one type of material within a matrix of another type of material provides a method of improving the strength of the resultant matrix. Historically, animal and vegetable fibres have been used to reinforce bricks, plaster and mud. Traditionally, straw has been used to increase the toughness of the unfired clay brick. In fact it was over the limited availability of straw that the first withdrawal of labour (by the Israelites) in the constructional industry was recorded. The traditional wattle and daub of the later years was a means of producing not only a stronger material but also a draught-proof screen. The weakness of lime-based plasters (**14.7.3**) was improved by the introduction of cow hair, as this reduced cracking and crazing, but the practice gave rise to problems associated with anthrax. Some examples of the properties of composite materials are given in **Table 9.5**.

Table 9.5 The main fibre reinforced composites

Fibre material	Strength	Matrix	Composite
Glass (normal)	Very brittle in tension	Plastics[a]	Glass reinforced polymer (plastic)
Glass	Very brittle in tension	Gypsum plaster[a]	(GRP)
Alkali resistant glass	Very brittle in tension	OPC	Glass reinforced gypsum (GRG)
Steel	Strong	OPC	Glass reinforced cement (GRC)
Polymers	Elastic	OPC, plastics	Reinforced concrete (RC) (**8.9.3**)
Asbestos[b]	Weak	OPC	

Notes: [a] used, for example, for floor and roof construction in the USA, BRE fire check doors, doubled skinned floor units, etc. GRG must not be wetted; [b] Asbestos (1.13) is resistant to alkali and so may be mixed with cement. It has good fire resistance (14.6.7) but constitutes a health hazard.

9.12 GLASS REINFORCED CEMENT AND GYPSUM

Portland cement and sand mixtures may be improved by reinforcement with steel, asbestos, glass, carbon fibres and plastics (nylon and polypropylene). Polymer fibres tend to increase the impact strength and offer a greater resistance to shattering (due to their low value of Young's modulus of elasticity, E). In order to restrict the unacceptable deflections, fibres such as glass are often added in order to increase the tensile strength.

Traditionally (within the last 70 to 80 years), asbestos has been used as a fibre reinforcement for cement and concrete. Asbestos is resistant to alkali and so may be mixed with cement. It has good fire resistance, but there is a danger of asbestosis from the 'blue form' of asbestos (Crocidolite). Experience gained in the formation of glass reinforced polymers (GRP) was utilised in the development of glass reinforced cement and concrete (GRC). However, in preliminary tests of GRC, the required stress improvements were not obtained. This was because OPC, which is very alkaline when hydrated (**7.17**), etched the glass surface, i.e., the glass was dissolved in contact with the alkali, with the result that the etched areas acted like small **stress raisers (9.6)**, causing the composite to fail. Pilkington Glass plc developed the alkali resistance glass CEMFIL™ (a borax glass, **9.3.1**), making the production of glass reinforced cement a possibility. Using the same alkali resistant glass, glass reinforced gypsum (GRG) could be produced to provide a tough fire resistant material suitable for sheets, doors and floors, replacing more combustible timber products.

Production methods for glass reinforced cement include

- **casting and injection moulding**, where a wet slurry of cement and admixture, with a water:cement ratio of about 0.5 (w/c = 0.5) is produced and 20 to 50 mm long chopped fibres are added and thoroughly mixed. This is poured into a mould and pressed to produce sheets or mouldings;
- **spray suction method**, where the slurry is sprayed through a nozzle into which the chopped fibres are introduced suspended in an airstream. The sprayed surface is then de-watered through a porous membrane;
- **layered method**, which produces a hand-layered sandwich of outer skins with the internal section being filled with organic fibres.

9.13 FIBRE REINFORCED PLASTICS

Fibre reinforced plastics have been widely used in the construction industry for 20 years for mainly non-structural applications (cladding, etc.). In recent times, more advanced forms of fibre reinforced plastics have been developed, mainly for the aerospace industry and the military. These materials are currently receiving attention for development in the construction industry[6,7] as sandwich panels, modular units, structural components, reinforcing bars, etc. Fibre reinforced plastics ordinarily comprise a **fibrous phase** dispersed in a continuous (resin based) **matrix phase**. The **resin phase** is composed either of thermoplastics or thermosetting plastics, with additives to improve fire resistance (incorporated in the resin itself or applied as a gel coat), mechanical properties, appearance and protection from the environment. Thermoplastic resins for fibre reinforced plastics include polyolefins, polyamides, vinyl polymers, polyacetals, polycarbonates, etc. Thermosetting resins for fibre reinforced plastics are usually either polyesters or epoxides. A wide range of amorphous and crystalline materials can be used for the fibrous phase, although the most common in the construction industry is **glass fibre (9.14.1)**. **Carbon fibre (9.14.2)** can be used separately or in addition the glass fibre to increase the stiffness of the structural member. **Aramid fibres (9.14.3)** (e.g, Kevlar^{TM}) can be used instead of glass fibre to again provide increased composite stiffness. These fibre types are considered below.

9.14 FIBRE TYPES

9.14.1 Glass Fibres

Glass fibres are of four types

- **E-glass** (of low alkali content) is widely employed, especially with polyester and epoxy resins;
- **AR-glass** (alkali resistant glass), developed for use in cementitious materials;
- **A-glass** (of high alkali content), now little used;
- **High strength glass fibre**, produced for extra high strength and high modulus applications (in aerospace industries).

Glass fibres for reinforcing thermosetting resins may be

- chopped to form milled fibres (30-3000 μm length), short chopped fibres (< 6 mm length) or long-chopped fibres (< 50 mm length);
- formed into chopped strand mats in which the chopped fibres are randomly orientated and loosely bonded with a resinous binder;
- formed into uni- or bidirectional woven rovings in which the fibres are orientated in one or two directions, giving the composite high directional strength properties;
- formed into a surface tissue comprising a thin glass fibre mat with a readily wetted medium for use when a resin rich surface is required or when the coarse fibre pattern of the chopped strand mat is to be concealed;
- formed into multi-axial, non-woven (stitched or warp-knitted) fabrics.

Figure 9.11a shows a schematic for the production of glass fibre, illustrating that the high surface area:mass ratio allows the glass fibre to rapidly cool and become toughened. ***Figure 9.11b*** illustrates the production of glass fibre mat by chopped strands, weaving and roving.

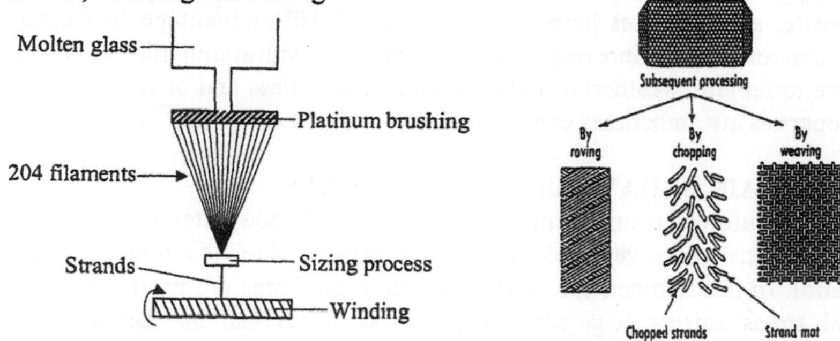

a) Glass fibre production *b) Glass fibre mat types*
Figure 9.11 *Manufacturing process for glass fibres*

Plastics are commonly strengthened by the addition of glass fibres; **Table 9.6** shows the effect on the strength of some plastics produced by the addition of the glass fibre.

Table 9.6 The effect on strength of fibre reinforcement of some plastics

Polymer	Young's modulus (x 10^9 N/m^2)		Tensile strength (x 10^6 N/m^2)	
	Unfilled	Filled	Unfilled	Filled
GRP (thermoset)	-	30	-	1200
Polypropylene	≈ 1.2	4	35	50
ABS + Glass	2	6	40	100
Polycarbonate	2	4	70	60

Table 9.6 illustrates that glass fibre is some 30 times stiffer than the polymer. As glass fibre is much cheaper than the plastic polymer, it is economic to design for high volume replacement. Fibre reinforced polymers are replacing more traditional materials in a variety of applications (e.g, GRP is replacing steel panels in boats and cars). A panel made of steel is some ten times stiffer and some two and a half times stronger in tension than the corresponding panel made from GRP. Hence, for a GRP panel to have the same stiffness as the steel panel, it would have to have increased thickness or be designed

with double or complex curvatures to increase the stiffness. Often other additional (inert) materials are also used to increase the stiffness of plastics (**13.11**).

9.14.2 Carbon Fibres

Carbon fibre is produced by heating polyacrylonitrile (PAN) fibre under tension in air at 250°C. During heating, the fibre will absorb oxygen, gain strength and change colour. Once it turns black, it is heated further in an inert atmosphere. By varying the processing conditions, mechanical property modifications can be obtained. There are three grades of carbon fibre

- **Type I**: the stiffest carbon fibre with the highest modulus of elasticity;
- **Type II**: the strongest carbon fibre;
- **Type III**: the least stiff carbon fibre with strength midway between Types I and II.

9.14.3 Aramid Fibres

Aramid fibres (e.g, Kevlar™) are aromatic polyamides. When used with polymers to form a composite, aramid fibres have up to a 35% and 10% advantage in weight compared to glass- and carbon fibre respectively. Composites with aramid fibres display good resistance to fatigue, weathering and chemical attack. Glass and/or carbon fibre composite properties are sometimes enhanced by the addition of Kevlar™.

9.15 STRESS-STRAIN BEHAVIOUR OF COMPOSITES

Modern day fibre reinforced resins are produced from a very brittle material (e.g, glass) whose strength is increased several fold by the resin matrix, which is a relatively low strength material. In this manner, the surfaces of the glass fibres are protected from environmental stress raisers (e.g, scratching). Glass fibres can be thermally or chemically toughened (**9.7.2**) to increase the mechanical strength, resistance to abrasion and to the formation of microcracks within the surface layers. Chemical toughening usually involves the treatment of the glass fibres with a tin compound which diffuses into the surface layers of the glass fibre to chemically toughen the fibre. Once the applied stress has overcome the compression stresses in the outer surfaces of a fibre (produced by thermal or chemical toughening), cracks will propagate only through a single fibre. However, glass is a very brittle material (**9.6.1**) and thus the use of glass fibre as reinforcement may seem surprising. To understand the principles of fibre reinforcement, a brief analysis of the theory of fibre reinforcement follows. The problem depends upon the relative volume fractions of the fibre and matrix (denoted V_f and V_m respectively), and upon Young's modulus elasticity (**2.8.1**) of the fibre and the matrix (denoted E_f, and E_m respectively) (***Figure 9.12***).

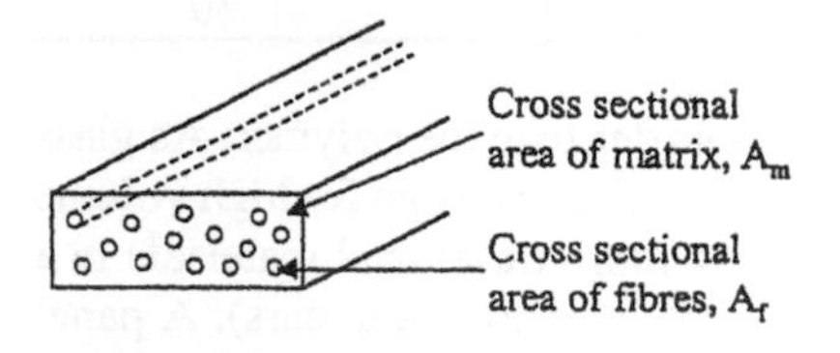

a) Schematic of fibre composite

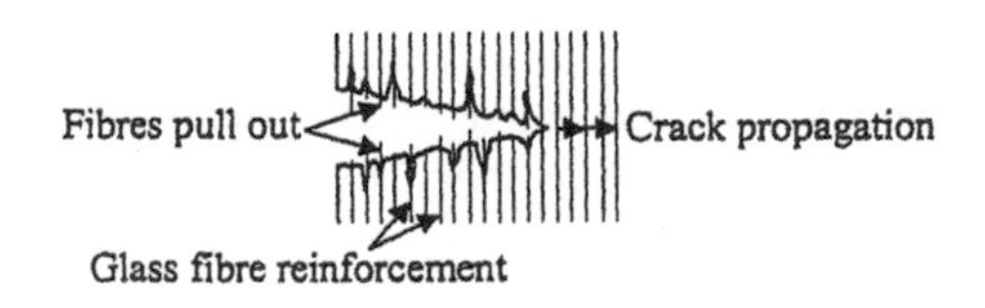

b) Crack propagation in fibre composite

Figure 9.12 *Glass reinforced plastics*

The general formula for a material under stress (Hookes law, **2.8.1**) is stress (σ) = Young's modulus (E) x strain (ϵ), where stress, σ = force/cross sectional area = F/A and strain, ϵ = increase in length/ original length. For a given strain, the total force is F. This force is carried on the composite by each of the fibres and the matrix making up that composite. Denoting the total force on the composite as F_c, then, since this force is carried by each component,

$$F_c = \sigma_f.A_f + \sigma_m.A_m \qquad ...(9.1)$$

where the suffix f and m refer to the fibre and the matrix respectively (***Figure 9.12***). The composite cross sectional area (A_c) is made up of the area of cross section of the fibres (A_f) and the matrix (A_m) (***Figure 9.12a***). The stress is carried equally by the fibres and the matrix provided there is a good bond across the interface. An added advantage is that when the composite breaks, work is done in sliding and pulling the fibre out of the matrix (***Figure 9.12b***).Thus, for the whole composite $F_c = \sigma_c.A_c$, which can be substituted into equation (9.1) to yield

$$\sigma_c.A_c = \sigma_f.A_f + \sigma_m.A_m \qquad ...(9.2)$$

Dividing each side by A_c yields

$$\sigma_c = \frac{\sigma_f.A_f}{A_c} + \frac{\sigma_m.A_m}{A_c} \qquad ...(9.3)$$

Given that $A_m = A_c - A_f$ ...(9.4)

then, substituting equation (9.4) into equation (9.2) yields

$$\sigma_c = \frac{\sigma_f.A_f}{A_c} + \frac{\sigma_m.(A_c - A_f)}{A_c} \qquad ...(9.5)$$

The volume fraction of the fibres (all aligned along the tensile axis) is

$$V_f = \frac{A_f}{A_c} \qquad ...(9.6)$$

Substituting equation (9.6) into equation (9.5) and rearranging yields

$$\sigma_c = \sigma_f.V_f + \sigma_m.(1 - V_f) \qquad ...(9.7)$$

This analysis assumes that the whole composite is at the same strain; this assumption is true **provided there is no loss of adhesion between fibre and matrix** (i.e., there is no **delamination**). Note, however, that for the fibre to stretch and elongate, the cross sectional area must decrease and, if there is **no delamination**, the cross sectional area of the fibre cannot decrease without deforming the matrix as well.

Assuming the whole composite is at the same strain, then $\epsilon_c = \epsilon_m = \epsilon_f = \epsilon$. Applying Young's law to both matrix and fibre, we have

$\sigma_f = E_f.\epsilon_f = E_f.\epsilon$ for the **fibres**
$\sigma_m = E_m.\epsilon_m = E_m.\epsilon$ for the **matrix** and
$\sigma_c = E_c.\epsilon_c = E_c.\epsilon$ for the **composite.**

Therefore, from equation (9.7)

$$E_c.\epsilon = E_f.V_f.\epsilon + E_m.(1 - V_f).\epsilon$$

thus

$$E_c = E_m + V_f.(E_f - E_m).$$

It is important to note that the fibres and matrix must be compatible. The analysis illustrates that, **for maximum effect,** $E_f >> E_m$. Therefore the reinforcing properties of the fibres depends upon the Young's modulus of the component materials, provided there is no separation of the fibre from the matrix (i.e., no delamination). One of the ways that delamination can occur is by shrinkage of the fibre relative to the matrix. During manufacture of the composite, therefore, it is most usual for the fibres to be solid and the matrix liquid, so that the liquid component will flow around and "bond" onto the solid fibres by the contraction occurring during the liquid to solid phase change. To eliminate delamination, the matrix should shrink onto the fibre. This is how the first plastic composite (Bakelite) **(13.5.2)** was produced in the 1920s. Here a thermosetting plastic (phenol formaldehyde) and fibres were heated together in a hydraulic press. Under the combined heat and pressure good adherence was obtained.

From the foregoing mathematical analysis, it can be seen that the most important aspect of increasing the strength of the composite is to have the fibres aligned along the major tensile axis. In this respect short fibres are very much less effective in acting as a reinforcement.

9.16 REFERENCES

1. BS EN 572: 1995. Glass in Building. Basic soda lime silicate glass products. Part 1. Definitions and general physical and mechanical properties.
2. BS EN 10002. British Standard for the testing of metallic materials. Part 1: 1990. Method of test at ambient temperatures.
3. BS EN ISO 3167: 1997. Plastics - Multipurpose-test specimens (= BS 2782: Part 9: Method 931A: 1993 = ISO 3167: 1993)
4. BS 6206: 1995. Performance requirements for flat safety glass and safety plastics for use in buildings.
5. BS 6262. Code of practice for glazing in buildings. Part 4: 1994. Safety related to human impact.
6. HALLIWELL, S.M. (1999) *Advanced polymer composites in construction.* BRE Information Paper IP/99. Garston, BRE.
7. HALLIWELL, S.M. (1999) *Architectural use of polymer composites.* BRE Digest 442. Garston, BRE.

10

METALS AND ALLOYS

10.1 INTRODUCTION

The structure of metals and alloys has been considered previously (**1.16**), as have stress-strain relationships (**3.2.2**) and hardness tests (**3.4.1**). **Chapter 10** focuses on the mechanisms and processes which govern strength, with particular emphasis on steels. We refer to **ferrous** metals and alloys (those containing iron, Fe), most importantly steels, and **nonferrous** metals and alloys (which contain no iron), e.g., brass (copper and zinc), solder (lead and tin), etc. Alloys have the same fundamental structure as metals (**1.16**); however, pure metals solidify at one temperature (the melting point) whilst alloys solidify over a temperature range. Metals and alloys have a number of properties that make them very useful as engineering materials

- They are generally strong in both tension and compression, usually having a high strength:weight ratio. They also conduct electricity;
- They may be readily **fabricated**, i.e., they can be readily worked and shaped, and they retain this new shape permanently. They may be **stiffened** by corrugating, curving or by making the metal into right angled pieces which increase the mechanical properties of the formed metallic product. In this form they resist **buckling** and **bending**. Using this method of stiffening, thin sheet steel can be made to have mechanical properties equivalent to a much thicker planar section. This arises because metals have the same mechanical properties in compression as they do in tension. Metals and alloys gernerally keep their formed shape up to fairly high temperatures, whereas other materials (such as thermoplastics) will revert to their original preformed shape (**13.6.1**). Metals acquire their 'new' shape under high stresses but with little 'spring-back' (**3.2.2**), a beneficial characteristic when forming complex panel shapes;
- They can be used in tension and undergo a strengthening process when deformed (**10.5.3**). Metals and alloys are the only materials which have this unique property;
- Metals resist the effects of **stress concentrations** (**10.14.1**) fairly well.
- Ferrous and nonferrous alloys cooled under **non equilibrium** conditions produce desireable engineering properties. As we will see later, ferrous alloy elements alter the position of the 'nose' of the non equilibrium TTT curve with respect to the cooling rate curves (**10.12**) whereas, in nonferrous alloys, non equilibrium cooling produces a mechanism to refine the precipitate dispersion (**10.5.2**).

The main problem with metals and alloys is that they may suffer corrosion (**Chapter 11**), creep (**3.5.4**), fatigue (**3.5.5**) and brittle fracture (**10.14.1**).

10.2 SOLIDIFICATION

The general process of solidification of a solid from a liquid is examined in **Chapter 2** (**2.6**). Pure metals grow by the formation of **dendrites** (**2.6.3**) (although these are not revealed in their microstructure, **10.3**), whilst alloys reveal dendritic growth only if the temperature drops so that equilibrium is not maintained (**10.6**; see ***Figure 10.3e***). The process is summarised in ***Figure 10.1***.

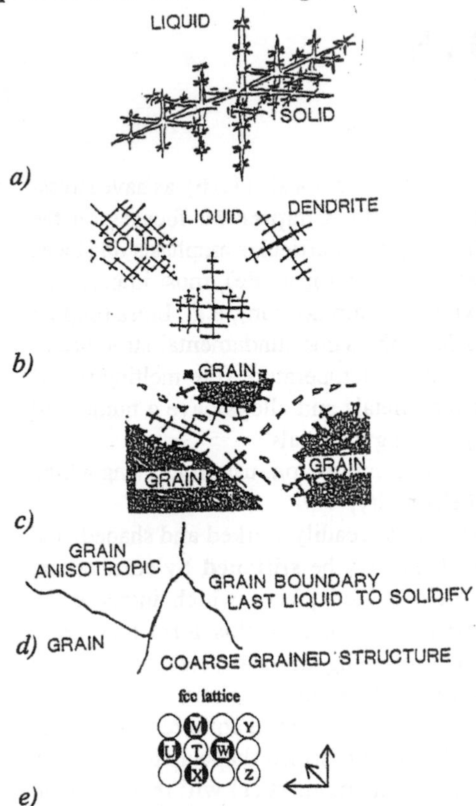

Figure 10.1a shows the first solid nucleus being formed within a liquid environment. This crucifix form of solidification is called a **dendrite** (***Figure 2.8b,c,d***). The solid grows in this form because the latent heat of solidification can be removed away from the solid phase along close packed directions (given by **UTW** and **VTX** in ***Figure 10.1e***).

Several nuclei (**2.6**) are formed with different orientations to one another. ***Figure 10.1b*** shows the different orientations of the dendrites formed at different nucleating sites.

Eventually, continued dendritic growth fills in the space between adjacent dendrites of the solid phase to form **grains** of the solid phase within the liquid phase ***Figure 10.1c***).

On complete cooling, the crystalline solid phase is formed, comprising closely packed grains of differing orientation (dependent on the orientation of the original nuclei). Within each grain, the atomic arrangement is regular (**1.15**). However, there is a **discontinuity** between adjacent grains at the **grain boundary** ***(Figure 10.1d***).

The grain boundary is important in restricting the movement of **dislocations** (which account for the transmission of shear stresses in metals, **10.4**). Dislocations only occur in crystalline materials.

***Figure 10.1** Solidification of a solid metal from a liquid*

10.3 MICROSTRUCTURE AND MACROSTRUCTURE

If it were possible to remove a grain from a metal, then the idealised grain would look similar to ***Figure 10.2a***. If the metal were all one orientation such that the grain constituted the **whole** of the metal, then the metal would be a **single crystal**. However, single crystals are rare and very difficult to produce (and are therefore very expensive). Normally, engineering metals consist of differently orientated regions or grains, i.e., they are **polycrystalline**, consisting of many grains each having a different crystalline orientation to its neighbour. It is possible to regrow and refine grains in an existing metal or alloy; this is the basis of strengthening metals, where small differently orientated grains are introduced, increasing the number of grain boundaries available to restrict the movement of dislocations. ***Figure 10.2b*** shows the grain boundaries within a sectioned and polished metal formed from equally sized grains. Note that the point of intersection **is a triple point** (a meeting of **four** boundaries at this point can never occur).

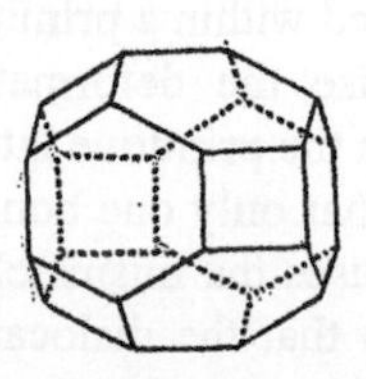

(grain completely removed from metal)

a) Grain shape

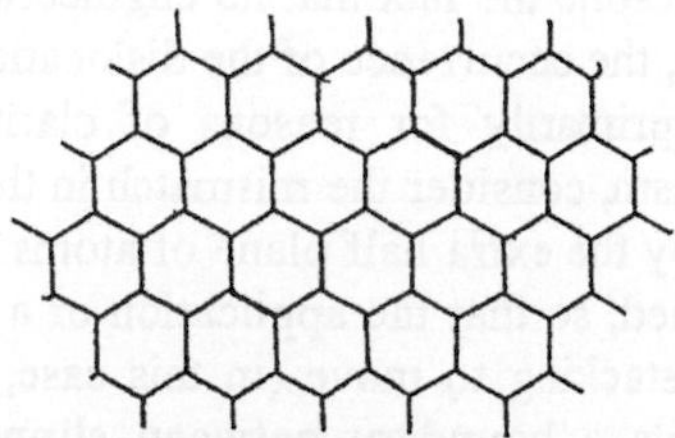

(equi-axed grains; idealised representation of full anneal)

b) Grain boundaries

Figure 10.2 *Idealised grain shape and structure of metals*

The structure of the metal or alloy is dependent upon a variety of factors, including composition, cooling, etc. These processes give rise to different structures which can be observed under the microscope by first obtaining **metallurgical sections**. ***Figure 10.3*** shows examples of metallurgical sections obtained in metals and alloys (the colour contrast is obtained as different phases etch and/or corrode differently). We will refer to the compounds noted throughout Chapter 10.

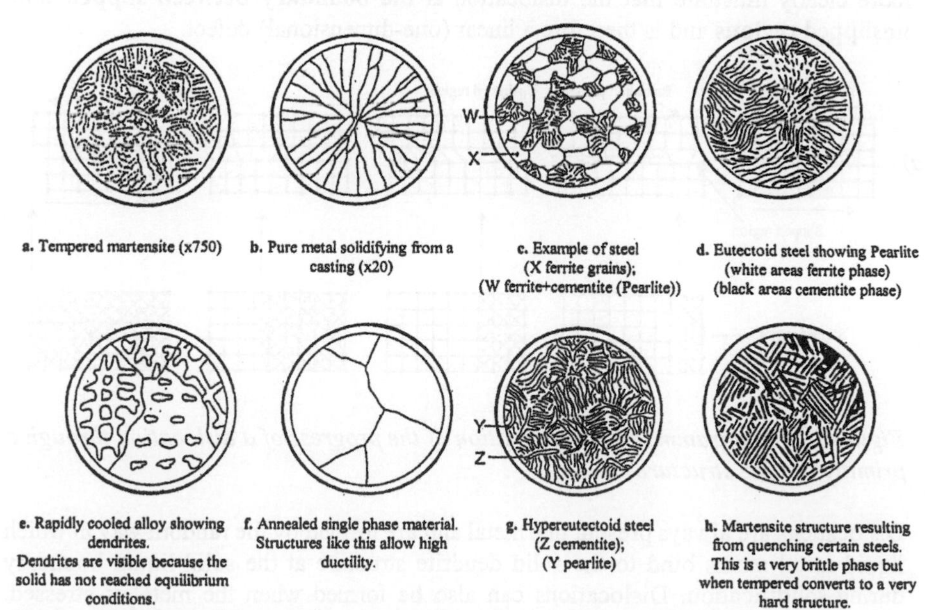

Figure 10.3 *Typical metallurgical sections observed through the microscope*

10.4 DEFORMATION

The process of deformation is modelled as the movement of dislocations, which are line defects caused by a mismatch in the atomic stacking in the metal lattice structure.

Dislocations are defined as the **boundaries between slipped areas of a crystal and unslipped regions**; they are **linear (one-dimensional) defects** which **cannot** cross the grain boundary. The **four** possible structures of a metal lattice are primitive (P), face centred cubic (fcc), body centred cubic (bcc) and hexagonal close packed (hcp) (***Figure 1.19***). Despite the fact that no engineering metals have the primitive lattice structure (**1.15.2**), the occurrence of the dislocation is most easily portrayed within a primitive lattice (primarily for reasons of clarity). In order to visualise the deformation mechanism, consider the mismatch in the atomic stacking within the primitive lattice shown by the extra half plane of atoms in ***Figure 10.4a***. Note that only one bond is unsatisfied, so that the application of a small shearing force causes the mismatch in atomic stacking to move (in this case, to the right). Note also that the dislocation represents a boundary **between slipped and unslipped regions**, conventionally represented by the symbol ⊥ (positive) or ⊤ (negative). The dislocation illustrated in the primitive lattice in ***Figure 10.4a*** is known as an **edge dislocation** and is viewed from the side. ***Figure 10.4a*** shows the stress field around the dislocation within the primitive lattice and how the atomic mismatch resulting from the extra half plane of atoms alters as the deformation passes across the lattice. This portrayal of the dislocation should be carefully interpreted as it appears that the dislocation is a two dimensional defect, whereas, as previously stated, it is one dimensional. Therefore, if we draw a 'bird's eye' view of the deformation, as in ***Figure 10.4b***, then the shaded area is the area where deformation has taken place. Seen together, ***Figures 10.4a*** and ***10.4b*** more clearly illustrate that the dislocation is **the boundary between slipped and unslipped regions** and is therefore a linear (one-dimensional) defect.

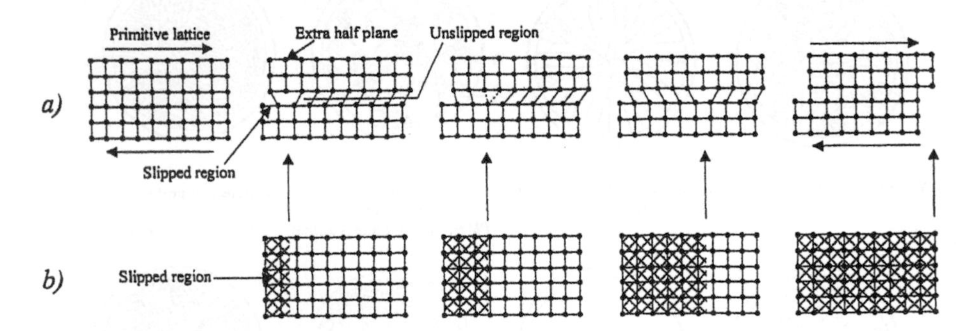

Figure 10.4 *Diagrammatic representation of the progress of a dislocation through a primitive lattice structure*

Dislocations are always present in a metal and are formed by the random way in which liquid metal atoms bind to the solid dendrite structure at the solid-liquid boundary during solidification. Dislocations can also be formed when the metal is stressed. Movement of dislocations through the lattice occurs only on certain atomic planes, known as **slip planes**. There are 8 slip planes in bcc lattice structure and 12 slip planes in the fcc lattice structure; in these structures, the slip planes intersect. In the hcp lattice structure there are only 3 slip planes and these do not interect; hcp metals are therefore relatively soft and ductile.

In the modelling of the movement of the dislocation line through a lattice structure, the dislocation is given a 'sense' (similar to the way that scientists give a 'sense' to a magnetic field). Hence we can refer to **positive and negative** edge dislocations, shown in ***Figure 10.5a*** and ***10.5b*** respectively. Comparing the result of the passage of positive and negative dislocations through the lattice structure, we can see that the net result of a positive and a negative edge dislocation is the same. If these dislocations occur in the same slip plane, they interact and annihilate one another (***Figure 10.5c***).

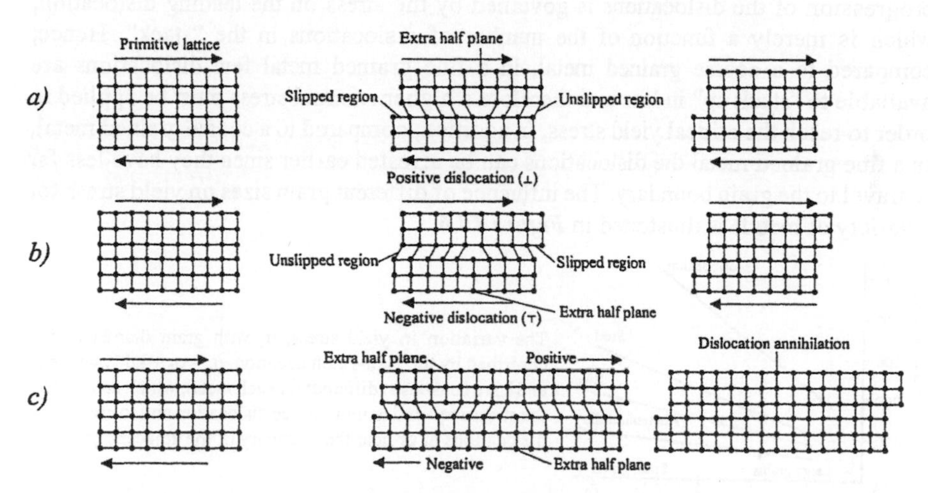

Figure 10.5 *Positive and negative edge dislocations and their interaction*

In metallic structures (such as the fcc and hcp lattice structures), dislocations are modelled in relation to their **Burgers vector** with respect to the sensed dislocation line. This is shown for the operation of the **Frank Read Source (FRS)** interpretation of the modelling of dislocations in alloys in section **10.8.**

10.5 MECHANICAL PROPERTIES

The engineering strength of a metallic material is dependent upon both **microscopic** and **macroscopic** structure **(10.3)** and upon **composition**. The structure of a metallic material and hence it's properties can be altered by several techniques, which can either be used singly or in combination. These techniques can basically be summarised as

- effect of **grain size** upon yield stress **(10.5.1)** and the effect that the alloy element has upon grain size and intermediate (intermetallic) second phase production;
- **alloying (by substitution** or **interstitial hardening) (10.5.2)**;
- **intermediate (intermetallic) second phase production (10.5.2)**;
- **strain (work) hardening (10.5.3)** by dislocation interaction **(10.4)** with or without **intermediate (intermetallic) second phases**;
- **heat treatment** and the effect on grain size of temperature **(10.5.4)**.

All these methods act to make dislocation movement more difficult, so increasing the yield stress **(3.2.2)** of the material. If the material requires a higher stress to produce yield, then the safe working stress is correspondingly increased.

10.5.1 Grain Size

It was noted above that within a single grain the atomic arrangement is regular and that at the grain boundary the arrangement becomes more irregular. In addition, it was noted that the dislocation cannot cross the grain boundary (until the shear stress reaches a critical value of sufficient magnitude to allow progression along a new slip plane in a new slip direction). Hence the movement of dislocations along the same slip plane is arrested at the grain boundary, where they "stack up" in line. The critical value for progression of the dislocations is governed by the stress on the leading dislocation, which is merely a function of the number of dislocations in the "stack". Hence, compared to a coarse grained metal, in a fine grained metal few dislocations are available to "stack up" in line and therefore a higher external stress must be applied in order to reach the critical yield stress. In addition, compared to a coarse grained metal, in a fine grained metal the dislocations can be arrested earlier since they have less far to travel to the grain boundary. The influence of different grain sizes on yield stress for a variety of metals is illustrated in ***Figure 10.6***.

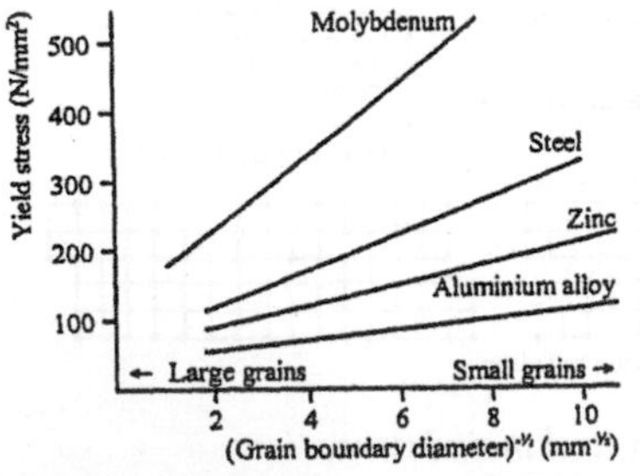

The variation in yield stress, σ_y with grain diameter, d is described by the Hall-Petch equation, $\sigma_y = \sigma_0 + kd^{-1/2}$, where σ_0 and k are constants, different for each metal, which account for the stress required to move a free dislocation and to create new dislocations at or near the boundary of the adjacent grain.

***Figure 10.6** Influence of different grain sizes on yield stress for a variety of metals*

For steels, **Table 10.3** gives a summary of how the factors depicted in ***Figure 10.6*** can be used to strengthen steel and **Table 10.4** gives the site of action of these factors and the effect on the properties of the steels. Note that all mechanisms by which metals can be strengthened (**10.5** above) fundamentally affect grain size and hence influence the yield stress, as shown by the relationships in ***Figure 10.6***.

10.5.2 Alloying and Second Phase Production

Metals and alloys all have their own structure (**1.16**) and this structure comprises similar sized atoms of similar chemistry. For simplicity, two metals are considered to form the alloy, the host metal **A** (the solvent, shown as white spheres in ***Figure 10.7***) and the foreign metal **B** (the solute, shown as black spheres in ***Figure 10.7***). The atoms of both metals are considered as hard spheres. The bond formed between one atom and its neighbour (the metallic bond, **1.10**) is generally indifferent to the types of metallic atoms forming the bond, provided

- the radius (**Table 1.5**) of the foreign metal **B** is not too different from the host metal **A**, so that there is no large scale distortion at the solute atomic sites or between the rows of atoms, whilst being just sufficient to cause an increase in the strength of the alloy by preventing one plane of atoms slipping over the adjacent plane (***Figures 10.4*** and ***10.5***) (an upper limit of 15% is the generally accepted maximum difference in radii of host and foreign metals);

- the structure of pure metal **A** and pure metal **B** are the same (i.e., **both** are either fcc or hcp or bcc);
- the solution resulting from dissolving metal **A** into metal **B** produces a **decrease in energy**.

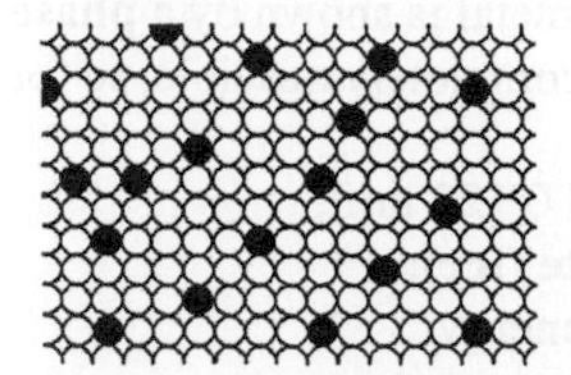
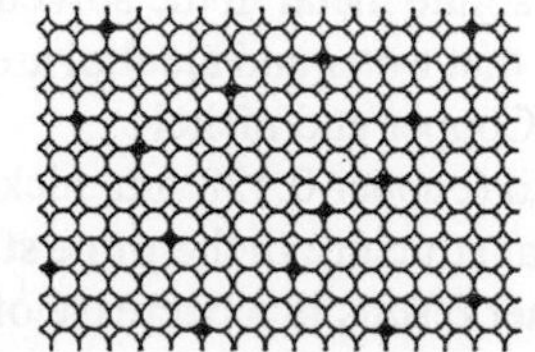
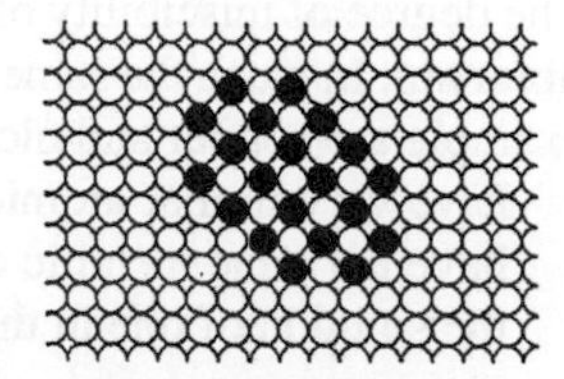

a) Solid substitutional hardening *b) Interstitial hardening* *c) Solid phase hardening*

***Figure 10.7** Alloying processes*

Where the atomic radii are very similar and both **A** and **B** have the same structure (e.g, fcc), then the metal atom **B** will form a random replacement of **A** atoms within metal **A**'s structure. This is called a **complete solid solution** (***Figure 10.7a***). When an atom of the foreign metal **B** replaces another in the host lattice of metal **A**, there is an energy imbalance which requires the structure of the host metal to change. The energy imbalance is primarily dependent upon the difference in the atomic radii of metal **A** and metal **B**. If the atomic radii are within 60% of one another, then the foreign atoms **B** go into the **free space (1.16)** within the structure of the host metal **A**. This is called an **interstitial solid solution** and is shown in ***Figure 10.7b*** (where the small black circles show interstitial solid solution of solute atoms). Since the commercially important metals range from cobalt (atomic radius 0.125 nm) to lead (atomic radius 0.175 nm), it can be seen that the atoms which can go in to interstitial solutions in these metals must have atomic radii less than 0.075 to 0.105 nm. This effectively limits the possibilities to the first six elements in the Periodic table (**Table 1.2**), hydrogen to sulfur. This range includes carbon, which forms particulary important interstitial solutions in iron to produce steels. In the iron-carbon (Fe-C) system, the very small carbon atoms are only 60% of the size of the iron atom and so are capable of filling the vacant sites between them. The iron-carbon system is considered in more detail in section **10.9**. If the difference in radii (and/or structure) between **A** and **B** is too great (>15% but <60%), then the solution formed has a new structure, called an **intermediate (intermetallic) second phase** and shown in ***Figure 10.7c*** (where the black solute atoms **B** are grouped together within the solute atoms **A**). The valency of the metals **A** and **B** is a good indicator of the propensity to form intermediate second phases. Where the valencies of the metals differ markedly, there will be a tendancy for form intermediate second phases. The second phase compounds are normally formed at, or near, the compositions corresponding to a simple ratio of components, such as **AB** or $\mathbf{AB_2}$ (similar to the complex ionic structures, **1.18.4**) and may exist over a compositional range around these simple ratios. Intermediate second phases often have a crystal structure which differes from that of either of the parent metals **A** or **B**.

Where the second phase has the same lattice spacing of the host structure, it is called a **coherent precipitate** (***Figure 10.19c***), but where it grows big enough to have its own lattice spacing it is called an **incoherent precipitate** (***Figure 10.19d***). The advantage of the incoherent precipitate over the coherent precipitate is the ability to arrest dislocations.

The degree of miscibility of the solute metal in the solvent metal is shown by a **phase diagram.** In much the same way that water and alcohol are completely miscible, so for example are copper and nickel. Copper and nickel

- have very similar atomic radii (copper 0.128 nm, nickel 0.125 nm);
- have the same metallic crystal structure in the **pure state (fcc)**;
- the solution of one in the other results in a decrease of energy.

Copper and nickel therefore form a **complete solid solution.**

In some metal solute-solvent systems, there is very little solution of one in the other (copper and zinc, copper and aluminium, lead and tin), in some cases despite the fact that both metals have the same crystal structure (e.g, both copper and aluminium are fcc). This arises primarily because of the differences in atomic radii of the metals (e.g, copper 0.128 nm and aluminium 0.143 nm) and leads to restricted areas of miscibility. These metals therefore form a **restricted solid solution**; the two metals "fight" over the structure to be achieved by the solution of one metal in the other. The result is that neither wins, except at low concentrations of one in the other, and a range of different competing crystal structures (intermediate second phases) are produced which are dependent upon the composition. As already noted, intermediate second phases contribute to the strength of the alloys formed.

Figure 10.8 compares the phase diagrams for a complete solid solution (e.g, pure copper and nickel) and for a restricted solid solution (say, a bcc and fcc metal). In these diagrams, the single phase regions are conventionally denoted by a Greek symbol (alpha **α**, beta **β**, gamma **γ**, etc.). If the composition is such that the single phase regions cannot be formed, then there exists a two-phase region (rather like the two phases of oil and water, which are immiscible). These intermediate two phase regions are designated (α + β), (β + γ), (δ + γ), etc. Two phase intermediate regions are very important because the second phases present act as "obstacles" to the movement of slip planes relative to each other, and hence to metallic deformation (**10.4**). Single phase structures are easily deformable. Two phase structures are not so easily deformed.

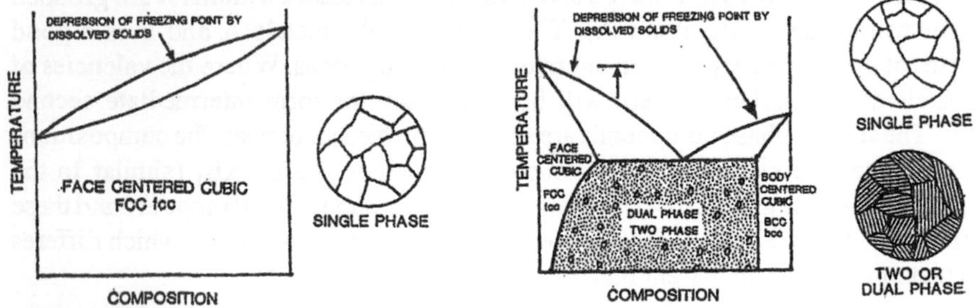

a) Complete solid solution (e.g, Cu-Ni) *b) Restricted solid solution (e.g, bcc-fcc)*

Figure 10.8 *Comparison of typical phase diagrams*

Figure 10.8a shows an example of a phase diagram showing **complete solid solution.** The solid solution region is represented as the **α** (alpha) structure. This is in fact the structure of pure copper and pure nickel. To form a complete solid solution, both metals have to have the same structure. In the fcc lattice of the pure copper and nickel metals, there is **random** replacement of the copper atoms by nickel (or vice versa). The microstructure is also shown for the range of compositions below the phase diagram.

Figure 10.8b shows an example of a phase diagram showing **restricted solid solution** and in this case the structure of the two metals is important. The shaded region represents the two phase region i.e., a mixture of **α** with a **fcc** structure and **β** with a **bcc** structure. The two phase region results because the structures of the two metals are different and each alloy element wants its own structure (i.e., they are **insoluble** in each other). The microstructures produced are also shown for the specific phases below the phase diagram, where the single phase and two phase (**duplex**) structures are illustrated. The lines on the single phase microstructures are grain boundaries. The parallel lines on the two phase structures are the **α** (white) and the **β** (black) lamellae. As single phase structures are easily deformable whereas two phase structures are not so easily deformed, alloys are usually heated up until a single phase region is achieved prior to rolling.

Other phase diagrams are more complex, as in the case of brass (an alloy of copper and zinc). The phase diagram for brass is shown in ***Figure 10.9***, and is a **restricted solid solution.**

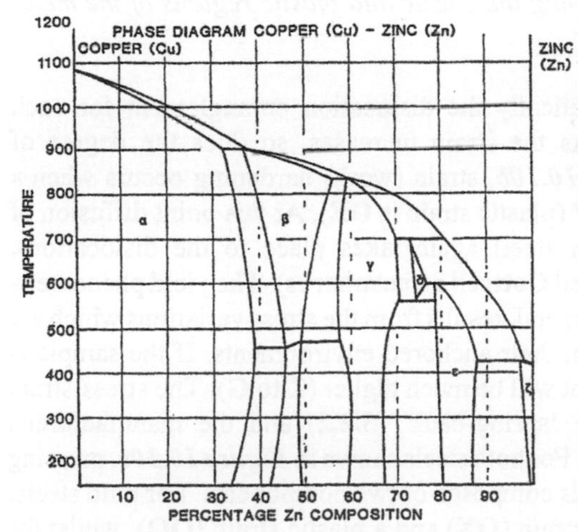

The structure of the two metals copper (Cu) and zinc (Zn) are different (fcc and hcp respectively). The single phase (one structure) regions, (alpha) **α**, (beta) **β**, etc., are shown in ***Figure 10.9***. Compositions between these single phase regions are two phase regions, e.g, (**α** + **β**), etc. Note that there are several single phases throughout the changes in composition, i.e., **α** (the structure of pure copper), **β**, **γ**, **δ**, **ε** and **ζ** (the structure of pure zinc); **β**, **γ**, **δ** and **ε** phases have different (complex) crystal structures to the metals from which they were formed. The effect of this can be seen, for example, with a brass alloy of composition 60% Cu, 40% Zn, which at room temperature has a two-phase structure (**α** + **β**) which is not easily deformed. It has to be heated up to 800°C before it achieves a single (**β**) phase whereupon it can be more easily worked.

Figure 10.9 Phase diagram for brass (the Cu-Zn system)

10.5.3 Strain (Work) Hardening

The typical stress-strain behaviour of metals is examined in section **3.2.2**, wherein it was noted that as a metal is extended its cross-sectional area decreases but the metal gets stronger. This phenomenon is known as **strain (work) hardening.** In bcc and fcc metals, slip planes frequently intersect one another so that dislocations in one slip plane act as a barrier to dislocations trying to move across them. Strain hardening occurs as a direct result of a dislocation source requiring a higher stress to operate in overcoming this barrier. This arises from the stress fields around the dislocation lines, which can be illustrated most easily in the case of the edge dislocation (***Figure 10.5b***), where the region around the extra half plane of atoms is in compression whilst the region below the extra half plane of atoms is in tension (this is shown in ***Figure 10.5b*** by the variation in interatomic distances). The complex interaction of dislocation lines places a "back stress" on the dislocation sources, which must be overcome in order for the dislocation to continue operating and thus produce further dislocations. This "back stress" increases with both **dislocation density** (i.e., the number of dislocation lines per unit volume) and the **degree of dislocation entanglement**, both of which increase with strain (***Figure 10a***). The dislocation density and degree of dislocation entanglement are related to the properties of the stress-strain curve in ***Figure 10.10***.

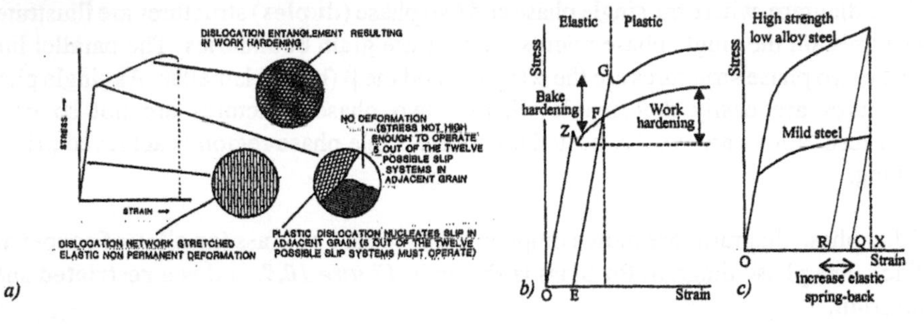

Figure 10.10 *Stress strain curve showing the elastic and plastic regions of the metal and the nature of the dislocations*

Figure 10.10a illustrates diagrammatically the dislocation entanglement for each section of the stress strain curve. As the strain increases, so does the degree of dislocation entanglement. In ***Figure 10.10b***, strain (work) hardening occurs when a metal sample is stressed to the point **F** (plastic strain is **OE**). At this point diffusion of solutes (e.g, carbon and nitrogen in steel) again takes place to the dislocations, anchoring them (these regions are called **Cottrell atmospheres**). The yield phenomena at **Z**, which is a characteristic of a bcc metal, results from the stress variations which are required to strip the dislocations from their anchored environments. If the sample is retested some time later, the yield point will be much higher (**Z to G**). The stress-strain curve can also be used to illustrate 'spring-back' (**3.2.2**) and the manufacturer's requirement for uniform quality steels. For both steels shown in ***Figure 10.10c***, pressing requires a plastic strain of **OX**, which is composed of two components. For mild steels, these two components are an elastic strain (**QX**) and a plastic strain (**OQ**), whilst for steels with a different yield strength, an elastic strain (**RX**) and a plastic strain (**OR**). The difference in the 'spring back' between the two steels is **RQ**, and this may give rise to problems, for example, in the alignment of components comprising these two steels.

10.5.4 Effect of Heat on a Strain (Work) Hardened Metal

In a strain (work) hardened metal, each dislocation produces a region of higher strain in the atomic lattice which is not thermodynamically stable. Consequently, comparatively little energy is required to cause a redistribution and cancellation of the dislocation arrays, destroying the beneficial strength effects of strain (work) hardening. This energy is most conveniently provided by heat, and the quantity of heat supplied influences the overall effect. The response of the strain hardened metal to heat is classified into three categories

- **recovery**;
- **recrystallisation**; or
- **annealing**.

The relationship between these categories as the temperature of a strain (work) hardened metal is raised and the recovery, recrystallisation and melting points of various metals and alloys is shown in ***Figure 10.11***.

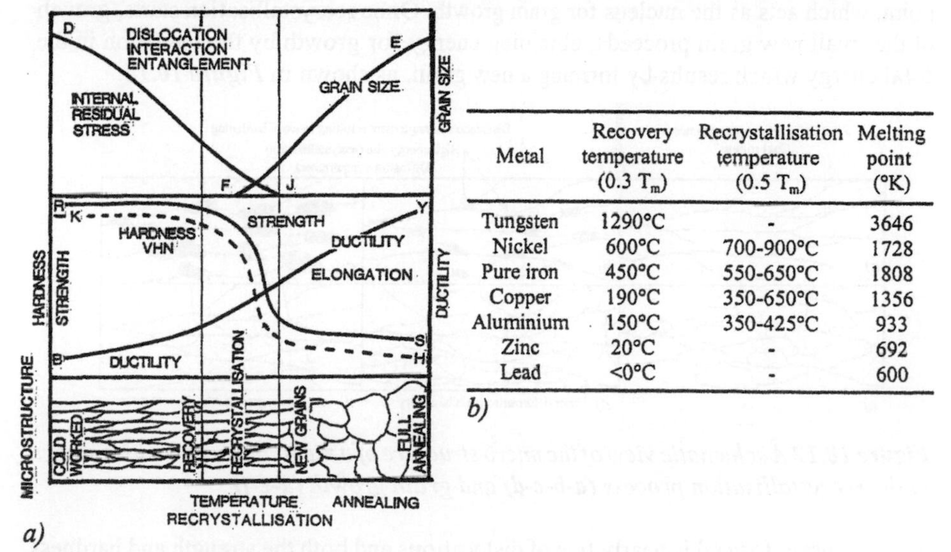

Metal	Recovery temperature ($0.3\ T_m$)	Recrystallisation temperature ($0.5\ T_m$)	Melting point (°K)
Tungsten	1290°C	-	3646
Nickel	600°C	700-900°C	1728
Pure iron	450°C	550-650°C	1808
Copper	190°C	350-650°C	1356
Aluminium	150°C	350-425°C	933
Zinc	20°C	-	692
Lead	<0°C	-	600

***Figure 10.11** a) Effect of temperature on properties of a strain (work) hardened metal; b) Recovery, recrystallisation and melting points of various metals and alloys*

Figure 10.11a illustrates that, as the temperature of a strain (work) hardened metal is raised, several of the material parameters alter. Dislocation entanglement, strength and hardness all decrease with increasing temperature, whereas grain size, ductility and elongation to fracture all increase with increasing temperature.

Recovery

Heating the strain hardened metal to above a certain temperature (about **$0.3\ T_m$**, where T_m is the melting point in °K) produces a process known as **recovery** in which the dislocation entanglements in certain areas take up a lower energy state. The degree of dislocation entanglement in certain areas therefore decreases and the structure becomes 'coarser'. However, the dislocation density remains relatively unchanged, so that the overall strength and hardness of the metal remain almost unchanged (***Figure 10.11***). In addition, the internal stresses are reduced whereas the grain size is unaltered by the recovery process. In consequence, recovery is an important process for releasing the internal stresses in forged, welded and fabricated equipment, for example, without lowering the strength acquired during **cold working (10.5.5)**. Recovery temperatures for common metals are shown in ***Figure 10.11b*** with respect to their melting points, T_m.

Recrystallisation

Heating the strain hardened metal above a certain temperature, called the **recrystallisation temperature** (approximately **0.5 T_m**), causes a further reduction in degree of dislocation entanglement but, more importantly, an alteration in the **grain boundary angles** caused by the formation of new grains. Therefore the alloy 'softens' and its hardness decreases as atomic diffusion is now possible. The structure **recrystallises** using the intersection of three grains (***Figure 10.2***), known as the triple point, which acts as the nucleus for grain growth. Once recrystallisation starts, growth of the small new grain proceeds, obtaining energy for growth by the reduction in the **total** energy which results by forming a new grain, as shown in ***Figure 10.12***.

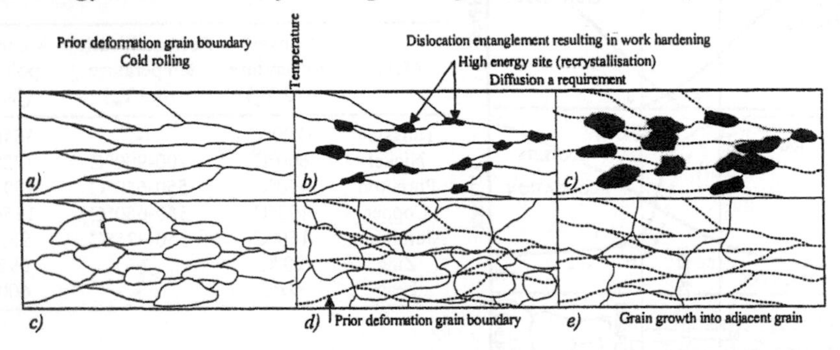

Figure 10.12** A schematic view of the micro structure of a metal through various stages in the recrystallisation process **(a-b-c-d)** and grain growth **(d-e-f)

The new grain formed is nearly free of dislocations and both the strength and hardness decrease during the process because there are fewer obstacles to dislocation movement. The recrystallisation temperatures for common metals are shown in ***Figure 10.11b*** with respect to their melting points, T_m.

Annealing

If the alloy is heated above the recrystallisation temperature, a process of **annealing** takes place. Annealing is a process which markedly continues to soften the alloy and allows an increase in the rate of grain growth, and results in a complete redistribution of dislocations such that some of the grains grow bigger whilst the stresses necessary for deformation to take place become less. Annealing, therefore, completely obliterates the prior history of the metal (except where insoluble phases are deposited, ***Figure 10.12f***). The process of annealing is thus an extension of the recrystallisation process as some of the smaller grains are engulfed into the larger grains.

10.5.5 Working and Manufacturing Processes

Figure 10.13 shows an abbreviated manufacturing route for British Steels products. These routes are typical manufacturing processes for the steels considered in this text.

dependence of the occurrence of one and two phase structures. As an example, the phase diagram for **solder**, an alloy of lead Pb and tin Sn, is illustrated in ***Figure 10.16b*** and the cooling curve for composition **X** in ***Figure 10.16a***. Plumber's solder (used for wiping joints) has a composition of 50% tin 50% lead (composition Z), whilst an electrician would use free-running solder (63% tin, 36% lead and 0.6% antimony Sb). Note that solders containing antimony (Sb) should not be used on zinc and galvanised components, because antimony and zinc form a brittle compound. Antimony is used to cheapen solders, reducing the amount of tin.

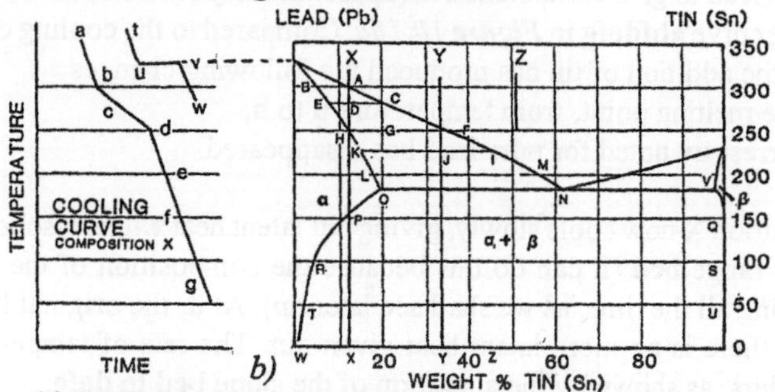

***Figure 10.16** The lead-tin phase diagram. a) Cooling curves for composition **X**; b) Lead-tin phase diagram*

Note that, in reading the phase diagram, letters have been used to highlight positions of interest (discussed below). These are upper case on the phase diagram (***Figure 10.16b***) and lower case on the cooling curve (***Figure 10.16a***). To highlight a **temperature** of interest, the phase diagram or the cooling curve can be used. For example, the temperature of 300°C is shown on the phase diagram (***Figure 10.16b***) as the horizontal line **AB** and on the cooling curve for composition **X** (***Figure 10.16a***) as point **b**. We can therefore denote the temperature ss T_A, T_B, T_{AB} or T_b. Similarly, we can refer to a temperature of 250°C, given on the phase diagram as the horizontal line **HGF** and on the cooling curve as point **d**, as T_H, T_G, T_F, T_{HGF} or T_d. To highlight a **composition**, the phase diagram is used. For example, the composition of 20% tin can be represented by the points **B, G** or **O**.

Also, note that the melting point of pure lead is 327°C, which is represented in the cooling curve (***Figure 10.16a***) by the horizontal line **uv**. Thus the melting point of pure lead can be denoted T_u, T_v or T_{uv}.

The lead-tin phase diagram is a restricted solid solution type (**10.5.2**) and shows

- two solid solution single phase regions, which are designated as **α** and **β**; and
- a single two phase region, designated (**α + β**), between the composition **R** and **S** at 100°C.

As the temperature decreases, the solubility of both tin in lead and lead in tin decreases. The restricted solid solution comes about because of the differences in both the atomic radii of tin (0.140 nm) and lead (0.175nm) (approximately a 25% difference in atomic radii) and structure of the two metals. The structure of tin is complex (it can form two crystal lattice structures, one a cubic lattice, the other a tetragonal lattice, **1.15.3**); lead has the fcc lattice structure.

The cooling curve for pure lead is shown by **tuvw** in ***Figure 10.16a***. A temperature arrest at the melting point of pure lead is shown by **uv**; the temperature would not decrease further until all the liquid phase had changed to the solid phase. It is the latent heat evolved which produces this horizontal line **uv** (**2.5**). Now consider the effect of adding some tin to the lead to give composition **X** (approximately 12% tin, 88% lead). This gives the cooling curve **abfdefg** in ***Figure 10.16a***. Compared to the cooling curve of pure lead (**tuvw**), the addition of tin has produced the following changes

- a depression in the melting point, from temperature **u** to **b**;
- the temperature arrest **uv** noted for pure lead has disappeared.

The solder of composition **X** now cools slowly, giving out latent heat whilst it solidifies over the temperature range **bcd** (it can do this because the composition of the solid being formed is altering all the time, as we shall see later on). At **d**, the original liquid has all solidified, so there is no more latent heat given out. The rate of temperature decrease therefore alters, as shown by the alteration of the slope **bcd** to **defg**.

Cooling curves of this type for a series of tin-lead alloys of different compositions are used to construct the lead-tin phase diagram. Points like **b** (the temperature at which solidifcation **starts**) and **d** (the temperature at which solidification is **complete**) are plotted on the temperature-composition plot. For example, for composition **X**, this gives points **A** and **H** on the lead-tin phase diagram. If all the temperatures at which solidification starts are joined up, this gives the **liquidus** line **ACFIMN**. Similarly, if the temperatures at which solidification is complete (i.e., the temperatures at which all the liquid has been changed to solid) are joined, this gives the **solidus** curve **BEHKLO**. The region between the line **VOW** and the left vertical temperature axis represents the solid solution of tin in lead, designated as the **α** phase, whereas the similar region on the right end of the diagram represents the solid solution of lead in tin, designated as the **β** phase. The limit of solubility of tin in lead is represented by the point **O** (approximately 19% tin) and that of lead in tin by the point **V** (approximately 96% tin or 4% lead). Since the solubility decreases with temperature, the composition of these two solid solutions will tend to approach that of the pure solvent metal at low temperatures, as indicated by the character of the lines **OW** and **VU**.

In order to examine how the phase diagram may be used to describe the cooling and subsequent solidification of a metal, consider the following scenario. Composition **X** (approximately 12% Sn, 88% Pb) will be cooled from the melt at high temperature to room temperature in specific temperature drops. As the temperature is reduced, there will be a difference in the rate of cooling as soon as the metal starts to solidify, as latent heat is given out. This takes place at temperature T_A and the first solid to solidify is of composition **B.** If the temperature now drops to T_D, the solid deposited is of composition **E** whilst the liquid is of composition **C**. The composition **E** of the alloy deposited at temperature T_D has the structure of lead, with random replacement of lead by tin. For the solid composition **E** to occur, the solid deposited at the higher temperature T_A (composition **B**) must **alter** and become richer in tin. Hence **diffusion** must take place. Scientists describe the **equilibrium state** as that state where the **solid and liquid have the compositions described by the phase diagram.** Equilibrium conditions therefore imply that the **temperature is considered held at the temperature points under discussion for a very long time** until diffusion is complete.

At this point it would appear that we are getting "extra" tin from somewhere, because now the liquid is richer in tin than when it started. This apparent anomaly is explained by the fact that the solid is **less** rich in tin than the liquid with which we started. We can determine the percentage quantities of solid and liquid using the **Lever rule (10.6.1)**. If the temperature now drops to T_H, all the solids previously deposited must change to become composition **H** and so, again, diffusion has taken place. All the remaining liquid at the reduced temperature T_H is of composition **F**. Again we can determine the relative amounts of the solid and liquid using the Lever rule. This procedure can be followed down to a temperature T_H (about 250°C), whereupon all the liquid has solidified. Below temperature T_H, the solid cools over the temperature range **HP** as a **single** phase (the **α** phase, a fcc structure with random replacement of lead atoms by tin atoms).

However, once temperature T_P (about 150°C) is reached, there exists a problem, as the decrease in temperature results in distortions in the crystal lattice due to the different radii between lead (Pb) and tin (Sn) and their different structures. At this temperature, the maximum solubility of the solid solution has been reached and further cooling will result in the formation of a supersaturated solution. However, since equilibrium conditions are maintained, the excess tin will precipitate as the **β** phase solid solution. Thus below the temperature T_P, the amount of tin which can exist in the fcc structure of lead decreases and so the **α** phase follows the composition line **PRT**. The excess tin is thrown out, **not** as **pure tin**, but as the **β** phase (with initial composition **Q**), which has the structure of tin with random replacement of tin by lead. Similarly, the **β** phase must follow the composition curve **QSU** and so throws out lead, **not** as **pure lead**, but as the **α** phase (with initial composition **P**), which has the structure of lead with random replacement of lead by tin). When the temperature falls to, say, 100°C, the compositions of the **α** and **β** phases when equilibrium has been reached correspond to the points **R** and **S** respectively, and there relative amounts can be calculated using the Lever rule. At room temperature, the compositions of the **α** and **β** phases are **W** and **U** respectively, and these are deposited as lamellae of **α** and **β** (which would etch differently).

The cooling curve for composition **Y** is not shown. It is suggested that a pencil cooling curve is made adjacent to the phase diagram on the tin side for this cooling curve. Briefly, if the composition **Y** is considered to cool from the melt, there will be a difference in the rate of cooling as soon as the metal starts to solidify due to the loss of latent heat. This takes place at temperature T_F and the first solid to solidify is of composition **H**. If the temperature now drops to T_J, the solid deposited is of composition **K** whilst the liquid is of composition **I**. Note that, as before, the solid deposited at the higher temperature (of composition **H**) must **alter** and become richer in tin, and hence diffusion must take place. The composition **K** of the alloy deposited at T_J has the structure of lead, with random replacement of lead by tin. The liquid follows the **liquidus** line **FIMN** and the solid follows the **solidus** line **HKLO.** If the temperature now drops to T_N, the solid deposited is of composition **O** whilst the liquid is of composition **N**. Note that there is a temperature arrest at this point (equivalent to that given by the line **uv** on the cooling curve for pure lead). The composition **N** (approximately 60% Sn) of the liquid is given the special name **eutectic**. The **eutectic** point is the lowest temperature at which the mixture of tin and lead will melt.

At temperatures below the eutectic point, the alloy will be composed of two crystalline solids, **α and β**, denoted (**α** + **β**) and deposited as lamellae. This solid is a complex structure of the **α** and **β** phases whose composition can be determined using the Lever rule. On further cooling the **α** phase follows the **solidus** line **OPRT** and the **β** phase follows the **solidus** line **VQSU**. For example, on cooling to a temperature of about 150°C, the **α** phase will have composition **P**; the excess tin is thrown out, **not** as **pure tin**, but as the **β** phase (composition **Q**, having the structure of tin with random replacement of tin by lead). Conversely, the **β** phase will have composition **Q**; the excess lead is thrown out, **not** as **pure lead**, but as the **α** phase (composition **P**, having the structure of lead with random replacement of lead by tin).

The relative quantities of the solid **α** and **β** phases is determined by the position of the **eutectic point**. For alloys with compositions to the left of the **eutectic composition N**, the predominant phase will be the **α** phase, which forms the **host matrix**. For alloys with compositions to the right of the **eutectic composition N**, the predominant phase will be the **β** phase, which forms the **host matrix**. Thus for composition **Y**, the host matrix is formed from the **α** phase and we refer to the **β** phase as the **second phase** (**10.5.2**). The second phases are the **insoluble** phases in the host matrix and act as agents which increase the strength of the alloy (**10.8**). This strengthening results by the formation of second phase solids which do not have the structure (or atomic spacing) of the metallic lattice of the host matrix.

Final cooling down to room temperature will result in the composition of the **α** and **β** phases gradually changing because the solubility of tin in lead and, to a lesser extent, lead in tin decreases with decreasing temperature, as shown by the lines **OW** and **VU** respectively. Thus at room temperature, the **α** solid solution will tend to become a pure metal (lead) and the **β** solid solution will tend to become a pure metal (tin), since their mutual solubilities become negligible. (In fact, for tin, this occurs at temperatures lower than about 175°C, since the solidus line **QSU** lies directly over the axis representing 100% tin). In practice, however, this is seldom achieved, because at lower temperatures the diffusion of atoms in the solid is so slow that an infinitely long time period would be needed to achieve complete equilibrium.

10.6.1 The Lever Rule

We can use the lead-tin phase diagram (discussed above) to illustrate the Lever Rule. At the temperature T_D, the alloy will consist of a mixture of a liquid of composition **E** and a solid of composition **C**. The fraction of tin in the alloy is **X** and the amount of the solid is (say) s, then the amount of tin in the solid phase is s**C** and the amount of tin in the liquid is $(1 - s)$**E**. Since the amount of tin in the alloy = amount of tin in the solid + amount of tin in the liquid

we can write $\mathbf{X} = s\mathbf{C} + (1 - s)\mathbf{E}$, so that the relative amount of tin in the solid is

$$s = \frac{\mathbf{X} - \mathbf{E}}{\mathbf{C} - \mathbf{E}}$$

Since **X** - **E** is represented by the line **ED** on the phase diagram and **C** - **E** by the line **EC**, the relative amount of the solid *s* will be equal to the length **ED** divided by the length **EC**, i.e.,

$$s = \frac{\mathbf{ED}}{\mathbf{EC}}$$

Similarly, the relative amount of the liquid, *l*, will be

$$l = \frac{\mathbf{C} - \mathbf{X}}{\mathbf{C} - \mathbf{E}} = \frac{\mathbf{CD}}{\mathbf{CE}}$$

10.7 PRECIPITATION HARDENING ALLOYS

The principle that a second phase, with a different lattice spacing and structure, is deposited within the host matrix solid is made use of in precipitation hardening alloys, which exhibit strength properties superior to those of the original host metal alone. **Aluminium alloys** are one example of precipitation hardened alloys, where the hardening process provides a means to obtain required engineering strengths whilst retaining the benefits of the aluminium host metal (e.g, corrosion resistance, high strength:weight ratio, etc.). To form a suitable second phase in the aluminium host metal, an alloying element was required that dissolved in aluminium at high temperatures, but was nearly insoluble in aluminium at room temperatures. The left-hand side of the phase diagram depicted in ***Figure 10.17*** was the type of phase diagram required (although the right-hand side would not give the hardening requirements). The element **copper** was found to exhibit the requisite properties, forming very small incoherent second phase precipitates of $CuAl_2$ which stop dislocation movement (**10.8**).

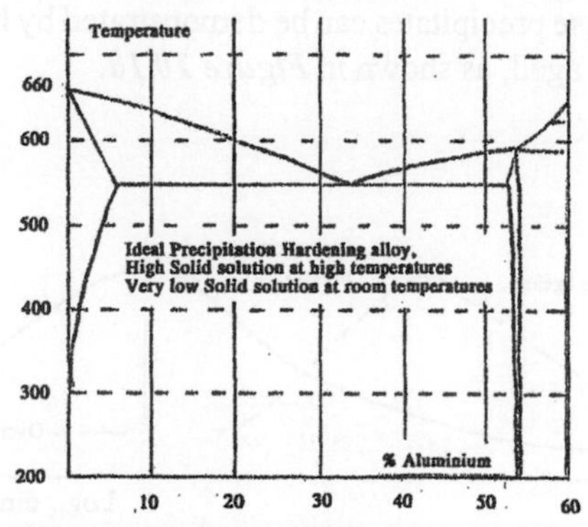

Figure 10.17 *Copper-aluminium phase diagram*

The maximum amount of copper allowed to form a good precipitation hardening alloy is 4.5%, which can be obtained at high temperatures. This has clear implications for the manufacturing process required

- the precipitation hardening process cannot be undertaken until all the rolling and forming processes (**10.5.5**) have taken place. Therefore prrecipitation hardening processes are undertaken on metals in the form of thin panel sheets;
- an alloy of copper and aluminium must be heated to a temperature at which all the copper must be in the solid phase, with no copper-rich phase remaining as a second phase. This requires a temperature of around 500-560°C, very close to that where solid and liquid phases coexist (>570°C).

Hence if the temperature control of the furnace poor or there is an uneven temperature distribution throughout the furnace, then the thin panel sheets could melt. For this reason, low alloy contents (2-4% copper) are used as these compositions give a much larger temperature range in which the sheets can be heated to obtain all the copper in the solid phase (i.e., the sheet is all in the **α** phase).

From these high temperatures, the alloy sheet is then **quenched**. Quenching is the rapid cooling of the solid solution, usually be immersion in water, of sufficient rapidity to ensure that the copper does not have chance to precipitate out and that equilibrium conditions are **not** achieved. The quenching process therefore effectively freezes in the **α** phase structure obtained at the higher temperatures. Once the quenching process is complete, the sheets can then be rolled in order to increase the dislocation density.

Thereafter, any shaping or cutting required is undertaken and the sheet is then **aged**. Ageing is a process of precipitation and cooling, which results in copper being deposited within the aluminium as an **inter-metallic** phase (an incoherent precipitate of $CuAl_2$) by heating it to specific ageing temperatures (sufficient to ensure that the alloy is still in the two phase region) over successive times. The nuclei for deposition of the inter-metallic phase are the dislocation entanglements. (If there were some undissolved $CuAl_2$ particles, then these would act as the nuclei and the full benefit of the precipitation hardening process would not be obtained). Subsequent cooling to room temperature ensures maximum strength and hardness. The formation and distribution of the inter-metallic phase precipitates can be demonstrated by hardness measurements (**3.4.1**) as the sample is aged, as shown in ***Figure 10.18***.

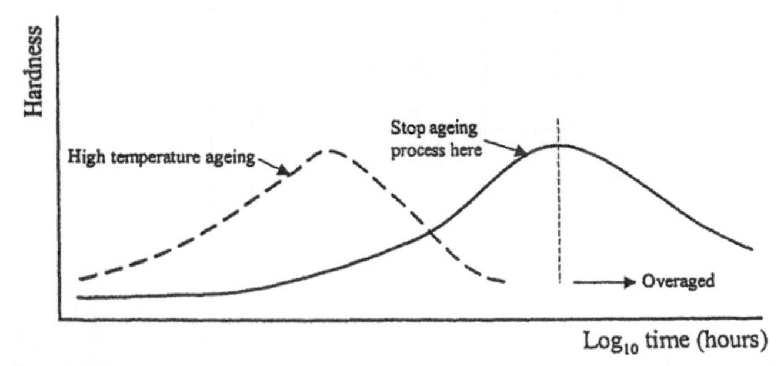

Figure 10.18 *The effect on hardness of ageing in copper-aluminium alloys*

To optimise strength and hardness, the second (inter-metallic) phase particles should be as small as possible and incoherent, but distributed uniformly throughout the host matrix (***Figure 10.19d***). These second phase incoherent particles act as obstacles to dislocation movement (**10.8**); each dislocation passing them creates a dislocation ring around the precipitate, making the precipitate **appear bigger** to the next dislocation. Thus the space between particles decreases as more dislocations pass and so the material rapidly achieves good **tensile strengths** with only a small percentage of an alloy element.

10.8 MODELLING DEFORMATION IN ALLOYS

Dislocation movement in alloys can be modelled by considering the operation of a **Frank Read Source (FRS)**. ***Figure 10.19*** illustrates the operation of a FRS, and shows some sequential stages (**1-6**) in the deformation process by considering what happens when a straight dislocation meets a non-deformable precipitate (or second phase)

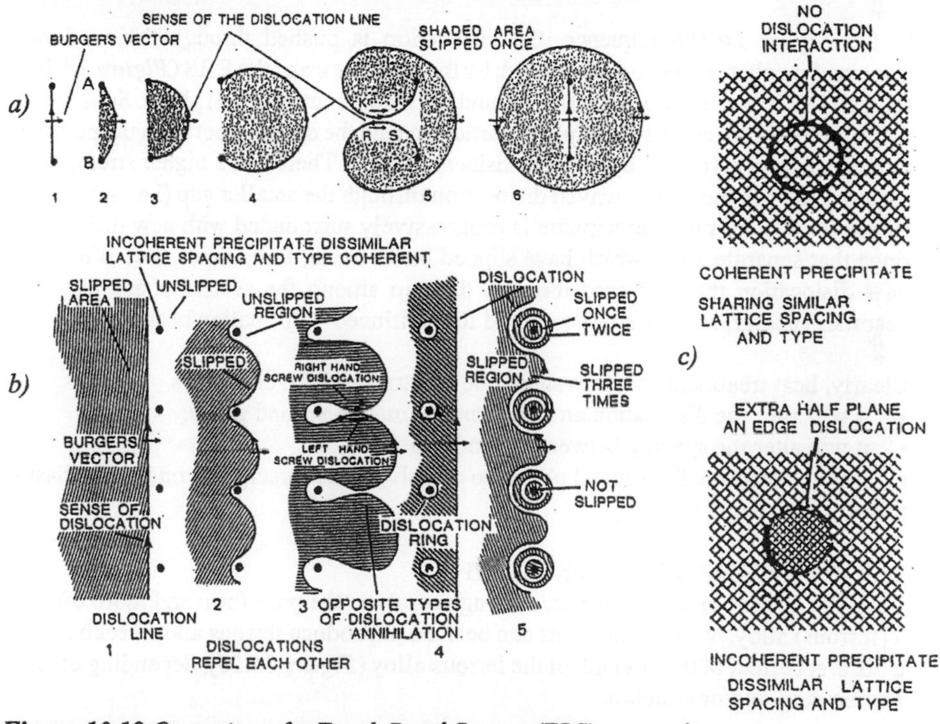

Figure 10.19 *Operation of a Frank Read Source (FRS)*

In ***Figure 10.19a***, stage 1 illustrates that a very high stress, σ is required to make a straight dislocation bend into a curve of radius r. Mathematically, this is expressed as $\sigma = k/r$, where k is a constant. The stress is always increasing until the radius of slip is equal to the distance between the two anchor points, whereupon the stress will decrease. The same phenomenon is observed in blowing soap bubbles (**5.4.2**); they are difficult to form at first, but when the radius becomes larger the increase in the blowing force required is smaller. Stages **2-5** show the bowing out of the dislocation line, and the respective dislocation sense and Burgers vector. The two dislocations at **KJ** and **RS** are opposite types, having the same Burgers vector but the opposite sense; **KJ** is righthand sided (**RHS**) and **RS** is lefthand sided (**LHS**). Therefore the dislocations **KJ** and **RS** will annihilate each other, forming a ring around the two anchor points (Stage **6**). Note that the **original dislocation line is retained**. This principle can be applied to a dislocation line meeting a row of precipitates (second phases), as shown in the sequences in ***Figure 10.19b***. This representation is valid provided we continue to view the precipitates (second phases) as non-deformable. Where the precipitates (second phases) have a crystal lattice structure and spacing different to the host (parent) metal, an **incoherent** phase (**10.5.2**) is formed (***Figure 10.19d***), which is then surrounded by a dislocation ring or sphere which cannot be broken into by an oncoming dislocation. Hence the model applies to incoherent precipitates. Where the precipitates (second phases) have a crystal lattice structure and spacing the same as the host (parent) metal, a **coherent** phase (**10.5.2**) is formed (***Figure 10.19c***), which is deformable.

In the ***Figure 10.19b*** sequence, the dislocation is pushed through the randomly dispersed incoherent second phases and, by the mechanism of the **FRS** (***Figure 10.19a***, stages **1- 6**), leaves a dislocation ring around the incoherent (second) phase. Subsequent dislocations arrive at the second phase particle to find the distance between the particles reduced, e.g, to r_1 ($r_1 < r$), by the new dislocation ring. Therefore a higher stress, σ_1, is required to push the newly arrived dislocation through the smaller gap (i.e., $k/r_1 > k/r$). Hence the second phase precipitate is progressively surrounded with new dislocation rings that separate areas which have slipped once, twice, three times, etc. With each **new** dislocation that is forced through, the gap among the second phase particles becomes smaller, and the stress required for continued deformation becomes higher.

Clearly, heat treatment of alloys should be undertaken with caution as
- it will alter the dislocation arrangement around the second phase;
- it will alter the spacing between second phases;
- it may cause smaller second phases to dissolve and reprecipitate on larger phases;

thereby weakening the alloy.

10.9 IRON-CARBON (Fe-C) SYSTEM

Iron (Fe) is used in several forms, although rarely in the pure form and more often as a (ferrous) alloy. Various elements can be used to produce ferrous alloys, each having a different effect of the strength of the ferrous alloy (***Figure 10.20***), depending on their atomic radii and/or structure.

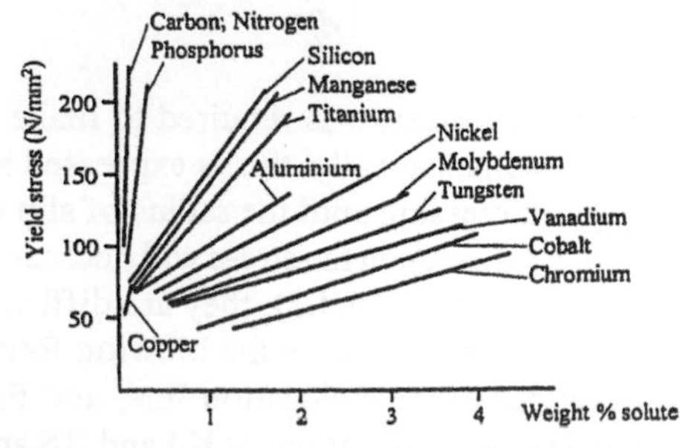

Figure 10.20 indicates that the addition of carbon causes there greatest increase in yield stress in ferrous alloys. Commonly, ferrous alloys used for engineering purposes are based on the iron-carbon system.

***Figure 10.20** Effect of alloying different elements and their quantity on yield stress*

10.9.1 Wrought Iron (WI)

Commercially pure iron is generally regarded as containing ≤ 0.008% carbon. Wrought iron is almost pure iron, containing less than 0.03% carbon. Wrought irons have high melting points and therefore do not flow well into moulds; they are generally shaped by **forging**. The forged article is usually stronger than the cast iron and does not suffer from **brittle fracture (10.14.1)**. This property allows the use of wrought iron for the couplings on articulated trailers, for example; wrought irons are commonly used for ornamental work (e.g, scrolls), as garden gates, etc.

10.9.2 Cast Iron (CI)

Cast iron usually contains more than 2% carbon, which can exist in two principal forms

- as **graphite** giving rise to the term **grey cast iron**; and
- as the **carbide** Fe_3C (cementite), giving rise to the term **white cast iron.**

The terms **white** and **grey** arise from the appearance of the fracture surface. The form that the carbon takes is governed by the cooling rate of the cast. **White cast irons** are used for fire grates, car cylinder heads, etc. whilst **grey cast irons** are used for piston rings, for example, where the graphite has lubricating properties.

10.9.3 Steels

Steels comprise a range of iron-carbon alloys where the carbon content ranges typically from 0.05% to 1.5%. The popularity of steels as engineering materials arises from the very wide range of strengths which can be achieved simply by the addition of carbon, leading to a wide variety of end uses **(Table 10.1)**.

Table 10.1 Applications of various iron-carbon steels related to carbon content

Carbon content (%)	Generic classification	Application
0.05-0.15	Low carbon mild steel	Nails
0.1 - 0.2	Mild steel	RSJ sections
0.2 - 0.3		Gears, forgings
0.3 - 0.4	Medium carbon	Axles, con rods
0.4 - 0.5		Crankshafts
0.5 - 0.6		Wire rope
0.6 - 0.7	High carbon	Set screws
0.7 - 0.8		Hammer heads
0.8 - 0.9		Cold chisels
0.9 - 1.0	Tool steels	Axe heads
1.0 - 1.1		Drills and taps
1.1 - 1.2		Ball bearings, lathe tools
1.2 - 1.3		Files and reamers
1.3 - 1.4		Razors, saws

The strength of steels can also be controlled by alloying (with additional elements) (**Table 10.3**) and by heat treatment (**Table 10.4**), both of which have effects on the iron-carbon phase diagram (**10.11**). The alloy element added is dependent upon the mechanical and corrosion resistant properties required. Where no alloy additions are added, the iron carbon system is referred to as **mild steel (10.10.3)**. It is incorrect to drop the *mild* from mild steel, although it is often done by engineers and other trades.

Steels are also described as **austenitic, ferritic** or **martensitic (10.10.7)**. These terms indicate how the alloy elements affect the room temperature structure e.g, cobalt (Co), nickel (Ni) and manganese (Mn) favour an austenitic steel (i.e., fcc structure at room temperature) whilst chromium (Cr), aluminium (Al), tungsten (W) and silicon (Si) favour a ferritic steel (bcc structure at room temperature). Varying the alloy element content of steels is often referred to as opening and closing the **austenite field (10.11.4)**.

Alloy steels, for example X2CrNi18-8 stainless steel (0.02% C, 18% Cr, 8% Ni), are formed by the **substitution** of iron in the host lattice structure with the nickel (Ni) and chromium (Cr). The carbon additions, however, are always **interstitial**. Their major contribution to the hardening and strengthening process is by the formation of insoluble carbides (**10.11**). These insoluble carbides are formed with both iron and the alloy additions. (Note that the **iron-carbon system** is more correctly referred to as the **iron-carbide system**, as it is the interaction of the iron with the relevant carbide, as opposed to carbon, that is described). The iron carbide of importance to the description of the phase diagram is called **cementite** (Fe_3C). In the presence of alloying elements, the alloy carbide is usually more complex, having the general formulae $M_{23}C_6$, M_6C, M_3C, etc. where M is iron (Fe), or molybdenum (Mo), or chromium (Cr), etc.

10.10 CLASSIFICATION OF STEELS

The classification of steels has recently been consolidated by BS EN 10027: Parts 1 and 2: 1992[1]. Part 1 designates steels by means of symbolic letters and numbers to express application and principal characteristics so as to provide an abbreviated identification of steels. Part 2 designates steels by means of a unique set of digits which represent the application and principal characterisitics of the various steels. Either designation system can be used to provide one unique steel name for all steels, although steel numbers more suitable for use for data processing.

A brief description of steel names (**10.10.1**) and steel numbers (**10.10.2**) is given below; however, for the purposes of this textbook, steel names will be used as they provide more information on the composition of the steel. In addition to the formal classification of steels given by the European Standard, steels are commonly referred to using traditional names, incluuding mild steel (**10.10.3**), bright drawn steel (**10.10.4**), high tensile steel (**10.10.5**), alloy steels (**10.10.6**) and stainless steels (**10.10.6**).

10.10.1 Steel Names

For the purposes of the European Standard designation, steel names are classified into two main groups

- Group 1: steels designated according to their application and mechanical or physical properties. Group 1 steel codings comprise of a principal letter (representing the end use of the steel), ordinarily followed by a number (representing the minimum or characteristic yield or tensile strength, depending on the end use). **Table 10.2a** summarises the coding for the main steels used in the construction industry.
- Group 2: steels designated according to their chemical composition and further divided into four subgroups. **Table 10.2b** summarises steel codings for Group 2.

Table 10.2a Summary of the classification for Group 1 steels (BS EN 10027: Part 1: 1992)

Principal letter coding and end use	Subsequent number
S Structural steels E Engineering steels	Specified minimum yield strength (N/mm^2) for smallest thickness range
B Steels of reinforced concrete	Characteristic yield strength (N/mm^2)
Y Steels for prestressing concrete	Specified minimum tensile strength (N/mm^2)
H Cold rolled flat products of high strength steels for cold forming T Cold rolled flat products of high strength steels for cold forming (tensile strength only specified)	Specified minimum yield strength (N/mm^2) Minimum specified tensile strength (N/mm^2)
D Flat products for cold forming (other than H and T above)	C Cold rolled products; or D Hot rolled products for direct cold forming; or X Products for which rolled condition unspecified; followed by two symbols characterising the steel and allocated by the appropriate body (BS EN 10027)

Table 10.2b Summary of the classification for Group 2 steels (BS EN 10027: Part 2: 1992)

Sub-group	Sub-group classification	Steel coding (successive symbols)			
2.1	• Non alloy steels (except free cutting steels) with Mn < 1%	C	100 x specified average percent carbon (C) content		
2.2	• Non alloy steels with Mn ≥ 1% • Non alloy free cutting steels and alloy steels (except high speed steels) where the content of each alloy element is < 5% by weight	100 x specified average percent carbon (C) content	Chemical symbols indicating alloy elements in decreasing order of content	Numbers, separated by hyphens, which indicate percent alloy element content multiplied by a factor[a] rounded to the nearest integer	
2.3	• Alloy steels (except high speed steels) where the content of at least one alloy element is ≥ 5% by weight	X	100 x specified average percent carbon (C) content	Chemical symbols indicating alloy elements in decreasing order of content	Numbers, separated by hyphens, which indicate percent alloy element content multiplied by a factor[a] rounded to the nearest integer
2.4	• High speed steels	HS	Numbers, separated by hyphens, which indicate percent alloy element content (in the order tungsten (W), molybdenum (Mo), vanadium (V), cobalt (Co)) rounded to the nearest integer		

Notes: [a] factors are: 4 (Cr, Co, Mn, Ni, Si, W); 10 (Al, Be, Cu, Mo, Nb, Pb, Ta, Ti, V, Zr); 1000 B (BS EN 10027: Part 1: Table 1).

10.10.2 Steel Numbers

The structure of steel numbers is set out as follows

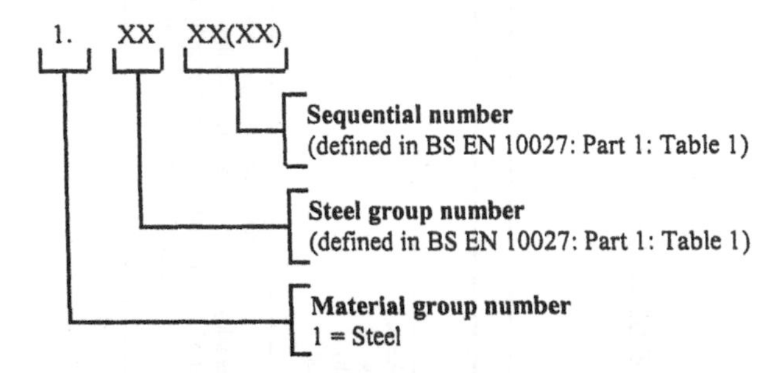

Although currently the numbering system is limited to steel, it is structured to permit extension to include other industrially produced materials, and this procedure may be applied to such materials during future ratifications of European Standards.

10.10.3 Mild Steel (MS)

Mild steels comprise principally iron, with up to 0.2% carbon. The carbon occupies the **interstitial** sites in the lattice structure of the host metal iron. The function of the carbon is to improve the mechanical properties of the steel by forming an **insoluble second phase** (**10.5.2**) with the iron. These insoluble second phases are known as **carbides** and they act as obstacles to dislocation movement (**10.8**). The heat treatment of mild steel is considered in a later section (**10.13**). Mild steel is used for a wide variety of applications (for example, girders, motor vehicle panels and other general engineering items) and, unlike higher carbon steels, is not adversely affected by welding.

10.10.4 Bright Drawn Steel (BDS)

Bright drawn steels are formed by cold drawing a two phase carbon steel. This technique produces a good finish, as atmospheric oxidation is eliminated. In addition, the mechanical properties of the product are significantly increased by the cold working, although the technique is used only where the tolerance and strength of the product offer a cost saving. BDS would not (normally) be specified when subsequent machining operations are required unless the enhanced mechanical properties were desired and the high cost was permissible.

10.10.5 High Tensile Steel (HTS)

High tensile steel is a high carbon steel used in prestressed concrete products (**8.9.3**).

10.10.6 Alloy Steels

Alloy steels are produced by introducing certain alloy elements (e.g, chromium or nickel) which substitute for the iron (Fe) in the iron lattice and modify the phase diagram (**10.11.4**). The addition of other elements (e.g, niobium, Nb) may result in the formation of different carbides (other than Fe_3C, cementite) as, compared to iron, they have a greater affinity for carbon. Alloy steels are usually tougher than the equivalent carbon steel and have different corrosion characteristics; good examples of this are the alloy steels known as **corrosion resistant steels** (including **stainless steels**) (**10.10.7**).

10.10.7 Corrosion Resistant Steels (Stainless Steels)

Corrosion resistant steels contain sufficient chromium (minimum 4-6%) to make them corrosion resistant. True stainless steels contain >12% Cr. Note that stainless steels are only "stainless" if they are used in oxygenated environments (**11.15.3**). Chromium on its own acts as a ferrite stabiliser (**Table 11.7**) and therefore corrosion resistant steels owe their resistance to corrosion to the strong adherent (but invisible) chromium oxide (Cr_2O_3) film which is formed on the surface in oxygenated environments. However, they are not corrosion resistant if they are put into a reducing atmosphere or where there is no oxygen available, as they are then unable to repair the tenacious chromium oxide film. An example of the requirement for an oxygenated environment can be demonstrated by the use of stainless steel for fermenting chambers in the preparation of beer. These chambers would soon leak, as the internal environment is predominantly carbon dioxide so that the stainless steel is unable to repair its protective oxide film. In addition, stainless steels are resistant to attack by air, weak mineral and organic acids and high temperatures.

Stainless steels are categorised by their chromium content or by their structure by BS EN 10088: Part 1: 1995[2] as

- **Ferritic stainless steels**, which have a carbon content limit of 0.08% and contain chromium (Cr) as the main alloying element (10.5% to 30%). The limited carbon content means that they do not display significant hardening after quenching (**10.13.6**). These stainless steels exhibit good weldability (provided the grain growth is reduced by the addition of nitrogen in the steel). This characteristic arises because there is no possibility of the **austenite phase** being formed (**10.11**). However, they possess poor ductility because they exhibit a bcc structure. Ferrite stainless steels are magnetic and must be hardened by **cold working (10.5.5)**. Unfortunately, ferritic stainless steels possess poor resistance to **crevice and pitting corrosion (11.8)** and their use is normally restricted to indoor situations (e.g, kitchen cutlery).
- **Martensitic stainless steels**, which have the highest carbon content (0.08% to over 1%) and contain 11.5% to 19% chromium. This results in a stainless steel which is **austenitic (10.11)** at the welding temperature. Rapid cooling from the welding temperature will produce the brittle phase **martensite (10.12.1)** in the **heat-affected zone (HAZ) (10.14.2)**, whilst slow cooling can cause some of the chromium to be deposited as the carbide (which is then not available to contribute to corrosion resistance). The mechanical properties of martensitic stainless steels can be increased considerably by quench heat treatment and the martensite structure obtained is magnetic and fragile; it must undergo **tempering (10.13.6)** treatment before use. Martensitic stainless steels are relatively cheap but are not usually suitable for building applications.
- **Austenitic stainless steels**, which contain the main alloying elements chromium (Cr) and nickel (Ni). These stainless steels are expensive (due to the high alloy content), are non-magnetic and highly corrosion resistant. However, they do suffer from **stress corrosion (11.15.3)** and **crevice corrosion (11.6.3)** in the presence of chloride ions and therefore effective design to limit the chloride ion concentration at key locations in the finished component is very important (e.g, leaking joints, boiler deposits, etc.). Austenitic stainless steels have good formability due to their

fcc structure. Whilst they do not display hardening after heat treatment, they are hardened both by cold working (and therefore suffer from 'spring back', **10.5.3**) and by nitrogen additions. Two of the most common austenitic stainless steel alloys are X2CrNi18-8 (i.e., 0.02% C, 18% Cr, 8% Ni) and X2CrNiMo18-10-3 alloy (i.e., 0.02% C, 18% Cr, 10% Ni, 3% Mo). Austenitic stainless steels are mostly used in the construction industry, particularly for external use where appearance is important; the X2CrNiMo18-10-3 alloy is particularly favoured in these conditions. Stainless steel can be safely used in contact with timber to provide protection from the acid nature of some timbers or where the timber has been treated with CCA (copper chrome arsenic) preservatives (CCA has been known to cause deposition of copper on to aluminium alloys and plain steels through galvanic couples, **11.3.2**). Due to their high resistance to corrosion, austenitic stainless steels are preferred to ferritic stainless steels for use as concrete reinforcement, particularly when concrete is contaminated with chlorides. Austenitic stainless steels for reinforcement are covered by BS 6744[3].

10.11 IRON-CARBON PHASE DIAGRAM

An important part of studying a ferrous alloy is to be able to interpret the **iron-carbon phase diagram**, given in ***Figure 10.21***. ***Figure 10.21*** includes

- the temperature bands where various metallurgical processes take place (e.g, spheroidising, stress relieving, homogenising, solution of the carbide network, full annealing, etc.). These processes are fully described in section **10.13**;
- some metallurgical sections formed for the temperatures and compositions indicated;
- the temperature dependance of the solubility of carbon in α-, γ- and δ iron, shown by the lines **QP** for α iron **(ferrite)**, **ES** for γ iron **(austenite)** and **AH** for δ iron;
- the structure of the single phases, γ iron (**austenite**, with a fcc structure), α iron (**ferrite**, with a bcc structure) and δ iron (with a bcc structure). A fourth solid phase, **cementite** (Fe_3C), has a complicated orthorhombic structure (**1.15.3**), with 12 iron atoms and 4 carbon atoms, and is not shown;
- the mixture of α iron (**ferrite**) and **cementite** (Fe_3C) phases are given the name **pearlite**, because the structure of the metallurgical section (after being cleaned with certain acids) resembles mother of pearl. Note that the phase diagram shows that γ iron (**austenite**) transforms to **pearlite** below 723°C. As examined in above (**10.5.3**), it is less energy consuming to roll steel whilst it is in the single (austenite) phase than when it has the two phase structure of pearlite. In the two phase (pearlite) state, very rapid hardening will take place.

The temperatures can be approximated by characteristic annealing temperature colours (emitted if Stephan's law, $\lambda T = \text{constant}$, is applied to the heated steel) (***Figure 10.22***). These colours can be used to give guidance on the temperature of the hot alloy. In addition, characterisitic tempering temperature colours (left from white light when interference takes place in the oxide film). These are equivalent to the interference colours caused by oil on water. The temper colours refer to the cooled metal and can be used to give guidance on the temperatures which have been used for tempering (**10.13.6**). As an example, Gillette razor blades are tempered to give a bright blue colour and then coated with a clear lacquer.

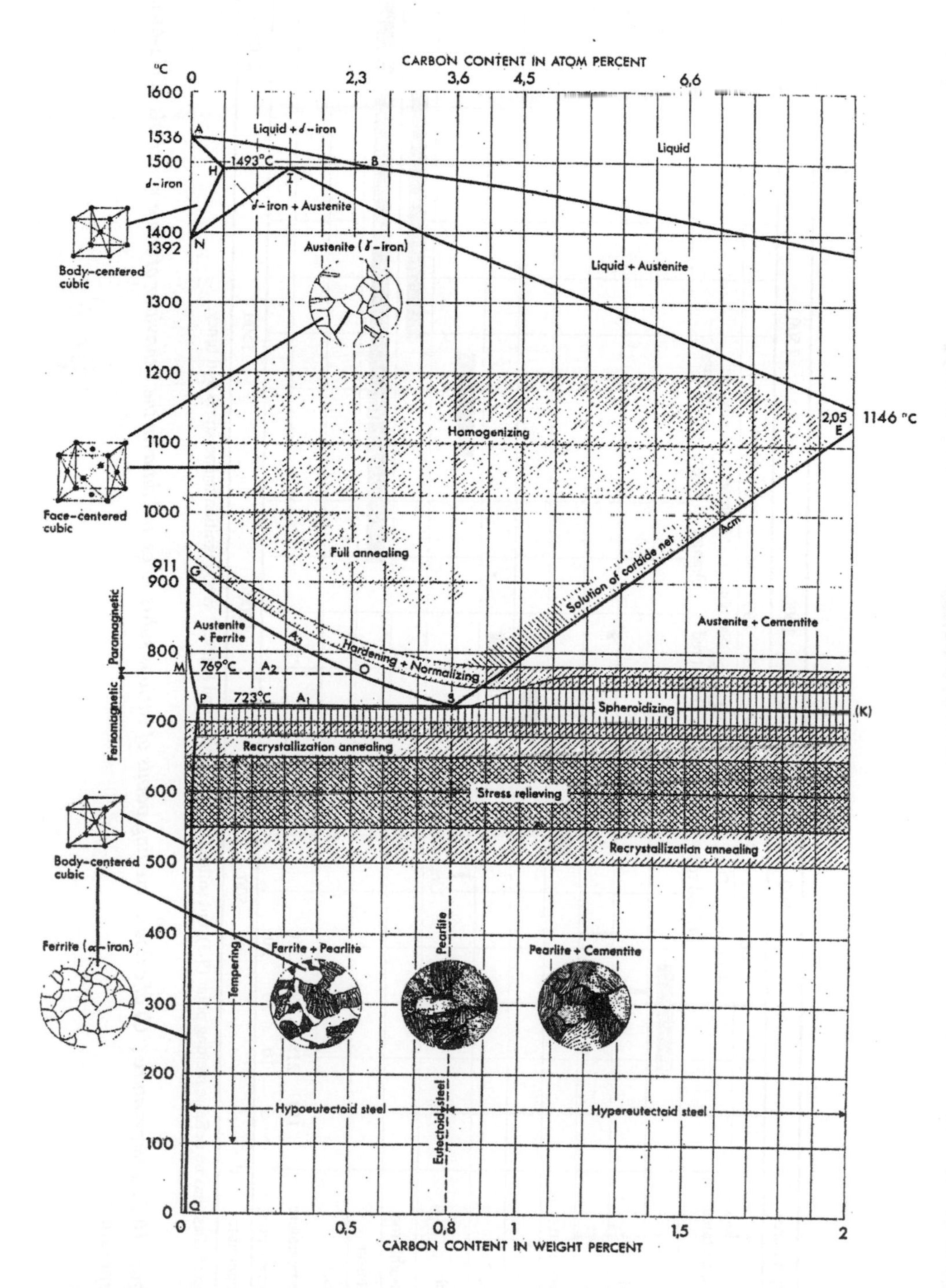

Figure 10.21 Iron-carbon phase diagram

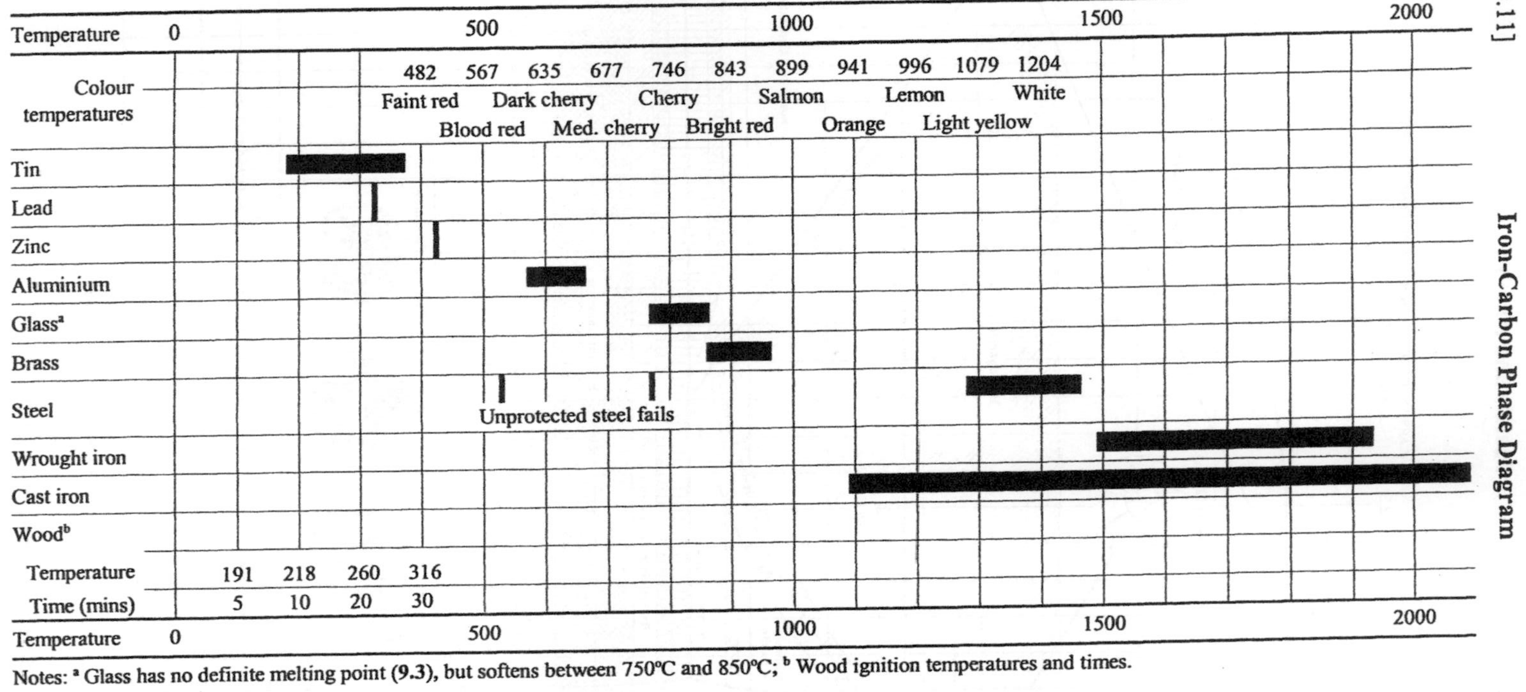

Notes: [a] Glass has no definite melting point (9.3), but softens between 750°C and 850°C; [b] Wood ignition temperatures and times.

Figure 10.22 *Temperature (°C) of colours, melting points of various substances and data on the behaviour of some materials at elevated temperatures*

The iron-carbon phase diagram (***Figure 10.21***) is complicated, but much of the phase diagram can be ignored for our purposes. For example, the phase diagram is shown only for carbon contents up to 2% and, although the phase diagram continues at higher carbon contents, most steels of practical importance contain < 2% carbon.

The line **ABC** is the **liquidus** line, above which only one liquid phase exists, consisting of iron and dissolved carbon. The line **AHIE** is the **solidus** line, below which the various iron-carbon compositions are completely solid. The regions between these two lines represent mixtures of solid and liquid phases. The solubility of carbon in steel is complex and structure dependent. The carbon goes into the 'free space' between the atoms to form an **interstitial solid solution (10.5.2)**.

The iron-carbon phase diagram is complicated by the different structures both **iron** and **carbon** can take at different temperatures. For **iron**, these structures, summarised in the iron-carbon phase diagram (***Figure 10.21***) and in ***Figure 10.23*** below, are

- the single α phase (**ferrite**) is stable between room temperature and 723°C when there is less than 0.025% carbon by weight and has the bcc structure. The interstitial carbon sites within the host (iron) lattice are shown in ***Figure 10.23a***;
- the single γ phase (**austenite**) is only stable between 723°C up to 1490°C and has the fcc structure. The interstitial carbon sites within the host (iron) lattice are shown in ***Figure 10.23b***;
- the single δ phase is stable between 1490°C and the melting point and again has the bcc structure.

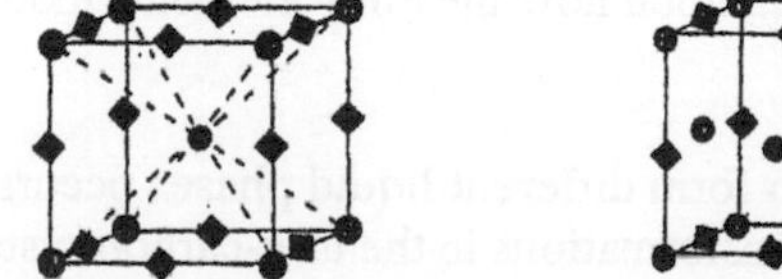

a) bcc α (ferrite) phase

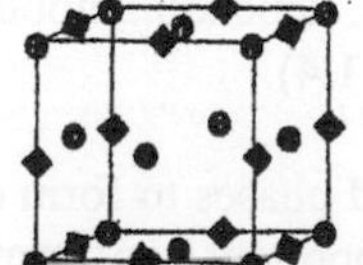

b) fcc γ (austenite) phase

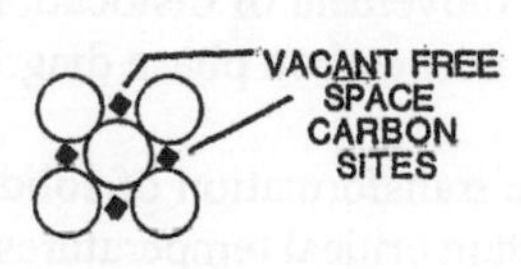

c) one face of the fcc lattice

Figure 10.23 *Interstitial carbon sites in the various phases*

The face of one FCC lattice is shown in ***Figure 10.23c***. The 'free space' in the lattice is **not** large enough to take in water but is can accommodate carbon and nitrogen atoms.

In the **carbon** rich side of the phase diagram, carbon can exist as the carbide **cementite** (Fe_3C) or as **graphite**. This affects the structure of the phase diagram (***Figure 10.24***).

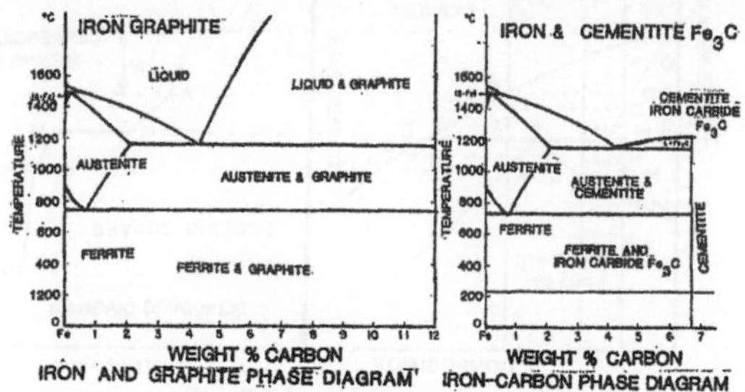

Figure 10.24 *The two phase fields involving graphite and iron carbide* (Fe_3C)

The α- (**ferrite**), γ (**austenite**) and δ phases are capable of dissolving various quantities of carbon, depending on the temperature. α iron (**ferrite**) can dissolve up to 0.025% carbon, as shown by point **P** on the phase diagram, γ iron (**austenite**) can dissolve up to 2.05% carbon, as denoted by point **E** (just off the right axis) and δ iron can dissolve up to 0.08% carbon, as denoted by point **H**. The temperature dependance of the solubility of carbon in α-, γ- and δ iron is shown on the phase diagram by the lines **QP** for α iron (**ferrite**), **ES** for γ iron (**austenite**) and **AH** for δ iron. The fourth solid phase existing in the phase diagram is **cementite** (Fe_3C) containing 6.69% carbon. Cementite is a hard and brittle compound, whereas ferrite is relatively soft and malleable.

The austenite region is the largest single phase region in the phase diagram and comes about because of the difference in the amount of 'free space' in the structure with elevated temperature (**1.16**), i.e.,

Interstitial void space of bcc structure = 32%
Interstitial void space of fcc structure = 26%

The structure of the carbide **cementite** is complex but can be considered as having a fixed ratio of Fe:C of 3:1, as suggested by the chemical formula (Fe_3C). When steel is alloyed with other elements (M) (**Tables 10.13** and **10.14**), different carbides are formed and these have the general formulae $M_{23}C_6$, M_6C, M_3C, where M is iron (Fe), or molybdenum (Mo), or chromium (Cr), etc. Carbides form **incoherent** structures (**10.5.2**) in both the austenitic and ferritic phases, and therefore constitute obstacles to the movement of dislocations (**10.8**), depending upon how the alloy elements modify the iron-carbon phase diagram (**10.11.4**).

The transformation of solid or liquid phases to form different liquid phases occurs at certain critical temperatures. Two important transformations in the iron-carbon system are known as the **peritectic** reaction and the **eutectoid** reaction; these reactions also occur in other alloy systems. These reactions are considered below.

10.11.1 The Peritectic Reaction

The peritectic reaction occurs at a temperature of 1493°C (given by the line **HIB** on the iron-carbon phase diagram, ***Figure 10.21***). In order to explain the peritectic reaction, the relevant proportion of the phase diagram is reproduced as ***Figure 10.25***.

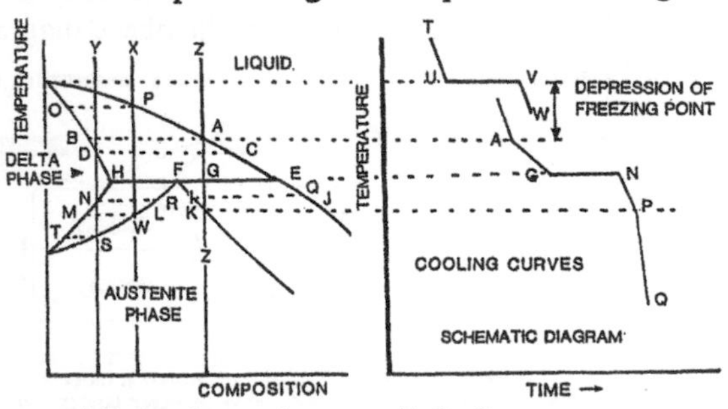

a) Peritectic reaction *b) Cooling curves*

***Figure 10.25** Enlarged area of the iron-carbon phase diagram (peritectic reaction)*

Figure 10.25a shows the peritectic reaction in the iron-carbon phase diagram and ***Figure 10.25b*** shows the cooling curves for pure iron (upper cooling curve) and for a iron-carbon alloy of composition **Z** (lower cooling curve). The peritectic reaction in the iron-carbon system can be described in the same way as the cooling curves in the lead-tin system (**10.6**). The cooling curve for **pure iron** would give a temperature arrest at the melting point of pure iron, as shown by the line **UV** (***Figure 10.25b***). The temperature would not decrease further until all the liquid had changed to the solid phase (***Figure 10.25a***) (for the same reasons as discussed for pure lead, **10.6**). However, if some carbon is added to the iron to give alloy composition **Z**, there is a depression in the melting point from temperature **U** to temperature **A**, even though the carbon is not going into **substitutional solid solution**. A similar depression in the melting point was noted in the lead-tin system (**10.6**). The iron-carbon alloy of composition **Z** cools slowly over the temperature range **AG**, whilst giving out latent heat. The liquid follows the **liquidus** line **ACE** whilst the solid (δ-) phase follows the **solidus** line **BDH** so that the first (δ phase) iron to be deposited will be of composition **B**. On cooling, diffusion will have to take place so that the carbon content can increase and follow the **solidus** line **BDH**. At temperature **HFE** the liquid undergoes a **peritectic** reaction. This reaction is accompanied by a temperature arrest **GN** (shown on the cooling curve ***Figure 10.25b***).

The **peritectic reaction** can be written for any alloy composition as

Liquid	+	Solid	→	Solid	+	Liquid	or	Solid
(composition **E**)		(composition **H**)		(composition **F**)		(composition **E**)		(composition **H**)
						(whichever is in excess)		

There will be excess liquid (composition **E**) from the peritectic reaction if the composition lies between points **F** and **E** or an excess solid (composition **H**) if the composition lies within the region **H** and **F**. For the alloy of composition **Z** (lying between points **F** and **E**), there will therefore be excess liquid of composition **E**. The amount of excess can be calculated from the Lever rule (**10.6.1**). If the composition line fell on the point **F** there would be **no** excess of either solid or liquid and the austenite phase of composition **F** would be formed. Point **F** is the **peritectic** composition.

Thus, for composition **Z**, the composition lies to the right of point **F** and the ratio of the liquid:solid is given by the ratio of line lengths **GE/GH**. In addition

- the first solid phase to deposit is the δ phase (of composition **B**), which follows the composition line **BDH**. All the δ phase is used up in the peritectic reaction;
- there will be an excess of liquid (of composition **E**), which follows the composition line **EQJ**. The γ (austenite) phase (of composition **F**) would cool and follow the composition line **FIK**;
- at temperature **K**, the last remaining liquid (of composition **J**) would have transformed to the γ (austenite) phase (of composition **K**). This would require diffusion of carbon;
- the liquid steel ends up as the γ (austenite) phase (of composition **K**), even though the first γ (austenite) phase deposited was of composition **F**;
- the whole structure is single γ (austenite) phase of composition **K**.

For a composition **X** (lying between points **H** and **F**)

- the γ (austenite) phase (of composition **F**) deposited from the peritectic reaction would transform to the δ phase, which follows the line **HNM**. The excess δ phase (of composition **H**) transforms to the γ (austenite) phase following the composition line **FRLWS**;
- at temperature **MW**, the last δ phase (of composition **M**) transforms to the γ (austenite) phase (of composition **W**);
- the whole structure is single γ (austenite) phase of composition **W**.

For a composition **Y** (lying to the left of point **H**)

- the first solid phase to deposit is the δ phase (of composition **O**), which follows the line **OBD**;
- at temperature T_D, all the liquid is used up and so the δ phase cools over the temperature range **DN** without undergoing any phase change;
- there would be **no** liquid phase left at the peritectic temperature T_{HFE}; in fact the last remaining liquid (which was of composition **C**) disappeared at temperature T_{DC} to produce the δ phase (of composition **D**);
- at the temperature T_N, the δ phase would have too much carbon in solution and so would deposit the γ (austenite) phase (of composition **R**), which would follow the composition curve **RLWS,** whilst the δ phase follows the curve **NMT**;
- the whole structure is single γ (austenite) phase of composition **S**.

Note that when the lines **FRLWS** or **FIK** intersect the original alloy composition line, the γ (austenite) phase stops altering in composition and cools as a single phase structure. This means that steels in the (single) γ **(austenite)** phase are very easy to work, as there are no second phases which obstruct the movement of dislocations (**10.8**).

The cooling curve of the system below the peritectic temperature T_{HGE} is shown in ***Figure 10.25b***. Latent heat is given out over a temperature range and so the rate of temperature decrease changes when there is a two-phase structure (liquid+solid). The kink shown at the point **P** is "artistic licence"; there is in fact very little alteration in the rate of cooling, but very accurate thermal expansion measurement techniques would detect volume (and energy) changes which would take place around this point.

Many textbooks omit the upper part of the iron-carbon system (showing the peritectic reaction) so that all steels can be considered from the simple phase diagram which concerns the **eutectoid reaction** only. The reason for this is that the peritectic reaction only affects solidification of steels with < 0.55% carbon (i.e., up to point **E** on the phase diagram in ***Figure 10.25b***), which all solidify to produce the γ (austenite) phase (it is the austenite phase which transforms to produce the obstacles to dislocation movement). Therefore, for alloys in the iron-carbon system, the upper part of the iron-carbon phase diagram is of little practical importance. With this omission, the remainder of the iron-carbon phase diagram can be interpreted in a similar way to the lead-tin system (**10.6**), except that the liquid phase (eutecti**c**) of the lead-tin system is now the solid γ (**austenite**) phase (eutect**oid**) of the iron-carbon system. The second important reaction in the iron-carbon system, the **eutectoid** reaction, is considered in **10.11.2**.

10.11.2 The Eutectoid Reaction

The eutectoid reaction occurs at a temperature of 723°C, given by the line **PSK** on the iron-carbon phase diagram (***Figure 10.21***). In order to explain the eutectoid reaction, the relevant proportion of the phase diagram is reproduced as ***Figure 10.26***.

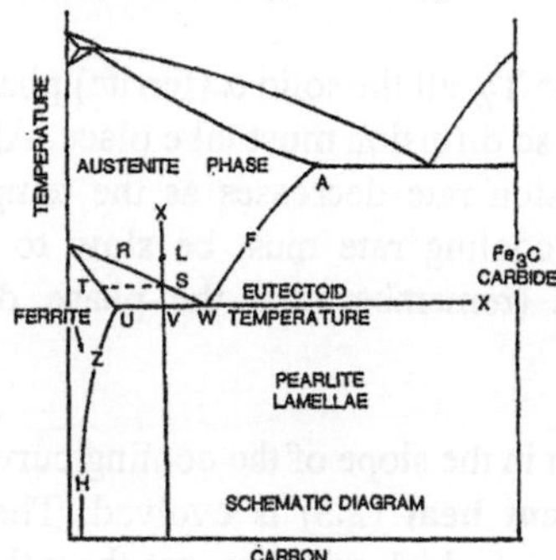

Figure 10.26 Enlarged area of the iron-carbon phase diagram (eutectoid reaction)

If an alloy of composition **X** is cooled from the γ (austenite) phase, there will be **no** differences in the cooling rates below the line **RSW**. At temperature T_S (point **S**), the alloy of composition **X** starts to transform to the α **(ferrite)** phase, which contains less carbon (composition **T**) than the original composition **X**, i.e., the first α (ferrite) phase deposited out of the solid γ (austenite) phase is of composition **T**. As the temperature drops, the α (ferrite) phase follows the composition line **TU**, thereby becoming richer in interstitial carbon, whilst the remaining γ (austenite) phase follows the composition line **SW**, also becoming richer in carbon. Both the solid deposited at the higher temperature (which was of composition T) and the remaining (solid) γ (austenite) phase **must alter** to become richer in carbon. Hence **diffusion** in the **solid state** must take place. At the temperature 723°C (termed the **eutectoid temperature**), the ratio of the untransformed γ (austenite) (now composition **W**) to α (ferrite) (now composition **U**), is given by the Lever rule **(10.6.1)**, i.e., γ (austenite)/α (ferrite) is **UV/VW**.

At the temperature T_S (the eutectoid temperature), the **eutectoid reaction** takes place and there is an imperceptible temperature arrest (detectable with very sensitive calorimetric equipment). Representing the eutectoid solid (γ (austenite) phase) of composition **W** by eutectoid (composition **W**), the eutectoid reaction is given by

Eutectoid (composition W)	→	Solid (composition U)	+	Solid (composition X)
γ (austenite) phase	→	α (ferrite) phase	+	cementite (Fe_3C)
(0.80% carbon)		(0.02% carbon)		(6.69% carbon)

indicating that diffusion of carbon is required. The product comprising the α (ferrite) phase and cementite formed by the eutectoid reaction is called **pearlite (α + Fe_3C)**. Pearlite appears as lathes, the α (ferrite) phase etching white and the cementite **(Fe_3C)** etching black (***Figure 10.3d***). If the composition of the steel is to the **left** of composition **W** (i.e., less than 0.8% carbon), the steel is termed a **hypoeutectoid** steel and some α (ferrite) phase will also be formed (***Figure 10.3c***). If the composition of the steel is to the **right** of composition **W** (i.e., more than 0.8% carbon), the steel is termed a **hypereutectoid** steel and there will be some cementite (**Fe_3C**) formed (***Figure 10.3g***). These phases are called **pre-eutectoid phases.** The transformations from one phase to another occurs at certain **critical temperatures** (discussed in section **10.11.3**).

If the temperature now drops to **T_Z**, all the solid α (ferrite) phase previously deposited must be of composition **Z** and so diffusion **must** take place. All the cementite remains as composition **X**. The diffusion rate decreases as the temperature decreases. For diffusion to take place, the cooling rate must be **slow** to allow for **equilibrium** conditions to be maintained (remember that the phase diagram **only** refers to **equilibrium** conditions).

Note that there is no alteration in the slope of the cooling curve in the solid part of the system (below **P**) as **no latent heat** (**2.5**) is evolved. The composition lines are determined by changes in volume which arise because there the phase change involves a change in structure, i.e., from a fcc structure (for the γ (austenite) phase) containing 26% 'free space' to a bcc structure (for the α (ferrite) phase) containing 32% 'free space' (**10.11**).

10.11.3 The Critical Temperatures

It is often easier to talk of the steel being **heated** up into the γ (austenite) phase. In order to generalise the iron-carbon phase diagram (***Figure 10.21***) for all carbon steels whatever their carbon content, the phase diagram includes specific **critical temperature lines** at which various transformations take place i.e., line **PSK** is denoted A_1, line **GS** is denoted A_3 and line **ES** is denoted A_{cm}. A_1 is known as the **lower critical temperature line** whilst A_3 and A_{cm} are known as **upper critical temperature lines.** From section **10.11.2** above, we can see that the γ (austenite) phase is not stable below A_3 and A_{cm}. In addition, for carbon contents below 0.8%, cooling the γ (austenite) phase results in the transformation to the α (ferrite) phase whilst the carbon content of the remaining γ (austenite) phase increases along the line A_3 until the point **S** (0.8% carbon) is reached. For carbon contents above 0.8%, cementite will separate out and the composition of the remaining γ (austenite) phase will vary along line A_{cm} until again point **S** (0.8% carbon) is reached. Point **S** is known as the **eutectoid** where all three phases (γ (austenite), α (ferrite) and cementite) coexist at equilibrium.

10.11.4 Effect of Alloying Elements on the Iron-Carbon Phase Diagram

Different alloy elements affect the iron-carbon phase diagram in different ways. These effects are summarised in ***Figure 10.27***, which portrays the effect of alloy elements upon the γ (austenite) phase. Alloying elements are subdivided into those that **open up** the austenite field by lowering the eutectoid temperature (producing a fcc structure at room temperature). These elements (e.g, cobalt, Co; nickel, Ni; manganese, Mn) give rise to **austenitic** stainless steels (**10.10.7**). Elements that **restrict** the austenite field by joining up the α (ferrite) and δ phases (e.g, chromium, Cr; aluminium, Al; tungsten, W; silicon, Si) form **ferritic** stainless steels (**10.10.7**). **Table 10.3** shows the effect of various alloying elements on the strength of steels obtained.

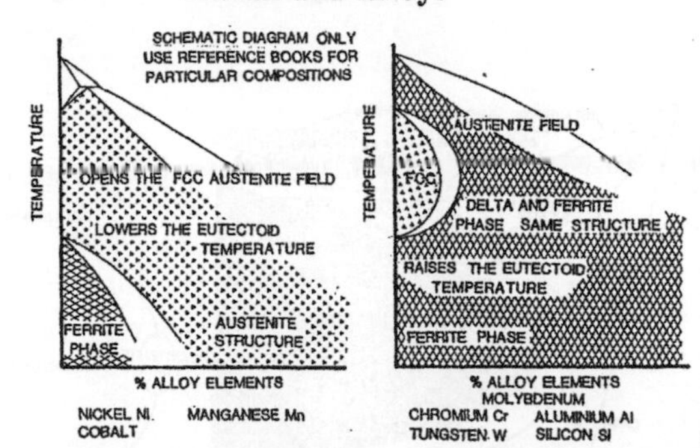

***Figure 10.27** Effect of alloying elements on the γ (austenite) phase in the iron-carbon phase diagram*

10.12 NON EQUILIBRIUM COOLING

In previous sections we have been concerned with interpreting phase diagrams obtained under **equilibrium** conditions. The phase diagram only models and tells us what phases are present if equilibrium conditions are achieved. This means that the temperature drop is small and the temperature is not allowed to fall until **all** the **necessary** diffusion has taken place, thus following equilibrium conditions. In practice, equilibrium conditions are seldom achieved because the rates of cooling are usually much faster than those required to maintain equilibrium. As the cooling rate increases, the transformation temperatures are lowered and new **metastable** (non equilibrium) structures may be formed which do not appear on the (equilibrium) phase diagram. For our purposes, these new structures are **martensite (10.12.1)** and **bainite (10.12.2)** (although **pearlite** **(α + Fe_3C), 10.11.2**, is also formed). These metastable structures may have properties that are different to those obtained under equilibrium conditions and have given rise to component failures, particularly when they are formed inadvertently (e.g, on welding, etc.). However, the factors which gave rise to these failures are now relatively well understood scientifically and understanding the difference between the structures formed from **equilibrium** and **non equilibrium** cooling of steel is essential to our understanding of ferrous metallurgy.

The intention in this section is to represent non equilibrium cooling behaviour in an easily conceivable way (after all, this is what science is doing for us all the time, even though some find the conceptual modelling more difficult than others). As the structure of the iron-carbon system is very sensitive to the rate of cooling **through the eutectoid temperature** (723°C), we will confine our discussions to this region.

Non equilibrium cooling is interpreted using **Temperature Time Transformation** curves (or **TTT** curves), which are shown for the iron-carbon system in ***Figures 10.28a*** and ***10.28b*** below. The curves shown in ***Figure 10.28*** are built up on the premise that the completely γ **(austenite)** phase is quenched very quickly to a new temperature (T) and held at that temperature for a very long period of time (hence the horizontal axis is Log(time)), whereupon the percentage of the austenite transforming at the variable time is calculated. These variables can be used to plot **TTT** curves, represented by the double lines in ***Figure 10.28a***, which have a characteristic "nose".

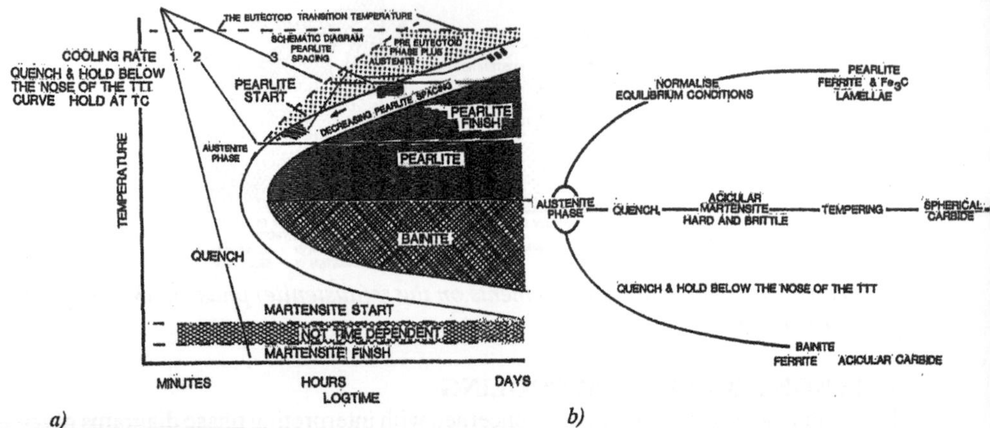

a) *b)*

Figure 10.28 *Non equilibrium behaviour in the iron-carbon system; a) Temperature-time-transformation (TTT) curves; b) transformations of the γ (austenite) phase depending on the rate of cooling*

In ***Figure 10.28a***, the **first (left) curve** of the "nose" represents the time at which the γ (austenite) phase **starts** transforming to either martensite, bainite or pearlite (α + Fe_3C), depending upon the temperature; and the **second (right) curve** of the "nose" represents the time at which these transformations **finish**. The **martensite** lines are shown **horizontally dotted**, as the formation of martensite is **not** time dependent. The **pre-eutectoid phase (10.11.2)** that would exist above the eutectoid temperature 723°C id the **dotted curve** that joins the "nose" of the TTT curve. The formation of **pearlite** and **bainite** are time dependent; pearlite ($\alpha + Fe_3C$) is formed above the "nose" of the TTT curve and bainite below the "nose". The spacing of the pre-eutectoid pearlite is also shown in ***Figure 10.28a***; the presence of pre-eutectoid pearlite does not alter the subsequent precipitation as it will act as nucleation sites for subsequent transformation.

The TTT curves are really masking the clear differences between equilibrium and non equilibrium conditions. The important aspect of the TTT curve concerns the temperature at which the "nose" occurs (for normal steels, the nose is at about 535 to 550°C). For most applications, it is important to appreciate that there is sufficient time for certain cooling rates **not** to cut the "nose" (if the cooling rate cuts the "nose", the efficacy of the heat treatment is reduced). It is also important to appreciate that inclusion of different alloying elements will alter the phase transformations and hence the TTT curves and the position of the "nose".

The TTT curves are not really amenable to the superposition of cooling curves (the usual method of interpretation). However, inclusion of cooling curves puts the TTT curve into perspective for new students of the subject. Three cooling curves have been used in ***Figure 10.28a***, denoted curve **1**, **2** and **3**. The slope of these cooling curves gives an idea of the rate of cooling. However, a cooling curve does not level out and become horizontal (as shown by curves **1** and **2** in ***Figure 10.28a***) unless heat energy is added (as can be achieved under controlled laboratory conditions but not ordinarily in practice). Hence transformations and interpretations based on (laboratory derived) TTT curves for application to engineering structures must be made with caution. Curves **1**, **2** and **3** refer to the very rapid cooling of a fully γ (austenitic) structure at the start of quenching (denoted, for example, by point **L** on ***Figure 10.26***).

Cooling curve 1

- the transformation of the γ (austenitic) phase to **martensite** takes place without the formation of any **pearlite** (α + Fe_3C);
- if cooling curve **1** was arrested just below the nose of the TTT curve (i.e., before crossing the martensite start lines), the γ (austenitic) phase would transform to **bainite** (a continuous α (ferrite) phase matrix, with fine, needle-like acicular carbides). The fine distributed carbides result in the higher strength of bainite compared to pearlite, whilst ductility is retained due to the continuous α (ferrite) phase matrix;
- two terms describe the nature of the temperature arrests that occur; **austempering** (higher temperature arrest) and **martempering** (lower temperature arrest) **(10.13.6)**; both produce **bainite.**

Cooling curve 2

- the γ (austenitic) phase does not start transforming until it has been held at the lower temperature for a long time (of the order hours);
- thereafter, the γ (austenitic) phase starts to transform to **fine pearlite (α + Fe_3C)** (denoted **W** in ***Figure 10.3c***). The **spacing** between the pearlite lamellae is very small (nearly unresolvable by optical microscopy) and this fine pearlite structure produces very good mechanical properties.

Cooling curve 3

- the γ (austenitic) phase takes a very long time to transform to **pearlite** (much longer than cooling curve **2**);
- in consequence, the γ (austenitic) phase transforms to **coarse pearlite (α + Fe_3C)**. The mechanical properties of the coarse pearlite structure are relatively poor and the steel would be very soft e.g, Vickers hardness number (VHN) **(3.4.1)** around 9.5.

Note that, for all the metal samples, when they have crossed the pearlite **finish** lines they can then safely cross the martensite **start** and **finish** lines without undergoing any **further** transformation.

These transformations are summarised in ***Figure 10.28b***, which illustrates schematically the various transformations (depending upon the rate of cooling) that the γ (**austenite**) phase can undergo, e.g, the equilibrium conditions to form pearlite lamellae, together with the non equilibrium quench and hold transformations to form martensite and bainite.

10.12.1 Martensite (a non equilibrium phase)

The microstructure of martensite is shown in ***Figure 10.3h*** and the crystal lattice structure is shown in ***Figure 10.29***.

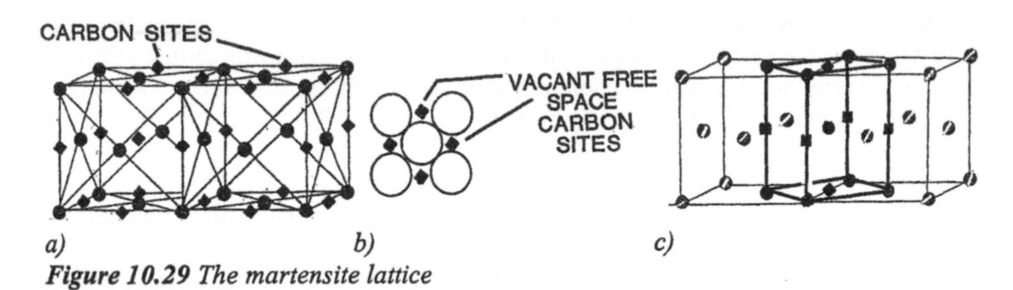

Figure 10.29 The martensite lattice

Figure 10.29a shows the positions of the carbon in martensite. The carbon atoms are located in the preferred **Z** crystallographic direction, in **non interstitial** positions causing a marked increase in the internal stresses. This structure is often described as a distorted bcc lattice i.e., a **tetragonal** lattice **(1.15.3)** (***Figure 10.29a*** and ***10.29c***). It is the position of carbon along the **Z** axis edges which gives rise to the distorted lattice. Note that the bcc lattice of **ferrite** would have carbon atoms along all edges and none in the face diagonal (as shown in ***Figure 10.23a***).

Martensite is an iron-carbon phase which is metastable and will slowly transform to more stable phases over time. As a metastable compound, martensite owes its (transient) existence to the fact that the diffusion of carbon is very slow at low temperatures whilst the formation of martensite is very rapid. In this respect, martensite is very much like other metastable compounds (**2.6.2**). Martensite is a very brittle phase and as such is **not** advantageous in the metallic structure. However, the metastable phase can be heat-treated (called **tempering, 10.13.6**) to produce very small second phase carbides that act as obstacles to dislocation movement (**10.8**); the resultant structure is **hardened and tempered**. The presence of certain alloy elements promote the existence of martensite; some produce a completely new phase diagram whilst others promote the formation of carbides (**Table 10.4**).

10.12.2 Bainite (a non equilibrium phase)

Bainite is formed when the γ (austenite) phase is rapidly cooled to a temperature below the "nose" of the TTT curve, but not so low as to produce martensite. This has the distinct advantage that the risk of cracking, which can occur in a martensitic structure, is reduced. The bainitic structure (α (ferrite) phase and acicular carbides) is **more ductile** than heat-treated (tempered) martensite.

10.13 PROCESSING TECHNIQUES

Steels are capable of very high strengths by the combination of alloying and the subsequent heat treatment of the steel. The effect on strengths obtained by different alloy elements are summarised in **Table 10.3** and the various hardening methods available are summarised in **Table 10.4**.

Table 10.3 Effect of alloy elements on steels

Alloy element	Where dissolved	Effect
Manganese	Dissolved in α (ferrite) phase	Strong effect upon strength
	Dissolved in γ (austenite) phase	Moderate effect upon hardness
	Undissolved carbide in γ (austenite) phase	Mild effect upon fine grain and toughness
	As oxide dispersion	Slight effect on toughness
Nickel	Dissolved in α (ferrite) phase	Mild effect upon strength; very strong effect upon toughness
	Dissolved in γ (austenite) phase	Moderate effect upon hardness
Chromium	Dissolved in α (ferrite) phase	Mild effect upon strength
	Dissolved in γ (austenite) phase	Moderate effect upon hardness
	Undissolved carbide in γ (austenite) phase	Strong effect upon fine grain and toughness
	Dispersed carbide (tempering, **10.13.6**)	Moderate effect on toughness
	As oxide dispersion	Slight effect on toughness
Molybdenum	Dissolved in α (ferrite) phase	Moderate effect upon strength
	Dissolved in γ (austenite) phase	Strong effect upon hardness
	Undissolved carbide in γ (austenite) phase	Strong effect upon fine grain and toughness
	Dispersed carbide (tempering, **10.13.6**)	Strong effect on toughness
Vanadium	Dissolved in α (ferrite) phase	Slight effect upon strength
	Dissolved in γ (austenite) phase	Very strong effect upon hardness
	Undissolved carbide in γ (austenite) phase	Very strong effect upon toughness
	Dispersed carbide (tempering, **10.13.6**)	Very strong effect upon toughness
	As oxide dispersion	Strong effect upon toughness

Table 10.4 Methods of hardening for steels[a]

Solid solution hardening (**10.5.2**)	Transformation hardening[b]	Ferrite grain refining (**10.11.4**)	Precipitation hardening (**10.7**)	Carbide promoters (**10.11**)
Phosphorus P	Carbon P	Aluminium P	Copper P	Chromium P
Silicon P	Manganese P	Vanadium P[d]		Vanadium S
Manganese P	Chromium P	Molybdenum S		Molybdenum S
Nickel P	Molybdenum P			Manganese M
Chromium M[c]	Vanadium P			
Molybdenum M[c]	Nickel S			
Vanadium M[c]	Silicon M			
Copper M				

Notes: P = Primary effect; S = Secondary effect; M = Minor effect. [a] Influence of alloy elements in decreasing order of intensity; [b] By various heat treatments; [c] Over and above the amount required to form carbides; [d] In conjunction with nitrogen.

In the following subsections, the various processes available are discussed with particular reference to the phase diagram (***Figure 10.21***) or TTT curve (***Figure 10.28***).

10.13.1 Forging

Forging is a mechanical deformation process undertaken whilst the steel is in the single γ (austenite) phase. The temperature at which the forging is undertaken is dependent upon the carbon content. An elongated grain structure (***Figure 10.14b***) is produced.

10.13.2 Homogenisation

Homogenisation of steel entails holding the steel for a long time in the γ (austenite) phase, which enables the carbide network to dissolve and diffuse into the interior of the grains, improving hardness (**3.4.1**). The aim of homogenisation is to remove all incipient nuclei as incompletely dissolved carbides (or α (ferrite) phases) act as nuclei for the production of second phases. Homogenisation therefore reduces the number of fine phases in favour of larger phases. The length of time at the homogenisation temperature (1035 to 1200°C) (***Figure 10.21***) is governed by the grain size of the γ (austenite) phase (smaller grains require less time for diffusion of carbon from the grain boundaries into the grain interiors). High holding temperatures promote better quality tool steels by dissolving more carbon, vanadium, tungsten, manganese, chromium, nickel, etc. Furthermore, as a general rule, the higher the concentration of alloy elements in the γ (austenite) phase, the less rapidly they transform (and therefore homogenisation promotes a deeper hardening effect).

10.13.3 Annealing

Annealing is the process where the steel is heat-treated to obtain maximum **softness** (as opposed to (VHN) hardness). The steel must be heated for a long time above the **eutectoid temperature** and slowly cooled through this temperature. Annealing

- improves the machinability of high carbon steels;
- produces an improved structure. Castings have large grains of varying size but within the grain boundary is a region of segregation of impurities (slag, oxides and other insoluble material), as the grain boundary is the last area to solidify. Impurities make cast materials shock sensitive and brittle, and they cannot therefore be deformed (rolled or bent);
- produces a new grain structure by a process of recrystallisation and grain growth (***Figures 10.11*** and ***10.12***), which removes the impurities from the grain boundaries;
- removes the **internal stresses** in cast materials produced by different sections cooling at different rates (***Figure 10.11***);
- produces large grains and a coarse pearlitic structure (***Figure 10.11***).

The **annealing temperature** varies with carbon content, as shown in ***Figure 10.21***, i.e

- with less than 0.8% carbon (i.e., a **hypoeutectoid** steel), a temperature of 25 to 50°C above the upper critical temperature line A_3 (about 900°C) is used, which represents the transformation temperature from the α (ferrite) phase (with a bcc structure) to the γ (austenite) phase (with a fcc structure). At these temperatures, the material must in the single γ (austenite) phase region;
- with more than 0.8% carbon (i.e., **hypereutectoid** steel), a temperature of about 50°C above the lower critical temperature line A_1 (= 723°C) is used and this is in a two-phase region, i.e., γ (austenite) phase + Fe_3C.

Soaking is the term which describes the time at the annealing temperature (generally one hour per 2.5 cm cross section). A typical structure is shown in ***Figure 10.3f*** for a single phase alloy, but not steel; for steel one would get a structure that looked like ***Figure 10.3d*** at very low magnifications.

10.13.4 Spheroidising (applicable to steels only)

Spheroidising is the formation of spheroids from cementite, which are embedded in a matrix of the α (ferrite) phase. Martensite, bainite and pearlite can be changed to this structure. Spheroidising results in considerable softening of the steel.

10.13.5 Normalising (applicable to steels only)

Annealing is an expensive process and so a cheaper process of normalising is often utilised. Normalising involves heating the steel into the γ (austenite) phase at the **normalising temperature** and removing the steel from the furnace to cool in still air. The normalising temperature is dependent upon the carbon content (***Figure 10.21***), i.e.,

- with less than 0.8% carbon (i.e., a **hypoeutectoid** steel), a temperature of about 60°C above the upper critical temperature line A_3 is used to produce a material in the single γ (austenite) phase region;
- with more than 0.8% carbon (i.e., **hypereutectoid** steel), a temperature of about 25 to 50°C above the lower critical temperature line A_1 (= 723°C) is used and this is in a two-phase region, i.e., γ (austenite) phase + Fe_3C.

Following heating and cooling in still air, the γ (austenite) phase transforms to **pearlite** (**α + Fe_3C**), which is formed as small lathes and etches to give a "mother-of-pearl" appearance. The spacing between the α phase and Fe_3C lamellae depends upon the cross sectional area of the steel section because this affects the rate of air cooling. A typical structure is shown in ***Figure 10.3d***.

10.13.6 Quenching and Tempering of Steels

Steels which have been correctly quenched and tempered give the best combined properties of strength and ductility. Quenching and tempering can occasionally take place **unintentionally**; for example, **chisels** and **screw drivers** cannot have their tips remade by a grindstone because the grindstone produces a lot of heat (making the tip glow red), which takes the tip into the γ (austenite) phase field. Although the tip is allowed to cool in air, the large mass of steel acts rapidly to quench the tip to produce a **martensitic** structure. Under these circumstances, the tip will shear off very readily. Similar brittleness arises when high strength steels are welded. The **heat-affected zone** (**10.14.2**) is very brittle. The (VHN) hardness (**3.4.1**) produced by various heat treatments is summarised in **Table 10.5**.

Table 10.5 Hardness and microstructure produced by various heat treatments

Heat treatments	Hardness (VHN)	Microstructure
Quenched from γ (austenite) phase	860	Martensite (**10.12.1**)
As above and tempered (**10.13.6**) 1 hr at 500°C	410	Precipitates of carbides in α (ferrite) phase
Air cooled from the γ (austenite) phase (normalised, **10.13.5**)	270	Pearlite (**10.12**)
Slow or furnace cooled from γ (austenite) phase	210	Coarse pearlite (**10.12**)
Heated for many hours at 690°C	160	Carbides completely spheroidised (**10.13.4**)

Quenching (applicable to all metal alloys and steels)

Quenching is a process that involves a rapid cool so that thermodynamic equilibrium conditions are not achieved. Quenching produces internal strains within a structure, but has very beneficial results if the temperature drop cuts a phase transformation line as, for example, in precipitation hardening alloys (**10.7**, ***Figure 10.28***). There is a slow transformation to the equilibrium conditions, a process called **ageing** (**10.13.7**). In **most** alloy systems (other than steels), quenching produces a supersaturated solution of the phase structure which is in equilibrium with the temperature of heating (**10.7**). To obtain the best results when quenching **steel** in water (or oil), the structure has to be in the γ (austenite) phase, but on quenching this transforms to **martensite** (***Figure 10.29***) if the quenching rate is sufficiently fast. A typical structure for martensite obtained from a quenched steel is shown in ***Figure 10.3h***. Martensite makes the steel a very brittle material. The internal stresses and brittleness formed by quenching steels can be relieved by a process of **tempering (10.13.6)**, which involves heating the steel up to about 100-200°C, whereupon carbon can diffuse within the structure to cause the precipitation of small carbides, thereby altering the structure of matensite (to that shown in ***Figure 10.3a***). Other alloy systems would show a single phase structure on quenching (***Figure 10.3f*** is the nearest) where the grain size would be much smaller.

Tempering (applicable to steels only)
Tempering is a low temperature heat treatment (***Figure 10.21***) that reduces the brittleness of martensite by producing another carbide in the steel (usually written as $\mathbf{M_{23}C_6}$, where M = Fe or any other element in the steel, e.g, Cr, Mo. etc.) (**10.11**). A typical structure for a tempered martensite steel is shown in ***Figure 10.3a***.

Austempering (applicable to steels only)
Austempering involves the rapid cooling of the steel to below the "nose" of the TTT curve, which transforms the γ (austenite) phase to **bainite**. This has distinct advantage that the risk of cracking (which can occur in a martensitic structure) is reduced because the bainitic structure is **more ductile** than tempered martensite.

Martempering (applicable to steels only)
Martempering is involves the rapid cooling of the γ (austenite) phase past the "nose" of the TTT curve to just above the **martensite** start temperature, whereupon the steel is allowed to stay at that temperature until the temperature gradients are minimal (a process called **soaking, 10.13.3**). The steel is then removed and allowed to cool to room temperature. The γ (austenite) phase transforms to **bainite** during this cooling period.

10.13.7 Ageing
The effect of ageing on the stress-strain curve is briefly discussed in section **3.2.2**. For **nonferrous alloys**, ageing is a result of the slow precipitation of the equilibrium phase from a quenched alloy giving rise to a supersaturated solution (e.g, precipitation hardened alloys, **10.7**). For **ferrous alloys**, such as in cold worked mild steel, ageing is due to the interstitial atoms (of carbon and nitrogen) diffusing to the edge dislocation (***Figures 10.4*** and ***10.5***), effectively locking the dislocation at that site. Thereafter a much higher stress (**FG** in ***Figure 10.10b***) is required to unlock the dislocation, resulting in the reappearance of the yield point (point **G** in ***Figure 10.10b***). This is often referred to as **strain ageing** and produces a decrease in the energy of fracture (***Figure 10.30***). Strain ageing is the mechanism at least partly responsible for brittle fracture after welding; the influence effect of strain ageing influencing brittle fracture can be removed by **stress relieving (10.13.8)**.

10.13.8 Stress Relieving
Stress relieving is a process carried out to remove the stresses that have been introduced

by welding or deformation. Stress relieving must be carried out within a precise temperature band (550 to 650°C, ***Figure 10.21***).

10.14 FAILURE MECHANISMS

In addition to mechanical deformation (**10.4**), metals and alloys are affected by corrosion (**Chapter 11**), creep (**3.5.4**), fatigue (**3.5.5**) and brittle fracture (**10.14.1**).

10.14.1 Brittle Fracture

The least expensive way of strengthening steel is by additions of carbon (C), nitrogen (N), phosphorus (P), manganese (Mn) and silicon (Si) (**10.11.4**). These metallic elements significantly raise the yield strength (**3.2.2**) of the steel (***Figure 10.20***). **Carbon** and **nitrogen** are particularly efficient in this respect. However, many metallic elements have a detrimental effect upon the transformation products which form when the temperature of the steel crosses the austenite-ferrite eutectoid (723°C). For example, yield strengths of steels are raised by the addition of **phosphorus**, but this alloy element also causes brittleness. In addition, 're-phosphorised' steels off-cuts cannot be mixed with normal scrap steel because phosphorus raises the **brittle transition temperature** (see below) of the steel. The brittle transition temperature is raised 5°C for every 0.01% phosphorous. Hence the scrap from "re-phosphorised" steel has very low value as phosphorous is considered a **very** deleterious alloy element for other steels to contain. Much less efficient in raising the yield strength is **silicon** and **manganese**. However, these elements also have a deleterious effect on the ductility of the steel and the way in which the steel behaves either under high stress rates or when tested at low temperatures. Atomic **hydrogen** (H^+) also effects the brittleness of steel (**1.8.8**).

Detrimental effects occur because these alloy elements promote the formation of the **bcc structure** (i.e., the α (ferrite) phase) at the eutectoid temperature. When steels with a bcc structure are tested at low temperatures (see ***Figure 10.30***), they gradually show a **decrease in ductility** and **toughness** and an **increase in tensile strength**. However, this increase in tensile strength is of no use to the designer because the steel can be broken in a brittle manner simply because the dislocations cannot interact to produce toughness and ductility (**10.4**). Steels which have the γ (austenitic) phase fcc structure do **not** show this brittleness with a decrease of temperature, despite the fact that both bcc and fcc structures (***Figure 10.4***) both have 12 slip systems. Scientists have investigated this loss of ductility by testing the steels at various temperatures below room temperature to produce curves similar to those reproduced in ***Figure 10.30***.

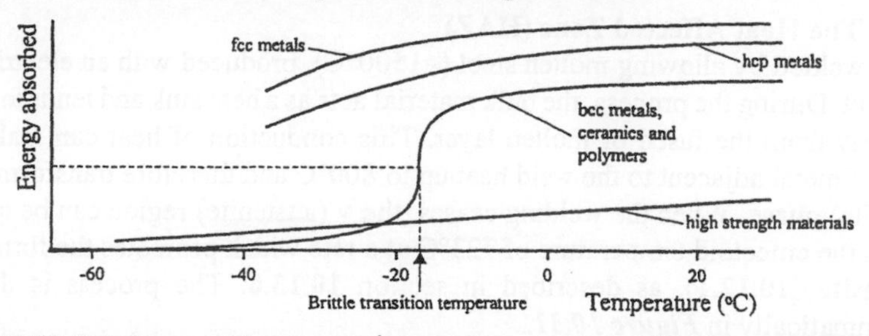

Figure 10.30 *Brittle transition temperatures for a variety of lattice structures*

The samples used in scientific tests (***Figure 10.30***) are usually **notched** to simulate the stress concentrations which may appear at small radii, at weld joints, etc. The characteristic of a metal is that it is not drastically affected by the presence of either notches forming a stress raiser or influenced by deformation at low temperatures. This is explained by the movement of dislocations which **strain (work) harden (10.5.3)** the region so that the subsequent deformation tends to 'blunt' the notch. A ductile fracture depends upon the steel being able to deform locally at the stress raising notch so as to blunt the effect of the stress raising notch. However, this effect is **not** exhibited by a **bcc** structure when the temperature at which the metal is tested decreases. The notch is not blunted at lower temperatures for steel, which undergoes a brittle transition at a temperature of about -18°C. This temperature is called the **brittle transition temperature** and is modelled as the average energy of absorption. Some metallurgical processes tend to shift this curve to higher temperatures so that the **brittle transition temperature** may occur at higher temperatures. Consequently, these metallurgical processes can be very damaging for structural units made from steels exhibiting the bcc structure. One important metallurgical process which causes this effect is **welding,** which produces a **heat affected zone (HAZ) (10.14.2).**

The **brittle transition temperature** is investigated using either the **Izod** or **Charpy** impact hammers **(3.4.1)**. The general rule is that fcc metals (and, to a lesser, extent hcp metals) do **not** show brittle fractures at low temperatures, whilst bcc metals, ceramics and polymers are very susceptible.

Laboratory testing (combined with experience of the analysis of failures which have taken place in industry) has shown that a good margin of safety should be used when designing structures from α (ferritic) steels (which have the bcc structure) which subsequently have to be welded. A good example of the analysis of failures which has been used to inform design decisions resulted from a change in the industrial process of rivetting to welding steel sections for use in ships. A number of ships which incorporated welded steel sections mysteriously disappeared without trace or warning. Investigations undertaken subsequently have shown that the welding process gave rise to metallurgical changes within a very small region of the weld (which is termed the **heat affected zone (HAZ) (10.14.2)**. Similar damage has been found in welded construction frames after the Kobe earthquake. The stresses induced by welding also gives rise to **strain ageing (10.13.7).**

10.14.2 The Heat Affected Zone (HAZ)

Steel is welded by allowing molten steel (>1500°C), produced with an electric arc, to fill a fillet. During the process, the bulk material acts as a heat sink and tends to conduct heat away from the fused or molten layer. This conduction of heat can make a thin region of metal adjacent to the weld heat up to 800°C and therefore transform to the γ (austenite) phase. When the welding ceases, the γ (austenite) region can be quenched through the eutectoid temperature of 723°C at a rate which promotes the formation of **martensite (10.12.1)**, as described in section **10.13.6**. The process is illustrated diagrammatically in ***Figure 10.31***.

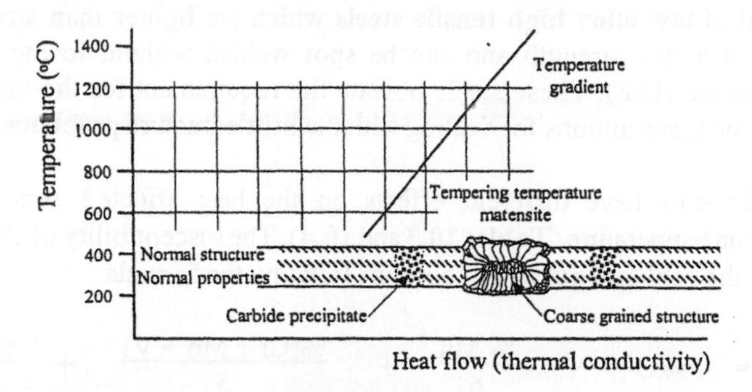

***Figure 10.31** Heat affected zone (HAZ) in welding*

The formation of martensite occurs when the rate of cooling is sufficiently high so that the cooling curve misses the "nose" of the TTT curve (shown by curve 1 in ***Figure 10.28a***). Adjacent areas will undergo different cooling rates such that they do not miss the "nose" of the TTT curve. Regions not taken in to the γ (austenite) field are unaffected. Therefore, if the steel is capable of undergoing martensitic transformation, this will occur in a very narrow zone of the weld (the **heat affected zone, HAZ**). Due to the brittle nature of martensite, the **HAZ** plays host to both the **initiation** and **propagation** of cracks.

Note that **strain (work) hardening (10.5.3)** of steel provides the designer with some margin of safety, which may be lost on welding due brittle fracture within the HAZ. For example, the first welded ships "broke their backs" by this form of brittle crack propagation within the heat affected zone. Problems exist today in welded steel framed structures which were thought to be more resistant to earthquakes. After some tremors, the buildings may remain standing, but the welds have shown signs of cracking along the HAZ.

Steps can be taken to eliminate the problem of brittle fracture in the HAZ, for example by preheating the metal prior to welding, by allowing the welded steel to cool at a much slower rate through the eutectoid temperature (723°C) (curve **2** in ***Figure 10.28a***) or by using steels with appropriate alloying elements. Some welded steels have to undergo a post-welding heat treatment in order to **temper** the martensite (**10.13.6**). However, this is an expensive method of overcoming the problem of the HAZ and would add to the price of welding.

In some circumstances, spot welding is used (e.g, in the manufacture of vehicle bodies, filing cabinets, some steel chairs, etc.), which has the advantage that it is a fast procedure. Preheating of the component to be spot welded could be used to overcome the problem of the HAZ, but the process would involve added expense and, furthermore, if there was a stoppage on the production line, there could be no guarantee that the "still to be welded" components would go back into the preheating ovens. Therefore, it is more practical to utilise steels which can be welded without having to go through preheating or post welding heat treatment processes. This has led to the development of **low alloy high tensile steels** which are lighter than structural mild steels, have a higher strength and can be spot welded without losing ductility or toughness (in the HAZ). These steels obviate the requirement for the expensive and often impractical precautions for dealing with the brittle fracture problems in the HAZ.

Different elements have different effects on the heat affected zone and brittle transformation temperature (**Tables 10.3** and **10.4**). The susceptibility of alloy elements is related to the **carbon equivalent content (CE)** by the formula

$$CE = \%C + \frac{\%Mn}{6} + \frac{\%(Cr + Mo + V)}{5} + \frac{\%(Ni + Cu)}{15}$$

If **CE** is below 0.25, there should be no significant detrimental effects, but if the **CE** is 0.5% or more then care should be taken in cooling the weld area (and the metal may require preheating before welding).

10.15 REFERENCES

1. BS EN 10027. Designation systems for steel. Part 1: 1992: Steel names, principal symbols. Part 2: Steel numbers.
2. BS EN 10088: Part 1: 1995. Stainless Steels. Part 1. List of Stainless Steels.
3. BS 6744: 1986. British Standard Specification for Austenitic stainless steel bars for reinforcement of concrete.

11

CORROSION

11.1 INTRODUCTION

Corrosion is a multi-disciplinary subject. Unfortunately, many half truths and myths exist concerning corrosion. These are perpetuated by having little or no knowledge of the science of the corrosion. A scientific approach is therefore adopted in this text. The subject matter is developed in such a way that the reader will gain confidence and eventually be able to recognise the many elementary misconceptions.

11.2 THE MYTHS?

Let us first of all discuss what is **not** the cause of corrosion.

- A metal is not porous, nor does it 'sweat' water. A metal is in fact a very dense substance and there is no room for water in its structure. Water appears on the surface of a metal because atmospheric water vapour condenses on the cold surface. Hygroscopic salts (e.g, road deicing salts) that may be present on the metal surface increase this condensation of water. The salts do not cause corrosion, they only accelerate it. In addition, salts are responsible for the water contained in paint blisters (**11.12.3**). The metal has not sweated this water from within its structure;
- Corrosion is not due to 'residual' corrosion from previously recycled steel. Once the metal has rusted, it must be heated up to the melting point (with carbon as the reducing agent) to reproduce the metallic state again. The oxide is very brittle and has low energy absorbing properties. Rust also contains water and harbours salts, thus aggravating the problem of corrosion (**11.15.1**);
- Some engineers have postulated that magnetic fields act as small generators to produce corrosion; this is not the cause of corrosion.

11.3 THE CORROSION REACTION

The word corrosion is derived from the Latin word *corrodo*, which means to gnaw away or to wear away. The dictionary definition is "a process in which a solid, especially a metal, is eaten away and changed by chemical action, in the presence of water, by an electrolytic process". Corrosion is thus an electrical phenomenon. It is equivalent to a "short circuited battery with latent potential difference" and is therefore rather like the dry cell battery (***Figure 11.1a***) which powers flashlights (***Figure 11.1c***), etc. Batteries have an internal resistance to the flow of charge that governs how the battery operates.

A knowledge of how internal resistance affects the operation of a battery is also important in understanding how the corrosion process operates. To illustrate this, two types of battery are compared in ***Figure 11.1***.

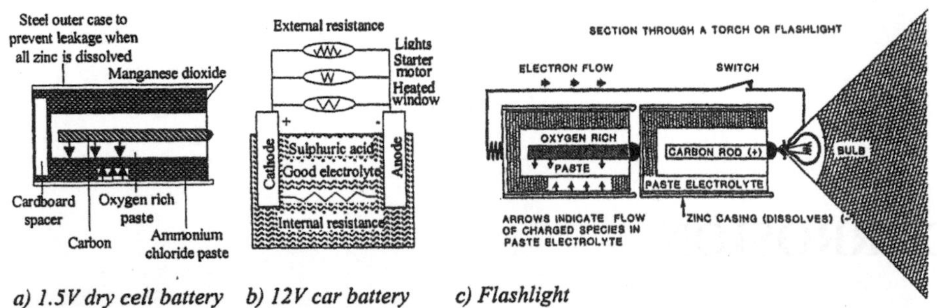

a) 1.5V dry cell battery b) 12V car battery c) Flashlight

***Figure 11.1** Corrosion reactions in a battery*

Figure 11.1a shows a cross section through a 1.5V dry cell battery. The battery produces electrons at the negative terminal, which causes the zinc outer case to dissolve. In order to make "leak-proof" batteries, the zinc case is given a layer of steel. The oxygen is provided from a chemical compound in the ammonium chloride paste which acts as the electrolyte. Several (8 x 1.5V = 12V) of these dry cell batteries can be joined up in series and produce 12V. However, this will **not** start a car because there is a large potential drop within the paste electrolyte due to the **high internal resistance** of the battery. ***Figure 11.1b*** shows the 12V car battery. This has a very **low internal resistance** which allows the battery to produce the very high starting current required to turn the starter motor over. If other items are also run off the battery at the same time that the starter motor is activated, the 12V supply available from the battery drops to sometimes unacceptable levels. Most motorists will have noticed the brightness of the headlights fading as the starter motor is engaged. The important point to understand is that the external current drawn must go through the **liquid (electrolyte) environment,** as well as the **metallic conductors**. These requirements apply to corrosion cells as well.

Figure 11.1c shows a cross section through a typical torch or flashlight. The electrons are shown flowing in the metal external circuit and the charged species in the paste electrolyte to provide the complete electrical circuit. The electrical energy for the light is produced by the zinc case dissolving. The zinc is termed the **anode** in corrosion technology and is the negative -ve terminal (or half cell) of the battery. The electrons are consumed at the carbon rod, which in corrosion technology is termed the **cathode** i.e., the positive +ve terminal (half cell) of the battery. The wasting away of the zinc case can be stopped by switching off the flow of electrons in the outer metallic circuit. Note that, unlike a battery, in many corrosion processes there is no switch to separate the cathode from the anode to prevent the continued wasting away of the anode.

A battery has a positive (+ve) and a negative (-ve) terminals (termed **electrodes**) at which specific chemical reactions take place. Similarly, a corrosion reaction has **cathodes** (positive +ve) and **anodes** (negative -ve) terminals at which specific chemical reactions take place. **A battery is a corrosion process which we want to make more**

efficient. A corrosion cell is an unwanted reaction which we want to stop or make less efficient. Table 11.1 shows a comparison of a battery with the corrosion process.

Table 11.1 Comparison of a battery with the corrosion process

Parameter	Battery	Corrosion
Potential difference obtained	Potential difference produced. Metallic circuit contains a switch and resistance (bulb, motor, etc.)	No potential difference produced. (because system is short circuited)
Efficiency	Want to increase efficiency.	Want to decrease efficiency.
Electrolyte	One or two distinct electrolytes for anodes and cathodes, separated by a semipermeable membrane[a]	Electrolyte common the anode and cathode (there can be many corrosion cells on one structure[b])
Corrosion terminals and environment	Positive (+ve) and negative (-ve). In some batteries these share the same electrolyte, but terminals are connected via a switch[c]	Cathode and anode. Share the same metallic substrate and the same electrolyte environment.
Current interruption	By switch in metallic circuit	Only by discarding the electrolyte
Chemical reactions	Similar chemical reactions in each half cell or electrode.	

Notes: [a] see, for example, ***Figure 11.10***; [b] see, for example, ***Figure 11.23***; [c] see, for example, ***Figure 11.1c***.

11.3.1 Anodic Reactions

The anode is **the electrode at which the current (positive charge) enters the electrolyte.** It is the (-ve) electrode of a 'corrosion battery'. There are a number of anode reactions; the most important in the context of the construction industry are considered below. In each case, the anode reaction **produces electrons.**

i. $$M \rightarrow M^{n+} + ne$$

where **n** is the **valency (1.5)** of the metal, **M** is any metal and **e** are the electrons. This is the most important anode reaction encountered, and expresses the fact that the **anode dissolves.** It should be remembered that this reaction can in fact proceed in either direction, depending on the conditions present **(11.4)**. In our battery, we prevent the (anode) reaction from proceeding from left to right by switching off the battery. Examples include

$$Zn_{(metallic\ state)} \rightarrow Zn^{2+}_{(liquid\ state)} + 2\ \text{electrons}$$

This reaction takes place in a battery (***Figure 11.1a***) and in the corrosion of zinc.

$$Fe_{(metallic\ state)} \rightarrow Fe^{2+}_{(liquid\ state)} + 2\ \text{electrons}$$

$$Mg_{(metallic\ state)} \rightarrow Mg^{2+}_{(liquid\ state)} + 2\ \text{electrons}$$

In each case, the electrons produced must stay in an electrical conductor (i.e., the zinc, iron or magnesium metal). It is expedient to leave out the subscripts (metal state) and (liquid state) and abbreviate electrons to e. The reaction of zinc can therefore be written

$$Zn \rightarrow Zn^{2+} + 2e$$

ii. $4OH^- \rightarrow (Air)\ O_2 + 2H_2O + 4e$

This reaction is important in the remediation of reinforced concrete using cathodic protection or prevention (**11.13.3**).

iii. $Cl^- + Cl^- \rightarrow Cl_2 + 4e$

This reaction is important in the remediation of chloride affected reinforced concrete, where the concrete cover surrounding the reinforcement contains chlorides (**8.21.5**).

The electrons formed at the anode must be used up by a cathodic reaction. Electrons can only move in a metallic conductor and therefore the anode and cathode regions must either be the same metal or be joined by electrical conductors.

11.3.2 Cathodic Reactions

The cathodic reaction is equivalent to the (+ve) electrode of the corrosion battery. **Cathodic reactions consume electrons**. There are several cathodic reactions, the most common of which are

i. $(Air)\ O_2 + 2H_2O + 4e \rightarrow 4OH^-$

This chemical reaction takes place in **most corrosion cells where oxygen is available** and produces OH^- ions (**hydroxyl ions**). Hydroxyl ions are **alkaline** and therefore this cathodic reaction results in the cathodic area becoming alkaline. The reaction is made to take place in the dry cell battery (***Figure 11.1a***) with great efficiency by using strong oxidising agents in the paste electrolyte.

ii. $H_2O \rightarrow H^+ + OH^-$

followed by $2H^+ + 2e \rightarrow H_2\ (gas)$

This cathodic reaction occurs on metals placed in acid electrolytes, provided the metal is **below hydrogen** in the electrochemical series (**Table 11.2**). For example, copper will not liberate hydrogen from an acid, but zinc will. For zinc placed in an acid, the reactions are

Anode reaction $Zn \rightarrow Zn^{2+} + 2e$

Cathode reaction $2H^+ + 2e \rightarrow H_2\ (gas)$

These reactions are sometimes shortened to

$Zn + 2H^+ \rightarrow H_2 + Zn^{2+}$

Note that the reaction $2H^+ + 2e \rightarrow H_2$ (gas) sometimes takes place where water is present without oxygen (**8.21.2**). Note that the H^+ ions (**hydrogen ions**) are derived from the water (or acid) electrolyte. Hydrogen ions are **acidic** and therefore, since hydrogen ions are **consumed**, this cathodic reaction also results in the cathodic area becoming alkaline.

iii. Cu^{2+} + 2e → $Cu_{(metal)}$

This reaction oan only take place when there are metallic ions (M^{n+}) in solution. In most corrosion reactions encountered in the building industry, there are no metallic ions (M^{n+}) in solution, and therefore this reaction is not usually important. For example, copper chrome arsenic (CCA) compounds are used as wood preservatives to protect timber (**12.25.4**). The copper must not be allowed to go into solution (as Cu^{2+} ions), since any iron or steel in contact with the wood (e.g, nails) would become the anode and corrode

Anode reaction $Zn \rightarrow Zn^{2+} + 2e$

and the copper ions in solution would form the cathode

Cathode reaction $Cu^{2+} + 2e \rightarrow Cu_{metal}$

i.e., the copper would be plated out on the steel and could then form a **galvanic** corrosion cell (**11.6.1**). This is prevented in the CCA wood preservative by ageing, which allows the chromium to chemically 'fix' the copper ions to the wood substrate, tying the copper up in a covalent structure. Note that, whilst relatively unimportant in the building industry, the cathode reaction ($M^{n+} + ne \rightarrow M_{(metal)}$) is important in **electroplating**, where metallic ions in solution are plated out as a metal e.g, zinc galvanised steel (**11.13.2**).

11.3.3 Combination of Anodic and Cathodic Products

The electrons formed at the anode must be used up by a cathodic reaction. Electrons can only move in a metallic conductor and therefore the anode and cathode regions must either be the same metal or be joined by electrical conductors. The **aqueous** by-products formed by the anode and cathode must react **wlthin** a liquid environment. Using the corrosion of steel (Fe) as an example

Fe^{2+} (anions) + $2OH^-$ (cations) → $Fe(OH)_2$

If the products of the anode and the cathode are unable to react, the corrosion process is inhibited. Adherent paint films (**11.12.1**) inhibit this reaction. If the products of the anode and the cathode are unable to react, the corrosion process is inhibited. Conversely, if we want a battery to produce a higher current, we must make the flow of these cations and anions faster (as explained in relation to dry cell and car batteries in section **11.3** above). The addition of more air (further **oxidation**) produces **rust**, Fe_2O_3, which is accompanied by an approximate 8-fold increase in volume compared to the original volume of the metal.

11.4 ELECTROCHEMICAL AND GALVANIC SERIES

As shown in **Chapter 1**, there are several metallic elements that comprise the Periodic Table (**1.2**). The metallic elements all have a different tendency to **ionise** (**1.12**). This is expressed as

$$M \rightleftharpoons M^{n+} + ne$$

Metals have an **equilibrium potential** (**11.4.1**) with respect to the **hydrogen electrode**

(**11.4.2**). These equilibrium potentials are tabulated in **Table 11.2**, which is known as the **electrochemical series** for pure metals. The electrochemical series gives the half cells (**11.4.1**) for most common metals in combination with their own ions;. each half cell reaction has been written so as to consume electrons. The oxygen electrode reaction is strictly not part of the electrochemical series; it has, however, been included as it is an important (cathode) reaction for the corrosion process. The more negative the electrode potential, the more stable the ionic form and the more unstable is the metallic state. As an example, sodium reacts with water to evolve hydrogen gas (**1.8.1**). The reaction is so vigorous that the hydrogen spontaneously ignites. If the anode and cathode reactions could be separated, a voltage of -2.7V could be measured.

Table 11.2 Part of the Electrochemical series (pure metals only)

Metal		Half cell reaction	Voltage (V)
Gold	'Noble end'	$Au^{3+} + 3e \rightarrow Au$	+1.4200
Oxygen electrode		O_2 (Air) + $2H_2O + 4e \rightarrow 4OH^-$	+1.4000 approx.
Silver		$Ag^+ + 1e \rightarrow Ag$	+0.7995
Copper		$Cu^{2+} + 2e \rightarrow Cu$	+0.3448
Hydrogen		$2H^+ + 2e \rightarrow H_2$ (gas)	0.0000 (by definition)
Lead		$Pb^{2+} + 2e \rightarrow Pb$	-0.1260
Tin		$Sn^{2+} + 2e \rightarrow Sn$	-0.1860
Nickel		$Ni^{2+} + 2e \rightarrow Ni$	-0.2500
Cadmium		$Cd^{2+} + 2e \rightarrow Cd$	-0.4020
Iron		$Fe^{2+} + 2e \rightarrow Fe$	-0.4400
Chromium		$Cr^{3+} + 3e \rightarrow Cr$	-0.7100
Zinc		$Zn^{2+} + 2e \rightarrow Zn$	-0.7620
Aluminium		$Al^{3+} + 3e \rightarrow Al$	-1.6700
Sodium		$Na^+ + 1e \rightarrow Na$	-2.7120
Magnesium	'Base end'	$Mg^{2+} + 2e \rightarrow Mg$	-2.8750

Pure metals are not very strong and so alloys are often used in engineering applications. The formation of an alloy alters the chemical potential of the metals forming the alloy. Alloys are arranged in order of their equilibrium potentials with respect to the hydrogen electrode in **Table 11.3**, which is known as the **galvanic series.**

Table 11.3 Part of the galvanic series (alloys and some pure metals)

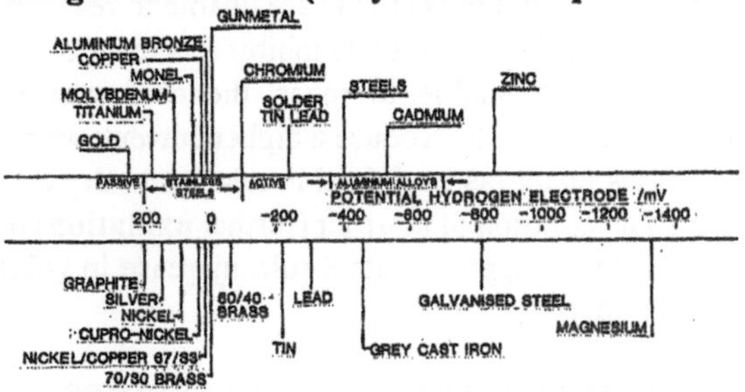

Most alloys are chosen because they form a protective oxide film which is chemically stable and adherent to the substrate metal. These oxides are also coherent (they do not fracture down to the metal surface) and so physically separate the underlying metal substrate from its environment. For this to happen in certain alloys (e.g, stainless steel, **11.15.3**) the alloy must be in an oxidising environment.

Table 11.4 illustrates that some engineering alloys suffer a **potential shift (11.4.5)** in different aqueous environments, e.g, 18/8 stainless steel (type 304) **(10.10.7)** suffers a shift of some 350 mV. Hence electrochemical series have to be interpreted with care as the equilibrium potentials are dependent of the nature of the **liquid environment.**

Table 11.4 Potentials of metals and alloys in different environments

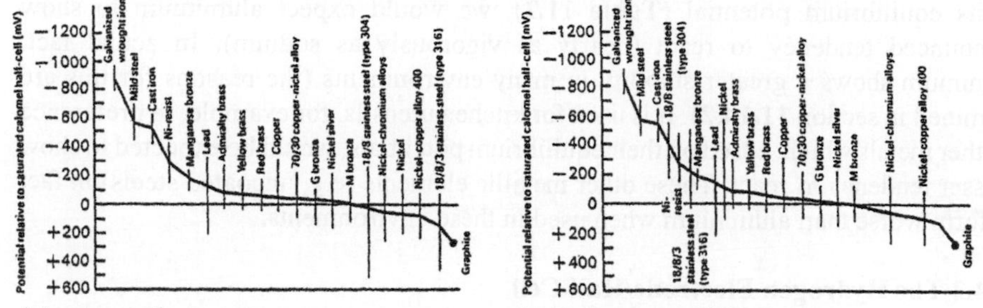

a) in aerated seawater *b) in de-aerated fresh water*

Notes: The reference potentials cited in **Table 11.4** are with respect to the **calomel electrode** (***Figure 11.4c***), rather than the hydrogen electrode.

11.4.1 Equilibrium Potential and Half Cells

Equilibrium potentials are best interpreted in relation to the electrochemical series, which expresses a **pure** metal in the solid state in equilibrium with the liquid state giving electrons. In section **11.3.1**, the equilibrium reaction for a number of metals is written as an anodic reaction ($M \rightarrow M^{n+} + ne$). We can generalise these reactions as

$$M \underset{(B)}{\overset{(A)}{\rightleftharpoons}} M^{n+} + ne$$

This reaction is called a **half cell.** There are many possible half cells, as shown in **Tables 11.2** to **11.4**. If the reaction proceeds in the direction **(A)** it is an **anodic** reaction, whilst if it proceeds in the direction **(B)** it would be a **cathodic** reaction (characteristic of an **electroplating process, 11.5.1**). If the rate of the reaction in direction **(A)** equals the rate of the reaction in direction **(B)**, there exists a **state of equilibrium.** Therefore an electrode is no longer at equilibrium when some current flows to or from its surface. Each corrosion reaction consists of two reactions: an **anodic** half cell and a **cathodic** half cell. The anodic reaction (the dissolving electrode) rises in potential when a current is drawn from the equilibrium potential, whilst the cathodic reaction falls in potential. This is shown in ***Figure 11.2.***

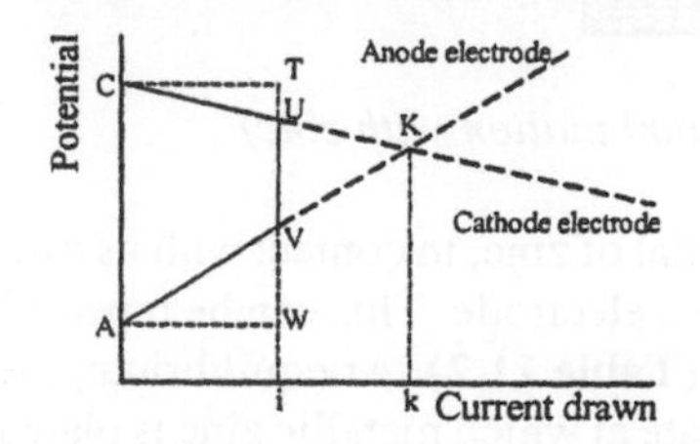

The potential difference between electrodes of a battery or corrosion cell falls with respect to the current drawn from the cell. This is modelled by considering a voltage drop across an internal resistance within the cells, due to the products of each electrode reaction remaining adjacent to the electrode. Representing the (open circuit) anodic and cathodic potential as A and C respectively, the external potential difference measured is AC if no current is drawn, but drops to UV if current 'i' is drawn. The open circuit potential AC has dropped due to a voltage drop across the cathode (TU) and anode (VW). These potential drops are referred to as **polarisation (11.4.8)**. Thus TU represents the cathodic polarisation and VW the anodic polarisation (see also ***Figure 11.9***).

Figure 11.2 The potential difference between the electrodes of a corrosion cell

The measured potential of these electrode/ion systems deviate from the equilibrium potential depending upon the magnitude of the current flowing. Therefore, in using the equilibrium potential of a metal in its environment (**Tables 11.3** and **11.4**), we are only able to obtain a **corrosion tendency** and cannot express a **corrosion rate (11.4.7)**. As corrosion rates are the main requirements for the designer and end user, it is particularly important that these equilibrium potentials are used with caution. For example, based on its equilibrium potential (**Table 11.2**), we would expect aluminium to show pronounced tendency to react (nearly as vigorously as sodium). In actual fact, aluminium shows a greater stability in many environments (the reasons for this are examined in section **11.14.2**). It is used for kitchen utensils, for example, in preference to other metals which, based on their equilibrium potentials, would be expected to show a lesser tendency to react. These other metallic elements (e.g, uncoated steels) in fact perform worse than aluminium when used in these environments.

11.4.2 The Hydrogen Electrode/Half Cell

There is a potential difference (voltage) associated with all half cells; to measure potential difference we either have to make a connection to another half cell or have some elaborate method of measuring the potential of the ions in solution. Commonly, we make a connection to another half cell, termed a **reference electrode (11.4.3)**. The reference electrode is usually given the position of arbitrary zero and must be fairly reproducible, so that the same measured potentials can be obtained anywhere in the world. A good analogy of the use of an arbitrary zero half cell is the ability to measure a distance. Historically, distances were first measured using a human's anatomy, an inch being the distance of the thumb finger nail to first joint and the yard being the average persons step. Obviously, these reference units changed from person to person and after several revisions throughout our scientific history, a standard unit of distance was introduced (the metre, defined as 1,650,763.73 of the wavelength, in vacua, of orange light emitted by Krypton, Kr). In corrosion, the main reference electrode was taken as the **hydrogen electrode**, which was given the status of being the reference arbitrary zero. The hydrogen electrode is shown in ***Figure 11.3***.

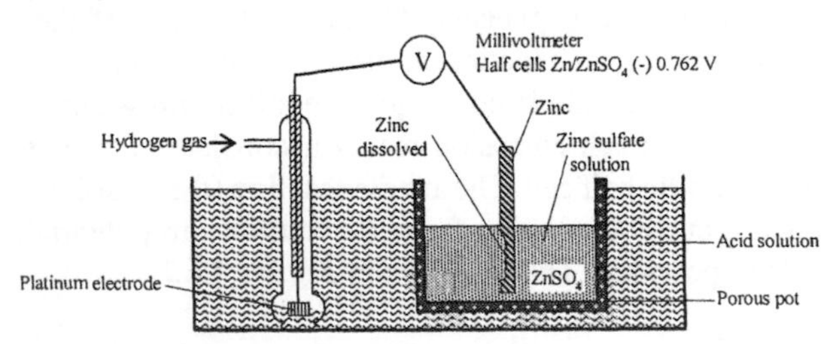

Figure 11.3 *The hydrogen reference electrode (in combination with zinc)*

Figure 11.3 illustrates a schematic view of the potential of zinc, in contact with its own ions, being determined against the hydrogen reference electrode. This can be repeated for all metals to produce the electrochemical series (**Table 11.2**). At equilibrium, the rate at which the metal zinc dissolves is equal to the rate at which metallic zinc is plated out from the zinc ions in solution.

11.4.3 Reference Electrodes/Half Cells

Currently, the hydrogen electrode is used predominantly for determining those metals which liberate hydrogen from an acid (metals more electronegative than the hydrogen electrode). More often, the hydrogen electrode is replaced by more practical reference electrodes (equivalent to the use of a tape measure in the measurement of distance), shown in ***Figure 11.4***. The standard reference electrodes are chosen because their potential is independent of the pH of the environment (see and compare with ***Figure 11.6***). The most used reference potentials (termed **practical reference electrodes**) all give a different half cell potential with respect to the hydrogen electrode (and also, therefore, to the metal measured against them).

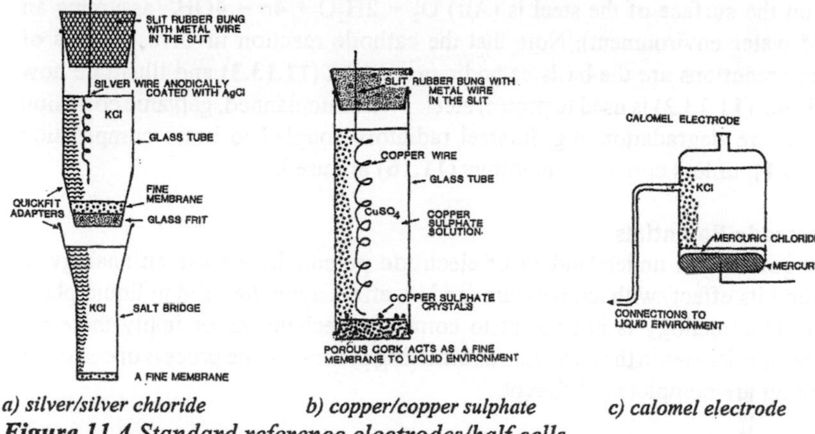

a) silver/silver chloride *b) copper/copper sulphate* *c) calomel electrode*

Figure 11.4 Standard reference electrodes/half cells

11.4.4 Coupling Different Half Cells

The way in which we shall use **Tables 11.2** to **11.4** is to consider two metals (Metal A and Metal B) in metallic contact in the same liquid environment (***Figure 11.5***).

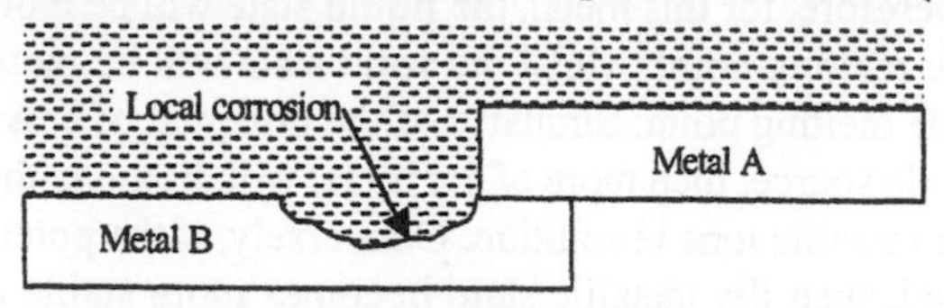

Figure 11.5 Galvanic corrosion of dissimilar metals

Metal B is more electronegative metal is more reactive (has a greater tendency to ionise) compared to Metal A (the more electropositive metal). Hence Metal B becomes the anode and corrodes ($B \rightarrow B^{n+} + ne$). The cathodic reaction occurring on the Metal A will depend upon the liquid environment in which the reactions occur. With sufficient oxygen, the cathode reaction will be (Air) $O_2 + 2H_2O + 4e \rightarrow 4OH^-$. If the liquid environment is de-oxygenated, the cathode reaction will be $H_2O \rightarrow H^+ + OH^-$, followed by $2H^+ + 2e \rightarrow H_2$ (gas). Note that the other cathode reaction that could take place is $A^{n+} + ne \rightarrow A$. However, for this reaction to take place, Metal A would have to be in a solution of its own ions (A^{n+}). If we assume that the liquid environment in ***Figure 11.5*** is water (the most usual case), then there are no Metal A ions in solution and the cathodic reaction $A^{n+} + ne \rightarrow A$ cannot take place. If we further assume that te water is oxygenated, the cathode reaction occurring on Metal A will be (Air) $O_2 + 2H_2O + 4e \rightarrow 4OH^-$; this is the most usual cathode reaction encountered in corrosion problems in the construction industry. Note that the cathodic reaction that actually takes place is dependent on the **reactivity of the metal** and the **type of liquid environment** in which it is placed.

In some cases, galvanic corrosion is designed for; if Metal B was zinc and Metal A was steel, zinc forms the anode and corrodes ($Zn \rightarrow Zn^{2+} + 2e$) whilst the cathode reaction occurring on the surface of the steel is (Air) $O_2 + 2H_2O + 4e \rightarrow 4OH^-$ (assuming an oxygenated water environment). Note that the cathode reaction involves no loss of metal. These reactions are the basis cathodic protection (**11.13.3**) and illustrate how galvanised zinc (**11.13.2**) is used to protect steel. Where unplanned, galvanic corrosion can cause severe degradation, e.g, in steel radiators coupled to brass compression fittings (**11.6.1**), unless corrosion inhibitors (**11.16**) are used.

11.4.5 Electrode Potentials

To aid the conceptual understanding of electrode potentials, we use an analogy of potential (and its effect) with temperature (and its effect upon the solid to liquid phase transition). This analogy is not meant to compare mechanisms or imply there are fundamental similarities in the corrosion and melting processes; the process of corrosion and of melting are completely different.

In ***Figure 11.6a***, the temperature scale is compared to the electrode potential of a metal. A metal above the melting point (m.pt.) will melt (at a rate determined by the supply of latent heat, **2.5**). Above the melting point, all the metal will be liquid, whilst below the melting point, all the metal will be solid. At the melting point the phase change (solid $\rightleftharpoons$ liquid) occurs. Therefore, for this metal, the liquid state will be more stable than the solid state above the melting point, whilst the solid state will be more stable than the liquid state below the melting point. Similarly, if the potential of a metal is artificially raised by some outside source, then more of the metal will dissolve, thereby increasing the concentration of metallic ions in solution. Conversely, if the potential of the metal is artificially lowered, then the metallic state becomes more stable and some of the metallic ions will plate out of the solution, i.e., electroplating (**11.5.1**).

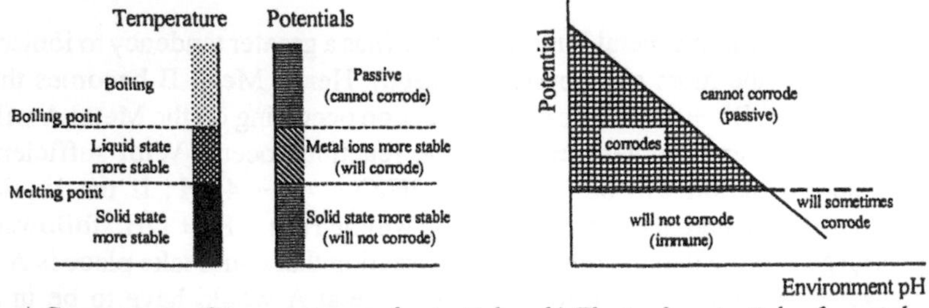

a) Comparison of temperature and potential *b) Electrode potentials of a metal*

Figure 11.6 *Comparison of the temperature scale with electropotentials of a metal*

If the temperature of the metal is raised further, the metal will boil. If the potential of the metal is raised, it often ceases to dissolve and becomes covered with an oxide film. The metal is said to be **passive**. The melting point of a metal is independent of the environment. Conversely, the ability of a metal to corrode is dependent upon the **pH** (**1.19.3**) of the environment, as shown by ***Figure 11.6b***. In the region of potential above the equilibrium potential (given for pure metals in **Table 11.2**), the metal **will** corrode as the ionic state is the preferred state. In the region of potential below the equilibrium potential, the metal **will not** corrode because it **cannot.** As shown in ***Figure 11.6b***, the potentials at which this takes place are dependent upon pH; ***Figure 11.6b*** is called a Pourbaix diagram (**11.4.6**). The Pourbaix diagram essentially plots the equilibrium potential (vertical axis) against pH (horizontal axis). In addition, for some metals, there is a region where the metal **cannot** corrode even above the equilibrium potential. This is due to the formation of a **passive oxide film**. Hence the environment (pH) affects the ease with which a metal dissolves but does not influence the melting point of a metal. In summary

Temperature	Melting	Potential	Effect upon corrosion
Below the melting point	All solid	Below the equilibrium potential	Can not corrode
Above the melting point	All liquid	Above the equilibrium potential	Will corrode
Far above melting point	Boiling starts	Far above equilibrium	Can not corrode; the metal is passive

11.4.6 Relationship of Potential and pH

A scientist called Pourbaix first examined the relationship of the pH of the environment and potential and plotted these for a variety of metals in an atlas of electrochemical equilibria in water. The general form of these plots is shown in ***Figure 11.7***. The potential-pH equilibria are a summary of thermodynamic data and so do **not** express a rate of corrosion. However, they do enable the scientist to determine whether a metal is capable of corroding or is immune (does not corrode) in specific pH environments.

The potential-pH diagrams for steel (iron) (Fe), zinc (Zn) and aluminium (Al), as postulated by Pourbaix, are shown in ***Figures 11.7a*** to ***11.7c*** respectively. The areas of corrosion, immunity and passivation are shown.

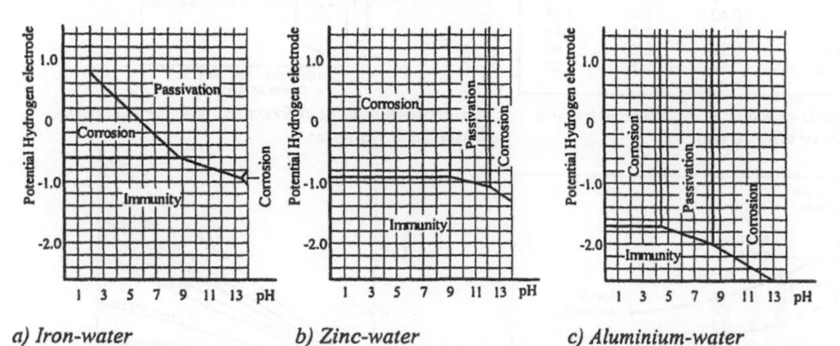

a) Iron-water *b) Zinc-water* *c) Aluminium-water*

Figure 11.7 Potential-pH diagrams for a variety of metal-water equilibria

Zinc and aluminium have been chosen because they are the only two metals widely used in the construction industry which dissolve in an alkaline medium. Aluminium is used, for example, for window frames, where it has to be set in wooden surrounds to protect it from the alkalinity of the mortar bed (**1.8.3**). The alkali dissolves the aluminium oxide film and prevents reformation of the oxide, so the aluminium metal beneath reacts fairly vigorously with water to produce hydrogen. This reaction is turned to good use to make aerated concrete products (**7.7.1**). Aluminium powder is mixed with a alkali cement, which strips the oxide film, allowing water to react with the aluminium metal beneath. Hydrogen (and steam) is evolved and these gases 'aerate' the concrete product. Zinc is used as galvanisation (**11.13.2**) and will etch in contact with an alkali.

Using the iron-water potential-pH diagram (***Figure 11.7a***), the metallic state of the metal is more stable at potentials below the line **EFH** (the region of immunity). If the potential is raised, rapid corrosion would result. With further raising of the potential, the steel would be protected by the passive oxide film (Fe_2O_3) as the steel is taken into the region **UKFH**. The Pourbaix diagrams shows that both zinc and aluminium will dissolve in (alkaline) hydrating cement (pH of about 9 to 13). Both metals show a region of passivation, between a pH of 9 and 12.5 (zinc) and a pH 4.5 and 8.5 (aluminium). Steel (Fe) is also shown to have a region of passivity at high pH. The regions of passivity are attributed to the formation of a **passive oxide film**, which is chemically protective to the metal substrate. For aluminium, the passive oxide film forms in neutral pH environments, whilst for steel (Fe), an alkaline environment is required. Hence steel reinforcement embedded in concrete is protected by the alkaline nature of the concrete matrix (**8.21.3**) (in the absence of chlorides, **8.21.5**). The nature of the passive protective oxide film on steel can be considered (as a first approximation) to be equivalent to the anodising process of aluminium (**11.14.2**).

11.4.7 Corrosion Rates

Factors influencing corrosion rates are shown in ***Figure 11.8***. ***Figure 11.8a*** shows the corrosion rates of various metals in three different environments.

Metal	Atmosphere		
	Industrial	Marine	Rural
Aluminium	0.81	0.71	0.025
Copper	1.19	1.32	0.58
Lead	0.43	1.41	0.48
Zinc	5.13	1.60	0.86
Mild steel	13.72	6.35	5.08
Weathering steel	2.54	3.81	1.27

a) Penetration rates of metals in different atmospheric conditions (μm/year after 10 yrs)

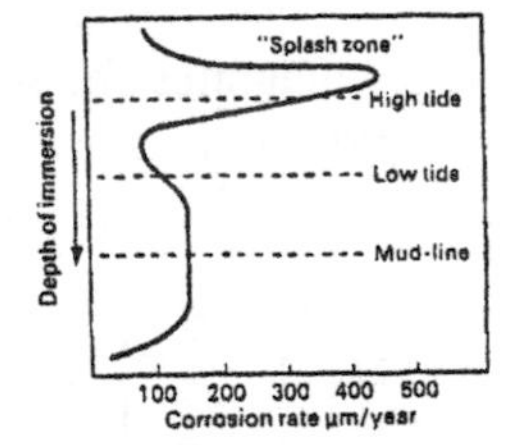

b) Corrosion rates at different positions for steel partly immersed in sea water

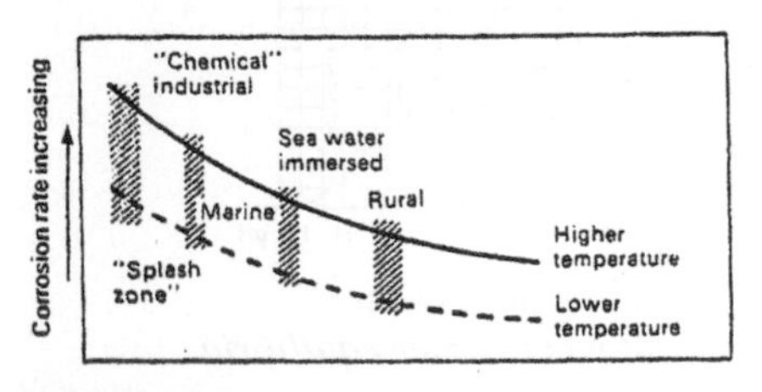

c) Temperature dependence of corrosion rates in different environments

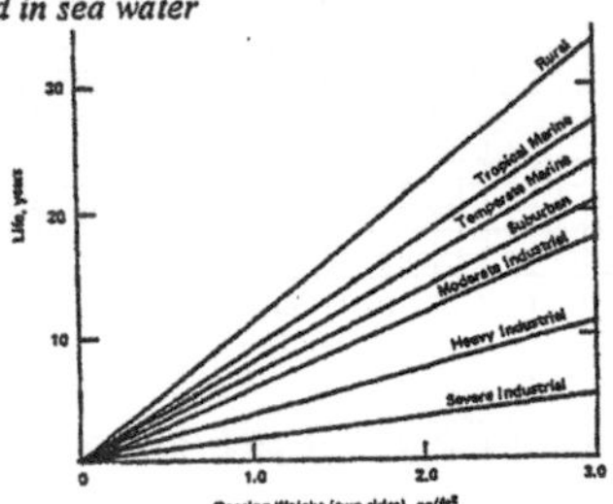

d) Life of zinc coatings on steel in different environments

Figure 11.8 *Factors affecting corrosion rates*

Industrial atmospheres contain sulphur dioxide and other ionic contaminants (e.g, water soluble salts, **6.7**), whilst marine atmospheres contain sodium chloride (to a first approximation, chemically the same material as road deicing salts). A rural atmosphere is considered as one devoid of contaminants and is the least corrosive atmosphere to the metals listed. Sea water is generally considered to be a very corrosive liquid. ***Figure 11.8b*** shows the corrosion rates for various depths of immersion in sea water and ***Figure 11.8c*** indicates the effect of temperature on the rate of corrosion in different environments (**11.6.4**). The splash zone is the most corrosive area. The automotive vehicle operates in this 'splash zone' during the winter period, and most of our roadside structures are subject to pulverised road deicing salts on dry days. The corrosive environment can depend upon how near a major arterial road the structure is sited. ***Figure 11.8d*** shows the life of zinc coatings on steel in different environments. The times are given for the first significant appearance of rust.

11.4.8 Polarisation

Potentials recorded in the electrochemical series are in fact **equilibrium potentials**, i.e., there is no current flowing and the circuit is an open circuit. The half cell electrodes are not at the equilibrium potential when some current flows either from or to the half cell electrode in the metal circuit. The deviation from equilibrium is known as **polarisation**. There are a number of processes which can cause polarisation; we can illustrate these processes by considering the corrosion of iron in a water electrolyte

- **Anodic polarisation**
  - Rate of production of Fe^{2+} ions;
  - Rate of removal of Fe^{2+} ions from the anodic region.
- **Cathodic polarisation**
  - Rate of consumption of electrons at the cathode;
  - Rate of diffusion of oxygen to the cathode area;
  - Rate of removal of OH^- ions from the cathodic region.
- **Mixed polarisation**
  - Ion combination rate to form iron hydroxide.

These processes occur at different rates and the slowest rate will determine the overall rate of corrosion. The degree of polarisation depends upon the current flowing, which can be represented graphically using **Evans curves** (***Figures 11.9b*** to ***11.9e***).

Figure 11.9 *Simplified Evans curves for different types of corrosion control*

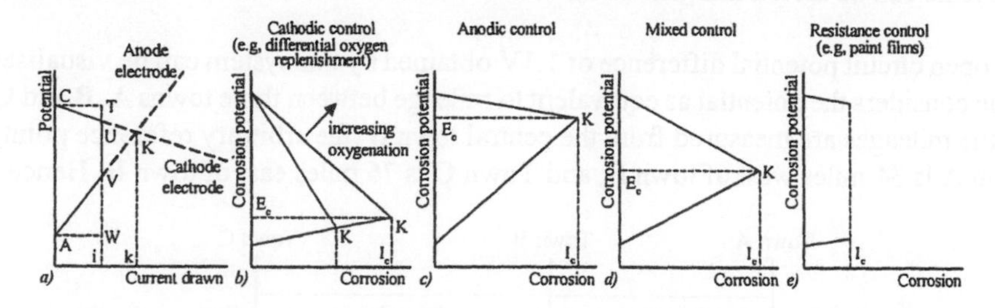

In section **11.4.1** we used ***Figure 11.2*** (reproduced here as ***Figure 11.9a***) to develop the general concept of polarisation at the electrode interface. Different types of polarisation lead to different types of corrosion control. The effect of anodic, cathodic and mixed polarisation on the cathodic and anodic potentials are shown in ***Figures 11.9b*** to ***11.9d*** respectively. Resistive control (***Figure 11.9e***) is characteristic of organic coating technology (**11.12**) using paint films. The corrosion potential, E_c, can be measured by a reference cell and is the basis of **potential mapping** of concrete structures (**8.24.2**).

11.5 EXAMPLES OF HALF CELLS

One of the earliest batteries (1836), attributed to Daniels, was made by separating the anodic and cathodic reactions, both electrically and ionically (***Figure 11.10a***). This was achieved by using copper (Cu) in copper sulphate ($CuSO_4$) as the **cathode** (+0.34V with respect to the hydrogen electrode, **Table 11.2**) coupled to zinc (Zn) in zinc sulphate ($ZnSO_4$) as the **anode** (-0.76V with respect to the hydrogen electrode). Thus the **chemical potential energy** of the zinc is being turned into **electrical energy**.

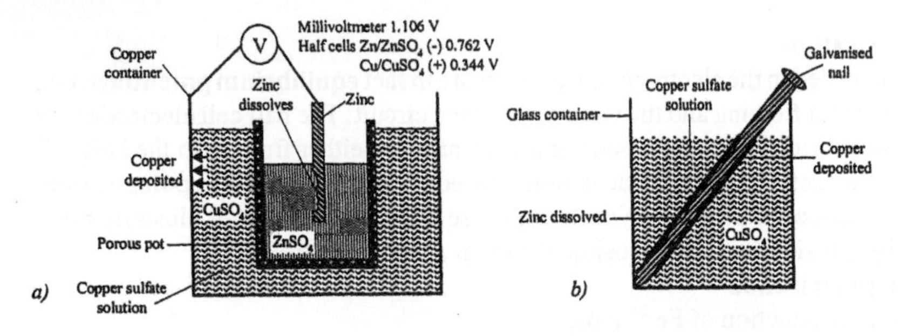

Figure 11.10 *Copper/copper sulfate in contact with zinc/zinc sulfate*

Zinc metal is in contact with its own ions (zinc sulphate). This is contained in a porous pot which allows hydrogen (H^+) ions to pass through to the solution of copper sulphate, which is in contact with copper metal. An 'open circuit' voltage (i.e., a circuit with **no** current drawn) of 1.1V is produced. The porous pot is required because if zinc is immersed in copper sulphate, a layer of copper is deposited at the same time as the zinc dissolves (as shown in ***Figure 11.10b*** for a galvanised nail placed in a copper sulfate solution) and the system is effectively short-circuited. This is why CCA (copper chrome arsenic) wood treatment compounds have to be aged before they can be used (**11.3.2**); the copper has to be 'fixed' (by covalent bonding) into the structure of the wood before steel nails can be driven into the wood.

The open circuit potential difference of 1.1V obtained by this system can be visualised if one considers the potential as equivalent to mileage between three towns **A, B** and **C**. All the mileages are measured from the central town **B** (the arbitrary reference point). Town **A** is 34 miles west of town **B**, and Town **C** is 76 miles east of town **B**. Hence

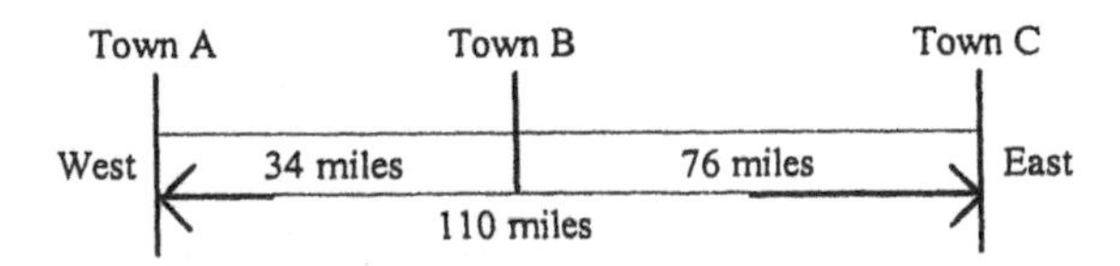

The distance between towns **A** and **C** is 34 + 76 = 110 miles and the voltage generated by our Daniel cell is (+ 0.34 - 0.76) = 1.10 volts (added as a difference irrespective of + or -). In the Daniels battery, zinc has the greatest electronegative potential and therefore, when connected to the copper metal contained in copper sulphate, the zinc will dissolve. Conversely, the copper is connected to a metallic/ion system which is lower than it's equilibrium reference potential and therefore copper ions will plate out of solution. The reactions are the zinc dissolving

Anode reaction $Zn \rightarrow Zn^{2+} + 2e$

and the copper plating out of solution, consuming the electrons produced at the anode

Cathode reaction $Cu^{2+} + 2e \rightarrow Cu_{metal}$

11.5.1 Electroplating

In the above example (***Figure 11.10a***), copper metal is being plated out of a solution of copper ions. This is **electroplating** of copper (albeit onto copper metal) and is described by the cathodic reaction

$$Cu^{2+} + 2e \rightarrow Cu_{metal}$$

However, it would be too expensive to use a zinc anode every time industry wanted to electroplate a metal, because the zinc metal is continually being lost into solution. Electroplating is therefore carried out by supplying a direct current (dc) source (supplying the electrons) across a plating bath. The normal supply of dc is from rectified alternating current (ac) power sources.

11.5.2 Effect of Impurities in the Electrolyte

The Daniel battery is able to maintain a voltage provided

- the two solutions ($ZnSO_4$ and $CuSO_4$) are separated (by the porous pot in the Daniel battery); and
- some of the zinc metal remains to provide a flow of electrons i.e., the battery can be switched off (by disconnecting the metallic contact between the zinc and the copper) to prevent complete loss of the zinc metal.

Sometimes impurities prevent a 'battery' from being switched off by forming local corrosion cells at the impurity site (which acts as the cathode). This is why distilled water has to be used in the 12V car battery system; domestic tap water would provide impurity sites on each electrode, effectively short circuiting the electrodes internally. A commercial battery does not have an infinite shelf life unless the metal and the electrolyte are absolutely pure.

11.6 MECHANISMS PRODUCING POTENTIAL DIFFERENCE

Differences in potential (and, in fact, reversals in polarity) can be produced by

- dissimilar metals
- differential deformation
- differential concentrations
- differential temperatures

11.6.1 Dissimilar Metals

Where two dissimilar metals are in good electrical contact in the same electrolyte environment, one metal will corrode in preference to the other (**11.4.4**; ***Figure 11.5***). This effect illustrates the importance of both the electrochemical series of pure metals (**Table 11.2**) and galvanic series of alloys (**Table 11.3**). For example, any alloy or metal more electropositive than mild steel will cause the mild steel to corrode. Other examples include

- copper water pipes in contact with brass compression fittings and steel radiators;
- the requirement for 'fixing' copper in CCA wood preservatives (**11.5**);
- lead sheets fixed with steel nails.

The more electropositive metal or alloy will be **cathodic** to the mild steel and so accelerate the corrosion of the mild steel. In each case, the cathode is the electrode at which electrons are consumed, given by the reaction

$$\text{(Air) } O_2 + 2H_2O + 4e \rightarrow 4OH^-$$

This reaction occurs on the copper water pipes and lead sheets respectively in the examples above. The mild steel dissolves in each case to produce electrons and ions in solution

$$Fe \rightarrow Fe^{2+} + 2e$$

In both examples, the dissimilar metal contact causes localised corrosion of the mild steel (Metal B in ***Figure 11.5***).

Overcoming Dissimilar Metal Contacts

The electrical contact can be removed to limit the effect of corrosion. This can be achieved by electrically insulating the two metals from one another to prevent the electrons from the more active electronegative metal from reaching the more electropositive metal. This can be achieved by plastic inserts which must completely isolate the two metals electrically (***Figure 11.11***) or by using adhesives, which have the advantage of chemically bonding the two metals together, whilst at the same time providing a gap filling role (**13.14**).

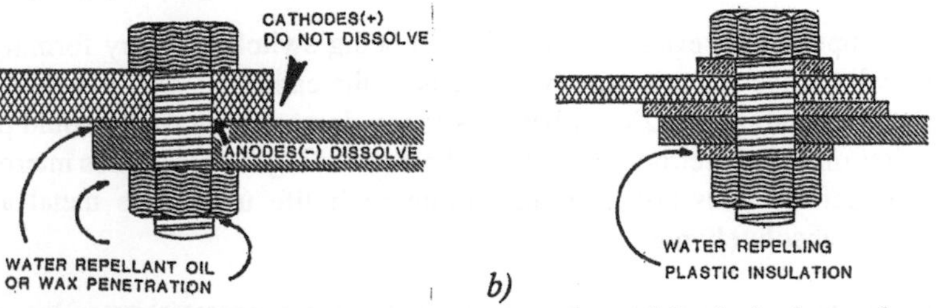

Figure 11.11 *Overcoming galvanic corrosion formed by bolted metal overlaps of dissimilar metals*

Making Use of Dissimilar Metal Contacts

Dissimilar metal contacts can be selected so that the more electropositive metal (the cathode) is protected whilst the more electronegative metal (the anode) dissolves. The anode is known as a **sacrificial anode** and the process is known as **cathodic protection**. These processes are considered in detail in section **11.13.3**.

11.6.2 Deformation Cells

The amount of deformation causes differences in electrochemical potentials. If a metal is bent, the more highly deformed area will anodic to the less deformed (cathodic) area, as shown in ***Figure 11.12***.

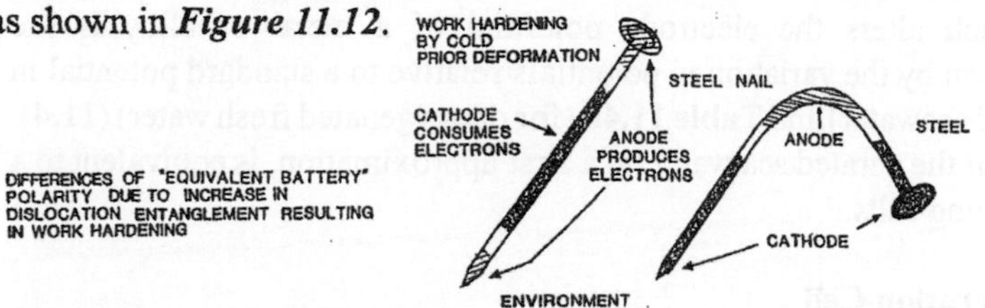

Figure 11.12 *The effect of differences in the amount of deformation*

A nail is made from cold rolled steel (**10.5.5**). The head is subsequently made by a process which involves deforming the head with a hammer (termed **upsetting**). The point is made by flying shears. A nail placed in water will therefore corrode with the head and point as the anodes whilst the less deformed shank will be the cathode. It is possible to reverse these potentials by bending the shank of the nail which produces more work (strain) hardening (**10.5.3**) in the shank compared to the head and point. The newly deformed shank becomes anodic. Similar effects in potential difference are observed with different **grain sizes** (***Figure 10.2***) or degrees of **work (strain) hardening** (**10.5.3**). This means that the same metal will have a different tendency to corrode depending upon its prior history and state of deformation. Overall, however, compared to other possible mechanisms of producing potential difference, the effect on corrosion of deformation cells is small.

11.6.3 Differential Concentration Cells

Differential concentration cells are cells that will produce a potential difference if there is a difference in the concentration of ions between two identical pieces of the same metal. Thus two pieces of copper metal in contact with solutions containing different concentrations of copper ions will produce an electric current (i.e., corrode) until both are in the same concentration of copper ions. This is shown in ***Figure 11.13a***.

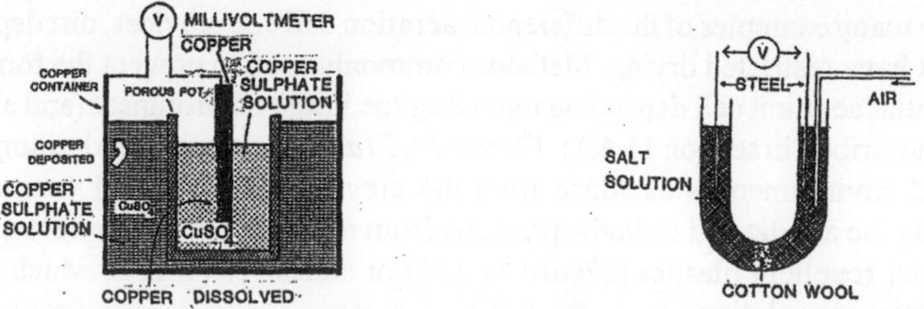

a) Copper in contact with solutions of copper ions of different concentrations *b) Experimental demonstration of the differential aeration cell*

Figure 11.13 *Differential concentration cells*

In ***Figure 11.13a***, two pieces of copper metal are immersed in two different concentrations of copper sulphate. For this system, a concentration potential will be produced. This can easily be demonstrated in the laboratory. There are other concentration cells more commonly encountered in the construction industry, in particular the concentration cell formed by the differences in oxygen replenishment within a liquid electrolyte i.e., a **differential aeration cell** (discussed below).

Another factor which alters the electrode potential of a metal or alloy is the environment, as shown by the variation in potentials relative to a standard potential in **Table 11.4a** (aerated seawater) and **Table 11.4b** (for deoxygenated fresh water) (**11.4**). It is worth noting that the aerated sea water, to a first approximation, is equivalent to a solution of road deicing salts.

The Differential Aeration Cell

The differential aeration cell is a specific type of concentration cell. The potential difference generated between two differentially aerated electrodes can be fairly large (~ 1.4V), although for most corrosion processes occurring as a result of differential aeration, the voltage cannot be detected because both anode and cathodes occur on the same metal surface i.e., they are ordinarily short-circuited (***Figure 11.2***). The fact that differential aeration produces a voltage can be demonstrated experimentally using a "U" tube. ***Figure 11.13b*** shows two pieces of mild steel immersed in a saline electrolyte, which is oxygenated differentially to show a potential difference. Both pieces of steel are connected to a digital voltmeter (DVM). The potential difference developed rises with increased oxygenation of one arm with respect to the adjacent arm. The cotton wool serves to prevent the oxygenated electrolyte in one arm of the "U" tube from mixing with the deoxygenated electrolyte in the other arm. Potential reversals of up to ± 90mV may be obtained, the sign (+ve or -ve) depending upon which arm is oxygenated (for comparison, a car battery produces 12V = 12,000mV). If the two arms are equally oxygenated and the steel is identical there is no potential difference; in this state, the system can be likened to a "run down" battery. Oxygenating one arm differentially to the other results in the steel in the oxygenated arm becoming the cathode (or +ve terminal of the "corrosion battery"), whilst the steel in the deoxygenated arm becomes the anode (or -ve terminal of the "corrosion battery") and dissolves. Phenolphthalein, an indicator which detects OH^- ions, will turn pink in the aerated arm.

There are many examples of the differential aeration cell e.g, crevices, dirt deposits and areas that have restricted drying. Methods commonly used to prevent the formation of a differential aeration cell depend on removing the liquid environment (and are similar to those described in section **11.6.1**). ***Figure 11.11a*** illustrates a typical example where the liquid environment is excluded from the crevice using oils and waxes, thereby preventing the anodic and cathodic products from reacting; a similar effect is obtained using water repellent plastics (***Figure 11.11b***) or adhesives (**13.14**), which provide a more permanent solution.

11.6.4 Differential Temperature Cells

A metallic component immersed in the same liquid will exhibit corrosion if there exists a temperature gradient along its length. The corrosion rate is increased with an increase in temperature (***Figure 11.8c***). Approximately, a 10°C rise in temperature will double the corrosion rate, whilst a 30°C rise will result in a 10 fold increase in the corrosion rate. In certain conditions, a rise in temperature may be beneficial; a higher temperature will also cause the corrosion microclimate to become dryer, changing the relative humidity. Remember, however, that when a liquid dries out, the dissolved salts are concentrated into the areas which are the last to dry, usually crevices.

11.7 ALKALINITY (CATHODIC AREA) AND ACIDITY (ANODIC AREA)

The corrosion of a metal results in cathodic areas becoming alkaline and anodic areas becoming acidic. These secondary effects play an important part in corrosion and are discussed separately below.

11.7.1 Alkalinity of the Cathodic Areas

Water is modelled as a covalently bonded molecule, H_2O. However, water also forms (H^+ and OH^-) ions as if it were ionically bonded; it is estimated that only one water molecule in every ten million are ionically bonded (**1.19.1**). Thus the following reaction is very biased towards the covalently bonded water molecule (i.e., towards the left)

$$H_2O \rightleftharpoons H^+ + OH^-$$

The cathodic reaction

$$\text{(Air) } O_2 + 2H_2O + 4e \rightarrow 4OH^-$$

produces more hydroxyl OH^- ions and these therefore mop up hydrogen H^+ ions to produce covalently bonded water. As pH is dependent upon the concentration of hydrogen H^+ ions (**1.29**), the cathodic area becomes very alkaline. Note that the alkaline regions of the cathode can sometimes cause "secondary" corrosion reactions to take place; for example, in the metals which occupy groups 2, 3 and 4 of the Periodic table (**1.8**), e.g, aluminium (Al) (group 3), zinc (Zn) (group 2), lead (Pb) (group 4) and tin (Sn) (group 4). These metals are **amphoteric** (**1.5**) and dissolve in both acids and alkalis. Of these metals, only zinc, lead and aluminium are used commercially in the construction industry. Zinc will therefore corrode in the presence of the by-products of the cathodic reaction, as it is one of the two metals which dissolve in an alkaline environment. Zinc is protected from this corrosion if carbon dioxide is available, as a basic carbonate film, $ZnCO_3.3Zn(OH)_2$, is formed which protects the underlying zinc metal. Where zinc products are painted, the basic carbonate will not form because the corrosion reaction is proceeding beneath the paint film. In general, zinc products should not be painted, unless special attention is paid to adherence of the paint film to the substrate; special chromate primer paints (**11.12.1**) should be used.

11.7.2 Acidity of the Anodic Areas

If an ion produced at the anode (e.g, Fe^{2+}) removes some of the hydroxyl OH^- ions from water to form the insoluble iron hydroxide, $Fe(OH_2)$, more covalently bonded water will dissociate to balance the hydroxyl OH^- ion deficiency (**1.19.1**). As hydroxyl OH^- ions are removed, the anodic areas become acidic due to the excess of hydrogen H^+ ions formed by the dissociation of water. Hence the water reaction

$$H_2O \rightleftharpoons H^+ + OH^-$$

is now biased to the right (with excess H^+ ions) and the anode environment becomes acidic. The rate of corrosion of steel in acidic solutions is much faster than in neutral solutions, due the excess of hydrogen H^+ ions. Hence crevices, pits and overlap joints (which form natural anodes due to the differential aeration of the crevice) become highly acidic as corrosion proceeds.

11.8 PITTING

Pitting results from the presence of a **large cathode and small anode** and is simply a manifestation of current density. ***Figure 11.14*** illustrates pitting corrosion caused by a water droplet around a dust particle (***Figure 11.14a***); ***Figure 11.14b*** shows a section through the water droplet. The same process occurs beneath dirt or dust particles on a metal surface (***Figure 11.15***).

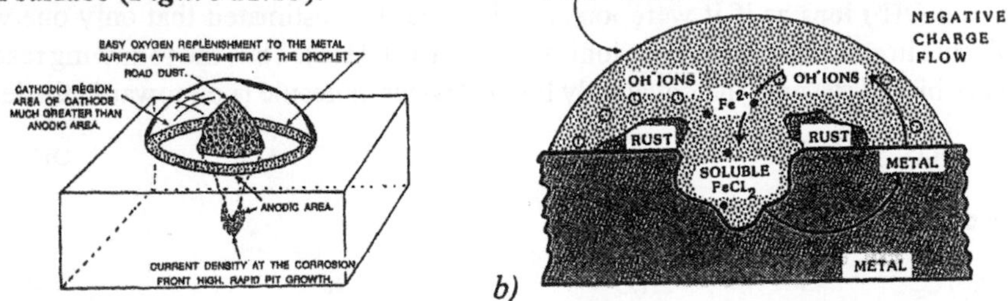

***Figure 11.14** Pitting corrosion formed by dirt particles and a water droplet and the transfer of negative charge*

Figure 11.14a illustrates pitting corrosion caused by a dirt particle (the water envelope has been extended to more clearly illustrate the process). The current density (current/area) is much higher at the anodic site (the centre of the water droplet) compared to the cathodic site (the perimeter of the water droplet). This large cathode/small anode leads to pitting of the metal.

It should also be appreciated that other factors promote pitting, including

- the acidic environment produced at the corrosion front (the pit) (**11.7.2**);
- increased differential aeration of the cathode compared to the anode as the corrosion front progresses into the metal, so favouring the differential aeration cell (**11.6.3**).

Figure 11.14b shows the circular transfer of negative charge, i.e.,

- production of negative charge (electrons) at the anode;
- transfer of negative charge (electrons) to the cathode, where the electron undergoes a reaction to convert the negative charge into hydroxyl OH^- ions;
- migration of hydroxyl OH^- ions to the anodic area, where they neutralise the positive ions produced at the anode (e.g, Fe^{2+} ions) at the metal/liquid interface.

The negative charge has gone round the whole circuit; conversely, the positive charge transfers in the opposite direction (i.e., the negative charge going anticlockwise round the circuit is equivalent to a positive charge going around clockwise). The rate of the corrosion reaction is governed by the slowest chemical reaction (**11.4.8**).

The differential aeration cell produced by a dirt particle and a water droplet responsible for pitting corrosion is shown in ***Figure 11.15***.

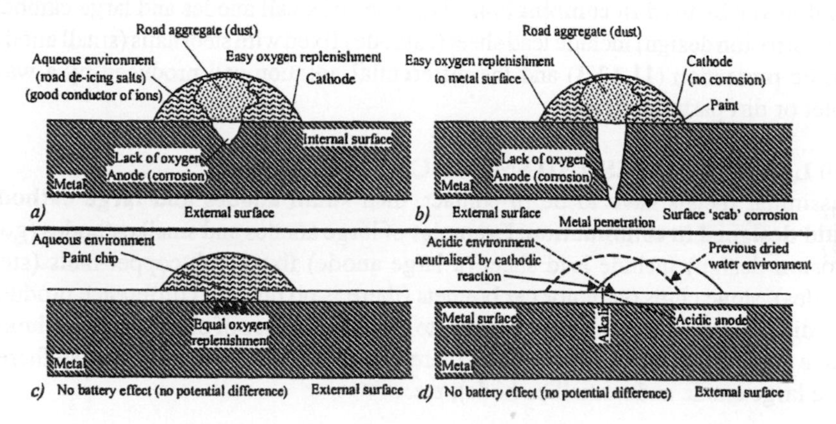

***Figure 11.15** Mechanisms responsible for pitting corrosion beneath dirt particles*

Figure 11.15a shows a schematic view of the hygroscopic road deicing salts/dirt particle on the surface of paint e.g, a car body or metallic soffit, which will absorb water from the atmosphere to form a differential aeration cell. The liquid environment is concentrated at the base of the salt/dirt particle upon drying, which helps fix the aggressive environment upon vertical surfaces of the metal. The short-circuited equivalent battery is generated by the difference in oxygen availability (***Figure 11.14b***). The oxygenated region is the cathode or (+ ve terminal of the 'corrosion battery') and the deoxygenated region in which the metal dissolves are the anodes or (- ve terminals of the 'corrosion battery'). ***Figure 11.15b*** shows diagrammatically the difference between internally initiated corrosion and metal penetration by pitting corrosion due to the lack of oxygen replenishment beneath these surface deposits and surface initiated corrosion or scab corrosion from stone chips. Note the cathodic electrode is much bigger than the anodic region (compare a paint chip, ***Figure 11.15c***). ***Figure 11.15c*** shows that a holiday in the paint work (for example, a stone chip on the bonnet of the car) does not cause very severe corrosion in the early stages of life. Both anodes and cathodes are equally oxygenated and dry rapidly; this is effectively a "run down" corrosion cell. ***Figure 11.15d*** highlights the fact that anodic areas can change into cathodic areas depending upon the location of the meniscus of the water droplet. For example, if the profile of the water droplet changed on two different days, anode and cathode areas change and effectively no corrosion will occur. The anodic area produced by the profile on day 1, changes into the cathode on day 2. A meniscus in the same position perpetually (caused, for example, by perpetual dust particles, bird droppings, etc.) would lead to pitting corrosion. If the two areas (anode and cathode) are of equal area and oxygenated to the same extent, there is no corrosion because there is no potential difference. Severe pitting corrosion results when large cathodes (perimeter of a water droplet) are in contact with small anodes (area beneath a dust particle) (**11.6.3**).

11.9 SMALL ANODES AND LARGE CATHODES

If dissimilar metals have to be in contact, then **small anodes and large cathodes should never be used in combination.** Examples of small anodes and large cathodes (poor corrosion design) include lead sheet (cathode) fixed with steel nails (small anode), cathodic protection (**11.13.3**) and the differential aeration cell produced by a water droplet or dirt particle (**11.6.3**).

11.10 LARGE ANODES AND SMALL CATHODES

If dissimilar metals have to be in contact, then **small anodes and large cathodes should designed in combination**. Examples of large anodes and small cathodes (good corrosion design) include lead sheet (a large anode) fixed with copper nails (small cathodes), stonė chips on motor car bonnets (there is no potential difference produced when different areas of steel are subjected to the same degree of oxygen replenishment, ***Figure 11.15d***; such paint chipping is therefore relatively harmless, because there is now a large anode and a small cathode), etc.

11.11 STOPPING THE CORROSION PROCESS

11.11.1 A Switch

Where a battery is used as a power source, the battery is switched off to conserve the battery. The switch breaks the metallic part of the circuit (***Figure 11.1c***). This method of stopping the corrosion process is not available in cases of unplanned corrosion because to use a switch we would have to cut the article up. The nearest we can get to this approach is to separate the two dissimilar half cells by an electrical insulator (as shown, for example, in ***Figure 11.11***) or by adhesive bonding (**13.14**).

11.11.2 Remove the Electrolyte

In the case of a 12 V car battery, if we throw away the electrolyte we will lose the 12V. For unplanned corrosion processes, we must

- exclude the liquid environment, for example by using waxes, plastic washers , etc (***Figure 11.11***) or adhesives (**13.14**); or
- isolate the two metals (***Figure 11.11***) or use one metal to protect the other, as in chrome plating, ***Figure 11.24***) and galvanisation (**11.13.2**); or
- isolate the metal surface from the agent causing the corrosion (using a physical barrier, such as a paint film, **11.12**). Note, however, that a paint film is permeable to both water and oxygen and so does not function when kept wet for long periods (**11.12.1**); or
- rely upon the passive oxide film that forms on some metals, as in the case of aluminium (**1.8.3**) and stainless steel (**10.10.7**). Thicker oxide films are placed on aluminium by **anodisation** (**11.14.2**), a special form of continuous coating (**11.14**).
- keep the surface free of dust particles, which form perpetual anode and cathode sites (***Figure 11.15***).

11.12 ORGANIC COATINGS (PAINT FILMS)

Paint is an organic coating that performs the primary function of providing a physical barrier, isolating the underlying metal from the environment and thus reducing corrosion. However, as shown later (**11.12.1**), this explanation is only a first approximation. Paint also

- provides a decorative coating;
- ensures that a metal surface is equally oxygenated (if the metal surface has an equal replenishment of oxygen over the whole surface, the 'corrosion battery' will produce zero volts and no corrosion will take place);
- adherent paints stop the flow of ions (from anode and cathode). Adhesion to the substrate metal must exist under dry, wet or corrosive conditions.

Paints contain a coloured pigment dispersed in a medium. In paint technology, the medium is called a **binder**. The binder initially produces a liquid on the metal surface that subsequently dries to form a solid film. The binder is the agent that determines the type of property possessed by the paint. The sole function of the volatile components of the paint (termed **thinners**) is to control the viscosity of the paint for ease of manufacture and application. The viscosity controls the application thickness that can be obtained before 'runs' and 'sags' are formed. Different types of paints require different thinners. One coat of paint will not possess all the properties required for corrosion protection and it is therefore necessary to have a two or three coat system. These are generally referred to a **primer, undercoat** and **top coat**.

Primer

A primer is the first coat of paint in contact with the metal and is put on to obtain good adhesion to the metal substrate. The primer usually has a high percentage of pigment in the binder, which means that it dries without lustre. The primer can be pigmented with anodic metallic elements (e.g, zinc particles) or with corrosion inhibitors. Hydrophilic additions usually form a high percentage of binder additives. As the primer is the first coat of paint put onto a metal surface, wettability is needed so that surface irregularities are filled rather than bridged. For better adhesion, mechanical abrasion (e.g, by emery cloths) or chemical keying (e.g, by phosphoric acid) is used (***Figure 11.16***). It is important that any surface irregularities are not larger than the thickness of paint film to be applied, otherwise these will "stick through" the paint film like "rocks out of the sea" and breakdown of the paint will result earlier than expected.

Undercoat

The undercoat is the coat often missed out but it does thicken up the paint film, making the water and oxygen diffusion path longer. Like the primer coat, it is usually high in pigment content and dries to a matt finish.

Top Coat

The top coat usually makes up for the matt finish of the primer. A top coat by itself does not have a good corrosion resistance, but does dry to give a bright lustre as well as giving the final colour to the article. The top coat has to possess stability to weathering, be resistant to chemical staining (e.g, by bird droppings, etc.) and possess abrasion resistance. It is impossible to apply a coat of paint and have it free from pin hole defects (termed **holidays** in paint technology). These defects mean that the protection produced by paints increases with film thickness.

11.12.1 Mechanisms of Paint Protection

Paint is an organic coating that provides a physical barrier, isolating the underlying metal from the environment and thus reducing corrosion. However, this mechanism is only a first approximation. In fact the paint film acts as a **semipermeable membrane,** allowing only one kind of molecule to pass through its thickness, namely uncharged ions, like water and oxygen. It is well established that a paint film will allow passage of

- the same amount of oxygen as is required for an uncoated metal to corrode;
- seven times the amount of water required for an uncoated steel to corrode. This means that the diffusion of water through a paint film is not the rate-determining step in the corrosion of steel.

However, a paint film is **impermeable to charged ions** (for example, Fe^{2+}, Cl^-, OH^-, Na^+, etc). The mechanism by which a paint film prevents corrosion lies in the origin of these charged ions, which could be

- components of airborne road deicing salts (Na^+ and Cl^-);
- aggressive atmospheres (sulfur dioxide and sulfates);
- products of the corrosion reaction itself (e.g, Fe^{2+} and OH^- ions).

The process of corrosion is stopped by arresting the lateral movement of ions produced by the corrosion reaction itself (e.g, Fe^{2+} and OH^- ions). These ions are prevented from spreading radially away from the micro-corrosion site at the metal/paint interface **only** if the paint film is **adherent** to the metal substrate. ***Figure 11.16*** illustrates the methods by which adherence of the paint film to the metal surface can be achieved.

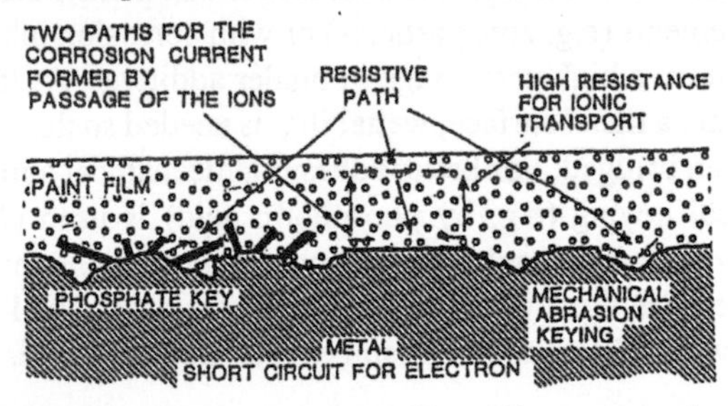

Figure 11.16 *Methods used to provide paint film adherence to the metal substrate*

A paint film stops lateral corrosion and the passage of the anions and cations from each corrosion half cell by providing an adherent film which increases the electrolyte resistance between anode and cathode. ***Figure 11.16*** illustrates the resistive path for a phosphate film, mechanical abrasion and a flat surface. Both phosphate keying and abrasion are methods of increasing the diffusion path of the electric circuit in the liquid environment. However, if there are any ionic contaminants at the metal paint interface, external water can diffuse through the paint film to form an electrolyte at the metal/paint interface. Ionic contaminants that could be present here include unwashed phosphoric acid, ferrous sulphate from industrial condensates or chlorides from road deicing salts. In the presence of ionic components at the paint/metal interface, the corrosion cell is complete. Surface cleanliness and preparation are therefore the most important aspects of preventing corrosion by the application of paint films. In addition

- a paint film is only as good as its surface preparation; therefore paint films are best applied on dry warm days so that there is no possibility of sulfurous acid condensates forming on the metallic surface. **Paint should never be applied when the temperature is falling** (for example, in late afternoon) due to the possibility of condensation. Condensates produce early breakdown of the film as a result of the catalytic action of sulfates on the corrosion reaction. A surface cleanliness test is described (**11.12.2**).
- the properties of a paint film are such that it is inadequate to protect the metal substrate in many corrosive environments. For example, a paint film would not be considered for an article submerged in hydrochloric acid. Paint films can only be used in mildly corrosive atmospheric environments because they are **permeable**.

11.12.2 A Test for Surface Contaminants

Filter paper soaked in potassium ferricyanide will turn blue in the presence of ferrous salts. This blue colour will indicate the areas of insufficient cleaning and where surface contaminants are present. The most expensive paint is wasted if the surfaces are contaminated by non-adherent impurities.

11.12.3 Mechanisms of Paint Failure

Mechanism 1: Paint blistering. The mechanism of paint breakdown is by a process of **osmosis**, which produces an osmotic (or solution) pressure (**6.6.4**) which can lead to paint blistering. Osmotic pressure operates in any situation where solutions of different concentrations are separated by a semipermeable membrane. Osmosis is the passage of water from a solution containing a low concentration of dissolved species, (eg pure water), to one containing a higher concentration of dissolved species, until stopped by the osmotic pressure. Thus an internal 'bursting' pressure can be obtained by the semi permeable membranes of both hydrating cement (**7.17**) and the product of the anodic and cathodic reaction within the semi permeable paint films. It can be demonstrated in the laboratory using a pig skin as a semi permeable membrane and a solution of cane sugar (***Figure 11.17a***). Eventually the head of water on the solute side, (osmotic pressure) will stop the water diffusing through the pig skin, or in the case of paint films, the osmotic pressure separates and 'bursts' the paint off the substrate steel.

Radioactive tracer techniques have shown that the permeation of ionic species increases with time of contact. In addition, unlike metals which are dimensionally stable in a moist environment, plastics and other non-metallic compounds are affected by moisture (**6.4**). The contact of a paint film with water vapour produces internal expansion stresses within the paint film, which can be demonstrated experimentally. Two pieces of thin section steel shim are coated with paint on one side only and suspended over water (***Figure 11.17b***) or submersed in water (***Figure 11.17c***). The results show that the one suspended in a moist environment is bowed due to the internal stresses developed within the paint film, whilst the shim immersed in water is not bent. There are several explanations to this phenomenon, but the most likely one is that moisture absorption has altered the glass transition temperature (**13.9.5**).

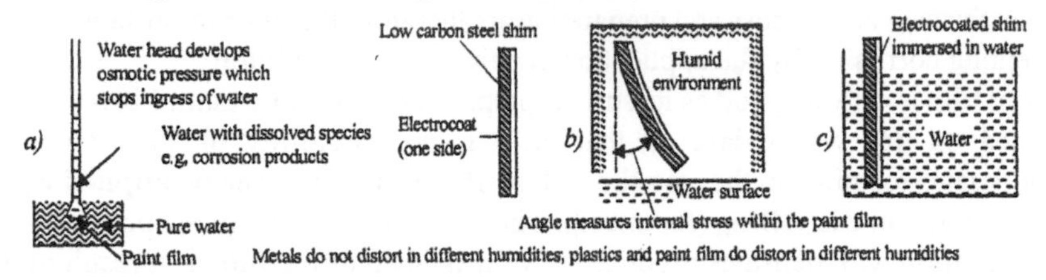

Figure 11.17 Demonstration of internal stress development in a paint film

The formation of a paint blister on a metal surface is illustrated in ***Figure 11.18***. Once a blister has formed, corrosion can proceed. Initially, the paint film provides similar oxygen replenishment to the whole metal surface and therefore there is **no** concentration cell developed (***Figure 11.18***). However dirt particles, which remain in contact with the paint film for long periods, or non-adherent paint films cause an unequal concentration of oxygen to develop and therefore provide the driving force for the corrosion of the substrate metal.

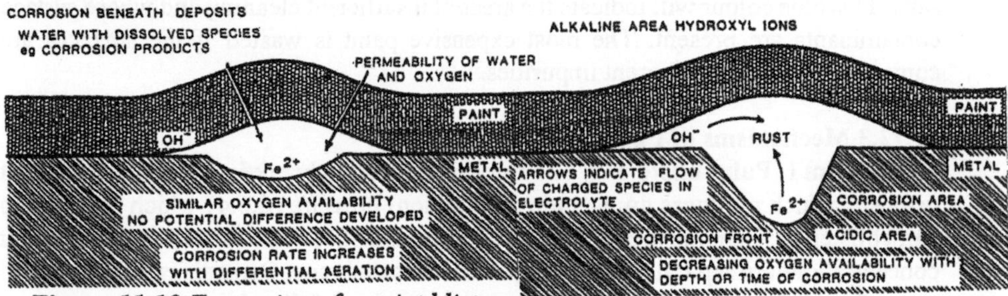

Figure 11.18 Formation of a paint blister

Corrosion proceeds as osmosis causes more water to be pumped into the gaps where the paint film has lost its adherence to the substrate metal. The process is rather like blowing up a balloon (the paint film) with water. Once corrosion is initiated, a difference in oxygen availability accelerates the corrosion rate. Pitting results as the metal is penetrated by the corrosion process as the corrosion front becomes further starved of oxygen and acidic in nature (**11.8**). The stripping of the paint film shown in ***Figure 11.18*** is further aided by the alkaline nature of the cathodic area, which is plentifully supplied with oxygen. As most paints are softened by alkalis (a process known as **saponification, 11.12.4**), the alkali formed is able to strip the paint from the underlying metal. It is also able to attack the pigment of the primer paint. Adhesion failure of the paint often precedes catastrophic corrosion.

Mechanism 2: Scab and filiform corrosion. Scab corrosion is where the paint film is damaged by external means so that it becomes non-adherent with the metal substrate but remains intact (coherent) within itself (***Figure 11.19a***). Filiform corrosion is characterised by threadlike filaments of the corrosive environment beneath the paint film. The corrosion arises as a result of identical processes to those described previously

(11.12.3). ***Figures 11.19b,c,d*** illustrate plan and sectional details of filiform corrosion.

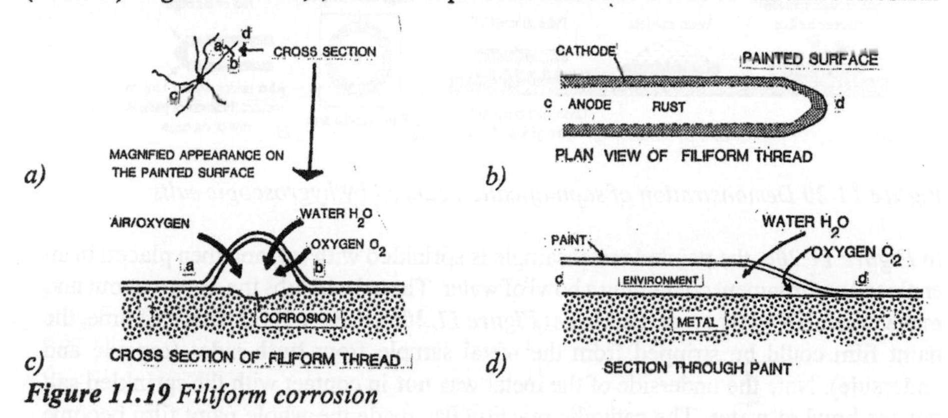

Figure 11.19 *Filiform corrosion*

Filiform corrosion starts from non-adherent paint formed by small paint 'chips' or by paint in contact with dirt or in crevices, which give a surface threadlike crack pattern similar to the shown in ***Figure 11.19a***. The corrosion process spreads out from the central area in a threadlike spider's web manner in plan, which resembles a tunnel in section (***Figure 11.19c*** and ***11.19d***). ***Figure 11.19b*** shows a magnified plan view looking down on the filiform **cd** (in ***Figure 11.19a***); ***Figure 11.19c*** shows a cross section of the filiform **ab** (in ***Figure 11.19a***); ***Figure 11.19d*** shows a cross section of the filiform **cd** (in ***Figure 11.19a***). The cathodic reaction takes place at the interface between the paint and metal surface (***Figure 11.19d***) because the diffusion path for oxygen is smallest at this point, whilst the anodic reaction takes place at the centre of the filiform 'tunnels' (***Figure 11.19c***). The corrosion path in filiform corrosion is not manifested as a total surface phenomenon, but is restricted to the filiform 'tubes' or 'threads' which are formed beneath the semi- non-adherent paint film. These 'tubes' or 'threads' are barely visible to the naked eye. The existence of filiform corrosion and the restriction of areas of corrosion to the very small filiform 'tubes' means that poorly prepared surfaces provide ideal sites for filiform corrosion beneath the paint film. What appears to be a good adherent paint may in fact contain a suitable environment for initiation of a corrosion cell. Therefore, in preparing a metal suffering from scab corrosion, it is wise to remove the old paint well back from the damaged areas. The problem with much defective paintwork is that most people are afraid to be drastic with the areas of scab and filiform corrosion and so rectification is often only temporary and the paint film soon fails again.

11.12.4 Paint Saponification

In previous sections we have shown how the cathodic product (hydroxyl OH^- ions) undermines the paint film and causes loss of adhesion to a metal surface. The cathodic areas become alkaline in nature and this will affect paints if they are not alkali resistant. It is widely acknowledged that alkali resistant paints are required for both marine conditions and for cementitious products where high pH conditions exist. ***Figure 11.20*** is a schematic view of an experiment carried out to discover how hygroscopic salts, e.g, road salts, which absorb atmospheric water vapour, would behave on a painted surface.

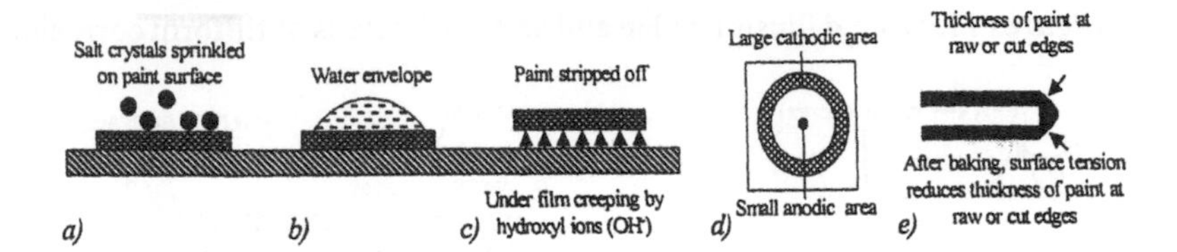

***Figure 11.20** Demonstration of saponification caused by hygroscopic salts*

In ***Figure 11.20a***, the painted metal sample is sprinkled with salt and then placed in an enclosed environment containing a bowl of water. The salt absorbs the water vapour and makes a pool of water around the salts (***Figure 11.20b***). After a short period of time, the paint film could be stripped from the metal sample from **both** sides (topside and underside). Note the underside of the metal was **not** in contact with the sprinkled salt nor the bowl of water. The cathodic reaction has made the whole paint film become non-adherent to the metallic surface. The corrosion cell formed is the result of a large cathode and a small anode (**11.9**), as shown in ***Figure 11.20d*** (which illustrates the result after about seven days). ***Figure 11.20e*** shows a section through the raw edge of a steel sheet, illustrating how the surface tension of the applied paint reduces the paint film thickness at sharp radii of curvature. The diffusion barrier provided by the paint film at raw edges is therefore less effective.

The saponification reaction can be written as

$$\underset{\text{(paint)}}{\mathbf{R_1\text{-}COOR_2}} + \underset{\text{(water)}}{\text{H-OH}} \xrightarrow{\text{saponification}} \underset{\text{(acid)}}{\mathbf{R_1}\text{-COOH}} + \underset{\text{(alcohol)}}{\mathbf{R_2}\text{-OH}}$$

(esters in an alkaline medium)

where $\mathbf{R_1}$ and $\mathbf{R_2}$ are any large organic hydrocarbon chain. If the hydrolysis is carried out in an alkaline environment (for example, in the presence of sodium hydroxide, NaOH), the sodium salt of the acid $\mathbf{R_1}$-COOH is formed. Since alkali salts of the higher acids are soaps, the alkaline hydrolysis reaction is known as **saponification** (derived from the Latin word meaning soap).

11.12.5 Summary

- a single paint film is not a good "barrier" as it contains micro-cracks and perforations (holidays). Good coatings rely upon a multi-coat system so that these perforations are not coincident;
- the paint film only acts as a barrier to the electrolyte of a mildly corrosive environment by preventing the lateral movement of the anodic and cathodic corrosion products. The paint film delays the onset of corrosion by preventing the anodic and cathodic products formed in small local sites (micro cells) from combining and therefore delays the onset of further corrosion;
- a paint film allows water and oxygen to pass through in sufficient quantities to allow corrosion to proceed. The common belief that a metal is porous and has 'sweated' water to produce the paint blister is incorrect. Osmosis causes water to be pumped to the metal/paint interface;
- a paint film does not allow the diffusion of charged (ionic) species, provided

intimate contact with the metal surface is maintained over a long period of time. If ionic species reach the metallic surface, a good electrolyte is formed which allows the products of corrosion from the micro cells to spread laterally and radially outwards;

- the alkali by-products of the cathode destroy the adhesion of the primer paint film, allowing the corrosion cell to spread. The cathodic reduction of oxygen and the subsequent saponification of the primer paint film is the main cause for the loss of adhesion and for the lateral growth of the micro cells to larger cells (viewable to the unaided eye);
- corrosion performance can be improved by better paint technology, for example by making a paint film more resistant to alkalis.

11.13 DISSIMILAR METAL CONTACTS

Dissimilar metal contacts can be designed whereby one metal (the **sacrificial anode**) is allowed to corrode to protect the other. Sacrificial metals which have been used include zinc and magnesium, Here we examine two applications of this theory, **coatings (11.13.1)**, in particular galvanised zinc coatings **(11.13.2)**, and **cathodic protection (11.13.3)**.

11.13.1 Sacrificial Metal Coatings

Figure 11.21 illustrates some well known methods of applying anodic metals to steel.

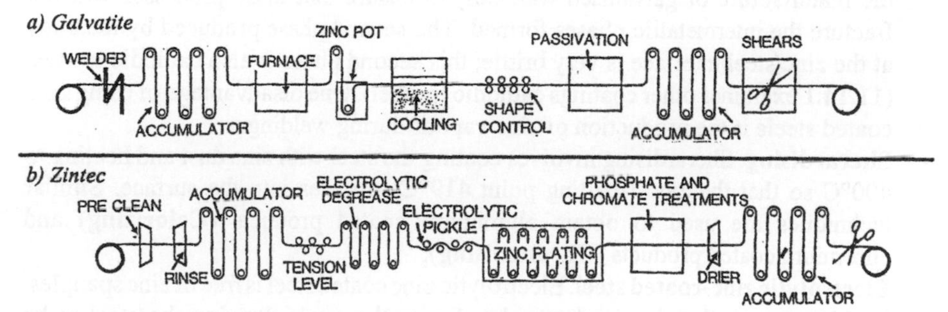

Figure 11.21 *Manufacturing processes: sacrificially protected British Steels products*

Figure 11.21a illustrates the manufacturing process for **Galvatite™**. Galvatite products are produced by a continuous zinc hot dipping process where the steel is dipped into molten zinc. Where an iron-zinc alloy is required, an extra heating oven is used prior to the cooling stage. ***Figure 11.21b*** illustrates the manufacturing process for **Zintec™**. Zintec is mild steel sheet coated with a thin film of electro-deposited zinc produced by British Steel. The coating process is capable of producing steel sheet with either one or both sides coated with zinc. Different thicknesses can be applied to the two opposite sides of the sheet. It should be noted that, just as the sacrificial anode does not protect the superstructure of a ship **(Table 11.5)**, so the zinc which is not in contact with the steel **through the electrolyte** will not protect the sawn edge of the metal sheet. This fact is not universally understood by the industry or manufacturers as there is too much reliance upon the zinc coated steels without providing adequate protection for the zinc.

More complex methods may be used to obtain coating of the metal substrate, e.g,

- **spraying** the molten metal onto the substrate surface from a high temperature gun (generally used for large structures *in situ*);
- **cementation**, where the substrate metal article is heated and tumbled with the metal powder and a flux to form the coating (generally used for small part production). Aluminium and zinc coatings can be applied this way;
- Coatings of the metal nickel may also be formed from a solution of the salt by a **chemical reduction process**, in contrast to the **electrical reduction process** of electroplating (**11.5.1**). Nickel coatings produced by chemical reduction are called **electroless nickel** to distinguish them from electroplated metal coatings.

11.13.2 Zinc Galvanising

Galvanising is one of the most promising corrosion preventive measures that can be applied to steel because it can be put on easily, commonly before the steel leaves the steel manufacturer. The durability of the total coating is dependent upon its thickness and the environment. There are several ways achieving zinc coating to steel

- **Hot dip galvanising**. The article is dipped in molten zinc and the sheet is therefore coated on both sides (***Figure 11.21a***). The process produces zinc 'spangles' (which are the large grains seen within the zinc coating), formed by the large grain growth when the zinc solidifies. The technique must be carefully controlled where there are any cold rolling and/or subsequent deformation steps envisaged for the product (e.g, the manufacture of galvanised wall ties) to ensure that these processes will not fracture the intermetallic phases formed. The second phase produced by the alloy at the zinc/steel interface is very brittle; this second phase is also cathodic to steel (**11.14.1** examines other coatings cathodic to steel). One disadvantage of using zinc coated steels is the production of zinc vapour during welding.
- **Sherardising**. Sherardising involves coating the steel with zinc dust and heating to 400°C so that the zinc (melting point 419°C) diffuses into the surface. Similar techniques are used to obtain aluminium coated products (**Calorizing**) and chromium coated products (**Chromatising**);
- **Electrolytic zinc-coated steel**. Electrolytic zinc coated steel is free of zinc spangles. In this process, the zinc is electroplated onto the steel, allowing the steel to be coated on one or both sides (***Figure 11.21b***);
- **By painting**. The metal, as zinc dust, is suspended in paint and (provided the zinc content is high enough) electrical contact will be made between the zinc particles and the steel to provide a sacrificial metallic coating. The zinc-rich paints are often referred to as **zinc rich primers**.

Note that the term **Galvanising** has come to be used as a general term for the deposition of the sacrificial metal zinc onto the surface of steel to provide corrosion protection. Technically, however, the term should only be used where the metal is deposited from a molten bath. Incorrect specifications may be written if the usage of the term is debased. Note also that the micro-structure arising from the hot dipping is completely different to that obtained in the Zintec process, which contains no second phase. Steels protected by the Zintec process are therefore widely used by the pre-coating Plastisol manufacturer, as bending and shaping processes are easier. Zintec products are also used in the automotive industry and by the steel manufacturers on plastic coated panels for shed roofs and in situations where corrosion has been a problem in the past (e.g, at sawn edges and overlap joints). Sometimes a zinc-iron alloy finish is required; this is achieved by passing the hot dipped steel strip through another furnace which is placed between the zinc pot and the cooling stage of the process (***Figure 11.21a***). The second phase formed is often brittle and potential shifts (**11.4.5**) can occur.

Advantages of zinc

The use of a metal more electronegative than steel is at first somewhat surprising, as one would expect the zinc to have a faster corrosion rate than steel. It is important to realise, however, that the position in the electrochemical series does **not** govern the **corrosion rate** (**11.4.7**). Zinc coatings on steel have gained in popularity because zinc has a much better corrosion resistance than does steel in most service conditions (***Figure 11.8a***) (exceptions are considered below). The normal atmospheric corrosion product from zinc is a colourless complex compound, the basic zinc carbonate $ZnCO_3.3Zn(OH)_2$, formed in the presence of atmospheric carbon dioxide. However, zinc coatings do not form this protective carbonate film when exposed to atmospheres containing small concentrations of sulphur dioxide, because the environment is too acidic to form the basic carbonate. There is also evidence to suggest that the corrosion rate of zinc in pure, atmospherically condensed water is fairly high and so zinc products usually suffer more when 'facing the ground' than when they 'face the sky'. One factor in this anomaly is the rate of evaporation from condensed droplets which have formed on sheltered surfaces, such as those which 'face the ground'. Zinc coatings are more resistant to chloride marine environments than they are to industrial environments which contain sulphur dioxide (***Figure 11.8a*** and ***11.8d***). Where zinc products are painted, the complex carbonate will not form because the corrosion reaction is proceeding beneath the paint film. In general, zinc products should not be painted, unless special attention is paid to adherence of the paint film to the substrate. Special chromate primer paints (**10.12.1**) should be used.

An additional advantage of zinc as a surface coating is at raw or sawn edges, where the zinc will sacrificially protect the steel provided the two metals are in aqueous contact. As the zinc coating applied to the steel is not heated, the brittle phase is not formed, therefore the sheared edges may also obtain a smearing of zinc on the exposed surfaces. The sacrificial protection obtained in these situations is dependent upon the conductivity (and resistance) (**1.19.1**) of the external liquid environment which coats the edges.

Disadvantages of zinc

Zinc is one of the few metals which dissolves in both alkaline and acidic environments (**11.4.6**). In section **11.7.1** we showed that the cathodic reaction produced an alkali, which would dissolve the zinc and so aid the undercutting of a paint film. Good examples of this can be seen where galvanised pedestrian railings and zinc coated roofing panels, etc., are painted. It is very difficult to keep a paint film on zinc coated (steel) products if there is little or no pretreatment (using, for example, chromate, phosphate or pre-ageing treatments, **11.15**) that allow the protective zinc carbonate ($ZnCO_3.3Zn(OH)_2$) coating to form.

11.13.3 Cathodic Protection

Cathodic Protection (CP) was first investigated by Sir Humphrey Davey in 1824, when it was noticed that some copper rivets in contact with steel made the steel corrode, whilst others did not. It was found that if the copper rivet did not make electrical contact with the steel (for example, by using an insulator, as in ***Figure 11.11***) there was little or no adverse reaction when copper and steel were placed together in a marine environment. Later it was found that some metals (e.g, zinc slugs) could be used for protecting the hulls of ships. The first instances of cathodic protection utilised **sacrificial metals** to protect iron. The sacrificial metal could be any metal more electronegative than iron in the electrochemical series (**Table 11.2**), although zinc was commonly used. However, the sacrificial metals for these CP systems had to be replaced periodically. Later, cathodic protection used impressed external current sources coupled to inert anodes, thus providing continuous, maintenance free protection. Since the beginning of the 20th century, CP by impressed current has been used to protect steel pipelines, maritime installations, reinforced concrete and steel processing plants. The first example of the application of cathodic protection (CP) to steel in concrete was to prestressed concrete pipelines in the 1950s. It was subsequently applied to underground tanks using conventional impressed cathodic protection techniques. However it was not applied to structures above ground until experiments were conducted in South Africa on bridge cross beams in 1966. In 1973, a bridge deck in the United States was protected with CP to counter problems caused by the application of high levels of road deicing salts to bridge decks. In the last 20 years, CP has been applied to a variety of concrete structures; for example, the M6 Gravelly Hill junction; M1 bridge decks; car parks subject to deicing salt contamination, high rise buildings and water storage towers. Cathodic protection for reinforced concrete is considered in detail in **Chapter 8 (8.25)**.

All cathodic protection systems are based on limiting the anodic corrosion reaction that would ordinarily take place in the steel. The anodic reaction is defined as the electrode at which **positively charged ions** (e.g, Fe^{2+}) leave the electrode in the liquid state and where **electrons** are produced at the metal surface and flow in the metallic conductor. The most usual way to produce electrons is for the metal to dissolve. The anode is therefore the negative electrode of a 'corrosion battery'. A common anodic reaction occurring in steel is

Fe	→	Fe^{2+}	+	2e
metal		metallic ions		electrons

The aim of cathodic protection is **to prevent or limit this anodic reaction from occurring** since it results in loss of the metal. Hence in all methods the steel reinforcement is polarised cathodically to various levels. A separate anode system is installed, normally on or very close to the concrete surface, either temporarily or permanently, depending on the technique. The type of anode installed for cathodic protection systems is therefore critical because the anode will tend to dissolve into ions and electrons (as shown by the reaction above). The steel and the installed anode are connected and it is imperative to ensure electrical continuity of the steel reinforcement within the system. There should also be no short-circuiting between the steel reinforcement and the anode and no insulating layers in the cover zone or surrounding the steel reinforcement.

A variety of cathodic protection systems have been used, either based on sacrificial metals or on impressed current. These two methods illustrated by the cathodic protection of underground pipes in ***Figure 11.22***. The most successful types of cathodic protection systems are generally of the impressed current rather than the sacrificial anode type.

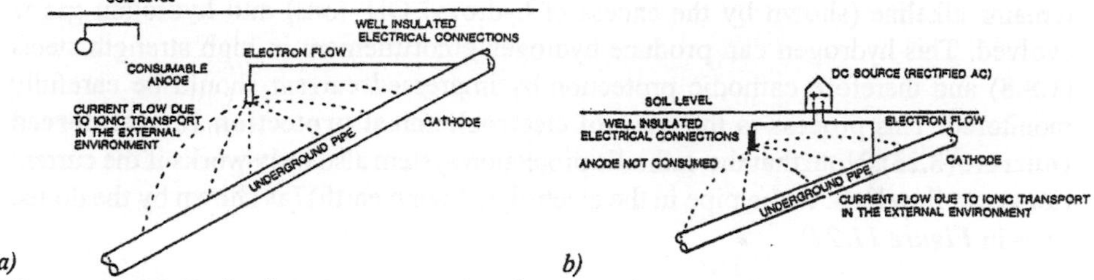

***Figure 11.22** Cathodic protection of underground pipework*

Underground steel pipes can be protected by wax wraps (not illustrated) or by cathodic protection. ***Figure 11.22*** illustrates cathodic protection of underground pipework using sacrificial metals (***Figure 11.22a***) and impressed current (***Figure 11.22b***). Cathodic protection works by depressing the steel pipe below its **equilibrium potential (11.4.5)** so that the steel **cannot** corrode. This can be done using

- a sacrificial anode (which is dissolved and has to be replaced) (***Figure 11.22a***); or
- a rectified direct current (dc) source connected to the pipe (***Figure 11.22b***).

Sacrificial metals usually used were zinc, aluminium and its alloys or magnesium. In this system, cathodic protection is achieved using the **chemical energy** of the anode i.e., $Zn \rightarrow Zn^{2+} + 2e$. The severe limitation on the use of sacrificial metals in most industrial applications is the maintenance required on the consumable anode, which must be periodically replaced. Other forms of cathodic protection based on impressed current are therefore commonly used. In these systems, the direct current (DC) required is obtained from a rectified alternating current (AC) source, usually from the mains supply. The anode of the rectified DC cell is a carbon rod which is buried in a specially constructed environment and does not dissolve. The fact that the steel pipe is forced below its equilibrium potential, and the cathode reaction which takes place on the underground pipes is to consume electrons, implies that any Fe^{2+} ions in the ground water will be plated out onto the steel pipe $Fe^{2+} + 2e \rightarrow Fe$. However there are no Fe^{2+} ions from the steel pipe available in the ground water, hence this cathodic reaction cannot take place. In the presence of adequate oxygen, the cathode reaction that will actually take place on the surface of the steel is $O_2 + 2H_2O + 4e \rightarrow 4\ OH^-$. Hence the steel structure will be protected by an alkaline environment produced and will not corrode **provided there is oxygen at the surface of the metal.** Note that this is a somewhat paradoxical situation. For successful cathodic protection, oxygen is required at the surface of the metal. Thus, for example, concrete cover to reinforcement must be permeable for successful cathodic protection of reinforced concrete (**8.25**); this permeability is the main factor in **reducing** concrete durability, since the ingress of aggressive acid gases (carbon dioxide, sulfur dioxide, etc.) is responsible for the carbonation of reinforced concrete (**8.21.4**). Note that if there is no oxygen available (for example, where the steel is submerged or buried underground) or when higher impressed current densities are used (**Table 8.22**), then another cathodic reaction must take place. This is the dissociation of covalently bonded water into hydrogen H^+ and hydroxyl OH^- ions, i.e., $H_2O \rightleftharpoons OH^- + H^+$ followed by $2H^+ + 2e \rightarrow H_2$. The whole reaction can be written as $2H_2O \rightarrow 2OH^- + 2H^+ + 2e \rightarrow H_2 + \mathbf{2OH^-}$, i.e., the cathode remains alkaline (shown by the excess of hydroxyl OH^- ions) and hydrogen gas is evolved. This hydrogen can produce hydrogen embrittlement in high strength steels (**1.8.8**) and therefore cathodic protection by impressed current should be carefully monitored. This process is the basis of **electrochemical protection for reinforced concrete (8.25)**. Note that the cathodic protection system also **only** works if the current (ions) can flow back to the pipe in the electrolyte (damp earth) (as shown by the dotted lines in ***Figure 11.22***).

Requirements for Cathodic Protection

The methods of cathodic protection using sacrificial metals and impressed currents are shrouded in misconceptions. In either case, for cathodic protection to be successful, the following conditions must apply

- **both** the anode and the metal to be protected must be in the **same liquid (electrolyte) environment**. With reference to the potential-pH diagram for the water-iron system (***Figure 11.7a***), the potential of the steel is not reduced from **A** to **D** merely by the attachment of the zinc metal. If attachment of the zinc metal was all that was required, we could protect the superstructure of ships (by putting the zinc on the mast head of a ship) and cars (by making a contact between any part of the car body and zinc, e.g, by putting the zinc in the windscreen washer bottle or underneath the bumper); any steel structure (by connecting a lump of zinc to steel columns immersed in the ground, rather than by galvanising the whole structure);
- the cathodic reaction (Air) $O_2 + 2H_2O + 4e \rightarrow 4OH^-$ has to take place on the steel, so that the electrons produced at the anode (e.g, zinc) can be consumed.

We can demonstrate these requirements by reference to the potential difference between the anode and cathode of the corrosion cell falls with respect to the current drawn from the cell, as in section **11.4.1**). In ***Figure 11.2***, the (open circuit) anodic potential and cathodic potential is represented by **A** and **C** respectively. Where sacrificial (zinc) anodes are used, the hydroxyl OH^- anions produced by the cathodic reaction must combine with the zinc Zn^{2+} cations produced by the anodic reaction ($Zn \rightarrow Zn^{2+} + 2e$) (***Figure 11.2***, point **A**). Following cathodic protection, the potential of the steel will be lowered from point **C** to point **U** for the corrosion current **AW**. Normally, both the steel and the zinc will be **galvanically coupled** at the same potential **K**, thus drawing the maximum current, **k**. We can also express this in relation to the Pourbaix diagram (***Figure 11.7a***); the potential of the steel (at the point, say, **A**, for an environmental pH of 7) is brought to a potential **D** below the line **EFH**, so that the steel enters the region of immunity. It is the dissolving zinc which produces the lowering of the potential of the steel with respect to the liquid environment.

Hence cathodic protection of any metal requires a **common electrolyte** and the **correct polarity**. The sacrificial metal or the impressed current will **not** protect any item which is not in contact with the common electrolyte. A summary of the requirements for the cathodic protection of various structures and key features of the cathodic protection system in each case is given in **Table 11.5**.

Table 11.5 Key requirements for cathodic protection of various structures

Structure	Requirements	Will not protect
Bridge decks and columns	Impressed anodes on top of bridge deck and columns. The electrolyte is the concrete matrix	Lamp posts, motor way crash barriers, railings, etc. (In fact these will suffer accelerated corrosion). Extra (electrical) insulation for these structures will be required in the acidic environments produced.
Ships at sea	Sacrificial metal (zinc/magnesium) in contact with ships hull. Only hull below water line is protected (same electrolyte environment).	Super-structure and ship's hull above the water line.
High rise buildings	Electrically conducting paint (to form the anode) over the surface of the concrete	Will not protect any other metal not under the throw of the impressed current. Will accelerate the corrosion of attached metals (e.g, TV aerials fixing, hand rails, etc.) which will require electrical insulation for protection.

For cathodic protection to work successfully, a return ionic current must flow from cathode to anode **over the whole metallic surface** i.e., the products of the anodic reaction must combine with the products of the impressed cathodic reaction. Thus any repair not passing a current to the anode will still allow localised corrosion to take place beneath the concrete repair. The electrically or ionically isolated areas of reinforcement can set up differential aeration corrosion cells (**11.6.3**) that allow corrosion to proceed in these areas. These concepts are illustrated diagrammatically in ***Figure 11.23***.

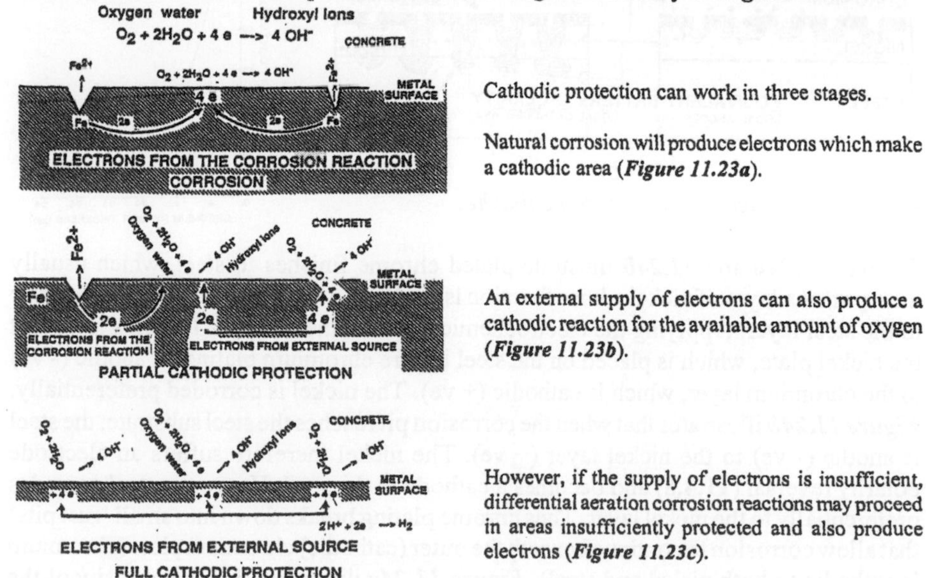

Cathodic protection can work in three stages.

Natural corrosion will produce electrons which make a cathodic area (***Figure 11.23a***).

An external supply of electrons can also produce a cathodic reaction for the available amount of oxygen (***Figure 11.23b***).

However, if the supply of electrons is insufficient, differential aeration corrosion reactions may proceed in areas insufficiently protected and also produce electrons (***Figure 11.23c***).

Figure 11.23 Diagrammatic representation of full and partial cathodic protection

Full cathodic protection is only achieved where the supply of electrons from an external source is able to react with all the oxygen. This concept of ionically isolated areas which may prevent corrosion protection by cathodic protection can be illustrated by the reference to the application of cathodic protection to the hull of a ship. The cathodic protection system applied to the hull of a ship will not protect the superstructure of the ship as there is no ionic transport contact between the cathodic and anodic areas of the superstructure. In fact the superstructure will only be protected when the ship sinks. Cathodic protection relies upon there being a good electrolytic path between the anodic and cathodic sites. In long term cathodic protection systems for reinforced concrete (**8.25.4**), this is provided by pore water within the concrete (**8.25**); in short term cathodic protection systems, the electrolyte is enhanced by introducing an external electrolyte (**8.25.2**).

11.14 CONTINUOUS COATINGS

11.14.1 Metals Cathodic to Steel

Metals cathodic to steel (copper, chromium, nickel, lead) are usually less corrodible and rely upon a continuous, pinhole- free coating to protect the underlying steel. In practice, this is never achieved by electroplating techniques due to the evolution of hydrogen. With chromium, for example, hydrogen reacts with the chromium to produce meta-stable chromium hydride which disrupts the film to produce micro-cracks (**holidays**). This is shown for chrome plating in ***Figures 11.24a*** and ***11.24b***. The porosity of the chromium plate relative to its thickness is shown in ***Figure 11.24c***.

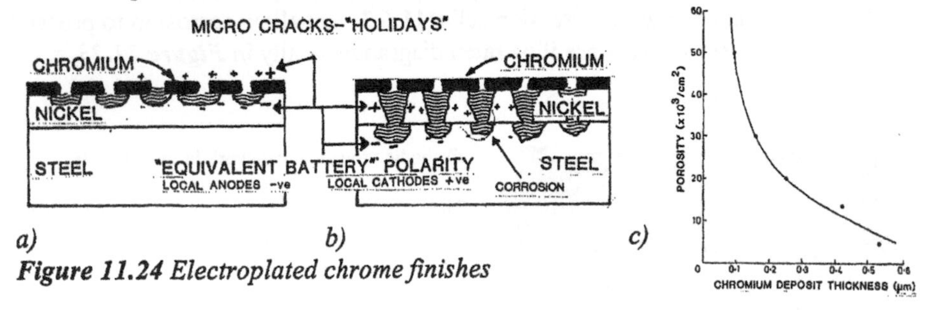

Figure 11.24 *Electroplated chrome finishes*

Figures 11.24a and ***11.24b*** illustrate plated chrome finishes to steel, which usually incorporate a layer of nickel whose function is to obtain adhesion of the chromium layer to the steel layer. Applying the electrochemical series, then ***Figure 11.24a*** shows that the nickel plate, which is placed on the steel before chromium plating, is anodic (- ve) to the chromium layer, which is cathodic (+ ve). The nickel is corroded preferentially. ***Figure 11.24b*** illustrates that when the corrosion pit reaches the steel substrate, the steel is anodic (- ve) to the nickel layer (+ ve). The nickel therefore suffers an electrode polarity reversal (**11.4.8**) and becomes a cathode to the steel. Hence the steel corrodes preferentially to the nickel layer. Thus chrome plating breaks down into small 'rust pits' that allow corrosion to continue beneath the outer (cathodic) chromium plate (chromium is cathodic to both nickel and steel). ***Figure 11.24c*** illustrates that the porosity of the microporous chrome plate decreases with increasing thickness, indicating that the number of holidays formed decreases with an increase of chrome plate thickness.

Specially alloyed corrosion resistant steels (**10.10.7**) (**weathering steels**) produce protective films on their surfaces. These protective films are often coloured (depending on the alloying metal used). For example, **Cor-ten™** steel (which contains various additions, including 0.25 to 0.55% Cu), develops a russet-coloured oxide film which darkens to purplish brown over time. These coloured protective films should be compared to the colourless (white) zinc corrosion product; the colourless result of the corrosion of zinc may be more aesthetically acceptable to the building designer.

Lead cannot be put onto steel like zinc, because it does not wet the surface. If tin is added (80 to 93% Pb, 20 to 7% Sn) and ammonium chloride used as a flux, wetting is improved. **Terne metal** (80.25% Pb, 1.75% Sb, 18% Sn) applied to steel is used for roofing material and for deep drawing manufacture; normally lead (Pb) alloy is the cathode and steel is the anode.

Tin plate is used to protect steel; normally tin (Sn) is the cathode and steel is the anode. Under normal conditions, tin plate corrodes at pin holes. The porosity of tin plate caused by these pinholes decreases with increased coating weight (as in ***Figure 11.24c***). Some (acidic) fruit juices produce a potential reversal (**11.4.8**) and therefore most tin cans are additionally protected by a lacquer.

11.14.2 Aluminium Anodised Oxide Films

Aluminium is a very reactive metal (**1.8.3**) and combines vigorously with oxygen to produce very high temperatures. This is most apparent by the white flashes present in fireworks produced by powdered aluminium. However, aluminium alloys are used in the construction industry, domestically for the manufacture of cooking utensils and in lightweight aircraft structures. In each case, the rapid combination with oxygen is not apparent, due to the presence of an oxide film (of approximately 10 nm thickness) which forms on the surface of the metal in normal environments and protects the metal from further attack by acting as a durable, coherent and adherent barrier (**1.8.3**).

Anodising is the process of forming a semi-porous oxide film on aluminium alloys. The aluminium metal is made the anode and placed in a suitable electrolyte (**Table 11.6**), whereupon the air-formed film breaks down and a much thicker oxide film (approximately 50 nm deep) is formed on the surface. Aluminium oxide is an extremely hard oxide, much stronger than the aluminium metal. The hardness and thickness of the anodic oxide film produced depends upon the composition of the electrolytic bath, as shown in **Table 11.6**.

Table 11.6 Conditions for anodising Aluminium alloys in various electrolytes

Process	Acid %	Current density (A/dm^2)	Voltage (V)	Temperature (°C)	Time (min)	Film Type
Chromic acid	3	0.4-0.6	(cycle)	40	60	Grey-white
	10	0.4-0.6	(cycle)	40	30	Hard Opaque
Sulphuric acid	14	0.8-1.0	15	10-15	30-40	Hard Translucent
	20	1.2-1.5	15	18-20	20	
Oxalic acid	5	0.5-1.5	50	20	30	Pale yellow
	3+ 2% Potassium oxalate	1.5	30-60	50	30-60	Thick Colourless

The aluminium oxide film is semi-porous (but impermeable to water) as a result of the central holes within each grain (***Figure 11.25***). These holes allow the oxide film to absorb a dye and so make coloured anodic oxide films. The holes also absorb ordinary household paint pigments and care should be taken when painting wooden surrounds to aluminium window frames to prevent splashing. It is estimated that there are 10^9 pores in each square millimetre of surface; the pores do not adversely affect the corrosion protection afforded by the oxide film, although they can be 'sealed' by exposing the surface to steam. This property is often utilised when aluminium articles are coloured (e.g, bronze coloured aluminium doors).

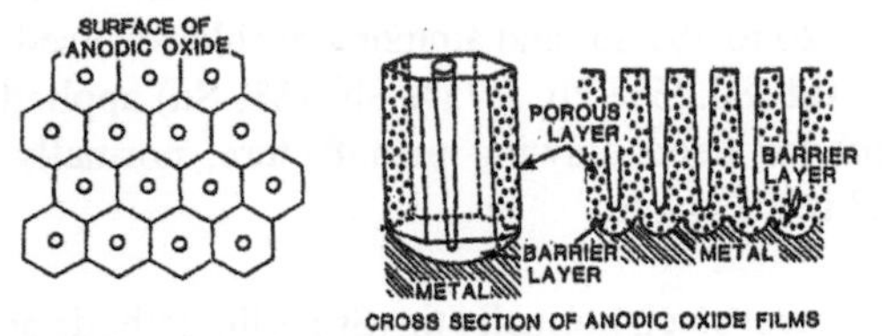

Figure 11.25 *Anodic oxide films on aluminium*

Bright anodising is a dual process. Initially the aluminium metal (or its alloys) is chemically polished to produce a shiny surface. This is followed by the production of the anodic oxide film. In this way, anodised aluminium trim has replaced chromium plated steel trim. It has the advantage that the oxide coat is free of pinholes. The anodic oxide film also acts as a good substrate for paint systems.

11.15 INORGANIC COATINGS

Inorganic coatings are used in the construction industry to provide corrosion protection. They are of two forms

i. **Inorganic phosphate films** (good for the protection of steel), where their function is two-fold
   - they must limit the area available for corrosion reactions by preventing the lateral spread of corrosion beneath the paint film (**11.12.1**);
   - they must increase the adhesion between the organic paint film and the metal in both dry and wet (aqueous) environments, preventing osmosis (**11.12.3**).

ii. **Inorganic oxides** (electrically produced on aluminium and its alloys).

Inorganic phosphate coatings are attacked and dissolved by both strong acids and alkalis, and are sensitive to changes in pH. If a phosphatised steel is galvanically connected to an unphosphatised steel plate and placed in a salt solution, then the phosphatised coating is anodic to the unphosphatised steel. However, after a period of time, the polarity of the couple is reversed (**11.4.8**) and the unphosphatised coating becomes anodic and dissolves. This will have relevance to paint repair systems where no **conversion coating** (**11.15.1**) is employed. Atmospheric exposure tests show that unpainted phosphatised steel sheets corrode at a faster rate than solvent wiped unpainted steel sheets. It is postulated that the phosphate ions accelerate the corrosion of the steel. This is important where conversion coatings are used, as these must be thoroughly removed by rinsing with demineralised (or distilled) water to prevent the trapping of foreign ions beneath the paint coating.

11.15.1 Conversion Coatings ('Rust Cures')

There are many preparations on the market that allow the repair of rusting paint defects with the minimum of surface preparation. Rust is a hydrated oxide, which can be written

Ferrous hydroxide	$Fe(OH)_2$ or $FeO.H_2O$	(Ferrous oxide hydrate)
Ferrous ferrite	Fe_3O_4 or $FeO.Fe_2O_3$	(Magnetite)
Ferric hydroxide	$2Fe(OH)_3$ or $Fe_2O_3.3H_2O$	(Ferric oxide hydrates)

This representation of rust highlights the fact that these oxides are compounds with water. The idea behind most rust cure preparations is to turn these oxides into the black oxide magnetite (Fe_3O_4) (containing no water). Most preparations depend upon the beneficial effect conferred on corrosion by the **tannate** molecule (present, for example, in timber, leather, fish oils, etc.). The benefits of these compounds were discovered very early when, for example, uncorroded steel and leather articles were pulled out of disused drinking wells and when it was observed that steel did not rust at sea where fish were handled. Tannins are added to phosphoric acid solutions and produce a black oxide on the surface of steel. However, the tannins are easily washed off steel and require some form of etch to produce satisfactory adhesion. Subsequent development of these solutions was undertaken during World War 2, where wash primers were developed for aluminium. These were found to be beneficial for steel and have since been produced industrially for the general public. A common conversion coating is bought as a single pack, typically comprising a resin pigment (7.2% proprietary vinylite, 6.9% basic zinc chromate, 1.1% talc, 48.7% isopropanol, 16.1% butanol) and a phosphator (3.65% phosphoric acid, 3.2 % water, 13.2% isopropanol).

11.15.2 High Temperature Oxidation Coatings

Oxidation is a corrosion process that occurs **without** an aqueous environment. In the high temperature oxidation coating process, metallic ions (Fe^{2+}) and their electrons can diffuse through the oxide film, causing growth in the thickness of the oxide film at the oxide/air interface (***Figure 11.26***).

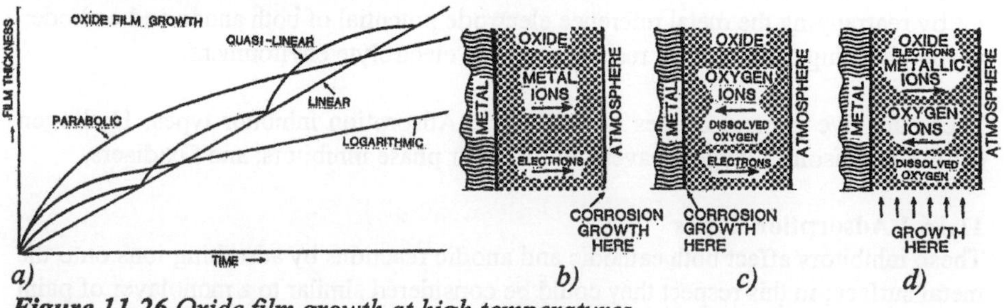

Figure 11.26 Oxide film growth at high temperatures

Growth in the oxide film thickness is dependent upon which species diffuses; growth with time may be linear, or may be best described by some other mathematical relationships (***Figure 11.26a***). For example, denoting y as the oxide film thickness and t as time, typical relationships are linear, $y = kt$; logarithmic, $y = \log_e (kt)$; parabolic, $dy/dt = k/t$, where dy is a small increase in thickness in the small time interval dt and k is a constant; quasi-linear (a combined linear and other form depending on where the oxide film fractures as a result of the increase in volume). During high temperature oxidation, oxygen must be oxidised ($O_2 + 4e \rightarrow 2O^{2-}$) to produce the oxide ion, O^{2-}. The electrons can diffuse through the oxide film because there are two states to iron, the ferrous state (Fe^{2+}) and the ferric state (Fe^{3+}), i.e., $Fe^{2+} \rightleftharpoons Fe^{3+} + e$; these ions can be considered as the means of transport (***Figure 11.26b***). An alternative mechanism has been postulated (***Figure 11.26c***), where the oxide ions diffuse through the oxide film, causing growth to take place at the metal/oxide interface. For high temperature corrosion resistance, the oxide film which is formed on the surface of the metal must protect the metal from further attack. This would produce a logarithmic or parabolic curve (***Figure 11.26a***); however, this is rarely observed, as the oxide film tends to fracture as a result of temperature fluctuations or volume differences.

11.15.3 Stainless Steel

Stainless steel is an alloy of iron with nickel and chromium (e.g, 18% Ni, 8% Cr) (**10.10.7**). The added metals produce a surface film of metal oxides which is impervious to water. Stainless steels owe their corrosion resistance to the tenacious coherent oxide film developed under atmospheric conditions. In aqueous conditions, stainless steel is not as good as its name implies, as it requires oxygen-containing mediums to keep it stainless. Stainless steels are not attacked by alkalis, hence their use as replacement wall ties. However, stainless steels are attacked by chlorides in alkaline mediums, leading to **stress corrosion cracking** (a disastrous form of corrosion failure, principally in alloys, that occurs along grain boundaries and is brought about by residual internal stresses and/or applied stresses in an aggressive environment). It is therefore important that heat treatment is carried out to relieve any residual stresses in the stainless steel.

11.16 CORROSION INHIBITORS

The methods by which corrosion inhibitors may work are

- by altering the nature of the corrosion product;
- by chemisorption onto the metal surface;
- by rearranging the metal reference electrode potential of both anode and cathode;
- by changing the electrical resistance of the electrolyte environment.

There are five general classes of inhibitors: Adsorption inhibitor types; Hydrogen evolution poisons; Oxygen scavengers; Vapour phase inhibitors; and Oxidisers.

11.16.1 Adsorption Types

These inhibitors affect both cathodic and anodic reactions by adsorbing ions onto the metal surface; in this respect they could be considered similar to a monolayer of paint (**11.12.1**). Typical of this class of inhibitor are the organic amines (***Figure 11.27***). Corrosion inhibitors based upon the amines produce chemically adsorbed monolayers adjacent to the metal surface. Amines depend upon their physical size to adsorb onto the substrate metal via the 'active end' of the molecule, which is through a positively charged nitrogen atom. The size of the amine unit is likened to the shield provided by an umbrella during a rain storm (***Figure 11.27c***). Adsorption is electrostatic in nature and the whole adsorbed molecule is thought to act as a barrier coating. Consequently a high molecular weight and spatial geometry of the molecule (i.e., balanced about the group 5 nitrogen atom) is required to aid the 'packing' of the 'umbrella' at the metal surface. Balance is required so that the umbrella sits flat to the surface rather than tilted; unbalanced molecular compounds have been found to be inferior to those which are balanced. Most inhibitors of this type are large oily macromolecules or ions.

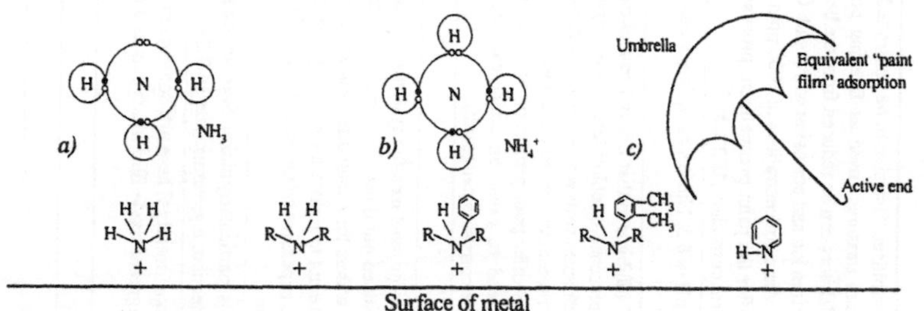

Figure 11.27 *Adsorption type corrosion inhibitors typified by organic amines*

11.16.2 Hydrogen Evolution Poisons

These inhibitors prevent the discharge of hydrogen on the cathodic areas of iron and thus reduce (eliminate) corrosion. They are thought to act as 'negative catalysts' to the discharge of hydrogen by forming a blanket over the metal surface. Hydrogen evolution poisons are therefore of little use in oxygenated waters and cannot be used outside acidic environments (pH 1 to 6). They are used primarily in the manufacturing and engineering trades for inhibiting phosphoric acid solutions where rusted steel is etched prior to painting.

11.16.3 Oxygen Scavengers

In cases where oxygen is involved in the cathodic reaction, e.g, $O_2 + 2H_2O + 2e \rightarrow 4OH^-$, oxygen scavengers will be effective in controlling corrosion. Hydrazine is an efficient oxygen scavenger and is oxidised to nitrogen and water, $N_2H_4 + O_2 \rightarrow N_2 + 2H_2O$. Such inhibitors are therefore consumed with time. In addition, oxygen scavengers are not suitable for use in acidic solutions.

11.16.4 Oxidisers

Inhibitors which are oxidisers include chromates, nitrates, nitrites and other oxidising salts which are used to inhibit the corrosion of iron and its alloys.

11.16.5 Vapour Phase Inhibitors (VPI)

VPIs are substances which possess a very low vapour pressure and readily transfer from their liquid formulation to the metallic surface by a process of **sublimation** (conversion of a solid substance by means of heat into vapour). VPIs are therefore only effective in enclosed areas. They are similar to the organic adsorption inhibitors (**11.16.1**) in their mode of operation. VPIs are usually incorporated into oily brown paper, which has a distinctive smell, used as a wrapping paper for metallic components (e.g, purchased spare parts in the motor industry). They are usually volatile nitrites, such as dicyclohexylammonium and di-isopropylamine nitrites. A range of inhibitors and their uses in neutral water are summarised in **Table 11.7**.

Table 11.7 General effectiveness of various inhibitors in the near neutral pH range

Corrosion inhibitor	Metal						Notes
	Mild steel	Cast iron	Zinc Zinc alloys	Copper Copper alloys	Aluminium Aluminium alloys	Lead-tin soldered joints	
Chromates	Effective	Effective	Effective	Effective	Effective	-	Chromium on its own acts as a ferrite stabiliser. Use should be controlled where mixed with organic compounds, e.g. ethylene glycol mixtures (used as freezing point depressants in antifreeze). In these cases, chromate inhibitors can be reduced from the hexavalent chromium ion to the non-inhibiting trivalent chromium ion and corrosion will result. Chromates can give rise to dermatitis. Used in car radiators where concentrations of 1000 ppm provide protection. Chromates are also used in the wash after the paint pretreatment process. These chromates prevent the formation of scab and filiform corrosion **(11.12.3)**
Nitrites	Effective	Effective	Ineffective	Partially effective	Partially effective	Aggressive	Optimum concentration 2,000 ppm at a pH of 8.5. Nitrites are a dangerous inhibitor if lead and lead solder are present.
Benzoates	Effective	Ineffective	Ineffective	Partially effective	Partially effective	Effective	Sodium benzoate is classed as a "safe" inhibitor in that it is not concentration dependent. Its efficiency is greatly influenced by the presence of chlorides. It was originally developed for use in radiators containing glycerol antifreezes with which it is compatible (cf. chromates). It is used in waxed wraps and temporary protectives based on rubber latex or bentonite pastes (it does not cause dermatitis). Benzoates offer poor protection with cast iron and zinc (used with galvanized products, benzoates tend to cause an anode-cathode reversal, the zinc becomes cathodic to the steel and severe corrosion can result).
Borates	Effective	Variable	Effective	Effective	Variable	-	
Phosphates	Effective	Effective	-	Effective	Variable	-	Phosphates exhibit average performance for most metallic systems, but the active component (the phosphate unit) tends to be precipitated out if dissolved in hard waters. Polyphosphates have been used as inhibitors in cooling waters for a long time. Additions of this inhibitor have been tried in car water washes in an attempt to reduce the effect of corrosion caused by road deicing salts. They were found to be too expensive when used on the scale necessary to reduce the aggressive effect of the chloride ion.
Silicates	Reasonably effective	Reasonably effective	Reasonably effective	Reasonably effective	Reasonably effective	Reasonably effective	Silicates are fairly good with most metals, particularly nickel. Sodium silicates exclude oxygen (see also their use in paints for fire protection, e.g, vermiculite **14.7.3)**
Tannins	Reasonably effective	Reasonably effective	Reasonably effective	Reasonably effective	Reasonably effective	Reasonably effective	Tannins are reasonably effective for most metals. These are conversion coatings **(11.15.1)**, where rust is converted into a black magnetite oxide film which acts as a base for subsequent paint films.
Molybdates	Molybdates are good for steel, used in conjunction with other inhibitors to obtain a synergistic effect.						
Triazoles	Triazoles are good for copper, brass, but not very effective with other metals.						
Thioureas	Thioureas are successful in pickling solutions, but can cause eye problems leading to temporary blindness.						

12

TIMBER AND TIMBER PRODUCTS

12.1 INTRODUCTION

Timber is a commercial raw material rather than a manufactured material; the only direct processing it receives is in the **conversion** of the tree to the sawn material, **seasoning** and **preservative treatment.** The attraction of timber as a construction material (both in terms of density and cost) lies in its high strength and stiffness to weight ratio, its workability and its thermal and sound insulating properties. The low weight of timber arises from its open boxlike (cellular) structure, whilst its ability to maintain high strength depends upon its durability, which in turn depends upon its biology and structure. Timber is an **organic carbohydrate** material, i.e., a hydrate of carbon, where the hydrogen and oxygen are in the same proportions as water, $C_x(H_2O)_y$. Thus wood is a food source for fungi and wood-boring beetles; it is **biodegradable**. To be used efficiently and effectively, knowledge of how the structure of timber affects strength, durability and resistance to biodegradation is required.

The structure of timber can be examined on three distinct levels: the **molecular**, **microscopic** and **macroscopic** level. The structure of timber at all three levels affects the durability and strength of timber and its products. At the **molecular** level, the arrangement of the **molecules** forming the **microstructure** is examined. At the **microscopic** level, the cellular structure of timber is observed with an optical microscope. At the **macrostructure** level, larger visual features observable with the ordinary eye (e.g, grain and knots) are examined.

Trees are classified as **hardwoods** or **softwoods**. These distinctions arise purely from botanical nomenclature and classification, and have nothing to do with the physical hardness or softness of the wood. For example, balsa wood is a hardwood, as is Lignum vitae (once used to make bowling balls). It is true, however, that commercial hardwoods are usually physically harder than most commercial softwoods. **Table 12.1** outlines the principal features associated with each type of tree.

Table 12.1 Definitions and terms associated with hardwoods and softwoods

	Hardwoods	Softwoods
Leaves	Broad leaves	Needle-like leaves
Type	Deciduous (leaves fall off in winter)	Coniferous, evergreen
Reproduction	Flowers (**angiosperm**). Enclosed or protected seeds	Cones (**gymnosperms**). Naked seeds
Examples	Temperate (Oak, Horse chestnut, etc.) Tropical (Teak, Mahogany, Sapele, etc.)	Pines, Firs, Spruces, Larch
Cost	Relatively expensive	Relatively cheap

12.2 MACROSTRUCTURE

A tree consists of three parts (***Figure 12.1a***), the **roots** (not commercially used, except for some particleboards), the **trunk** (commercially used) and the **crown**, consisting of the branches and leaves (not commercially used).

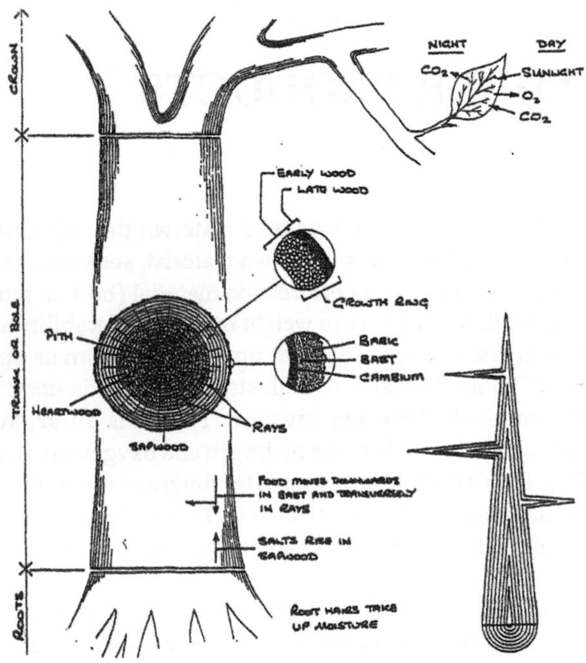

a) Growth of the tree *b) Continuity of growth*

***Figure 12.1** The growth of the tree*

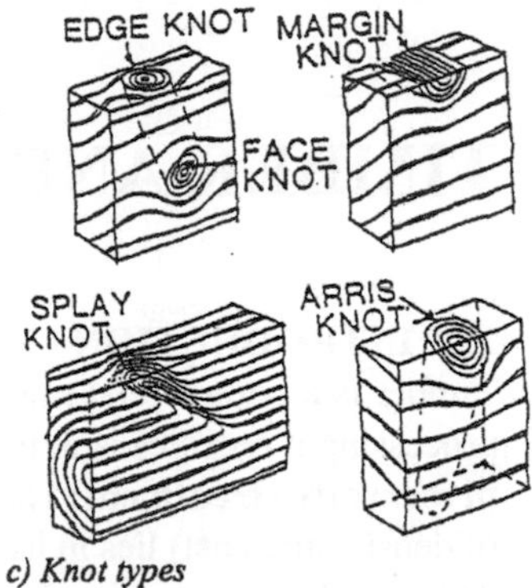

c) Knot types

Growth of the tree (in height and girth) is dependent upon two main factors

- the ability of the roots to extract minerals (as dissolved salts) from the soil. Some plants (like strawberries) put out new runners, called **stolons,** to tap new ground for the minerals required to sustain growth; these plants are capable of living forever;
- the ability of the leaves to synthesise **carbohydrates (12.4.1)** from carbon dioxide and water.

The **crown** of the tree comprises the upper branches and leaves. The leaves contain a green substance called **chlorophyll** which, under the action of sunlight, acts as a catalyst to chemically bind carbon dioxide (from the air) and water (from the roots) into a new substance called **glucose (12.4.1)**. This process is called **photosynthesis**. The crown is also the region responsible for reproduction (i.e., the production of seeds). The **trunk** of the tree has three physical functions to perform

- it must **support** the crown;
- it must **conduct** the minerals absorbed by the roots upwards (to the **cambium layer**);
- it must **transport and store** the photosynthesised glucose until required.

As will be described later (**12.3**), these tasks are performed by different types of cells. The entire cross-section of the trunk fulfils the function of support (increasing crown diameter is matched with increasing trunk diameter). However, conduction and storage are restricted to the outer region of the trunk known as the **sapwood** (the region in which the cells no longer fulfil these functions is termed the **heartwood**). A tree differs from other perennial plants (like bracken) in its ability to stand erect after completing its growth period, even though the interior is **biologically dead**. This means that the leaves are able to reach the forest canopy and utilise the sunlight for producing glucose by photosynthesis. The glucose is changed to **cellulose** (**12.4.1**) in the **cambium layer**.

The **heartwood** is the centre portion of the tree and is biologically dead. The function of the heartwood is to supply rigid support for the tree. The heartwood of most trees is darker in colour than the sapwood due to the higher content of **extractives**, principally **tannins**, which provide increased resistance to both fungal and insect attack. In addition, the heartwood contains no free glucose or starch, a potential food source for wood-destroying organisms. The heartwood of hardwoods usually contains more tannin than the heartwood of softwoods, and is therefore more durable. This darker colour does not occur in all woods; for example, Spruce, Hemlock, Fir, Poplar and Obeche do not, as a rule, have a darker heartwood and there is no colour differentiation between the heartwood and the sapwood because there are no extractives in the heartwood. These timbers therefore have lower durability than those where the heartwood contains extractives. In general, the darker the heartwood, the more extractives there are and the greater the durability of the wood (**12.25.2**).

The **sapwood** (the last few years of growth) is also biologically dead, and is the part of the tree responsible for transporting water and minerals up the trunk and glucose down from the crown. Cellulose is produced in the **cambium layer**, which is situated beneath the **bark**; it is the only living part of the tree. The soft centre portion of the trunk is the **pith** and the outer section is the **bark**, which protects the cambium layer from physical damage. The cambium is the 'growth factory' of the tree; in temperate zones, it lies dormant during the winter period and manufactures new wood cells during the growing season (spring to autumn) at the outer girth of the tree. Any damage to the cambium layer can kill the tree (e.g, Dutch Elm disease). During the growing season, the cambium cells divide to form additional cells, some of which will remain as cambium cells, whilst those on the outside of the cambium zone form the bark and those on the inside form the sapwood. The newly formed sapwood is more porous than the inner heartwood. Therefore, although the sapwood is more liable to biodegradation than the heartwood, it is also more easy to protect, as fungicides and insecticides are able to penetrate more easily. The width of the sapwood varies with the species and age of the tree, but it is usually less than one third of the total trunk radius.

Growth may be continuous throughout the year in some parts of the world and the wood formed tends to be uniform in structure. In temperature regions, growth is seasonal, resulting in the formation of **growth rings**. In the UK, there is a defined growth period each year, resulting in the formation of **annual growth rings**. Wide growth rings indicate rapid growth, giving rise to thin-walled cells and less summer wood. There is a loss of strength associated with wide growth rings; although now withdrawn, BS 1186[1] required at least 8 growth rings within 25 mm for joinery (a European preStandard dealing with the classification of timber quality for joinery timber, prEN 942[2], exists but has not been formally adopted by European member states).

With increasing radial growth of the trunk, increases in crown size occur, resulting in the enlargement of existing branches and the formation of new ones. Radial growth of the trunk must accommodate existing branches, and this is achieved by the structure known as a **knot**. If the cambium of the branch is still alive at the point where it fuses with the cambium of the trunk, continuity of growth will occur (***Figure 12.1b***). The structure formed is a **green** or **live knot**. If, however, the cambium of the branch is dead (e.g, lower branches), there will be no continuity of growth in the cambium layer as the trunk will grow around the branch, often complete with its bark. This structure is a **black** or **dead** knot, frequently dropping out of planks on sawing. Both types of knot produce distortion of the grain and therefore a reduction in strength in sawn timber (**12.9**). Knot types are shown in ***Figure 12.1c***.

12.3 MICROSTRUCTURE

The identification of a timber species is made by examining the microstructure of the timber. The microstructure for a typical timber is shown in ***Figure 12.2***.

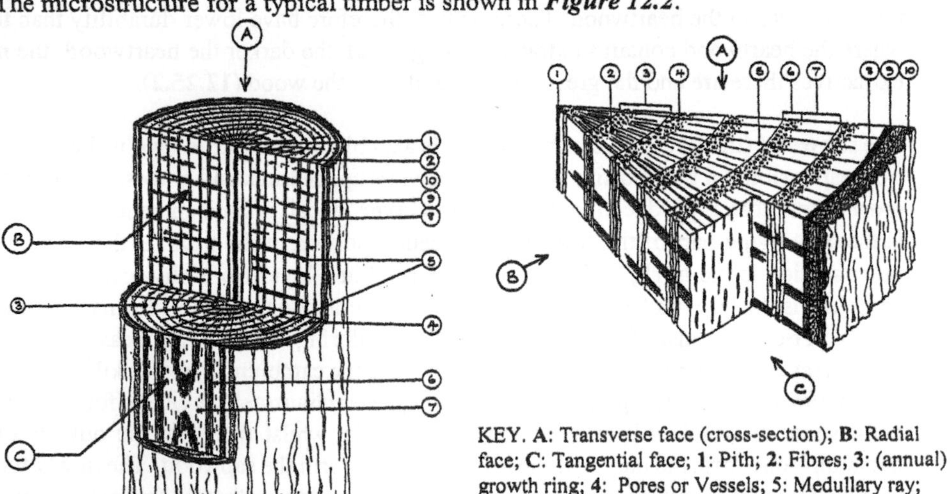

KEY. **A**: Transverse face (cross-section); **B**: Radial face; **C**: Tangential face; **1**: Pith; **2**: Fibres; **3**: (annual) growth ring; **4**: Pores or Vessels; **5**: Medullary ray; **6**: Earlywood; **7**: Latewood; **8**: Cambium; **9**: Inner bark (or bast); **10**: Outer bark

***Figure 12.2** Generalised microstructure of the tree (hardwood)*

During the growing period, new cells produced in the cambium layer are stacked up end to end. Approximately 90-95% of the cells are orientated longitudinally within the tree, whilst the remaining 5-10% are present in radially orientated bands (known as **medullary rays**). This variation in cell orientation is a major reason why timber exhibits **anisotropic** strength and moisture movement characteristics (**12.4.3**). Within a few weeks, the cells undergo changes, notably the formation a secondary cell wall and the **lignification** of the inside of the cell wall. Lignification is the deposition of **lignin** ribs of various patterns which gives the cell added strength and prevents its wall from caving in. Lignification also renders the cell walls impermeable to water and solutes, and so the cell dies. The lignified walls are perforated by numerous **pits**, where lignin fails to be deposited and only the cell wall remains. Pits allow the passage of water and solutes both vertically and horizontally. In hardwoods, the horizontal cell walls of adjacent cells gradually break down, leaving adjacent cells in open communication with one another (essentially forming a long hollow tube, ideally suited to the passage of water). In softwoods, the cell walls are tapered at both ends and passage of water occurs through pits (known as **bordered pits**). The pit structure comprises a **torus** (which acts like a diaphragm) suspended by **margo** strands (***Figure 12.3***).

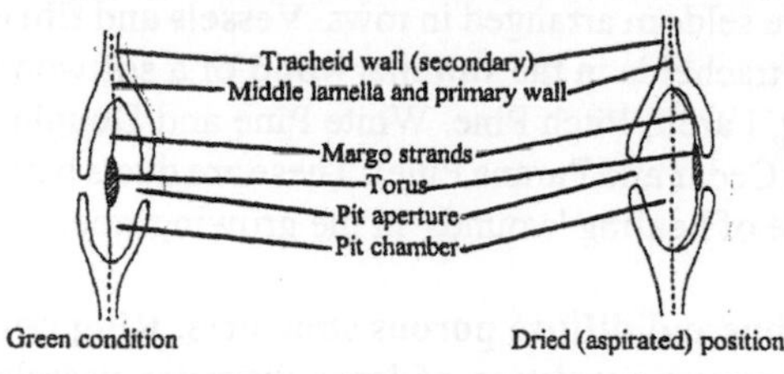

Differential water pressure between adjacent tracheids causes the torus to move against the pit aperture, effectively closing the cell to the passage of water. Pits have important implications in the preservation treatment of timber (**12.24**), as the preservative formulation must be carefully selected to prevent pit closure to maximise penetration of the preservative.

***Figure 12.3** Schematic representation of the structure of a pit*

An additional change that occurs in the new cells formed is the transformation into one of four basic cell types (**Table 12.2**).

Table 12.2 The functions and wall thicknesses of the various cells

Cell	Notes	Hardwoods	Softwoods	Primary function	Schematic
Tracheids	Length 2-5 mm; Length:diameter ratio 100:1	Scarce	Present	Mechanical support. Conduction	
Parenchyma	Length 0.2 mm; Length:diameter ratio 8:1	Present	Present	Storage (also conduction of glucose to the outside of the tree)	
Fibres	Length 1-2 mm; Length:diameter ratio 100+:1	Present	Absent	Support	
Vessels/pores	Length 1 mm; Length:diameter ratio 10:1	Present	Absent[a]	Conduction	

Notes: [a] Some softwood conifers contain resin ducts (see text). Care needs to be taken to distinguish these structures from vessels/pores.

When seasonal growth commences, the dominant function is conduction, while in the latter part of the year it is support. This change in function is seen in softwoods with the presence of thin-walled, large diameter tracheids in the early part of the growing season (when the tree has to take up a large amount of water), known as the **earlywood** (or **springwood**), and thick-walled, slightly longer tracheids of a smaller diameter in the latter part of the season (when less water is available), known as **latewood** (or **summerwood**). The latewood is much denser and stronger than the earlywood (***Figure 12.4a***).

Tracheid cells are present in softwoods and to a limited extent in hardwoods. In softwoods, tracheids form about 90% of the wood and are entirely responsible for conducting minerals in solution from the roots up the trunk. In addition, in softwoods, they are responsible for support (in hardwoods this function is performed by **fibres**). **Vessels** (or **pores**) are found in hardwoods and are thin-walled, short, wide cells arranged together to form long 'hose pipes' running the length of the trunk. They are the equivalent of **tracheids** in the **springwood** of softwoods. Vessels conduct minerals in solution from the roots up the trunk. **Fibres** only appear in hardwoods and are similar in appearance to tracheids in softwoods, although they do not conduct water and are seldom arranged in rows. **Vessels** and **fibres** in hardwoods together are equivalent of the tracheids in the **summerwood** of a softwood. In some conifers, **resin ducts** are found (e.g, Larch, Pitch Pine, White Pine and Douglas fir, but not in Western Hemlock, Western Red Cedar and Parana Pine). These are ducts between cells which contain a resin for the purpose of healing 'wounds' in the growing tree.

Hardwoods are categorised into **ring porous** and **diffuse porous** structures. **Ring porous** hardwoods are characterised by the earlywood consisting of large diameter **vessels** (or **pores**) with very few **fibres**, whilst the latewood has more fibres and much smaller **vessels**. However, the majority of hardwoods do not show this differentiation, and in these cases the **vessels** are distributed evenly throughout the annual ring; these hardwoods are called **diffuse porous**. Both ring porous and diffuse porous hardwoods have **medullary rays** similar to the softwoods; whereas the rays in softwoods are difficult to see, those in hardwoods are very much bigger and more defined, giving the attractive **figuring** which characterises radial sections of hardwoods. The microstructure of a typical softwood, ring porous hardwood and diffuse porous hardwood are given in ***Figure 12.4***.

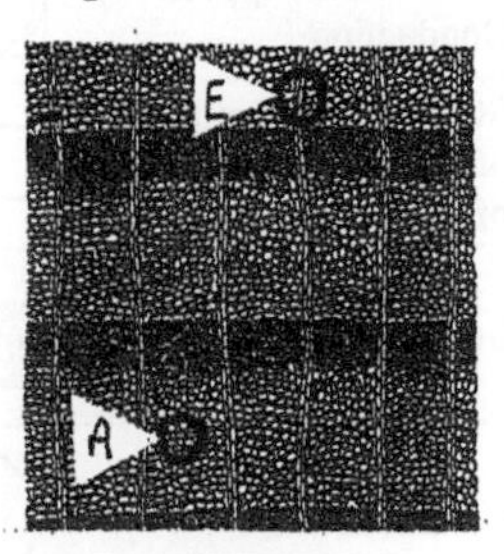

Section 1

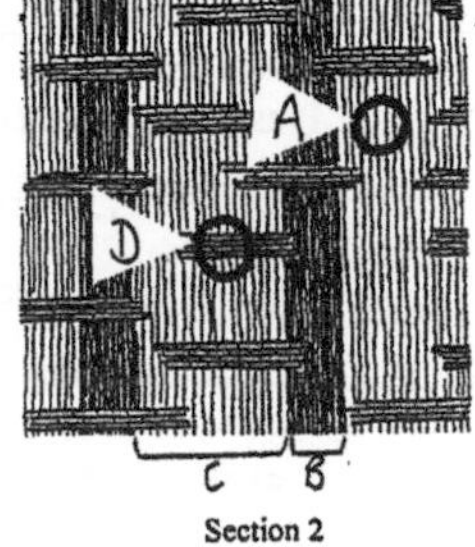

Section 2

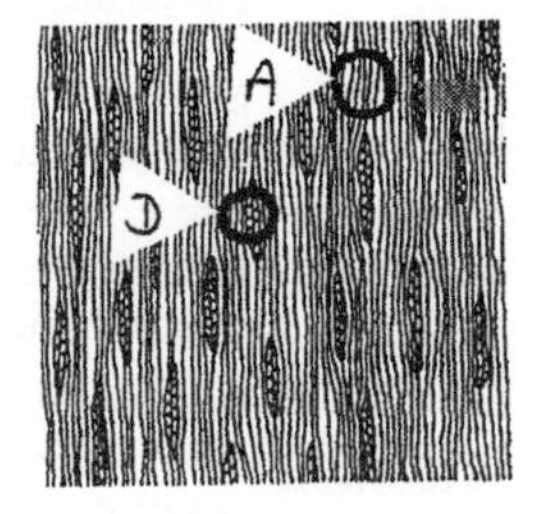

Section 3

Figure 12.4 *Generalised microstructure of the tree a) Softwood*

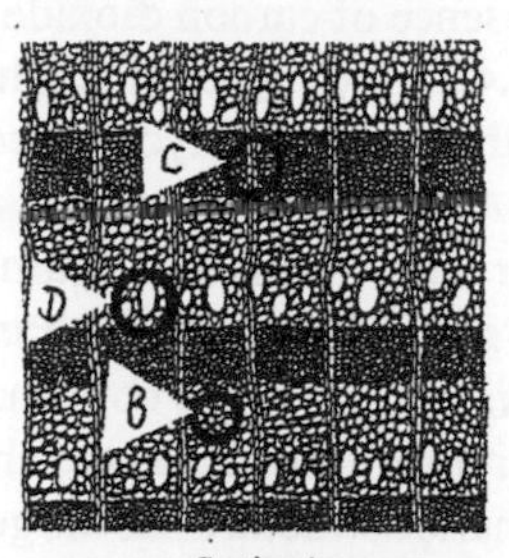

Section 4

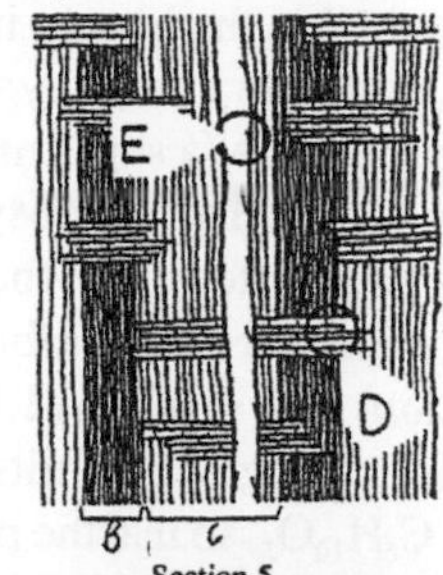

Section 5

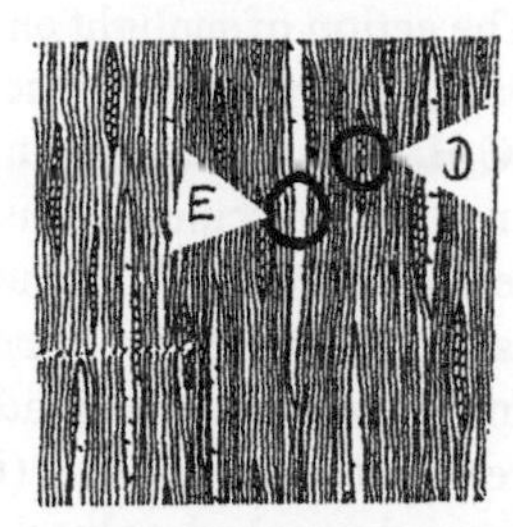

Section 6

b) Ring porous hardwood

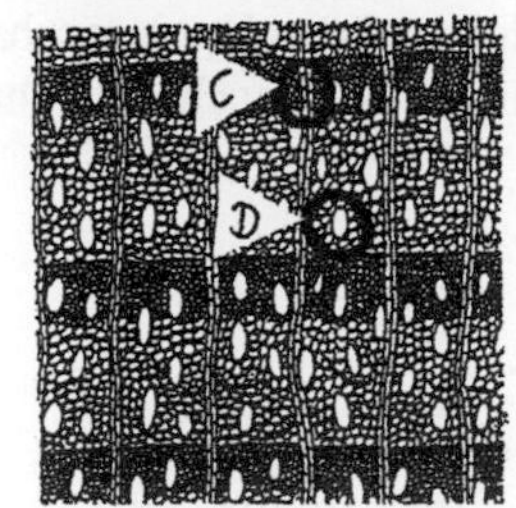

Section 7

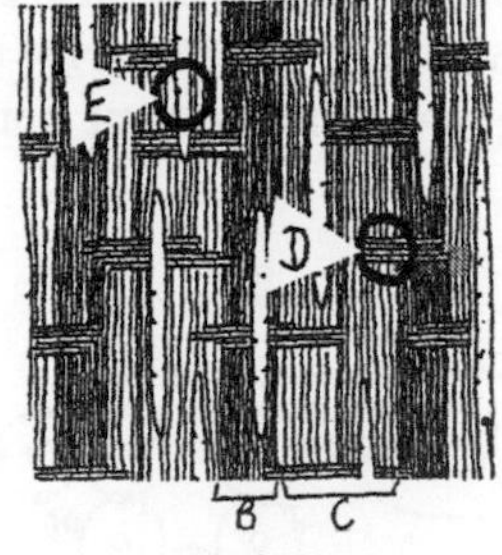

Section 8

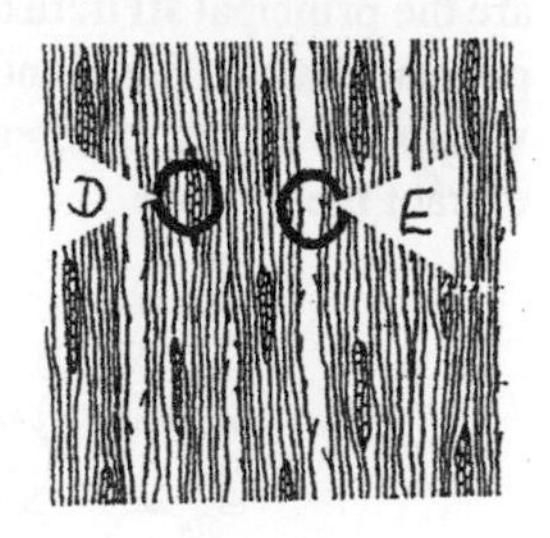

Section 9

c) Diffuse porous hardwood

Notes: There are nine sections here, showing the typical microstructure of a softwood, ring porous hardwood and diffuse porous hardwood in the three principal direction: transverse, radial and tangential. From ***Figure 12.2***, try and identify the sections and the items labeled in each section. The answers are given at the end of this Chapter.

***Figure 12.4** Generalised microstructure of the tree*

12.4 MOLECULAR STRUCTURE

Chemical analysis of timber reveals the existence of four chemical constituents and provides information on their relative proportions (**Table 12.3**).

Table 12.3 Chemical composition of timber

Constituent[a]	% (by weight)	Molecular derivatives	Polymeric state
Cellulose $(C_5H_{10}O_5)_n$	40-50	Glucose	Crystalline, highly orientated[b], large 3D molecule. Cellulose chain length (degree of polymerisation, **13.5**), n = 7000-10000. Anhydroglucose $(C_5H_{10}O_5)_n$, covalently linked in filament form.
Hemicellulose $(C_5H_{10}O_5)_n$	20-35	Galactose Mannose	Semi-crystalline, smaller molecule. Cellulose chain length n = 500-1000. Equivalent to the adhesive material between crystals.
Lignin	15-35	Phenyl propane	Amorphous, large 3D molecule. Resistant to fungi.
Extractives	2-10	e.g, terpenes, polyphenols	Responsible for the colour and flame in wood. Resistant to fungal attack.
Water	varies		

Notes: [a] Most timber also contains trace amounts of inorganic minerals (about 0.1-1.0%); [b] Cellulose consists of crystalline regions (cellulose chains are arranged in an orderly 3D crystal lattice) and amorphous regions (cellulose chains show much less orientation). However, compared to lignin, cellulose can be considered crystalline.

The action of sunlight on chlorophyll in the leaves, in the presence of carbon dioxide from the air and water from the roots, produces a carbohydrate (**12.4.1**), the sugar called **glucose** ($C_6H_{12}O_6$). Glucose is a **monosaccharide** (a sugar molecule that cannot be hydrolysed into smaller molecules). This process is called **photosynthesis** and requires sunlight and a temperature of >6°C. Glucose is transported down the **bast** (or **inner bark**) to the **cambium** layer, where they are joined to give longer chain carbohydrate molecules (**polysaccharides**) in a polymerisation (**condensation**) type reaction (**13.5.2**) (***Figure 12.5***). The condensation reaction removes water (H_2O) from the glucose units and therefore each unit in the chain is termed an anhydroglucose unit, $C_6H_{10}O_5$, so that the polysaccharide structure has the general formula $(C_6H_{10}O_5)_n$. In wood, **cellulose** (n = 7000-10000) and **hemicellulose** (n = 500-1000) are the principal **structural** polysaccharides and **starch** (n = 100+) is the principal **storage** polysaccharide. Fungi and wood-boring insects derive nourishment from all polysaccharides within the timber by breaking down the covalent bonding in the polysaccharide chain to extract the glucose.

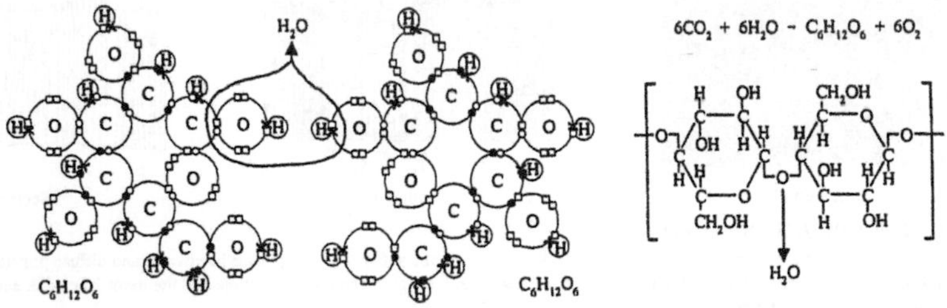

***Figure 12.5** Molecular structure of glucose and cellulose (showing condensation reaction)*

12.4.1 Carbohydrates

Carbohydrates are substances with the general formula $C_x(H_2O)_y$, and are, as the general formula suggests, hydrates of carbon possessing the same ratio of hydrogen to oxygen as exists in water. Not all organic compounds that have this general formula are carbohydrates, e.g, formaldehyde (HCHO) and acetic acid CH_3COOH $C_2(H_2O)_2$ are not carbohydrates. Carbohydrates are either

- linear chains of polyhydroxyl aldehydes (**CHO**.CHOH.CHOH.CHOH.CHOH.CH_2OH);
- or ketones (**C=O**.CHOH.CHOH.CHOH.CHOH.CH_2OH).

(see **Table 13.7** for nomenclature in bold text). Note that the many OH groups give rise to the term **polyhydroxyl** or produce a cyclic ring compound shown in ***Figure 12.5***. Carbohydrates are further subdivided into two main classes, **monosaccharides** ('basic' sugars, such as glucose) and **polysaccharides** (longer chain carbohydrates). Sugars (**trivial name**) are crystalline substances with a sweet taste that are soluble in water, can be hydrolysed to produce 'basic' monosaccharide sugars (no further breakdown of 'basic' sugars is possible by hydrolysis) and have a low melting point (and therefore melt when heated). Examples of sugars include glucose ($C_6H_{12}O_6$) and cane sugar ($C_{12}H_{24}O_{12}$); cane sugar (sucrose, a **disaccharide**) will hydrolyse to produce two 'basic' sugars with the same chemical formula but different optical activity (produced by rotation in the plane of polarised light; (+) *dextro* clockwise; (-) *levo* counterclockwise. Glucose ($C_6H_{12}O_6$) is a well known 'basic' sugar present in, e.g, urine of diabetics, human blood and sweet grapes, and is one of the sugars produced by the tree.

Polysaccharides on the other hand are more complex 'sugars' which do not taste sweet, are insoluble in water, are not fully crystalline nor optically active; however, they can be hydrolysed to produce sugar (in fact wood has been used to produce sugar for human consumption). Polysaccharides are represented by the formula $(C_6H_{10}O_5)_n$; they have a higher melting point than 'basic' sugars (due to the high value of n in their general formula). Everyday examples include starches (**trivial name**) occurring in potatoes, rice, wheat etc; chemically, starches comprise two chemical compounds, 20% α amylose (which is soluble in water) and 80% β amylose (which is insoluble in water). By boiling starches we tenderise them by dissolving the 20% α amylose.

Carbohydrates (sugars) are made by all living vegetation by the action of sunlight and chlorophyl (**12.4.1**) $CO_2 + H_2O \rightarrow HCHO + O_2$ (HCHO is formaldehyde, which can be polymerised by calcium hydroxide to form the sugar $C_6H_{12}O_6$). Polymerisation takes place in the leaves; further polymerisation can take place in the cambium layer to form a polysaccharide $(C_6H_{10}O_5)_n$; it is the length of the $(C_6H_{10}O_5)_n$ chains that dictates the name of the polysaccharide formed (**Table 12.3**).

The most common polysaccharide present in the tree is cellulose. From the structure of cellulose (***Figure 12.5***), a number of important points can be noted

- a single water molecule is removed from a pair of monosaccharide (glucose) molecules during the condensation reaction;
- adjacent anhydroglucose units are joined by covalent bonding through an oxygen atom. Covalent bonds within and between the monosaccharide units are relatively strong, thus the cellulose chain (and therefore timber) exhibits relatively high axial tensile strength;
- cellulose is a relatively straight chained molecule (unlike starch, which is a helical molecule). In cellulose, the way the anhydroglucose molecules are orientated means that - OH groups stick outwards from the chain in all directions (***Figure 12.5***). These - OH groups can form van der Waals bonds (**1.14**) with adjacent chains, producing a kind of three dimensional lattice, so that cellulose forms a relatively strong, rubbery structure. In starch, the anhydroglucose units are orientated such that the - OH groups potentially capable of forming hydrogen bonds project into the helix, so that starch exhibits no cross linkages. This is one reason why starch lacks the structural properties of cellulose;
- The - OH groups that give rise to van der Waals bonding also have a high affinity for water, thus explaining the high affinity of cellulose for water.

The degree of crystallinity of cellulose varies between regions of high crystallinity and regions of non-crystallinity. The individual chains of cellulose pass through regions of high crystallinity (known as crystallites or micelles) and through regions of low crystallinity (in which the cellulose chains are only in loose association with one another). Collectively, the chains of cellulose are grouped into structures known as **microfibrils (12.4.3)** which exhibit variations in crystallinity along their length. The hemicelluloses are usually intimately associated with the cellulose chains, effectively binding the microfibrils together.

12.4.2 Lignin

Lignin is non-crystalline and comprises complex aromatic compounds composed primarily of phenyl groups (**Table 13.7**) whose exact composition is indeterminate. About 25% of the total lignin content of the wood is found in the thin middle lamella between adjacent cells (***Figure 12.6***), whilst the majority is found within the cell wall (deposited during the lignification of the cell, **12.3**). One prime function of the lignin is to protect the hydrophilic cellulose and hemicelluloses, which are mechanically weak when wet.

12.4.3 Molecular Structure of the Microfibril

From the analysis of the main chemical constituents of timber, the fibre composite nature of the material can be seen, where the cellulosic microfibrils act as the fibres within a matrix of hemicelluloses and lignin. The cellulosic microfibrils impart high tensile strength on the timber as a consequence of the strong covalent bonds both within and between adjacent anhydroglucose units, whilst the hemicellulose-lignin matrix cements microfibrils together and imparts shear resistance and stiffness to the timber. The location and orientation of the microfibrils within the timber governs, to a large degree, the resultant strength properties of the timber. After a cambium cell divides to form two new cells, each cell produces a secondary cell wall whose thickness will depend on the function that the cell will perform (**12.3**). The intercellular region is known as the **middle lamellae**. Studies indicate that the secondary cell wall can be divided into three layers (***Figure 12.6***), denoted S_1, S_2 and S_3, based upon the orientation of the microfibrils within each region.

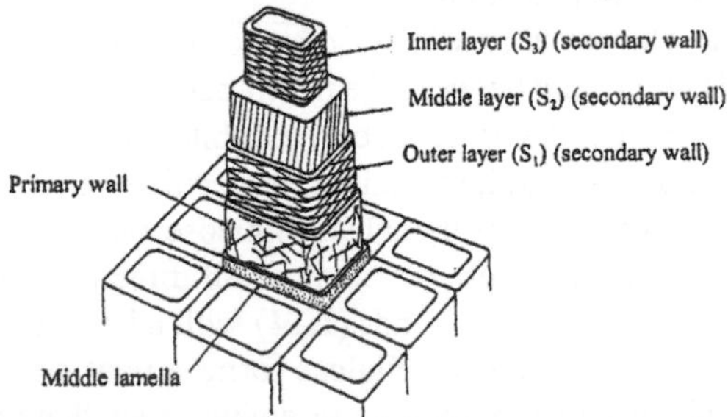

Figure 12.6 *Simplified structure of the cell wall showing orientation of microfibrils*

In the majority of timber, the S_2 layer comprises about three-quarters of the cell wall width, and therefore the orientation of the microfibrils within this layer (about 10 to 30° to the longitudinal axis) profoundly influence the tensile strength, stiffness and shrinkage behaviour of the timber. To illustrate this, we can consider strength (**12.8**) and moisture movement properties (**12.6**) of timber in relation to the **grain** direction (the general arrangement of the cells vertically aligned in the longitudinal direction; note that the majority of cells within the tree are orientated in this direction, **12.3**). Since the majority of the microfibrils are located in the S_2 layer and are orientated at only a small angle to the longitudinal axis, the timber exhibits high tensile and compressive strength **along the grain** but relatively low tensile and compressive strength **across the grain** (**Table 12.4**). In addition, moisture movement across the grain is relatively high compared to moisture movement along the grain, since the hydroxyl (- OH) groups in the cellulose chain capable of binding water are orientated predominantly perpendicular to the longitudinal axis (note that water cannot penetrate the crystalline regions of the cellulose chains since the hygroscopic hydroxyl groups are mutually satisfied by hydrogen bonding, as noted earlier; water can, however, combine with free hydroxyl groups in both the hemicelluloses and the amorphous cellulose regions). The variation in properties along and across the grain indicates that timber is an **anisotropic** material; unlike isotropic materials (such as metals and concrete), the properties of timber vary according to the orientation under consideration.

Table 12.4 Anisotropic properties of timber (Douglas fir)

Tensile strength[a]			Compressive strength[a]			Moisture movement[b] (%) (max)		
Along grain (N/mm²)	Across grain (N/mm²)	Along / Across	Along grain (N/mm²)	Across grain (N/mm²)	Along / Across	Across grain	Along grain	Across / Along
138	2.90	47.6	49.6	6.90	7.19	0.8[c]	<0.1	8.0

Notes: [a] 12% moisture content; [b] at equilibrium, on increasing the RH from 35% to 85%; [c] radial direction.

12.5 CONVERSION

The method by which the trunk is sawn (conversion) can dramatically affect its performance as a building material. Cuts in timber are **tangential** (or plain, flat or slash sawn timber) where the growth rings meet the cut face at **less than 45°** or **radial** (or quarter, rift or comb grained timber) where the growth rings meet the cut face at **not less than 45°** (***Figure 12.7***).

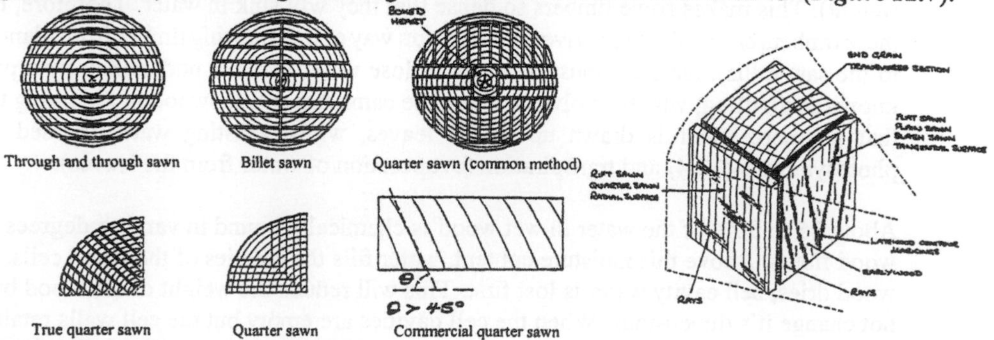

Figure 12.7 Conversion of timber

The variables affected by different orientation of cuts are

- strength; timber is stronger when loaded parallel to the grain compared to when loaded perpendicular to the grain. The strength of timber is also dependent upon the moisture content below the fibre saturation point (***Figure 12.9***);
- the wearing quality of timber;
- the resistance to shrinkage;
- the exposure of the grain or 'figuring' which occurs in the timber.

12.5.1 Conversion Defects

Distortion is caused by the difference in shrinkage in the tangential direction compared with that radially, coupled with the fact that the grain of a piece of timber rarely runs true. Thus, changes in moisture content below the fibre saturation point can result in bowing or twisting of studs or cupping of floor or cladding boards. Differential movement associated with the loss of water from longitudinally and radially orientated cells causes defects known as heart shakes and radial shakes. Wane is the loss of the square edge of the cut timber at the curved surface of the trunk. Some typical examples of conversion defects are shown in ***Figure 12.8***.

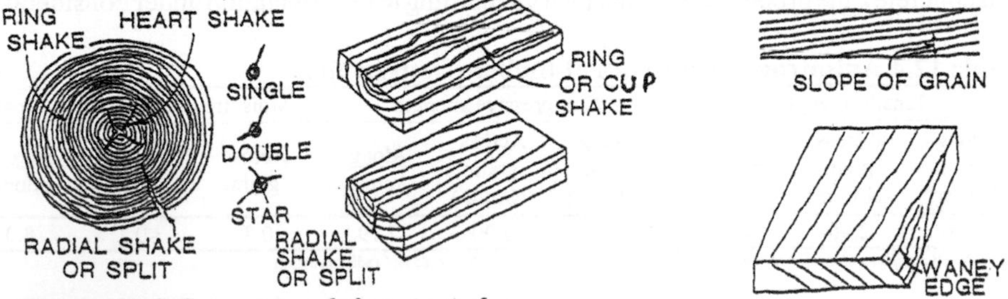

Figure 12.8 *Conversion defects in timber*

12.6 MOISTURE IN TIMBER

Wood is porous and, when growing, the hollow cells are filled with water (sap). The water content of living trees and freshly felled timber can be up to 200% (based on the oven dried weight). This makes some timbers so dense that they will sink in water. Therefore, before these timbers can be floated in rivers (a common way of transporting timber from plantations to the saw mills), the trees must be made to lose water. This is undertaken by a process known as **girdling**, which involves scoring the cambium and sapwood in the living tree so that no more water is drawn up to the leaves, whilst existing water is used up by photosynthesis (**12.4**) and transpiration (evaporation of water from the leaves).

About 25 to 30% of the water in wet wood is chemically bound in varying degrees to the wood fibres. Above this moisture content, water fills the cavities of the wood cells. When wood dries, cell cavity water is lost first. This will reduce the weight of the wood but will not change it's dimensions. When the cell cavities are empty but the cell walls retain their bound water, the wood is said to be at **fibre saturation point** (***Figure 12.11***). Further drying below the fibre saturation point results in shrinkage of the wood as the cell walls lose water. In theory, the fibre saturation point is reached when there is no free water in the cell cavities while the cell walls have lost none of their chemically bound water. In practice, this situation rarely exists (free water may be present when bound water has been lost) and therefore the fibre saturation point is more correctly a moisture content range over which the transition occurs. Below the fibre saturation point, timber is hygroscopic, absorbing moisture from the atmosphere if dry and yielding moisture to the atmosphere if wet. Timber therefore reaches an **equilibrium** with the water vapour conditions of the surrounding atmosphere. Hence for any combination of ambient temperature and relative humidity, there is a corresponding moisture content of timber such that there will be no inward or outward diffusion of water vapour; this timber moisture content is known as the **equilibrium moisture content.**

12.6.1 Moisture Movement in Timber

Moisture movement through wood occurs by the following mechanisms

- capillary movement through tracheids and vessels (or pores). Timber can be dried from the green state to the fibre saturation point without shrinkage. The loss of this water has the greatest effect upon the density and durability of the wood.
- hygroscopic or physically bound water, i.e., water which is weakly bonded to the hydroxyl (- OH) groups in the cellulose molecule. Removal of this water (i.e., below the fibre saturation point to a condition where the wood is in equilibrium with the humidity of the environment) causes shrinkage (of up to 9%);
- water vapour from the atmosphere condenses into cells (**5.7**). The vapour pressure depends on the radius of the cell wall, and the vapour pressure gradient will determine the rate of water movement through the timber.

12.6.2 Influence of Moisture on Timber Strength

Loss of water from the cell cavities (above the fibre saturation point) has no effect on the strength of the timber. However, below the fibre saturation point, chemically bound water from the cell wall is lost, causing it to shrink. Loss of water below the fibre saturation point occurs principally from the non-crystalline regions and hemicelluloses, with the net effect that microfibrils are forced closer together, increasing interfibrillar bonding and strength (***Figure 12.9***). These effects are almost completely reversible.

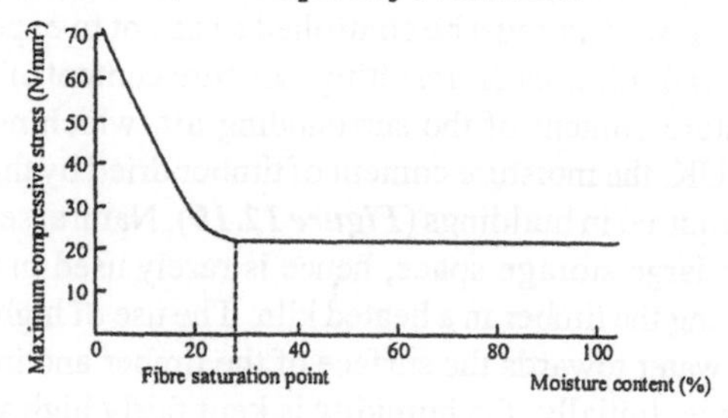

***Figure 12.9** Relationship between longitudinal compressive strength and moisture content*

12.7 SEASONING

To avoid shrinkage of a timber article after fabrication, it is essential that the timber is dried to a moisture content which is in equilibrium with the relative humidity of the atmosphere in which the article is to be located (***Figure 12.10***). Seasoning refers to the controlled loss of moisture from green timber to an appropriate moisture content for a given end use and is achieved by natural seasoning or kiln drying. A certain tolerance can be allowed in, for example, timber frames and roof trusses, but in the production of window frames, flooring, etc. it is essential that timber is seasoned to expected service conditions. In practice this usually means seasoning timber to 12% moisture content for regular intermittently heated buildings and 10% in buildings with central heating. If this seasoning is omitted, shrinkage in service will occur with loosening of joints, crazing of paint films, buckling and delaminating of laminates, etc. An indication of the moisture content of timber for different environments is given in ***Figure 12.10***; further information is given in Eurocode 5 (**12.10**).

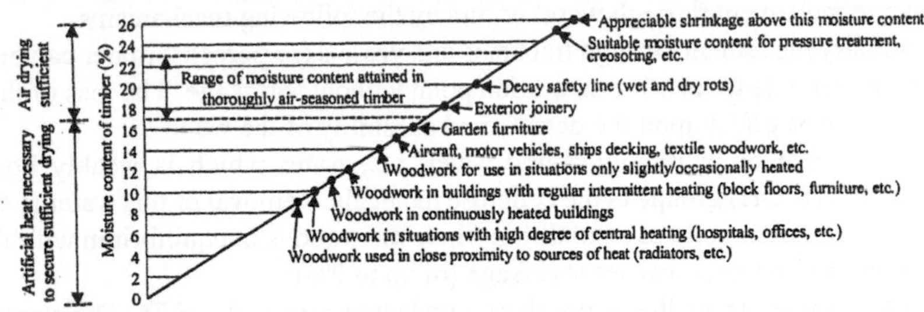

***Figure 12.10** Equilibrium moisture content of timber in various environments*

Natural seasoning relies on the drying action of naturally circulating air (with no assistance from artificial heat or ventilation). The sawn timber is stacked horizontally with piling sticks between each plank. The timber is protected from the external environment by a cover and the transverse cut ends of the timber are usually coated (e.g, with bitumen) to prevent excess drying. A 'rough and ready' rule commonly applied in the past was to allow a year for each 25 mm of thickness[3]. The general principle is to remove moisture from the surface of the sawn timber at approximately the same rate as it is lost from the centre. If the surface dries out more quickly than the centre, the surface layers will shrink and split. In practice this means that the flow of air must be controlled so as not to expose the surface layers to unduly rapid drying. In addition, as the resulting moisture content of the wood will be in equilibrium with the moisture content of the surrounding air, which never usually drops below 18% outside in the UK, the moisture content of timber dried by this method can never be reduced to the levels required in buildings (***Figure 12.10***). Natural seasoning is also time consuming and requires a large storage space, hence is rarely used in the UK today. **Kiln seasoning** involves stacking the timber in a heated kiln. The use of higher temperatures accelerates the movement of water towards the surface of the timber and increases the rate of evaporation from the surface. Initially, the humidity is kept fairly high and the temperature moderately low. Thereafter, as the moisture content of the wood begins to fall, humidity is reduced and temperature raised until the wood reaches the desired moisture content. A 'rough and ready' rule used traditionally was one day kiln seasoning per 25 mm section. Kiln seasoning has the added advantage of killing any fungi or beetle infestation which may be present.

12.7.1 Seasoning Defects

Seasoned timber is **dimensionally unstable** and moisture movement is **anisotropic**. This anisotropy occurs because moisture causes the timber cell walls to expand predominantly perpendicular to the cell axis **(12.6)** and, since the majority of cells are orientated longitudinally but a small proportion of the cells are orientated radially (the medullary rays), dimensional moisture movement in timber varies in each principal axis. For most practical purposes, the effects of moisture on the dimensional stability of timber can be summarised (***Figure 12.11***) as

- timber does not shrink or swell lengthwise along the grain;
- shrinkage starts as the timber dries below the fibre saturation point (about 25-30% moisture content);
- timber shrinks almost twice as much across the width of a flat sawn board (i.e., in the tangential direction) as it does across a quarter sawn board (i.e., radially) (***Figure 12.11***);
- tangential shrinkage (or swelling) can be estimated as roughly 1% for every 3% change in moisture content below about 30%. Radial shrinkage (or swelling) is about half this.

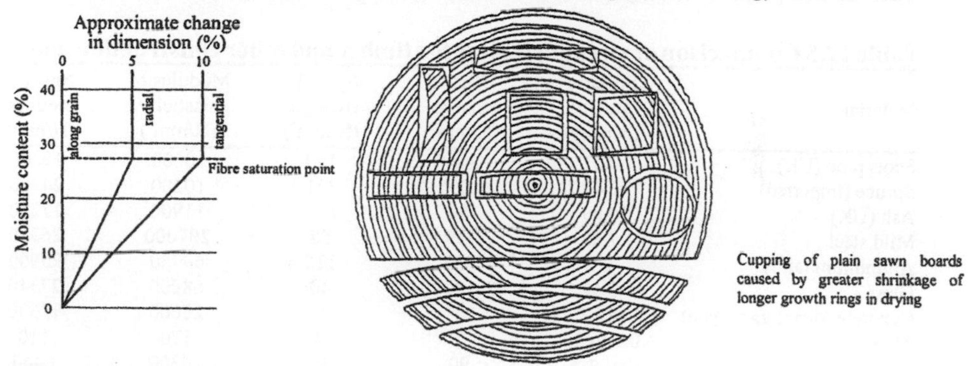

Figure 12.11 *Equilibrium moisture content values and shrinkage defects*

12.8 PHYSICAL PROPERTIES

Compared to the density of other materials e.g, metal 2640-11373 kg/m^3, stone 2082-3200 kg/m^3, plastics 900-1400 kg/m^3, timber has a relatively low density, averaging between 380- 840 kg/m^3 for seasoned timber (there are exceptions, e.g, Balsa 160 kg/m^3 and Lignum vitae 1208 kg/m^3). Density varies with species and moisture content. Timber density is a function of the ratio of cell wall thickness to cell diameter. Consequently, increased density will result in increased strength and stiffness of the timber cell and hence overall strength and stiffness of the timber (***Figure 12.12***). Generally, timber density (specific gravity) is an excellent predictor of timber strength.

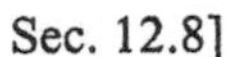

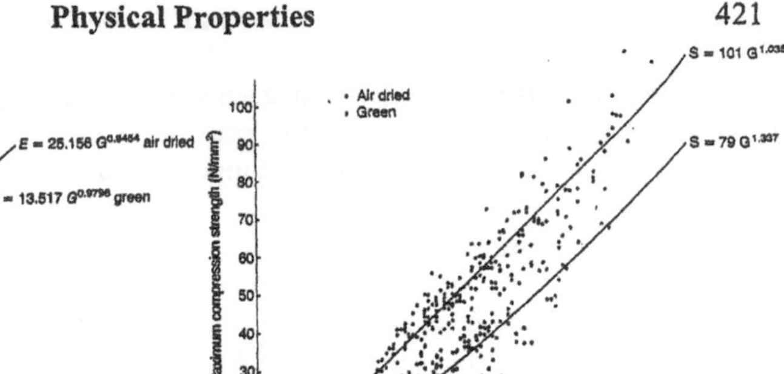

a) modulus of elasticity *b) maximum compressive strength*

Notes: Most experimental relationships obtained for strength properties and timber density take the form S = kG^n, where S is any strength property, G is the specific gravity, k is a proportionality constant differing for each strength property and n is an exponent that defines the shape of the curve.

***Figure 12.12** Effect of density on mechanical properties of timber*

As previously noted (**12.4**), timber strength properties are anisotropic. Short term tensile strength in relation to density compares favourably with other materials (**Table 12.5**), whereas compressive strength is much lower, due to the tendency of timber cells to buckle.

Table 12.5 Comparison of strength indices for timber and other construction materials

Material	Specific gravity (density)	Tensile strength (N/mm²)	**Specific strength (N/mm²)**	Modulus of elasticity (N/mm²)	**Specific modulus (N/mm²)**
Scots pine (UK)	0.51	92[a]	180	10000	19600
Spruce (imported)	0.42	139[a]	331	10200	24290
Ash (UK)	0.69	180[a]	261	11900	17250
Mild steel	7.85	459	58	207000	26370
Aluminium (alloy)	2.7	310	115	69950	25900
Glass	2.52	100	40	68900	27340
Concrete (dense aggregate)	2.32	4	2	28600	12330
Thermoplastics	0.9 to 2.3	7	4	170	110
Thermosetting plastics	av. 1.6	90	56	10300	6440
Glass fibre	3.5	2400	686	72400	20690
Epoxy resin	1.8	100	56	5000	2780

Notes: The strength of wood becomes more apparent if its properties are compared with other materials in terms of **specific strength** (strength/specific gravity) and **specific modulus** (elastic modulus/specific gravity); [a] Tensile strength parallel to the grain.

Timber is particularly useful for bending applications where a significant proportion of the permanent load is attributed to the self-weight of the structural component, as in roofing and domestic flooring. Problems in utilising timber efficiently for structural applications include

- the strength of identical ('clear') specimens of the same species (and even from the same tree) are very variable. Coefficients of variation (standard deviation/mean strength) over 20% are not uncommon;
- long term strengths are typically 40% lower than short term strengths, due to the tendency of timber to creep (**3.5.4**);

- strengths of timber specimens are greatly reduced by the presence of inherent defects (such as knots and fissures) and by factors such as the rate of growth, ratio of earlywood to latewood, moisture content and slope of grain.

These problems are overcome by **grading** timber according to its likely performance.

12.9 GRADING TIMBER

All structural timber used in the UK is strength graded. Grading is carried out visually or by machine. **Visual grading** of timber must be carried out in accordance with rules which comply with BS EN 518[4]. This standard is not a grading standard; it simply sets down the criteria which visual grading standards have to meet. In the UK, BS 4978[5] (softwoods) and BS 5756[6] (hardwoods) conform to the requirements of BS EN 518 and are therefore used to visually grade timber. Under these standards, the following grades are defined:

For **softwoods** to BS 4978, two strength grades are defined

- **General structural** (GS) grade, estimated to have a strength of between 30 to 50% of that for clear (perfect) timber of the same species;
- **Special structural** (SS) grade, estimated to have a strength of between 50 to 60% of that for clear (perfect) timber of the same species

and each strength grade is assessed at two levels of moisture content

- **dry graded timber** is assessed when the batch of timber has an average moisture content of 20% or less, with no reading exceeding 24%. Dry graded timber should be specified for use in Hazard Class 1 (**12.25.1**) (timber >100 mm target thickness is exempt from this requirement);
- **wet graded timber** is assessed at a moisture content above 20%. Wet graded timber should be specified for use in contact with water i.e., Hazard Classes 2 to 5 (**12.25.1**).

For **hardwoods** to BS 5756, five visual grades are defined

- **Tropical hardwood** (HS)
- **General structural temperate hardwood** (TH1 and TH2) for timber of a cross-sectional area less than 20000 mm^2 and a thickness of less than 10 mm;
- **Heavy structural temperate hardwood** (THA and THB) for timber of a cross-sectional area greater than 20000 mm^2 and a thickness of 100 mm or more.

Visual grading involves assessing the strength of the timber visually by recording those observable features which affect strength, as defined by BS 4978 (e.g, size and distribution of knots, slope of grain, rate of growth, presence of fissures, etc.). These requirements are summarised in **Table 12.6**. Knots have a tendency to distort growth and hence weaken timber. Hence a key parameter is the **'knot area ratio'** (KAR), defined as the sum of the projected cross-sectional areas of all the knots at a cross-section in terms of the cross-sectional area of the test piece. Ordinarily, this ratio can be readily determined. However, the capacity of knots to reduce strength depends primarily upon their location. For example, the presence of live ('tight') knots (**12.2**) in the central portion of a given cross-section have little detrimental strength effects whilst those at the edges of the timber will be more highly stressed under load, thus their affect is more critical. Under compression, the presence of 'tight' knots at the edge will cause little reduction in strength but under tension the reduction in strength is broadly equivalent to a hole of the size of the knot, irrespective of whether the knot is 'tight' or 'loose'. Under these circumstances, determination of the knot area ratio is more time consuming. To account for these strength reductions, BS 4978 defines a 'margin' condition (MKAR) as being present when more than half the top or bottom quarter of any section is occupied by knots. The term TKAR is then used to distinguish the overall KAR from the MKAR. The key TKAR values are 1/2, 1/3 and 1/5. Hence the TKAR for the largest knot cluster for each piece must be placed within one of the following ranges TKAR < 1/5, 1/5 < TKAR < 1/3, 1/3 < TKAR < 1/2, TKAR > 1/2. Knot area ratios in excess of 1/3 (margin condition) or 1/2 (no margin condition) lead to rejection of that piece. The knot area ratio values for overlapping knots are added.

Table 12.6 Properties and requirements for visually graded timber (BS 4978)

Property	Type	SS	GS
Knots (see text)	Margin condition (MKAR > 1/2)	TKAR ≤ 1/5	KAR 1/5 to 1/3
	No margin condition (MKAR < 1/2)	TKAR ≤ 1/3	TKAR ½ to 1/3
Fissures	Not through thickness	Not longer than half the length of the piece	Unlimited
	Through the piece	Not longer than twice the width of the piece	Not longer than 600 mm on any running metre
Slope of grain	-	Must not exceed 1 in 10	Must not exceed 1 in 6
Wane	-	Wane shall not reduce the full edge and face dimensions to < 2/3 of the dimensions of the piece. Length of wane is unlimited.	
Rate of growth	-	Average width of annual rings ≯6 mm	Average width of annual rings ≯10 mm
Distortion (over 2 m length)	Bow	≯ 10 mm	≯20 mm
	Spring	≯ 8 mm	≯12 mm
	Twist	≯ 1 mm per 25 mm width	≯2 mm per 25 mm width
	Cup	Unlimited	Unlimited

Machine grading is undertaken in accordance with BS EN 519[7], which recognises two types of strength grading machine, **output controlled** and **machine controlled. Output controlled** machines apply sufficient load to the piece of timber to induce a defined deflection. The load required indicates the grade. Most machines operating in the UK are **machine controlled** and operate by applying a defined load to each piece of timber as it passes through the machine. The resulting deflection indicates the grade, which is then marked on the piece. Under the European Standards, machine grading is undertaken directly to the strength class limits of the strength classes defined in BS EN 338[8]. The timber is marked according to its grade along the length at one metre intervals and at the end with its lowest grade. Dry machine graded timber and wet machine graded timber are defined in the same way as for visually graded timber.

12.10 STRUCTURAL TIMBER

The use of structural timber in construction in the UK is governed by the requirements of the Building Regulations. The regulations and associated documentation refer to relevant codes for design, materials and workmanship as evidence of compliance with the regulations for general structural timber. Currently, changes in the standards governing structural timber design (from British Standards to European Standards) are being implemented. These changes represent a significant deviation in the nature of design from a **permissible stress** design approach in the current British Standards (e.g, BS 5268: Part 2[9] for timber) to a **limit states** design approach in the new European Standards (governed by Eurocode 5[10]). This brings the structural design of timber into line with that for other materials (steel, concrete, etc.) which are already governed by Standards which utilise the limit state approach. Currently, these two systems operate side by side, and the designer can opt to follow either code (although it is recommended that one code is adopted for each job). The UK permissible stress codes will be formally withdrawn in January 2004 and thereafter only the European limit states design approach will be acceptable[11,12].

Limit states are checked by comparing effects on a structure and its components of loads and imposed deformations with the material properties of the components concerned and their capacity to resist these effects. This is achieved using **characteristic values** of both loads and material properties, which are modified by specified **partial factors** which reflect the reliability of the characteristic values. Partial factors increase the values of loads and decrease the values of material properties. Characteristic values are generally fifth percentile values derived directly from **strength grading** tests (**12.9**); they do not take into account safety and other factors included in the design process which are the responsibility of the designer. Characteristic values are therefore considerably higher than the **grade stress** values in BS 5268: Part 2 used for permissible stress design, as these are obtained from long term loading, are mean values and include a safety factor. In principle, Eurocode 5 contains all the rules necessary for the design of timber structures (it does not, however, contain the material properties, fastener loads and other design information found in BS 5268: Part 2). To find, for example, the mechanical properties of the 15 European Strength Classes of solid timber, the designer must have reference to other European Standards (BS EN 338 in this case), enter the various properties of the timber members into formulae which give the different loads at which possible modes of failure can occur.

12.10.1 Strength Class

Under the original British Standard system of stress grading timber, no account was taken of the fact that timber strength is species related; thus the same grade in different species would have different strengths. The designer could not refer merely to 'SS' grade timber, as different timber species have different stress capacities at a given grade. Strength classes were therefore introduced in the 1991 edition of BS 5268 to simplify design, where species/grade combinations of similar strength were grouped together in **strength classes**. Ten strength classes (denoted SC) were defined for solid timber. Of these, strength classes SC3 and SC4 were common as they corresponded to European redwood and whitewood in the GS and SS stress grades respectively. Timber design tables (e.g, in the Building Regulations) for simple structures, such as domestic floors, were based on these classes. Strength classes for timber are now defined in BS EN 338 (solid timber), BS EN 1194[13] (glued laminated timber, Glulam) and prEN 12369[14] (wood-based panel products). The strength class numbering system relates to the characteristic bending strength for that class, prefixed by a letter, i.e., softwood (C), hardwoods and Poplar (a softwood) (D), Glulam (GL). Summaries of the characteristic values for strength and stiffness properties which define strength classes are included in **Table 12.8** (solid timber and Glulam) and **Table 12.9** (wood-based panel products). Visually and machine graded timber (**12.9**) is acceptable within European Standard design procedures. Advice on the equivalence of the British Standard strength class and grading system with the new European system is given by TRADA[15] and summarised in **Table 12.7**.

Table 12.7 Commonly used timber grades assigned to BS EN strength classes

Species	Strength class to BS EN 338 (nearest 'SC' equivalent)								
	Softwoods					Hardwoods			
	C14 (SC2)	C16 (SC3)	C18	C22	C24 (SC4)	D30	D40	D50	D70
UK grown									
Douglas fir	GS		SS						
British pine	GS			SS					
British spruce	GS			SS					
Larch		GS			SS				
Oak						TH1, THB	THA		
Imported									
Redwood		GS			SS				
Whitewood		GS			SS				
Sitka spruce	GS		SS						
Hem fir		GS			SS				
Parana pine		GS			SS				
Western red cedar	GS		SS						
Greenheart									HS
Iroko							HS		
Opepe								HS	
Teak							HS		

Notes: Strength class C27 is equivalent to SC5; Visually grading only shown (softwoods to BS 4978 and hardwoods to BS 5756). Machine grading results incorporated directly to strength class limits.

Table 12.8 Characteristic values for solid timber (BS EN 338) and Glulam (BS EN 1194) strength classes

		Softwoods (BS EN 338)									Hardwoods (BS EN 338)						Glulam timber (BS EN 1194)				
Strength class		C14	C16	C18	C22	C24	C27	C30	C35	C40	D30	D35	D40	D50	D60	D70	GL20	GL24	GL28	GL32	GL36
Characteristic strength and stiffness properties (N/mm²)																					
Bending	$f_{m,k}$	14	16	18	22	24	27	30	35	40	30	35	40	50	60	70	20	24	28	32	36
Tension parallel to grain	$f_{t,0,k}$	8	10	11	13	14	16	18	21	24	18	21	24	30	36	42	15	18	21	24	27
Tension perpendicular to grain	$f_{t,90,k}$	-	-	-	-	-	-	-	-	-	-	-	-	-	-	-	0.35	0.35	0.45	0.45	0.45
Compression parallel to grain	$f_{c,0,k}$	16	17	18	20	21	22	23	25	26	23	25	26	29	32	34	21	24	27	29	31
Compression perpendicular to grain	$f_{c,90,k}$	4.3	4.6	4.8	5.1	5.3	5.6	5.7	6.0	6.3	8.0	8.4	8.8	9.7	10.5	13.5	5.0	5.5	6.0	6.0	6.3
Shear	$f_{v,k}$	1.7	1.8	2.0	2.4	2.5	2.8	3.0	3.4	3.8	3.0	3.4	3.8	4.6	5.3	6.0	2.8	2.8	3.0	3.5	3.5
Mean modulus of elasticity parallel to grain	$E_{0,mean}$	7000	8000	9000	10000	11000	12000	12000	13000	14000	10000	10000	11000	14000	17000	20000	10000	11000	1200[illegible]	13500	14500
5%ile modulus of elasticity parallel to grain	$E_{0.05}$	4700	5400	6000	6700	7400	8000	8000	8700	9400	8000	8700	9400	11800	14300	16800	8000	8800	9600	10800	11600
Mean shear modulus[a]	G_{mean}	440	500	560	630	690	750	750	810	880	600	650	700	880	1060	1250	625	688	750	844	906
Density (kg/m³)																					
Characteristic density	ρ_k	290	310	320	340	350	370	380	400	420	530	560	590	650	700	900	360	380	410	440	480
Average density	ρ_{mean}	350	370	380	410	420	450	460	480	500	640	670	700	780	840	1080	390	410	440	480	520

Examples of lay-ups in Glulam timber	GL20	GL24	GL28	GL32	GL36
Homogeneous Glulam (all laminations)	C18	C22	C27	C35	C40
Combined Glulam (outer laminations)[c]	C22	C24	C30	C35	C40
Combined Glulam (inner laminations)	C16	C18	C22	C27	C35

Notes: Eurocode 5 adopts the following conventions: f = strength of a material; E = modulus of elasticity; G = permanent action; Q = variable action; ρ = mass density and the subscripts k = characterisitic value; v = shear; m = bending; c = compression; 0 and 90 = relevant directions in relation to grain direction;
[a] $G_{mean} = E_{0,mean}/16$; [b] based on the values of ρ_k assuming a 5% coefficient of variation; [c] the requirements apply to the extreme one sixth of the depth on both sides.

Table 12.9 Characteristic values for wood-based panels (prEN 12369) strength classes

		Particleboards (BS EN 312)				Fibreboards (BS EN 622)				Plywood (3 plies)		Plywood (5 plies)	
Board classification		P4	P5	P6	P7	HB.HLA2	MBH.LA2	MDF.LA	MDF.HLS	P30[a]	P30[a]	P30[a]	P30[a]
Thickness (mm)	t_{nom}	>13-20	>13-20	>13-20	>13-20	>3.5-5.5	>10	>12-19	>12-19	7.0	9.0	12.0	15.0
Density (kg/m³)	ρ_k	600	600	600	600	850	600	600	600	>420	>420	>420	>420
Characterristic strength properties (N/mm²)													
Bending	$f_{m,0,k}$	12.5	13.3	15.0	16.7	35.0	15.0	21.0	22.0	27.8	27.8	23.0	23.0
Bending perpendicular to outer layer grain	$f_{m,90,k}$	-	-	-	-	-	-	-	-	4.9	7.2	11.4	12.0
Tension	$f_{t,0,k}$	7.9	8.5	9.5	10.6	26.0	8.0	12.5	16.5	16.7	16.7	15.0	15.0
Compression	$f_{c,0,k}$	11.1	11.8	13.3	14.7	27.0	8.0	12.5	16.5	16.7	16.7	15.0	15.0
Tension and compression perpendicular to outer layer grain	$f_{t,90,k}$ $f_{c,90,k}$	-	-	-	-	-	-	-	-	9.9	9.9	12.0	12.0
Panel shear	$f_{v,k}$	6.1	6.5	7.3	8.1	18.0	4.5	6.5	8.5	2.9	2.9	2.9	2.9
Planar shear	$f_{r,k}$	1.6	1.7	1.7	2.2	3.0	0.25	-	-	0.9	0.9	0.9	0.9
Mean stiffness values (N/mm²)													
Bending	$E_{m,0,mean}$	2900	3300	4100	4200	4800	2900	3000	3000	11000	11000	9200	9200
Bending perpendicular to outer layer grain	$E_{m,90,mean}$	-	-	-	-	-	-	-	-	2000	2900	4600	4800
Tension and compression	$E_{t,0,mean}$ $E_{c,0,mean}$	1700	1900	2400	2500	4800	2900	2700	2800	8000	8000	7200	7200
Tension and compression perpendicular to outer layer grain	$E_{t,90,mean}$ $E_{c,90,mean}$	-	-	-	-	-	-	-	-	4000	4000	4800	4800
Panel shear	$G_{v,mean}$	830	930	1150	1200	2000	1200	800	1000	600	600	600	600

Notes: Eurocode 5 adopts the following conventions: f = strength of a material; E = modulus of elasticity; G = permanent action; Q = variable action; ρ = mass density and the subscripts.
k = characterisitic value; v = shear; m = bending; c = compression; 0 and 90 = relevant directions in relation to grain direction.
Particleboard and fibreboard types (including cement-bonded particleboards and OSB) not shown in this Table have yet to be included in prEN 12369.
pr EN 12369 includes further information on Swedish, Finnish, US, Canadian, German, Italian and French plywoods, including ply, layer and thickness variations and calculation methods.
[a] Swedish designation, spruce, unsanded

12.11 TIMBER BOARD (PANEL) PRODUCTS

Timber has a number of inherent weaknesses as a construction material, for example it possesses a high degree of variability, is strongly anisotropic in both strength and moisture movement, is dimensionally unstable in changing environmental humidity conditions and is available in only limited widths. Such deficiencies are reduced by processing timber into small units and subsequently reconstituting it, usually in the form of large flat sheets. The size and orientation of these smaller units and the method by which they are bonded together determines the degree of increased dimensional stability and reduced anisotropy of the board compared to the solid timber. Hence, in comparison with timber, board materials possess a lower degree of variability, a lower degree of anisotropy, higher dimensional stability and availability in large and variable dimensions. There are a large variety of wood based panel products available, including Plywood (including veneer plywoods and core plywoods, such as blockboard and laminboard); Particleboards, including wood chipboard, flake boards (e.g, oriented strand board (OSB) and wafer board) and cement bonded particleboard; Fibre building boards; Compressed straw building slabs; Wood wool cement slabs.

12.11.1 Timber Board Product Standards

Recently (1997), new European Standards have been introduced to replace existing British Standards. Until January 2004, the existing UK system will run alongside the new European system. European Standards are in general performance based and therefore wood based panel products are now classified in terms of their intended use conditions. Appropriate test methods are laid down in seperate standards which, in some cases, differ from previous British Standard tests. Thus the grades of panel products and their designations are different from those defined in previous British Standards.

With the exception of plywood, all wood-based panel products are designated by levels of performance for various mechanical and physical properties in appropriate standards. Characteristic values strength and stiffness properties for established wood-based panel products (listed in prEN 12369) are summarised in **Table 12.9** and can be compared with values for solid timber and Glulam in **Table 12.8 (12.10.1)**.

Table 12.10 gives the levels of performance required for the various mechanical, physical and moisture related properties of particleboards and fibreboards and includes typical values, for selected boards of given nominal thickness, in parentheses. **Table 12.10** includes the voluntary colour coding system (defined in BS EN 622: Part 1) used to define the different grades and types of wood-based panel products, which comprises two coloured stripes. The **first colour** denotes the suitability of the board for particular applications. Either one or two stripes of this colour is used; white (**W**) for general purpose boards and/or yellow (**Y**) for load-bearing boards; the **second colour** denotes the environmental conditions for which the board is suitable; blue (**B**) for dry conditions, green (**G**) for humid conditions or brown (**Br**) for exterior conditions.

Table 12.10 Mechanical, physical and moisture related properties required for each type of particleboard and fibreboard

Board type	Bending strength	Modulus of elasticity	Internal bond	Moisture content	Swelling in thickness	Dimensional changes	Density	Surface soundness	Sand content	Moisture resistance		formaldehyde content
	BS EN 310[16]	BS EN 310	BS EN 319[17]	BS EN 322[18]	BS EN 317[19]	BS EN 318[20]	BS EN 323[21]	BS EN 311[22]	ISO 3340[23]	BS EN 1087[24] BS EN 321[25]		BS EN 120[26]
										Bending strength[a]	Internal bond strength[b]	
	(N/mm²)	(N/mm²)	(N/mm²)	(%)	(%, 24hr)			(N/mm²)		(N/mm²)	(N/mm²)	
Particleboards (BS EN 312) (required properties: >13 - 20 mm nominal thickness boards)												
P2 (WWB)	✓ (11.5)	✗	✓ (0.24)	✓ (5-13)	+	+	+	+	✗	✗	✗	✓
P3 (WB)	✓ (13.0)	✓ (1600)	✓ (0.35)	✓ (5-13)	+	+	+	✓ (0.8)	+	✗	✗	✓
P4 (YYB)	✓ (15.0)	✓ (2150)	✓ (0.35)	✓ (5-13)	✓ (15.0)	+	+	+	✗	✗	✗	✓
P5 (YYG)	✓ (16.0)	✓ (2400)	✓ (0.45)	✓ (5-13)	✓ (10.0)	+	+	+	✗	✗	✓ (0.22)[e]	✓
P6 (YB)	✓ (18.0)	✓ (3000)	✓ (0.50)	✓ (5-13)	✓ (14.0)	+	+	+	✗	✗	✗	✓
P7 (YG)	✓ (20.0)	✓ (3100)	✓ (0.70)	✓ (5-13)	✓ (8.0)	+	+	+	✗	✗	✓ (0.36)[e]	✓
Oriented strand board (OSB) (BS EN 300) (required properties: >13 - <20 mm nominal thickness boards)												
OSB 1 (WB)	✓ (18.0)[f]	✓ (2500)[f]	✓ (0.28)	✓ (2-12)	✓ (25.0)	+	+	✗	✗	✗	✗	✓
OSB 2 (YYB)	✓ (20.0)[f]	✓ (3500)[f]	✓ (0.32)	✓ (2-12)	✓ (20.0)	+	+	✗	✗	✗	✗	✓
OSB 3 (YYG)	✓ (20.0)[f]	✓ (3500)[f]	✓ (0.32)	✓ (2-12)	✓ (15.0)	+	+	✗	✗	✓ (8.0)[f]	✓ (0.15)[e]	✓
OSB 4 (YG)	✓ (28.0)[f]	✓ (4800)[f]	✓ (0.45)	✓ (2-12)	✓ (12.0)	+	+	✗	✗	✓ (14.0)[f]	✓ (0.17)[e]	✓
Cement-bonded particleboard (BS EN 634) (required properties: all thicknesses)												
Class 1 (WWBr)	✓ (9.0)	✓ (4500)	✓ (0.50)	✓ (6-12)	✓ (1.5)	+	✓ (1000)	✗	✗	✗	✓ (0.30)	✗
Class 2 (WWBr)	✓ (9.0)	✓ (4000)	✓ (0.50)	✓ (6-12)	✓ (1.5)	+	✓ (1000)	✗	✗	✗	✓ (0.30)	✗
Fibreboards: Hardboards (BS EN 622: Part 2) (required properties: > 3.5 - 5.5 mm nominal thickness boards)												
HB (WWB)	✓ (30.0)	✗	✓ (0.50)	✓ (4-9)	✓ (30)	+	✗	✗	+	✗	✗	✗
HB.LA (YYB)	✓ (32.0)	✓ (2500)	✓ (0.60)	✓ (4-9)	✓ (30)	+	✗	✗	+	✗	✗	✗
HB.H (WWG)	✓ (32.0)	✗	✓ (0.60)	✓ (4-9)	✓ (20)	+	✗	✗	+	✗	✓ (0.30)	✗
HB.HLA1 (YYG)	✓ (36.0)	✓ (3600)	✓ (0.70)	✓ (4-9)	✓ (13)	+	✗	✗	+	✗	✓ (0.42)	✗
HB.HLA2 (YG)	✓ (42.0)	✓ (4300)	✓ (0.70)	✓ (4-9)	✓ (13)	+	✗	✗	+	✓ (16.0)	✓ (0.42)	✗
HB.E (WWBr)	✓ (35.0)	✓ (3100)	✓ (0.60)	✓ (4-9)	✓ (10)	+	✗	✗	+	✗	✓ (0.42)	✗

Notes: ✓ Requirement; ✗ no requirement; + additional requirement; [a] wet, after boil test; [b] after boil test; [c] BS EN 622: Part 4 notes that special provisions exist for softboards used exclusively as an insulation material; [d] boards over 4 mm thick; [e] tested in accordance with BS EN 321; [f] major axis.

Table 12.10 Mechanical, physical and moisture related properties required for each type of particleboard and fibreboard (contd.)

Board type	Bending strength BS EN 310[16] (N/mm²)	Modulus of elasticity BS EN 310 (N/mm²)	Internal bond BS EN 319[17] (N/mm²)	Moisture content BS EN 322[18] (%)	Swelling in thickness BS EN 317[19] (%, 24hr)	Dimensional changes BS EN 318[20]	Density BS EN 323[21]	Surface soundness BS EN 311[22] (N/mm²)	Sand content ISO 3340[23]	Moisture resistance BS EN 1087[24] BS EN 321[25]: Bending strength[a] (N/mm²)	Moisture resistance: Internal bond strength[b] (N/mm²)	formaldehyde content BS EN 120[26]
Fibreboards: Medium boards (BS EN 622: Part 3) (required properties: > 10 mm nominal thickness boards)												
MBL (WWB)	✓ (8.0)	✗	✗	✓ (4-9)	✓ (20)	+	✗	✗	+	✗	✗	✗
MBH (WWB)	✓ (12.0)	✗	✓ (0.10)	✓ (4-9)	✓ (15)	+	✗	✗	+	✗	✗	✗
MBH.LA1 (YYB)	✓ (15.0)	✓ (1600)	✓ (0.10)	✓ (4-9)	✓ (15)	+	✗	✗	+	✗	✗	✗
MBH.LA2 (YB)	✓ (18.0)	✓ (2500)	✓ (0.20)	✓ (4-9)	✓ (15)	+	✗	✗	+	✗	✗	✗
MBL.H (WWG)	✓ (10.0)	✗	✗	✓ (4-9)	✓ (15)	+	✗	✗	+	✗	✗	✗
MBH.H (WWG)	✓ (15.0)	✗	✓ (0.30)	✓ (4-9)	✓ (10)	+	✗	✗	+	✓ (5.0)	✗	✗
MBH.HLS1 (YYG)	✓ (22.0)	✓ (2100)	✓ (0.40)	✓ (4-9)	✓ (7)	+	✗	✗	+	✗	✓ (0.20)	✗
MBH.HLS2 (YG)	✓ (25.0)	✓ (2800)	✓ (0.40)	✓ (4-9)	✓ (7)	+	✗	✗	+	✓ (9.0)	✓ (0.20)	✗
MBL.E (WWBr)	✓ (12.0)	✗	✗	✓ (4-9)	✓ (9)	+	✗	✗	+	✗	✗	✗
MBH.E (WWBr)	✓ (18.0)	✓ (2200)	✓ (0.30)	✓ (4-9)	✓ (6)	+	✗	✗	+	✓ (6.0)	✗	✗
Fibreboards: Softboards (density > 230kg/m³) (BS EN 622: Part 4)[c] (required properties: > 10 - 19 mm nominal thickness boards)												
SB (WWB)	✓ (0.8)	✗	✗	✓ (4-9)	✓ (10)	✗	✗	✗	✗	✗	✗	✗
SB.LS (YYB)	✓ (1.1)	✓ (130)	✗	✓ (4-9)	✓ (8)	✗	✗	✗	✗	✗	✗	✗
SB.H (WWG)	✓ (1.0)	✗	✗	✓ (4-9)	✓ (7)	✗	✗	✗	✗	✗	✗	✗
SB.HLS (YYG)	✓ (1.2)	✓ (140)	✗	✓ (4-9)	✓ (6)	✗	✗	✗	✗	✗	✗	✗
SB.E (WWBr)	✓ (1.1)	✗	✗	✓ (4-9)	✓ (6)	✗	✗	✗	✗	✗	✗	✗
Fibreboards: Dry process boards (BS EN 622: Part 5) (required properties: > 12 - 19 mm nominal thickness boards)												
MDF (WWB)	✓ (20.0)	✓[d] (2200)	✓ (0.55)	✓ (4-11)	✓ (12)	+	✓	+	+	✗	✗	✓
MDF.LA (YYB)	✓ (25.0)	✓ (2500)	✓ (0.60)	✓ (4-11)	✓ (12)	+	✓	+	+	✗	✗	✓
MDF.H (WWG)	✓ (26.0)	✓ (2500)	✓ (0.80)	✓ (4-11)	✓ (10)	+	✓	+	+	✗	✓ (0.20)	✓
MDF.HLS (YYG)	✓ (30.0)	✓ (2700)	✓ (0.75)	✓ (4-11)	✓ (8)	+	✓	+	+	✗	✓ (0.20)	✓

Notes: ✓ Requirement; ✗ no requirement; + additional requirement; [a] wet, after boil test; [b] after boil test; [c] BS EN 622: Part 4 notes that special provisions exist for softboards used exclusively as an insulation material; [d] boards over 4 mm thick; [e] tested in accordance with BS EN 321; [f] major axis.

12.12 PLYWOOD

Veneer plywood (which is the correct name for what is commonly called 'plywood'), is 'plywood in which all the layers are made of veneers orientated parallel to the plane of the panel' (BS EN 313[27]). The term plywood includes core plywoods, such as blockboard and laminboard, and is a 'wood based panel product consisting of an assembly of layers bonded together with the direction of the grain in adjacent layers usually at right angles', with the outer and inner plies placed symmetrically on each side of a central ply or core. However, as long as the veneer plywood is balanced about its centre line, plies may consist of two adjacent veneers bonded with their grain parallel. British Standards do not recognise unbalanced construction as plywood (although some overseas standards do). Both softwoods and hardwoods within the density range 400-700 kg/m^3 are utilised.

Core plywood (Blockboard and laminboard)

Blockboards are composite boards having a core made up of strips of wood, each not more than 25 mm wide, laid separately and glued (or otherwise joined) together to form a panel, to each face of which is glued one or more veneers with the direction of the grain of the core strips running at right angles to that of the adjacent veneers. **Laminboards** are similar to blockboards, except the core strips are narrow (up to 7 mm) and, unlike some blockboards, they are continuously glued (laminboards are thus more free from surface distortion and provide an excellent base for high quality veneers or for high gloss paint). Although dated, BS 3444[28] is frequently used by manufacturers as a production aid and for quality control.

BS EN 313: Part 1: 1996[29] provides a number of sets of classification criteria for plywoods, including general appearance, construction (e.g, veneer plywood and wood core boards, such as blockboards and laminboards), form and shape (flat, moulded, etc.), principal characteristics, durability (for use in exterior, humid or dry conditions), mechanical properties, surface appearance, surface condition (e.g, sanded, unsanded, prefinished, overlaid, etc.) and user requirements. BS EN 315[30] specifies tolerances for plywoods at a moisture content of 10 ± 2%. Commonly adopted criteria include:

Conditions of use (BS EN 636), defined in accordance with the parameters laid down for hazard classes (**12.25.1**) in separate parts of BS EN 636

- Dry conditions (for Hazard Class 1) (BS EN 636: Part 1)
- Humid conditions (for Hazard Class 2) (BS EN 636: Part 2)
- Exterior conditions (for Hazard Class 3) (BS EN 636: Part 3)

Bond quality (BS EN 314[31]) is a fundamental parameter in determining whether a particular plywood is suitable for a given hazard class. Three bonding classes are defined in BS EN 314 which relate to the hazard classes for the end use conditions (**12.25.1**)

- Bonding class 1: Dry conditions (for Hazard Class 1)
- Bonding class 2: Humid conditions (for Hazard Class 2)
- Bonding class 3: Exterior applications (for Hazard Class 3)

Appropriate test procedures are outlined in BS EN 314: Part 1 which classifies the plywood into one of the three bonding classes according to two criteria (mean shear strength and mean apparent cohesive wood failure) in response to various climatic conditions. These tests are essentially similar to those for wood adhesives for non-loadbearing applications (BS EN 204:1991) (**Table 13.15**).

Plywood is ordinarily made with three types of glues which confer varying durability characteristics on the plywood. These glues (along with other adhesives) are considered in **13.20** and are summarised (using existing British Standard nomenclature) below.

- **Urea formaldehyde (UF)** is used in **Moisture Resistant (MR)** and **Interior (INT)** grade plywood. **MR** plywoods will survive full external exposure to weather for limited periods only. They will withstand cold water for a long period and hot water for a limited time, but fail when exposed to boiling water. **INT** grades are required to be strong and durable in dry weather and to be resistant to cold weather.
- **Phenol formaldehyde (PF)** is used in **Weather and Boil Proof (WBP)** plywoods for exterior use, where the glue bond is required to be highly resistant to the weather, micro-organisms, cold water, boiling water and wet and dry heat.
- **Melamine urea formaldehyde (MUF)** is used in **Cyclical Boil Resistant (CBR)** plywoods forming a bond with intermediate resistance to moisture and the weather and producing a plywood intermediate between **WBP** and **MR** grades. Note CBR is essentially equivalent to BR in the adhesives standards (**13.20**).

Plywood for interior uses will normally be produced from non-durable species whilst plywood for external use will be produced from durable timbers or permeable non-durable species which have been preservative treated.

The **durability** of plywood (BS EN 335: Part 3) is affected by the wood species used in the plies, ply thickness and the gluelines. Risks of biological attack are given in BS EN 335: Part 3 (**12.25.1**) in relation to Hazard Classes. Use of plywood in Hazard classes 4 and 5 is noted as being appropriate only if the inherent properties of the boards are adequate. Guidance on the use of plywood under different environmental conditions (hazard classes) is given in DD ENV 1099[32]. For fungal attack, the durability of the wood used in the plies is related to the hazard class in which the plywood is to be used, and recommendations as to whether natural durability is sufficient or whether preservative treatment is required is included. Ratings of the resistance of plywood to insects is also included (**12.25.2**).

Formaldehyde can cause irritation of the respiratory tract in humans when inhaled at low concentrations. Although longer term tests for carcinogenicity are not conclusive, many regulatory authorities have classified formaldehyde as a 'questionable human carcinogen'. In view of health concerns, three **formaldehyde release** classes are defined by BS EN 1084[33] which can be used to restrict the use of formaldehyde-emitting plywoods in interior environments

- Class A: ≤ 3.5 mg formaldehyde/m^2.h
- Class B: > 3.5, ≤ 8 mg formaldehyde/m^2.h
- Class C: > 8 mg formaldehyde/m^2.h

Unlike other wood based panel products, the properties of the plywood depend not only on the species of timber selected, but also on the type of adhesive used. It is not possible to examine the strength properties of plywood in general terms because there are different strength properties in different grain directions; in addition, strength is affected by the configuration of the plywood in terms of the number, thickness, orientation and quality of the veneers. The approach adopted is therefore to utilise the characteristic values for mechanical and physical properties from prEN 12369 (**Table 12.9**). Generally, plywood can be said to possess high strength and stiffness (especially when expressed in terms of specific gravity). Since the principal characteristic of plywood is the crossed plies which distribute the longitudinal wood strength, plywood is particularly strong in shear; hence its use in the manufacture of ply-web beams[34].

12.13 PARTICLEBOARDS

BS EN 309[35] defines particleboards as manufactured from particles of wood (flakes, chips, shavings, sawdust, wafers, strands, etc.) and/or other ligno-cellulosic materials (such as flax, hemp, etc.) under heat and pressure with an adhesive. Particleboards include chipboard, flakeboards such as waferboard and oriented strand board (OSB) and cement-bonded particleboard. **Chipboard** is the common name for a particleboard made primarily of small softwood particles and a binder (a synthetic resin). Boards are available typically from 3 to 50 mm thick and may be of uniform construction through their thickness, of graded density or of distinct 3 or 5 layered construction.

Particleboards can be classified in a number of ways; those defined in BS EN 309 are

- manufacturing process (e.g, pressed or extruded);
- surface finish (e.g, unsanded, sanded, planed, coated or surfaced with a solid material, such as veneers);
- shape (flat, profiled surfaces or edges);
- size and shape of particles (chipboard, waferboard, oriented strand board, etc.);
- board structure (single layer, multi-layer, graduated, extruded, etc.);
- use categories, as defined in various parts of BS EN 312[36] for particleboards, in BS EN 300[37] for oriented strand board and BS EN 633[38] and BS EN 634[39] for cement-bonded particleboards.

Particleboards are separated on the basis of their performance in differing environmental situations. Three environmental conditions are specified and these relate to the Hazard Class conditions defined in Eurocode 5 (**12.25.1**)

- **Dry**: Moisture content in the board material corresponding to a temperature of 20°C and an ambient relative humidity exceeding 65% for a few weeks each year. Boards of this

type are only suitable for use in Hazard Class 1 of BS EN 335: Part 3;
- **Humid**: Moisture content in the board material corresponding to a temperature of 20°C and an ambient relative humidity exceeding 85% for a few weeks each year. Boards of this type are suitable for use in Hazard Class 1 or 2 of BS EN 335: Part 3;
- **Exterior**: Moisture content in the board material higher than that in humid conditions.

BS EN 312, Parts 2 to 7, specify six grades of particleboard (P2 to P7) suitable for use in different environmental conditions, as follows
- P2: general purpose boards for use in dry conditions;
- P3: boards for interior fitments (including furniture) for use in dry conditions;
- P4: load-bearing boards for use in dry conditions;
- P5: load-bearing boards for use in humid conditions;
- P6: heavy duty load-bearing boards for use in dry conditions;
- P7: heavy duty boards for use in humid conditions.

Particleboards suitable for use in each Hazard Class are summarised in **Table 12.14**. The risks of biological attack of particleboards are summarised in **Table 12.15**.

The properties required for the manufacture of particleboards are summarised in **Table 12.10** and typical characteristic values are given in **Table 12.9**. Strength and rigidity tend to increase with density. All particleboards are particularly affected by moisture, resulting in a marked reduction in mechanical performance. The principal difference in performance between different particleboard grades is that PF and MUF boards are capable of considerable recovery on drying after wetting, while UF boards show very little recovery. In addition, the permanent loss of strength of UF boards in the presence of moisture increases markedly at raised temperatures.

12.14 FLAKEBOARDS

Flakeboard is the generic name for two types of wood-based panel, **wafer board** and **OSB (oriented strand board)**. Both panel types are outwardly similar products manufactured from large flakes of wood bonded together with a synthetic resin. This type of manufacture has two principal advantages
- as 'fines' are not included, the total surface area of the constituent particles will be lower compared to other wood-based panel products (e.g, chipboard), and less adhesive is required to achieve the same board strength;
- the strength and stiffness properties of the board are enhanced by the use of large flake-like particles (wafer board and OSB differ in the size, shape and orientation of the flakes and this results in differences in their properties).

Wafer board is a type of wood particleboard made from wood wafers or flakes which are mostly square, or nearly square, in shape (approximately 75 x 75 mm) and 0.4-0.6 mm thick. The flakes are produced from ribbons of wood cut tangentially from the timber. The plane of the flakes run parallel with that of the board but are otherwise randomly orientated. **OSB** is made from large wood strands having a length of at least twice their width, which are orientated in predetermined directions in each layer to simulate some of the characteristics of plywood. Boards are assembled as three (or, more rarely, five) layered boards. Both types of board can be made from various species of wood although wafer board tends to be mainly hardwood (aspen) and OSB either hardwood (aspen) or softwood (pine). Wafer board is not currently used in the UK but, when first imported in the late 1970's, it was used as sheathing in timber frame housing. It has been widely used in North America as a flooring board, for temporary shuttering, for pitched roof sarking and flat roof decking.

BS EN 300 defines four types of OSB for which mechanical and physical properties are prescribed (for manufacturing purposes rather than design values)

- **OSB1**. General purpose boards and boards for interior fitments, including furniture, for use in dry conditions;
- **OSB2**. Load-bearing boards for use in dry conditions;
- **OSB3**. Load-bearing boards for use in humid conditions;
- **OSB4**. Heavy duty load-bearing boards for use in humid conditions.

OSB can be used for pitched roof sarking and flat roof decking (except in warm deck flat roofs to insulated and heated buildings), soffits, some sheathing applications, light duty (e.g, domestic) suspended floors and overlay to structural floors. OSBs are not suitable for conditions where sustained exposure to water and/or high humidity is expected.

The properties required for the manufacture of OSB are summarised in **Table 12.10** whilst characteristic values are not yet available. Overseas Standards are, however, available for both wafer board and OSB; for example, Canadian Standard CAN3-0437.0-M85[40] describes the basic properties and characteristics of both products whilst Canadian Standard CAN3-0437.1M85[41] describes appropriate test methods. OSB, due to particle orientation, has higher strength and stiffness values in the long (machine) direction and lower values in the transverse direction than wafer board. Alignment of the core flakes in three layer boards at right angles to those in the surface layers in both board types was a design adopted to emulate (and compete with) plywood. However, although OSB strength and stiffness properties compare favourably with particleboards (**Table 12.10**), both wafer board and OSB fall short of the levels found in plywood. In addition, although OSBs are bonded with adhesives which are inherently moisture resistant, durability with respect to fungi attack of the timber strands used in the boards would be classified as 'slightly durable' or 'non-durable' by BS EN 350 (**12.25.1**). Thus both board types are expected to have limited durability in conditions of prolonged wetting, limiting their use to 'protected' environments.

12.15 CEMENT BONDED PARTICLEBOARD

Cement bonded particleboard is made from small particles of wood, bonded with Ordinary Portland Cement (OPC), formed and cured into panels. In the past, magnesite cement binders have been used (to give boards suitable for internal (dry) conditions), although these have now been withdrawn from the UK market. The constituents of the particleboard are approximately cement 60%, wood 20% and water 20% by weight.

BS EN 633 provides a number of sets of classification criteria for cement-bonded particleboards based on

- the type of binder (OPC, magnesium based cements, e.g, magnesite);
- the state of the surfaces (pressed, sanded, coated, surface sheeted, etc.);
- colouration (integrally coloured, no added colouration, etc.);
- shape (e.g, flat surfaces and square edges, profiles surfaces, profiled edges, etc.).

BS EN 634 distinguishes Class 1 and Class 2 cement-bonded particleboards on the basis of the required modulus of elasticity (**Table 12.10**). Currently, only one plant in the UK produces cement bonded particleboard (product trade name **Pyrok 5 Star**), although a number of brands (e.g, **Cemboard, Duripanel, Viroc**) are imported. Cement bonded particleboards have been used successfully for cladding, internal wall lining and flooring where high resistance to either fire and/or moisture is specified, and in sound insulation barriers. In the UK, the principal use has been for internal walls and partitions, due both to its good resistance to fire and sound transmission, and to its superior impact resistance compared to gypsum board. Other uses include pitched roof sarking and flat roof decking (including high humidity locations), cladding, fascias, soffits, sheathing, domestic flooring and overlay to structural non-domestic flooring.

The claimed advantages of cement bonded particleboard over other panel products are

- superior dimensional stability in wet or humid conditions compared to resin-bonded particleboards and retention of a smooth surface;
- superior behaviour in fire (for fire propagation and surface spread of flame to BS 476 (**14.9**), cement bonded particleboard is rated Class O in accordance with the Building Regulations);
- OPC confers high alkalinity producing high resistance to fungi. The product has performed well in tests against insect attack[42] and in accelerated ageing tests, suggesting good resistance to natural weathering;
- good sound insulating properties resulting from high density.

12.16 FIBREBOARDS

Fibreboard is the generic name for a spectrum of board types of differing density and hardness, made from wood in the fibrous state. Fibreboards are defined in BS EN 316 as 'panels with a thickness of 1.5 mm and greater, manufactured from lignocellulosic fibres with application of heat and/or pressure'. The bond is derived either from the felting of the fibres and their inherent adhesive properties or from a synthetic binder. Other additives may be included. Their properties (**Table 12.10**) and performance in service depend mainly on the extent of pressing of the fibrous mat in manufacture and on the incorporation of chemical additives. BS EN 316 distinguishes fibreboards by their production process and their density and BS EN 622 defines 25 types of panel covering a wide range of densities and suitable for a variety of end use environments.

Modification of the pressure applied in the final pressing produces a wide variety of boards of different densities (e.g, soft boards <400 kg/m^3, hardboard ≥900 kg/m^3). Fibre building boards can therefore be subdivided on the basis of density, although within these groups there is a wide variation in strength and moisture resistance. BS EN 622 should be referred to for more information.

Wet process fibreboards have a moisture content of >20% at the stage of forming. The following types are differentiated, according to their density

- **Hardboards** (density >900 kg/m^3) (BS EN 622: Part 2). All hardboards are denoted by the symbol **HB**, and are further distinguished by their intended use (general purpose, load-bearing or heavy duty load-bearing) and expected service environment (dry, humid or exterior), as follows
  - **General purpose boards. HB** (the standard grade with no additives for use in dry conditions), **HB.H** (humid conditions) and **HB.E** (exterior conditions).
  - **Load-bearing boards. HB.LA** (the standard load-bearing grade with no additives for use in dry conditions) and **HB.HLA1** (humid conditions).
  - **Heavy duty load-bearing boards. HB.HLA2** (humid conditions).

  General purpose boards are typically used for ceilings, internal wall linings, overlay to structural floors, etc. and load-bearing and heavy duty load-bearing boards for box and I-beams, external cladding, formwork, overlay to structural floors, roof sarking, sheathing, soffits and fascias.

- **Medium boards** (>400 kg/m^3 to <900 kg/m^3) (BS EN 622: Part 3). All medium boards are denoted by the symbol **MB**, and are further distinguished by their density range, intended use (general purpose or load-bearing) and expected service environment (dry, humid or exterior), as follows
  - **Low density medium boards** (>400 kg/m^3 to <560 kg/m^3) **MBL**;
  - **High density medium boards** (>560 kg/m^3 to <900 kg/m^3) **MBH** ('panelboard').

  Both low- and high density medium boards can be used in dry conditions (**MBL** and **MBH** respectively), humid conditions (**MBL.H** and **MBH.H** respectively) and in exterior conditions (**MBL.E** and **MBH.E** respectively). Additionally, high density medium boards can be used for load-bearing applications in dry (**MBH.LA1**) and humid (**MBH.HLS1**) conditions and for heavy load-bearing applications in dry (**MBH.LA2**) and humid (**MBH.HLS2**) conditions. Medium boards can be used for a wide range of applications; appropriate grades of low density medium board are commonly used in ceilings, internal wall linings, roof sarking, whilst high density medium boards are commonly used for external cladding, soffits and facias (dry and humid types) and for external cladding, sheathing, soffits and facias (exterior types).

- **Softboards** (>230 kg/m^3 to <400 kg/m^3) (BS EN 622: Part 4)
  Softboards have thermal and acoustic properties, and can be given additional properties by the addition of appropriate additives (e.g, to improve fire resistance, moisture resistance, etc.). All softboards are denoted by the symbol **SB**, and are further distinguished by their intended use (general purpose or load-bearing) and expected service environment (dry, humid or exterior), as follows
  - **General purpose boards. SB** (the standard grade with no additives for use in dry conditions), **SB.H** (humid conditions) and **SB.E** (exterior conditions).
  - **Load-bearing boards. SB.LS** (the standard load-bearing grade with no additives for use in dry conditions, **SB.HLS** (humid conditions).

  Softboards suitable for humid and exterior conditions are generally bitumen-impregnated; type **SB.E** has the higher bitumen loading and lower water uptake. All softboards have a thermal conductivity of the order ≤0.055 W/mK. The standard **SB** and **SB.LS** grades are used in ceilings, internal wall lining, overlay to structural timber floors, etc. The **SB.H** and **SB.HLS** grades are suitable for flat roof insulation overlay, movement-joint fillers, overlay to structural floors, roof sarking, sheathing. The **SB.E** grade can be used in roof sarking and sheathing.

Dry process fibreboards have a moisture content of <20% at the stage of forming, and have a density of ≥600 kg/m^3. Dry process boards conforming to these criteria are termed '**medium density fibreboards**' (**MDF**) and are produced with a synthetic adhesive under heat and pressure. They can be given additional properties (e.g, moisture resistance, fire retardence, biodegradation resistance to biological attack, etc.) by the application of appropriate additives. Medium density fibreboards are denoted by the symbol **MDF**. Uses for MDF include architectural mouldings, staircases, fully sealed window boards, etc. BS EN 622: Part 5 further distinguishes boards by their intended use (general purpose or load-bearing) and expected service environment (dry or humid), as follows

- **General purpose boards. MDF** (the standard grade with no additives for use in dry conditions) and **MDF.H** (humid conditions).
- **Load-bearing boards. MDF.LA** (the standard load-bearing grade with no additives for use in dry conditions), **MDF.HLS** (humid conditions).

Table 12.10 gives the mechanical and physical requirements for compliance with BS EN 622 for selected board thicknesses. The high strength and stiffness requirements for hardboard (particularly the load-bearing grades) are roughly comparable with plywood and superior to most other wood-based panels. However, the relative performance of fibreboards exposed to moisture is poor (as shown by the 'swelling in thickness' criteria, which is larger than for most other wood-based panels and exceeds that for plywood).

12.17 COMPRESSED STRAW BUILDING SLABS

Compressed straw building slabs consist of natural straws compressed by heat and pressure, restrained in, and firmly bonded to, facings of paper or other materials to form rectangular slabs. BS 4046[43] gives methods of testing and guidance on the use, which is primarily for thermal insulation, including roof linings and ceilings, as partitioning or as internal linings to exterior walls. The thermal conductivity of compressed straw building slabs must not be more than 0.108 W/mK (BS 4046). Typically, slabs are 50 mm thick, range in density from 16.5 to 22.5 kg/m^3, have a bending strengths of 0.70 N/mm^2 and 1.2 N/mm^2 parallel and perpendicular to the plane of the board respectively and a modulus of elasticity of 200 N/mm^2 and 600 N/mm^2 in these directions. Slabs, like other board materials, are prone to moisture movement; typical values are about 0.3% as a result of a change in the relative humidity from 35% to 85%. Consequently, compressed straw slabs should not be used where permanently damp or persistent high humidity conditions exist. When used for internal wall or roof lining, a grade of slab with a shower proof facing should be used with the facing adjacent to the cavity between the slab and the roof or wall. The low strength and stiffness properties of these slabs means they should not be self-supporting (as required in some roof decking applications).

12.18 WOOD WOOL CEMENT SLABS

Wood wool cement slabs are made of long wood shavings, coated with cement and compressed but leaving a high proportion of thermal insulating voids. Specification for wood wool cement slabs up to 125 mm thick is given by BS 1105[44], which is now obsolete, although a European preStandard exists (prEN 13168[45]) which is yet to be formally adopted. BS 1105 defines three grades of panel according to their mechanical performance. Wood wool cement slabs are used primarily for thermal and sound insulation, including flat roof decking, ceilings, as partitioning or as internal linings to exterior walls (Types B and SB except in conditions of high humidity). As roof decking the 51 mm thick slab can span joists at 610 mm centres. The open texture is a good base for plasters and renderings although the natural surface, which can be decorated by spraying with paint, provides good sound absorption. The thermal conductivity of slabs 25 mm or 38 mm thick must not be more than 0.1 W/mK and the thermal conductivity of slabs equal to or greater than 50 mm thick must not exceed 0.093 W/mK (BS 1105). The Standard also gives requirements for minimum sound absorption coefficients for different nominal thicknesses and methods of mounting. Wood wool cement slabs are resistant to biodegradation. They are combustible but not readily ignited (BS 476: Part 4) and the surface spread of flame is low, i.e., Class 1 (**14.9**).

12.19 GLUED LAMINATED TIMBER (GLULAM)

The most common form of reconstituted solid structural timber is **glued laminated timber**, commonly known as **Glulam.** It consists of timber or veneer laminations bonded with an adhesive (urea formaldehyde or resorcinol formaldehyde depending whether the material is required for internal or external use) so that the grains of all laminations are parallel with the long direction. Ordinarily, a minimum of four laminations within the member are used in order to benefit from increased stress factors. In deriving strength grade values for structural timber, allowance must be made for strength-reducing effects of natural characteristics, such as knots, fissures and slope of grain (**12.9**). Thus the strength of a clear, straight-grained portion of a length of timber of a given grade is not being utilised to its full potential; for example, a softwood beam containing a knot at a critical position may fail at a value as low as 40% of the clear wood value. This can be overcome by cutting up the piece and reconstituting it, thus dispersing the strength reducing characteristics. This will permit a higher overall stress level in the member and will ensure more efficient use of the material. This is the basis for the use of Glulam. Reconstitution is also able to produce sizes of members not available in the original tree. Laminated construction is used in such diverse items as tennis racquets, skis, hulls of wooden ships and arched beams in halls and stadiums.

Design using Glulam under European Standards is governed by Eurocode 5 (**12.10**). Compared to the design approach adopted in the previous governing design standard (BS 5268: Part 2: 1991), Glulam beams designed using Eurocode 5 will carry approximately 5% more load if the bending strength governs the design or from 5% to 20% more load if deflection governs the design (as is normally the case)[46]. Characteristic values for Glulam properties for design to Eurocode 5 are given in BS EN 1194 (selected grades are reproduced in **Table 12.7**). In the European system, the number in each strength class refers to the characteristic bending strength of each class; thus GL20 has a characteristic bending strength of 20 N/mm^2.

Glulam has higher design strength values than other timber products resulting from lower variability. It also has a number of other advantages over single solid timber sections

- members of large cross-section can be produced easily;
- the relatively small section laminations are quick and cheap to kiln-dry;
- laminations can be bent to produce pre-cambered or curved profiles;
- end-jointing of laminations using finger joints enables long members to be produced;
- reductions in distortion and splitting;
- no single defect can be thicker than the individual lamination thickness.

Compared to steel sections, Glulam also has a number of advantages

- it works much better than unprotected steel in cases of fire and requires no encasement to ensure adequate fire performance;
- lighter foundations are required;
- ease of handling and on site fixing.

12.20 TIMBER DEGRADATION

Agencies which attack wood are varied, but include fungi (**12.21**); wood boring insects (**12.22**); some chemicals (which attack the structure of the wood) (**12.23**); heat (which tends to make the wood brittle) (**14.6.2**); and fire (an advanced form of chemical attack) (**14.6.2**).

12.21 FUNGI

Trees feed on simple inorganic materials which they build up into complex organic compounds (**autotrophic nutrition**). In contrast, wood-rotting fungi do not contain chlorophyll and obtain their food by breaking down wood cell walls into soluble organic compounds (**saprophytic nutrition**). Where this occurs in structural timber, significant loss of strength may occur. Fungi need water and oxygen to grow but, unlike other plants, do not require sunlight to grow (although they do require light to produce spores). Fungi cannot utilise wood at or below 20% moisture content (or wood that is thoroughly wet so that oxygen is excluded). Since the moisture content of building timber is usually in the range 9 to 19%[47] (**12.6**), fungi do not occur in internal building timbers unless poor design or building faults have led to water entry or condensation. For significant growth, fungi require

- a temperature of 3 to 23°C (*Serpula lacrymens* (dry rot) is particularly sensitive to temperatures above 25°C);
- a wood moisture content of 20% or more;
- a supply of oxygen (air);
- a supply of organic compounds (i.e., timber).

The basis of timber preservation (**12.25** and **12.26**) is to treat the timber with a toxic chemical in order to poison the food supply and thereby eradicate the fungi. The life cycle of different wood destroying fungi is essentially the same, as typified by ***Figure 12.13***.

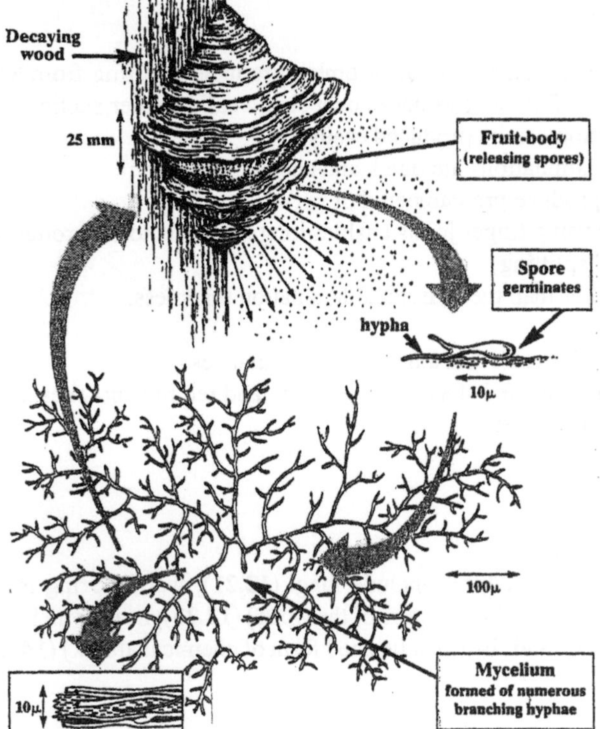

Fungal growth does not require light, and usually occurs out of sight until such time as it becomes exposed through crevices and joints. In contact with sunlight, reproduction of the fungi occurs by the production of **spores** from a fruiting body (known as a **sporophore**). These spores are released and carried in the air, to land on new timber food sources. Fugal spores are ubiquitous. Although numbers vary seasonally, about 50000 spores/m^3 occur in outdoor air in the summer in the UK compared to about 500 spores/m^3 in indoor air (indoor spore counts show large variations; e.g, vacuuming will increase suspended spore counts; in mould affected dwellings, spore levels may reach 3000-7000 spores/m^3)..

Figure 12.13 *Life cycle of a typical wood-rotting fungus*

Under suitable temperature conditions, spores can germinate on wood when the moisture content is greater than about 26 to 30%. Although this range is required for germination, spores can remain dormant in the absence of water for several years, germinating when the correct conditions occur. On germination, fine strands called **hyphae** (singular **hypha**) are produced, analogous to the roots of a plant in that they spread out through the wood structure in search of new food supplies. During the saprophytic nutrition, the hyphae use **oxygen** and produce **carbon dioxide** and **water**. Indeed '*lacrymens*' (as in the correct name for dry rot, *Serpula lacrymens*) is Greek for *tears*, named for the visible presence of water droplets on the hyphae of this fungus. It has been suggested that *Serpula lacrymens* utilises evolved water to moisten dry wood to above the moisture content required for growth, although this ability will clearly be dependent on the ambient evaporation conditions.

The masses of hyphae collect together to form a mat-like structure known as a **mycelium** (Greek = mushroom). The hyphae making up the mycelium continue to penetrate the wood, breaking down the wood cell walls and feeding on them. Eventually, in the presence of sunlight, the mycelium again produces a **sporophore**, and the fungal life cycle resumes again. ***Figure 12.14*** highlights the main types of fungi found in buildings.

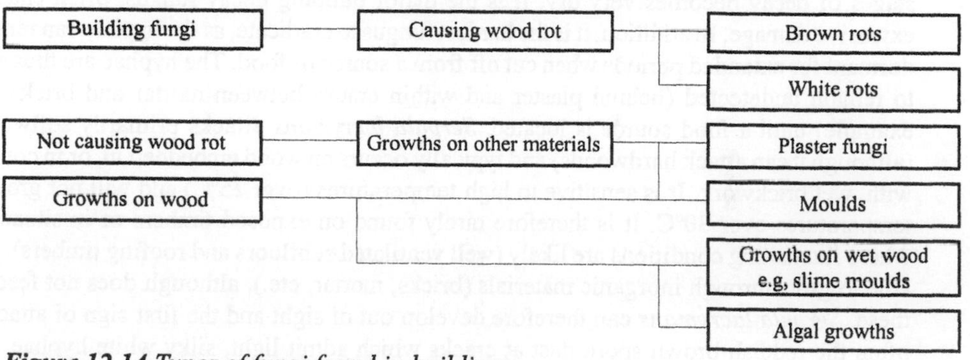

Figure 12.14 *Types of fungi found in buildings*

As noted in ***Figure 12.14***, not all of the fungi found in buildings cause wood rot. However, the presence of any fungal growths in buildings indicates that damp conditions exist which are suitable for the development of wood-rotting species. Wood-rotting species found in buildings feed on the structure of the tracheid, pore or ray cell walls and can be divided into two major groups, according to which part of the timber is extracted as food.

- **White rots** thrive on all the constituents of the wood, including the lignin, causing the wood to become lighter in colour and fibrous in texture without cross-cracking. Hardwoods are particularly susceptible and some, such as beech and Ash, show characteristic dark zone lines. Formerly, white rot fungi were considered to attack external timber only, but they are now recognised as a major cause of joinery decay. Woods containing these fungi rarely shrink or collapse. All white rots are wet rots.

- **Brown rots**, which cause the wood to become darker in colour by extracting the polysaccharides (**12.4.1**) from the cellulose of the wood cell wall but leaving the darker lignin (**12.4.2**). Cracking takes place along and across the grain. Even when decay is only slight, the strength can be affected. Permeability is also affected and exposed woodwork wets more quickly and takes longer to dry. When the timber effected by brown rots is moist it is soft and easily breaks up and when dry will crumble to dust (this is especially true of very decayed wood). Partly because of this softening, decayed timber is often more susceptible to insect attack. Many common wet rots are brown rots; dry rot (*Serpula lacrymens*) is also in this group.

Except for one brown rot, *Serpula lacrymans* (commonly called **dry rot**), **all** brown rots and white rots are commonly referred to as **wet rots**. Note, however, that under the international rules of botanical nomenclature, the terms 'dry' and 'wet' rot are discouraged. 'Dry rot' is a misleading term to apply to fungal decay which requires moist timber.

12.21.1 Dry Rot (*Serpula lacrymans*) (formerly *Merulius lacrymans*)[48]

Serpula lacrymans is a **brown rot.** It is often called 'dry rot' because the wood in the final stages of decay becomes very dry. It is the major building decay fungus, often causing extensive damage. In addition, it is the hardest fungus to eradicate, as the hyphae can remain dormant for extended periods when cut off from a source of food. The hyphae are thus able to remain undetected (behind plaster and within cracks between mortar and bricks, for example) until a food source is located. *Serpula lacrymans* attacks primarily softwoods (although it can attack hardwoods) and typically occurs on wood embedded in, or in contact with, wet brickwork. It is sensitive to high temperatures (over 25°C) and will not grow at temperatures over 30°C. It is therefore rarely found on exposed timbers or in situations where fluctuating conditions are likely (well ventilated subfloors and roofing timbers). It is able to grow through inorganic materials (bricks, mortar, etc.), although does not feed on these. *Serpula lacrymans* can therefore develop out of sight and the first sign of attack is often the reddish brown spore dust at cracks which admit light, silky white hyphae or a characteristic 'mushroom' or 'musty' smell.

Successful **remedial treatment** of *Serpula lacrymans* is usually more involved and elaborate than for other rots because of its ability to spread extensively behind plaster, through wall materials, etc. Remedial measures will involve[49,50,51]

Primary measures

- establishing the size and significance of the attack (e.g, determine whether structural timbers are affected and whether structural repairs are necessary);
- locating and eliminating moisture sources (**5.2**) and promoting the rapid drying of the structure[52];
- removing all rotted wood, cutting away timber approximately 300 to 450 mm from the last evidence of fungus or rot.

Secondary measures

- containing the fungus within the wall using preservatives (where drying is delayed);
- in replacement work, preservative treated wood should be used (remember that newly exposed saw cuts are not protected by the preservative treatment, and additional treatment will be required);
- treating the remaining sound timbers potentially at risk (minimum two brush coats);
- introducing measures to support the preservative treatment (such as increased ventilation between sound timber and wet brickwork, etc.);
- inspections periodically to ensure that there is no re-infestation.

12.21.2 'Wet' Rots

Wet rots are much more common in buildings than *Serpula lacrymens* ('dry' rot). There are many fungal species causing wet rot and, since the same remedial measures are required for all of them, the only requirement for identification is that they are distinguished from dry rot (remedial measures will be less destructive to the building fabric than the treatment for dry rot). The most common wet rot fungi are the **brown rots** *Coniophora puteana* (formerly *Coniophora cerebella*), sometimes called 'cellar fungus', and *Coniophora marmorata*. These fungi grow best on wood slightly wetter than that required by *Serpula lacrymans* (40 to 50% moisture content) and often attack timbers in roofs, cellars or other situations where water leaks or excessive condensation occurs. Occasionally, the decay is internal and the surface of the timber appears intact. **Soft rots** can be regarded as a superficial form of wet rot, and occur in timber in very wet conditions (e.g, timber in ground contact). They are remedially treated in the same way as wet rot fungi. The importance of this group of fungi was only recognised in the 1950's. Since that time, it has been discovered that the soft rot fungi attack a much broader range of timbers than do brown or white rot fungi. Hardwoods are especially susceptible. Normally, the timbers attacked by these fungi retain their original shape, but the surface becomes discoloured, softened and possibly eroded. When dry, the hyphae-ridden surface is crumbly and shows many fine cracks and fissures. **Remedial treatment** of wet rot differs from that for *Serpula lacrymens* ('dry' rot) in the secondary measures which may be required. Treatment will involve[53,54]

Primary measures

- locating and eliminating sources of moisture and promoting rapid drying of the structure

Secondary measures

- establishing the size and significance of the attack;
- removing rotted wood (consider localised application of preservative treatment *in situ*);
- using preservative treated replacement timbers;
- introducing supporting measures

12.21.3 Non Wood-rotting Fungi

Many types of **mould** are prevalent in buildings and building materials, where they colonise surfaces of damp wood, plaster, wallpaper, paint, etc. Moulds feed on the **contents** of the timber cells (the free sugars or sap) or on surface detritus, **not** on the wood structure itself. They are only important where the wood is left in its natural state (without pretreatment, **12.25**). They do not affect the mechanical properties of the timber, although they may make the timber absorb more water and so aid the degradation by other fungi. Moulds may also be aesthetically unacceptable. **Remedial treatment** for moulds[55] involves drying the area and increasing ventilation to disperse the mould spores. Surface spores can be removed with a damp cloth; surfaces should then be cleaned with a **fungicidal wash** to prevent a regrowth. Heating and ventilation of the area should be continued and where damp conditions are likely to prevail, only effective fungicidal paints should be used when redecorating.

Different mould species produce different colour staining (bluish, yellow, orange and green colours are common). **Blue stain in service**[56] is a fairly common mould causing disfigurement of wood, especially with clear finishes. It is often confused with **'sap stain'** or **'blue stain'** of freshly felled logs or fresh sawn timbers, which are caused by a different species but still produce a blue colour which is detectable in service. Blue stain in service causes early failure of surface coatings by rupture caused by the fruiting bodies. **Remedial treatment** will involve removing the damaged surface coating, drying the timber and treating with a preservative specifically active against these fungi prior to redecoration.

Plaster fungi occur in damp brickwork or plaster and feed on surface detritus or on organic material included in walls (e.g, bitumenised felt DPC's, the keratin set retarded in Class B gypsum plaster, hair contained in old plasters). They invariably produce large fruiting bodies or mycelial growths which may be mistaken for those of wood-rotting fungi. **Remedial treatment** will involve the location and removal of sources of dampness. **Slime moulds** attack damp wood, usually exterior joinery, where they feed on the bacteria within the wood, becoming visible when they produce fruiting bodies on the surface. **Remedial treatment** will involve locating and removing sources of dampness and any surface growths.

Algal growths[57] occur primarily on external damp surfaces (masonry walls, roof coverings, etc.). They require sunlight for growth, which is encouraged by rough-textured, damp surfaces which allow the accumulation of organic, nutrient-rich detritus. Initially, algal growths have a bright green colour which subsequently thickens and changes to slime green. Drying causes the growth to die. In the long term, growths may progress to the lichen stage, where crust-like patches of varying colours develop. Localised growth indicates water flow, which must be rectified (e.g, leaking rainwater goods or overflows, absence of drip channels in window cills, etc.). Generalised growth over large surface areas indicates high levels of exposure to driving rain, a porous surface or one with an inadequate slope to ensure rapid drainage. Periodic removal of growths by wire-brushing and application of a biocide to restrict re-colonisation is required. Growths on wall surfaces indicate possible fungal decay problems in associated internal timbers (e.g, joist ends).

12.22 WOOD BORING INSECTS

A number of insects (mainly beetles and their larvae) are able to use wood as a food source and some of them can cause serious damage to building timbers. Wood boring insects are able to break down cellulose in the wood cell wall to its constituent monosaccharides (**12.4.1**), which in turn provide the food source for the insect. Interestingly, the enzyme required to do this, **cellulase**, is relatively rare in the animal kingdom. For example, humans cannot produce cellulase and are therefore unable to digest cellulose. In most herbivores, it is the presence of micro-organisms in the gut of the animal which produce the cellulase enzyme and enable the animal to digest cellulose. In the insect kingdom, a good example of this symbiotic relationship is seen with the flagellate *Trichonympha*, which lives in the intestine of termites. Termites are responsible for the largest destruction of construction timber worldwide and yet cannot themselves produce cellulase! It is the *Trichonympha* flagellate which produces cellulase and digests the timber and, since not all the digestive products are used by the flagellate, the remainder are absorbed by the termite.

As a general rule, wood boring insects attack the sapwood preferentially to the heartwood because of the higher starch (**12.4.1**) content and lack of toxic extractives. Timber with a moisture content of <20% will not be attacked by fungi, but can be infested by wood boring insects. In UK wood boring insects, **the majority of damage is caused by the larvae of the insect rather than the adult**. In other parts of the world, damage to constructional timber by termites is a major problem; it is the adult termite that consumes the majority of the timber. UK wood boring insects all have fairly similar life cycles, as shown in ***Figure 12.15***, although there are variations in the length of each stage of the life cycle, the type of wood attacked and the extent and type of damage caused.

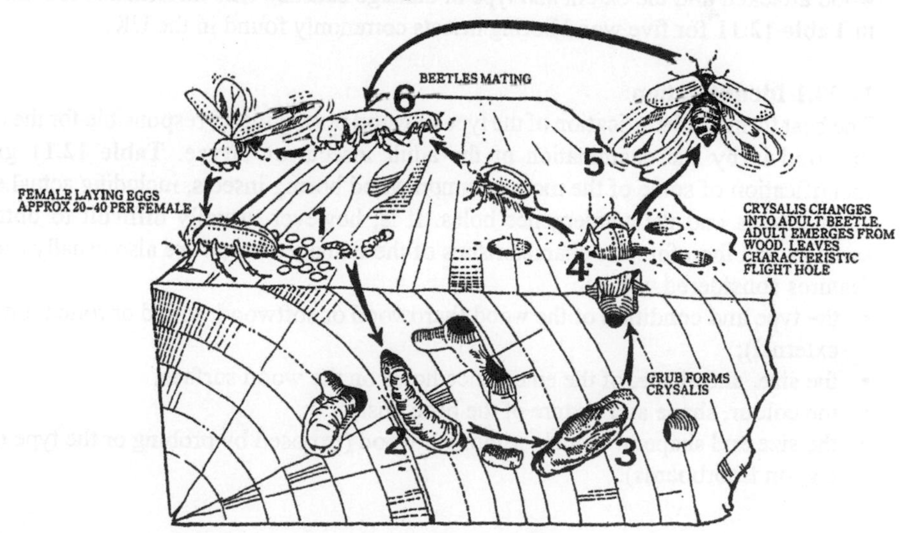

Figure 12.15 *Life cycle of a typical wood-boring beetle*

The **adult** female beetles lay eggs (about 20 to 40, maximum 80) on the wood surface (***Figure 12.15(1)***), in splits or in bark (they do not bore into the wood). The eggs hatch (within about 2 to 4 weeks) into active, grub-like **larvae** (***Figure 12.15(2)***) which (over the next 1 to 5 years) eat their way into the wood, creating tunnels. It is the feeding and tunneling of the larvae which is largely responsible for the damage in timber.

The larvae of most wood-boring insects fill the tunnels with excreted wood pellets known as **bore dust** (or **frass**), which can be used as an aid to identification of the insect (**12.22.1**). The larva produces a chamber just beneath the surface of the wood, where it forms a cocoon or chrysalis and **pupates** (***Figure 12.15(3)***). It does not eat during pupation. During pupation, the insect slowly transforms from a larvae into an adult beetle (a process which takes 2 to 3 weeks, maximum 6 weeks). The beetle eats it way out of the pupal chamber and emerges from the surface of the timber (***Figure 12.15(4)***), leaving characteristically sized and shaped emergence holes, which can also be used as an aid to the identification of the insect (**12.22.1**).

The newly emerged adult lives for only 10 to 14 days from emergence (as it is doubtful whether they can eat and digest the wood). During this period, the prime objective it to find a mate (***Figure 12.15(6)***). Some insects are capable of flying (***Figure 12.15(5)***) to infest other timber. Alternatively, adults may simply walk and re-infest the same piece of timber. After mating, the life cycle recommences with egg-laying.

As noted above, there are variations in the length of each stage of the life cycle, the type of wood attacked and the extent and type of damage caused. This information is summarised in **Table 12.11** for five wood boring insects commonly found in the UK.

12.22.1 Identification

The best form of identification of the type of wood boring insect responsible for the damage in wood is by the examination of the adult insects or larvae. **Table 12.11** gives an identification of some of the more common wood boring insects, including actual sizes of adult insects and their emergence holes. It is, however, usually difficult to obtain live samples and therefore the characteristics of the damaged wood are also usually used. The features considered will be

- the type and condition of the wood (hardwood or softwood, sound or rotted, internal or external);
- the size, and shape, of the emergence holes on the wood surface;
- the colour, shape and texture of the bore dust;
- the size and shape of tunnels within the wood (exposed by probing or the type of wear, e.g, on floorboards).

Table 12.11 Comparison of five common UK wood boring insects

Name	Furniture Beetle *Anobium punctatum*	Death Watch Beetle *Xestobium rufovillosum*	Lyctus (Powder Post) Beetles (various species)	House Longhorn *Hylotrupes bajulus*	Weevils *Pentarthrum huttoni* *Euophryum confine*
Adult Beetle					
Larva					
Damage Category	A	A	A	A	B
Type of wood attacked	Softwoods and European hardwoods (tropical hardwoods rarely attacked). Sapwood only attacked, unless wood rot present. Modern plywoods immune, though may attack older plywoods with animal-based glues.	Sapwood and heartwood of partially decayed hardwoods. Softwoods may be attacked if adjacent to infected hardwood.	Attack tropical and coarse-pored European hardwoods with high starch contents e.g, oak, elm. Softwoods immune. Plywoods, blockboards, etc., are susceptible.	Softwoods only, principally the sapwood, commonly in roofing timbers. Wood-based panel products not usually attacked (but occasionally found in softwood plywood.	Decayed softwoods and hardwoods in damp conditions are attacked, including plywood etc.
Occurrence	Attack most common in damp situations, rare in very dry situations (<12% moisture content).	Common in the south of England; rarer in the north. Damp conditions essential for establishment of infestation. If damage severe, survey structural timbers.		Larval tunnels may be present under a surface veneer of intact timber. Found mainly in SW England, where the Building Regulations require pretreatment of roofing timbers.	Damage secondary to fungal decay.
Complete Life cycle	2 or more years	Usually 4-5 years, can be up to 10 years	1-2 years, may be less in hot buildings	3-11 years or more	7-9 months
Length of pupal stage	4-8 weeks	2-4 weeks	Up to 1 month	About 3 weeks	6-8 weeks
Adults emerge	May-September	March-June	May-September	July-September	No fixed time
Size of exit hole	2 mm	3 mm	2 mm	6-10 mm	Channels on surface and some holes 5 mm
Boredust ('frass')	Roughly 'egg' shaped	Bun-shaped pellets	Fine Floury dust	Short, compact cylinders and powdery dust	Similar to *Anobium* but smaller
Number of eggs laid	Usually 20-40, but up to 80	40-70 or more	70-220	Up to 200	About 25 laid singly
Eggs hatch in	4-5 weeks	2-8 weeks	2-3 weeks	2 or more weeks	About 2 weeks

As the presence of damage by wood-boring insects does not always indicate the need for remedial treatment, the BRE[58] classifies wood-boring insects according to the **types of damage** found in building timbers to enable the correct remedial treatment to be specified. This information is summarised in **Table 12.12**.

Table 12.12 Classification of wood boring insects according to damage caused

Damage category[a]	Remedial measures required	Insect
A	**Insecticidal treatment usually needed** These beetles' activities are normally restricted to the sapwood, although in timbers where there is no colour difference between heartwood and sapwood (e.g, spruce, beech) or where the wood is slightly rotted. Both the heartwood and sapwood may be attacked. The deathwatch beetles, which infests slightly rotted wood, attacks both heartwood and sapwood. With the exception of the deathwatch beetle, these beetles attack sound, dry wood. Some cause structural damage, (e.g, the death watch beetle and the house longhorn beetles). Under the Building Regulations[59,60], all softwood roof timbers in specified areas in the south of England should be treated with a preservative against the House Longhorn beetle.	Common furniture beetles (*Anobium punctatum*); Ptilinus (*Ptilinus pectinicornis*); House longhorn beetle (*Hylotrupes bajulus*); Powder-Post beetles (*Lyctus brunneus*); Imported tropical Powder-Post beetles (*Bostrychids*); Death Watch beetles (*Xestobium rufovillosum*)
B	**Treatment is necessary only to control associated wood rot** These insects attack only damp fungi rotted wood (they cannot infest sound dry timber). They usually infest sapwood but, where rot has extended into the heartwood, it can also be attacked.	Wood-boring weevils; Wharf borers; (*Nacerdes melanura*); Leaf cutter bees and solitary wasps; Tenebrionid (mealworm) beetles; Stag beetles
C	**No treatment needed** Damage by insects which attack green or partially dry timber may be incorporated into buildings, but as the insects have usually been killed during timber drying, no remedial treatment is necessary.	Bark (waney edge) borer beetles (*Ernobius mollis*); Pinhole borer beetles (ambrosia/shothole beetle); Wood wasps (*Sirex* sp); Forest longhorn (*Cerambycid* sp); Shipworm or marine borers (*Teredo* sp); Sawfly (*Ametastegia glabrata*)

Notes: Damage categories for common UK wood boring insects given in **Table 12.11**.

The **remedial treatment** of damage by wood boring insects will involve

- identifying the insect responsible;
- if the damage is caused by those species which fall into **Damage Category A (Table 12.12)**, determine whether the infestation is active or not;
- if damage is associated with fungal decay (i.e., **Damage Category B** insects and deathwatch beetles), check for dampness and carry out necessary remedial measures for dealing with fungal decay;
- if the infestation is thought to be active, estimate the significance of the damage (more than 20 emergence holes per 100 mm run of timber can be considered as a severe attack, but fewer holes may be significant for house longhorn and deathwatch beetles);
- if structural timbers are affected, carry out a full structural survey and ensure structural integrity is maintained;
- decide upon the type and extent of remedial treatment most suitable for the particular insect and type of component damaged.

12.23 CHEMICAL DEGRADATION

Acid or alkaline chemicals from industrial atmospheric pollution, leaking flues, etc. can cause damage to timber which may be mistaken for attack by wood rotting fungi. Acids act principally on the cellulose, breaking the oxygen bonds between adjacent monosaccharide units (**12.4.1**), i.e., $(C_6H_{10}O_5)_n + nH_2O \rightarrow nC_6H_{10}O_5$, causing the timber to become brittle and fibrous in appearance. Whereas acids have little effect on the lignin constituent, alkalis attack both lignin and hemicelluloses, dissolving the cellulose (this is a standard chemical processing technique to make viscose rayon and wood pulp); the timber softens and loses strength. One form of chemical attack in building timbers results from sulfates leached from clay roof tiles, which can attack the lignin in roof timbers, making it soluble in water; the wood becomes defibrated and may become covered with powdery crystals (not unlike fungi mycelium) that crumble when touched. Attack is usually superficial but may be widespread. Wood may also suffer from photodegradation by UV light (in a similar way to plastics, **13.12.1**), which effects the lignin constituent. Photodegradation products of lignin can be leached out, leaving a loose matrix of partially modified cellulose fibres, so that the affected timber may look similar to that attacked by some fungi. However, the limited depth of penetration of UV light into timber (0.05 to 0.5 mm) restricts damage to the surface.

12.24 TIMBER PRESERVATION

Appopriate timber preservation depends on the timber species, the size and condition of the timber, the effectiveness of the preservative, the depth of penetration and amount retained. Timber preservatives contain a toxic chemical **pesticide** (a **fungicide** or an **insecticide**) for eradication of the wood destroying organism, carried into the timber by a liquid **carrier fluid** (e.g, an organic solvent or water), and other additives (to increase water repellency, prevent deposition of pesticide crystals on the surface following evaporation of the carrier fluid, etc.). For timber at risk from both wood boring insects and fungi, **'dual purpose'** preservatives, comprising compatible insecticides and fungicides, can be used. Preservatives ideally combine high toxicity to fungi or insects; low mammalian toxicity and odour; high permanence; non-toxicity to plants; good penetration; non-flammability (in the longer term); no adverse effects on polishes, paints or adhesives; chemical stability; stability in sunlight; non-corrosive to metals[61,62]; low cost. Types of timber preservatives and methods of application vary according to whether timber is being **pretreated** (prior to use in the building) or **remedially** treated (*in situ* within a building). Almost all green timber for construction use will be treated for **sap stain** (or **blue stain**) fungi at the saw mill.

12.25 PRETREATMENT

The objective of pretreatment is to protect the timber against possible future infestation and attack by fungi or wood boring insects, given the expected service conditions and level of performance required of the timber, by creating a persistent layer of pesticide in the zone penetrated sufficient to prevent an initial infestation and attack. Recent changes to the specification of pretreated timber in the UK have been implemented within Europe. Previously, UK timber pretreatment was by a **schedule based** approach to defining required pretreatment quality; wood preservatives were specified according to their active ingredients and were applied to the timber by a specified process to suit the given end use requirements. Compliance with these procedures was considered to afford an appropriate degree of protection for a given commodity. In the new system, a **results based** approach is adopted, implemented by five new European Standards which essentially define the environments to which timber may be exposed in service (the **hazard class**) (BS EN 335[63]) (**12.25.1**); the natural durability (**12.25.2**) and treatability (**12.25.3**) of timber (BS EN 460[64] and BS EN 350[65]); a classification of retention and penetration for pretreatments (BS EN 351[66]) (**12.25.4**); and performance criteria for wood preservatives (BS EN 599[67]) (**12.25.4**). To comply with the new European Standards, it is no longer appropriate to adhere rigidly to any particular active ingredient and treatment process. ***Figure 12.16*** summarises the approach for selecting timber for a given end use environment.

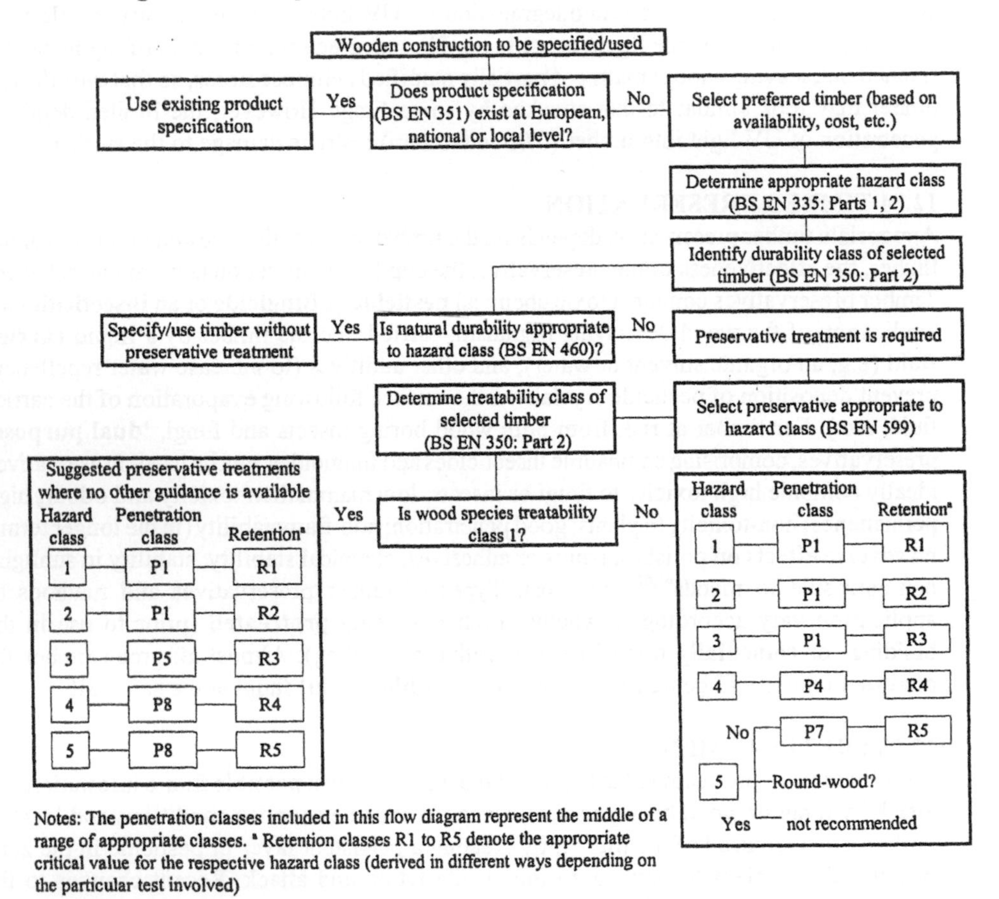

***Figure 12.16** General approach for selection of timber appropriate to hazard class of use*

12.25.1 Hazard Classes

Table 12.13 summarises the five hazard classes (BS EN 335), which represent the various service environments in which timber is used. As service conditions are defined in relation to the moisture conditions, hazard classes essentially define the risk of biodegradation.

Table 12.13 Summary of hazard classes for wood and wood-based products

Hazard class	Typical moisture content	Typical service examples	Occurrence of biological agency						
			Fungi						
			All	Basidiomycetes	Soft rot	Blue stain (Mould)	Beetles[a]	Termites	Marine borers
1. Above ground, covered, permanently dry.	always < 20%	All timbers in normal pitched roofs **except** tiling battens and valley gutter members. Floor boards, internal joinery, skirtings. All timbers in upper floor not built into solid external walls	S✗	P✗ B✗ O✗ F✗	P✗ B✗ O✗ F✗	P✗ B✗ O✗ F✗	SU PU B✗ O✗ F✗	SL PL BL OL FL	S✗ P✗ B✗ O✗ F✗
2. Above ground, covered, risk of high humidity and occasional wetting	occasionally > 20%	Tiling battens, timbers frames in timber frame housing, timber in pitched roofs with high condensation risk, ground floor joists, sole plates above dpc	SU	PU BU OU FU	P✗ B✗ O✗ F✗	PU BU OU FU	SU PU B✗ O✗ F✗	SL PL BL OL FL	S✗ P✗ B✗ O✗ F✗
3. Above ground, not covered, risk of frequent wetting	frequently > 20%	External joinery including roof soffits and facias, bargeboards, etc., cladding, valley gutter timbers, fence rails, boards and gates, etc.	SU	PU BU OU FU	P✗ B✗ O✗ F✗	PU BU OU FU	SU PU B✗ O✗ F✗	SL PL BL OL FL	S✗ P✗ B✗ O✗ F✗
4. In contact with ground or fresh water, exposure permanently wet	permanently > 20%	Fence posts, gravel boards, poles, sleepers, sole plates below dpc, cooling tower packing	SU	PU B[b] O[b] FU	PU B[b] O[b] FU	PU B[b] O[b] FU	SU PU B[b] O[b] F✗	SL PL B[b] O[b] FL	S✗ P✗ B[b] O[b] F✗
5. In salt water, exposure permanently wet	permanently > 20%	Marine piling, piers and jetties, dock gates, sea defences, ships hulls, etc.	SU	PU B[b] O[b] F[b]	PU B[b] O[b] F[b]	PU B[b] O[b] F[b]	SU PU B[b] O[b] F[b]	SL PL BL O[b] F[b]	SU P✗ B[b] O[b] F[b]

Notes: The risk of attack of cement-bonded particleboard (12.15) by biodegradation agencies is insignificant in all hazard classes. S Solid timber; P Plywood (12.12); B Particleboard (12.13); O Oriented strand board (12.14); F Fibreboard (12.16); U Biodegradation agency Universally present in Europe; L Biodegradation agency Locally present within Europe; ✗ Biodegradation agency not present; [a] Risk of attack can be insignificant according to specific service conditions. [b] No suitable boards of these types are manufactured for use in this hazard class.

It is recognised that different wood and wood based products are likely to attain different equilibrium moisture contents in a given hazard class and specific guidance for each material is given in Parts 2 and 3 of BS EN 335. Advice is also given on the selection of a suitable material for use in the respective hazard class (**Table 12.14** for wood-based panel products) and on the use of preservative treatments. The main advantage of BS EN 335 is that it allows the end-use of the wood product to be assigned to one of the five hazard classes across the whole of Europe. Once placed within a hazard class, appropriate decisions on the preservative treatment required or the selection of an appropriate level of natural durability can be made.

Table 12.14 Wood-based panel products for use in Hazard Classes

Hazard Class	Board type			
	Particleboards	Fibreboards	Oriented strand board	Cement-bonded particleboards
1	P2 to P7	All	All	Suitable
2	P5, P7	HB.H types, HB.E MB.H types, MB.E types SB.H types, SB.E MDF.H types	OSB 3, OSB 4	Suitable
3	No products recommended	HB.E MB.E types SB.E	Only certain products in which the wood species imparts a degree of durability are suitable	Suitable
4	No products suitable	Certain boards appropriate if properties adequate	No products listed as suitable	Suitable
5	No products suitable	No products listed as suitable	No products listed as suitable	Suitable

Notes: BS EN 335: Part 3 notes that there is no risk of attack of modern panel products from insects except termites (**Table 12.13**)

12.25.2 Natural Durability Class

BS EN 350 and BS EN 460 document the principles of testing natural durability of timbers, provide a guide to the natural durability and treatability of selected timbers and relate the natural durability characteristics to the hazard classes. Natural durability ratings for wood destroying fungi, beetles and termites (BS EN 350) are summarised in **Table 12.15.**

Table 12.15 Natural durability ratings for wood destroying fungi, beetles and termites

Fungi (EN 252[68] and 113[69])[a]		Beetles (EN 46[70], EN 49[71], EN 20[72])[a]		Termites (EN 118[73])[a]
Class 1	Very durable	D	Durable	D Durable
Class 2	Durable	S	Susceptible	M Moderately durable
Class 3	Moderately durable	SH	where heartwood is known to be susceptible to the insect under consideration	S Susceptible
Class 4	Slightly durable			
Class 5	Not durable			

Notes: Natural durability ratings for marine borers are excluded (although are essentially the same as for termites); [a] Tested in accordance with Standard noted.

12.25.3 Treatability Class

Treatability classes are included in BS EN 350: Part 2 based on observations of vacuum pressure treatment processes (**Table 12.16**). Treatability cannot be exactly defined, thus some overlap between Classes will occur (particularly between Class 2 and 3).

Table 12.16 Treatability classes

Class	Description[a]	Explanation
1	Easy to treat	Complete penetration of sawn timber by pressure impregnation
2	Moderately easy to treat	Usually complete penetration not possible, but after 2-3 hours pressure treatment > 6 mm lateral penetration can be achieved in softwoods and a large proportion of the vessels of hardwoods will be penetrated
3	Difficult to treat	3-4 hours pressure treatment may not result in > 3-6 mm lateral penetration
4	Extremely difficult to treat	Virtually impervious to treatment. Little preservative absorbed even after 3-4 hours pressure treatment. Both lateral and longitudinal penetration minimal

Notes: [a] Historically, treatability data may refer to other descriptive terms[74] which approximate to the treatability classes (Class 1: Permeable; Class 2: Moderately resistant; Class 3: Resistant; Class 4: Extremely resistant).

Natural durability and treatability for selected timber species is summarised in **Table 12.17**.

Table 12.17 Natural durability and treatability of selected timber

Timber		Natural durability				Treatability	
Common name	Scientific name	Fungi	*Hylotrupes* sp.	*Anobium* sp.	Termites	Heartwood	Sapwood
Softwoods							
Fir	*Abies alba*	4	SH	SH	S	2-3	2
Parana pine	*Araucaria angustifolia*	4-5	D	S	S	2	1
Yellow cedar	*Chamaexyparis nootkatensis*	2-3	S	S	S	3	1
Larch	*Larix decidua*	3-4	S	S	S	4	2
Norway spruce	*Picea abies*	4	SH	SH	S	3-4	3
Sitka spruce	*Picea sitchensis*	4-5	S	SH	S	3	2-3
Scots Pine	*Pinus sylvestris*	3-4	S	S	S	3-4	1
Douglas fir (European)	*Pseudotsuga menziesii*	3-4	S	S	S	4	2-3
Western red cedar	*Thuja plicata*	2	S	S	S	3-4	3
Hardwoods							
Norway maple	*Acer platanoides*	5	-	S	S	1	1
Beech	*Fagus sylvatica*	5	-	S	S	1	1
Ash	*Fraxinus excelsior*	5	-	S	S	2	2
Walnut	*Juglans regia*	3	-	S	S	3	1
Mahogany	*Khaya* sp.	3	-	n/a	S	4	2
Iroko	*Milicia excelsa*	1-2	-	n/a	D	4	1
Oak	*Quercus robur*	3	-	n/a	M	4	1
Obeche	*Triplochiton scleroxylon*	5	-	n/a	S	3	1

Notes: [a] see **Table 12.15**; [b] see **Table 12.16**.

BS EN 460 gives guidance on the minimum natural durability classification acceptable for use in each hazard class, together with recommendations as to whether preservative treatment is required, as summarised in **Table 12.18**.

Table 12.18 Durability classes of wood species for use in hazard classes

Hazard class	Durability class				
	1	2	3	4	5
1	N	N	N	N	N
2	N	N	N	(N)	(N)
3	N	N	(N)	(N) - (P)	(N) - (P)
4	N	(N)	(P)	P	P
5	N	(P)	(P)	P	P

Notes: N Natural durability sufficient; (N) Natural durability normally sufficient, but for certain end uses treatment may be advisable; (N) - (P) Natural durability may be sufficient but (depending on wood species, permeability and end use) preservative treatment may be necessary; (P) Preservative treatment normally advisable, but (for certain end uses) natural durability may be sufficient; P Preservative treatment necessary.

12.25.4 Pretreatment Preservatives and Treatment

BS EN 351: Part 1 specifies wood preservative treatments for solid wood products, including glued laminated timber (Glulam, **12.19**). Treatment is defined in terms of penetration and retention of the wood preservative active ingredients in the treated timber. BS EN 351: Part 2 gives guidance on the general procedures for sampling treated wood to determine penetration and retention.

Penetration of preservative into wood is defined in terms of nine classes ranging from P1 (no penetration requirement) to P9 (penetration requirement full sapwood and 6 mm into exposed heartwood). If the boundary between the heartwood and the sapwood cannot be clearly distinguished then the penetration requirement is for the sapwood. **Retention** requirements are defined as the loading of the wood preservative in the 'analytical zone' (usually the same as the required minimum lateral penetration). The level of retention is defined as the 'critical value' (i.e., the minimum amount of wood preservative product required for effectiveness for that hazard class) derived from tests carried out in accordance with BS EN 559: Part 1 (see below). BS EN 351 also specifies procedures for two production control systems available for measuring penetration and retention requirements

- **direct testing,** where penetration and retention are measured on representative samples;
- **indirect testing** allows a specific treatment process to be used if a safe relationship has been established between penetration and/or retention and measurable features of the treating process. Indirect testing is therefore similar to the long established UK practice of using schedules for process and quality control (e.g, BS 5268: Part 5 and BS 5589).

Performance of Preservatives

BS EN 599: Part 1 covers the performance of preservatives and defines the performance that any wood preservative is required to achieve in any laboratory or field tests in order to be acceptable for a given hazard class. The selection of appropriate tests to demonstrate this takes into account the relevant hazard class; the method of application; the type of wood;

the insect species to protect against; and the relevance of painting or other coating. Detailed tables define which of the various efficacy tests must have been carried out before a wood preservative is deemed acceptable for use in a given Hazard Class. Two types of preservative pretreatment are recognised

- superficial treatments (see below), such as brushing, spraying and dipping, which are only suitable for timber to be used in Hazard Class 1 to 3;
- penetrating treatments (see below), such as pressure impregnation, which may be used for timber to be used in all Hazard Classes.

A maximum application rate for both types of treatment is specified (this will frequently correspond to a manufacturers proposed recommended loading). The outcome of the range of tests is a series of Biological Reference Values which are the amount of preservative (g/m^2 for superficial treatments and kg/m^3 for penetrating treatments) found to be effective in each individual test in preventing attack by the biological agency tested. The Critical Value is the highest biological reference value obtained from all the required tests and represents the minimum quantity of preservative product required for effectiveness in that Hazard Class. The Critical Value is then used in BS EN 351: Part 1 to define the actual amounts of the product needed for efficacy in service. Additional advice is available from relevant trade associations (e.g, the BWPDA[75]).

BS EN 599: Part 2 specifies requirements for classifying and marking of wood preservative products according to their performance and suitability for use. Marking includes product name, appropriate Hazard Class(es), application processes, need for coating, type of wood, additional biological efficacy, critical value and recommended application rate.

Types of Preservative

BS 1282[76] classifies timber preservatives into three main types

- **Type TO.** Tar oil type;
- **Type OS.** Organic solvent type;
- **Type WB.** Water-borne type.

Tar oil type (TO) consist of a complex mixture of distillate oils from coal tar; a common example is **creosote.** Wood preservation using coal tar creosotes is currently governed by BS 144[77]. Creosote is suitable for most outdoor applications, being both fairly resistant to leaching by rain, etc. and having a degree of water repellency. Creosote is not corrosive to metals and not readily flammable but should be avoided for internal use, due primarily to its characteristic odour, which can be picked up by foodstuffs in the vicinity of (but not in actual contact with) the treated wood. Creosote will bleed through surface coatings and these should be avoided. Porous materials will be stained by creosote. Penetration is limited when applied by brush; it is normally applied under pressure in, for example, the pretreatment of railway sleepers and telegraph poles. Coal tar creosotes are phytotoxic (toxic to plant life) whereas modern tar oils from the petroleum industry are less so.

Organic solvent type (OS) timber preservatives consist of solutions of one (or more) fungicides or insecticides in an organic solvent carrier (e.g, white spirit). Other additives, such as water repellents (waxes or resins) and pigments, may be included. Organic solvent type wood preservatives are currently governed by BS 5707[78]. The solvent will evaporate within a few days, leaving the pesticide active ingredient, which is responsible for killing the timber-degrading organism and can be any one of a number of different chemicals. Currently, concern is being expressed over the high mammalian toxicity and environmental persistence of older fungicides (e.g, pentachlorophenol and tributyltin oxide) and insecticides (e.g, dieldrin and lindane), and these are being withdrawn in favour of more environmentally acceptable chemicals with equivalent pesticide action (i.e., fungicides based on organoborons, metallic naphthenates[79] and soaps, and insecticides such as permethrin). Organic solvent type preservatives are suitable for internal and external use and are resistant to leaching by water (although evaporative losses may occur). Organic solvents types are generally not corrosive to metals but they should not be allowed to come into contact with PVC cables as they cause embrittlement by leaching of the plasticiser. Once the solvent has evaporated, timber can usually be painted and glued. The solvents are readily flammable and the timber will remain an increased fire risk until the solvent has evaporated. Organic solvent types are non-swelling to wood.

Water-borne type (WB) preservatives consist of a mixture of inorganic salts together with a 'fixing' agent responsible for chemically binding the active ingredients to the cell wall. The most common water-borne preservative is a mixture of copper sulphate (the fungicide), sodium dichromate (the 'fixing' agent) and arsenic pentoxide (the insecticide), commonly known as CCA (currently governed by BS 4072). The fixation of the wood preservative renders it resistant to leaching by water and also 'safe' for human exposure. The addition of water to the timber may cause swelling and distortion, and may necessitate re-drying. Contact with rubber and aluminium should be avoided. CCA is non-flammable and will not stain adjacent materials. **Celcure** is the Rentokil proprietary name for its CCA formulations.

Application Methods

Surfaces should be free from extraneous materials e.g, water, bark, mud and surface finishes. All pretreatments (except diffusion) require the timber to be seasoned to a moisture content of below 28%. For diffusion treatments, the moisture content must be greater than 50%. All cutting, drilling, etc. of the timber should be carried out before pretreatment since the penetration of the timber by the preservative may not be sufficient to totally protect the entire cross-section.

Superficial treatments

Deluging involves feeding timber through a tunnel on a conveyer system as an organic solvent type preservative is sprayed and flooded over it. As the wood is in contact with the preservative for only a few seconds, the treatment effectively equates to a short dip (10 seconds) or a 'flood' brush application. **Immersion** involves submerging the timber in a bath of preservative liquid for a period which can range from a few seconds to several minutes (usually called 'dipping') to several hours or days ('soaking' or 'steeping'). Organic solvent preservatives are widely used and creosote is used occasionally. The degree of penetration obtained depends on the preservative fluid used, the species of timber and the period of immersion. For example, the degree of penetration obtained in Pine sapwood in 3 minutes is about double that after 10 seconds; double again requires a period of longer than 30 minutes. **Hot and cold open tank processes** involve submerging the timber in cold preservative (usually tar oil types), which is then heated to 85 to 95°C and held at this temperature for one to three hours. This results in much of the air in the wood being expelled and also lowers the viscosity of the tar oils, increasing penetration. The timber remains in the preservative during cooling, when the residual air in the wood contracts and the fluid is drawn into the wood. Most of the absorption takes place during the cooling period, which may last many hours before ambient temperature is reached.

For **unseasoned timber, diffusion** processes can also be used. This method consists of applying a concentrated solution of a water-borne preservative (a highly soluble boron salt is widely employed) to the surface of the wood and then stacking the timber for several weeks under cover to restrict drying. This allows the salt to diffuse into the wood and it is possible to obtain a through-and-through penetration of the preservative even in timbers which are very difficult to impregnate using any other method.

Penetrating treatments

In the **double vacuum process**, timber is placed in a treating cylinder and a partial vacuum is created and held for a few minutes before flooding with preservative. The vacuum is then released and the timber allowed to remain in the preservative fluid for up to one hour, either under atmospheric pressure or a small excess pressure. After the preservative is emptied from the cylinder, a final vacuum is applied to the timber to recover some of the preservative solution from the wood and to leave a dry surface. Organic solvent type preservatives are used. The treatment gives approximately two to three times the penetration obtained from a 3 minute dip. **Pressure impregnation** is either by

- the **full cell process**, where an initial vacuum is applied to the timber in the cylinder before filling with preservative, and then pressure is applied. After about one to six hours, the pressure is released, the cylinder emptied and a final vacuum applied to dry the surface of the wood. Creosote (at 65 to 95°C) or water-borne CCA are commonly used; organic solvent types are not applied by this process; or
- the **empty cell process**, where no initial vacuum is employed. Instead, air at atmospheric pressure (or a small excess pressure) is present in the cylinder before and during the introduction of the preservative. After the pressure period, the cylinder is emptied and a final vacuum is applied which can recover a large proportion of the absorbed preservative, leaving the wood cell cavities only partially filled with preservative. The actual depth of penetration is, however, similar to that achieved by the full cell process. Creosote is commonly applied by this method.

Protim (protimised) and Tanalith (tanalised) are trade names, the former referring to a pressure impregnation formulation (**protimised** has become almost a generic term for pressure-treated timber) and the latter referring to pressure pretreatment chemicals and the treatment process of **tanalising**, which includes various products/services, but is best known for CCA formulations.

12.26 REMEDIAL (*IN SITU)* TREATMENT

Once biological degradation of the timbers of an existing building has been recognised, it is important to establish the cause of the damage by accurately identifying the fungus (**12.21**) or wood boring insect (**12.22.1**) responsible to avoid unnecessary and costly remedial treatment. Once the need has been established, various types of treatment are available.

12.26.1 Remedial Treatments (Insects and/or Fungi)

Liquid treatments are most commonly used remedially. The treatment strategy is

- to kill any insects or fungi already within the timber by
  - penetrating the timber and killing insect larvae or fungal hyphae within the zone penetrated (initial kill);
  - leaving a persistent layer of pesticide active ingredient in the zone penetrated which kills either successive generations of emerging adults developed from surviving larvae or hyphae emerging from the untreated zone;
- to prevent further infestation and so protect timbers against future attack by leaving a persistent layer of pesticide active ingredient in the zone penetrated to kill either any eggs or hatching larvae or any spores or hyphae which encroach from elsewhere.

The effectiveness of liquid treatments in achieving these objectives depends on the formulation, the size and permeability of the timber and the thoroughness of the application.

Organic solvent type liquid preservatives are commonly used for remedial treatment and are similar to those described for pretreatment (above). When used remedially, organic solvent type preservatives are most commonly applied by brush or spray, at recommended rates of 200 to 300 ml/m^2 (in practice, this usually means until 'run-off'). With sapwood of permeable timbers penetration of up to 10 mm can be expected. Larvae of wood boring insects are killed immediately within the treated zone by the combined effects of the solvent and the insecticide. In less permeable timbers, penetration of the treatment fluid may be improved by injection under pressure through one-way valve inserts or slow 'irrigation' into drilled holes.

Emulsion based liquids consist of an emulsified mixture of water and organic solvent, where the pesticide active ingredient is carried in the organic solvent phase. The ratio of oil:water varies considerably between products but is typically about 1:10. Formulations based entirely on water as a carrier cannot be produced because the pesticide active ingredients currently used in remedial treatments are insoluble in water (unlike water-borne type pretreatment preservatives, above). Emulsion-based liquids are applied by spray or brush (not injection, as this would cause swelling of the timber) and achieve shallow penetrations compared with organic solvent treatments, which may mean that attack can continue in the large, untreated inner zone.

Paste formulations (bodied emulsions), also called **'mayonnaise pastes'**, consist of the same pesticide active ingredients which are incorporated into liquid formulations but are carried in a gelatinous emulsion paste with a high oil-solvent content. Application rates are between 600 to 800 ml/m^2. They adhere to the wood surface and are thus able to deliver larger quantities of pesticide active ingredient to the timber than pure liquid formulations, which run off. Pastes are applied by caulking gun or pallet knife, which is very time consuming, and tend to leave a waxy deposit on the wood surface which can stain adjacent plaster.

Painting provides a decorative coating and also a protective barrier against the external environment. Paint will therefore protect the timber from fungal attack, both by the exclusion of water and by the physical barrier between the fungal spore and its food source, and from insect attack since the timber surface is protected from the egg-laying activities of the adult. The effectiveness of the paint system in preventing timber decay is dependent upon maintaining the physical barrier between the timber and the external environment; regular maintenance is therefore required (minimum 2 year cycle).

Temperatures of >30°C for prolonged periods and between 52 to 55°C for 30 to 60 minutes is lethal to dry rot (*Serpula lacrymans*) and the wood boring insects found in the UK. In the UK, **heat sterilisation** is used commercially only in the kiln sterilisation of sawn hardwoods which have become infested with Lyctus powder post beetle during storage. In Europe, a hot-air process is used to sterilise roof space insect infestations. In the UK, blow lamps have been used to sterilise masonry walls infected with *Serpula lacrymans*; however, calculations have shown that even when the surface temperature of the wall is raised to 1100°C it would take one hour to sterilise the wall to a depth of 150 mm, rising to five hours for a depth of 230 mm. These calculations, coupled with the increased fire risk, imply that heat sterilisation against *Serpula lacrymans* is probably impracticable for large areas of construction timber.

12.26.2 Remedial Treatments (Insects)

Smoke treatments consist of pyrotechnic canisters which are used to produce clouds of insecticidal particles in the form of a smoke. When released, the smoke leaves a fine deposit of insecticide on all internal surfaces. These particles kill any insect that comes into contact with the surface, provided the application rate is adequate. The smoke only kills beetles, eggs and hatching larvae (it cannot affect larvae in the wood). The method is useful on impermeable timbers (where inadequate penetration of preservative liquids would occur) and in situations where there is a large quantity of inaccessible timber. As the surface deposits are only effective for a few weeks, repeated annual treatments are required, timed to coincide with the emergence of the adults. Larval activity will continue within the wood until eradication by annual treatments is complete, during which time considerable damage may be caused. In addition, the smoke deposit may affect non-timber surfaces and building occupants and thus extensive cleaning operations are required after each treatment.

Volatile vapours of certain insecticides are effective against many insects, including wood boring species. **Dichlorvos** is one such insecticide, which has been marketed as 'slow-release' strips (principally for domestic fly control), and is effective against wood boring insects at aerial concentrations recommended for domestic fly control. Treatment has the same drawbacks as for smoke treatments (above), but its low cost made it attractive in certain situations (e.g, against small infestations in roof voids which are semi-enclosed with sarking felt and where ventilation rates are low). Recently, concerns that dichlorvos may be a genotoxic carcinogen (causing cancer by damaging DNA) have led to the suspension of sales of dichlorvos products in the UK.

Gas fumigation, usually using methyl bromide gas, is effective if the infested timbers can be temporarily enclosed in an atmosphere of the gas. The insects are killed by direct toxic action of the gas which diffuses into the damaged timber. Again, however, the method provides no residual protection against re-infestation. Whole building gas fumigations against termite infestations are carried out in some countries (USA, Australia, etc.), but in the UK the extensive sealing required and the toxic hazard make gas fumigation impracticable except for specialised uses (e.g, treatment of museum pieces in gas-tight chambers); even in these cases, stringent codes of practice must be adhered to.

12.26.3 Remedial Treatments (Fungi)

Borate rods, pastes or tablets have been used for fungal attack and rely on diffusion to carry the active ingredient throughout the timber component. In this respect, treatment is equivalent to the diffusion application process for the pretreatment of timber (**12.25.4**). The use of borates relies upon the availability of moisture in the wood, and therefore this method can only be used where the timber to be treated will maintain a moisture content of >50%. The borate reacts with the moisture in the wood to form boric acid, having pronounced fungicidal activity, which diffuses throughout the moist timber. Care must be taken to ensure that the borates not leached out of the wood by transient water. Borate rods are widely used in the protection of window joinery. The boron diffusion process is called '**Timborising**'.

12.27 ANSWERS TO QUESTIONS. Section **12.3**; ***Figure 12.4***.

Section 1 Transverse section of a Softwood. **E** medullary ray, **A** tracheids.

Section 2 Radial section of a Softwood (because there are no vessels). **D** medullary ray, **A** tracheids, **C** spring wood, **B** summer wood

Section 3 Tangential section of a Softwood (because there are no vessels). **D** medullary ray, **A** tracheids.

Section 4 Transverse section of a Ring porous hardwood. **C** medullary ray, **B** tracheids,

D pores or vessels.
Section 5 Radial section of a Ring porous hardwood (because there are only vessels in the spring wood). **D** medullary ray, **E** vessels or pores, Item **C** spring wood, **B** summer wood.
Section 6 (and 9) Tangential section of a hardwood (to differentiate between diffuse and ring porous hardwoods, you would have to look for the presence of vessels in both the springwood and summerwood). **D** medullary ray, **E** vessels or pores.
Section 7 Transverse section of a Diffuse porous hardwood. **C** medullary ray, **D** pores or vessels.
Section 8 Radial section of a Diffuse porous hardwood (because there are vessels in both the spring wood and summer wood). **D** medullary ray, **E** vessels or pores, **C** spring wood, **B** summer wood.

12.28 REFERENCES

1. BS 1186. Timber for and workmanship in joinery. Part 1: 1991. Specification for timber.
2. prEN 942: 1992. Timber joinery. Classification of timber quality.
3. MAUN, K.W. and CODAY, A.E. (1997) *Timber drying manual. BRE Report 321.* Garston, BRE.
4. BS EN 518: 1995. Structural timber. Grading. Requirements for visual strength grading standards.
5. BS 4978: 1996. Specification for visual strength grading of softwood.
6. BS 5756: 1997. Specification for visual strength grading of hardwood.
7. BS EN 519: 1995. Structural timber. Grading. Requirements for machine strength graded timber and grading machines.
8. BS EN 338: 1995. Structural timber. Strength classes.
9. BS 5268. Structural use of timber. Part 2: 1996 Code of practice for permissible stress design, materials and workmanship.
10. BS ENV 1995. Design of timber structures. Eurocode 5. Part 1-1. General rules, and rules for buildings (together with United Kingdom National Application Document (NAD).
11. TRADA (1995) *TRADA Wood Information Sheet No. 1-37. Eurocode 5. A introduction.* High Wycombe, TRADA.
12. BRE (1997) *BRE Digest 423. The structural use of wood-based panels. A commentary on the changes ahead with European standardisation.* Garston, BRE.
13. BS EN 1194: 1995. Timber structures. Glued laminated timber. Strength classes and the determination of characteristic values.
14. prEN 12369: 1999. Wood-based panels. Characteristic value for structural design.
15. TRADA (1997) *TRADA Wood Information Sheet No. 4-21. European strength classes and strength grading.* High Wycombe, TRADA.
16. BS EN 310: 1993. Wood-based panels. Determination of modulus of elasticity in bending and of bending strength.
17. BS EN 319: 1993. Particleboards and fibreboards. Determination of tensile strength perpendicular to the plane of the board.
18. BS EN 322: 1993. Wood-based panels. Determination of moisture content.
19. BS EN 317: 1993. Particleboards and fibreboards. Determination of swelling in thickness after immersion in water.
20. BS EN 318: 2002. Fibreboards. Determination of dimensional changes associated with changes in relative humidity.
21. BS EN 323: 1993. Wood-based panel products. Determination of density.
22. BS EN 311. Particleboards. Surface soundness. Test methods.
23. ISO 3340: 1995. Fibre building boards. Determination of sand content.
24. BS EN 1087. Particleboards and fibreboards. Moisture resistance. Part 1: 1995. Boil test.

25. BS EN 321: 1993. Fibreboards. Cyclic tests in humid conditions.
26. BS EN 120: 1992. Wood-based panels. Determination of formaldehyde content. Extraction method called the perforator method.
27. BS EN 313. Plywood. Classification and terminology. Part 2: 1995. Terminology.
28. BS 3444: 1972. Specification for blockboard and laminboard.
29. BS EN 313. Plywood: Classification and terminology. Part 1: 1996. Classification.
30. BS EN 315: 1993. Plywood: Tolerances for dimensions.
31. BS EN 314. Plywood. Bonding quality. Part 1: 1993. Test methods. Part 2: 1993. Requirements.
32. DD ENV 1099: 1999. Plywood. Biological durability. Guidance for the assessment of plywood for use in different hazard classes.
33. BS EN 1084: 1995. Plywood. Formaldehyde release classes related to the gas analysis method.
34. BRE (1988) *BRE Information Paper IP7/88. The design and manufacture of ply-web beams.* Garston, BRE.
35. BS EN 309: 1992. Wood particleboards. Definition and classification.
36. BS EN 312. Particleboards. Specifications. Part 1: 1997. General requirements for all board types. Part 2: 1997. Requirements for general purpose boards for use in dry conditions. Part 3: 1997. Requirements for boards for interior fitments (including furniture) for use in dry conditions. Part 4: 1997. Requirements for loadbearing boards for dry conditions. Part 5: 1997. Requirements for load-bearing boards for use in humid conditions. Part 6: 1997. Requirements for heavy duty load-bearing boards for use in dry conditions. Part 7: 1997. Requirements for heavy duty load-bearing boards for use in humid conditions.
37. BS EN 300: 1997. Oriented strand boards (OSB). Definitions, classification and specifications.
38. BS EN 633: 1994. Cement-bonded particleboards. Definition and classification.
39. BS EN 634. Cement-bonded particleboards. Part 1: 1994. General requirements. Part 2: 1995. Requirements for OPC-bonded particleboards for use in dry, humid and exterior conditions.
40. Canadian Standard CAN3-0437.0-M85 (1985) Wafer board and strand board.
41. Canadian Standard CAN3-0437.1-M85 (1985) Test methods for wafer board and strand board.
42. BRE (1981) *BRE Information Paper IP9/81. The termite resistance of board materials.* Garston, BRE.
43. BS 4046 (1991) Compressed straw building slabs.
44. BS 1105 (1994). Specification for wood wool cement slabs up to 125 mm thick.
45. prEN 13168 (1998) Thermal insulation products for buildings: Factory made wood wool products: Specification.
46. TRADA (1995) *Glued laminated timber: European standards. Wood Information Sheet WIS 1 Sheet 38.* High Wycombe, TRADA.
47. BRE (1991) *Design of timber floors to prevent decay. BRE Digest 364.* Garston, BRE.
48. JENNING, D.H. and BRAVERY, A.F. (eds.) (1991) *Serpula lacrymans. Fundamental biology and control strategies.* Chichester, Wiley and Sons.
49. BERRY, R.W. and TAKENS-MILNE, E. (1993) Remedial treatment of wood rot and insect attack in buildings. Garston, BRE.
50. BRAVERY, A.F.; BERRY, R.W.; CAREY, J.K and COOPER, D.E. (1992) *Recognising wood rot and insect damage in buildings. BRE Report 232.* Garston, BRE.
51. BRE (1989) *Dry rot: its recognition and control. BRE Digest 299.* Garston, BRE.
52. BRE (1988) *House inspection for dampness: a first step to remedial treatment for wood rot. BRE Information Paper IP19/88.* Garston, BRE.
53. BRE (1989) *Wet rots: Recognition and control. BRE Digest 345.* Garston, BRE.
54. BRE (1985) *Preventing decay in external joinery. BRE Digest 304.* Garston, BRE.
55. BRAVERY, A.F. (1985) *Mould and its control. BRE Information Paper IP11/85.* Garston, BRE.
56. CAREY, J.K. (1991) *Blue staining of timber in service: its cause, prevention and treatment. BRE Information Paper IP9/91.* Garston, BRE.
57. BRE (1992) *Control of lichens, moulds and similar growths. BRE Digest 370.* Garston, BRE.
58. BRE (1992) *Identifying damage by wood-boring beetles. BRE Digest 307.* Garston, BRE.
59. BRE (1994) *House longhorn beetles: geographical distribution and pest status in the UK. Information Paper IP8/94.* Garston, BRE.

60. Building Regulations Approved Document to support Regulation 7: Materials and Workmanship (1992 ed.), paragraph 1.9. London, HMSO, 1991.
61. BRE (1985) *Corrosion of metals by wood. BRE Digest 301.* Garston, BRE.
62. BS 4072: 1999. Copper/chrome/arsenic preparations for wood preservation.
63. BS EN 335. Hazard classes of wood and wood based products against biological attack. Part 1: 1992. Classification of hazard classes. Part 2: 1992. Guide to the application of hazard classes to solid wood. Part 3: 1996. Application to wood based panels.
64. BS EN 460. Durability of wood and wood based panel products. Natural durability of solid wood. Guide to durability requirements for wood to be used in hazard classes.
65. BS EN 350. Durability of wood and wood based products: Natural durability of solid wood. Part 1: 1994. Guide to the principles of testing and classification of the natural durability of wood. Part 2: 1994. Guide to the natural durability and treatability of selected wood species of importance in Europe.
66. BS EN 351. Durability of wood and wood based products. Preservative treated solid wood. Part 1: 1996. Classification of preservative penetration and retention. Part 2: 1996. Guidance on sampling for the analysis of preservative treated wood.
67. BS EN 599. Durability of wood and wood based products. Performance of preventive wood preservatives by biological test. Part 1: 1997. Specification according to hazard class. Part 2: 1997. Classification and labelling.
68. EN 252: 1989. Field test method for determining the relative protective effectiveness of a wood preservative in ground contact (= BS 7282: 1990).
69. EN 113: 1980. Wood preservatives. Determination of the toxic values against wood destroying *Basidiomycetes* cultures on an agar medium (= BS 6009: 1982).
70. EN 46: 1988. Wood preservatives. Determination of preventive action against recently hatched larvae of *Hylotrupes bajulus (Linnaeus)* (laboratory method) (= BS 5434: 1989).
71. EN 49. Wood preservatives. Determination of the protective effectiveness against *Anobium punctatum (De Geer)* by egg laying and larval survival. Part 1: 1992. Application by surface treatment (laboratory method).
72. EN 20. Wood preservatives. Determination of the protective effectiveness against *Lyctus brunneus (Stephens)*. Part 1: 1992. Application by surface treatment (laboratory method).
73. EN 118: 1990. Wood preservatives. Determination of preventive action against *Reticulitermes santonensis de Feytaud* (laboratory method) (= BS 6240: 1990).
74. BRE (1985) *Timbers: their natural durability and resistance to preservative treatment. BRE Digest 296.* Garston, BRE.
75. BWPDA (1999) *BWPDA Manual.* London, British Wood Preserving and Damp-proofing Association.
76. BS 1282: 1975. Guide to the choice, use and application of wood preservatives.
77. BS 144. Wood preservation using coal tar creosotes. Part 1: 1990. Specification for preservation. Part 2: 1990. Methods for timber treatment.
78. BS 5707. Solution of wood preservatives in organic solvents. Part 1: 1979. Specifications for general purpose applications, including timber that is to be painted. Part 2: 1979. Specification for pentachlorophenol wood preservative solution for timber that is not required to be painted. Part 3. 1980. Methods of treatment.
79. BS 3769: 1993. Specification for copper naphthenate and copper naphthenate concentrates.

13

PLASTICS AND ADHESIVES

13.1 INTRODUCTION

The rubber industry was well established before the plastics industry and was based on naturally occurring rubber (initially derived from the *Hevea brasiliensis* tree cultivated in large plantations around the Malay peninsula area). Developments in rubber technology have concentrated on producing materials which have similar properties to natural rubber, notably high elasticity. Initially, natural rubbers were not perceived to be **polymeric** materials. Today, natural rubbers are used alongside synthetic rubbers and both are often referred to as **elastomers** (**13.10**). Although the raw products from which plastics are made are naturally occurring, techniques are required to extract these in a useable form. The plastics industry has developed alongside the growing petrochemical industry and considerable research and development work was required to exploit these extracted compounds. Plastics, elastomers, paint coatings, sealants and adhesives all use the similar "basic" chemical units and associated chemical technology.

Plastics are a group of materials which have a wide range of mechanical, functional and processing properties. These properties stem from the ability of the carbon atom to bond in various ways and with various other elements and compounds, and the directionality of the carbon to carbon bond (**1.11.2**). Plastics generally have low density, have high strength:weight ratios, are relatively inert (they do not rot, degrade, absorb water) and have good resistance to low temperatures, pollutants, dilute acids and alkalis. Plastics are also good thermal insulators, particularly when foamed, and have high thermal capacity. The principal advantage of plastics over more traditional materials (timber, metals, etc.) is their ability to be processed, shaped and moulded to form complex products with reproduceable properties. There are several good examples e.g, intricate moulding of camera bodies, nylon gear wheels in food mixers, plastic water pipes, guttering, water traps, etc. However, plastics are combustible (some are highly flammable, whilst others are more difficult to burn, **14.6.6**), have high coefficients of thermal expansion (**2.8.2**), have lower rigidity and stiffness than virtually all other building materials (**3.2.3**) and are liable to creep (**3.5.4**). Most of these adverse properties can be modified in various ways (e.g, by altering the base structure of the polymer, the additional of side groups to the polymer chain, the addition of additives, etc.), as considered in more detail in section **13.9**.

As scientists continue to understand more about the mechanisms by which these properties are achieved, more use will be made of plastics in the field of engineering and construction, especially for easily moulded items. For example, plastics can be made rigid (grease free bearings) or springy (and hence development is being successfully undertaken to replace heavy steel springs by lightweight plastic units); plastic springs (e.g, LITEFLEX™ General Motors) have been used successfully in the automotive industry.

13.2 FUNDAMENTAL PROPERTIES

Fundamentally, a plastic

- is **organic**, i.e., containing carbon, although some plastics may be partially inorganic in nature;
- has **high molecular weight** (commonly over 25,000);
- is **plastic**, i.e., able to change shape during manufacture.

Plastics are covalently bonded materials, which are bad conductors of electricity (under low potential differences) because the bond electrons are 'fixed' between each atom and cannot move away. Covalently bonded materials are therefore **insulators**. Note that, under high potentials, this electrical insulation can be broken down as the electrons are stripped off the bond. The covalent bond formed between adjacent carbon atoms will allow rotation (**1.11.2**), which gives rise to the large variation in mechanical properties exhibited by plastics.

13.3 BONDING

Organic chemistry (**1.2.2**) is characterised by the **covalent bond (1.11)** and **van der Waals bond (1.14)**. **Ionic bonds (1.12)** may occur in organic chemistry as weak acids; acetic acid (vinegar), CH_3COOH (**1.19**) is an example. Acetic acid ionises as follows

$$\underset{\text{Covalent}}{CH_3COOH} \rightarrow \underset{\text{ionic}}{CH_3COO^-} + \underset{\text{acid part}}{H^+} \qquad \textbf{(Table 1.15)}$$

The slight acidity (due to the 'excess' hydrogen ions) gives vinegar a sharp taste.

The main principle behind the covalent bond is that the atom carbon (C) does not need to get too big (e.g, by forming C^{4-} ions) or too small (e.g, by forming C^{4+} ions). This is achieved by covalent bonding by sharing electrons (**1.11**) (as opposed to donating electrons as in the ionic bond, **1.12**). Each atom in the plastic molecule will want to get the inert gas electron configuration for chemical stability. Thus, in the simplest case, four atoms of the element hydrogen combine with one atom of carbon by a mutual sharing of electrons to produce the simplest molecule, methane CH_4 (***Figure 13.1***).

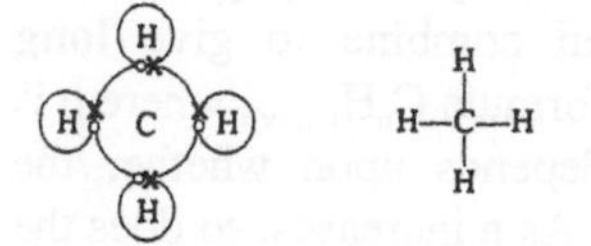

The sharing of the electrons produces a methane **molecule**, which has two electrons in each of the hydrogen atom's outer orbit and 8 electrons in the carbons outer orbit. This, therefore, produces a **very directional bond** across the shared electrons which are **'fixed'** at the point of sharing. There are no further electrons available to bond one CH_4 molecule to another CH_4 molecule and therefore methane CH_4 exists in the gaseous phase. The sharing of two electrons is normally referred to as a **single bond** and is represented by a single line. Being covalent, the bond is fairly difficult to break.

Figure 13.1 *Structure of the methane molecule,* CH_4

13.4 FORMATION

13.4.1 Monomers

The fundamental building blocks of plastics are called **monomers**. Monomers are derived in the first instance from crude oil, coal and natural gas which occur naturally beneath the earth's crust. Crude oil is a dark viscous liquid formed as a result of the bacterial decomposition of early (in geological time) marine life. Crude oil is a mixture of **hydrocarbons** which are a product of hydrogen (Group 1 of the Periodic Table) and carbon (Group 4 of the Periodic Table) (**Table 1.2**). These hydrocarbons comprise molecules of differing chain lengths and boiling points. The first requirement of the manufacturing process is therefore to separate crude oil into its constituent parts (by chain length and boiling point) by a process called **fractional distillation (Table 13.1)**

Table 13.1 Fractional distillation of crude oil showing the relationship between molecular chain length and boiling point

Fraction		Carbon atoms /molecule	Boiling point	Calorific value[1]		Name and use
				MJ/kg	mJ/m^3	
Gases and volatile liquids		1 to 5	< 40°C			Fuel gases solvents
General formula	$C_nH_{(2n+2)}$					
e.g,	CH_4	1	- 161°C	37.78		Natural gas (methane)
	C_2H_6	2	- 90°C			Ethane
	C_3H_8	3	- 42°C	50	66.07	Propane gas (bought in cylinders)
	C_4H_{10}	4	0°C	41.3	121.8	Butane (cigarette lighter fuels)
	$C_{10}H_{22}$	10	+ 174°C			Liquid solvent
Gasoline		5 to 10	40 to 190°C			Petrol, naphtha, etc.
Kerosine		11 to 14	190 to 260°C			Paraffin oil, jet fuel
Light fuel oils		15 to 19	260 to 330°C			Diesel oil
Light lubricating oils		19 to 30	330 to 400°C			Medicinal paraffin oil to wax, solids
Bitumen		> 30	> 400°C			Heavy fuel oils, tars for road asphalt
Wood[2]	$C_6H_{10}O_5)_n$			Pine 21.3 Oak 19.3		

Notes: [1] see **14.10.1**; [2] Wood is not a product of distillation, but is included here for comparison.

Note that 30 carbon atoms represents a relatively short chain length compared to that of a plastic, which may contain over 1,000 CH_2 units (**13.9.1**).

Table 13.1 indicates that the boiling point and melting points of the hydrocarbons obtained are dependent upon the chain length of the hydrocarbon compound. The boiling points and melting points of a substance occur at temperatures at which the thermal energy is capable of breaking the weakest chemical bonds that form the substance. Therefore, the thermally dependent mechanical properties of plastics increase with increases in chain length.

The hydrocarbons compounds produced by fractional distillation are subdivided into **aliphatic** (chain-like materials) and **aromatic** (ring or cyclic) compounds (***Figure 13.2***).

- In **aliphatic compounds**, the carbon and hydrogen combine to give **long hydrocarbon chains** (***Figure 13.2a***) with the general formula $C_nH_{(2n+x)}$, where n is the number of carbon atoms in the chain and x depends upon whether the hydrocarbon is saturated (x = 2) or unsaturated (x = 1). As n increases, so does the melting and boiling point of the hydrocarbon (**Table 13.1**);
- In **aromatic compounds** the carbon and hydrogen are arranged in **rings**, with each

ring containing six carbon and hydrogen atoms (***Figure 13.2c***); ordinarily, these rings are very stable. Several rings can be joined together (***Figure 13.1d***) and are then usually referred to as **cyclic compounds** (where only one ring is involved) or **dicyclic compounds** (if two rings are joined together). The general formula for one ring compounds is C_nH_n and for dicyclic ring compounds $C_nH_{(n-2)}$, where n is the number of carbon atoms in the ring. The mechanical properties of the hydrocarbon improve as the number of rings increases.

The mechanical properties of the hydrocarbon improve as the number of rings increases. For the same molecular weight, aromatic hydrocarbons are usually solids (or not very volatile liquids), whereas aliphatic hydrocarbons are liquids that are more volatile (and the chains are more easily broken). Note that, in writing complex chemical compounds which may partly comprise aliphatic or aromatic compounds, these compounds alone are often abbreviated R, R' etc.

In ***Figure 13.2***, the structure of representative aliphatic and aromatic hydrocarbons is shown (the simplest **aliphatic** hydrocarbon, methane CH_4 is shown in ***Figure 13.1***).

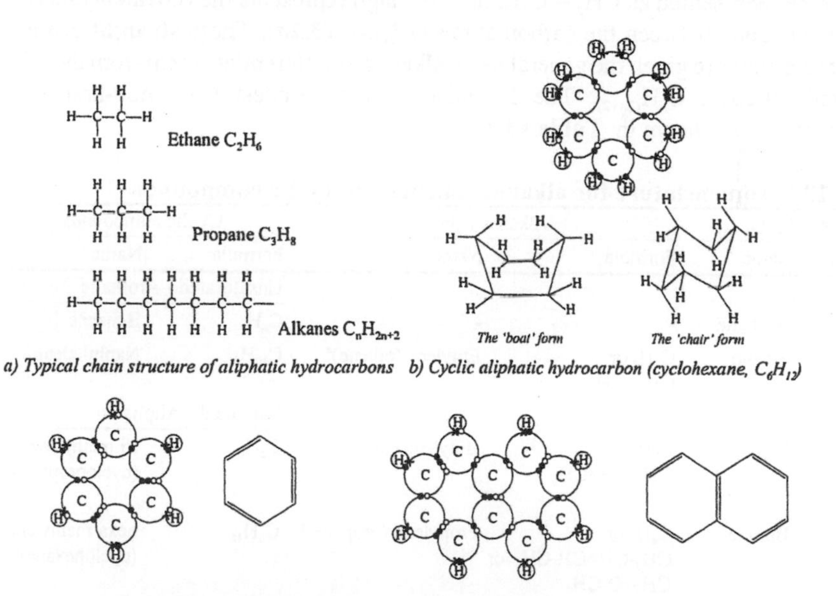

Figure 13.2 Structure of some representative aliphatic and aromatic hydrocarbons

Figure 13.2a illustrates the structure of some aliphatic alkane molecules showing the typical chain-like structure. Aliphatic hydrocarbons can also form **cyclic** compounds; ***Figure 13.2b*** shows the **fully saturated** (no double bond) cyclic **aliphatic** compound, cyclohexane (C_6H_{12}) found in oils. Thus the hydrocarbons are all categorised according to their general formula and the presence of double bonds in the cyclic ring rather than their structural appearance. In both aliphatic and aromatic rings the spatial orientation of the carbon atom (***Figure 1.11***) has to be fully satisfied and therefore the rings are 'puckered' (giving the two forms for cyclohexane, the 'boat' and the 'chair' form). ***Figure 13.2c*** shows benzene (C_6H_6), the most usual form of aromatic (cyclic) compound. Also shown is the scientists' shorthand representation of benzene. ***Figure 13.2d*** shows an **aromatic** compound based on (here) two benzene rings (naphthalene, $C_{10}H_8$), together with the appropriate scientists' shorthand representation.

Long chain **aliphatic** hydrocarbons are usually referred to as **alkanes** when they are **fully saturated** (i.e., when the hydrocarbons have the formula $C_nH_{(2n+x)}$ and x = 2). When x is not equal to two, then

- **either** the hydrocarbon is aromatic and is composed of cyclic compounds ***(Figures 13.2c*** and ***13.2d***);
- **or** double bonds are present in the molecule (***Figure 13.2a***).

The simplest hydrocarbon with a **double** bond in the molecule is ethene C_2H_4, which can also be represented as $CH_2 = CH_2$, the " = " sign represents the (covalent) sharing of **four** electrons between the carbon atoms (***Figure 13.2a***). These **straight** chained C_nH_{2n} molecules are given the general name alk**enes** to differentiate them from the fully saturated alk**anes**, $C_nH_{(2n+2)}$. The formulae of the simplest (i.e., non-branched) hydrocarbons are shown in **Table 13.2**.

Table 13.2 Nomenclature for alkanes, alkenes and cyclic compounds

Alkanes $C_nH_{(2n+2)}$		Alkenes C_nH_{2n}		Cyclic compounds	
Formula	Name	Formula	Name	Formula	Name
				Unsaturated - Aromatic	
CH_4	Methane	-	-	C_6H_6	Benzene
C_2H_6	Ethane	C_2H_4 or $CH_2{=}CH_2$	Ethylene (ethene)[a]	$C_{10}H_8$	Naphthalene
				Saturated - Aliphatic	
C_3H_8	Propane	C_3H_6 or $CH_3\text{-}CH{=}CH_2$	Butylene (butene)[a]	C_3H_6	tri methylene (cyclopropane)[a]
C_4H_{10}	Butane	C_4H_8 or $CH_3\text{-}CH{=}CH\text{-}CH_3$ or $CH_2{=}C(CH_3)\text{-}CH_3$	Propylene (propene)[a]	C_6H_{12}	hexa methylene (cyclohexane)[a]

Notes: [a] Common names have been used in **Table 13.2** (standard IUPAC nomenclature in parentheses).

The fractional distillation process only separates the hydrocarbons into the components of the crude oil according to their boiling points. Often there are hydrocarbons in the crude fractions which are not as commercially saleable as some other hydrocarbons of shorter chain length. In order therefore, to reduce the chain length of the hydrocarbon, a process called **cracking** is used to produce the more saleable hydrocarbon(s). Catalytic cracking therefore reduces the length of the carbon chain.

13.4.2 Polymers

Polymers (plastics) are formed by reacting a number of monomers together, a process known as **polymerisation** (or chain formation) (**13.5**). As the monomer chain grows, the number of links in the chain (the **degree of polymerisation**) increases. When the polymerisation reactions produce many hundreds or thousands of repeating units, a **macromolecule** is formed of suitable molecular weight (25,000+) and the polymerisation process is terminated. Plastics are not composed of chains having the same length or degree of polymerisation and therefore comprise areas of different molecular weight. The distribution of molecular weights throughout the plastic will affect both processability and functional performance.

13.5 POLYMERISATION

Polymerisation reactions are of principal two kinds

- **additional polymerisation** (normally referred to as **polymerisation**). Here monomer A is polymerised to polymer A with no by-product being formed, i.e., monomer A (e.g, ethene) → polymer A (e.g, polyethene). The polymers formed through polymerisation are **thermoplastic** (**13.6.1**); the most commonly used include members of the olefin, vinyl and styrene families (**13.9.4**) (styrene is vinyl benzene, or phenylethylene, $C_6H_5CH{=}CH_2$).
- **condensation polymerisation** (normally referred to as **condensation**). Here two or more co-monomers combine to form the desired polymer **and** a by-product (usually **water**), i.e., monomer A + monomer B → polymer C + by-product. Polymers formed through condensation can be **thermoplastic** (**13.6.1**) or **thermosetting** (**13.6.2**); all thermosets are based on the condensation reaction. Examples of thermoplastics produced using the condensation reaction include thermoplastic polyesters (PET) and polyamides (Nylons), which both generate water as the by-product, and thermoplastic polycarbonate (PC), which yields HCl as the by-product.

13.5.1 Additional Polymerisation

Additional polymerisation is the reverse process of **cracking** and is the process by which the length of the hydrocarbon chain is increased to make a longer hydrocarbon chain molecule. Of great importance to the polymerisation process is the **double bond** in the **aliphatic** hydrocarbon compounds. The double bond represents an **active site** (***Figure 13.3***) which may be chemically broken to allow the addition of more monomers. Note that joining up monomers to form a polymer in this way is a characteristic of aliphatic molecules with double bonds and **not** of the **aromatic** double bond (as in benzene, ***Figure 13.2c***). The polymerisation process can be illustrated with reference to the structure of the **alkenes**, typified by the **ethylene** (ethene) molecule, C_2H_4. Ethylene is a gas at room temperature. When pressurised (200 N/mm^2) at temperatures of about 220°C in the presence of a catalyst, the double bond between the two carbon atoms of ethylene is opened up or **activated** (***Figure 13.3***). When in the activated state, the ethylene molecule will join up with other separate ethylene **monomers** to form a **polymer**. This chemical joining up occurs through the covalent bond between adjacent carbon atoms. Note that there is never a double bond between the carbon and hydrogen, nor can the chains be joined through a hydrogen atom of the ethylene (ethene).

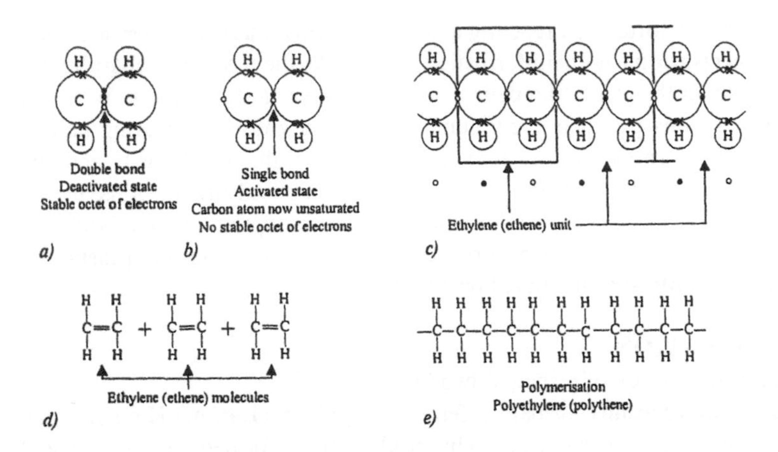

Figure 13.3 Polymerisation of ethylene (ethene)

The resulting polymer is polyethylene (polythene, abbreviated PE). The molecular chain comprises covalent bonds whilst between each chain there are intermolecular (van der Waals) forces which hold the chains together (***Figure 13.4***)

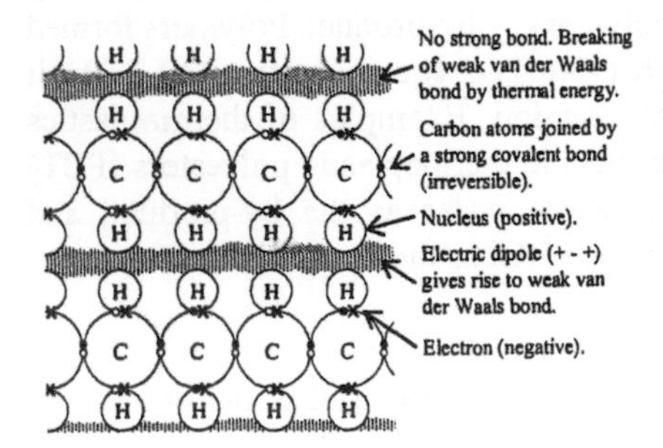

Figure 13.4 shows the intermolecular forces between the chains of the polymer. The centre of each hydrogen atom is positively charged and the negative charge is fixed between the carbon and hydrogen atoms. This creates the van der Waals forces ('bonds') which are, in effect, electric dipoles, (+ - +) (**1.14**). Although these forces are individually weak, their effect over the extremely large surface areas of adjacent chain is sufficient to produce a coherent solid.

Figure 13.4 Intermolecular forces between adjacent molecular chains

PE is solid, stable, impermeable to water, unaffected by salt water and resistant to bacteria and acids. As a result of these properties, PE is used as a damp proof membrane in the building industry. In addition, Polythene (PE) is an electrical insulator (in solid and molten states) because the electrons are fixed by the strong covalent bond into the molecule. PE has a melting point of between 110 to 140°C (depending upon the chain length and perfection of the structure) and fairly readily catches fire. When PE burns, it burns like a "super paraffin wax". If the oxygen supply is restricted, it burns with a smoky flame. This sooty flame arises from the fact that the **hydrogen** has 'first call' upon any oxygen (**14.3**). PE is degraded by the ultraviolet radiation from the sun which turns polythene products into rather brittle products. PE can be stabilised to UV radiation by the addition of carbon black (this is mixed into the molten PE plastic rather like the ingredients of a pudding are mixed). Note that, unlike metals, there are no solid solutions or dislocations involved in plastics.

Generally, the result of polymerisation of different monomers produces a class of plastics which can be reshaped several times by heating and moulding to shape and cooling to "freeze" them into the new shape. These plastics are the **thermoplastics (13.6.1)**. If, however, the plastic molecule is bonded in three dimensions by the covalent bond so that each hydrocarbon chain is linked through another carbon atom, then the resultant plastics **cannot** be reheated to reset them in another configuration; they can only be heat-set **once and once only**. These materials are called **thermosetting plastics** (or **thermosets**) **(13.6.2)** and are formed by **condensation polymerisation (13.5.2)**. Condensation reactions may "cross link" the hydrocarbon chains through atoms other than carbon (e.g, sulfur during **vulcanisation, 13.10**) or by including additional monomers within the reaction. Commonly, these additional monomers are produced from an **aldehyde** group (***Figure 13.5***), such as formaldehyde.

13.5.2 Condensation Polymerisation

If the joining up of the hydrocarbon chain involves the production of water, the reaction is called a **condensation reaction**. The transformation of glucose into starch is an example of a condensation reaction in plant life (***Figure 12.5***), i.e.,

$$nC_6H_{12}O_6 \rightarrow (C_6H_{10}O_5)_n + nH_20$$

Glucose (carbohydrate) → Cellulose + Water

Another example is the formation of phenol formaldehyde (Bakelite™, one of the first plastics to be produced). This reaction is shown in ***Figure 13.5a***, which illustrates two hydrocarbon chains being covalently bonded together with the production of water.

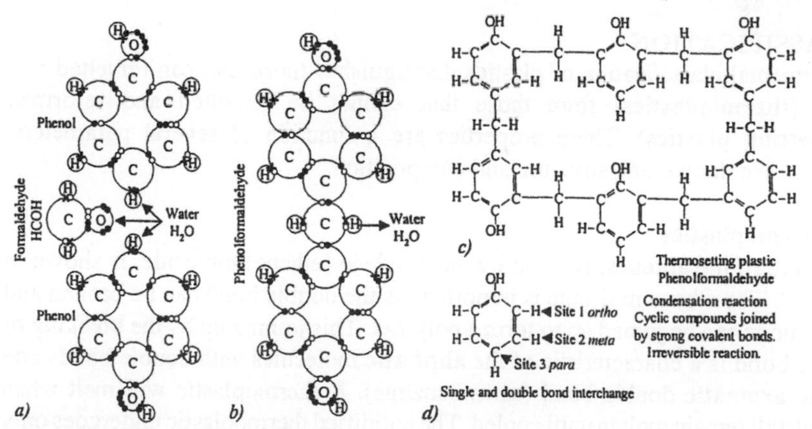

***Figure 13.5** Condensation reaction of phenol and formaldehyde monomers*

Figures 13.5a and ***13.5b*** show the bonding of two cyclic phenol rings through a strong covalent bond through the - CH_2 - i.e., the (H - CHO) group from the formaldehyde (***Figure 13.5d***). In fact condensation reactions usually require an **aldehyde** unit (ending in - CHO) which provides the **oxygen** and a (- CH_2 -) group so that the two adjacent chains may be linked by the **covalent** bond. The condensation reaction shown in ***Figure 13.5*** produces water (***Figure 13.5b***). The water arises from the two hydrogens (derived from the cyclic hydrocarbon phenol, melting point 43 °C) when combined with the oxygen (from formaldehyde gas, boiling point -21 °C), with the result that the (-CH_2) group from the aldehyde can covalently bond into the adjacent chains, thereby producing the rigidly linked three-dimensional structure (**13.9.7**). The product of this condensation reaction, phenol formaldehyde, has niether a melting point nor a glass transition temperature, **13.9.5** (it is used, for example, for ashtrays). Certain wood glues also use formaldehyde (H - CHO) as the "activator" in order to cross link the glue and form a load-bearing bond (**13.16.3**). The phenol groups (denoted 1 to 6 in ***Figure 13.5c***) can be further linked via more formaldehyde to other phenol units through the 'active sites' 1 and 3 in the phenol ring (***Figure 13.5d***). The technical name for these active sites are the **ortho-** and **para-** positions. Cross-linking therefore occurs in three dimensions. ***Figure 13.5a*** and ***13.5b*** show the electron configuration produced by this cross-linking. The three dimensional cross-linking produced by condensation reactions forms polymers with no **glass transition temperature** (**13.9.5**). Aliphatic chains (even if they contain double bonds) **cannot** be cross-linked with formaldehyde, but may be cross linked with other agents, such as sulfur. This cross-linking is particular to the 'double bond' of the **benzene** ring (***Figure 13.2c***). Cross-linking is also used to strengthen elastomers (**13.9.7**), such as in the repair of bicycle tire punctures, where it is known as **vulcanisation**. The crosslinked (or **vulcanised**) repair patch will not come off the tire even if the temperature of the tyre rises because it is bonded predominantly with the strong covalent bond. However, a bicycle patch repair also makes use of the weak van der Waals bond which weakens as the temperature rises. This form of repair is therefore illegal on the motor vehicle tyre which becomes fairly hot at high speeds.

13.6 CLASSIFICATION

The fundamental classification of plastics distinguishes those that can remelted and reformed (**thermoplastics**) from those that cannot be remelted and reformed (**thermosetting plastics**). These properties are a function of several parameters, including molecular weight, structure and composition.

13.6.1 Thermoplastics

The structure of the **alkenes**, typified by the ethylene (ethene) molecule, is shown in ***Figure 13.3***. The ethene molecule is important as the double bond can be broken and the ethene units can be joined up to form a polymer. This joining up by the breaking of the double bond is a characteristic of the **aliphatic** molecules with double bonds and **not** of the **aromatic** double bond (as in benzene). A thermoplastic will melt when heated and will remain molten until cooled. The solidified thermoplastic undergoes only a physical change on cooling. If the solidified plastic product was unacceptable, it could be ground and recycled. This is one of the main advantages of thermoplastics in that they can be recycled many times, often with minimal loss of properties. An example of a thermoplastic-type process is water freezing into an ice cube mould. When fully solidified, the water has undergone only physical changes (and no chemical changes). If the shape of the ice cube was not acceptable, it could be easily ground, remelted and moulded again and again. Some common examples of thermoplastics and their use, with particular reference to the construction industry, are summarised in **Table 13.3**.

Table 13.3 Some examples of common thermoplastics and their use

Example plastic and structure	Properties and uses
Acrylonitrile butadiene styrene (ABS) $[-CH_2-CH(CN)-]$ $[-CH_2-CH=CH-CH_2-]$ $[-CH_2-CH(C_6H_5)-]$ Acrylonitrile Butadiene Styrene	**Properties:** Opaque. Fairly stiff. Fairly hard. Good surface finish. Good impact resistance even at low temperatures. Fairly easily cut. Smooth edges. Rather like PVC in appearance (but higher softening point). Burns readily. Normally not self-extinguishing. Orange-yellow flame with black soot. Odours similar to polystyrene mixed with a bitter odour (due to acrylonitrile) and a rubbery odour (due to butadiene). Suffers from creep (3.5.4). Poor surface weathering characteristics (but may be coated or metallised). **Uses:** Water supply pipes and fittings, internal sink and bath wastes, inspection chambers, ventilator pipes and grilles.
Poly oxymethylene (POM) $[-CH_2-O-]$	**Properties:** Opaque. Stiff. Hard. Fairly difficult to cut. Smooth edges. Burns readily. Drips (these do not usually continue burning). Almost invisible pale blue flame. Very unpleasant pungent odour of formaldehyde. **Uses:** Taps and pipe fittings; light beams and bearings.
Polymethyl methacrylate (PMMA) (Perspex) $[-CH_2-C(CH_3)(COOCH_3)-]$	**Properties:** Can be crystal clear with usually a good surface gloss. Stiff. Hard. Rigid. Somewhat brittle. Splinters on cutting. Good electrical insulation. Burns readily. Not self-extinguishing. Molten droplets continue to burn. Yellow flame with clear edges. Not unpleasant, fruity odour. Good weathering behaviour. **Uses:** Baths, basins, sinks, tap heads, glass replacement (light fittings, roof and dome lights, wall and window glazing), door furniture.
Cellulose acetate (CA) 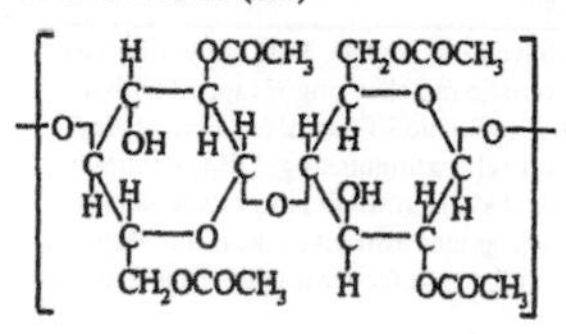	**Properties:** Transparent (unless fillers or pigments have been added). May be flexible (depending on plasticiser content). Fairly hard. Easily cut. Smooth edges. Burns readily. May or may not be self-extinguishing. Dark yellow flame with some smoke. Odour of acetic acid (vinegar-like) or the really rancid smell of perspiration (butyric acid).
Cellulose nitrate (CN) (R is NO_2 or H)	**Properties:** Transparent (unless filler or pigments have been added). May be flexible (depending on plasticiser content). Fairly hard (first plastic developed and used as billiard balls). Easily cut. smooth edges. Extremely flammable and very fast burning. Yellow flame with some smoke. Smell of moth balls from the camphor (used as plasticiser).
Polytetrafluoroethylene (PTFE) (Teflon) $[-CF_2-CF_2-]$	**Properties:** Opaque, usually white as solid piece. Smooth and slippery to the touch. Fairly soft. Easily cut. smooth edges. Will not ignite in an ordinary flame. Behaves rather like a thermosetting material in that it will not melt very easily. Chemically and thermally very inert. **Uses:** Low friction materials, plumber's tape, bridge bearings.
Polyamides (PA) (e.g. Nylon 6)[a] $[-NH-(CH_2)_5-CO-]$	**Properties:** Opaque except as thin film. Stiff, mechanically strong, with a high softening point. Tough and hard material. Good surface finish. Good resistance to creep. Low friction. Fairly easy to cut. Smooth edges. Difficult to ignite. Not self-extinguishing. Melts to a free flowing liquid which drips, carrying the flame with it. Blue flame with yellow top, characteristic odour of burning hair. Affected by sunlight, relatively expensive. **Uses:** Hinges and window fittings, sliding door runners. Used in quiet running gears where oil based products would be a disadvantage e.g. food mixers. Many forms of nylon
Polycarbonate (PC) $[-C_6H_4-C(CH_3)_2-C_6H_4-O-CO-O-]$	**Properties:** Transparent glass like forms and coloured forms available. Stiff with a high softening point (110°C). Hard and tough with good impact resistance. Difficult to cut. Self extinguishing or burns only very slowly, with a smoke flame. Good weathering properties, relatively expensive. Uses: Light fittings, vandal proof lights.
Polyethylene terephthate (PETP) (Polyester film) (Do not confuse with polyester resin, which is a thermosetting material) (see note [c])	**Properties:** Transparent. Stiffer than many other plastics films. Fairly hard. Fairly easy cut. Smooth edges (but tearing can leave jagged edges). Burns readily. Not self extinguishing. Tendency to drip. Yellow, smoky flame, odour said to resemble that of burnt raspberry jam.

Table 13.3 Some examples of common thermoplastics and their use (contd.)

Example plastic and structure	Properties and uses
Low density polyethylene (LDPE) $\left[-\overset{H}{\underset{H}{C}}-\overset{H}{\underset{H}{C}}- \right]$	**Properties:** Transparent only as thin film, translucent in thin sections. Fairly flexible (see HDPE). Soft "waxy" feel. Easily and smoothly cut. Not self-extinguishing. Molten droplets, which usually go out on reaching the bench or floor. Blue flame with yellow tip and little smoke. Smell of burning candle when flame is extinguished. Loses strength in sunlight due to UV light. More resistant to UV when carbon black is inserted as a filler. Otherwise good chemical resistance. Large thermal movement. Not very resistant to oil based jointing compounds which cause chemical stress cracking within the polythene. **Uses:** Many uses, including damp proof courses and membranes, site protection, etc.
High density polyethylene (HDPE) $\left[-\overset{H}{\underset{H}{C}}-\overset{H}{\underset{H}{C}}- \right]$	**Properties:** Transparent only as thin film, translucent in thicker sections. Fairly stiff (see LDPE). Fairly hard. Easily cut, smooth edges. Not self-extinguishing. Molten droplets, which usually go out on reaching bench or floor, Blue flames with a yellow tip and little smoke. Smell of burning candle when flame is extinguished. **Uses:** Many uses, including pressure pipes for water and gas, cold water cisterns, sink and bath wastes, cable insulation, etc.
Poly propylene (PP) 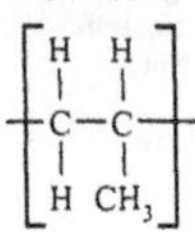	**Properties:** Transparent only as thin film, translucent in thicker sections. Stiffer than HDPE and more expensive. Hard, Easily cut. Fairly smooth edges. Not self-extinguishing. Molten droplets, which usually go out on reaching the bench. Flame mainly yellow with a trace of clear blue at the bottom. Smell of burning candle when the flame is extinguished, but not identical to that from polyethylene. Not very good in exterior uses. Good electrical insulation. Large thermal expansion. Resistant to most materials except chlorinated solvents. **Uses:** Many uses, including cold water and WC cisterns, overflow tanks, sink and bath wastes, pressure pipes for water and gas, wall tiles, etc.
Polystyrene (PS) 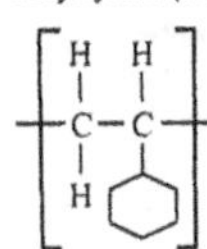	**Properties:** Transparent (unless fillers or pigments have been added). Stiff but with a low softening point. Hard and sometimes brittle. Characteristic metallic ring if tapped with a pencil. Good surface finish. Fairly difficult to cut (but some modified polystyrene grades can be cut to leave smooth" edges). Burns readily. Not self-extinguishing. Drips continue to burn. Orange-yellow flame with black, sooty smuts. Odour in small amounts is said to resemble that of marigolds (some modified polystyrene grades also give rise to rubbery odours). Inexpensive plastic, capable of being made in the cellular form where it is used as a cavity wall in-fill. **Uses:** WC cisterns, wall tiles, light fittings. Expanded polystyrene (EPS) has been used for various insulating applications (walls, floors, flat roofs, etc.) and disposable thermal drinking cups. Impact sound absorbing sheets. Cheapest thermoplastic. Attached by white spirit based solvents and plasticiser leached from e.g. unplasticised PVC.
Polyvinyl chloride, unplasticised (PVC-u) 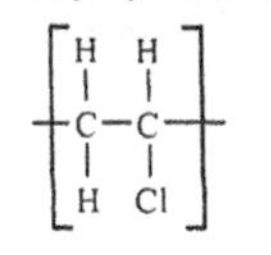	**Properties:** Transparent, (unless fillers or pigments have been added). Stiff. Hard. Good surface finish. Fairly easy to cut. Smooth edges. Burns with difficulty. Self-extinguishing. Yellow flame, blue-green at bottom edges. Unpleasant, acrid odours of hydrochloric acid. Notch sensitive when unplasticised. Suitable for external use. Easily jointed by solvent welding. **Uses:** Many uses, including structural uses, such as window frames and external cladding sheets. Rainwater goods, soil and waste goods, electrical conduit, wall skirting, weatherboarding, etc. Also as Post chlorinated PVC (CPVC) (which raises the softening point, making CPVC harder to mould or extrude than PVC) for hot water and central heating pipes, internal sink wastes
Polyvinyl chloride, plasticised (PVC) $\left[-\overset{H}{\underset{H}{C}}-\overset{H}{\underset{Cl}{C}}- \right]$	**Properties:** Unless fillers or pigments have been added, transparent. Flexible. Soft. Very easily cut. Smooth edges. Burning behaviour will depend on the plasticiser, but generally not self-extinguishing. May drip. Yellow, smoke flame. acrid odour of hydrochloric acid, with an aromatic odour due to the plasticiser. **Uses:** Many uses, including flexible substitutes for leather (but loss of plasticiser causes embrittlement), electrical cable covering and insulation, water stops, roof sarking, coatings on board and sheet material (Plastisol is vinyl chloride acetate copolymer and vinyl chloride resins)[b]

Notes: [a] The numbers of carbon atoms between each nitrogen atom pair govern the type of polyamide produced and the resultant properties.; [b]The similarity between **polymethyl methacrylate** (PMMA) and **polyvinyl acetate** (PVA) can be seen by comparing the following representation of the two polymers

PVA is a chain of

$$(-CH_2-\underset{\displaystyle CO-O-CH_3}{\underset{|}{CH}}-)_n$$

and PMMA is a chain of

$$(-CH_2-\overset{\displaystyle CH_3}{\overset{|}{\underset{\displaystyle CO-O-CH_3}{\underset{|}{C}}}}-)_n$$

[c] Esters are examined in section 6.10.4. A **polyester** contains the characteristic (- O - CO -) unit. A polyester is a chain of these esters and they are written in general terms as

$$(-O-R-O-CO-R'-CO-)_n$$

13.6.2 Thermosetting Plastics (Thermosets)

Thermosetting plastics are formed when smaller or shorter hydrocarbon chains are irreversibly **crosslinked** into a colossal three-dimensional molecule (a macromolecule) by the joining up of the individual molecules into a thermally stable network. One method of achieving this cross linking is by the **condensation reaction (13.5.2)** (which should be contrasted with **additional polymerisation, 13.5.1**, to form a thermoplastic, which is thermally reversible). A thermosetting plastic develops its functional properties as it chemically and physically changes during processing. The thermosetting plastic 'cures' (or crosslinks, **13.9.7**), a process analogous to vulcanisation of an elastomer (**13.10**). The thermosetting plastic solidifies as a result of chemical reactions. Some thermosetting plastics are cast as liquids and cured at room temperature, whilst others require external heating. Thermosetting plastics are formed through condensation reactions (**13.5.2**) normally achieved using relatively inexpensive operations. However, unlike thermoplastic materials, since the thermosetting scrap cannot be reformed again economically, thermosetting plastics are considered relatively expensive. An example of a thermosetting-type process is hard boiling an egg. Initially the egg is liquid, but when exposed to heat it will undergo a physical and chemical change to solidify within the mould (the egg shell). If the shape of the hard boiled egg is unacceptable, it cannot be reformed on heating. (Another example of a thermosetting-type reaction is the production of cured concrete).

There are no van der Waals bonds of any consequence in a thermosetting unit and so there is **no** glass transition temperature and **no** softening on heating. Thermosetting plastics can only be heat set **once** and once **only**. They do **not** burn because there are **no** short chain hydrocarbons to come off in a vapour and ignite (**14.6.6**); thermosetting plastics just char producing **carbon**. Examples of thermosetting plastics are phenol formaldehyde (PF), melamine formaldehyde (MF), urea formaldehyde (UF). Note that all contain the **-aldehyde** derivative in their names. Materials which use other mechanisms to cross link include epoxy resins (EP), polyesters (UP) (from the name unsaturated polyester), polyurethane resins (PUR) and polyurethane foam (PUF). Some common examples of thermosets and their use, with particular reference to the construction industry, are summarised in **Table 13.4**.

Table 13.4 Some examples of common thermosets and their use

Example plastic and structure	Properties and uses
Epoxy resins and epoxides (see note *)	**Properties:** good bonding properties. excellent resistance to alkalis. hard and difficult to cut. usually rather brittle. does not melt, but burns fairly readily to produce a 'fishy' odour. **Uses:** Usually used as an adhesive (e.g. 'Araldite') or in repair work. Heavy duty flooring compositions, resin concretes, glass-reinforced composites. The high cost of epoxy resins in offset by good performance.
Urea formaldehyde (UF) 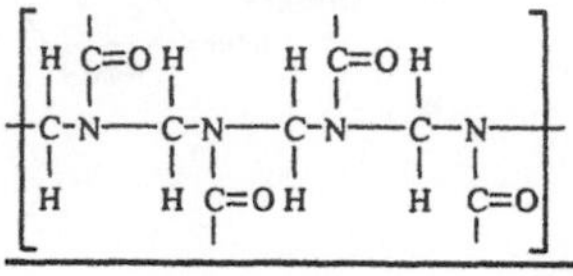 H C=O H H C=O H –C-N—C-N—C-N—C—N– H H C=O H H C=O	**Properties:** Opaque usually light in colour. Stiff, hard material with a "solid" feel. May be brittle in some cases. Flakes on cutting. Burns with great difficulty. Self-extinguishing. Swells, cracks and turns white at edges of burnt portion. Pale yellow flame with light blue-green edges. Pungent odour of formaldehyde and also, sometimes, a fish-like smell. Relatively inexpensive, with poor weathering characteristics. **Uses:** Cavity wall infilling when foamed, electrical fittings, adhesives and binders for wood particleboards.

Table 13.4 Some examples of common thermosets and their use (contd.)

Example plastic and structure	Properties and uses
Melamine formaldehyde (MF) Triazine ring with three $N(CH_2-O-CH_3)_2$ groups: CH_3-O-CH_2, CH_2-O-CH_3	**Properties:** Opaque, usually light in colour. Stiff. Hard. 'Solid' feel. Flakes on cutting. Burns with great difficulty. Self extinguishing. Sample cracks and turns white at edges of burnt portion. Pale yellow flame with light blue-green edges. Pungent odour of formaldehyde and also, sometimes, a fish-like smell. Relatively expensive. **Uses:** Used as the outer surface of decorative heat resisting laminates (of phenol formaldehyde).
Phenol formaldehyde (PF) $OH-C_6H_4-CH_2-C_6H_4-OH$	**Properties:** Opaque, dark coloured. Stiff. Hard. 'Solid' feel. Flakes on cutting. Burns with difficulty. Self-extinguishing. Odour of carbolic acid. **Uses:** Used as the backing on heat resisting laminates. Brown light fittings. WC seats. Composite panels of foam (on aluminium/wood substrates) as thermal insulation (e.g. in caravans).
Polyester resins (PER) (e.g, styrene-polyester coopolymer) $-CH_2CH_2-O-C(=O)-CH-CH-C(=O)-O-CH_2CH_2-$ cross-linked through Styrene (CH_2, CH, phenyl) to $-CH_2CH_2-O-C(=O)-CH-CH-C(=O)-O-CH_2CH_2-$	(Do not confuse with the thermoplastic material polyester film in the finished state). **Properties:** Stiff. Hard. 'Solid' feel. Difficult to cut. Brittle. Burns readily. Often formulated to be self extinguishing. Smoky flame. Fruity odour. As GRP, normally the glass fibre reinforcement can be seen; good mechanical properties, especially high tensile strength though relatively low stiffness for structural applications; burns readily (low flammability grades available but weathering behaviour may be adversely affected by the incorporation of flame retardants); suffers moisture loss and shrinkage (surface of the glass fibres is protected to prevent admission of water) **Uses:** As GRP, for a wide variety of precast moulded units, including cold water cisterns, baths, basins, roof lighting sheets and domes, etc., external cladding panels, fascias, miscellaneous architectural features.

Notes: * Only those compounds where an oxygen atom is cyclically linked between two carbon atoms are prefixed by the term **epoxy**. Thus, ethylene oxide is epoxy ethane. Epoxide resins are

$$\overset{O}{\overbrace{CH_2 - CH_2}} \qquad -\overset{O}{\overbrace{CH_2 - CH} }- CH_2 - \left[O - C_6H_4 - \underset{CH_3}{\overset{CH_3}{C}} - C_6H_4 - O - CH_2 - \overset{OH}{CH} - CH_2 - \right]_n O - C_6H_4 - \underset{CH_3}{\overset{CH_3}{C}} - C_6H_4 - O - CH_2 - \overset{O}{\overbrace{CH - CH_2}} -$$

Ethylene oxide (epoxy ethane) — Epoxide resins

13.7 FABRICATION PROCESSES

Plastics can be fabricated using a wide variety of processes, of which only the most common are examined here (***Figure 13.6***)

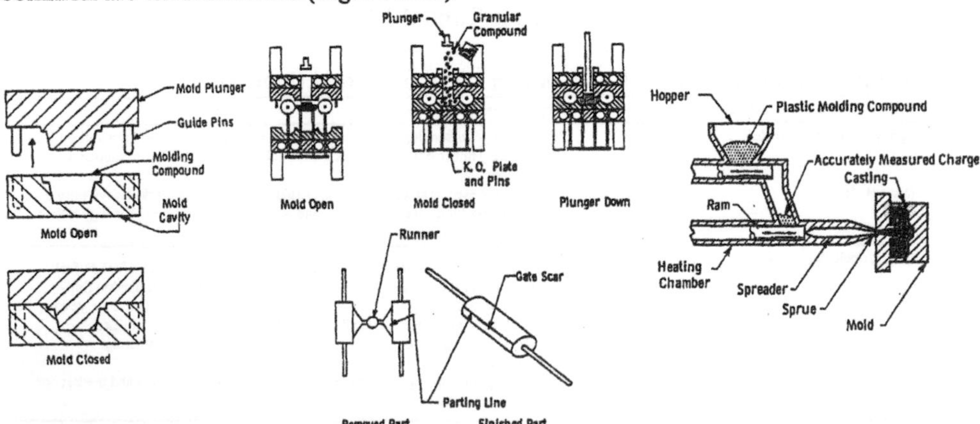

a) Compression moulding *b) Transfer moulding* *c) Injection moulding*

Figure 13.6 *Diagrammatic illustration of fabrication processes*

Compression moulding is the most common process for forming thermosetting raw materials (and some thermoplastics) and is shown diagrammatically in ***Figure 13.6a.*** A given quantity of the raw material is added to the open mould. The mould plattens are heated and closed under pressure, causing the material to run into the shape of the mould cavity. In a similar process known as **transfer moulding**, the raw material is placed into a heated pot, separate from the mould cavity. The molten material is then passed under pressure from the pot to the closed cavity of the mould (***Figure 13.6b***). This system has the advantage that the mould is closed at the time the raw material enters, thus parting lines (which may appear in compression moulding) are minimised.

Injection moulding most common processes for moulding thermoplastic materials. The raw material is added to a heated reservoir and, once molten, is forced into a cool mould. The mould is opened as soon as the material is sufficiently rigid to hold its shape on demoulding (***Figure 13.6c***). Compared to compression moulding of thermosets, greater moulding cycle speeds and lower moulding costs are obtained with injection moulding. This is the main reason why thermoplastics are relatively cheap compared to thermosetting materials. Recent advances in thermosetting additives have enabled thermosetting materials to be injection moulded by screw-injection, allowing significant cost savings. **Thermoforming** is the process by which a plastic sheet is heated and forced in to a mould, either by differential air pressure or by mechanical means. Initial sheet thickness and the applied force controls the thickness of the formed product. **Calendering** is a thermoforming operation in which granulated raw materials are run through a series of heated rollers to produce sheet materials. The process allows formation of laminates (from many sheets) and inclusion of reinforcement material. **Extrusion** involves forcing the melted and heated raw material through the opening in a die with the required cross section to produce the finished item. The process is usually used for thermoplastic materials and is particularly useful for forming continuous sheets, pipe, wire jacketing, etc. **Pultrusion** is an increasingly used technique for pulling resin soaked fibres through an orifice as it generates significant strength improvements.

13.7.1 Blends

Due to the high costs and the long time periods involved in the development of new polymers, the trend today is to introduce a **polymeric blend** based on existing coploymers with proven and predictable service. Simply, the blend is designed so that the benefits of polymer A will offset the limitations of polymer B, and vice versa. Examples include (ABS + PC), (ABS + PVC), (ABS + PU) and (PVC + PMMA).

13.8 CELLULAR PLASTICS

Foamed plastics (having an open or closed cell structure) and expanded plastics (having substantially a closed cell structure) have been widely used in construction for insulation (acoustic, impact and heat). Cellular plastics are available as rigid or flexible boards, granules or as foams. **Table 13.5** summarises the main types, applications and properties of cellular plastics used in construction.

Table 13.5 Main types, applications and properties of cellular plastics used in construction

Material	Description	Applications	Properties
Expanded polystyrene ('bead board')	White rigid material formed by the fusion of expanded beads of PS. Also obtainable in a pre-compressed form that is more flexible and resilient, and in granular form as unfused expanded beads	Lining of walls and skin roofs, insulation of flat roofs and concrete floors. In floating floor construction, to absorb impact sound. Wall and ceiling tiles. Shaped formwork for concrete. Cavity wall and roof insulation.	• attractive appearance • good thermal insulation. Thermal conductivity 0.035-0.038 W/m°C (bead board) 0.036 W/m°C (extruded) 0.030 W/m°C (extruded with surface skin) • relatively low softening point • attacked by most organic solvents • loss of plasticiser from plasticised PVC can cause embrittlement • burns fairly rapidly. Often flames and collapses. (Flame retardant grades are available)
Extruded expanded polystyrene	White or light-coloured material formed by a simultaneous extrusion and foaming process		
Expanded polystyrene beads/granules	White material formed by granulating bead board or as unfused expanded beads		
Expanded polyvinyl chloride	Yellow to deep brown rigid material formed by foaming PVC while in the plastic state, and cooling to solidify	Lining of walls and skin roofs, semi-structural sandwich panels	• relatively strong • good thermal insulation. Thermal conductivity 0.035-0.047 W/m°C • low water vapour transmission • low softening point • burns with difficulty. Collapses.
Foamed urea formaldehyde resins	White friable material, usually formed by foaming an aqueous dispersion of resin and curing it in this condition	In situ filling of cavity walls for insulation	• poor physical and mechanical properties • good thermal insulation. Thermal conductivity 0.035 W/m°C • releases formaldehyde gas • resistant to ignition
Foamed phenol formaldehyde resins	Pink or deep red rigid material, somewhat friable, made by foaming and subsequently curing phenolic resin	Lining of walls and flat roofs and as core material for sandwich panels	• good fire and high temperature characteristics • good thermal insulation. Thermal conductivity 0.043 W/m°C • relatively poor physical and mechanical properties
Foamed polyurethane	Colourless to deep brown material, sometimes artificially coloured. Prepared from two liquid components which are sometimes pre-expanded before complete chemical interaction and curing occurs	Flexible foamed PU used for shaped insulation for pipework, acoustic applications, sealing and joining. Rigid foamed PU used for in situ insulation in roofs, as insulating boards for use in roofs, walls, underfloor applications. Core material for sandwich panels. Insulating and stabilising cavity walls. Sprayed foam has been used externally with a weather-resistant coating for walls and roofs.	• Foamed PUs generally burn rapidly, producing dark thick smoke. (Flame retardant grades are available) • flexible foamed PU has relatively high water vapour transmission and thermal insulation. Thermal conductivity 0.037 W/m°C • rigid foamed PU has fairly good high temperature and water vapour transmitting characteristics and excellent thermal insulation. Thermal conductivity 0.03-0.07 W/m°C (CO_2 blown) 0.01-0.03 W/m°C (fluorinated hydrocarbon blown)

13.9 STRUCTURAL CHARACTERISTICS AND PROPERTIES

The structural characteristics of plastics determines their functional characteristics and processability and include

- moleoular weight (**13.9.1**);
- chain structure (**13.9.2**);
- homogeneity (**13.9.3**);
- the number and composition of side groups (functional groups) (**13.9.4**);
- the glass transition temperature, T_g with respect to the operating temperature, T_r (**13.9.5**);
- the degree of crystallisation (**13.9.6**);
- the degree of cross linking (particularly important in thermosetting adhesive compounds) (**13.9.7**).

Unlike metals, there are **no** dislocations (**9.4**) present in plastics. The mechanical deformation of the plastic is modelled by the slipping of the weekly bonded molecular chains past one another as the van der Waals forces ('bonds') are broken (**13.9.5**). The structural characteristics of plastics (listed above) will all influence the ease with which the molecular chains slip relative to one another, and will therefore influence the ultimate strength of the plastic.

The properties of a plastic can be scientifically modelled as several compartments of a triangle, as shown in ***Figure 13.7***, which illustrates the magnitude (or absence) of the glass transition temperature. The Figure brings together the various models and relates the models to the ultimate strengths. The strength properties can be achieved by a mixture of the processes outlined. Hence a plastic with some cross-linking and some weak van der Waals forces ('bonds') will be fairly stiff, pliable and heat resistant. However, it will soften on heating but would not be expected to lose its shape completely. A plastic completely cross-linked would remain rigid right up to the onset of chemical decomposition and the onset of charring.

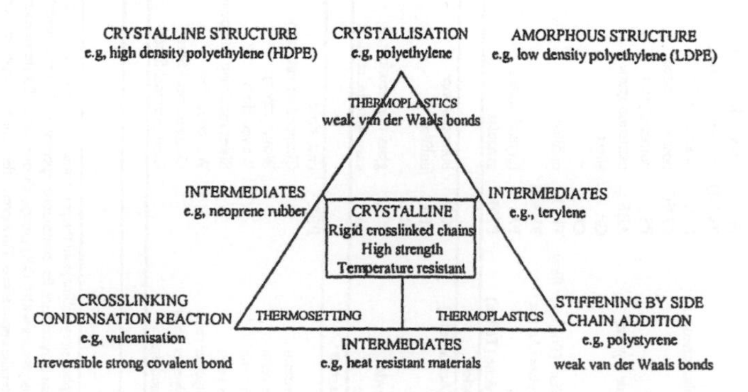

***Figure 13.7** Inter-relationship of the functional properties of plastic materials*

Some functional properties of common thermoplastics and thermosetting plastics are summarised in **Table 13.6**.

Table 13.6 Some functional properties of common thermoplastics and thermosetting plastics

	Abbrev.	Form	am/cry	T_g	T_m	T_{max}	SG	H_2O (%)	TS	EL (%)	TM	Izod	α ($\times 10^{-6}$)
Thermoplastics													
Low density polyethylene	LDPE	branched homopolymer	40/60	98-115	-120	80	0.91-0.93	<0.01	7-31	100-650	0.1-0.3	NB	160-200
High density polyethylene	HDPE	homopolymer	10/90	130-138	-120	104	0.94-0.96	<0.01	21-38	10-1200	0.4-1.3	0.5-5.4	110-140
Polycarbonate	PC	unfilled moulding and extrusion resin			150	110	1.2	0.15	63-72	110-120	2.4	16.3-24.4	60-70
Polypropylene	PP	homopolymer, unfilled	40/60	160-176	-10	120	0.90-0.91	0.01-0.03	29-38	100-600	1.1-1.6	0.5-1.9	80-110
Polyvinyl chloride	PVC-U	unplasticised, rigid	90/10	212	75-105	65	1.30-1.58	0.04-0.4	41-52	40-80	2.4-4.1	0.5-30	50
Polyvinyl chloride (plasticised)	PVC	flexible, unfilled	90/10	212	-35	40-65	1.16-1.35	0.15-0.75	10-24	200-450			70
Polyvinyl chloride	CPVC	post-chlorinated	90/10		110	100	1.49-1.58	0.02-0.15	47-62	4.0-100	2.4-3.3	1.4-7.6	70
Polystyrene	PS	homopolymer	100/0		74-105		1.04-1.05	0.01-0.03	36-52	1.2-2.5	2.3-3.3	0.5-0.6	60-80
Acrylonitrile butadiene styrene	ABS	extrusion grade	100/0		88-120	90	1.02-1.08	0.20-0.45	7-55	20-100	0.9-2.9	2.0-16.3	60-100
Cellulose acetate	CA	sheet	100/0		80		1.28-1.32	1.70-4.40	31-55	20-50	2.1-4.1	2.7-11.5	
Cellulose nitrate	CN						1.35-1.40	1.0-2.0	48-55	40-45	1.3-1.5	6.8-9.5	
Polyethylene terephthate (Polyester film)	PETP	unfilled	60/40	212-265	68-80		1.10-1.46	0.1-0.6	41-90	30-300	2.1-4.4	0.3-1.0	
Poly methyl methacrylate (Acrylic)	PMMA		90/10		85-105	80	1.17-1.20	0.1-0.4	48-76	2.0-5.5	2.2-3.2	0.3-0.5	50-90
Polyamide (Nylon 6)	PA	(crystallisable)	0/100	210-225		70	1.12-1.14	1.3-1.9	65-185	30-100	1.4-3.1	0.8-3.0	80-130
Poly tetra fluoro ethylene (Teflon)	PTFE	granular	15/85	327	-120		2.14-2.20	<0.01	14-35	200-400	0.4-0.6	4.1	
Thermosetting plastics													
Poly oxymethylene (Acetal resin)	POM	homopolymer	0/100	172-183	-76	80	1.43	0.24-1.0	67-69	10-75	2.8-3.6	1.5-3.1	80
Epoxy resin	PER	unfilled		thermoset			1.11-1.40	0.08-0.15	28-90	3-6	2.41	0.3-1.4	39
Urea formaldehyde	UF			thermoset			1.56	0.4-0.8	38-90	<1	6.9-10.3	0.3-24.4	27
Melamine formaldehyde	MF	glass fibre reinforced		thermoset			1.5-2.0	0.09-1.3	34-72	0.6	11.0-16.6	0.8-24.4	30-55
Phenol formaldehyde	PF	casting resin, unfilled		thermoset			1.24-1.32	0.1-0.36	34-62	1.5-2	2.8-4.8	0.3-0.5	30-55
Fibre reinforced plastics													
Polyester resin	PER	cast, rigid		thermoset			1.04-1.46	0.15-0.6	4-90	<2.6	2.1-4.4	0.3-0.5	20-35
Glass fibre. Polyester resin		Chopped strand mat		thermoset		175	1.65-2.30	0.06-0.28	20-69	<1	6.9-17.2	2.0-21.6	30
Glass fibre. Polyester resin		Woven roving		thermoset			1.35-2.30	0.01-1.0	103-207	1-5	5.5-13.8	2.7-27.1	15
Glass fibre. Polyester resin		Woven cloth		thermoset			1.50-2.10	0.05-0.5	207-345	1-2	10.3-31.0	6.8-40.5	
Glass fibre. Polyester resin		Sheet moulding compound				60	1.65-2.60	0.1-0.25	48-172	3	9.7-17.2	9.5-29.7	
Polyaramid fibre. Polyester resin		Woven roving		thermoset					390		24		
Polyaramid fibre. Polyester resin*		Unidirectional fibres		thermoset					1380		76		57
Carbon fibre. High modulus*		Unidirectional fibres							1260		200		25
Metals													
Mild steel						50			450		207		12
Aluminium alloy						25			300		70		24

Notes: **am/cry** (amorphous/crystalline) and the percentage (where known/useful). Reflects previous thermal history; rapid cooling (amorphous structure), slow cooling (crystallinity). T_g (glass transition temperature), T_m (melting point). T_{max} (maximum temperature for continuous operation). T_m only available for the most crystalline polymers (amorphous materials have no melting point); **SG** (specific gravity). **H_2O (%)** (weight % water absorbed in 24 hours). Amorphous plastics with (-OH) groups absorb water. **TS** (tensile strength, MN/m^2). In plastics, refers to force per unit of original cross-section on elongating to rupture, thus not really an important property as plastics rarely used in purely tensile applications. However, other plastic properties correlate well with tensile strength (wear and tear resistance, resilience, cut resistance, creep and, in some polymers, e.g, neoprene, ozone resistance). **EL (%)** (elongation at break). Maximum extension at rupture; a plastic with <100% elongation will usually break if doubled over on itself. Most plastics undergo considerable elongation and necking (**3.2.1**). **TM** (Tensile modulus, GN/m^2). Force per unit of original cross section to a specific extension (usually 300%); shows stiffness of the plastic. It is the ratio of the stress/strain, but differs from metals (not a Young's modulus stress-strain type curve). Stress-strain values extremely low for slight extensions but increase logarithmically with increased extensions (e.g, natural rubber is extensible to about 10x its original length). **Izod** impact (energy absorbed (J) on fracture; 3.2 mm specimens). **NB** (no break). α (per °C) Coefficient of thermal expansion (**Table 2.3**). * perpendicular to reinforcement.

13.9.1 Molecular Weight

The length of the plastic molecule is usually determined by the molecular weight. The thermoplastic may have a molecular weight of 20,000 to 50,000. The length of the molecular chain can be estimated by the fact that each carbon atom (atomic weight = 12) is associated with two hydrogen atoms (atomic weight = 2 x 1). Thus one - CH_2 - unit has a molecular weight of 14 and therefore a molecular weight of 20,000 is equivalent to a chain with 1430 - CH_2 - units. This length of chain is far longer than the hydrocarbons given in **Table 13.1** (derived from the fractional distillation of crude oil).

13.9.2 Chain Structure

The structure of the plastic depends on the nature of the chain structure. Four basic chain structures can be produced

- **linear** i.e., long chains that can either be **aliphatic** or **aromatic**;
- **branched** i.e., long polymer "backbones" with side chain additions;
- **crosslinked** i.e., linear chains connected to one another by short, low molecular weight groups, forming a complex three-dimensional network;
- **ladder** i.e., a highly developed network of ring structures, tightly connected at multiple sites.

It is possible for a plastic to have more than one structure. A good example is polyethylene which, depending upon how it is polymerised, can be either linear (as in HDPE for injection moulding) or branched (as in LDPE for extruded films). In addition PE can be crosslinked for enhanced properties.

13.9.3 Homogeneity

Plastics can be based on one or more monomers. A **homopolymer** is based on one starting monomer (e.g, ethylene polymerises to polyethylene). Plastics based on two monomers (called **co-monomers**) include impact resistant polystyrene (styrene and butadiene co-monomers) and styrene acrylonitrile (SAN), and an example of a plastic based on three co-monomers is acrylonitrile butadiene styrene (ABS). Co-polymers themselves have various structures. Denoting the various co-monomers as A, B, ..., etc., these structures are

- **random** (not usually used commercially) i.e.,
  ABAABAAABBAAABABBA
- **block**, containing large sections or blocks of each co-monomer i.e.,
  AAAAAAAAAA-BBBBBBBBBB-AAAAAAAAAA
  (e.g, ethylene-propylene copolymer). Crystalline block polymers form an important group of adhesives (**13.16.2**);
- **graft**, consisting of large molecules of one co-monomer grafted on to the backbone of the second co-monomer i.e.

```
-BBBBBBBBBBBBBBBBBBBBBBBBBBBB-
      A      A      A      A
      A      A      A      A
      A      A      A      A
      A      A      A      A
      A      A      A      A
```

(e.g, rigid styrene grafted on to the flexible butadiene backbone). Graft polymers can be chemically engineered to overcome brittleness and low impact resistance (**13.9.6**);

- **alternating** (typical of fluoro copolymers and new generation copolymers, e.g, aliphatic polyketones) i.e. ABABABABABABABABABABABABABA

Note that the macro molecular structure of copolymers governs the properties and end uses of the resultant plastic. For example, the copolymer based on styrene and butadiene can either be the graft or block structure, and the properties of the copolymer produced by each different arrangement are greatly different.

13.9.4 Chemical Grouping

The composition of both the polymer chain and of the side groups (**functional groups**) can be varied to produce polymers with different properties. **Aliphatic** groups, based on repeating $-CH_2-$ units, provide flexibility to the polymer, although they are prone to oxidative attack during processing at elevated temperatures. **Aromatic** groups, based on the large benzene ring, C_6H_5, contribute stiffness, rigidity, hardness, and creep and heat resistance, but at the expense of impact resistance.

The various functional groups that are commonly employed in polymers are summarised in **Table 13.7**. These functional groups characterise the chemistry of the polymer and give rise to their nomenclature.

Table 13.7 Functional groups

Group	Group Name	Group	Group Name
R - CH_3	Methyl	R - OH	Alcohol
R - C_2H_5	Ethyl	R'- OH	Phenol (R' is an aromatic compound)
R - NH_2	Amine	R - Cl	Chloride (or chloro)
R' 〉C=O R	Ketone	R' 〉C=O H	Aldehyde
R - C ≡ N	Nitrile	R - C_6H_5	Phenyl
R - C〈 O, OH	Carboxyl	R - O - R'	Ether
R - $C_nH_{(2n+1)}$	Alkyl	R - C〈 O, O - R'	Ester[a]

Notes: Standard IUPAC nomenclature and abbreviations are adopted. R and R' represent different chemical groupings depending on the functional group and may or may not be the same. [a] An ester is formed by reacting an alcohol with an acid (the inorganic equivalent of an acid plus base reaction). It is therefore the **alkyl** salt of the **acid**. The fundamental unit of an ester is the (- O - CO -) group. In general terms it would be written as R - CO - OR' where R is from the acid and R' is from the alcohol. Thus ethyl acetate is CH_3 - CO - OC_2H_5 (i.e., R = CH_3 and R' = C_2H_5).

These functional groups provide the nomenclature for organic compounds. Hence, for example,

CH_4	is called	methane (a gas)
CH_3 - Cl	is called	methyl chloride (or chloro methane)
CH_3 - OH	is called	methyl alcohol (or methanol) (the poisonous alcohol)
C_2H_5 - OH	is called	ethyl alcohol (or ethanol) (alcohol in beers, etc.)
H - CHO	is called	formaldehyde (or methanal) (preservation fluid)
CH_3 - CHO	is called	acetaldehyde (or ethanal)
H - COOH	is called	formic acid (or methanoic acid)
CH_3 - COOH	is called	acetic acid (or ethanoic acid (vinegar)

As a point of interest, the alcohol C_2H_5 - OH oxidises in the human body to produce acetaldehyde CH_3 - CHO, which is responsible for the 'hangover' if too much alcohol is consumed. Methyl alcohol CH_3 - OH ('methelated spirits') will oxidise to give formaldehyde, which is the liquid used for preserving (human) organs. Domestically this is referred to as pickling. Hence this reaction occurring in the human brain is very dangerous (and results in a "pickled" brain).

These functional groups can also be combined with one another. For example,

CH_3 - CO - C_2H_5	becomes	ethyl methyl ketone
CH_3 - O - C_2H_5	becomes	ethyl methyl ether

and so forth. Hence organic chemistry builds up compounds by combining various "units" together. The terminology of the compound arises from the combined fundamental units. Some compounds have popular usage terms which are a direct derivation of their chemical name, e.g, polythene is derived from polyethylene (or polythene). As a final example, copolymers may take on the name of the two polymers producing the copolymer. Ethylene-vinyl acetate copolymers would be chains of

$$(-CH_2-CH_2-)_n \; -(CH_2-\underset{\displaystyle O-CO-CH_3}{\underset{|}{CH}}-)_n$$

Ethylene : Vinyl acetate : Methyl group

Subtle changes in the polymer structure will dramatically affect processability and functional behaviour of the plastics. Side chain "grafting" is one of the best methods for increasing the mechanical properties of the plastic. This not only strengthens the plastic but raises the glass transition temperature. To illustrate this, we consider examples based on the simplest organic group, vinyl, comprised of repeating units of

$$-(-CH_2-\underset{\displaystyle X}{\underset{|}{CY}}-)-$$

For the purpose of this comparison, Y = hydrogen (H) and X represents the functional group varied to produce plastics of differing properties. Resulting properties of vinyl polymers are summarised in **Table 13.8**.

Table 13.8 Properties of vinyl polymers

X is:	Monomer is:	Polymer (plastic) is:	Properties
H	H H | | C = C | | H H ethylene	H H H H H H H H | | | | | | | | - C - C - C - C - C - C - C - C - | | | | | | | | H H H H H H H H Polyethylene (polythene) (PE)	flexible, soft, opaque, chemically resistant, burns, T_g = -120°C PE is extruded into wire and cable jacketing, blown film and moulded bottles
CH_3	H H | | C = C | | H CH_3 propylene	H H H H H H H H | | | | | | | | - C - C - C - C - C - C - C - C - | | | | | | | | H | H | H | H | CH_3 CH_3 CH_3 CH_3 Polypropylene (PP)	stiff, rigid, opaque, heat resistant, burns, poor low temperature flexibility, T_g = -10°C PP is moulded into automotive components and extruded into packaging films and textile fibres

These plastics are members of the olefin family (which consist of aliphatic groups) and are semi-crystalline as they are based on carefully aligned, closely packed, repeating units containing only carbon and hydrogen.

X is:	Monomer is:	Polymer (plastic) is:	Properties
C_6H_5	H H | | C = C | | H C_6H_5 styrene	H H H H H H H H | | | | | | | | - C - C - C - C - C - C - C - C - | | | | | | | | H | H | H | H | C_6H_5 C_6H_5 C_6H_3 C_6H_3 Polystyrene (PS)	clear, strong, burns, poor chemical resistance, poor impact resistance, T_g = 74-105°C PS is extruded into film and sheeting and moulded into many consumer products, including pens, disposable glasses, CD and cassette boxes, etc.

Although this polymer also contains only carbon and hydrogen, its properties vary dramatically from the olefins. The aromatic benzene group contributes stiffness, rigidity, flatness, hardness and good resistance to heat and creep. However, the size, shape and arrangement of the repeating units do not allow the polymeric chains to pack closely together. Consequently, crystallinity is not possible, and this material is referred to as amorphous (without structure) and hence transparent (like glass)

X is:	Monomer is:	Polymer (plastic) is:	Properties
Cl	H H | | C = C | | H Cl vinyl chloride	H H H H H H H H | | | | | | | | - C - C - C - C - C - C - C - C - | | | | | | | | H | H | H | H | Cl Cl Cl Cl Poly vinyl chloride (PVC)	transparent, resistance to chemicals, brittle, slow burning, marginal heat resistance, T_g = 75-105°C PVC is an extremely versatile material, based on how it is chemically modified. It can be extruded into food packaging films, blow moulded into containers, etc.

The addition of a third element, chlorine (Cl), significantly modifies the processability and performance of the resultant polymer. Chlorine is the most commonly used halogen for contributing flame-retardant properties (**14.7.4**).

Note that, in **Table 13.8**, we have assumed Y = hydrogen (H). In practice, Y can also be changed. For example, in poly methyl methacrylate (Perspex), X = CH_3 and Y = COO.CH_3.

13.9.5 Glass Transition Temperature

The hydrocarbon chain is held together through the carbon atoms by the strong covalent bond. A very weak "residual" force is left to "bond" one chain to the adjacent chain. This is the **van der Waals 'bond'** (**1.14**). The glass transition temperature T_g is the temperature at which bond rotation and 'slip' between the weakly bonded molecular chains can take place, as a result of thermal vibration, and is the temperature at which the plastic softens (this can be measured using the Vicat apparatus, **3.4.1**). If the plastic is used at a temperature T_r (e.g, room temperature) and $T_r < T_g$, this relative motion is "frozen out" and the polymer is fairly rigid. In addition

if $T_g = T_r$ the plastic is **elastic or rubbery**
if $T_g << T_r$ the plastic is **rubbery or liquid**
if $T_g >> T_r$ the plastic is **stiff, hard (glass like)**, and maybe **brittle**

Hence the mechanical properties of the polymer are dependent upon the difference between the glass transition temperature T_g and the temperature of use T_r, as shown in ***Figure 13.8a***.

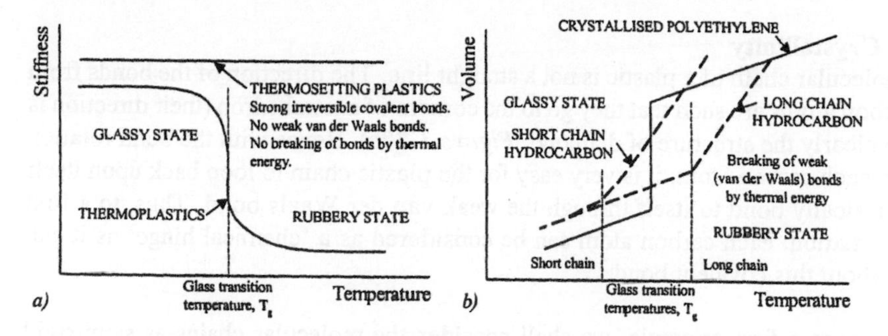

***Figure 13.8** a)Polymer stiffness in relation to the glass transition temperature, T_g; b) The effect of temperature on the volume of a thermoplastic*

There is no glass transition temperature associated with covalently bonded cross-linking. Hence all covalently cross linked structures are rigid and glass-like. Elasticity or rubbery features are conferred on the plastic if the adjacent chains are held together by a weak van der Waals forces instead of covalently bonded chains. The extent of this elasticity or rubbery state is dependent on the glass transition temperature. The glass transition temperature increases with increases in both the length of the hydrocarbon chain and the existence of (and length or size of) the side chains attached to the 'straight' hydrocarbon chain. These side chains give rise to a different naming of the hydrocarbon (**13.9.4**).

If a plastic is heated through the glass transition temperature, it will increase in volume. The weakest van der Waal bonds are the first to break. There will not be many of these and so there is no great increase in volume. As the temperature is increased still further, the next strongest bonds are broken corresponding to this temperature. At a certain temperature, where the majority of van der Waals bonds are broken, there will be a marked increase in volume (***Figure 13.8b***). This temperature corresponds to the **glass transition temperature** T_g. Naturally this temperature will be higher if the hydrocarbon chain is longer. It is the breaking of the van der Waals bond by thermal energy that creates the thermoplastic and the temperature at which this occurs is the glass transition temperature. ***Figure 13.8b*** illustrates that the change in the volume of the thermoplastic with temperature is chain-length dependent; also illustrated is the volume-temperature relationship for a fully crystalline plastic.

Above the glass transition temperature, the material is a very viscous liquid. The number of bonds broken is smaller, but their **strength** is **greater** than those broken at lower temperatures and so the volume increase is large (***Figure 13.8b***). This is accompanied by a movement and rotation of the chain about the tetrahedral bond (***Figure 1.11***), which must require co-operative movement or a certain free volume. If this free volume is less than **2.5%**, the co-operative movement is denied and the plastic is glassy. An analogy is people in a lift. If there is spare floor space, the people can move about, but if they are packed in with very little space, they cannot move without co-operation. Hence, a very dense plastic has little or no glass transition temperature (***Figure 13.8b***). The free space available can be modified by the addition of modifiers (**13.11**).

13.9.6 Crystallinity

The molecular chain of a plastic is not a straight line. The direction of the bonds from the carbon atoms are such that they go to the corners of a tetrahedron (their direction is shown clearly the structure of diamond, ***Figure 1.20b***). Hence with the bond rotation around each carbon atom, it is very easy for the plastic chain to loop back upon itself and physically bond to itself through the weak van der Waals bond. Thus, to a first approximation, each carbon atom can be considered as a 'chemical hinge' as it can rotate about this covalent bond.

However, as a first principle, we shall consider the molecular chains as semi-rigid "matchsticks" in order to discuss the effect of thermal energy upon the structure. ***Figure 13.8a*** illustrates the effect on specific volume ($1/\rho$) of the plastic with temperature as the plastic is slowly heated. Over the whole temperature range, the weak van der Waals bonds are broken at the temperature corresponding to their strength. Note that the covalent intra-molecular bonds (i.e., the C - C or C - H bonds) are **not** affected. For the purposes of this illustration, we assume that the plastic as it cooled down from the molten state. In the molten state, the molecular chain matchsticks will have a lot of vibrational energy, which will allow them to rotate. However, owing to their length there will be little translational movement or diffusion. The strongest van der Waal bonds at this temperature will tend to align the matchsticks to get the best packing. If these matchsticks are tipped out onto the desk (analogous to a rapid or fast cool) there will be several sticks which cross each other and there will be no obvious packing order. This lack of packing order is called an **amorphous** structure and is the structure of the "frozen liquid state". If the plastic could have been slowly cooled, the chains would have rotated and fitted into each other. This is analogous to gently vibrating the table onto which the match sticks have fallen, which would produce alignment in regions of complete order. This is called a **crystalline region**. If the ordered regions ran throughout the whole solid plastic, it would be fully **crystalline**. Due to the size of these polymeric "matchstick" molecules, complete rearrangement to get full crystallinity is **not** possible. In practice, the structure of a "crystalline" plastic can be viewed as composed of crystalline regions (called **spherulites**) within an amorphous matrix (comprising regions of complete disorder).

Figure 13.9 schematically represents the structure of polythene (40% amorphous, 60% crystalline).

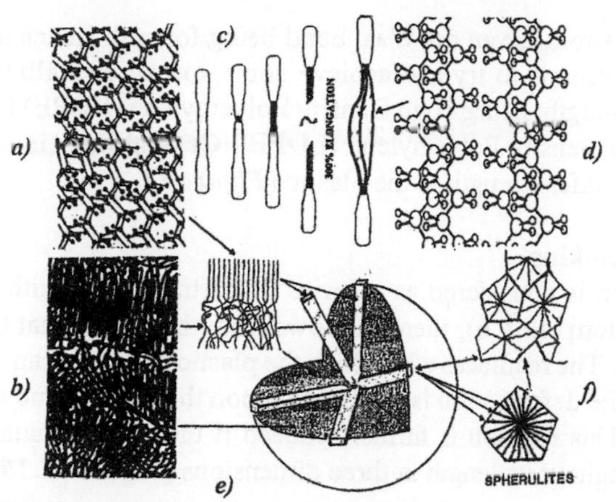

***Figure 13.9** Schematic representation of polythene*

The crystalline region is depicted in ***Figure 13.9a*** and the amorphous region in ***Figure 13.9b***. When polythene is pulled, test samples show elongations of around 300%, with considerable loss in cross-sectional area (***Figure 13.9c***). The deformation is initially concentrated in one area which, if pulling continues, broadens to include previously non-deformed areas, so increasing the cross-sectional area affected. The deformation causes the amorphous regions (in which the molecular chains are orientated randomly) to break down, i.e., the hydrocarbon chains effectively 'unzip' by breaking the van der Waals bonds between adjacent chains. The load orientates the molecular chains parallel to one another (***Figure 13.9d***) so that the load is carried by the covalent bond within the chain. Aligned chains can easily be separated (by simply pushing a fingernail through the plastic, ***Figure 13.9c***). The structure of the crystalline regions of polythene (the spherulites) are illustrated in ***Figure 13.9e***, which appears similar to the (3 dimensional) spokes in a wheel. The spokes are sub-microscopic regions of crystallinity (called **lamellae**) (in fact the lamellae appear twisted, but this is omitted from ***Figure 13.9*** for clarity). The lamellae have the crystalline structure shown in ***Figure 13.9a***, whilst the region between the lamellae is amorphous (having the amorphous structure of ***Figure 13.9b***). ***Figure 13.9f*** illustrates schematically the lamellae (observed by X-ray diffraction and rotation of the plane of polarisation in optical microscopy). Note that there are **no dislocations** in a plastic, even in the crystalline region. This means that plastics do not **strain (work) harden** like metals (**10.5.3**).

Effect of side chain addition on crystallinity
If the hydrocarbon molecular chain has chemical side chains along its length, the matchsticks are no longer straight but can be considered **barbed**, causing the formation of crystalline regions to become **more** difficult. This means that organic thermoplastics with side chains do not crystallise so well. Furthermore, these barbed chains cannot move easily relative to each other. Therefore, compared to unbarbed molecular chains, barbed molecular chains have higher glass transition temperatures, T_g, and stronger mechanical properties at all temperatures. The amorphous structure confers a "glass-like transparency" to the "frozen" plastic, whilst a crystalline structure tends to give opacity or milkiness to the plastic. If in our analogy the matchsticks all align with each other,

there is more chance of a stronger van der Waal bond being formed. Hence one of the parameters used by scientists is to try and achieve some sort of crystallinity in the structure to give added strength; hence High Density Polyethylene (HDPE) is stronger and more rigid than Low Density Polyethylene (LDPE). **Crystallinity** is **one** of the many parameters which confer strength to the plastic (***Figure 13.7***).

13.9.7 Degree of Cross-linking

If the plastic hydrocarbon is considered as a series of partial hinges with complete rotation at each carbon atom (**13.9.6**), then it follows from this model that the plastic will be easily deformable. The readiness with which the plastic molecule can "unhinge" and rotate to accommodate deformation is dependent upon the number and size of the side branches (**13.9.4**). This rotation is further reduced if the plastic chain is **cross-linked** periodically throughout its length in three dimensions (***Figure 13.10***).

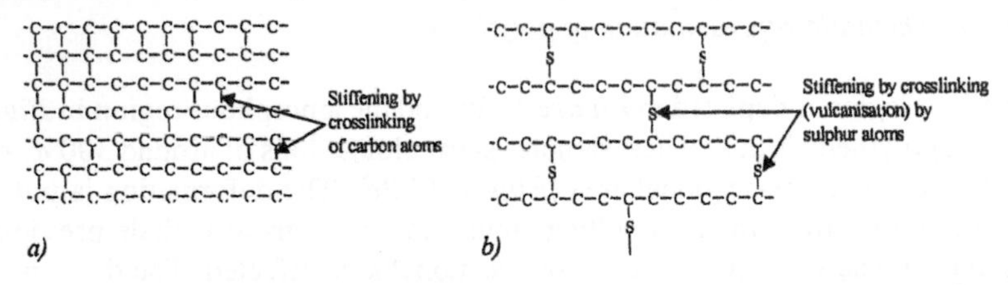

Figure 13.10 *Model of a cross-linked structure is depicted as a three-dimensional array of covalently bonded carbon atoms*

As noted in section **13.6.2**, the actual compounds used in thermosetting reactions are a mixture of aliphatic and aromatic hydrocarbons. This produces a very complex three dimensional cross-linking. In ***Figure 13.10***, for simplicity,

- the array is shown without accompanying hydrogen atoms (which are bonded to those carbon atoms which do not have four nearest covalently bonded neighbours);
- without the cross bonding of - CH_2 - units (as in ***Figure 13.5c***).

In thermoplastics, the linear chains can be cross-linked (like the rungs of a ladder) by sulphur in a process called **vulcanisation** (***Figure 13.10b***). Vulcanisation changes the thermoplastic in to a thermosetting plastic. Furthermore, the reverse process can be undertaken (e.g, thermosetting rubber → thermoplastic rubber) by opening up the cross-links so that the rubber can be reclaimed.

13.10 ELASTOMERS

Elastomers are macromolecules which, when deformed by a weak applied stress, will return rapidly to approximately their initial shape and dimensions after the stress has been removed. There are a great number of elastomers, including natural rubber and synthetic polymers. Elastomers can be thermoplastic (e.g, styrenes, olefins, polyesters, polyamides and polyurethanes) or vulcanised (**13.9.7**). Vulcanisation provides excellent functional properties but prevents subsequent remelting. As with moulded and extruded plastics, rubber and elastomers are most often used after the base polymer has been modified by additives (**13.11**). Some common elastomers are considered in the following subsections; functional properties are summarised in **Table 13.9**.

Table 13.9 Some functional properties of common elastomers

Material	Abbrev.	Vulcanisation agent	T_s	T_b	T_{max}	T_{min}	Flame resistance	SG	Water resistance	Weathering (sunlight ageing)	Oxidation	TS (MN/m^2)	EL (%)	Crystallisation on stretching	α (per °C x 10^{-6})
Natural rubber	NR	Sulfur	-29 to -45	-62	82	-54	D	0.92	A	D	B	31	650	B	37
Isoprene rubber	IR		-29 to -45	-62			D	0.91	A	E	B	31	650		37
Neoprene	CR	MgO or ZnO	-12 to -45	-65	100	-54	B-A	1.25	B	B	A	28	600	C	34
Nitrile rubber	NBR	Sulfur	0 to -45	-40	121	-54	D	1.0	B-A	D	B	28	650	E	39
Polysulphide	PTR		-23 to -43	-56			D	1.35	B	B	B	10	450		42
Urethane	PU	Diisocyanates	-23 to -34	-51 to -129	100	-54	D	1.25	C-B	A	B	34	750	B	27
Silicone	SI	Peroxides	-51 to -118	-68 to -118	237	-84	A	1.1-1.6	A	A	A	10	900	D	45

Notes: Physical properties shown in this Table apply to the general class of base polymers. A - Excellent; B - Good; C - Fair, D - Poor (use with caution); E - Very poor (not recommended).
T_s (stiffening temperature), T_b (brittle point temperature), T_{max} (maximum service temperature), T_{min} (minimum service temperature). SG (specific gravity). TS (tensile strength, MN/m^2). In elastomers, tensile strength refers to the force per unit of original cross-section on elongating to rupture, thus not really an important property as elastomers rarely used in purely tensile applications. However, other elastomer properties correlate well with tensile strength (wear and tear resistance, resilience, cut resistance, creep and, in some elastomers (e.g, neoprene) ozone resistance. EL (%) (elongation at break). The maximum extension at the moment of rupture, i.e., an elastomer with <100% elongation will usually break if doubled over on itself. Most elastomers undergo considerable elongation and necking (3.2.1). α (per °C) Coefficient of thermal expansion (**Table 2.3**).

13.10.1 Natural Rubber

Natural rubber was the original basis of electrical insulation. In comparison with modern synthetic elastomers, natural rubber possesses high tensile strength, resilience, resistance to wear and tear and to acids and bases (except oxidising agents). Resilience is the main superior quality compared to synthetic elastomers. However, natural rubber does not perform as well for sunlight, oxygen and ozone resistance (although additives offset these disadvantages to some extent), or in resistance to oils from petroleum, vegetable or animal origin. In addition, natural rubbers will soften over time, whereas synthetic elastomers gradually harden. Despite these disadvantages, natural rubbers are used in tires and various mechanical goods; they have also been used to mount buildings to provide vibration protection (e.g, from underground trains).

13.10.2 Isoprene Rubber (IR)

Isoprene rubber is the synthetic equivalent of natural rubber and was developed during World War 2 when natural rubber supplies were scarce. Isoprene rubber is formed from a polymerisation of the monomer **isoprene**, which is $CH_2 = C(CH_3) - CH = CH_2$. The scientist represents the isoprene rubber polymer as the breaking of the two double bonds at either end as $(- CH_2 - C(CH_3) = CH - CH_2 -)_n$; note that one double bond per monomer is lost through polymerisation. Isoprene rubber has the advantage over natural rubber of uniform molecular weight (giving uniform processing qualities) and odourlessness.

13.10.3 Neoprene (CR)

Neoprene is made from the monomer **chloroprene,** and therefore the scientists name for neoprene is **polychloroprene** (CR). Neoprene consists of chains of

$$- (CH_2 - \underset{\displaystyle Cl}{\overset{}{\underset{|}{C}}} = CH - CH_2 -)_n$$

Neoprene, developed by du Pont in the 1920s, was the first commercial synthetic rubber. It has inherently high tensile strength and elongation properties (comparable with natural rubber) as it crystallises on stretching (**Table 13.8**). Wear properties are also good. It has excellent flame resistance (and is in fact self-extinguishing) and therefore has been used in areas where fire is a potential hazard (e.g, coal-mining). Neoprene is also resistant to oxidative chemicals and has good weather- and ozone-resisting properties, making it an ideal material for sealing strips for glazing units, 'O' ring seals and gaskets. Neoprene is also used in **contact adhesives (13.16.1**).

13.10.4 Nitrile rubber (NBR)

Nitrile rubber is mostly an alternating copolymer of butadiene and acrylonitrile. Butadiene is derived from the butane monomer C_4H_6 or $CH_2 = CH - CH = CH_2$ (butane is lighter fuel C_4H_{10} or $CH_3 - CH_2 - CH_2 - CH_3$). The butadiene polymer derived from the C_4H_6 monomer is depicted as $(- CH_2 - CH = CH - CH_2 -)_n$, where n stands for chains of $(- CH_2 - CH = CH - CH_2 -)$. Note again that one double bond per monomer is lost through polymerisation. Nitrile rubber consists of chains of

$-(CH_2-CH=CH-CH_2)$	$[-CH-CH_2-]_n$ (with CN attached to CH)
Butadiene	Acrylonitrile

This method of writing down the formula is a scientists representation of the way the copolymer adds on. Nitrile rubbers are noted for their oil resistance and are extensively used as petrol hoses and engine mounts. They are also used in the production of adhesives. A high impact plastic can be made by adding another copolymer (styrene) to give acrylonitrile butadiene styrene (ABS); the styrene addition stiffens up the chain.

13.10.5 Polysulphides

Polysulphides consist of the unit - C - S - S - . They have good resistance to ozone and ultraviolet light, and are used in the automotive and construction industries as sealants.

13.10.6 Urethanes

Polyurethane consists of the fundamental unit - O - CO - NH -, whilst polyurethane rubber is chains of $(-R-O-CO-NH-R'-NH-CO-O-R''-)_n$, where R and R" are aliphatic chains like $-CH_3$, $-C_2H_5$, $-C_4H_{10}$, etc. and R' is a cyclic compound like $-C_6H_5$. The number of carbon atoms in either R, R' or R'' is dependent upon type of rubber and the physical properties required from the compound. Urethanes cross link and undergo chain extension to produce a wide variety of compounds. They can be cured to form tough, elastic solid rubbers with outstanding load-bearing properties, abrasion resistance, resilience and tensile strength (two to three times that of natural rubber). They have good resistance to oils, solvent, oxidation and ozone, but poor resistance to hot water (they are not recommended for temperatures above 80°C). Foamed polyurethane (available as foam or granules) is widely used in the construction industry as both a heat and acoustic insulation material (cavity wall insulation, subfloor insulation, etc.) (**13.8**), having excellent low thermal conductivity at low temperatures, low vapour permeability and the ability to be foamed *in situ* (by mixing a volatile inert organic liquid with a resin mixture to produce a foamed polyurethane which expands after placing and is adherent to backing materials). It is therefore possible to surround complex series of pipes, for example, to provide thermal insulation or to fill complex cavities. It should be noted that voids in the filling and water bridging caused by unseen obstacles within the cavity may lead to aggravated moisture penetration of the cavity wall at these positions. In addition, a great deal of skill is required to correctly apply the foam on building sites (compared to controlled laboratory conditions), where adverse weather conditions and wide variations in the wetness of surfaces occur.

13.10.7 Silicones

Silicones comprise a polymer backbone of silicon and oxygen atoms, rather than the carbon structure of all other elastomers. Silicon is in the same chemical group as carbon and, since it is a more stable element, scientists predicted that more stable compounds could be obtained if carbon was replaced with silicon in the molecular chain structure. Silicones have high resistance to oxidation, ozone and weathering and are commonly used as sealants in the construction industry.

13.11 POLYMER MODIFIERS AND ADDITIVES

In order to alter the properties of the plastics, certain additions, summarised in **Table 13.10**, may be made.

Table 13.10 Typical polymer additives

Additive	Description
Blowing agents	Blowing agents can be either physical or chemical types. Physical blowing agents change from a liquid to a gas when heated, i.e., they only change form. Expanded polystyrene (EPS) is a typical product based on a physical blowing agent. A chemical blowing agent decomposes when exposed to a critical temperature and the gaseous product expands the molten polymer. Structural foams are commonly moulded using chemical blowing agents. Different types and concentrations of chemical blowing agents result in very different physical properties, even at the same density reduction.
Colorants	Colorants can either be chemical dyes or organic and inorganic pigments. There are very specific rules for colorant selection based on resins used, amounts and end use requirements.
Fillers/reinforcements	A filler can be either an extender type, a functional filler or a reinforcement. Extender type fillers are used primarily to lower the overall product cost. These are usually a cheap inorganic additive whose function is to "pad out" and stiffen the plastic and therefore reduce the cost of the product when used in mouldings. Functional fillers are also inexpensive, but also contribute one or more desirable performance properties (typically tensile modulus, impact resistance and/or heat resistance). Examples include carbon black (conferring stability to UV light, electrical conduction and colour), chalk, Fullers earth etc. A reinforcement is usually more expensive than the base resin, but does contribute to a significant enhancement of tensile modulus, tensile and flexural strengths and upper use temperature. Fillers tend to be fine powders whilst reinforcements possess greater physical size (e.g, fibres, honeycombs, etc.)
Flame retardants	Flame retardants act to slow down the combustion rate of the polymer, often by generating a heavy smoke or moisture to cool the fire. Flame retardant are often used with smoke suppressants. Many flame retardants adversely affect other polymer properties. Note that the charring in the thermosetting plastics (**14.6.6**) is a natural fire retardant, as carbon does not vaporise and therefore does not produce a flame. For further consideration of flame retardant additives, see **14.7.3**.
Plasticisers	The primary functions of a plasticiser are to enhance processibility by reducing melt temperatures and to improve basic polymer properties (such as impact behaviour, low temperature flexibility and softness) without adversely effecting dimensional stability. Unfortunately, plasticisers adversely affect modulus, strength, upper use temperature (by reducing the glass transition temperature) and chemical resistance. Plasticisers used to impart flexibility are usually large compatible molecules which act like ball bearings. For example, plasticisers used in PVC are saturated, long chain paraffins of high melting point and low volatility. Unplasticised PVC (PVC-U) is a rigid material used for guttering, window frames, etc. Introduction of a plasticiser increases the flexibility of the product to produce products that imitate the properties of leather, for example (such as the "plastic" seat coverings of motor vehicles, chairs, etc.). Over a long period of time this plasticiser is driven out of the PVC, as can be seen in motor vehicles exposed to high temperatures when left locked in the midday sun, where these paraffinic plasticisers are deposited on the windscreen giving the waxy film on the inner window surfaces. The PVC seat material becomes "harsher" and loses the ability to bend back on itself and so cracks.
Ultraviolet absorbers	Ultraviolet absorbers are added in small quantities to absorb the harmful, high energy UV rays that will attack polymers. Unlike some additives that are used to protect the polymer during high temperature processing, UV absorbers are required to have a long-term function.
Antioxidant	Antioxidants are used to ensure thermal stability against oxidation during high temperature processing.
Antiozonant	Antiozonants are used to ensure protection against ozone attack, especially in rubber-modified plastics.
Antistatic agents	Antistatic agents are used to protect the polymer from excessive static charge accumulation due to processing.
Impact modifiers	Impact modifiers enhance impact behaviour but at the expense of reduced modulus, strength and heat resistance.
Lubricants	Lubricants (release agents) are added either during compounding or during the actual melt processing. They prevent the hot polymer from 'wetting' (sticking) to the metal mould making it easier to remove components after manufacture.
Thermal stabilisers	The primary function of thermal stabilisers is to ensure short term stability during processing at elevated temperatures. Thermal stabilisers are a critical additive in any vinyl compound where they operate to remove the released HCl generated during processing. If not properly controlled, excess HCl will cause the vinyl polymer to degrade (called chain scission).

13.12 DEGRADATION OF POLYMERS

In general, thermoplastics are combustible, whilst thermosetting materials are not (although thermosetting plastics will give off obnoxious vapours when heated and before charring, **14.8.2**). Thus thermoplastics should be used with caution, particularly when the temperature of the service environment is accurately known. Generally, plastics do not rot or corrode and offer good resistance to moisture, some oils, acids and alkalis. However, they do deteriorate through a wide variety of mechanisms. In most cases, these mechanisms are extremely complex and, even now, are not fully understood. Degradation of plastic materials can broadly be described as irreversible changes in some of their properties which are detrimental to their usefulness and performance. Changes may occur during processing and during exposure to the service environment. Conditions during processing are relatively severe but are of short duration and usually involve the action of oxygen. Although plastics are usually well protected against thermal oxidation during processing, impurities and some additives can initiate or accelerate degradation in service.

Plastics represent a relatively new class of materials in the construction industry and therefore long-term 'in service' tests on the degradation mechanisms are not readily available. As most plastics do not readily degrade, deterioration of plastics is commonly assessed in accelerated tests. Degradation mechanisms in plastics (which, once elucidated, may lead to novel applications for plastic products) will continue to be found as the accumulated data from laboratory and long term 'in service' tests becomes available. Degradation in service usually starts at the outer surface and penetrates gradually into the bulk of the material, and may take the form of discolouration, pitting, reduction in gloss, crazing or cracking, erosion, leaching (e.g, the loss of plasticiser in plasticised PVC, **13.11**), embrittlement, fibre prominence (in glass fibre reinforced plastics) and deterioration of mechanical properties. Plastics can deteriorate in service through chemical processes (principally under the combined action of UV light and oxygen) and physical processes (under the effects of environmental water and heat).

13.12.1 Chemical Degradation Processes

The deleterious effects of weathering on plastic materials is generally ascribed to a complex set of processes in which the combined action of UV light and oxygen are most important (termed **photodegradation**). To undergo a photodegradation reaction, the material must absorb light (radiation wavelengths of greater than approximately 290 nm, since these are the wavelengths that normally reaches the earths surface; shorter wavelength radiation is absorbed by the ozone layer). When a polymer molecule absorbs UV radiation it attains an electronically excited state. The excited molecule may either release its absorbed energy by re-emitting it at longer wavelengths (as fluorescence, phosphorescence or heat) or it may transfer it to another molecule during a collision. In either case, the excited molecule will return to the unexcited state. When absorbed energy is not dissipated by any of these processes, it will initiate photodegradation. Generally, degradation is dependent of the wavelengths to which the polymer is exposed and upon the impurities and additives incorporated in the polymer.

The absorption of UV light causes the dissociation of bonds (mostly C - C and C - H) within the molecules of one or more constituents of the polymer, leading to one or more of the following chemical changes

- cleavage into smaller molecules (chain scission) (e.g, the embrittlement of weathered PE, PTFE, PMMA and cellulosics);
- crosslinking (e.g, the embrittlement of PE, PP, PS and certain grades of PVC);
- elimination of small molecules;
- formation of double bonds in the main chain (e.g, oxidative reactions in PVC cause dehydrochlorination, leading to colour change and loss of light transmission, particularly in warm weather conditions);
- depolymerisation;
- photohydrolysis.

These mechanisms are together termed **ageing**. Chain growth and crosslinking generally result in the formation of a hard, rigid material, whilst chain scission and depolymerisation result in a plastic or resinous mass. In addition to UV light, heat, oxygen, stress with atmospheric ozone and atmospheric nitrous oxide promote polymer degradation. A well known example is the ageing of polythene and some cheap rubbers, which lose flexibility over time. This occurs by the crosslinking of adjacent chains through oxygen (or, if present, sulfur) atoms under the catalytic action of UV light, though this is often limited by including an anti-oxidant (e.g, phenol) and an opaque filler (e.g, carbon black) to exclude UV light.

13.12.2 Physical Degradation Processes

Moisture exerts an important influence on the degradation of most plastics by causing them to shrink and swell (**13.6.1**). In addition, free moisture may leach out soluble products of photodegradation and certain additives. Water may also enter secondary photodegradation reactions (e.g, hydrolysis of photactivated polyesters and polyamides).

Thermal effects during weathering are determined by the actual temperature of the material and particularly the temperature at the exposed surface (where most of the degradation takes place). The difference between the exposed surface of the material and the ambient air temperature can be relatively large (temperature differences greater than 30°C are not uncommon in sheet materials on summer days and higher temperature differences are likely where the material is dark in colour).

The combined effect of moisture and temperature cycles on plastics can cause severe degradation in the form of surface cracks. This degradation is greatly accelerated by the action of solar radiation. The effects are particularly severe in polycarbonate and glass fibre reinforced polyester. For these materials, cyclic variation in humidity causes absorption and desorption of moisture, resulting in alternate swelling and shrinkage of the surface, and cyclic variations in temperature induces alternate volume expansion and contraction. Cyclic dimensional changes are not uniform and thus produce variable, nonuniform stresses that result in stress fatigue (**3.5.5**). The effect of physically induced stress fatigue is particularly important in composites (e.g, GRP), where dissimilarities in the properties of the main components (matrix and reinforcement) are large. Cracks in the glass-resin interface occur relatively quickly on exposure to cyclic moisture and thermal cycles and prolonged exposure leads to gradual fragmentation of the resin matrix. UV radiation accelerates surface degradation by rendering the matrix more brittle as a consequence of UV induced crosslinking. Eventually the glass fibre filaments become delaminated completely, resulting in a large number of fibres becoming exposed at the surface. After long term weathering, the exposed surface develops randomly orientated, single micro cracks which propagate gradually to form a network. This occurs because the exposed surface of the GRP sheet undergoes gradual shrinkage as a result of the crosslinking caused by UV light. The crosslinking causes the matrix to become more rigid, so that the permanent tensile stresses produced by the shrinkage can no longer be reversibly accommodated and the material fractures.

13.13 IDENTIFICATION OF COMMON PLASTICS

If identification of the common plastics used in the construction industry is required, the following tests can be carried out. Based on these simple tests, ***Figure 13.11*** gives the procedure for identifying a particular plastic. Reference should be made to **Tables 13.3 and 13.4** which summarise the basic properties of thermoplastics, thermosets and elastomers, and to **Tables 13.6 and 13.9** which compare functional properties for these polymer groups.

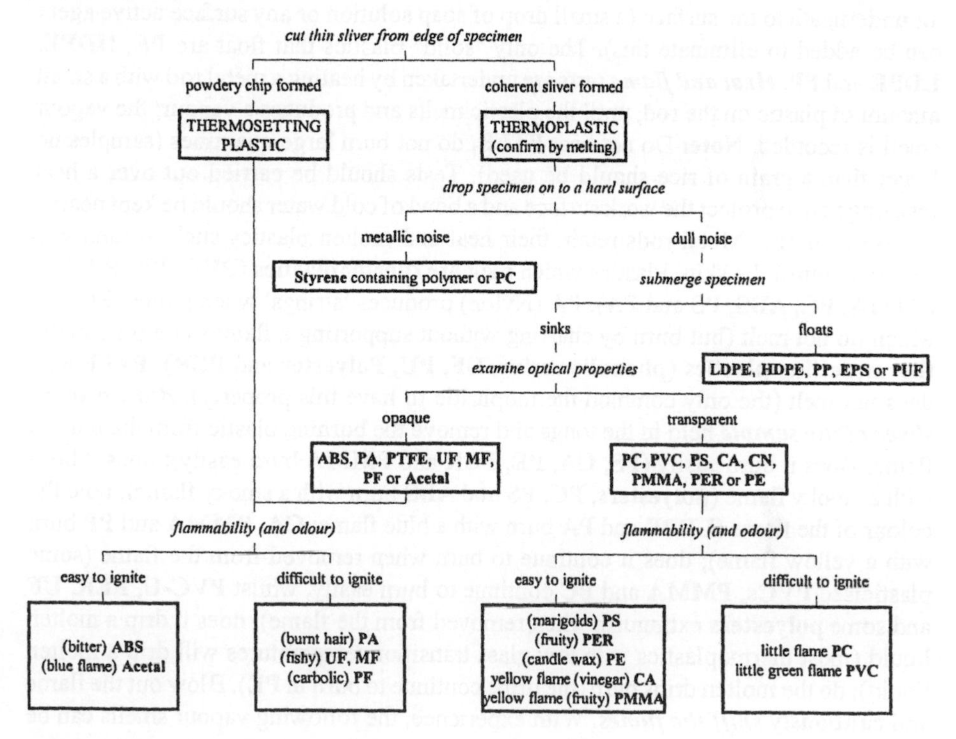

Figure 13.11 *Identification of common plastics*

The identification process will include ***Visual inspection***: is it transparent or opaque; is it filled (chalk calcium carbonate, calcium silicate and carbon black will make a transparent plastic opaque; however, an opaque plastic cannot be made transparent. Opaque plastics include **UF** and **PF**). ***Surface examination***: scratch the plastic with a thumb nail (cellular plastics); does it bend easily; is it flexible for its physical dimensions; is it greasy (waxy materials include **PE, PP, PA** and **PTFE**); does it melt if a hot rod is pushed into it (a **thermoplastic** will give a thin fibre of the plastic hydrocarbon chain when the rod is withdrawn from the molten environment whilst a **thermoset** will not react to the hot rod). ***Cut the sample*** using a razor blade or sharp knife: note the appearance of the sliver (a jagged or a smooth edge); ease of cut (the addition of fillers may make the plastic more difficult to cut whilst the addition of plasticisers may make cutting easier); examine the cut surface optically for evidence of aeration (**13.8**). The result of these tests will indicate whether the plastic is a **thermoplastic** (cut easily to produce a sliver, if not filled too much) or a **thermosetting** plastic (produces a jagged cut without a sliver); the hot rod test will tend to differentiate these.

The ***relative density*** with respect to water helps classify the plastic. If the plastic sinks, it has a greater density than water (1000 kg/m^3), whereas if it floats it is less dense. However, care should be taken in interpreting this test because fillers make the plastic more dense, aeration (foam filled) makes the plastic less dense, greasy plastics can trap air underneath to the surface (a small drop of soap solution or any surface active agent can be added to eliminate this). The only "solid" plastics that float are **PE, HDPE, LDPE** and **PP**. ***Heat and flame tests*** are undertaken by heating a metal rod with a small amount of plastic on the rod, until the plastic melts and produces a vapour; the vapour smell is recorded. **Note:** Do not heat **PTFE**, do not burn large quantities (samples no larger than a grain of rice should be used). Tests should be carried out over a heat resisting mat to protect the work surface and a bowl of cold water should be kept nearby in case of burns. Metal rods retain their heat and molten plastics stick to (and will seriously burn) the skin. Plastics which melt are **thermoplastics (PVC, PE, PP, PA, PMMA, PC, ABS, PS** and **PA). PA** (Nylon) produces 'strings' when pulled. Plastics which do not melt (but burn by charring without supporting a flame))are principally **thermosetting plastics (phenolic resins, UF, PU, Polyester** and **PER). PTFE** also does not melt (the only common thermoplastic to have this property). ***Burn a small sliver of the sample*** held in the tongs and remove the burning plastic from the bunsen flame. Does it burn easily (**PS, CA, PE, PER** and **PMMA** burn easily); does it burn with a smoky flame (**polyesters, PC, PS** and **ABS** burn with a smoky flame); note the colour of the flame (**LDPE** and **PA** burn with a blue flame; **CA, PMMA** and **PP** burn with a yellow flame); does it continue to burn when removed from the flame (some plasticised **PVCs, PMMA** and **PC** continue to burn easily, whilst **PVC-U, PER, UF** and some **polyesters** extinguish when removed from the flame); does it drip a molten liquid (most thermoplastics with low glass transition temperatures will drip a molten liquid); do the molten drips burn (the drips continue to burn in **PE**). Blow out the flame and **cautiously *sniff the fumes***. With experience, the following vapour smells can be identified

Burning candle	Fruity	Acrid	Fishy	Hair	Formaldehyde	Styrene	Antiseptic
PE PP	PMMA	PVC PU	U MF PER	PA	POM (PF, MF, UF)[a]	PS PE ABS	PC PF

Notes: [a] smell likely to be masked

13.14 ADHESIVES

Adhesives are a group of chemical compounds which are capable of sticking two or more components to form a new complete entity. Some adhesives allow the bonding together of materials which are impracticable (if not impossible) to join by any other means. Adhesives are even used in the aerospace industries and, as such, adhesive technology is advancing very rapidly. Adhesives can be used to stick a wide range of materials, ranging from ceramics and glass reinforced polyesters (GRP) to metal, motor vehicle plastic trim to a glossy paint surface and very thin non-weldable alloys to metals.

Advantages of adhesives include

- The adhesive bond has an **amorphous** structure that provides high vibration damping and good performance in high fatigue environments. For this reason helicopter rotor blades, and the tail section and some 80% of the aerodynamic surfaces of a jet liners wing, use adhesive bonding. Metallic coupling of rubber to metal in automotive suspension systems is through a thin film of copper, which has to be plated onto the steel before the adhesive will bond. It is thought that this bonding takes place through the porosity of the electroplate, where the copper-sulphur chemical bond formed promotes adhesion. Brake linings are adhesively bonded to brake shoes (replacing localised fixing provided by rivets);
- The adhesive bond will not affect the overall properties of metallic components (as would occur in welding processes, **10.14.1**). Other methods of joining (e.g, welding, brazing and soldering) can cause distortion and warping due to differential thermal expansion and subsequent contraction and/or colour changes in the metal substrate. The maximum temperature required for adhesive bonding is 65°C so that distortion through thermal movement is minimised;
- Corrosion problems resulting from residual fluxes (from welding or brazing processes) are eliminated;
- The adhesive does not rely upon drilling, countersinking and other processes and so is labour saving and cosmetically attractive (as adhesives will fill gaps). The stresses associated with drilled holes (see, e.g, **9.6**) are eliminated;
- Adhesives spread the load over the total area of contact and therefore the stress concentration effects associated with other processes of joining (tack welds, rivets and screws, etc.) is therefore eliminated. This is important in engineering where parts undergo high stress cyclic loading (e.g, aircraft uses);
- Adhesives make use of the total area of joining to produce a load-bearing joint, producing in one operation a vapour tight, crevice free joint (***Figure 11.11***);
- Adhesives can be very successfully used where **two chemically dissimilar metals** are joined together. In these circumstances, galvanic bimetallic couples (**11.6.1**) could cause accelerated corrosion if there were metallic contact. The adhesive serves to form an electrically insulating layer between the two metals and also eliminates the crevice. Hence sealants or adhesives are used as inter-weld mastics in automotive applications. Certain inter-weld mastics and jointing compounds are known as **adhesive sealants.**
- Adhesives may be used to close joints where the surfaces to be joined may not be too close or in continuous contact (**gap-filling adhesives**). Gap-filling adhesives are commonly synthetic resins. The limit for the glue line is about 1.3 mm thickness, above which gap-filling adhesives should not be used unless otherwise stated by the manufacturer.

Disadvantages of adhesives include the common requirement for expensive jigging and heating operations. There is no universal (chemical) adhesive.

13.15 PROPERTIES OF ADHESIVES

The adhesive bond is developed by a process of **adhesion** and **cohesion**. The adhesive has to be applied as a **liquid** and it has to stick to **two different substances** with sometimes completely differing chemical bonds (i.e., **adhesion**). The adhesive in its final state has to be a **solid** to allow high loads to be carried (i.e., **cohesion**). The fundamental requirement for all adhesives is **permanence** and **durability**. Therefore, for an adhesive to work satisfactorily, it must possess good adhesion (stickability to the surfaces) and cohesion (within itself) so that it remains intact when subjected to stress. The cohesive forces develop in the curing process and are a mixture of covalent and van der Waals bonds. The cohesive forces are therefore a property of

- the van der Waals bond (**1.14**);
- the degree of cross-linking between the hydrocarbon chains (**13.9.7, 13.16.3**);
- the number and size of the side chains (**13.9.6**).

There are several ways in which the requirements of cohesion and adhesion can be satisfied by different adhesive formulations; these are examined below.

13.15.1 Adhesion

Adhesion is the ability of the adhesive to stick to the surfaces of the two adherends (the material being stuck is termed the **adherend**), and can be generated in two ways

- by **specific bonding**. Specific bonding is a term used to describe the process by which the adhesive reacts chemically with the adherend, giving rise to a property called **tackiness**. Certain hydrocarbons have a natural affinity for a surface and are tacky to the touch. These naturally occurring materials (e.g, resins from trees, gum arabic, etc.) have been used for thousands of years as adhesives and are still used (in conjunction with modern polymers) to obtain adhesion and strength;
- by **mechanical interlocking**. Mechanical interlocking is a term used to describe the process of producing a joint by a physical cementing process, where the adhesive penetrates the surface (pores, microscopic pores, holes, scratches, etc.) of the materials to be jointed. Note that many apparently smooth materials have sufficient surface roughness (**13.17.1**) to allow mechanical interlocking adhesion.

In order to generate adhesion, adhesives have the following fundamental properties

- they are thin, low viscosity liquids. As a liquid, adhesives are predominantly bonded with weak van der Waals bonds (so that the adhesive will spread easily). However, these weak bonds **must be capable of transforming to strong bonds** (in a short time) when adhesion and cohesion is required. In addition, this property must not develop over a period of time (otherwise the adhesive will sets slowly in the container). 'Shelf life' is very important property of adhesives (**13.17.2**);
- the "tacky" materials must have relatively short molecular chains (so they can flow to follow the surface contours of the adherends easily). Note that many materials are rough on the microscopic scale (**13.17.1**) but mountainous when compared with the dimensions of the interatomic bond of carbon (the main constituent of an adhesive).

13.15.2 Cohesion

Cohesion is the ability of the adhesive to remain intact when subject to stress. Cohesion forces develop in the curing and hardening process (the classification of adhesives, **13.20**, is by their chemical mechanism of hardening). Cohesive forces are a function of

- cross linking (by covalent bonding) within the adhesive mass (**13.9.7**); and

- the number of side chains on the long chain molecule, which act in a similar way as the barbs in tangled barbed wire (**13.9.2**).

13.16 CURING OF ADHESIVES

The mechanisms by which the liquid adhesive transforms to a solid can be

- by evaporation of a solvent (**13.16.1**);
- by formation of regions of crystallinity (**13.16.2**);
- by crosslinking (**13.16.3**);
- by excluding oxygen from the interface (**13.16.4**).

13.16.1 Evaporation of a Solvent

The adhesive (normally rubber compounds, polybutylene, polybutadiene, etc.) is a high viscosity, very tacky, high molecular weight polymer, which is carried in a solvent (either an organic solvent or water). The solvent evaporates (or is absorbed in the adherend) leaving the adhesive to carry the load and to provide the "stick". These types of adhesives are often helped by leaving the applied surfaces exposed to the air to aid evaporation. Once tacky, the applied surfaces are brought and held together to form the bond. These adhesives are known as **contact adhesives** and require a close contact joint (less than 0.15 mm). Joints for which the bond thickness is greater than 0.15 mm but less than 1.3 mm (**gap joints**) ordinarily require a gap-filling adhesive (**13.14**). Note that if adjacent surfaces were brought together before the solvent had evaporated, the stresses generated by the volume shrinkage (of some 70-80%) would self-destroy the adhesive joint. This is particularly important for non-absorbent surfaces. Some adhesives are allowed to completely dry before bonding and in these cases the surfaces must be reactivated (either by heating with infra-red lamps or a solvent wipe).

Contact adhesives are usually categorised by the solvent as either water borne adhesives (WBA) or solvent borne adhesive (SBA). Water borne adhesives are emulsions of the polymer particles which coalesce when the water is adsorbed by the porous adherend. The emulsion has the appearance of a milky liquid. When bonding nonporous materials, WBAs take longer to reach full strength than the solvent-based derivatives. Note that **organic solvents** are flammable, usually undesirable to inhale and frequently greenhouse gases. Increasingly therefore, procedures which reduce the use of organic solvents are being investigated and introduced.

13.16.2 Formation of Regions of Crystallinity

Some polymers have sufficient structural symmetry that crystalline areas are produced in an otherwise amorphous matrix (**13.9.6**). These crystalline areas act in the same way as cross-linking (**13.9.7**) by tieing the polymer chain together, thus preventing movement. The areas of crystallinity also act as the **hard** segments within a polymer adhesive which act to restrict flow and promote good creep resistance. Some adhesives of this type are made from two polymers with different properties (e.g, one polymer may be glassy whilst the other is rubbery). The two polymers have the property of being immiscible with each other and it is the chemists job to join these polymers together into one chain so that each exists as a **block** (**13.9.3**). The immiscible blocks separate in the polymer solid, but because they are chemically linked together, they do not fully separate. This mechanism is the basis of **crystalline block polymers**; a good example is the styrene/butadiene/styrene block copolymer. These materials will flow when the temperature is raised above the glass transition temperature T_g of the glassy block.

Adhesives which form areas of crystallinity can be

- **Cold-setting**, comprising a synthetic resin adhesive intended to set and harden satisfactorily at temperatures between 10°C and 50°C. Cold-setting adhesives may, however, be heated to accelerate setting if required;
- **Warm (or hot) setting**, comprising a synthetic resin adhesive which sets and hardens satisfactorily at temperatures between 25°C and 90°C. Note that some warm-setting adhesives may require a minimum temperature above 25°C. Many hot setting adhesives are **thermoplastic** and so they melt to produce a liquid at elevated temperatures, which allows the adhesive to run into the rough surface and be easily spread. On cooling they revert to their "solid state" and produce strong bonds. The highest temperature of heating is around 200°C (above which the carbon chain will start to char and colour the adhesive);
- **Hot melt (setting)** adhesives require temperatures above 100°C to set and harden. If temperatures above 90°C are to be employed, it is essential that the suitability of the adherend for heating to such temperatures, and the method to be employed, be confirmed by the manufacturer.

13.16.3 Crosslinking

Reactive adhesives stiffen by cross linking. Polymer chains can be made more rigid and less mobile by crosslinking the chains in certain areas to produce a three dimensional structural material with a very high (or even absent) glass transition temperature. Resultant polymers are glassy, have very high creep resistance and are capable of transmitting very high loads. Such polymers are the basis of high strength structural adhesives. Unfortunately, the disadvantage of glassy polymers is their characteristic brittleness. These adhesives are therefore formulated as **copolymer (13.9.3)** compounds of differing glass transition temperatures (brittle/glassy components with high T_g and rubbery materials with low T_g). Resultant polymers are resistant to impact loads without losing the high strength characteristics of the glassy adhesive (***Figure 13.12***). The rubbery domains stop the crack propagating in the adhesive. Most reactive adhesives are formulated as copolymer compounds.

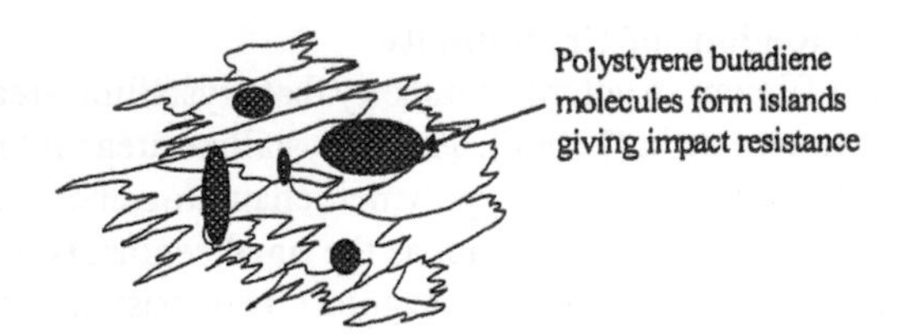

Figure 13.12 *Formulation of brittle adhesives to absorb sudden impact loads*

Some modern adhesives are a formulation in which a rubbery polymer is reacted **into** the organic chain (rather than blended into the adhesive). Materials such as polymethyl methacrylate (PMMA) can be **graft** copolymerised (**13.9.3**) with the rubbery polymer to produce a 'rubberised Perspex'. In this case the rubber forms a separate phase (as particles of rubber of about 0.1 μm in diameter) which are chemically bound into the acrylic hydrocarbon chain (or phase). These adhesives provide for the applied stress to be uniformly distributed over the whole contact area with the ability to reduce stress concentration effects. This is achieved because the cracks within the harder phase are being arrested by the rubber microspheres.

In some reactive adhesives, condensation reactions (**13.5.2**) promote crosslinking and pressure must be applied to stop non-absorbent adherends from being blown apart. Cross-linkable, solvent-based, reactive adhesives are used to bond many varied items (from inflatable dinghies to shoes). The crosslinking stiffening process is brought about

- by **heating**. Some adhesives cure at temperatures as low as 120°C and so they are 'unstable' at room temperatures and have to be stored in refrigerators (i.e., frozen). Adhesives of this type are very attractive for certain industries (e.g, the motor manufacturer) as the adhesive can be cured at the same time as the paint coatings are cured. In other (one-part) adhesives, crosslinking is initiated by heat, UV or electron beam energy. Once initiated, the reaction proceeds at room temperatures (and therefore caution is required when storing the adhesive, **13.17.2**).
- using a **surface catalyst**, which can be
  - a metallic ion; or
  - surface absorbed water (most materials have a layer of adsorbed water on their surface) e.g, "super-glues" use absorbed surface water as a catalyst and this is why they stick the skin so well; or
  - atmospheric water e.g, **silicone bath sealants** are triggered by atmospheric water vapour (causing polymerisation and the evolution of acetic acid); or
  - the absence of oxygen e.g, **anaerobic adhesives** (**13.16.4**) are triggered by a metallic ion but only set if oxygen is excluded from the surface.
- using an adhesive composed of **two reactive parts**, which are mixed prior to use, or by spreading a reactive component on one of the surfaces to be stuck and the adhesive on the adjacent surface.
- using a **combination of methods**, e.g, some solvent-based adhesives have a reactive component in them in order to gain better adhesive strengths. Examples include cycle tire repair adhesives based on natural rubber, which can be made to have very high strength and wear resistance if it is heated with sulphur to **vulcanise** the rubber (a process of cross linking to join up the molecular chains to produce a more rigid three-dimensional structure, **13.10**). Cycle tire repair adhesives still use contact adhesives with relative low glass transition temperatures.

13.16.4 Exclusion of Oxygen from the Interface

Adhesives that cure when oxygen is excluded from the interface are called **anaerobic adhesives** (e.g, Loctite™). They are not suitable where tensile forces are possible. Anaerobic adhesives have to be sold in large bottles with a small amount of liquid present to ensure that there is sufficient oxygen available to prevent setting.

13.17 ADHESIVE PERFORMANCE

13.17.1 Surface Topography and Chemical Bonding

One of the most difficult factors to achieve is the intimate contact of two surfaces which have **different surface topographies**. Polished steel has microscopic troughs and peaks at the surface, formed by the movement of dislocations when the metal is deformed (**10.4**) to produce slip planes. The troughs and peaks differ in height by about 0.3 μm. The slip steps produced when dislocations reach the surfaces of rolled metals also produce an undulating surface. Typical roughness values for various engineering surfaces are illustrated in **Table 13.11**.

Table 13.11 Surface roughness of various engineering surfaces

Surface Profile	Roughness values (from an average centre line) (μm)
Milled	1.6 - 6.3
Bored or turned	0.4 - 6.3
Ground	0.1 - 1.6
Lapped and polished	0.05 - 0.4

The roughness values appear small but are far greater than the carbon-carbon bond length (about 5 nm). When a metal is deformed, some surfaces are extruded (with a clearly defined slip plane) whilst adjacent surfaces can be intruded. An intrusion is a crystallographic 'valley' (***Figure 13.13a***). The adhesive is a long hydrocarbon chain and is very viscous. It follows that there can be difficulty for the adhesive to enter troughs between the ridges to make good contact over the whole surface area (***Figure 13.13b***). The combined effect of viscosity and bond direction favours the adhesive just sticking to the ridges and spanning the troughs (like a suspension bridge).

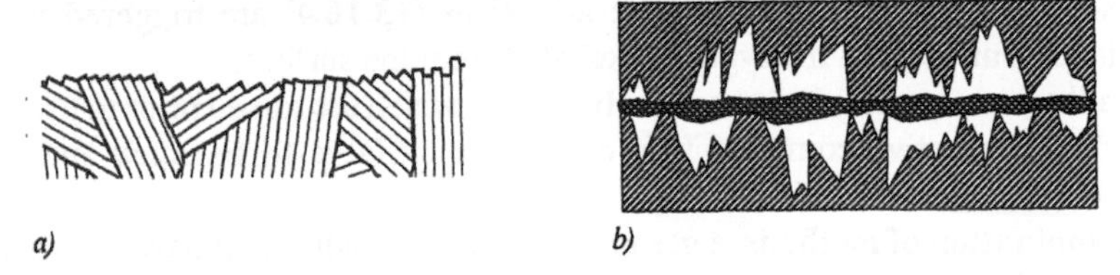

***Figure 13.13** Surface roughness and the effect on adhesive contact*

To investigate this effect further, we must look at the van der Waals forces (**1.14**) responsible for holding a great variety of materials together (e.g, the bond between chains in thermoplastics, the forces giving cohesion to liquids, etc.). The van der Waals forces are residual bonds which operate over very short distances. They are considered weak bonds but are in fact the **only** bonds which **can operate at the surfaces of all dissimilar materials**; van der Waals forces are therefore the chemical bonds in adhesives which need to do the sticking. In consequence,

- as the van der Waals forces ('bonds') operate only over a short range, the adhesive and the adherend surface must be in **intimate contact**. This means that the surfaces to be stuck must be clean and be capable of being **wetted** by the adhesive solution (i.e., the adhesive must form a low **contact angle** (**5.4.3**) with the adherent substrate;

- adhesives must be capable of **spreading** to maximise the surface area of contact (i.e., they must provide coherent coverage). The adhesive must have a low glass transition temperature so that the weak van der Waals bonds can be readily broken to allow the fluid to be mobile. Alternatively, fluid mobility can be achieved using **solvent bases** to carry the adhesive (**13.16.1**).

13.17.2 Shelf Life

Broadly, a 10°C rise in temperature doubles the rate of chemical reaction. Therefore a temperature rise from 20°C to 100°C represents a 2^8 fold increase in the rate of a reaction, i.e., the reaction rate at 100°C is about 256 times the room temperature reaction rate. Hence an adhesive which takes 24 hrs to cure at 100°C will take about 7 months to cure at room temperature. This simple calculation illustrates why single pack adhesives initiated by external energy sources generally have restricted shelf lives.

13.17.3 Adherends

A wide variety of materials are used in the construction industry, and therefore a large number of possible adherends exist (e.g, wood and allied materials, natural fibrous materials, metals and alloys, etc.). These materials have a wide range of surface properties necessitating a wide variety of adhesives types. Where the two adherends are completely different chemically, it is often not possible to use the same adhesive to stick both surfaces. Therefore chemical **conversion processes** are used to change one surface into a more suitable "chemical state" so that the one adhesive will stick and **adhere** to both surfaces. Such conversion processes include

- the use of 'primers' (e.g, organo-silanes for glass);
- anodising (of aluminium) (**11.14.2**);
- acid etching by chromic acid for chlorinated plastics.

Adhesives do not stick oily metals or treated timber. In certain industries, such as the automotive industry, pressing oils are used to lubricate the metal in the pressing operations. Large quantities of steel are used directly from the pressing shop and it would be too expensive to clean these oils off before the body was welded up. To solve this problem, advanced adhesives have been formulated to 'absorb' these oils into discrete islands and, although there is a loss of bond strength due to the loss of area being stuck (the oil islands), these **oil absorbing adhesives** perform adequately.

13.17.4 Compatibility

There must be compatibility between the adhesive and the adherend. Compatibility issues may arise, for example, when

- an acid-based adhesive is applied to a metal substrate (which may then suffer from enhanced corrosion);
- the migration of a plasticizer from a (thin) flexible plastic into the adhesive, causing accelerated degradation (e.g, sticking of plasticised PVC). The vapours from the adhesive can also affect the properties of the adherend;
- adhesives are used to bond treated timber (i.e., timber treated either with fire retardant material or with wood preservation). Very few adhesives will adequately bond treated timber; manufacturers advice is required in these circumstances.

13.17.5 Environmental Temperature and Moisture

Most adhesives are temperature and moisture sensitive. As an illustration of this, special self-adhesive labels have to be used on freezer packs as ordinary self-adhesive labels would fall off as their adhesive strength is very temperature dependent.

13.18 MECHANICAL PROPERTIES OF ADHESIVES

The strength of the adhesive bond depends upon

- adhesive thickness. The adhesive may be applied by solution spraying, trowelling or by extrusion from a tube. The film thickness **must** be carefully controlled as thick films give rise to peel forces whilst thin films produce only localised bonding (spottiness), which may be inadequate to support required loads;
- type of solvent;
- the service environment (heat, moisture, light, etc.).

The mechanical properties of the hardened adhesive are dependent upon the chemical constituents and design of the joint. If the molecular chains of these chemical structures are capable of bending, rotation and slip then the adhesive will be rubbery. If the adhesive has a high glass transition temperature it will be brittle. The mechanical properties can be controlled by the addition of fillers which 'pad out' the adhesive and reduce the degree of elasticity. Too much filler will make the adhesive very brittle and liable to crumble under low shear stresses. Mechanical properties are also dependent on the chemical bonds formed during curing, as the different bonds are associated with intrinsic bond energies, as summarised in **Table 13.12**.

Table 13.12 Bond energies and lengths of chemical bonds

Bond Type	Bond Energy (kJ/mole)	Bond length ($x10^{-10}$ m)	Bond Saturation
Ionic	600-1200	2-4	Saturated complete bonds
Covalent	60-800	1-2	Saturated complete bonds
Metallic	100-350	2-6	Saturated complete bonds
Dipole	20	2-4	Van Der Waals forces/bonds
Hydrogen	60	3	Van Der Waals forces/ bonds
Physical absorption	60	10	Van Der Waals forces/ bonds

13.19 DESIGN REQUIREMENTS FOR ADHESIVES

The design requirements for adhesives are

- to maximise the surface area over which the bonding occurs;
- to avoid peeling or cleavage stresses, where adhesives are very weak;
- to design in shear (i.e., so that the force is in the plane of the adhesive).

Adhesives work best where the stresses that act on them are in **compression** (or, to a lesser extent, in **shear**) so that the deformation forces are in the plane of the adhesive (***Figure 13.14a***). The tensile strength of an adhesive is only passable in pure tension where the forces of deformation are at right angles to the plane of the adhesive (***Figure 13.14b***). Some shear stresses in a simple overlap joint produce tensile forces at the end of the adhesive, causing the joint to peel (***Figure13.14c***) or cleave (***Figure 13.14d***). The peel strength of the adhesive is only some 10% of the shear strength. ***Figures 13.14e*** to

13.14h illustrate some failure modes for adhesives.

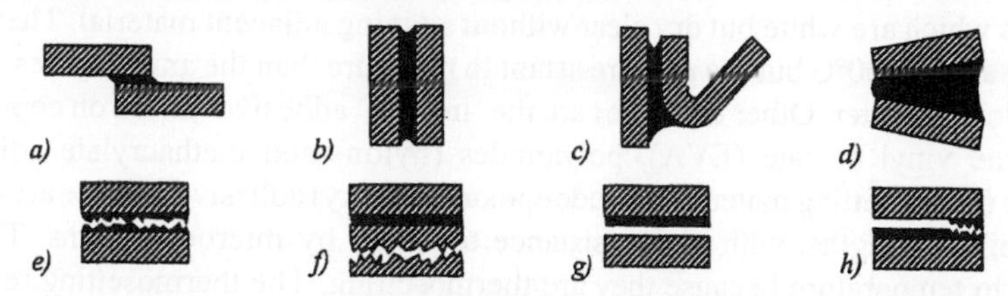

***Figure 13.14** Adhesive joint design and modes of failure*

13.20 CLASSIFICATION OF ADHESIVES

Annex A of BS EN 923:1998[1] classifies the types of adhesives available (by the physical nature of the adhesive) and their main components (BS EN 923 provides a useful glossary of terms used in adhesive technology). The classification is summarised in **Table 13.13** (over).

Table 13.13 Classification of adhesives (BS EN 923:1998)

Adhesive (physical nature)	Binder	Number of components[a]	Setting[a]	Notes
Glue[b, d]	Starch, dextrin Cellulose ethers Casein Poly acrylic acid esters Poly vinyl alcohol	1	Cold	Volatile solvents (mainly water) evaporate during setting
	Glutin	1	Hot	
Paste[b, d]	Starch Cellulose ether, acrylic derivatives	1	Cold	
Solvent and dispersion adhesives[b,e]	Natural rubber, some synthetic rubbers Acrylonitrile-butadiene copolymers Polyurethane Poly vinyl acetate Poly vinyl chloride Polyester Polystyrene	1, (2)	Cold, hot	Volatile solvents or dispersion agents evaporating largely before or during setting
	Poly vinyl chloride Plastisols	1	Hot	No volatile solvents or dispersing agents
Hot melt adhesives[b,e]	Styrene-butadiene and styrene-isoprene rubber Ethylene vinyl acetate Polyamides Polyester	1	Hot	
Reactive adhesives[c,e]	Soluble silicates	1, (2)	Cold (hot)	Volatile solvents or water evaporate with conversion products formed during setting
	Hydraulic cement	1, (2)	Cold	
	PF-, MF-, UF-resins, water soluble	2, (1)	Hot, Cold	
	Phenolic resin with poly vinyl acetals or nitrile rubber	1, 2	Hot	Volatile conversion products evaporate during setting
	Polyamides	2	Hot	
	Silicone resins, moisture cure	1	Cold	
	Unsaturated polyester resins, vinyl- and acrylic derivatives	2	Cold	Volatile, reactive components are chemically incorporated into the bond
	Dimethyl acrylic acid, anaerobic setting	1	Cold	No volatile components
	Polyisocyanates, moisture cure	1	Cold	
	Epoxy/Poly(amido)amines	2	Cold (hot)	
	Epoxy/Acid anhydride	2	Hot	
	Special epoxy resins	1	Hot	

Notes: [a] Alternatives are listed in brackets; [b] Physically or physically and chemically setting; [c] Chemically setting; [d] Natural raw materials of animal origin set slowly, are sensitive to water and are attacked by microorganisms above about 80% relative humidity. Examples of these adhesives are albumin, animal glues, casein, shellac, beeswax, etc. Natural raw materials of vegetable origin include cellulose (from wood) pastes for wall paper pastes, starch/flour mixes, gum arabic, linseed oil based glues, Canada balsam and water/alkali mixes for interior plywood.; [e] Natural raw materials of mineral origin include bituminous adhesives derived from coal tar. They are used for bonding/laying wood tiles and roofing felt and are used hot or as emulsions. Generally, they have poor resistance to heat, but good resistance to moisture. Examples include sodium silicate adhesives (with a SiO_2:Na_2O ratio of between 2 and 3.5), magnesium and zinc oxychlorides (which are known for their chemical resistance, hence their use in fungicidal treatment of porous brickwork).

Synthetic thermoplastic materials (for example, polyvinyl acetate, PVA) are emulsion adhesives which are white but dry clear without staining adjacent material. They suffer softening above 200°C but are more resistant to moisture than the animal glues. Setting is by the loss of water. Other examples are the 'iron on' adhesives based on copolymers of ethylene vinyl acetate (EVA), polyamides (nylon) and methacrylate adhesives. Synthetic thermosetting materials include epoxide (Epoxy) adhesives. These are capable of very high strengths, with high resistance to decay by microorganisms. They are resistant to temperature because they are thermosetting. The thermosetting resins are based upon the formaldehyde resin (e.g, urea formaldehyde and phenol formaldehyde resins); these are resistant to boiling.

Traditionally in the UK, phenolic and amino plastic contact adhesives for wood have been categorised according to BS 1204: 1979. As these categories are widely used today, they are explained below

- **Type WBP** (Waterproof and boil-proof). Adhesives of the type which, by systematic tests and by their records in service over many years, have been proved to make joints which are highly resistant to weather, microorganisms, cold and boiling water and dry heat. At present, only certain phenolic and amino plastic resin adhesives have been shown to meet these requirements. Examples include some phenol formaldehyde adhesives.
- **Type BR** (Boil resistant). Joints made with these adhesives have good resistance to weather and to boiling water, but, unlike Type WBP adhesives, will fail under very long weathering exposure. Joints will withstand cold water for many years and are highly resistant to attack by microorganisms. Examples include melamine formaldehyde adhesives.
- **Type MR** (Moisture resistant and moderately weather resistant). Joints made with these adhesives will survive full exposure to the weather for only a few years. They will withstand cold water for a long period and hot water for a limited time, but fail when exposed to boiling water. They are resistant to attack by microorganisms. Examples include urea formaldehyde adhesives, but inclusion of melamine in the hardener (fortified UF adhesives) brings the durability up to Type BR.
- **Type INT** (Interior). Joints made with these adhesives are resistant to cold water but neef not withstand attack by microorganisms. Examples include phenol-based adhesives and resorcinol formaldehyde adhesives.

Currently, the BS specification Type MR phenolic and amino plastic grades for wood is retained in BS 1204:1993[2], whilst other grades are reclassified according to whether the adhesive is required for loadbearing or non-loadbearing applications; loadbearing phenolic and amino plastic grades for wood (BS EN 301:1992[3] and BS EN 302:1992[4]) and non-loadbearing adhesives (BS EN 204:2001[5], BS EN 12765: 2001[6] and BS EN 205:2001[7]).

For loadbearing applications, adhesives for wood are classified by BS EN 301:1992 according to their suitability for use in specified climatic conditions. The classification is summarised in **Table 13.14** (over).

Table 13.14 Classification of phenolic and aminoplastic adhesives for loadbearing wood structures

Classification	Temperature	Climatic equivalent to[a]	Service examples
Type I	> 50°C	Not specified	Prolonged exposure to high temperature
	≤ 50°C	> 85% RH at 20°C	Full exposure to the weather
Type II	≤ 50°C	≤ 85% RH at 20°C	Heated and ventilated buildings. Exterior protected from the weather. Short periods of exposure to the weather.

Notes: [a] 85% RH at 20°C will result in a moisture content of approximately 20% in softwoods and most hardwoods (**12.7**), and a somewhat lower moisture content in wood-based panel products.

For non-loadbearing applications, adhesives for wood are classified by BS EN 204:2001 according to their bond durability (assessed as minimum adhesive strength) in response to various climatic conditions (measured in accordance with BS EN 205) (**Table 13.15**).

Table 13.15 Classification of adhesives for non-loadbearing wood structures

Durability class		Minimum values of adhesive strength (N/mm^2)					
	Exposure conditions[a]	A	ABA	AC	ACA	ADE	ADEA
	Examples of climatic conditions for durability classes						
D1	Interior, in which the temperature only occasionally exceeds 50°C for a short time and the moisture content of the wood is 15% maximum	≥ 10	NT	NT	NT	NT	NT
D2	Interior with occasional short-term exposure to running or condensed water and/or to occasional high humidity provided the moisture content of the wood does not exceed 18%	≥ 10	≥ 8	NT	NT	NT	NT
D3	Interior with frequent short-term exposure to running or condensed water and/or to heavy exposure to high humidity. Exterior not exposed to weather.	≥ 10	NT	≥ 2	≥ 8	NT	NT
D4	Interior with frequent long-term exposure to running or condensed water. Exterior exposed to weather but with adequate protection by a surface coating.	≥ 10	NT	≥ 4	NT	≥ 4	≥ 8

Notes: NT No test required; [a] **A** - 7 days in standard atmosphere (23 ± 2°C and 50 ± 5% RH or 20 ± 2°C and 65 ± 5% RH); **B** - 3 hours in cold water; **C** - 4 days in cold water; **D** - 6 hours in boiling water; **E** - 2 hours in cold water.

13.21 SEALANTS

A sealant is the same chemical compound as an adhesive except that it

- is formulated to fill gaps and so is **not formulated** to have a high bond strength
- must have the ability to **lose tackiness** after use
- must possess gap filling properties, which are often achieved by the addition of fillers (e.g, chalk, clay, etc.). There is a physical limit to the size of the gap which can be filled, as the curing process seals off further activating species (which can, for example, be atmospheric water vapour or oxygen). A sealant with a good **slump resistance** is required if a large gap is to be filled as it would be detrimental to find the gap filling sealant sliding down the gap under gravitational forces

Sealants are used widely in the construction industry e.g, to seal open seams to exclude water and dust from compartmental areas. BS 6213: 1982[8] gives guidance on the types of constructional sealants available on and their selection and correct application, as summarised in **Table 13.16** (over); BS EN 26927[9] provides a useful glossary of terms used in adhesive technology.

Table 13.16 Characteristics of various sealant types (adapted from BS 6213: 1982)

Sealant type (service life)	Characteristics
Oleo-resinous (up to 10 years)	• packaged in cartridges and bulk containers. • develops a flexible surface skin by oxidation within 24 hours. • skin prevents the mass beneath the surface from rapid oxidation and hardening (painting will also prolong the effective service life of the sealant). • have basic plastic properties (only the surface skin is flexible). • will bond with most common clean and dry building materials (no primer required).
Bitumen based (up to 10 years)	• packaged in cartridges and bulk containers. • thermoplastic. Retains a degree of flexibility. • gun applied forms contain solvent and are therefore subject to shrinkage.
Rubber/bitumen based (up to 10 years)	• packaged in cartridges and bulk containers. • bitumen modified with rubbers to achieve better flexibility and elasticity. • develop non-tacky surface 24-48 hours after application (but does not skin). • essentially plastoelastic in nature. • cartridge grade for gun application contains solvent and therefore subject to small shrinkage.
Butyl rubber (up to 10 years)	• packaged in cartridges and bulk containers. • available in a wide variety of formulations. Those based on solid butyl rubber have to be heavily plasticised and solvated to be suitable for gun applications and are therefore subject to shrinkage. • predominantly plastic.
Butyl rubber (cross-linked) (up to 10 years)	• packaged in cartridges and bulk containers. • butyl rubber can be partially vulcanised and then formulated into a sealant with higher elasticity. • normally contain solvent and are therefore subject to shrinkage. • have greater resistance to ageing and weathering than butyl rubber sealants. • predominantly plasto-elastic. • generally used without a primer.
Acrylic resin (solvent types) (up to 15 years)	• packaged in cartridges and bulk containers. • good adherence without a primer. • shrinkage occurs with loss of solvent. • thermoplastic with plastoelastic properties.
Acrylic resin (emulsion types) (up to 15 years)	• packaged in cartridges and bulk containers. • forms a skin as soon as exposed to atmosphere. • shrinkage due to loss of water in conjunction with development of plastoelastic properties after a few days.
Flexible epoxide (up to 20 years)	• packaged in pre-proportioned bulk containers. • based on epoxy resins. • degrees of flexibility imparted by addition of other polymers and extenders. • generally of a two (or more) part system and require mixing prior to application. • cure at room temperature. • predominantly plastic properties (although can show limited flexibility at room temperatures and above, becoming more rigid at low temperatures)
Polysulphide (1 component) (up to 20 years)	• packaged in cartridges and bulk containers. • cure on exposure to atmospheric moisture (cure time is much longer than for two component systems). • surface skin thickens progressively through the depth of the sealant until complete cure is achieved. • speed of cure dependent on rate of penetration of atmospheric moisture (becoming progressively slower with increasing thickness of the skin). • when fully cured, exhibit elastic properties.
Polyurethane (1 component) (up to 20 years)	• packaged in cartridges and bulk containers. • cure on exposure to atmospheric moisture. • diversity of polymers enables formulation of sealants with widely varying properties. • rate of cure is faster than for one component polysulphide sealants. • curing reaction with moisture increases sealant viscosity. • some emit small quantities of CO_2 during cure, causing swelling. • when fully cured, exhibit elastoplastic properties.
Silicone (up to 20 years)	• packaged in cartridges packs for gun application. • diversity of polymers enables formulation of sealants with widely varying properties. • when fully cured, exhibit fully elastic properties. • cure on exposure to atmospheric moisture to form a rubber-like surface skin. • cure proceeds rapidly through depth of sealant. Tooling (if required) must take place soon after application. • although cured product generally non-corrosive, by-products from the cure reaction of some formulations may be corrosive to concrete and metals.
Polysulphide (2 component) High modulus sealants (up to 20 years)	• packaged in two packs requiring mixing prior to application. • diversity of polymers enables formulation of sealants with widely varying properties. • cures chemically, enabling optimum properties to be achieved more rapidly than from one component polysulphide sealants. • two component polysulphide sealants available in non-sag and self levelling grades. • when fully cured, exhibit elastoplastic properties.
Polyurethane (2 component) (non-sag type) (up to 20 years)	• packaged in two packs requiring mixing prior to application. • diversity of polymers enables formulation of sealants with widely varying properties. • cures chemically, enabling optimum properties to be achieved rapidly. • when fully cured, exhibit elastic properties.

Table 13.16 Characteristics of various sealant types (contd.)

Sealant type (service life)	Characteristics
Polysulphide (2 component) Polyurethane (2 component) (up to 20 years)	• packaged in two packs requiring mixing prior to application. • used for cold-poured joints for concrete pavements. • can be modified to achieve resistance to liquid fuel, etc. if required. • diversity of polymers enables formulation of sealants with widely varying properties. • cured sealants based on polysulphide liquid polymer are predominantly elastoplastic while those based on polyurethane polymer are predominantly elastic.
Sealing strips	• packaged in a variety of sizes and sections, in pre-cut lengths, reels and boxes. • made from preformed extruded materials based on non-drying oils, polymer modified bitumens or synthetic polymers and resins, resulting in sealants with widely varying properties. • are either non-loadbearing (essentially plastic in nature) or loadbearing (predominantly plasto-elastic).
Rubber bitumen hot poured	• supplied as solid products requiring melting and pouring in accordance with manufacturers instructions. • formulated to seal joints in concrete pavements. • essentially elastoplastic in nature.
Setting putty	• packaged in kegs and buckets. • hand- or knife- applied, drying oil based products for the traditional face glazing of glass. • available either as linseed oil putty for use with primed wood frames or metal casement putty for steel frames. • set to a hard consistency and require the continuous protection (e.g. of a complete paint system).
Non-setting glazing material	• packaged in kegs and buckets. • hand-applied oil based compounds for bead glazing. • usually polymer modified to accommodate limited movement. • dry to form a surface skin beneath which the product remains plastic. • always applied to non-absorbent surfaces (otherwise loss of oil medium results in premature hardening). Porous surfaces must be made non-absorbent by application of a surface sealer. • gradual hardening by oxidation and polymerisation of oil medium occurs throughout the life of the compound. • plastic in nature (therefore unsuitable for face glazing. In bead glazing essential that provision is made to prevent displacement under load).
Bedding material	• packaged in kegs and buckets. • hand-applied oil based materials for bedding metal and timber frames into subframes or prepared openings. • must be applied to non-absorbent (or sealed) surfaces. • not recommended for exposure to weather (but may be protected by pointing with gun applied sealant or painting).
Glazing material (2 component)	• packaged in packs requiring mixing prior to application. • usually hand-applied. • cure chemically to a soft rubber-like compound. • cure is temperature dependent (can be made sufficiently rapid for use in factory glazing techniques). • primer required to ensure adhesion to all surfaces. • not suitable for over-painting.

13.22 CLASSIFICATION OF SEALANTS

BS ISO 11600:1993[8] specifies the types and classes of building construction sealants according to their applications and performance characteristics (**Table 13.17**).

Table 13.17 Classification of sealants (BS ISO 11600:1993)

Application		% movement capability	Tensile moduls (N/mm²)				Elastic recovery	
Application	Code		High	Code	Low	Code	(%)	Code
Glazing sealants	G	25 20	> 0.4 (23°C) and/or > 0.6 (-20°C)	HM	(other than high)	LM	-	-
Construction sealants[a]	F	25 20	> 0.4 (23°C) and/or > 0.6 (-20°C)	HM	(other than high)	LM	-	-
		12.5	-	-	-	-	≥ 40% < 40%	E (Elastic) P (Plastic)
		7.5	-	-	-	-	-	-

Notes: Sealants of classes 25, 20 and 12.5E are called elastic sealants whilst those of classes 12.5P and 7.5 are called plastic sealants. [a] for use in building joints other than glazing;

Using this classification, a construction sealant (F) having a movement capability of at least 12.5% and an elastic recovery of less than 40% (12.5P) is designated "Sealant ISO 11600 - F - 12.5P". BS ISO 11600:1993 gives further requirements for type G and F sealants in respect of

- maximum flow (3 mm for all sealants);
- elastic recovery;
- tensile modulus at 23°C, -20°C, at specified extensions and at maintained extension;
- cohesion/adhesion properties at constant and variable temperature, after exposure to artificial light and after maintained extension following water immersion;
- resistance to compression; and
- loss of volume.

The classification is summarised in ***Figure 13.15***.

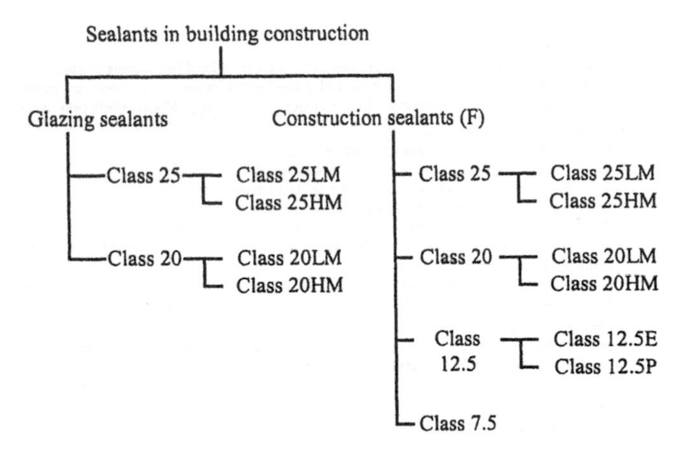

Figure 13.15 *Diagram of the classification of sealants in building construction*

13.23 FAILURE OF SEALANTS

Common types of sealant failure and their possible causes are outlined in **Table 13.18**.

Table 13.18 Failure mechanisms in sealants

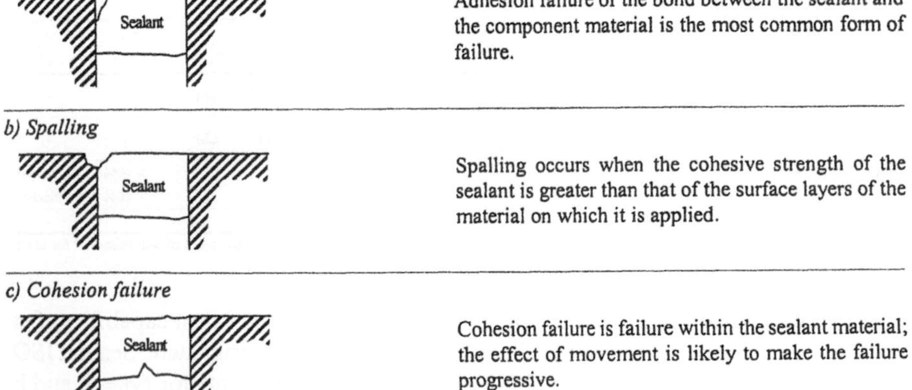

a) Adhesion failure	Adhesion failure of the bond between the sealant and the component material is the most common form of failure.
b) Spalling	Spalling occurs when the cohesive strength of the sealant is greater than that of the surface layers of the material on which it is applied.
c) Cohesion failure	Cohesion failure is failure within the sealant material; the effect of movement is likely to make the failure progressive.

Table 13.18 Failure mechanisms in sealants (contd.)

d) Folding

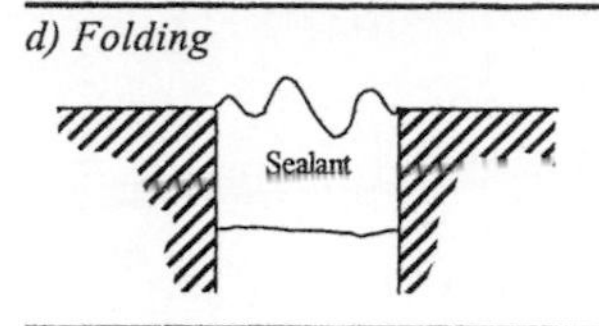

Folding occurs mainly with plastic and elastoplastic sealants under excessive movements, which result in permanent set. Subsequent movement leads to folding and to the progressive breakdown of the seal due to intrusion, removal of material (e.g, by traffic) or adhesion failure.

e) Intrusion

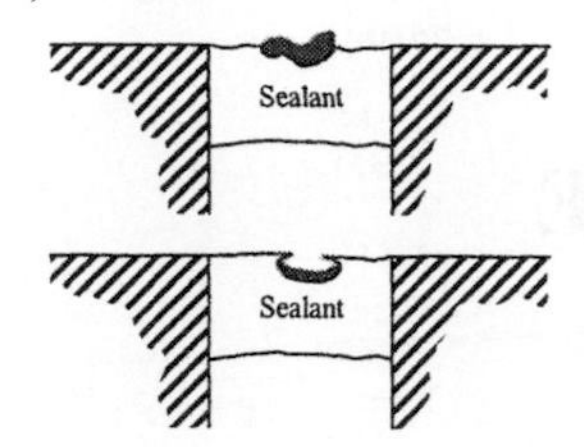

Intrusion is common in horizontal joints in traffic areas and occurs when the sealant surface is penetrated by grit and debris. On subsequent closing, the grit is not ejected, causing abrasion and eventual failure of the sealant. In extreme cases, the functioning of the joint is impaired.

f) excessive extrusion, which is ejection from the joint of part of the sealant, which becomes vulnerable to mechanical damage and impairs the future ability of the sealant remaining in the joint to perform its function;

g) slumping, which occurs when the sealant flows from its intended position owing to its own selfweight. This can commonly occur with joint which are excessively wide in relation to the properties of the sealant. Excessive slumping constitutes failure;

h) chemical attack, which occurs mainly in specific aggressive environments, but also from the use of some cleaning materials and some decorative or other finishes;

i) displacement (or removal of the sealant), which may occur by the scouring action of fast turbulent water or by hydraulic pressure.

13.24 REFERENCES

1 BS EN 923: 1998. Adhesives. Terms and definitions.
2 BS 1204:1993. Specification for Type MR phenolic and aminoplastic synthetic resin adhesives for wood.
3 BS EN 301:1992. Adhesives, phenolic and aminoplastic, for load-bearing timber structures: classification and performance requirements.
4 BS EN 302:1992. Adhesives for load-bearing timber structures: test methods.
5 BS EN 204:2001. Classification of thermoplastic wood adhesives non-structural applications.
6 BS EN 12765: 2001. Classification of thermosetting wood adhesives non-structural applications.
7 BS EN 205:2001. Test methods for wood adhesives for non-structural applications. Determination of tensile shear strength of lap joints.
8 BS 6213: 1982. British Standard Guide to Selection of Constructional Sealants.
9 BS EN 26927: 1991. Building Construction. Jointing products. Sealants. Vocabulary.
10 BS ISO 11600: 1993. Building Construction - Sealants - Classification and requirements.

14

FIRE AND FIRE RESISTANCE

14.1 INTRODUCTION

(To aid interpretation of the terms associated with fire and fire resistance, some useful definitions have been included in **14.10**).

Combustion is an **exothermic**, self-sustaining chemical reaction involving a solid-, liquid- and/or gas-phase fuel. Combustion accompanied by smoke or a flame or both, which spreads uncontrolled in time and space, is commonly called a **fire**. Most combustion processes are associated with **oxidation** of a fuel by (atmospheric) oxygen. In order to produce a flame, the combustion reaction must involve a fuel in the gas-phase, and therefore solid- and liquid-phase fuels must be capable of producing flammable vapours to produce a flame. However, combustion may proceed without producing a flame, by glowing or smouldering. This mode of combustion occurs with certain porous materials in the solid-phase that can form a carbonaceous char when heated, and is characteristic of materials in the final stages of a fire decay, or where a material is treated to suppress the production of flammable vapours and prevent flaming, for example tobacco, intumescent coatings (**14.7.2**).

There are three basic requirements for combustion, often represented as 'the triangle of fire' (***Figure 14.1***).

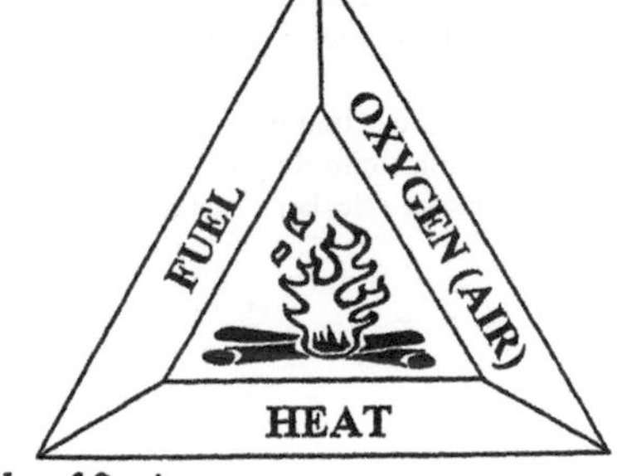

Figure 14.1 *The 'triangle of fire'*

To reduce or prevent burning (i.e., fire extinguishing), any one requirement within the **triangle of fire** must be removed. Three methods of removal can be employed

- remove the **fuel** , which is something that burns **exothermically** **(14.10.7)**;
- remove the **air** by smothering, i.e., remove the oxygen required to sustain the exothermic reaction);
- remove **heat** by cooling, i.e., providing an **endothermic** **(14.10.7)** reaction adjacent to the fire. Heat is removed by conduction, convection and radiation **(14.4.1)**. Combustion products and unburnt gases (e.g, nitrogen) remove a lot of heat **(14.4.2)**.

14.2 FUELS

Fuels include numerous materials which, due to their chemistry, can thermally degrade to yield relatively stable products (such as carbon dioxide and water) and heat. Most fuels of practical significance contain a high proportion of carbon and hydrogen, and may be gases or vapours, liquids or solids **(Table 12.1)**.

14.2.1 Gas- and Vapour-phase Fuels

Gas- and vapour-phase fuels can either be pre-mixed intimately with air before ignition or may involve diffusion of flammable vapours from a solid or liquid surface, where combustion takes place in the region where the gas- or vapour-phase fuel and oxygen are mixing. Controlled combustion of pre-mixed fuel-air mixtures can be achieved by controlling the rate at which the fuel and oxygen are mixed; this is achieved in domestic gas supplies by limiting the flow rate of the methane fuel. Unregulated pre-mixed combustion may result in a rapid expansion of gaseous products resulting in a sudden pressure increase, giving rise to an **explosion** **(14.3.1)**. Diffusion of flammable gas- or vapour-phase products from solid- and liquid-phase fuels is of primary importance in accidental fires in buildings. Combustion of solid building materials and contents occurs along thin flame sheets, which separate regions rich in gas or vapour fuel (derived from the solid) from regions rich in oxygen (the air in the volume of the room). Gas or fuel vapour and oxygen diffuse towards each other from opposite sides of the flame, where they combine to produce combustion products and heat, which in turn diffuse away from the flame. For ignition of gas- or vapour-phase fuel-air mixtures, the fuel concentration must lie within certain **flammable (explosive) limits** **(14.2.4)**. Once ignition of pre-mixed fuel-air mixtures has occurred, a flame will propagate through the unburnt mixture until it is entirely consumed.

14.2.2 Liquid-phase Fuels

Combustion of liquids depends on the ability of the liquid to produce a flammable vapour, which requires thermal (heat) input **(14.4.1)**. Many flammable liquids produce vapours below room temperature which can be ignited by a pilot ignition source (termed the **flash point** temperature, **14.2.4**, **14.4.1**), whilst others require an external heat source to raise their temperature sufficiently to produce vapours and reach the **ignition temperature** **(14.4.1)**.

14.2.3 Solid-phase Fuels

Solid-phase fuels undergo thermal degradation at elevated temperatures by processes involving **oxidation** (in the presence of oxygen in the reaction zone), **pyrolysis** (in the absence of oxygen in the reaction zone) or both. For flame initiation, thermal (heat) energy must be supplied to convert a sufficient quantity of the fuel to a vapour, thus creating a flammable vapour-air mixture in the vicinity of the surface. A good example of this is a solid wax candle which melts to produce a liquid. The liquid is drawn up the wick whereupon thermal energy from the flame converts the liquid to a vapour which then supports the flame. Building materials and contents (furnishings, etc.) constitute fuels for fires in buildings (**14.5**). Pure carbon has very high melting and boiling points (**1.11.2**) and so does not form a vapour (thus cannot support a flame). Carbon reacts with oxygen (air) exothermically (**14.3**).

14.2.4 Flammable (Explosive) Limits

The flammable (explosive) limits for a gas- or vapour-phase fuel-air mixture are the extreme fuel-to-air concentration limits beyond which propagation of a flame will not occur in the presence of a pilot ignition source (**14.4.1**). Limits can be demonstrated using a basin with varying amounts of petrol in an bowl and ignited with an electrical discharge. When the fuel concentration is too high, ignition is not possible; this is why a car is so difficult to start when the engine is flooded with petrol. Over-chocked petrol engines require the accelerator to be completely depressed to start the car (this effect is particularly noticeable on short journeys). If the fuel concentration is too low (e.g, because the petrol is too cold), the car will not start (a choke is required to enrich the mixture). Flammable (explosive) limits are shown for selected gas and liquid fuels in **Table 14.1**, where the upper and lower limits are denoted UFL and LFL respectively.

Table 14.1 Flammable (explosive) limits for various gaseous fuels in air (data from[1])

Gases		Flammability limits (% by volume in air)		
		LFL	UFL	UFL-LFL
Carbon monoxide	(gas)	12.5	74.95	62.45
Hydrogen	(gas)	4.0	75.0	71.0
Acetylene (used in welding)	(gas)	2.5	100	97.5
Methane (natural gas)	(gas)	5.0	15.0	10.0
Petroleum spirit	(liquid)	1.3	6.0	4.7
White spirit	(liquid)	1.1	6.0	4.9

Note volatile liquids generally have lower ranges than gases. In addition, methane, used for domestic gas supplies, has a very small range of flammable concentrations compared to other gases, so that, in the event of a leak, there is a limited chance of explosion compared to other gas-air mixtures. Flammable (explosive) limits are effected both by ambient pressure and temperature. When the mixture temperature is increased, the flammability range widens whereas when the mixture temperature is decreased, the flammability range narrows (***Figure 14.2a***).

For a liquid fuel in equilibrium with its vapours in air, the vapour concentration is dependent upon temperature and pressure. There are two maximum temperatures of interest

- the maximum temperature **below** which there is insufficient vapour to form a flammable vapour-air mixture, i.e., the mixture is too lean (the **lower flash point**);
- the maximum temperature **above** which the vapour concentration is too great to propagate a flame, i.e., the mixture is too rich (the **upper flash point**).

Upper and lower **flash points** are shown in ***Figure 14.2a***. The general use of the term 'flash point' refers to the lower flash point temperature.

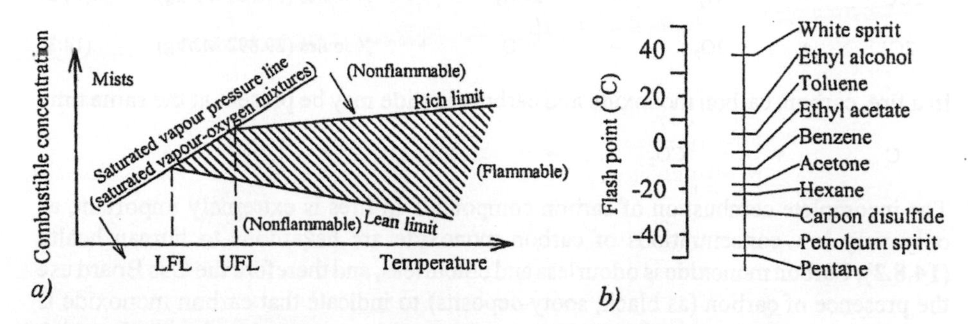

***Figure 14.2** a) Flammability (explosive) limits (methane in oxygen) as a function of temperature[2]. b) Flash point temperatures for various liquid fuels[3]*

Test procedures for obtaining flammability limits and flash points for petroleum and its products are given in BS 2000[4], for paints and varnishes in BS 3900[5], BS 6664[6] and BS EN 456[7] and for adhesives in BS EN 924[8]. Flash points for selected liquid fuels are shown in ***Figure 14.2b***.

14.3 OXYGEN

Oxygen required for combustion is most commonly supplied by the ambient air. However, certain chemicals are powerful oxidising agents which, when mixed intimately with a solid or liquid fuel, produce a highly reactive mixture (**14.3.1**). The requirement for oxygen can be illustrated by chemical equations for the **incomplete and complete combustion** of carbon compounds (C). Chemical equations can tell us how much oxygen (and air) is required for complete combustion, as well how much water is produced. This is important in terms of the ventilation requirements of the room, which must be adequate both to supply sufficient oxygen and to remove water produced by gas, oil and solid fuel heating or cooking systems. These requirements are reflected in Building Regulations (Part F. Ventilation), for which Approved Document F recommends for kitchens in domestic buildings

- an openable window of any area to provide rapid ventilation (or, if windows are not present, mechanical ventilation providing 30 litres/second adjacent to cooking hobs and 60 litres/second elsewhere);
- a controllable ventilation opening of 4000 mm^2 (e.g, 'trickle' ventilators) to provide background ventilation, or continuous mechanical ventilation equivalent to 1 air change rate per hour.

The nature of combustion reactions for carbon compounds depends on the quantity of oxygen (and air) available. If insufficient oxygen is available, **incomplete** combustion occurs (without a flame)

$$2C + O_2 \rightarrow 2CO + X \text{ Joules (22.8 MJ/kg)} \quad ...(14.1)$$

In the presence of additional oxygen, carbon monoxide (CO) will react further to produce carbon dioxide (CO_2), i.e., **complete combustion**

$$2CO + O_2 \rightarrow 2CO_2 + \text{X Joules (14.184 MJ/kg)} \quad ...(14.2)$$

$$2C + 2O_2 \rightarrow 2CO_2 + \text{X Joules (29.892 MJ/kg)} \quad ...(14.3)$$

In a fire, carbon, carbon monoxide and carbon dioxide may be present at the same time

$$C + CO_2 \rightleftharpoons 2CO$$

The incomplete combustion of carbon compounds in fires is extremely important, as only very low concentrations of carbon monoxide are hazardous to human health (**14.8.2**). Carbon monoxide is odourless and colourless, and therefore the Gas Board use the presence of carbon (as black, sooty deposits) to indicate that carbon monoxide is being produced (***Figure 14.3***). Relatively cheap carbon monoxide detectors, which change colour in the presence of carbon monoxide, are available (e.g, 'Safety Spot' carbon monoxide detector).

Advice given is to check for signs of the reaction $2CO \rightarrow C_{\text{causing stains}} + CO_2$ by looking for

- staining of front panels above radiants;
- staining of wall or ceiling above the gas fire;
- scorching of outer gas fire case;
- peeling wall paper above gas fire;
- soot, or debris, appearing beneath the gas fire.

***Figure 14.3** Typical carbon monoxide awareness public information notification*

Combustion reactions for hydrocarbon fuels can be written (using methane or natural gas, CH_4, as an example) as

Incomplete combustion:

$$2CH_4 + 3O_2 \rightarrow 2CO + 4H_2O + \text{X Joules} \quad ...(14.4)$$

Complete combustion:

$$CH_4 + 2O_2 \rightarrow CO_2 + 2H_2O + \text{X Joules (37.78 MJ/kg)} \quad .....(14.5)$$

If equation (14.5) is rewritten for the same volume of methane as equation (14.4), then

Complete combustion:

$$2CH_4 + 4O_2 \rightarrow 2CO_2 + 4H_2O + \text{X Joules} \quad ...(14.6)$$

1 volume of air only contains 1/5 volume of oxygen (i.e., only 1 volume of oxygen is contained in 5 volumes of air, **1.8.5**). Comparing the volume of oxygen in equations (14.4) and (14.6), incomplete combustion of methane requires 15 volumes of air

$$2CH_4 + 15\ AIR \rightarrow 2CO + 4H_2O + \text{X Joules} \quad ...(14.7)$$

whereas complete combustion requires 20 volumes of air

$$CH_4 + 20\ AIR \rightarrow CO_2 + 2H_2O + \text{X Joules} \quad ...(14.8)$$

Therefore, if the amount of air is restricted to a natural gas burner, incomplete combustion will result, with the formation of the poisonous carbon monoxide (CO). Hence the need for adequate ventilation in gas appliances, central heating boilers, etc. (we can actually calculate the gas volumes involved, because the molecular weight of a gas is contained in a volume of 22.4 litres). In the above equations, water is produced as a result of the oxidation of the **hydrogen** containing compound (methane). Indeed water is produced by oxidation of any hydrogen containing compound; the hydrogen has 'first call' on the oxygen (in preference to the carbon in the compound)

$$2H_2 + O_2 \rightarrow 2H_2O + \text{X Joules (37.66 MJ/kg)}$$

Note that sulphur and phosphorous containing compounds also produce their oxides, which are obnoxious to humans (**14.8.2**).

14.3.1 Explosions

In certain circumstances, abrupt oxidation or decomposition of a fuel produces a large volume of flammable gases or vapours and a simultaneous increase in temperature, pressure or both (an explosion). The extent depends upon the rate at which heat energy is liberated. The rate of chemical reactions approximately double for every 10°C rise in temperature. Hence exothermic combustion reactions can produce a sudden release of thermal energy which increases the volume of liberated gases at such a rate that a pressure wave spreads out (approximately seven times) faster than the velocity of sound. The liberation of gases can be illustrated using the combustion reaction for gunpowder, which contains the fuel carbon and the oxidising agent potassium nitrate (KNO_3).

$4KNO_3$	+	$5C$	→	$2K_2CO_3$	+	$2\ N_2$	+	$3\ CO_2$	+	X Joules
Potassium nitrate				Potassium carbonate						
(solid)		(solid)		(solid)		(gas)		(gas)		
Gunpowder										

i.e., a nominal 5:1 increase in the volume of gas:solid. Note that the oxygen comes from the **solid** potassium nitrate (KNO_3). Remember also that the interatomic distances within a gas are roughly 12 times those of a liquid (**2.2.4**), and the increase in volume is some

60 times for the solid → gas transformation. This volume increase excludes any effect due to a temperature increase, which must also be added: this is about 1/273 V added for each 1°C rise (i.e., equivalent to a doubling of the volume for every 273°C rise).

A more efficient gunpowder can be made by ensuring that the oxygen does not stay locked up in the solid reaction product (potassium carbonate, K_2CO_3). This is achieved by addition of more fuel (sulphur).

$4KNO_3$	+	6C	+	S	→	$2K_2S$	+	$2N_2$	+	$6CO_2$	+	X Joules
(solid)		(solid)		(solid)		(solid)		(gas)		(gas)		

i.e., an nominal 8:1 increase in the volume of gas:solid. The volume of gas is further increased by the increase in temperature which arises from the exothermic reaction, as described above. Thus certain chemicals should **not** be stored adjacent to each other, as in the case of the constituents of gunpowder, since the oxygen required for combustion is already present in the solid oxidising agent potassium nitrate (KNO_3).

14.4 HEAT

Thermal energy is **required** for combustion reactions to proceed, both to produce flammable vapours and gas (by vaporisation of liquids or decomposition of solids) and to initiate the oxidation reaction. Thermal energy is also **produced** by combustion reactions when the new products of combustion are formed.

14.4.1 Heat Input

The binding of molecules to each other to form new molecules of another reaction product requires energy. A system containing these molecules is therefore associated with a certain intrinsic binding energy. When fuel molecules react with oxygen, the atoms rearrange and the bonding energy changes. This reaction will only occur if the final energy is less than the starting energy, i.e., the reaction is **exothermic** and the products of the reaction are more stable than the reacting molecules (***Figure 14.4a***).

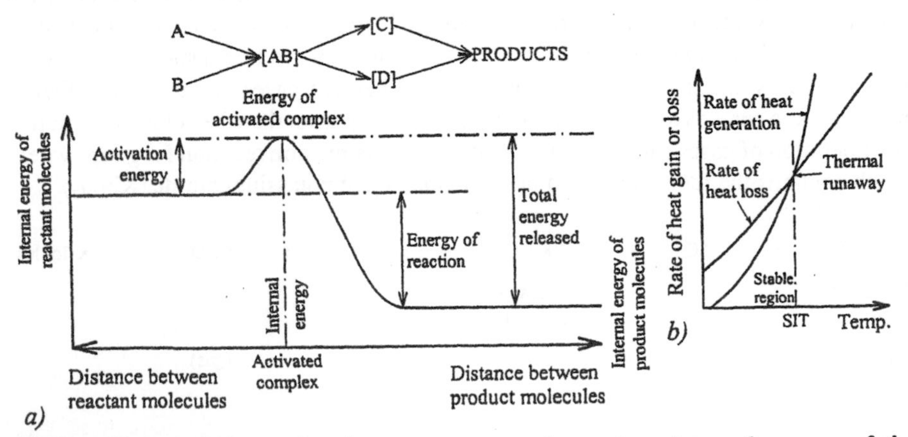

Figure 14.4 *a) Relationship between energy of reaction, internal energy of the molecules and intermolecular distances*[9]. *b) Heat gain/loss and 'thermal runaway'*[10]

The two gas-phase molecules in ***Figure 14.4a***, denoted [A] (a fuel, e.g, methane, CH_4) and [B] (oxygen, O_2), behave similarly to hard spheres, simply bouncing away from one another when they collide. In order for the oxidation reaction to proceed, the molecules [A] and [B] must first be excited to some activated state to overcome the initial repulsion forces and approach one another sufficiently closely to become mutually attracted. The excitation energy needed to overcome the repulsive forces is termed the **activation energy** (e.g, the spark energy for ignition). After interacting chemically, the fuel and oxygen form an 'activated complex' [AB] which is unstable and breaks down through the free radicals [C] and [D], which react further with fuel and oxygen molecules to produce a stable product. Heat energy is evolved during the process. Stable products produced by the methane fuel and oxygen include carbon dioxide (CO_2) and water (H_2O). Carbon monoxide (CO) may also be formed, but may combine further with excess oxygen to form carbon dioxide with the release of thermal energy.

Combustion reactions release energy, which must be dissipated. A proportion of this energy is 'lost' to non-reactive molecules (e.g, nitrogen and the combustion products), whilst some is absorbed by the unreacted fuel and oxygen, so increasing their chance of reacting. In the presence of a **pilot ignition source** (such as a flame or spark), the temperature at which a fuel generates sufficient vapours to form a **momentary** ignition, flash or explosion when mixed with oxidant is termed the **flash point**. The minimum temperature of a fuel at which **sustained combustion** is initiated in the presence of a pilot ignition source is termed the **ignition temperature**. The fuel must be heated to its ignition temperature before it can be ignited or support flame spread.

As the temperature of the fuel is raised, less and less activation energy is required for ignition. At low temperatures, the rate at which heat energy is produced by the combustion reaction is less than the rate at which it is lost (see ***Figure 14.4b***). As the temperature rises, the number of molecules reacting increases rapidly until eventually the rate at which energy is released by the combustion reaction balances the rate at which energy is lost. At this point the reaction becomes self-sustaining in the absence of a pilot ignition source and the onset of combustion may become spontaneous. The minimum temperature at which a material will ignite spontaneously is termed the **spontaneous (or auto-) ignition temperature**; this is responsible for **flashover** in the development of building fires (**14.5.2**). In general, self-sustained ignition will only occur in conditions capable of supporting self-sustained combustion; thus if either the ambient pressure or the oxidant concentration is insufficient for sustaining combustion, it will also be insufficient for ignition. At even higher temperatures, the rate of release of energy is even greater and the temperature of the gas mixture increases rapidly ('thermal runaway'). Combustion will continue until all the available fuel or oxidant has been consumed or until the heat source is removed (e.g, the flames are extinguished), i.e., removal of one or more of the requirements of combustion from the 'triangle of fire'.

Reported ignition and spontaneous ignition temperatures (and flash points, **14.2.4**) depend somewhat on the specific test conditions. For example, ignition temperatures of solids and liquids can be influenced by the rate of oxygen flow and the rate of heating, whilst those for flammable gas-air mixtures are affected by composition, ambient pressure and the size, shape and composition of the vessel in which the measurements are made. Test methods are given in BS EN ISO 11114-3[11] for gases and BS ISO 871[12] for plastics. Ignition temperatures and spontaneous ignition temperatures for certain fuels (common building materials and some hydrocarbons fuels) are shown in ***Figure 14.5***. These values should be considered indicative only.

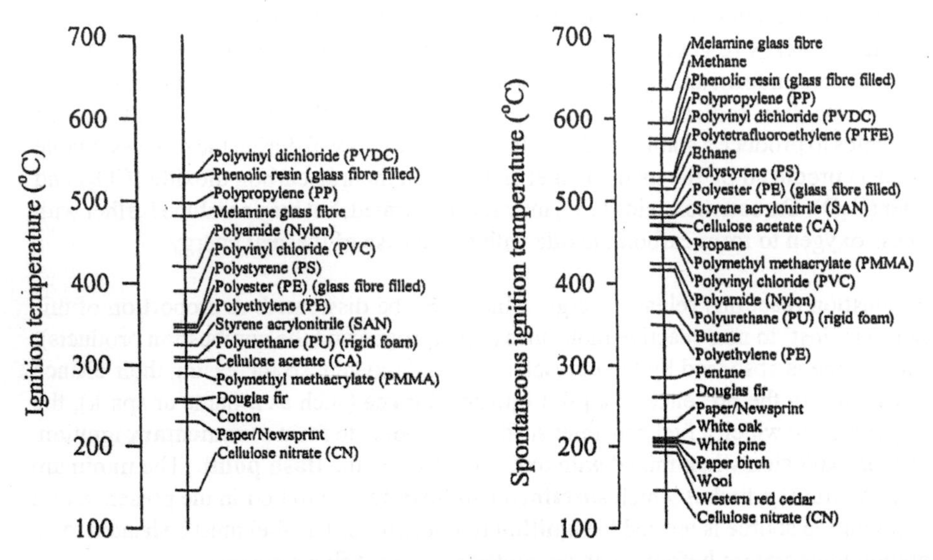

***Figure 14.5** Ignition and spontaneous ignition temperatures for various materials*[12,13]

Within a given homologous series of hydrocarbon fuels, ignition and spontaneous ignition temperatures decrease as molecular weight (or carbon chain length) increases, all other factors being equal, e.g, compare spontaneous ignition temperatures given in ***Figure 14.5*** for the alkanes methane, propane, butane, pentane (**Table 13.2**).

Requirements for ignition

The ability of a material to ignite (**ignitability**) is controlled by its resistance to heating (thermal inertia) and the temperature rise required for combustion to initiate. Ignitability is therefore inversely proportional to the time taken for a given applied heat flux to raise the surface temperature to the **ignition temperature (14.4.1)**. Whether a material becomes hot enough for ignition, and the time taken for it to ignite, depends upon

i. the rate at which the material is heated;
ii. the ability of the material to absorb heat, which is influenced by
   - thermal capacity (**14.10.5**) and thermal inertia;
   - surface area:mass ratio. A thin sheet of material (e.g, a piece of paper) requires less heat for ignition than a solid block of the same material to raise it to a given temperature. Dusts (e.g, wood and coal), can form explosive mixtures;
   - the juxtaposition of the material. If a material is in close proximity to another which absorbs heat readily, it becomes very difficult to ignite (e.g, wall paper stuck to a plaster wall);

- the orientation of the material. A material generally ignites more easily on an edge, particularly a lower one than on a plane surface.

Although a flammable gas-air mixture is readily ignited in the flammable range, it is relatively more difficult to ignite near the flammable limits (**14.2.4**) and more easy to ignite somewhat above the stoichiometric concentration (**14.4.2**). Minimum energies required for ignition will depend on the energy source; the spark energy required for the ignition of a **methane-air mixture** is shown in ***Figure 14.6***.

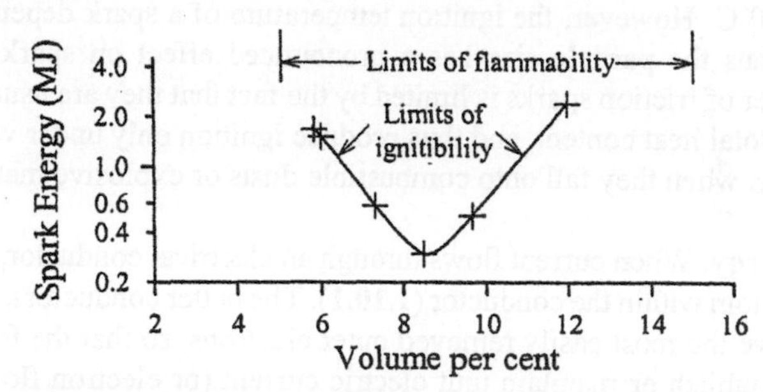

Figure 14.6 *Concentration dependence of the spark energy required for ignition of a methane-air mixture*[14]

Sources of heat energy for ignition
The principal sources of heat energy which can act as ignition sources include hot surfaces, flames, friction sparks, electrical energy and adiabatic compression. In addition, certain fuels undergo decomposition or exothermic oxidation reactions which may produce sufficient heat for ignition or spontaneous ignition.

Hot surfaces. Hot surfaces can be ignition sources if they are sufficiently large and hot enough. The potential for hot surfaces to result in ignition of a fuel depends on the ability of the surface to transfer heat, which can be by one or more of three methods

- **Conduction** (transfer of thermal energy by direct contact from one body to another);
- **Convection** (transfer of thermal energy by a circulating gas or a liquid medium);
- **Radiation** (transfer of thermal energy across a space or through materials as electromagnetic waves. The radiation emitted is dependent upon the temperature of combustion).

Flames. Flames are a zone of combustion in the gaseous phase from which light and heat is emitted. They are unfailing sources of ignition for flammable gas-air mixtures that are within the flammable range. Flames must be capable of heating the flammable mixture to its ignition temperature in order to be a source of ignition. For some liquid- and solid-phase fuels, it is necessary for the flame to be of sufficient duration for heat to evaporate the fuel and to ignite the released vapours. Once ignited, heat radiated from the burning vapours perpetuates the combustion process.

Friction sparks. Friction sparks include the sparks that result from impact of two hard surfaces, at least one of which is usually a metal. Heat energy, generated by impact or friction, initially heats up the particle. The maximum temperature reached is usually determined by the lowest melting point of the materials involved. For some metals, however, the freshly exposed surface of the particle may oxidise at elevated temperature, with the heat of oxidation increasing the temperature further until the particle is incandescent. Although temperatures necessary for incandescence vary for different metals, in most cases they are well above the ignition temperatures of flammable materials. For example, the spark temperature from a steel tool approaches 1400°C, and sparks from copper-nickel alloys containing small amounts of iron may be well above 300°C. However, the ignition temperature of a spark depends on its total heat output; thus the particle size has a pronounced effect on spark ignition. The practical danger of friction sparks is limited by the fact that they are usually very small and have low total heat content, and thus produce ignition only under very favourable conditions (e.g, when they fall onto combustible dusts or explosive materials).

Electrical energy. When current flows through an electrical conductor, electrons pass from atom to atom within the conductor (**1.10.1**). The better conductors, such as copper and silver, have the most easily removed outer electrons, so that the force or voltage required to establish or maintain unit electric current (or electron flow) through the conductor is less for substances composed of more tightly bound electrons. Thus the electrical resistance of any substance depends upon atomic and molecular characteristics; the electrical resistance is proportional to the energy required to move unit quantity of electrons through the substance against the forces of electron capture and collision. This energy expenditure appears in the form of heat.

Adiabatic compression. Controlled adiabatic compression is the basis of diesel engine operation. The air is compressed so quickly that it heats up to a temperature greater than the spontaneous ignition temperature of the diesel fuel (about 210°C) . When the piston is near the top dead centre, diesel fuel is injected into the engine so that the flammable diesel vapour-air mixture spontaneously ignites.

Exothermic oxidation reactions. If exposed to the atmosphere, almost all organic substances capable of combination with oxygen will oxidise, with the evolution of heat. The rate of oxidation at normal temperatures is usually so slow that it is imperceptible (e.g, the perishing of rubber over time). Under these circumstances, the heat released is transferred to the environment as rapidly as it is formed, so that no appreciable temperature rise in the combustible material occurs. However, certain materials undergo oxidation reactions at normal temperatures that generate heat more rapidly than it can be dissipated, which may raise the temperature of the material to the ignition or spontaneous ignition temperature. For spontaneous ignition to occur, very precise conditions are required to supply sufficient oxygen to permit oxidation but not so much as to carry heat away by convection as it is formed. Suitable conditions have been known to produce spontaneous ignition of oil-soaked rags and straw bales and hayricks.

Decomposition reactions. Certain compounds, formed by endothermic reactions, are intrinsically unstable. When decomposition of these compounds is started (e.g, by heating above a critical temperature) decomposition continues on removal of the heat source with the liberation of heat. Acetylene and cellulose nitrate are known to decompose in this way with the liberation of dangerous quantities of heat. Other unstable compounds form the basis commercial and military explosives.

14.4.2 Heat Output

The quantity of heat produced by the **complete combustion (14.3)** of unit mass of a fuel in oxygen is termed the **calorific value**. The calorific value of a material depends upon the kinds, numbers and arrangement of the atoms in the molecule and is constant for a given material. Calorific values for various materials and fuels are shown in **Table 14.2**.

Table 14.2 Calorific values for building materials and hydrocarbon fuels[15]

Material	State	Calorific value (kJ/kg)
Carbon monoxide	gas	14184
Hydrogen	gas	24394
Methane	gas	25586
Propane	gas	175195
Wood sawdust (oak)	solid	19755
Wood sawdust (pine)	solid	22506
Newspaper	solid	18336
Petroleum coke	solid	36751
Asphalt	solid	39910
Oil (paraffin)	liquid	41031
Polypropylene	solid	42000-46000
Polyvinyl chloride	solid	20000-30000

Calorific values can be used in the calculation of **fire loads and densities** and **flame temperatures**, but do not necessarily indicate relative fire hazard, since fire hazard depends on **the rate of burning**. Also, heat is produced by **incomplete combustion (14.3)**, which occurs at some stage in almost all accidental fires and in spontaneous combustion by exothermic oxidation reactions **(14.4.1)**.

Solid organic materials fall into two broad classes, hydrocarbon based and carbohydrate (cellulose) based. Hydrocarbons are derivatives of the unoxidised hydrocarbon building blocks: $-CH_2-$ or -CH-. Carbohydrates are based on the partly oxidised carbon unit: -CH(OH)-. In a sense, carbohydrate-based materials are already partly burnt in their natural state, so that when the two classes are combusted to produce carbon dioxide and water, carbohydrates consume less oxygen and produce less heat. For equivalent amounts of oxide formed, the hydrocarbon-based materials consume 50% more oxygen and thus produce about 50% more heat. For almost all hydrocarbon and carbohydrate based materials, heat produced by oxidation, whether complete or incomplete, depends upon the amount of oxygen consumed. For this reason, the heat produced in a fire or by spontaneous combustion by exothermic oxidation is often limited by the air (oxygen) supply. For common hydrocarbon and carbohydrate materials (e.g, coal, natural gas, common plastics, oils, woods, cotton, sugar, etc.), the heat of oxidation in air is approximately 3 kJ/g (3700 kJ/m^3) regardless of the completeness of combustion.

Rate of burning. The rate of burning is a function of how fast the oxidation reaction occurs, as well as the speed at which the vaporised fuel and oxygen are delivered to the combustion zone. In premixed flames (**14.2.1**), the burning rate is controlled by the inherent rate at which the substances combine. This is generally quite fast; flames propagate under premixed conditions at about 1 m/s. It is for this reason that the contact of air and flammable gases is so dangerous; the process, once started, is virtually impossible to interrupt. A more common mode of burning is where the vaporised fuel mixes with oxygen in the combustion zone (**14.2.1**). The rate of burning is essentially controlled by the rate at which the two components arrive in the heated combustion zone. Once they arrive, combustion is, by comparison, instantaneous. Since gases mix with one another readily, the burning of a gaseous fuel in air is a rapid process. However, the burning of a liquid or solid requires first that the fuel be converted to a the gaseous state (**14.2.3**). This process requires an appreciable amount of heat energy, often from the fire itself, and is almost always slow compared to the rate of oxidation. The rate of volatilisation therefore strongly affects its rate of burning.

Fire loads and densities. Fire load is a measure of the maximum heat that would be released if all the combustible material in the fire area is burned. Maximum heat release is the product of the weight of each combustible material multiplied by its **calorific value** (**14.10.1**). Fire density is the product of the fire load divided by the floor area. As a rough indication, 25 kg/m^2 of combustible material will radiate approximately 84 kJ/m^2 (i.e., 84 W). In a typical building, fire load includes combustible contents, finishes and structural elements.

Flame temperatures. The volumes of fuel and oxygen derived from the balanced chemical equations for complete combustion are such that neither is in excess; the concentration of the mixture is termed the **stoichiometric concentration.** Under these conditions, the energy generated per unit volume will be a maximum. For a methane-oxygen mixture, the balanced chemical equation for complete combustion is given by equation (14.5) (**14.3**) and the stoichiometric concentrations (v/v) are methane 33.33% and oxygen 66.66%. The maximum flame temperature occurs at the stoichiometric concentration. As the concentrations depart further from the stoichiometric, so the flame temperatures fall until the limits of flammability are reached (***Figure 14.7***).

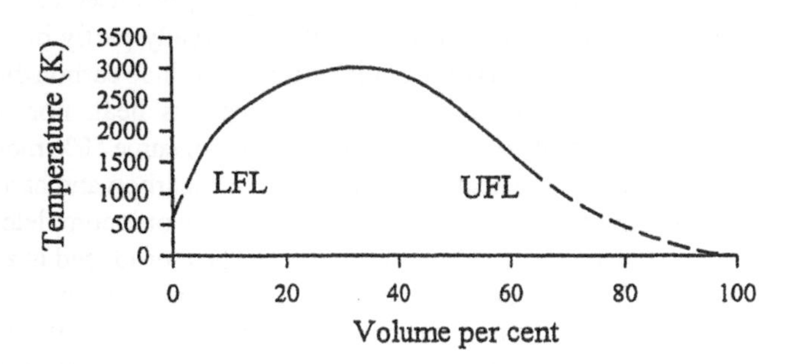

***Figure 14.7** Flame temperatures produced by a methane-oxygen mixture*[16]

The maximum flame temperature of methane in oxygen is approximately 2700°C. Both the upper limit and the lower limit concentrations generate flames at approximately 1200°C. This indicates that the combustion reaction no longer becomes self-sustaining when the resulting flame temperature drops below approximately 1200°C.

Flame temperatures can be calculated using the balanced complete combustion equations. For illustrative purposes, the flame temperature produced following the complete combustion of acetylene in pure oxygen and in air is calculated below. These combinations have been chosen to illustrate the dramatic reduction in flame temperature when a fuel is burnt in air compared to when burnt in pure oxygen. This reduction is caused by the presence of noncombustible nitrogen in air.

The balanced equation for the complete combustion of acetylene in oxygen is

$$2C_2H_2 \quad + \quad 5O_2 \quad \rightarrow \quad 4CO_2 \quad + \quad 2H_2O \quad + \quad \text{X Joules}$$

Each product of combustion ($4CO_2 + 2H_2O$) removes heat from the combustion zone. The heat removed by mass (M) of each product of combustion is (M x c_p x ΔT), where c_p is the specific heat capacity of each product of combustion (834 J/kg.°C for CO_2 and 2016 J/kg.°C for H_2O (steam)) and ΔT is the temperature rise (flame temperature). The quantity of heat produced by complete combustion of unit mass of acetylene (i.e., the calorific value of acetylene) is 74203 kJ/kg. We can therefore equate the heat input and heat output to determine the unknown flame temperature ΔT. The relevant atomic weights are C = 12, H = 1 and O = 16, from which the molecular weights are calculated

acetylene (C_2H_2)	=	(2 x 12) + (2 x 1)	=	26
carbon dioxide (CO_2)	=	(1 x 12) + (2 x 16)	=	44
water (H_2O)	=	(2 x 1) + 16	=	18

We can therefore determine the masses involved as follows

	$2C_2H_2$	+	$5O_2$	$\rightarrow$	$4CO_2$	+	$2H_2O$
Mass (M)	(2 x 26) 52 kg				(4 x 44) 176 kg		(2 x 18) 36 kg

Note that there is no need to determine the mass of oxygen involved because the calorific values of fuels are obtained by burning the fuel in oxygen (**14.10.1**). Equating the heat input and the heat output (kJ) gives

Heat input   Heat output

$$(52 \times 74203) = (176 \times 0.834 \times \Delta T) + (36 \times 2.016 \times \Delta T)$$

$$\Delta T = \frac{(52 \times 74203)}{(176 \times 0.834) + (36 \times 2.016)}$$

$$= 17{,}590°C$$

The flame temperature produced when acetylene is burnt in oxygen at the stoichiometric concentration is 17,590°C. The complete combustion of acetylene in air produces a flame temperature far lower than that produced in pure oxygen, as heat is also removed by the nitrogen in the air. 1 volume of air contains 1/5 volume of oxygen and 4/5 volume of nitrogen. Although the nitrogen takes no part in the combustion process, it must still be heated to the flame temperature and this heat absorption reduces the resultant flame temperature.

From the balanced equation for the combustion of acetylene in oxygen (above), 5 volumes of oxygen are required for the complete combustion of acetylene. Since 1 volume of air contains 1/5 volume of oxygen, 25 volumes of air are required to provide the 5 volumes of oxygen required for complete combustion. Note that 1 volume of air contains 4/5 volumes of nitrogen, so that the 25 volumes of air required for complete combustion of the fuel contains 20 volumes of nitrogen. The balanced equations for the complete combustion of acetylene in air can therefore be written as

$$2C_2H_2 + 5O_2 + 20N_2 \rightarrow 4CO_2 + 2H_2O + 20N_2$$

$$2C_2H_2 + 25\text{Air} \rightarrow 4CO_2 + 2H_2O + 20N_2$$

The presence of $20N_2$ removes heat from the site of combustion. The quantity of heat removed is added to that removed by the products of combustion, calculated above. Thus, with the atomic weight of nitrogen (N = 14) and the specific heat capacity of nitrogen (c_{N2} = 1040 J/kg.°C),

	$2C_2H_2$	+ 25Air →	$4CO_2$	+	$2H_2O$	+	$20N_2$
Mass (M)	(52 kg)		(176 kg)		(36 kg)		(560 kg)

Heat input (kJ) — Heat output (kJ)

$$(52 \times 74203) = (176 \times 0.834 \times \Delta T) + (36 \times 2.016 \times \Delta T) + (560 \times 1.04 \times \Delta T)$$

$$\Delta T = \frac{(52 \times 74203)}{(176 \times 0.834) + (36 \times 2.016) + (560 \times 1.04 \times \Delta T)}$$

$$= 4{,}813°C$$

The temperature of the oxy-acetylene flame is 17,590°C, and is capable of cutting through steel. Temperature of acetylene burning in air is a theoretical 4,813°C. This temperature is not reached in practice because there is insufficient oxygen available to burn all the carbon. A very smoky flame is produced; in the same way, a domestic wax candle (approximately $C_{40}H_{82}$) will burn with a characteristic yellow smoky flame. The carbon can be made to deposit upon any cold surface. If a natural gas flame is yellow, it means there is insufficient oxygen to burn all the carbon. This unburnt carbon will reduce the carbon dioxide to carbon monoxide (**14.3**).

14.5 FIRES IN BUILDINGS

Fires in buildings are very often the source of significant risks of occupant injury and loss of life. Costs of direct property damage as a result of all serious fires (those resulting in fatalities or losses in excess of £50,000) in the UK are produced by the Fire Protection Association. For the period 1992-1995, annual losses were of the order of £200 to £300 million (1991 £417 million)[17].

14.5.1 Fire Class

BS EN 2[18] defines four categories of fire in terms of the materials consumed

Class A: fires involving solid materials, usually of an organic origin, in which combustion normally takes place with the formation of glowing embers;
Class B: fires involving liquids or liquefiable solids;
Class C: fires involving gases;
Class D: fires involving metals.

As the classification is based on materials consumed, fires involving electrical risks are not separately classified, although special precautions will be required in the selection of fire extinguishers (**14.5.4**) where electrical hazards are present.

14.5.2 Fire Growth in a Room

A room fire normally starts as a small fire caused by the ignition of a material by a heat source in the presence of oxygen in the ambient air (**14.1**). The growth of a fire in a room has four distinct phases, initiation, growth, the fully developed phase and decay (see ***Figure 14.8***).

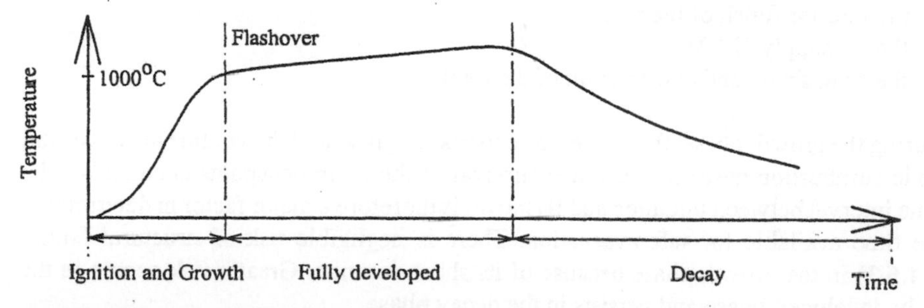

Note: The scales for the two axes of the graph have been deliberately omitted because the rate of development and the severity of fires differs greatly, although the general relationship varies very much less.

***Figure 14.8** General relationship between temperature and time, and the stages of a typical uncontrolled fire in a compartment*[19]

Initiation (initial ignition) phase

Potential ignition sources are considered in **14.4.1**. In buildings, the most common ignition sources are **flames**, e.g, from matches or gas fires, **smouldering sources** e.g, cigarette ends and **hot surfaces**, e.g, ovens, irons, hot electrical wires, etc. Chip pan fires are also common as the spontaneous ignition temperature of domestic cooking oil is low (around 310°C). Structural fabric and fittings account for less than 7% of materials first ignited in fires in occupied buildings; more commonly, room contents (furniture and furnishings) are first ignited. The major factors influencing the initiation phase are

- the amount and duration of the heat flux;
- the surface area of the material exposed to the ignition source;
- the ignitability of the material (**14.4.1**).

Growth phase

The growth phase is characterised by a rapid increase in temperature as the less dense, and most suitably orientated, materials ignite and burn exothermically. This is termed **flashover**, occurring when the vapours produced by the materials in the room undergo spontaneous ignition. A notional **flashover temperature** (of about 500°C) is usually quoted. Most building materials and room contents have ignition and spontaneous ignition temperatures below this temperature (***Figure 14.5***).

The growth phase is characteristically short (≤ 10 minutes). During the growth phase, the fire spread to unaffected materials is essentially by a series of ignitions. When flames are small, flaming will spread to materials which are either in contact with, or very close to, the flames. The fire may self-extinguish if

- there are no materials sufficiently close to allow flame spread; or
- oxygen supply is limited; or
- heat output is insufficient to raise the fuel temperature to the ignition temperature,

otherwise the fire will continue to grow. The main factors which influence the likelihood of spread to full room involvement in a fire are

- the fuel load and its distribution (**14.4.2**);
- the interior finish of the room;
- the air supply (**14.3**);
- the size, shape and construction of the room.

During the growth phase, the temperature rise is so quick, and the volume of smoke and toxic combustion gases so great, that survival of the room occupants is unlikely. The time interval between initiation and flashover is therefore a major factor in determining the time available for safe evacuation. There is negligible risk of structural failure (**14.5.3**) in the growth phase because of its short duration. Greater risk occurs in the fully developed phase and persists in the decay phase.

Steady state (fully developed) phase

The material set alight in previous phases will continue to burn, but the temperature rise is less as the materials become coated with carbon (char) formed by incomplete combustion due to lack of oxygen. Occupants still present in the room will be dead. Structural failure (**14.5.3**) during this phase will be of concern to fire-fighters. The intensity (rate of heat production) of a fully developed fire is determined either by

- the fuel surface area available to participate in the combustion reaction; or
- the amount of oxygen available for combustion.

The maximum rate of heat transfer occurs at the point where the ventilation is just sufficient so that combustion is controlled at the fuel surface. At higher ventilation rates, more heat is removed from the fire by the excess air. At lower ventilation rates, the combustion heat release rate is less, and more unburnt pyrolysis products and fuel particles are vented to outside the fire area.

Decay phase

The decay phase is characterised by declining temperatures caused by the thickness of carbon (char) layer, which limits flaming, allows a temperature gradient to form from the outer surface to the inner (unburnt) material and also restricts the accessibility of oxygen to the flammable vapours. The temperature may be reduced to such a degree that insufficient vapours are produced to support a flame and/or the flammable vapour-air mixture which is produced may be below the ignition temperature. Oxygen (air), however, is able to enter the shrinkage cracks which occur in charcoal, allowing localised surface oxidation. The temperature of the room falls, both by loss of heat due to convection, conduction and radiation, and through providing latent heat for the liquid to vapour phase change (**14.5.6**). Structural failure (**14.5.3**) during this phase will be of concern to fire-fighters.

14.5.3 Structural Failure

Failure of structural elements in a building allows fire to spread to adjacent compartments by

- convection, conduction and radiation of heat;
- movement of the burning fuel itself (e.g, due to collapse of a floor, etc.)

Elements of structure will only act as effective fire breaks if they have the necessary degree of **fire resistance**. The degree of fire resistance necessary will generally vary from 30 minutes to several hours, depending on the occupation of the building, as reflected in the Building Regulations. The three criteria of fire resistance are

- **Insulation** (the ability of an element of the structure to resist the passage of heat through it by conduction. If heat is readily conducted through a wall, for instance, combustible materials on the far side of the wall may be ignited);
- **Integrity** (the ability of an element of structure to prevent the passage of flames and hot gases through it. If gaps through which pipes pass are not made good, or the structure cracks during a fire, there will be little resistance to the spread of fire gases through the element);
- **Stability** (the ability of an element of structure to resist collapse and, if it has a loadbearing function, to continue to support its load).

14.5.4 Fire Extinguishers

BS 4422[20] defines 5 types of fire extinguishers based on the extinguishing medium used

Group 1: Water extinguishers (either soda/acid or water gas pressure). Oil floats on water and so water extinguishers should not be used to put out oil-based fires. The action of water on fires is considered in **14.5.6**.

Group 2: Carbon dioxide to BS EN 25923[21] (Pressurised carbon dioxide which exists in the liquid phase under pressure). These can only be used for small fires as the carbon dioxide is dispersed by the draught of bigger fires.

Group 3: Vaporising liquid extinguishers based on halogenated compounds to BS EN 27201[22] (halons) (**14.5.5**).

Group 4: Dry powder extinguishers to BS EN 615[23]. Dry powder is propelled into the fire by carbon dioxide.

Group 5: Foam solution extinguishers to prEN 1568[24]. These are used for extinguishing burning oils, as the froth floats on the oil and prevents the ingress of oxygen. Generally, an acid is mixed with a bicarbonate when the containers are broken to release a jet of carbon dioxide gas. Foam types include **aluminium sulfate,** $Al_2(SO_4)_3.18H_2O$, which is mixed with sodium bicarbonate, $NaHCO_3$, when the two containers containing them are broken, so as to form a froth containing carbon dioxide. The froth is stabilised by the aluminium hydroxide, $Al(OH)_3$.

$$Al_2(SO_4)_3.18H_2O \quad + \quad 6\,NaHCO_3 \quad \rightarrow \quad 2Al(OH)_3 \quad + \quad 2Na_2SO_4 \quad + \quad 6CO_2$$

Table 14.3 gives recommended fire extinguisher types for each fire class (**14.5.1**).

Table 14.3 Comparison of different fire extinguishers

Fire Class	Water	Foam	CO_2	Dry powder	Halons
Electrical	NO	NO	YES	YES	YES
	(conductors)		(good non-conductor of electricity)		
A	YES (water cools and saturates)	YES	small fires only (convection currents remove gas/powder)		
B	NO (depends upon density of oil)	YES (floats)	YES	YES	YES
			(smothering gas/powder cloud produced excluding the necessary oxygen)		
C	YES (water cools the container)	NO	NO	NO	NO
			(shut off gas supply and seek fire service help)		
D	NO (chemical reaction to produce hydrogen)	NO	YES	YES	NO
		(limited applications)			

14.5.5 Halogenated Compounds (Halons)

Halogenated compounds (halons) used as the extinguishing medium in Class 3 fire extinguishers (**14.5.4**) are named using a simplified system of nomenclature which describes the chemical composition of the materials without the use of chemical names. In this system, complex chemical compounds are referred to as 'Halon' and a 'halon number'. The halon number identifies the chemical composition of the compound

- the **first** digit represents the number of **carbon** atoms in the compound;
- the **second** digit represents the number of **fluorine** atoms in the compound;
- the **third** digit represents the number of **chlorine** atoms in the compound;
- the **fourth** digit represents the number of **bromine** atoms in the compound;
- the **fifth** digit represents the number of **iodine** atoms in the compound (if any). If the fifth digit is zero, it is not expressed.

Hence bromotrifluoromethane (BTM, $BrCF_3$) is referred to a Halon 1301, bromochlorodifluoromethane (BCF, CF_2BrCl) as Halon 1211, dibromotetrafluoroethane ($C_2F_4Br_2$) as Halon 2402, etc.

Chlorine and, in particular, bromine damage ozone. The chemical stability of halons means that they can be carried into the stratosphere, where they deplete ozone. Weight for weight, halons have a much greater ozone-depleting potential than CFCs (although the former are used in smaller quantities). Halons (including Halon 1211 and Halon 1301) are listed in the Montreal Protocol on substances that deplete the ozone layer. The UK (and other EC states) are signatories to the Protocol and agreed to phase out the use of halons by 1st January 1994.

14.5.6 Effect of Water on Fires

The action of water on fires is two fold, **turning water to steam** and **reacting endothermically with carbon compounds**. Heat energy removed by these two processes is **additional** to that removed by products of combustion and unburnt gases.

The order of magnitude of the heat energy removed by the products of combustion and the unburnt gases, the heat removed for every 1°C rise in the flame temperature can be calculated from the product of the mass multiplied by the specific heat capacities of the combustion products. Complete combustion of acetylene in air **(14.4.2)** is assumed and the **specific heat capacities (14.10.2)** for the products are 834 J/kg.°C for carbon dioxide, 2016 J/kg.°C for water (steam) and 1040 J/kg.°C for nitrogen.

	$2C_2H_2$ + 25Air →	$4CO_2$	+	$2H_2O$	+	$20N_2$
Mass (M)		(176 kg)		(36 kg)		(560 kg)
Heat removed per 1°C rise in flame temperature	=	(176 x 0.834)	+	(36 x 2.016)	+	(560 x 1.04)
	=	146.78 kJ	+	72.58 kJ	+	582.40 kJ
	=	801.76 kJ				

Turning water to steam

The heat required to raise a mass of water (M) to boiling point is (M x c_{pw} x ΔT), where c_{pw} is the specific heat capacity of water (= 4190 J/kg.°C) and ΔT is the change in temperature of water to the boiling point. The heat required to undertake the phase transformation from liquid to vapour (steam) is (M x L), where L is the latent heat of vaporisation of water (= 2260 kJ/kg) **(2.5)**. The steam can also be heated up to above 100°C. The heat required to do this is (M x c_{ps} x ΔT), where c_{ps} is the specific heat capacity of steam (= 2016 J/kg.°C) and ΔT is the change in temperature of the steam. The total heat taken out of the fire is therefore

(M x c_{pw} x ΔT)	+	(M x L)	+	(M x c_{ps} x ΔT)
Heat required to bring water to the boil		Latent heat Liquid → vapour		Heat required to 'super heat' the steam

As an example, it is assumed water at 20°C is used to extinguish a fire. As a consequence, steam is produced at a temperature of 1000°C. The notional total quantity of heat (in kJ) removed from the fire per kg of water is determined as follows

Heat required to bring water to the boil		Latent heat Liquid → vapour		Heat required to 'super heat' the steam
(M x c_{pw} x ΔT)	+	(M x L)	+	(M x c_{ps} x ΔT)
4.190 x (100 - 20)	+	2260	+	2.016 x (1000 - 100)
335.2 kJ/kg	+	2260 kJ/kg	+	1814.4 kJ/kg

Note that the greatest heat removed from the fire is by the liquid → vapour phase transformation. In firefighting operations, fine water sprays are often used, which have the dual function of producing a high water surface area:mass ratio to facilitate the liquid → vapour phase transformation (thereby removing large quantities of heat) and providing a fire curtain to protect firefighters from radiant heat. Note also that the heat required to 'superheat' the steam (1814.4 kJ/kg) is **additional** to that removed by the products of combustion (determined above).

Endothermic reaction with carbon compounds

Steam reacts **endothermically** with carbon to produce hydrogen and carbon monoxide

$$C + H_2O \rightarrow H_2 + CO - \text{Y Joules}$$

This is called the water-gas reaction and was used to make 'town gas' before the advent of natural gas. The reaction produces two combustible gases and is the cause of the flames leaping into the sky from a fire and/or spreading elsewhere, as the carbon monoxide, CO, and hydrogen, H_2, burn exothermically. Carbon can react with air to produce carbon monoxide (CO)

$$C + \text{Air} \rightarrow CO_2 + \text{X Joules}$$

or, if starved of oxygen,

$$C + \text{Air} \rightarrow CO + \text{X Joules}$$

These pockets of carbon monoxide are the cause of deaths amongst firemen and are also responsible for the 'fire walls' which are often produced when doors are opened, as this action lets in air which then allows the carbon monoxide (and hydrogen) to burn.

14.6 FIRE CHARACTERISTICS OF BUILDING MATERIALS

All structural and nonstructural materials used in construction are adversely affected by the elevated temperatures experienced in fires. The degree and significance of this adverse behaviour depends primarily on the function of the element and on the degree of protection afforded. At elevated temperatures, mechanical properties are reduced and other adverse effects, such as thermal expansion and accelerated creep, develop. The effects vary depending on the material under consideration. ***Figure 14.9*** compares the reduction in strength of four common materials (steel, aluminium, concrete and timber) resulting from elevated temperatures experienced in a fire.

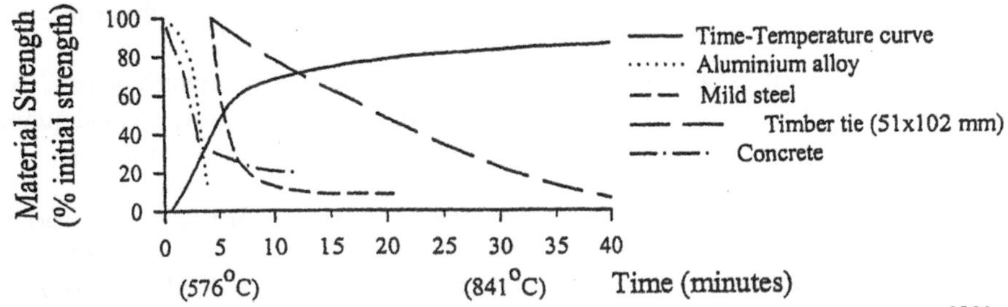

***Figure 14.9** Strength loss of timber, mild steel, aluminium and concrete in a fire*[25,26]

14.6.1 Metals

Metals are generally considered noncombustible in massive solid form. Combustibility depends primarily on the surface area:mass ratio (**14.4.1**); for example, finely divided metals oxidise readily (e.g, magnesium, aluminium, steel wool and iron filings). In addition, the presence of alloying elements influence combustibility. Some metals melt at relatively low temperatures e.g, aluminium (660°C), zinc (419°C) and lead (327°C). On exposure to increasing temperatures, metals exhibit characteristic colours, illustrated for the Fe-C system in ***Figure 10.22***.

Laboratory tests at elevated temperatures to assess metal beams and columns yield critical temperatures within the metal at which failure of the test specimen occurs. These temperatures may differ from those at which failure occurs in buildings, since this will additionally depend on

- the load and support conditions of the member;
- the dimensions of the member;
- the geometry of the member;
- the variation in temperature both along the length of the member and over a single cross-section.

Cast iron

Although the use of cast iron as a structural material has all but ceased, it possesses high compressive strength and has traditionally been used in structural columns (e.g, in shops in the UK). Cast iron generally has good fire resistance properties, as illustrated by extensive use in fire grates and boilers. However, cast iron exposed to temperatures in excess of 426°C (well below the melting point) will deteriorate and fracture if quenched by water, for example, during fire-fighting operations.

Steel

Different grades of steel have a wide range of strengths, but both tensile and yield strengths of all steel grades are similarly effected by temperatures typically encountered in building fires. The range of critical temperatures for structural failure are around 540°C to 760°C (***Figure 14.9***). Steels have high **thermal conductivity (14.10.4)**, and heat transfer away from a local heat source can be rapid. This property, in conjunction with a relatively high **thermal capacity (14.10.5)**, means that steels can act as heat sinks, re-radiating heat to unaffected zones. Where steels are able to transfer heat from high to low temperature regions, it can take a relatively long time for critical temperatures to be attained. Steel has a relatively high coefficient of thermal expansion ($\alpha = 12 \times 10^{-6}$) (**2.8.2**). Where a structural steel member is axially restrained, thermal expansion causes thermal stresses to be induced in the member that may contribute to normal loading and cause more rapid failure. Where unrestrained, thermal expansion causes lateral movement, and may lead to sufficient eccentric loading or pressure on adjacent components to cause collapse.

For structural use in buildings, steel is protected from exposure to heat using an appropriate thickness of insulating material (although this may be difficult to introduce in some locations, for example roof trusses). Insulation may be achieved using suitable materials (including concrete (**14.6.3**), brickwork (**14.6.7**), plasterboards and various sprayed materials exhibiting high fire resistance qualities) or through appropriate detailing (for example, suspended ceilings forming part of a floor-ceiling assembly). All insulating materials and methods create a large temperature gradient between the exposed surface and the steel, which at least slows down the rate at which the structural element reaches a critical failure temperature.

Aluminium

Aluminium has generally one third of the strength of mild steel, but may approach the strength of mild steel in alloyed (with copper) form (**10.7**). However, the critical temperature for yield is about half that for mild steel, and is rapidly followed by melting. Aluminium alloys require approximately three times the thickness of cover compared to mild steel. Aluminium has about twice the coefficient of linear expansion of mild steel (**2.8.2**) and approximately three times the thermal conductivity. Properties vary considerably with the type of aluminium alloy used structurally (aluminium is used in the pure form only for roofing details). Aluminium is not a good fire-resisting material (***Figure 14.9***) and its selection for use in building should be given careful consideration.

14.6.2 Timber

When timber is exposed to a radiant heat source, the surface temperature increases, causing gases (principally steam) to be emitted as the boiling point of water is reached. Over time, the drying zone penetrates deeper into the timber. As the surface temperature increases, thermal degradation of the surface occurs by pyrolysis, and darkens in colour. During pyrolysis, combustible vapours are emitted and the timber surface reverts to carbonaceous char (***Figure 14.10***). The pyrolysis zone penetrates deeper into the timber as the heating process continues. Soon after active pyrolysis commences, combustible vapours evolve sufficiently quickly to support gas-phase combustion in the presence of a **pilot ignition source** (**14.4.1**). At this stage, the **ignition temperature** (**14.4.1**) of the volatiles is reached. The volatile vapours, in moving to the surface, cool the char zone, and are subsequently ejected to the boundary layer where they block incoming convective heat (known as **'transpirational cooling'**). In the absence of a pilot ignition source, the temperature of the timber surface must rise (to the **spontaneous ignition temperature** (**14.4.1**), about 500°C) before the volatile vapours spontaneously ignite.

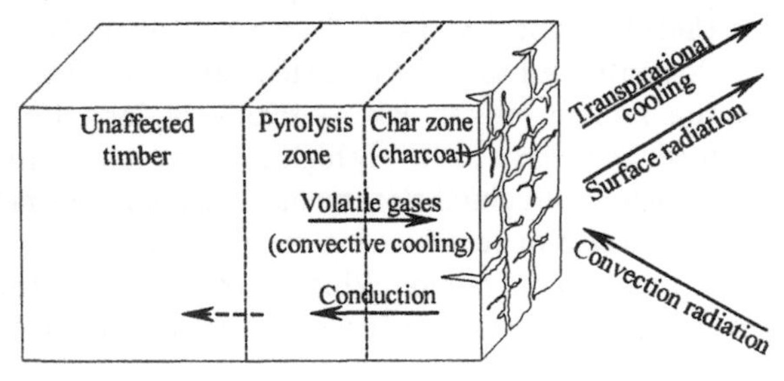

Figure 14.10 *Diagrammatic representation of the thermal decomposition of timber*[27]

Once ignition occurs, a flame rapidly covers the pyrolysing surface and shields it from direct contact with oxygen. The flame continues to heat the surface of the timber and causes an increase in the rate of pyrolysis. If the original radiant heat source is withdrawn at ignition, the flame may go out if the timber loses too much heat by thermal radiation and conduction to its interior; the heat lost depends on the thickness of the timber and the proximity of adjacent materials (which can reflect the radiated heat lost back to the pyrolysing surface of the timber). As heating continues, the charred surface emits heat by radiation. The surface layers crack both along and across the grain, and the surface material is slowly but continually lost. The char zone gradually progresses into the depth of the timber at a rate dependent upon the density of the timber (approximately 0.6-0.75 mm/min for untreated structural softwoods, and approximately 0.5 mm/min for untreated structural hardwoods)[28]. The char zone is a good thermal insulator, restricting the flow of heat from the surface to the depth of the timber, and the char itself produces little or no volatile vapours. The char zone therefore acts to protect the depth of the timber from thermal degradation and may result in a core of unaffected timber. Failure of a timber component will only occur when the cross-sectional area of the unburnt timber becomes too small to support the load. By increasing the dimensions to above those required for structural support, it is possible to design for the continued structural integrity of a timber component for a given period of fire exposure. The ability of timber to retain its strength with increasing temperature and time exposure compares favourably with other materials, notably metals (***Figure 14.9***). When the rate of pyrolysis decreases to a point where the production of volatile vapours is insufficient to sustain gas-phase combustion, the flame will extinguish. If oxygen remains in the system in contact with the char, glowing combustion may continue.

14.6.3 Concrete

Concrete is virtually noncombustible, has low thermal conductivity (**14.10.4**) and thermal capacity (**14.10.5**), and thus provides effective cover for steel reinforcement (**14.6.4**). Building fires lasting one to two hours only moderately damage concrete walls and collapse of concrete structures is comparatively rare. Concrete is, however, adversely effected at sustained elevated temperatures and loss in strength, spalling and other deleterious effects do occur. A change in the natural grey colour of concrete to pink or brown is indicative of heating to temperatures in excess of about 230°C.

Concrete behaves better in actual fires than it does in laboratory experiments, due to the increase in strength with the age of concrete (**8.13.2**). When heated to 100°C, concrete made from OPC cement starts to lose strength, but the strength reduction is not significant until the temperature rises to 300°C. At 600°C there is little strength left (***Figure 14.9***). However, the poor thermal conductivity of concrete means that the core takes a long time to reach these high temperatures. The most significant factors determining the performance of concrete at elevated temperature are

- **Type of aggregate**. Flint, quartz and granite aggregates tend to explode at around 400°C. Crushed sandstones and limestones offer a greater resistance. Lightweight concrete formed using aggregates of low thermal conductivity (which are often produced by a sintering process) e.g, pumice, expanded slags and shales, vermiculite (**14.7.3**) and perlite, provide increased protection. For refractory concrete and slabs for fireplaces, a crushed firebrick and HAC (**7.14.5**) mix is often used.

- **Moisture content.** A considerable quantity of the heat energy of a fire is expended in vapouring the absorbed and capillary moisture in concrete (**14.5.6**). In horizontal members heated from below, water vapour is driven upward and maintains a temperature at the top of the member of 100°C until the water has been driven off. This increases the fire endurance of the member as it maintains the temperature on the unexposed side below the failure temperature. Voids caused by the evaporation of water contribute to shrinkage and a decrease in concrete strength.

14.6.4 Reinforced Concrete

The low thermal conductivity and thermal capacity of concrete makes it an ideal protective covering for other materials, notably steel. These characteristics result in large temperature gradients through the concrete from the exposed to the unexposed surface[26]. Thus for monolithic reinforced concrete, considerable fire exposure times are required before steel reinforcement is adversely affected, as shown in **Table 14.4**. The level of stress in a reinforced concrete member exposed to elevated temperatures has significant influence on endurance[29]. The reduction is attributed primarily to the reduction in the mechanical properties of steel and concrete at elevated temperatures.

Table 14.4 Nominal cover to all reinforcement (including links) to meet specified periods of fire resistance[30,31]

Fire resistance (hours)	Nominal cover (mm) for reinforced concrete (prestressed concrete in parentheses)						
	Beams[a]		Floors		Ribs		Columns[a]
	Simply supported	Continuous	Simply supported	Continuous	Simply supported	Continuous	
0.5	20[b] (20[b])	20[b] (20[b])	20[b] (20)	20[b] (20)	20[b] (20)	20[b] (20)	20[b]
1	20[b] (20)	20[b] (20[b])	20 (25)	20 (20)	20 (35)	20[b] (20)	20[b]
1.5	20 (35)	20[b] (20)	25 (30)	20 (25)	35 (45)	20 (35)	20
2	40 (60)	30 (35)	35 (40)	25 (35)	45 (55)	35 (45)	25
3	60 (70)	40 (60)	45 (55)	35 (45)	55 (65)	45 (55)	25
4	70 (80)	50 (70)	55 (65)	45 (55)	65 (75)	55 (65)	25

Notes: The nominal covers given relate specifically to the minimum member dimensions for normal reinforcement given in BS 8110: Part 1: Figure 3.2. Guidance on increased covers necessary if smaller members are used is given in BS 8110: Part 2: Section 4. Cases that lie below the dashed line (reinforced concrete) or the dotted line (prestressed concrete) may require additional measures necessary to reduce the risks of spalling (BS 8110: Part 2: Section 4). [a] For the purposes of assessing the nominal cover for beams and columns, the cover to the main bars which would have been obtained from BS 8110: Part 2: Tables 4.2 and 4.3 has been reduced by a notional allowance for stirrups of 10 mm to cover the range 8 mm to 12 mm; [b] These covers may be reduced to 15 mm provided that the nominal maximum size of aggregate does not exceed 15 mm.

14.6.5 Prestressed Concrete

Concrete used for prestressed concrete is of a higher strength than that used for ordinary concrete and the overall fire resistance is better. However, prestressed concrete subjected to elevated temperatures usually performs worse than ordinary reinforced concrete, for two principal reasons

- prestressed concrete has an increased tendency to spall, exposing prestressing steel;
- the type of steel used for prestressing (cold drawn, high carbon) is more susceptible to strength loss at elevated temperatures than normal reinforcement (hot rolled, low carbon). For example, normal reinforcement and prestressing steel retain 50% of their ambient strength at about 550°C and about 450°C respectively. The modulus of elasticity of prestressing steel is reduced by about 20% when temperatures reach

about 300°C to 320°C. In addition, unlike hot rolled reinforcement, cold drawn prestressing steels do not regain strength on cooling and are permanently weakened at temperatures in excess of 420°C.

For these reasons, prestressed steel wires and tendons require a thicker protective concrete cover compared to conventional reinforcement, as shown in **Table 14.4**.

14.6.6 Plastics

When a plastic is exposed to a fire and it's temperature rises, several phenomena may occur. In **thermoplastics (13.6.1)**, the hydrocarbon chains acquire vibrational energy, so that the thermal energy breaks the van der Waals forces (**1.14**). This change occurs at the **glass transition temperature (13.9.5)**. As the plastic is heated still further, depolymerisation may occur where the vibrationally excited bonds begin to break, eventually yielding fragments with sufficient vapour pressure to be lost into the gaseous phase. Gases lost can support a flame. In **thermosets (13.6.2)** crosslinked by covalent bonds (**13.9.7**), no softening occurs. High temperatures lead to some simple fragmentation, but the reactions are generally highly complex, leading to the formation of a carbonaceous char. The resulting porous residue is generally less readily volatile than the original polymer, hence loss of flammable vapour is limited; it also acts as a diffusion barrier to volatiles leaving the plastic and to oxygen reaching the reactive surface. The thermal insulation properties of the carbonaceous char reduce the transfer of heat to the surface, further reducing degradation. Char-forming thermosets are commonly selected for fire-safety applications, and additives (**13.11**) promoting char formation are used as components of many commercial polymer formulations (**14.7.1**).

14.6.7 Other Building Materials

Glass, used for glazing in doors and windows, cracks quickly because of differential temperatures between surfaces (**9.6.1**) and thus has little capacity to resist fire. **Fibreglass** is non-combustible and no flammable vapour is given off on heating. However, fibreglass is often covered with a resin binder which is combustible and will spread flame, albeit relatively slowly. In glassfibre reinforced plastic (GFRP) products (e.g, translucent window panels, bathroom units, etc.) the fibreglass acts as reinforcement for a thermosetting (**13.6.2**) resin. The resin is combustible and, as it usually comprises over 50% of the product, renders GFRP products combustible.

Clay bricks (**7.3**) are exposed to temperatures in excess of 1000°C during production, temperatures sufficient to remove any carbonaceous material present in the clay raw material. In a fire, clay bricks are noncombustible and retain their strength, although some distortion is possible when heating under load. Spalling may occur following sustained exposure to elevated temperatures and is a definite risk where hot bricks are drenched with water during fire-fighting operations. Reinforcing steel embedded in the centre of a clay brick wall would be protected by a minimum of 75 to 100 mm of brick.

Emulsion paints contain PVA (**Table 13.3**) which chars and lifts from the surface to produce a secondary surface, providing some protection to the substrate.

Asbestos is a mineral fibre that has been used extensively for fire protection in buildings, although its use has ceased in recent years as asbestos fibres are a health hazard (**9.12**). Asbestos has been used

- in combination with cementitious binders and sprayed onto structural members;
- in combination with Portland cement to make asbestos cement products;
- in combination with other materials to produce e.g, asbestos insulation board, asbestos wood, etc.

14.7 FIRE/FLAME RETARDANT TREATMENTS

A fire/flame retardant treatment interferes with the chemistry and physics of combustion in such a way as to change the flammability of a material, making ignition more difficult and/or reducing the rate of flame spread across the surface. Fire/flame retardant treatments are available for cellulosic materials, textiles, plastics and to cover internal surfaces (e.g, steel, board materials, etc.).

14.7.1 Methods of Treatment

There are four methods of treatment: **chemical changes** (substitutions, admixtures), **impregnation** (saturation, absorption), **pressure impregnation** and **coating**. Chemical changes and pressure impregnation are limited to manufacturing processes and are not adaptable to field use (impregnation and coating methods are used for *in situ* treatments). Generally, methods employed in manufacturing processes are preferred since they provide greater uniformity, permanence and dependability.

Chemical changes

Fire/flame retardant chemical changes are primarily effective with plastics and synthetic fibres, because different chemical elements and compounds can be chosen to alter the burning characteristics of the final product. These elements and compounds are either blended into the polymer mix during processing (**additives**) or chemically bound to the substrate during polymerisation in a separate step (**reactants**). For wood, new chemical grafting techniques have combined bromine (**14.7.3**) with the lignin components, phosphates (**14.7.3**) associated with nitrogen compounds, such as urea, $CO(NH_2)_2$ with the cellulosic component, or, on heating, by polymerisation of fire/flame retardant monomers inside the wood cells.

Impregnation

Impregnation involves the saturation of the material to be treated in a solvent (usually water) in which the fire/flame retardant chemicals are dissolved or dispersed. The solution may be applied by brushing, spraying or immersion. In the simple water-soluble salt impregnation process, the result is merely the deposition of minute crystals of the salts within or on the surface of the material. The process is physical and no chemical reactions are involved. Typical water soluble salts include aluminium trihydrate and potash alums (**14.7.3**). Certain cellulosic-based products, such as paper and wood-based composites, incorporate a wet pulp stage during the manufacturing process. It is possible to add fire/flame retardant chemicals at the wet pulp stage, resulting in an even distribution of chemicals throughout the entire mass of the treated product.

Pressure impregnation
Fire/flame retardant pressure impregnation is used for treating relatively dense, non-absorbent materials (e.g, wood). The process replaces air in the material with fire/flame retardant chemicals, whioh are deposited as the solutions dry.

Coating
There are several types of fire/flame retardant coatings that can be used for treating a variety of materials. Coatings may be applied at any stage, from manufacture to use, and may either actively inhibit flame spread to some degree or present a non-combustible surface over which the flame cannot spread. They are used predominantly on non-absorbent building materials that cannot be treated by other means. The effectiveness of the coating depends on

- the chemical and physical properties of the material to which the coating is applied;
- the effectiveness of the coating on this material;
- the workmanship and thoroughness of the treatment.

Many impregnation and pressure impregnation fire/flame retardant systems for wood incorporate salt mixtures which become increasingly hygroscopic at humidities in excess of about 75%. Prolonged exposure may result in salt efflorescence and/or migration. These treatments are classified by the British Wood Preserving and Damp-proofing Association (BWPDA) based on the properties which limit or recommend their use in specific circumstances

- Interior type: BWPDA Type A: For use where the relative humidity is less than 75%;
- Humidity resistant type: BWPDA Type B: Better able to withstand intermittent superficial wetting and humidities in excess of 75%. Not suitable for exterior use where weather exposure or leaching is likely;
- Leach resistant type: BWPDA Type C: Based on a polymeric resin system, where chemical fixation (**14.7.1**) occurs as a result of polymerisation of organophosphate resins induced by heat treatment at the end of the kiln-drying process.

14.7.2 Mechanism of Operation

There are four general mechanisms by which fire/flame retardant materials retard the flaming or glowing of combustible materials, **thermal, coatings, gases** and **chemical**. For most fire/flame retardants, more than one mechanism is involved.

Thermal
Fire/flame retardants may reduce thermal buildup in the treated material by

- increasing thermal conductivity to dissipate the heat of combustion;
- increasing thermal absorption to reduce the amount of heat available for pyrolysis;
- providing thermal insulation to reduce heat access to the substrate.

Coatings
Some fire/flame retardants melt or fuse at relatively low temperatures to form an insulating glaze over the treated material (***Figure 14.11a***), acting to exclude oxygen and inhibit the escape of flammable gases. More effective retardants exhibit a bubbling or foaming action (**intumescence**), creating an insulating barrier (***Figure 14.11b***). An

effective, coherent intumescent coating is generated in several steps. Formulations are specific to a particular application. Several classes of compound are common to many formulations, including

- **latent acids**: ammonium phosphate and zinc borate (**14.7.3**), which are activated at high temperatures;
- **residue-forming carbonaceous materials**: pentaerythritol, alkyd resins, polyols;
- **blowing agents** (a substance capable of producing a cellular structure in a plastic) e.g, urea, melamine or dicyandiamide, which typically contain nitrogen;
- **fire/flame retardant additives**: alkyl halophosphates or organohalogens (**14.7.3**), which are compatible with the other ingredients and with the coating;
- **a vehicle**, which is usually a lightly cross-linked thermoset resin or a thermoplastic. The vehicle must be a good film-former.

Intumescent coatings have been used to increase the endurance of structural steel. However, beneficial effects persist only for as long as the coating remains intact. As sustained elevated temperatures tend to destroy the coating, intumescent coatings are primarily used for non-exposed structural steel.

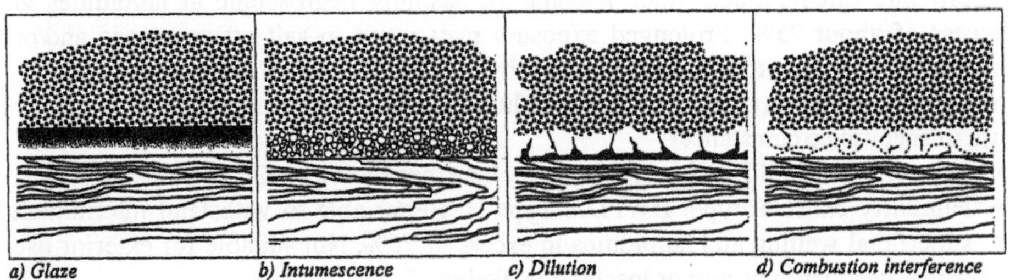

***Figure 14.11** The fire/flame retardant action of coatings*

Gases

Some fire/flame retardants function by releasing non-flammable gases (such as water, ammonia and carbon dioxide) following exposure to heat. These gases dilute the combustible gases sufficiently to render the mixture non-flammable in air (***Figure 14.11c***). Other retardants catalytically inhibit free radical chain reactions (**14.4.1**) that occur in the flame (***Figure 14.11d***).

Chemical

Combustion of certain untreated materials (e.g, wood) results in breakdown into liquids and a smaller amount of solids. The liquids in turn breakdown into flammable gases and a little char, whilst the solids decompose into char and gases, some of which are flammable. The flammable gases derived from the liquids combine with those from the solids to cause flaming and the emission of additional heat, which continues the reaction. When chemically treated, few liquids and more solids are formed, and more of the solids become char. The reduced amounts of liquids create some char and some flammable gases, but overall the quantities of flammable gases from both the liquid- and solid-phases is greatly reduced. Thus less and less heat is produced and a degree of flame resistance is achieved.

14.7.3 Fire/Flame Retardant Compounds

Historically, the development of fire/flame retardant coatings started in the 1300's when King John decreed, after the fires in London, Bath, and York, that lime (**7.10**) plaster (see below) had a good 'fire protective value' or resistance and incorporated its use into the byelaws. However, these byelaws were never properly enforced. During the 17th century, a lot of softwoods were imported from Scandinavia. These timbers were easier to work and lighter than the home-grown hardwoods, but were also less fire resistant than the hardwoods. In 1666, a bake-house in London caught fire and 13,000 houses were burnt down in 2 days. This led again to the Rebuilding Act (1667) where the poor fire resistance of the softwoods was counteracted by the use of (lime) plastered ceilings. The next milestone in fire prevention came with the multi-storey buildings during the Industrial Revolution. For these buildings there was a requirement that the fire had to be contained not only within the building, but also within the floor in which it started. With the rapid increase in the use of steel in industrial buildings, added structural integrity was achieved by encasing steel stanchions within a gypsum (see below) plaster cover. Currently, many compounds are known to have fire/flame retardant qualities, the most important of which in terms of fire protection in buildings are considered below.

Lime plaster

The benefits of lime plasters in a fire result from the low thermal conductivity of the product, which produces high temperature gradients through the material so that, although exposed surfaces may gain high temperatures, considerable exposure times are required to endanger the backing material. Temperatures in excess of the calcination temperature obtained during manufacture(**7.10.1**) are required to decompose calcium carbonate. Any uncarbonated lime, present as calcium hydroxide, $Ca(OH)_2$ ($CaO.H_2O$) (**7.10.2**), only loses water depending upon the relative humidity; the reactions are very temperature sensitive (**2.2.7**). For example, calcium hydroxide loses no water at 100°C, but at elevated temperatures water is expelled; at 400°C about 30% water is expelled, and at 450°C nearly all water is driven off. These effects result from the requirement for the water vapour driven off to be in equilibrium with the saturated vapour pressure of water at that temperature (**2.2.6**).

Gypsum plaster

Gypsum for building purposes is made from the naturally occurring mineral gypsum ($CaSO_4.2H_2O$) or calcium sulfate anhydrite ($CaSO_4$), or is obtained as a by-product from the chemical industry. It is manufactured by crushing and rolling the raw material to produce a white powder which is calcined (heated to about 107 to 170°C remove the water of crystallisation) to produce gypsum hemihydrate, $CaSO_4.½H_2O$

$$\underset{\text{Gypsum}}{CaSO_4.2H_2O} \xrightarrow{\text{Heating}} \underset{\text{Gypsum hemihydrate}}{CaSO_4.\tfrac{1}{2}H_2O} + \underset{\text{Free water}}{1\tfrac{1}{2}H_2O} - \text{Y Joules}$$

If the calcination temperature is increased (to about 220 to 340°C), the remaining water of crystallisation is driven off to leave anhydrous calcium sulfate ($CaSO_4$). When the powdered plaster is mixed with sufficient water, needle-like crystals of $CaSO_4.2H_2O$ are reformed within a short time as the plaster first sets (i.e., a change of flow properties) and then hardens (reverts to calcium sulfate dihydrate, i.e., gypsum). Excess water evaporates on drying and is responsible for the porosity of gypsum plaster (the greater the proportion of excess water, the greater the porosity). Retarded hemihydrate plasters set rapidly once the setting process has begun; retarded anhydrous plasters, however, set more slowly, where the slow continuous setting action helps to produce a hard finish if plenty of water is used and the coat is re-trowelled before final hardening (e.g, Class C plasters for squash courts). Typically, the retarder is the colloid (**2.9.2**) keratin. Thus the setting and hardening of gypsum plasters is **not** by the evaporation of water or by carbonation (as for lime plasters), but by the formation of interlocking crystals of calcium sulfate dihydrate. This recrystallisation is too fast for plastering work and has to be slowed down by chemical retarders, as described above. The powdered gypsum plaster crystals ($CaSO_4.½H_2O$) contain 6% of water of crystallisation, but when hardened the gypsum plaster crystals ($CaSO_4.2H_2O$) contain nearly 21% of water of crystallisation. Hence gypsum plaster sets to form $CaSO_4.2H_2O$ with an increase in volume and with the evolution of heat.

$$\underset{\text{Gypsum hemihydrate}}{CaSO_4.½H_2O} + \underset{\text{Free water}}{1½H_2O} \rightarrow \underset{\text{Gypsum}}{CaSO_4.2H_2O} + \text{X Joules}$$

This heat evolution helps provide a degree of cold weather working resistance and helps with the setting reaction. The strength of gypsum means that it can act like an adhesive. It can be mixed with sand, vermiculite or perlite to produce lightweight plasters. Gypsum plasters are completely free from all the disadvantages associated with lime plasters, e.g, contraction on setting and a slow chemical set which cannot be accelerated. However, if excess water is used, setting will not take place, but hydration will, resulting in shrinkage cracking of the surface on drying. As the setting process takes place, the needle-like crystals orientate themselves, resulting in 'setting expansion'. Expansion increases as the mix water decreases. The advantages of this expansion include a reduction in the possibility of cracking; enables production of casts of fine detail; ideal for repairs; the finishing coat may be in a state of compression. However, excessive expansion may reduce adhesion or cause buckling of backing materials (e.g, 'wavy' plasterboard ceilings). BS 1191 limits the linear expansion of Class B plaster to $\leq 0.2\%$ after one day where low expansion plaster is required.

BS 1191[32] classifies gypsum plasters in to five classes: Class A and B are based on calcium sulfate hemihydrate ($CaSO_4.½H_2O$) plaster; Class C and D are based on anhydrous calcium sulfate ($CaSO_4$) plaster; Class E is based on calcium anhydrite ($CaSO_4$). ***Figure 14.12*** shows the interrelationships between these different gypsum plasters.

Class A plaster (Plaster of Paris). Plaster of Paris is marketed as a fine powder, which is usually white but may be pink or grey in colour depending on the purity of the product. These colours arise from the iron oxide content of the raw material, an impurity commonly associated with natural mineral gypsum. When Class A gypsum plaster is mixed with water the mass quickly sets hard. The time available between mixing the plaster with water and the initial stiffening of the mix is too short for application by common building plastering methods employed in the UK, but it may be used for small patching repairs when a rapid set is advantageous (this property if clearly of importance in other areas where a rapid set is required, such as setting broken bones). Class A gypsum plaster is sometimes added to lime plaster to impart a set and to reduce shrinkage cracking ('crazing'); the overall product is referred to as 'gauging plaster'.

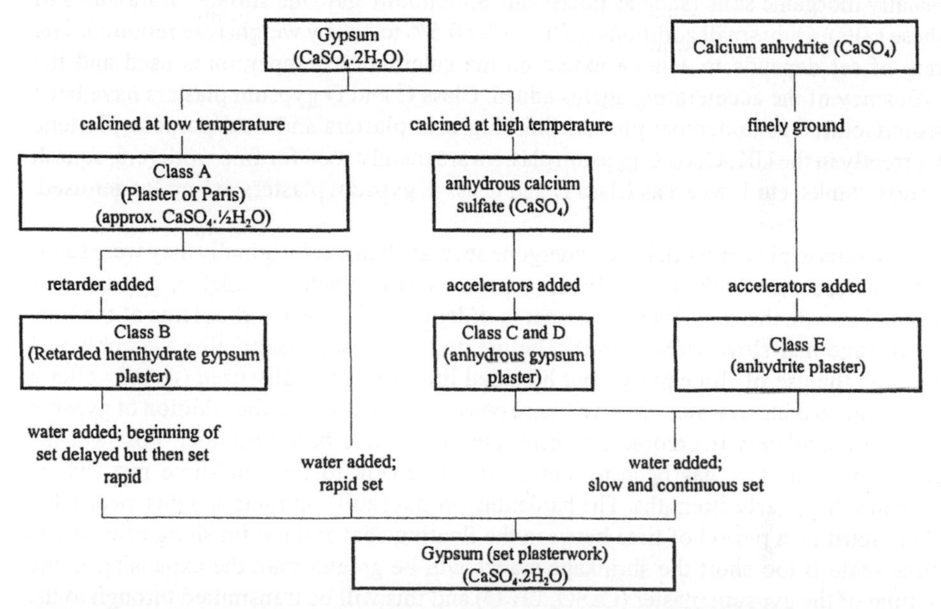

Figure 14.12 Interrelationships between gypsum and anhydrite plasters

Class B plaster (retarded hemihydrate gypsum plaster). To convert Class A gypsum plaster into a suitable material for general building work, it is necessary to lengthen the time interval between the addition of water and the start of the setting process. This is done at the manufacture stage by adding a retarder, usually a colloidal **(2.9.2)** material (commonly based on keratin, a natural protein present in animal hoof and horn material). The retarder is added in relatively small amounts and functions to delay the start of the setting process to an extent varying with the quantity added. The proportion is adjusted to suit the type of plaster and the use to which the product is put. Thus plaster for undercoat work are more heavily retarded than that for finishing coats. The presence of the keratin retarder can provide a source of nutrients for plaster fungi **(12.21.3)**. Class B gypsum plasters are marketed as

- Undercoat plasters, having a setting time of about 120 minutes;
- Finishing coat plasters, having a setting time of about 25 minutes;
- Dual-purpose plasters, having a setting time of about 60 minutes.

Special finishing coat plasters are available for single coat work on boards or on concrete and are characterised by low thermal expansion on setting.

Other plaster classes. Class C gypsum plaster (anhydrous gypsum plaster, Sirapite), Class D gypsum plaster (Keenes or Parian plaster) and Class E gypsum plaster (Anhydrite plaster) have much the same appearance as Class A plasters. Generally, the higher the calcination temperatures used for these plasters, the more unreactive they are. These plasters are too slow in hardening to be used as building plasters without modification and accelerator additions have to be made to increase their reactivity, usually inorganic salts (such as potassium, aluminium and zinc sulfate, or mixtures of these salts). Only small additions (of the order 0.5% to 1% by weight) are required. The rate of set depends to a large extent on the calcination temperatures used and the efficiency of the accelerating agents added. Class C and D gypsum plasters have been manufactured as undercoat plaster, finishing coat plasters and dual-purpose plasters. Currently in the UK, Class C gypsum plasters are mainly used for fine work (e.g, squash courts, banks, etc.), whereas Class D and Class E gypsum plasters are no longer used.

Undercoats to plaster work have undergone several changes. Originally they were sand-lime undercoats which shrank. When gypsum became readily available, gypsum was gauged in with the sand-lime undercoat in order to overcome the problems of the lime (shrinkage and slow carbonation). The greatest increased workability was obtained through the use of 'lime putty', but hydrated lime was generally used (usually after a period of soaking in water) as it was more convenient. However, the addition of gypsum did not completely overcome the shrinkage of the lime-based product. Post-war the practice in the UK has been to add Portland cement to the sand-lime mix, which provides high early strengths. The hardening product still contracts and this means that there must be a period of time between the floating coat and the finishing coat. If this time scale is too short the shrinkage cracks will be greater than the expansion in the setting of the gypsum plaster ($CaSO_4.2H_2O$) and this will be transmitted through to the finishing coat. Gypsum plaster should never be used as an addition to Portland cements due to the problem of sulfate attack (**8.19.2**). A finishing coat containing gypsum plaster may, however, be safely applied over cementitious substrates provided the substrate has had adequate time to dry and subsequent conditions are not persistently damp.

When exposed to elevated temperatures, the response of gypsum plasters is the reverse process to that during their manufacture. Above about 90°C, gypsum loses water of crystallisation; this reaction is strongly endothermic, consumes thermal and radiant energy from the flame, and slows the rate of pyrolysis of the substrate. In addition, the vaporised water acts as an inert diluent and cools the flame, reducing the effective heat flux to the substrate surface. If a sufficient quantity of gypsum is present, the heat flux and rate of flammable gas generation becomes inadequate for continued flame propagation and the sample self-extinguishes. This process occurs in all materials which contain water of crystallisation (e.g, aluminium trihydrate), but is inefficient when compared to chemical inhibition. High loadings of the material are required for satisfactory flame retardance. However, gypsum is relatively inexpensive and, as it does not promote incomplete combustion, does not significantly increase (and may even reduce) smoke and toxic gas emission.

Boron compounds
Boron compounds, principally boric acid (H_3BO_3) and sodium borate ('Borax') ($Na_2B_4O_7.10H_2O$) mixtures, are used for impregnation and pressure impregnation of cellulosics (cotton, paper, wood, etc.), polymers and in intumescent paints (**14.7.2**). When heated, the solids (melting points 171°C and 75°C respectively) fuse and form a glass-like glaze on the surface of the material. Further heating dehydrates the borax, which causes the coating to intumesce, increasing the effectiveness of the protective layer. The endothermic dehydration consumes thermal energy from the flame and releases water vapour. The boron coating interacts with the substrate forming char and creates another protective layer between the substrate and the flame. For polymers, zinc borate is also used, usually in combination with another fire/flame retardant, such as halogenated compounds, antimony oxide or aluminium trihydrate (see below). The requirement for a co-retardant often limits the application of borates to those that can be processed at low temperatures (e.g, plasticised PVC or unsaturated polyester resins).

Aluminium trihydrate (ATH) and Potash alums
These compounds contain a high proportion of water of crystallisation (**1.18.5**); for example, ATH ($Al_2O_3.3H_2O$) (**1.8.3**) contains 34.62% water of crystallisation by weight and potash alum ($K_2SO_4.Al_2(SO_4)_3.24H_2O$) contains 45.57% water of crystallisation by weight (determined from the atomic weights, **1.3**). When exposed to elevated temperatures (above about 250°C), these compounds lose water of crystallisation (see gypsum plasters, above). ATH is used in polymers in applications where high loadings are required, such as carpets and upholstery back coatings, filled thermoset polymers. Potash alums are impregnated in to textile materials, fire curtains, etc.

Phosphorous-containing compounds
For example, monoammonium phosphate, $NH_4H_2PO_4$ and diammonium phosphate, $(NH_4)_2HPO_4$. The pyrolysis of pure cellulose involves two competing processes; one leads to dehydration and formation of H_2O, CO_2 and a carbonaceous residue, and the other leads to the formation of volatile gases by way of depolymerisation (**14.6.2**). Phosphorous compounds enhance char formation in cellulose by inhibiting the depolymerisation reaction, which thus reduces the quantity of flammable gases produced. The principal commercial applications for phosphorous-based fire/flame retardants are in cellulosics, flexible and rigid polyurethane foams, and vinyls. Phosphorous-based fire/flame retardants used as **reactants** (**14.7.1**) are employed in rigid and flexible urethanes, cellulosics, thermosets, polyester resins, adhesives and engineering thermoplastics (including polyvinyl chloride, PVC)

Antimony oxides
Antimony trioxide (Sb_2O_3) and antimony pentoxide (Sb_2O_5) are among the most widely used commercial fire/flame retardants. Their application is based on the synergistic interaction between antimony oxides and halogenated compounds. In the absence of a halogen, antimony oxides function only as inert fillers. The combined system is always more effective than the halogen alone. The effectiveness of the combined system is due to the formation of volatile antimony halides and oxyhalides which reduce the heat generated by combustion reactions, resulting in flame inhibition. Antimony oxides can

be applied to both additive and reactive halogenated compounds for all classes of polymer substrates. In additive systems, the antimony oxide is applied together with the halogen to ensure thorough blending. In reactive systems, the oxide may be added during or after polymerisation. The ratio of halogen to antimony is between 1:1 and 5:1.

Halogenated compounds

Halogenated compounds (**14.5.5**) contain the halogens methane, bromine, fluorine, chlorine or iodine, which have been substituted for a hydrogen atom in the hydrocarbon chain. Whilst compounds containing all these elements have been found to be very effective fire/flame retardants, bromine, fluorine and chlorine are most commonly employed (iodine lacks thermal and photolytic stability for most applications). The halogen influences the relevant properties of the compound in the following manner

- **fluorine** imparts stability, reduces toxicity, reduces the boiling point and increases thermal stability;
- **chlorine** imparts fire extinguishing effectiveness, increases the boiling point, increases toxicity and reduces thermal stability;
- **bromine** has the same effects as chlorine, but to a greater degree.

The action of halogenated compounds is to break into the combustion reaction (**14.4.1**) and convert the highly reactive intermediate products ('free radicals' [C] and [D]) to a stable form. This results in destabilisation of the flame and a reduction in the thermal feedback to the substrate surface, which in turn reduces the rate of fuel gas formation which further destabilises the flame. If sufficient quantity of the halogenated compound is introduced into the flame, it becomes self-extinguishing.

Halogenated compounds are used in plastics as **additives** and **reactants (13.11)**. In practice, however, chlorine and bromine are seldom used in without a synergist (e.g, antimony oxide), which allows lower concentrations of total additives to achieve the required resistance. Additive halogenated compounds are primarily employed in thermoplastics and as an ingredient in coating formulations, and include **chlorinated paraffin** (halogen concentration 20-40%: intumescent coatings and polyolefins for paints, canvasses and adhesives), **chlorinated cycloaliphatics** (halogen concentration ≤40%: polyolefins, nylons, thermosets and thermoplastic polyesters), **brominated aromatics** (halogen concentration ≤20%: high impact polystyrene, polycarbonate, polyolefins, polyester thermoplastics and polyester resins) and **brominated polyaromatics** (halogen concentration ≤20%: engineering resins, nylon and thermoplastic polyesters). Reactant halogenated compounds contain an additional chemical functionality that allows them to be permanently incorporated into the substrate structure, and include chlorendic acid (C_5Cl_6.2CH.2COOH) and chlorendic anhydride (C_5Cl_6.CHCO-O-CHCO) (halogen concentration ≤25%: epoxy resins).

Vermiculite

Vermiculite is a hydrated, laminar mineral that is soft and flaky, and consists mainly of silicates (SiO_2, 39 to 45%) and oxides of aluminium (Al_2O_3, 14 to 20%), iron (Fe_2O_3-FeO, 6 to 11%) and magnesium (MgO, 15 to 20%). When vermiculite is heated to between 700°C and 1000°C under controlled conditions, the flakes exfoliate (expand) to between 7 and 15 times their original volume, resulting in a highly insulating material of very low density. Vermiculite is used in combination with powdered carbon for the fire protection of structural steel (e.g, sprayed Limpet™ Vermiculite).

Glass and mineral wool

Glass and mineral wools are non-combustible (**14.6.7**) and are incorporated in products for the fire protection of structural steel and concrete, external walls and ceilings (e.g, sprayed Limpet™ Mineral Wool, HT grade).

14.8 EFFECTS OF FIRES ON BUILDING OCCUPANTS

Fire atmospheres have a number of characteristics which pose hazard risks to occupants. These are invariably present together in fires, and include

- high temperatures;
- the presence of toxic and narcotic gases (i.e., those that induce unconsciousness);
- low oxygen concentrations;
- the presence of particulates and smoke;
- the presence of irritants.

Low concentrations of fire gases may cause behavioural changes and incapacitation, so reducing the chances of escape. High concentrations may bring about rapid unconsciousness and death within a few minutes. The relative importance of incapacitation and acute effects depends on the duration and conditions of exposure. Generally, carbon monoxide is the most important toxic product in fires, but other products, particularly those derived from modern synthetic polymers, contribute to the overall toxicity and irritancy. Pathological studies of fatalities of domestic fires carried out in the Strathclyde region of Scotland[33] indicate that high alcohol levels are commonly associated with fire deaths, although alcohol levels were significantly lower in findings for the rest of the UK[34].

14.8.1 Effects of High Temperatures

The body dissipates heat by evaporative cooling (perspiration) and by circulation of the blood. If the total heat energy reacting with the body surpasses the capability of the physiological defence processes to compensate, a series of events will occur, ranging from minor injury to death. In dry air temperatures below about 60°C, the body can remove heat successfully by evaporative cooling. The time for skin temperatures to increase depends upon the exposure temperature which, in most fires, increases very quickly (**14.5.2**). Thus temperature may increase faster than the defence mechanisms respond. Thermal data for unprotected skin of humans at rest[35] suggests a limit of about 121°C, above which considerable pain would occur; a maximum survivable breathing temperature of 140°C has been determined[36]. Adverse effects are increased by moisture in the atmosphere (derived from ambient humidity, from the combustion process itself or from fire fighting operations). With higher moisture content, the transfer of heat energy is more efficient and the body is less able to rid itself of the excess heat burden. Skin tissue burns are commonly classified as first-, second- or third-degree burns

- **first-degree burns** involve only the outer layer of skin and are characterised by abnormal redness, pain and sometimes a small accumulation of fluid under the skin;

- **second-degree burns** penetrate more deeply. The burned skin area is pink, moist and blistered. Second-degree burns occur at skin surface temperatures as low as 71°C maintained for about 1 minute[37];
- **third-degree burns** are usually dry, charred or pearly white. If a large percentage of the body skin tissue is effected, post-exposure consequences are extremely critical.

14.8.2 Combustion Product and Oxygen Vitiation Effects

Combustion products have direct and indirect toxic effects, and may cause visual obscuration by smoke and particulates, impeding escape. Toxic effects depend on the compound, dosage, duration of dosage and on the storage capability of the body. The combustion chemicals produced all have some effect on occupants, but determination of the products of combustion from fires is extremely complex due to the wide variation in material types and localised heating regimes involved, as illustrated in **Table 14.5.**

Table 14.5 The complexity of combustion products from polymeric materials[38]

Elemental composition	Examples	Products	
		Pyrolysis	Oxidative decomposition
CH	Polyethylene, polypropylene	hydrocarbons	CO, hydrocarbons, oxygenated species
CHO	wood, polyesters	CO, hydrocarbons, oxygenated species	CO, hydrocarbons, oxygenated species
CHN	polyacrylonitrile	hydrocarbons, amines, cyanides	CO, hydrocarbons, amines, cyanides
CHNO	polyurethanes	CO, hydrocarbons, cyanides, oxygenated species	CO, hydrocarbons, cyanides, oxygenated species
CHCl	polyvinyl chloride	hydrocarbons, hydrogen chloride	CO, hydrocarbons, hydrogen chloride
phosphorous and antimony	various polymeric substances	phosphorous and antimony containing products	

From the point of view of fire safety, the most important products are those which attain concentrations injurious to health in a rapid period of time following commencement of a fire. Chemical studies of undiluted combustion products over time for fires involving simple fuels have been undertaken by the BRE, for example ***Figure 14.13*** for a room fire involving 40 kg of wood with low ventilation.

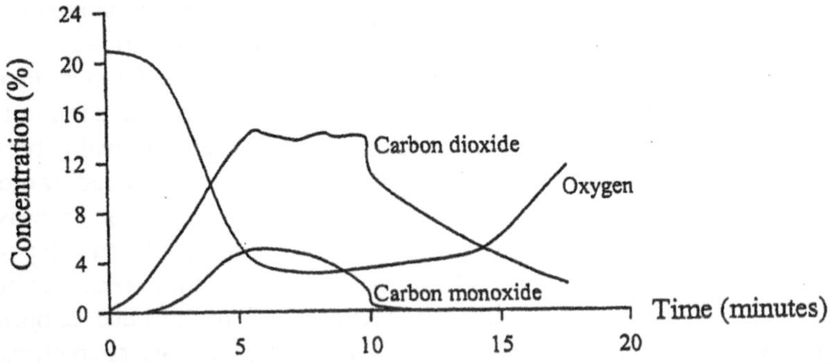

Figure 14.13 *Combustion product (CO_2, CO) and O_2 concentrations in wood fires[39]*

Movement of combustion gases away from the combustion zone in a real fire would act to dilute the concentrations indicated in ***Figure 14.13***. The physiological effects of the most important combustion products are summarised in ***Figure 14.14a*** and ***14.14b***. Carbon monoxide and hydrogen cyanide are considered toxic, whereas hydrogen chloride and acrolein are primarily irritants at concentrations typical of those encountered in fires. The essential difference between the toxicity of the products of combustion of the many different natural and synthetic materials are small when compared on the basis of the mass of product consumed. The major problems arise because of the considerable differences in the rates at which different products burn; for example, compared to rigid materials, foams burn readily and consequently produce large quantities of smoke and gases. In the short term, hazards from smoke and toxic gases can be decreased by reducing the ignitability of materials, thus slowing their rate of burning (**14.7.2**), rather than by attempting to reduce their toxicity.

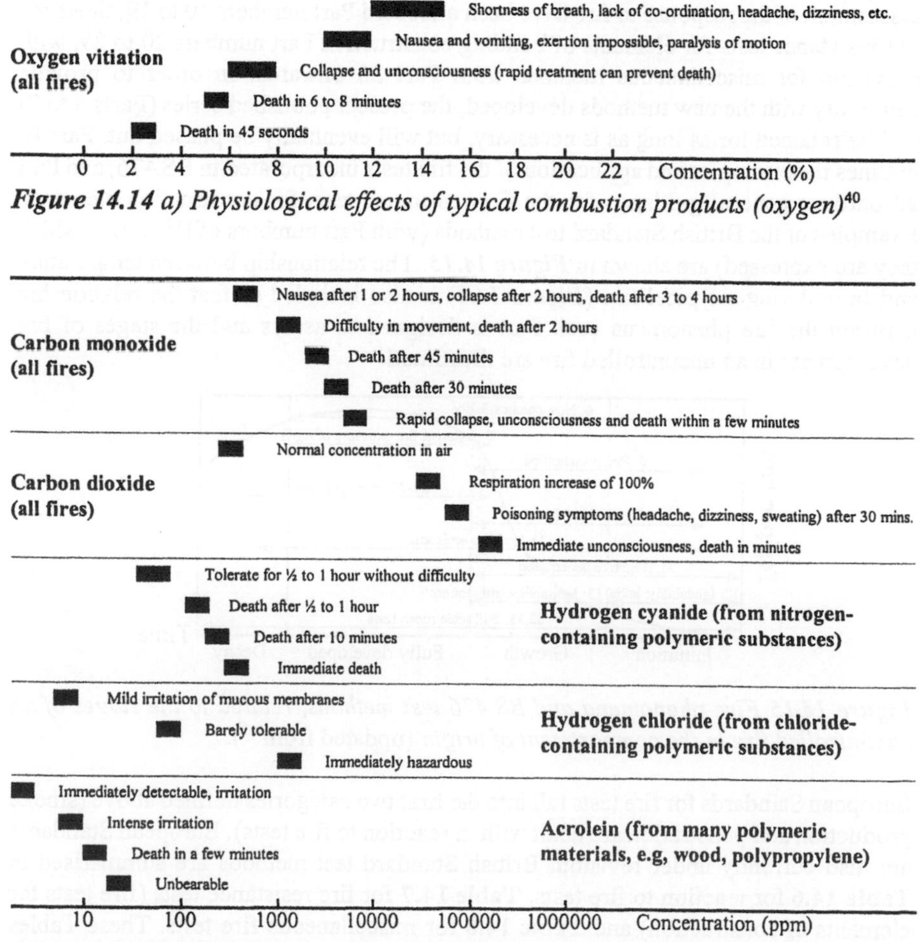

***Figure 14.14** a) Physiological effects of typical combustion products (oxygen)*[40]

***Figure 14.14** b) Physiological effects of typical combustion products (product gases)*

14.9 STANDARD FIRE TESTS

Standard fire tests aim to simulate as closely as possible the conditions that arise in typical fires so that the performance of materials, components, elements of structure and assemblies can be assessed and compared. The standards fall into three categories

- Reaction to fire tests (fire tests for products) related to initiation and early stages of fire growth;
- Fire resistance tests (fire tests for elements of construction) related to the behaviour of building elements exposed to fully-developed fires;
- Miscellaneous fire tests (including smoke penetration).

The British Standard for fire tests is BS 476. The various Parts of this British Standard deal with specific tests. The British Standards Institution is currently moving away from a practical approach to fire testing based on experience and limited technical data to a more rational approach, involving the revision of existing British Standards and the preparation of new methods. As part of this process, British Standards specific to the assessment of the response to fire have been allocated Part numbers 10 to 19, those for fire resistance tests for elements of building construction Part numbers 20 to 29, with provision for miscellaneous methods from Part 30 onwards. In order to provide continuity with the new methods developed, the present published series (Parts 3 to 7) will be retained for as long as is necessary, but will eventually be phased out. Part 10 outlines the principles and application of the fire tests incorporated in BS 476, and Part 20 outlines general guidance for the fire resistance tests for construction elements. Examples of the British Standard test methods (with Part numbers of BS 476 in which they are expressed) are shown in ***Figure 14.15***. The relationship between temperature and time during a typical fire (***Figure 14.8***) is also included so that the relationship between the fire phenomena each Part is designed to assess and the stages of fire development in an uncontrolled fire are illustrated.

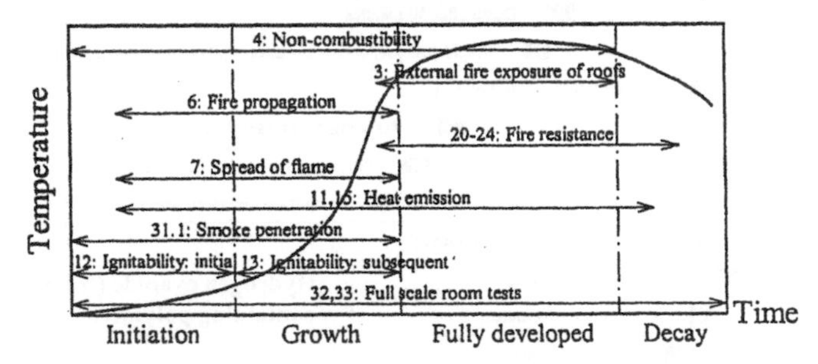

Figure 14.15 *Fire phenomena and BS 476 test methods, related to the stages of an uncontrolled fire in the compartment of origin* (updated from [41]).

European Standards for fire tests fall into the first two categories defined above (smoke production and penetration are dealt with in reaction to fire tests). European Standards are also currently under revision. British Standard test methods are summarised in **Table 14.6** for reaction to fire tests, **Table 14.7** for fire resistance tests (fire tests for elements of construction) and **Table 14.8** for miscellaneous fire tests. These Tables include equivalent standards, where present.

Table 14.6 Reaction to fire tests (for products). BS 476.

Part 4: 1983. Non-combustibility test for materials.	
Purpose	Method of test for determining whether building materials are non-combustible, within the meaning of the definition.
Applicable to	Materials, whether coated or not.
Not applicable to	The coating alone.
Method	Three specimens are tested, each 40 mm x 40 mm x 50 mm high. Each specimen is exposed for 20 minutes in a furnace preheated to 750°C.
Outcome	The material is deemed non-combustible if, during the test, none of the three specimens either • cause the temperature reading from either the specimen or the furnace to rise by 50°C or more above the initial furnace temperature; or • is observed to flame continuously for 10 seconds or more inside the furnace.
Equivalent standards	
Notes	Part 4 will be retained for as long as necessary but will eventually be phased out and replaced.

Part 6: 1989. Method of test for fire propagation of products.	
Purpose	Test method that provides a comparative measure of the contribution to the growth of fire made by an essentially flat material, composite or assembly. It is primarily intended for the assessment of the performance of internal wall and ceiling linings.
Applicable to	Essentially flat materials, composite or assembly.
Not applicable to	
Method	For each face to be tested, a minimum of three and a maximum of five samples are selected to give each test sample. The dimensions of the specimens are 225 mm square and of the normal thickness of the material (50 mm maximum). The specimens are exposed separately in a small combustion chamber for 20 minutes to a specified heating regime of increasing intensity.
Outcome	The performance is expressed as a numerical index (the fire propagation index) where low values indicate a low rate of heat release. The fire propagation index, $I = i_1 + i_2 + i_3$, where i_1 is derived from the first 3 minutes of the test, i_2 from the next 7 minutes and i_3 from the final 10 minutes. The fire propagation index is based on three valid test results. If four or five test specimens have been tested to achieve the three valid test results, a suffix 'R' is added to indicate variability in test results.
Equivalent standards	
Notes	Part 6 will be retained for as long as is necessary but will eventually be phased out and replaced.

Part 7: 1997. Method of test to determine the classification of the surface spread of flame of products.	
Purpose	A test method for measuring the lateral spread of flame along the surface of a specimen or a product orientated in the vertical position, and a classification system based on the rate and extent of flame spread. It provides data suitable for comparing the performance of flat materials, composites or assemblies which are used primarily as the exposed surfaces of walls or ceilings.
Applicable to	Flat materials, composites or assemblies which are used primarily as the exposed surfaces of walls or ceilings.
Not applicable to	
Method	The test sample comprises a minimum of six and a maximum of nine specimens for each face to be tested. The dimensions of the specimens are 885 ± 5 mm x 270 ± 5 mm and of the normal thickness of the material (50 mm maximum). The length of the specimen may be reduced to 250 mm where only Class 1 classification is being sought. Each specimen is exposed for 10 minutes to a 850 mm square radiant panel, with the specimen held with its long axis horizontal and its face vertical. A pilot flame is applied to the bottom corner near the radiant panel for the first minute of the test.
Outcome	The extent of flame spread after 1.5 minutes and at the end of the test is used to classify products into 4 classes. Class 1 represents the best performance. If seven, eight or nine specimens have to be tested to achieve six valid test results, a suffix 'R' is added to indicate variability in test results. Where softening and/or other behaviour which may affect flame spread occurs, a suffix 'Y' is added. The standard also imposes limits on surface irregularities of products; a prefix 'D' is added when a product not meeting these limits is tested in a modified form (e.g. corrugated sheeting tested flat).
Equivalent standards	
Notes	Class 0 is not a classification identified in a British Standard method of test. It is a term defined in connection with the Building Regulations, which makes use of BS performance ratings to limit both the spread of flame across the surface of a material or product (BS 476: Part 7) and the rate at which heat is released from it (BS 476: Part 6). Such materials or products are generally required to be non-combustible or to have a Class 1 surface spread of flame with fire propagation indices $i \leq 12$, $i_1 \leq 6$. Part 7 will be retained for as long as is necessary but will eventually be phased out and replaced.

Part 10: 1983. Guide to the principles and application of fire testing.	
Purpose	Description of the procedure for a laboratory fire test of building products, components and elements of construction.
Applicable to	Essentially flat products, composites or assemblies which are used primarily as the exposed surfaces of walls or ceilings.
Not applicable to	
Method	(dependent on test - see Parts 11-19)
Outcome	Test results provide information on the whether building products, components and elements of construction will ignite so easily as to significantly increase the chance of a fire occurring and, given that a fire has started and recognising that the contents of the building are usually the prime source of fuel for the fire, whether the product significantly increases the speed and extent of fire spread and/or contributes significantly to the heat and smoke generated by a fire. Results also provide information on whether, once a fire has become established, the elements of construction remain stable and barriers, such as floors, walls and doors, successfully contain the heat and smoke.
Equivalent standards	
Notes	Part 10 provides guidance on Reaction to fire tests (fire tests for products) comprising Parts 11-19. It performs a similar function to Part 20 for Fire resistance tests (fire tests for elements of construction) comprising Parts 21-29.

Part 11: 1982. Method for assessing the heat emission from building materials.	
Purpose	Method for assessing the heat emission from building materials when inserted into a furnace at 750°C. The method is applicable to simple materials or mixtures of materials, either manufactured or naturally occurring, that are reasonably homogeneous and from which it is possible to obtain specimens representative of the material as a whole. It is also applicable to non-homogeneous materials providing that irregularities within the material (such as density gradient, particle size, voids, etc.) are not disproportionately large compared with the size of the test specimen.

Table 14.6 Reaction to fire tests (for products). BS 476. (contd.)

Part 11: 1982. Method for assessing the heat emission from building materials. (contd.)	
Applicable to	
Not applicable to	This method is not normally suitable for assessing combinations of materials, such as those that are surface coated, veneered or faced or that contain discrete layers of materials that have been fixed or glued together as laminates. However, providing that sufficiently representative specimens can be produced, the individual discrete materials may be assessed separately.
Method	Five cylindrical specimens are prepared for testing, each with a diameter of 45 mm and a height of 50 mm. Materials thicker than 52 mm are reduced accordingly; for those less than 48 mm thick, the specimen is made from two or more layers. The specimens are exposed for a maximum of 120 minutes in a preheated cylindrical furnace and note taken of any rise in the temperature of both the furnace and the specimen, and of any sustained flaming.
Outcome	From these results, the mean furnace temperature, the mean specimen temperature rise and the mean duration of sustained flaming for each of the specimens tested is calculated.
Equivalent standards	
Part 12: 1991. Method of test for ignitability of products by direct flame impingement.	
Purpose	Method of test for the determination of the ignitability of materials, composites and assemblies subjected to direct impingement of flames of different size and intensity but without impressed irradiance.
Applicable to	Essentially flat products, composites or assemblies which are used primarily as the exposed surfaces of walls or ceilings.
Not applicable to	
Method	Vertically held specimens are exposed to specified flames of different sizes and intensities and their ignition behaviour is observed. The flame is applied to the surface and/or a bottom edge on different specimens. Seven ignition sources, representing a range of flame heights and intensities are specified. A minimum of six full thickness representative specimens for each ignition source to be applied and each flame application time (three for the surface ignition and three for the bottom edge ignition). Specimen sizes range from 100 mm x 150 mm to 500 mm x 1200 mm and flame application times range from 1 second to 180 seconds, and are dependent on the ignition source type. For each ignition source and for each flame application time, ignition characteristics (sustained or transient ignition, separation of debris, extent of flaming or glowing) are recorded.
Outcome	Results can be used to assess the performance of a product in the early stages of a fire. Results cannot be considered or used by themselves for describing or appraising the fire hazard of a material, composite or assembly under actual fire conditions. A future part of the standard will address these aspects.
Equivalent standards	
Part 13: 1987. Method of measuring the ignitability of products subjected to thermal irradiance.	
Purpose	Method for examining the ignition characteristics of the exposed surfaces of specimens of essentially flat materials, composites or assemblies not exceeding 70 mm in thickness, when placed horizontally and subjected to specified levels of thermal irradiance.
Applicable to	Intended to provide information which can be used in the evaluation of wall and ceiling linings, flooring systems, external claddings and duct insulation materials.
Not applicable to	
Method	Five specimens are tested at each level of irradiance selected and for each exposed surface. The dimensions of the specimens are 165 mm square and of the normal thickness of the material. Products thicker than 70 mm are reduced accordingly. The upper surface of each specimen is exposed for up to 15 minutes to selected levels of constant radiation within the range 10 kW/m^2 to 50 kW/m^2. A pilot flame is applied at 4 second intervals to a position 10 mm above the centre of the specimen to ignite any volatile gases given off.
Outcome	At each irradiance, the time to sustained surface ignition is reported for each of specimen tested.
Equivalent standards	
Part 14. Method for measuring the rate of flame spread on surfaces of products.	
Notes	(in course of preparation)
Part 15:1993. Method for measuring the rate of heat release of products).	
Purpose	Method for assessment of the heat release rate of essentially flat products exposed to controlled levels of radiant heating with or without an external igniter.
Applicable to	Essentially flat products, composites or assemblies used primarily as the exposed surfaces of walls or ceilings. Products with surface irregularities can be tested according to specific requirements.
Not applicable to	
Method	The rate of heat release is determined by measurement of the oxygen consumption derived from the oxygen concentration and the flow rate in the combustion product stream. The time to ignition (sustained flaming) is also measured.
Outcome	The results are derived from a particular aspect of the potential fire situation typified by a radiant heat source and a spark and therefore cannot alone provide any direct guidance on behaviour or safety in fire. Results may, however, be used for comparative purposes or to ensure the existence of a certain quality of performance in relation to heat release rates considered to have a bearing on fire performance generally.
Notes	'Heat release rate' is defined as the calorific energy released per unit time by a material during combustion under specified test conditions.
Equivalent standards	ISO 5660-1:1993. Fire tests - Reaction to fire - Part 1: Method of measuring the rate of heat release of products (cone calorimeter method).
Part 16. Method of measuring the smoke release (obscuration) of products.	
Notes	(in course of preparation)
Parts 17 to 19. Other methods of test relating to products.	
Notes	(to be issued when required).

Table 14.7 Fire resistance tests (elements of construction). BS 476.

Part 3: 1975. External fire exposure roof tests.	
Purpose	Measurement of • the capacity of a representative section of a roof to resist penetration by fire when the external surface is exposed to radiation and flame; • the distance of the spread of flame on the outer surface of the roof covering under certain conditions.
Applicable to	Flat and sloping roofs
Not applicable to	
Method	Seven specimens 840 mm square are tested, one for the preliminary ignition test using a pilot flame and (if successful) three each for separate tests for penetration and spread of flame where the specimen is exposed to heat radiated from gas fired panels before the flame is applied.
Outcome	Roofs are graded according to the time for which they resist penetration by fire and the distance of superficial spread of flame on their external surface. Constructions are designated by two letters, the first letter indicates the time to penetration and the second letter the distance of flame spread, with AA representing the best performance. A suffix 'EXT F' signifies flat roofs and 'EXT S' sloping roofs. A suffix 'X' is added where dripping from the underside of the specimen, any mechanical failure or any development of holes has taken place.
Equivalent standards	EN 1187. External fire exposure to roofs. Part 1: 1993. Method of test simulating exposure to burning brands, without wind and supplementary radiant heat. Part 2: 1994. Method of test simulating exposure to burning brands, with wind and supplementary radiant heat.
Notes	
Part 20:1990. Method of determination of the fire resistance of elements of construction (General principles)	
Purpose	Description of the procedure for a laboratory test for the determination of fire resistance of elements of construction.
Applicable to	
Not applicable to	
Method	A representative sample of the element is exposed to a specified regime of heating and the performance of the test sample is monitored on the basis of criteria described in the standard. Fire resistance of the test element is expressed as the time for which the appropriate criteria have been satisfied. The times so obtained are a measure of the adequacy of the construction in a fire but have no direct relationship with the duration of a real fire. Heating conditions are specified based on the temperature time curve which furnaces are required to follow. Only one specimen is normally tested and it should be either full size, or where this is not possible, it should have the following minimum dimensions • non-separating elements: vertical 3 m high; horizontal 4 m span • separating elements: vertical 3 m high x 3 m wide; horizontal 4 m span x 3 m wide Specimens are normally heated to simulate their exposure in a fire, e.g. walls are heated from one side, floors from beneath, and columns from all sides, and their performance is expressed as the time (in minutes) to failure of one or more of the following criteria • Loadbearing capacity: resistance to collapse or excessive deflection. • Integrity; resistance to penetration of flame and hot gases • Insulation: resistance to excessive temperature rise on the unheated face.
Outcome	The test data can be used directly to show compliance with fire resistance requirements in regulations. the tests can also be used to study the behaviour of constructions at high temperatures and obtain guidance on the effects of design features on fire resistance. It is intended to extend the application of the test data by agreed procedures for interpolation and extrapolation. Due to restrictions of size and the absence of surrounding construction, the laboratory test cannot reproduce the actual behaviour pattern of an element in a fire. However, test data can provide a basis for making engineering evaluations. A future part of the standard will address these aspects.
Equivalent standards	EN 1363. Fire resistance tests on elements of building construction. Part 1: 1993. Alternative and additional procedures.
Notes	'Separating element' is defined as 'an element that is required to satisfy the criteria of integrity and insulation, in addition to loadbearing capacity, if applicable, for the purposes of maintaining separation between two adjacent areas of the building in the event of a fire'. Part 20 provides guidance on Fire resistance tests (fire tests for elements of construction) comprising Parts 21-29. It performs a similar function to Part 10 for Reaction to fire tests (fire tests for products) comprising Parts 11-19.
Part 21:1987. Methods for determination of the fire resistance of loadbearing elements of construction.	
Purpose	Procedures for determining the fire resistance of loadbearing elements of building construction when subjected to the heating and pressure conditions specified in the Part 20. This part provides requirements for the specimen selection and/or its design and construction, loading and restraint conditions, equipment, conditions and criteria as they apply to loadbearing specimens.
Applicable to	The methods are applicable to beams, columns, floors, flat roofs and walls.
Not applicable to	The methods described are not applicable to suspended ceilings protecting steel beams (BS 476:Part 23). Owing to limitations imposed both by the furnaces and their method of control, structures employing water cooling techniques are not capable of being evaluated realistically by these procedures. The methods described are not applicable to assemblies of elements e.g. wall/floor combinations, although some guidance is provided in appendix A for tests of this type.
Method	(see Part 20)
Outcome	(see Part 20)
Equivalent standards	
Notes	EN 1365. Fire resistance tests on loadbearing elements in buildings. Part 1: 1994. Internal walls. Part 2: 1994. External walls. Part 3: 1993. Floor construction. Part 4: 1993. Roof constructions heated from the underside. Part 5: 1993. Beams.
Part 22:1987. Methods of determination of the fire resistance of non-loadbearing elements of construction.	
Purpose	Procedures for determining the fire resistance of non-loadbearing elements of building construction when subjected to the heating and pressure conditions specified in Part 20. The methods described are appropriate to normal combinations of these elements.
Applicable to	This part is applicable to vertical partitions, to fully insulated, partially insulated and uninsulated vertical doorsets and shutter assemblies, to ceiling membranes and to glazed elements.
Not applicable to	Fire dampers incorporated in ducts.

Table 14.7 Fire resistance tests (elements of construction). BS 476. (contd.)

Part 22:1987. Methods of determination of the fire resistance of non-loadbearing elements of construction.	
Outcome	(see Part 20)
Equivalent standards	EN 1364. Fire resistance tests on non-loadbearing elements in buildings. Part 1: 1993. Partitions. Part 2: 1994. External walls. Part 3: 1994. Ceilings. Part 4: :1997. Curtain walling. EN 1634. Fire testing of door and shutter assemblies. Part 1: 1994. Method of test for fire resistance of fire doors and shutters. Part 3: 1996. Smoke control doors and shutters.
Notes	
Part 23:1998. Methods for determination of the contribution of components to the fire resistance of a structure.	
Purpose	Describes procedures for determining the contribution made by components to the total fire resistance of a structure or other elements used in conjunction with them when subjected to the relevant heating and pressure conditions specified in Part 20.
Applicable to	The methods are applicable to suspended ceilings protecting steel beams and intumescent seals for use in conjunction with single-acting, latched timber fire resisting door assemblies.
Not applicable to	
Method	(see Part 20)
Outcome	This part provides the requirements for specimen selection and/or its design and construction, the edge conditions, the equipment (including any special apparatus or instrumentation) and the procedures and criteria as they apply to components which make a contribution to the fire resistance of a complete assembly.
Equivalent standards	EN (number not yet known). Part 1. Membrane protection, horizontal. Part 2. Membrane protection, vertical. Part 3. Concrete elements. Part 4. Steel elements. Part 5. Flat concrete/profiled sheet composite materials. Part 6. Concrete-filled hollow steel columns. Part 7. Timber elements.
Notes	It is expected that Part 23 will be extended by tests for the contribution made by other components as such tests are developed.
Part 24:1987. Method for determination of the fire resistance of ventilation ducts.	
Purpose	Method of test and criteria for the determination of the fire resistance of vertical and horizontal ventilation ducts under standardised fire conditions.
Applicable to	It is applicable to horizontal and vertical ducts, with or without branches, taking into account joints, air supply and exhaust openings, as well as suspension devices, etc.
Not applicable to	The standard in not applicable to ducts above fire resisting suspended ceilings (horizontal membranes) in those cases where the ducts rely for their fire resistance on the performance of the ceiling ducts containing fire dampers at points where they pass through fire separations The standard is not appropriate for the following ducts, unless the further criteria described in the annex are established to the satisfaction of the appropriate authority ducts of materials which are extremely sensitive to thermal shock smoke outlet ducts ducts lined on the inside with combustible material or which in practice may accumulate combustible deposits on their inside face (e.g. kitchen extract ducts)
Method	(see Part 20)
Outcome	Measurements of the ability of a representative duct or duct assembly to resist the spread of fire from one fire compartment to another without the aid of fire dampers.
Equivalent standards	EN 1366. Fire resistance tests on service installations in buildings. Part 1: 1994. Fire-resisting ducts. Part 2: 1994. Dampers. Part 3: 1993. Penetration seals. Part 4: 1998. Linear joint seals. Part 5: 1998. Service ducts and shafts. Part 6: 1996. Smoke extraction.
Notes	The standard does not take into consideration the effect of impact shock loading on ducts due to collapse of supporting or adjacent structural members of other components, or of impact, or of thermal shock loading resulting from the application of a water (hose) stream.
Parts 25 to 29. Other methods related to the determination of fire resistance.	
Notes	(to be issued when required).

Table 14.8 Miscellaneous fire tests. BS 476.

Part 31. Methods for measuring smoke penetration through doorsets and shutter assemblies.	
Purpose	Method for measuring smoke penetration, as represented by the measurement of air leakage rate, through doorsets and vertically orientated shutter assemblies under ambient temperature conditions. Used to evaluate the performance of a fire door used for smoke control purposes but gives no information on the fire resistance of a fire door (for which the methods described in Part 22 are applicable).
Applicable to	
Not applicable to	
Method	One full size specimen is tested when fitted to form one side of an air leakage chamber. The rate of air leakage is established by measuring the airflow for increasing pressure differentials between the two faces of the specimen, up to the maximum pressure differential for which information is required.
Outcome	Smoke penetration, as represented by air leakage rates. Results are given for the adjusted rate of air leakage of the specimen (in m^3/h) for each pressure differential and face of the specimen.
Equivalent standards	
Part 32:1989. Guide to full scale fire tests within buildings.	
Purpose	Recommendations on the conduct of full scale experiments simulating fires in buildings. The guide can be used to establish, as far as is reasonably possible, laboratory conditions that represent a specific set of fire conditions, with an emphasis on the pre-flashover behaviour and contribution to fire growth of the product(s) under consideration.

Table 14.8 Miscellaneous fire tests. BS 476. (contd.)

Part 32:1989. Guide to full scale fire tests within buildings. (contd.)	
Applicable to	
Not applicable to	
Method	Tests can be used to evaluate the behaviour of building components, assemblies or constructions and/or contents under actual fire conditions. Tests are carried out in normal storey height (2.3 m) minimum test rigs with uncontrolled ventilation. In the range of tests covered by the guide, specimens may vary from a room lining material applied to a standard support structure to a full room/corridor, as well as complete free standing assemblies. Various measurements can be taken depending on the aims of the assessment. These may include assessment of ignitability and critical exposure of products, flame spread, heat output, smoke and noxious gas production, etc.
Outcome	Depending on the objective of the fire experiment, and may include • a comparison of the fire performance of different materials; • a comparison of theory and experiment; • a simulation experiment; • measurement of the fire behaviour of composites, assemblies and finished products; • investigation of the interaction of the components within the system; • authentication of fire properties measured in small scale tests.
Equivalent standards	
Part 33: Full-scale room test for surface products.	
Purpose	Specification of a test method that simulates a fire that, under well ventilated, controlled conditions, starts in a corner of a small room with a single open doorway. The method is intended to evaluate the contribution to fire growth provided by a surface product using a specified ignition source. The test provides data for the early stages of a fire from ignition up to flashover.
Applicable to	Particularly suited to products that for some reason cannot be tested in a small laboratory scale (e.g. thermoplastic materials, the effects of an insulating substrate, joints and surfaces with great irregularity.
Not applicable to	
Method	The tests are undertaken in a room of dimensions 3.6 m x 2.4 m x 2.4 m high, which includes a 0.8 m x 2.0 m high door in one wall.
Outcome	The potential for fire spread to other objects in the room, remote from the ignition source, is evaluated by measurements of the total heat flux incident on a heat flux meter located in the centre of the floor. The potential for fire spread to objects outside the room of origin is evaluated by measurement of the total rate of heat release from the fire. An indication of the toxic hazard is provided by the measurement of certain toxic gases. The hazard of reduced visibility is estimated by measurement of light-obscuring smoke.
Equivalent standards	ISO 9705: 1993. Fire tests. Full-scale room test for surface products.
Notes	

14.10 DEFINITIONS

14.10.1 Calorific Value

The calorific value of a fuel is the amount of heat which would be released by complete combustion of unit quantity of the fuel under specified conditions of temperature and pressure, and has units J/g or J/m^3 (**Table 14.2**). Complete combustion is achievable with gaseous fuels at atmospheric pressure, but solid and liquid fuels require higher pressures. Since the heat released at high pressure differs from that which would be obtained with complete combustion at atmospheric pressure, conditions for solid and liquid fuels are defined in such a way that the difference can be evaluated and kept small enough to be ignored in most circumstances. Calorific value is the term used by the solid fuel and gas industries. An alternative term, used in the oil industry, meaning the same is 'heat of combustion'. Calorific values are determined by burning a sample of the fuel in oxygen under defined pressure in a bomb calorimeter (BS 7420:1991[42]). This is the only satisfactory method for the determination of the calorific value of solid and liquid fuels.

14.10.2 Specific Heat Capacity

The specific heat capacity is the amount of energy required to raise unit mass of the substance by 1°C, and has units J/g.°C. Specific heat capacity therefore defines the amount of heat absorbed by a substance as its temperature rises. Specific heat capacities of materials are important parameters for fire protection because they indicate the relative quantity of heat needed to raise the temperature to danger levels, or that must be removed to cool a substance to a safe temperature. One reason why liquid water (specific heat capacity 4.19 J/g.°C) is an effective fire fighting agent is that its specific heat capacity is generally higher than other materials.

14.10.3 Latent Heat

Latent heat is the quantity of energy required to initiate a change of phase of a material (e.g, solid → liquid, liquid → gas or gas → liquid, liquid → solid) and has units J/g. For water at normal pressure, the latent heat of fusion (liquid → solid) (**2.5**) at the melting/freezing point (336 J/g) and the latent heat of vaporisation (liquid → gas) at the boiling point (2260 J/g) are generally greater than those for other materials. The greater latent heat of vaporisation of water is another reason for the effectiveness of water as a fire extinguishing agent. Heat absorbed by water converts the liquid to steam; this heat is removed from the combustion system so that it cannot be used to propagate flame further by vaporising more liquid fuel or pyrolysing more solid fuel.

14.10.4 Thermal Conductivity

Thermal conductivity is the rate of heat transfer through unit area of a slab of a uniform, homogenous material of unit thickness, when unit temperature difference is maintained between two opposite surfaces (the other surfaces being insulated), and has units W/m.°C. Some materials (e.g, metals) have high thermal conductivites (e.g, aluminium 164 W/m.°C) and are therefore able to transmit heat energy effectively. Other materials (e.g, cementitious materials and masonry) have low thermal conductivites (e.g, cast concrete 1.4 W/m.°C and clay facing bricks 0.84 W/m.°C), and may therefore provide a contribution to the thermal resistance of building components and structures.

14.10.5 Thermal Capacity

The thermal behaviour of a material or component is much influenced by its ability to store heat. The amount of heat stored in a material depends largely on the density of the material and its specific heat capacity. More dense materials can absorb more heat and will therefore take longer to heat up or to cool. The thermal capacity of a material is therefore a measure of the amount of heat necessary to raise unit volume of that material by unit temperature. Thermal capacity is derived as the product of the density of the material and its specific heat capacity, and has units $J/m^3.°C$.

14.10.6 Gas and Vapour

There is a very subtle distinction between a gas and a vapour. A gas is a vapour **above** its critical temperature, whilst a vapour is a substance in the gaseous state at a temperature **below** its critical temperature. **Critical temperature,** T_c is the temperature above which a substance in the gaseous state **cannot** be liquefied (**Table 14.9**), however large the applied pressure. Above the critical temperature, the kinetic energy of the gas molecules are too high to allow the intermolecular forces to hold the molecules together as a liquid. **Critical pressure,** P_c is the pressure required to liquefy the gas at the critical temperature. Below the critical pressure, the gaseous state **cannot** be liquefied because the pressure is not high enough to allow the intermolecular forces to hold the molecules together as a liquid. Some liquids which are normally used quite safely can be taken to temperatures above the critical temperature during the course of a fire, therefore causing explosions.

Table 14.9 Critical pressures and temperatures for common compounds[43]

Material	Carbon dioxide (CO_2)	Ammonia (NH_3)	Water (H_2O)	Propane (C_3H_8)	Acetylene (C_2H_2)	Freon
Critical temperature, T_c	31°C	133°C	374°C	97°C	36°C	96°C
Critical pressure, $P_c \times 10^5$	73 Pa	111 Pa	218 Pa	42 Pa	63 Pa	50 Pa
Uses	Fire extinguisher	Refrigerant	Fire extinguisher	Gas cylinders	Welding	Refrigerant

14.10.7 Exothermic and endothermic reactions

Exothermic chemical reactions **give out** heat as they proceed (usually represented as +X Joules on the right hand side of the chemical equation). Endothermic chemical reactions **require** heat to proceed (usually represented as -Y Joules on the right hand side of the chemical equation). Usually, therefore, if this heat is not supplied (from an external source), endothermic reactions will stop.

14.11 REFERENCES

1. FPA (1988) *The Physics and Chemistry of fire.* Fire Safety Data Sheet NB 1. London, Fire Protection Association. p. 3.
2. DRYSDALE, D.D. (1991) Section 1. Chapter 4. Chemistry and Physics of Fire. In: COTE, A.E. and LINVILLE, J.L. (eds.) *Fire Protection Handbook, 17th ed.* Quincy, Massachusetts, National Fire Protection Association. p. 1-46. Figure 1-4A.
3. ROSE, J.W. and COOPER, J.R. (eds.) (1977) *Technical Data on Fuel, 7th ed.* London, British National Committee of the World Energy Conference.
4. BS 2000. Methods of test for petroleum and its products. Part 35:1993. Determination of open flash and fire point. Pensky-Martens method. Part 170:1995. Determination of flash point. Abel closed cup method. Part 403. Determination of flash and fire points. Cleveland open cup method (= BS EN 22592: 1994). Part 404: 1994. Determination of flash point. Pensky-Martens closed cup method (= BS EN 22719: 1994).
5. BS 3900. Methods of test for paints Group A: Tests on liquid paints (excluding chemical tests). Part A8: 1986. Methods of test for flash/no flash (closed cup equilibrium method) (= BS 6664. Part 1: 1986). Part A9: 1986. Determination of flash point (closed cup equilibrium method) (= BS 6664. Part 2: 1986). Part A13: 1986. Methods of test for flash/no flash (rapid equilibrium method) (= BS 6664. Part 3: 1986).
6. BS 6664. Flashpoint of petroleum and related products. Part 4: 1991. Method for determination of flashpoint (rapid equilibrium method).
7. BS EN 456: 1991. Method for determination of flashpoint of paints, varnishes and related products by the rapid equilibrium method.
8. BS EN 924: 1995. Adhesives. Solvent-borne and solvent-free adhesives. Determination of flashpoint.
9. FPA (1985) *The Combustion Process.* Fire Safety Data Sheet NB 4. London, Fire Protection Association. Dec. 1985. p. 2. Figure 3.
10. FPA (1985) *The Combustion Process.* Fire Safety Data Sheet NB 4. London, Fire Protection Association. Dec. 1985. p. 3. Figure 4.
11. BS EN ISO 11114. Transportable gas cylinders. Compatibility of cylinder and valve materials with gas contents. Part 3: 1998. Autogenous ignition test in oxygen atmosphere.
12. BS ISO 871: 1996. Plastics. Determination of ignition temperatures using a hot-air furnace.
13. BS 5345. Selection, installation and maintenance of electrical apparatus for use in potentially explosive atmospheres (other than mining applications or explosives). Part 1: 1993. General recommendations (= BS EN 60079. Part 14: 1997).
14. CRUICE, W.J. (1991) Section 1. Chapter 5. Explosions. In: COTE, A.E. and LINVILLE, J.L. (eds.) *Fire Protection Handbook, 17th ed.* Quincy, Massachusetts, National Fire Protection Association. p. 1-62.

15. BEALS, H.O. (1991) Section 3. Chapter 3. Wood and wood-based products. In: COTE, A.E. and LINVILLE, J.L. (eds.) *Fire Protection Handbook, 17th ed.* Quincy, Massachusetts, National Fire Protection Association. p. 3-28. Table 5-3E.
16. FPA (1985) *The Combustion Process.* Fire Safety Data Sheet NB 4. London, Fire Protection Association. Dec. 1985. p. 3. Figure 6.
17. SCOONES, K. (1997) Serious fires in historic buildings. *FPA Records, Vol. 303 (Oct. 1997).* p. 38.
18. BS EN 2: 1992. Classification of Fires
19. BS 476. Fire tests on building materials and structures. Part 10: 1983. Guide to the principles and application of fire testing. London, British Standards Institution. p. 4. Figure 1.
20. BS 4422. Glossary of terms associated with fire. Part 4: 1994. Fire extinguishing equipment
21. BS EN 25923: 1994 Fire Protection. Fire Extinguishing Media. Carbon dioxide (= ISO 5923:1994) (replaces BS 6535: 1990)
22. BS EN 27201. Fire Protection. Fire Extinguishing Media. Halogenated Hydrocarbons. Part 1: 1994. Specifications for Halon 1211 and Halon 1301 (= ISO 7201. Part 1: 1989) (replaces BS 6535. Part 2. Section 2.1: 1990). Part 2: 1994. Code of practice for safe handling and transfer procedures (= ISO 7201. Part 2: 1991) (replaces BS 6535. Part 2. Section 2.2: 1989).
23. BS EN 615: 1995 Fire Protection. Fire Extinguishing Media. Specifications for powders (other than class D powders) (replaces BS 6535. Part 3: 1989)
24. prEN 1568: 1999 Fire Extinguishing Media. Foam Concentrates (= ISO 7203: 1995)
25. WILCOX, W.W; BOTSAI, E.E. and KUBLER, H. (1991) *Wood as a Building Material: A guide for designers and builders.* New York, John Wiley and Sons. p. 111. Figure 7.3.
26. BENJAMIN, I.A. (1961) Fire Resistance of Reinforced Concrete. *Symposium on Fire Resistance of Concrete, American Concrete Institute.* Detroit, MI, American Concrete Institute. pp. 29, 30.
27. ROSS, R.B. (1980) *Metallic Materials Specification Handbook, 3rd ed.* London, E&FN Spon.
28. BS 5268. Structural Use of Timber. Part 4. Fire resistance of timber structures. Section 4.1: 1978. Recommendations for calculating fire resistance of timber members. p. 1. Table 1.
29. NBS (1961) *Fire tests of columns protected with gypsum. NBS Research Paper No. 563.* Washington DC, National Bureau of Standards.
30. BS 8110. Structural use of concrete. Part 1: 1997. Code of practice for design and construction. Section 3. Design and detailing. Reinforced concrete. p.17. Table 3.4.
31. BS 8110. Structural use of concrete. Part 1: 1997. Code of practice for design and construction. Section 4. Design and detailing. Prestressed concrete. p.85. Table 4.9.
32. BS 1191:1973. Gypsum building plasters.
33. HARLAND, W.A. and WOOLLEY, W.D. (1979) *Fire Fatality Study,* BRE Current Paper CP 18/79. Garston, BRE.
34. BRE (1985) *Toxic effects of fires.* BRE Digest 300. Garston, BRE. p. 3.
35. PURSER, D.A. (1988) Section 1. Chapter 4. Toxicity assessment of combustion products. In: *The SFPE Handbook of Fire Protection Engineering,* Quincy, MA, US National Fire Protection Association. pp. 200.
36. SHORTER, G.W.; McGUIRE, J.H.; HUTCHEON, N.B. and LEGGET, R.F. (1960) The St. Lawrence Burns. *NFPA Quarterly, Vol. 53(4). April 1960.* pp. 300-316.
37. WOODSON, W.E. (1981) *Human Factors Design Handbook.* New York, McGraw-Hill. p. 812.
38. BRE (1985) *Toxic effects of fires.* BRE Digest 300. Garston, BRE. p. 3. Table 1.
39. BRE (1985) *Toxic effects of fires.* BRE Digest 300. Garston, BRE. p. 4. Figure 2.
40. BRE (1985) *Toxic effects of fires.* BRE Digest 300. Garston, BRE. p. 5. Table 2.
41. BS 476. Fire tests on building materials and structures. Part 10: 1983. Guide to the principles and application of fire testing. London, British Standards Institution. p. 5. Figure 2.
42. BS 7420: 1991. Guide for the determination of calorific values of solid, liquid and gaseous fuels (including definitions).
43. TENNENT, R. M. (ed.) (1997) *Science Data Book,* Harlow, Oliver and Boyd. p. 63. Table 14.

INDEX

Printed and bound by CPI Group (UK) Ltd, Croydon, CR0 4YY
03/10/2024
01040437-0016